T0183253

Lecture Notes in Computer Science   10177

Commenced Publication in 1973
Founding and Former Series Editors:
Gerhard Goos, Juris Hartmanis, and Jan van Leeuwen

More information about this series at http://www.springer.com/series/7409

Selçuk Candan · Lei Chen
Torben Bach Pedersen · Lijun Chang
Wen Hua (Eds.)

Database Systems for Advanced Applications

22nd International Conference, DASFAA 2017
Suzhou, China, March 27–30, 2017
Proceedings, Part I

 Springer

Editors
Selçuk Candan
Arizona State University
Tempe - Phoenix, AZ
USA

Lijun Chang
University of New South Wales
Sydney, NSW
Australia

Lei Chen
Hong Kong University of Science
and Technology
Hong Kong
China

Wen Hua
The University of Queensland
Brisbane, QLD
Australia

Torben Bach Pedersen
Aalborg University
Aalborg
Denmark

ISSN 0302-9743   ISSN 1611-3349 (electronic)
Lecture Notes in Computer Science
ISBN 978-3-319-55752-6   ISBN 978-3-319-55753-3 (eBook)
DOI 10.1007/978-3-319-55753-3

Library of Congress Control Number: 2017934640

LNCS Sublibrary: SL3 – Information Systems and Applications, incl. Internet/Web, and HCI

Printed on acid-free paper

This Springer imprint is published by Springer Nature
The registered company is Springer International Publishing AG
The registered company address is: Gewerbestrasse 11, 6330 Cham, Switzerland

Preface

It is our great pleasure to welcome you to DASFAA 2017, the 22nd edition of the International Conference on Database Systems for Advanced Applications (DASFAA), which was held in Suzhou, China, during March 27–30, 2017.

The long history of Suzhou City has left behind many attractive scenic spots and historical sites with beautiful and interesting legends. The elegant classical gardens, the old-fashioned houses and delicate bridges hanging over flowing waters in the drizzling rain, the beautiful lakes with undulating hills in lush green, and the exquisite arts and crafts, among many other attractions, have made Suzhou a renowned historical and cultural city full of eternal and poetic charm. Suzhou is best known for its gardens: the Humble Administrator's Garden, the Lingering Garden, the Surging Wave Pavilion, and the Master of Nets Garden. These gardens weave together the best of traditional Chinese architecture, painting, and arts. Suzhou is also known as the "Venice of the East." The city is sandwiched between Taihu Lake and Grand Canal. A network of channels, criss-crossed with hump-backed bridges, give Suzhou an image of the city on the water.

We were delighted to offer an exciting technical program, including three keynote talks by Divesh Srivastava (AT&T Research), Christian S. Jensen (Aalborg University), and Victor Chang (Xi'an Jiaotong University and Liverpool University), one 10-year best paper award presentation; a demo session with four demonstrations; two industry sessions with nine paper presentations; three tutorial sessions; and of course a superb set of research papers. This year, we received 300 submissions to the research track, each of which went through a rigorous review process. Specifically, each paper was reviewed by at least three Program Committee (PC) members, followed by a discussion led by the PC co-chairs. Several papers went through a shepherding process. Finally, DASFA 2017 accepted 73 full research papers, yielding an acceptance ratio of 24.3%.

Four workshops were selected by the workshop co-chairs to be held in conjunction with DASFAA 2017. They are the 4th International Workshop on Big Data Management and Service (BDMS 2017), the Second Workshop on Big Data Quality Management (BDQM 2017), the 4th International Workshop on Semantic Computing and Personalization (SeCoP), and the First International Workshop on Data Management and Mining on MOOCs (DMMOOC 2017).

The workshop papers are included in a separate volume of the proceedings also published by Springer in its *Lecture Notes in Computer Science* series.

The conference received generous financial support from Soochow University. The conference organizers also received extensive help and logistic support from the DASFAA Steering Committee and the Conference Management Toolkit Support Team at Microsoft.

We are grateful to the general chairs, Karl Aberer, EPFL, Switzerland, Peter Scheuermann, Northwestern University, USA, and Kai Zheng, Soochow University, China, the members of the Organizing Committee, and many volunteers, for their great support in the conference organization. Special thanks also go to the DASFAA 2017 local Organizing Committee: Zhixu Li and Jiajie Xu, both from Soochow University, China. Finally, we would like to take this opportunity to thank the authors who submitted their papers to the conference and the PC members and external reviewers for their expertise and help in evaluating the submissions.

February 2017

K. Selçuk Candan
Lei Chen
Torben Bach Pedersen

Organization

General Co-chairs

Karl Aberer   EPFL, Switzerland
Peter Scheuermann   Northwestern University, USA
Kai Zheng   Soochow University, China

Program Committee Co-chairs

Selçuk Candan   Arizona State University, USA
Lei Chen   HKUST, Hong Kong, SAR China
Torben Bach Pedersen   Aalborg University, Denmark

Workshops Co-chairs

Zhifeng Bao   RMIT, Australia
Goce Trajcevski   Northwestern University, USA

Industrial/Practitioners Track Co-chairs

Nicholas Jing Yuan   Microsoft, China
Georgia Koutrika   HP Labs, USA

Demo Track Co-chairs

Meihui Zhang   Singapore University of Technology and Design, Singapore
Wook-Shin Han   POSTECH University, South Korea

Tutorial Co-chairs

Katja Hose   Aalborg University, Denmark
Huiping Cao   New Mexico State University, USA

Panel Chair

Xuemin Lin   University of New South Wales, Australia

Proceedings Co-chairs

Lijun Chang   University of New South Wales, Australia
Wen Hua   University of Queensland, Australia

Publicity Co-chairs

Bin Yang   Aalborg University, Denmark
Xiang Lian   University of Texas-Pan American, USA
Maria Luisa Sapino   University of Turin, Italy

Local Organization Co-chairs

Zhixu Li   Soochow University, China
Jiajie Xu   Soochow University, China

Steering Committee Liaison

Xiaofang Zhou   University of Queensland, Australia

Conference Secretary

Yan Zhao   Soochow University, China

Webmaster

Yang Li   Soochow University, China

Program Committee

Amr El Abbadi   University of California at Santa Barbara, USA
Alberto Abello   UPC Barcelona, Spain
Divyakant Agrawal   University of California at Santa Barbara, USA
Marco Aldinucci   University of Turin, Italy
Ira Assent   Aarhus University, Denmark
Rafael Berlanga Llavori   University Jaume I, Spain
Francescho Bonchi   ISI Foundation, Italy
Selcuk Candan   Arizona State University, USA
Huiping Cao   New Mexico State University, USA
Barbara Catania   Università di Genova, Italy
Qun Chen   Northwestern Polytechnical University, China
Reynold Cheng   University of Hong Kong, SAR China
Wonik Choi   Inha University, South Korea
Gao Cong   Nanyang Technological University, Singapore
Bin Cui   Peiking University, China
Lars Dannecker   SAP, Germany
Hasan Davulcu   Arizona State University, USA
Ugur Demiryurek   University of Southern California, USA
Francesco Di Mauro   University of Turin, Italy
Curtis Dyreson   Utah State University, USA
Hakan Ferhatosmanoglu   Bilkent University, Turkey

Elena Ferrari	Università dell'Insubria, Italy
Johann Gamper	Free University of Bozen-Bolzano, Italy
Hong Gao	Harbin Institute of Technology, China
Yunjun Gao	Zhejiang University, China
Yash Garg	Arizona State University, USA
Lukasz Golab	University of Waterloo, Canada
Le Gruenwald	University of Oklahoma, USA
Ismail Hakki Toroslu	Middle East Technical University, Turkey
Jingrui He	Arizona State University, USA
Yoshiharu Ishikawa	Nagoya University, Japan
Linnan Jiang	HKUST, SAR China
Cheqing Jin	East China Normal University, China
Alekh Jindal	MIT, USA
Sungwon Jung Jung	Sogang University, South Korea
Arijit Khan	Nanyang Technological University, Singapore
Latifur Khan	University of Texas at Dallas, USA
Deok-Hwan Kim	Inha University, South Korea
Jinho Kim	Kangwon National University, South Korea
Jong Wook Kim	Sangmyung University, South Korea
Jung Hyun Kim	Arizona State University, USA
Sang-Wook Kim	Hanyang University, South Korea
Peer Kroger	LMU Munich, Germany
Anne Laurent	Université de Montpellier II, France
Wookey Lee	Inha University, South Korea
Young-Koo Lee	Kyung Hee University, South Korea
Chengkai Li	University of Texas at Arlington, USA
Feifei Li	University of Utah, USA
Guoliang Li	Tsinghua University, China
Zhixu Li	Soochow University, China
Xiang Lian	Kent State University, USA
Eric Lo	Chinese University of Hong Kong, SAR China
Woong-Kee Loh	Gachon University, South Korea
Hua Lu	Aalborg University, Denmark
Nikos Mamoulis	University of Ioannina, Greece/University of Hong Kong, SAR China
Ioana Manolescu	Inria, France
Rui Meng	HKUST, SAR China
Parth Nagarkar	Arizona State University, USA
Yunmook Nah	Dankook University, South Korea
Kjetil Norvag	Norwegian University of Science and Technology, Norway
Sarana Yi Nutanong	City University of Hong Kong, SAR China
Vincent Oria	New Jersey Institute of Technology, USA
Paolo Papotti	Arizona State University, USA
Torben Bach Pedersen	Aalborg University, Denmark
Ruggero Pensa	University of Turin, Italy

Dieter Pfoser	George Mason University, USA
Evaggelia Pitoura	University of Ioannina, Greece
Silvestro Poccia	University of Turin, Italy
Weixiong Rao	Tongji University, China
Matthias Renz	George Mason University, USA
Oscar Romero	UPC Barcelona, Spain
Florin Rusu	University of California Merced, USA
Simonas Saltenis	Aalborg University, Denmark
Maria Luisa Sapino	University of Turin, Italy
Claudio Schifanella	University of Turin, Italy
Cyrus Shahabi	University of Southern California, USA
Jieying She	HKUST, SAR China
Hengtao Shen	University of Queensland, China
Yanyan Shen	Shanghai Jiao Tong University, China
Alkis Simitsis	HP Labs, USA
Shaoxu Song	Tsinghua University, China
Yangqiu Song	HKUST, SAR China
Xiaoshuai Sun	University of Queensland, Australia
Letizia Tanca	Politecnico di Milano, Italy
Nan Tang	Qatar Computing Research Institute, Qatar
Egemen Tanin	University of Melbourne, Australia
Junichi Tatemura	Google, USA
Christian Thomsen	Aalborg University, Denmark
Hanghang Tong	Arizona State University, USA
Yongxin Tong	Beihang University, China
Panos Vassiliadis	University of Ioannina, Greece
Sabrina De Capitani Vimercati	University of Milan, Italy
Bin Wang	Northeastern University, China
Wei Wang	National University of Singapore, Singapore
Xin Wang	Tianjin University, China
John Wu	Lawrence Berkeley Lab, USA
Xiaokui Xiao	Nanyang Technological University, Singapore
Xike Xie	University of Science and Technology of China, China
Jianliang Xu	Hong Kong Bapatist University, SAR China
Jeffrey Xu Yu	Chinese University of Hong Kong, China
Xiaochun Yang	Northeastern University, China
Bin Yao	Shanghai Jiao Tong University, China
Hongzhi Yin	University of Queensland, Australia
Man Lung Yiu	Hong Kong Polytechnic, SAR China
Yi Yu	National Institute of Informatics, Japan
Ye Yuan	Northeastern University, China
Meihui Zhang	Singapore University of Technology and Design, Singapore
Wenjie Zhang	University of New South Wales, Australia
Ying Zhang	University of Technology Sydney, Australia

Zhengjie Zhang   Advanced Digital Sciences Center, Singapore
Xiangmin Zhou   RMIT, Australia
Yongluan Zhou   University of Southern Denmark, Denmark
Lei Zhu   University of Queensland, Australia
Esteban Zimanyi   Université Libre de Bruxelles, Belgium
Andreas Zufle   George Mason University, USA

Industry Track Program Committee

Akhil Arora   Xerox Research Centre, India
Jie Bao   Microsoft, China
Senjuti Basu Roy   UW Tacoma, USA
Neil Zhenqiang Gong   Iowa State University, USA
Defu Lian   University of Electronic Science and Technology,
                              China
Qi Liu   University of Science and Technology of China, China
Alkis Simitsis   HP Labs, USA
Kostas Stefanidis   University of Tampere, Finland
Lu-An Tang   NEC Lab, USA
Fuzheng Zhang   Microsoft, China
Hengshu Zhu   Baidu, China

Demo Track Program Committee

Jinha Kim   Oracle Labs, USA
Xuan Liu   Baidu, China
Yanyan Shen   Shanghai Jiao Tong University, China
Yongxin Tong   Beihang University, China

Additional Reviewers

Jinpeng Chen   Beihang University, China
Xilun Chen   Arizona State University, USA
Yu Cheng   Turn Inc., USA
Alexander Crosdale   UC Merced, USA
Tiziano De Matteis   University of Pisa, Italy
Vasilis Efthymiou   University of Crete, Greece
Roberto Esposito   University of Turin, Italy
Yixiang Fang   The University of Hong Kong, SAR China
Christian Frey   LMU Munich, Germany
Yash Garg   Arizona State University, USA
Concorde Habineza   George Mason University, USA
Jiafeng Hu   The University of Hong Kong, SAR China
Zhiyi Huang   UC Merced, USA
Zhipeng Huang   The University of Hong Kong, SAR China
Shengyu Huang   Arizona State University, USA

Contents – Part I

Trajectory and Time Series Data Processing

Data Mining

Query Processing and Optimization (I)

Text Mining

Recommendation

Security, Privacy, Senor and Cloud

Social Network Analytics (I)

Tutorials

Contents – Part II

Big Data (Industrial)

Social Networks and Graphs (Industrial)

Demos

Semantic Web and Knowledge Management

A General Fine-Grained Truth Discovery Approach for Crowdsourced Data Aggregation

Yang Du[1]([✉]), Hongli Xu[1], Yu-E Sun[2], and Liusheng Huang[1]

[1] School of Computer Science and Technology,
University of Science and Technology of China, Hefei, China
jannr@mail.ustc.edu.cn, {xuhongli,lshuang}@ustc.edu.cn
[2] School of Urban Rail Transportation, Soochow University, Suzhou, China
sunye12@suda.edu.cn

Abstract. Crowdsourcing has been proven to be an efficient tool to collect large-scale datasets. Answers provided by the crowds are often noisy and conflicted, which makes aggregating them to infer ground truth a critical challenge. Existing fine-grained truth discovery methods solve this problem by exploring the correlation between source reliability and task topics or answers. However, they can only work on limited tasks, which results in the incompatibility with Writing tasks and Transcription tasks, along with the insufficient utilization of the global dataset. To maintain compatibility, we consider the existence of clusters in both tasks and sources, then propose a general fine-grained method. The proposed approach contains two integral components: kl-means and Pattern-based Truth Discovery (PTD). With the aid of ground truth data, kl-means directly employs a co-clustering reliability model on the correctness matrix to learn the patterns. Then PTD conducts the answer aggregation by incorporating captured patterns, producing a more accurate estimation. Therefore, our approach is compatible with all tasks and can better demonstrate the correlation among tasks and sources. Experimental results show that our method can produce a more precise estimation than other general truth discovery methods due to its ability to learn and utilize the patterns of both tasks and sources.

1 Introduction

In recent years, producing large-scale datasets has been vital for both research and industry. Traditional strategies, mostly hiring motivated experts, are not competent enough when collecting datasets like ImageNet [1] and TinyImages [2] which contain millions of samples. On the other hand, crowdsourcing services, like AMT (Amazon Mechanical Turk) [3] or CrowdFlower [4], offer a convenient way to collect datasets by distributing tasks to workers (sources) all over the world.

For its efficiency and effectiveness in producing large-scale datasets, crowdsourcing has become increasingly popular. As sources are not necessarily experts, tasks are often distributed more than once, which results in the noisy and conflicted answers. Therefore, how to aggregate answers and infer the ground truth has been a critical challenge in crowdsourcing.

© Springer International Publishing AG 2017
S. Candan et al. (Eds.): DASFAA 2017, Part I, LNCS 10177, pp. 3–18, 2017.
DOI: 10.1007/978-3-319-55753-3_1

An intuitive solution to this challenge is majority voting, which selects the answer with most votes as the final output. However, it overlooks the rugged reliability of sources and shows poor performance when the number of low-quality sources exceeds that of high-quality ones. To solve this problem, a set of weighted voting approaches was proposed [5–7]. Regardless of the differences in their models, their principles are assigning larger weights to sources with higher reliability and outputting the answers with largest weights. Nevertheless, these approaches assume that each source's reliability is consistent with all tasks, which is not true, as everyone has his areas of expertise.

Given the shortcomings of above approaches, current efforts try to utilize sources' fine-grained reliability. Some of them divide tasks into topical-level clusters and estimate source's topical reliability, like FaitCrowd [8], which employs a topic model on task description and divides tasks into groups. Others attempt to model source's reliability for different answers [9,10]. However, a common drawback of these methods is their limited compatibility with tasks.

Table 1. Crowdsourcing task templates in AMT

Task form	Task description		Candidate answers	
	task-specific	non-specific	task-specific	non-specific
Image tagging/Moderation		✓		✓
Writing		✓	✓	
Data collection	✓		✓	
Transcription		✓	✓	
Sentiment wizard	✓			✓
Categorization wizard	✓			✓

To better demonstrate the incompatibility issues of existing fine-grained approaches, we consider the task templates in AMT and propose a rough classification for their *task description* and *candidate answers*. As Table 1 shows, we divide each feature into two groups: *task-specific* or *non-specific*. For example, the *task description* of Data Collection tasks, like *"What is the height of the Mount Everest?"*, is labeled as *task-specific* for that each description is targeted at a specific task. In contrast, the *task description* of Image Tagging tasks, like *"Is there a duck in the picture?"*, which can be used for multiple images, is labeled as *non-specific*. The classification process of *candidate answer* is similar. Consider a typical Image Tagging task which asks workers to answer whether they see birds in the picture. We label the *candidate answers* of this task as *non-specific* since they can be shared among a set of Image Tagging tasks.

We find out that current fine-grained approaches are only competent for limited tasks in AMT. For instance, the approaches which divide tasks into topical-level groups, like FaitCrowd, can only work with the tasks whose description can reflect their topics. Therefore, those approaches can only perform well on tasks with the *task-specific* description but are not competent for those with *non-specific* description. Other approaches, which model fine-grained reliability for different answers, limit the tasks to those whose candidate answers are *non-specific*. As consequences, (1) none of these approaches can deal with the Writing tasks or the Transcription tasks, and (2) they can only utilize a subset of the global dataset to help infer the ground truth. Therefore, it is crucial to design a general fine-grained truth discovery approach.

To tackle with challenges above, we consider the clusters in both tasks and sources. It has been observed that members of the same task cluster or source cluster share identical patterns. For instance, the pattern shared in a source cluster is sources' rugged reliability levels for different task clusters. More specifically, sources can have different reliability levels for various task clusters, but for a particular task cluster, sources coming from the same cluster show an equal reliability level. The patterns of task clusters are similar to that of source clusters. Therefore, we can first learn the patterns of task clusters and source clusters, then derive correct answers for target tasks based on the learned patterns.

In this paper, we propose a general fine-grained truth discovery approach to aggregate crowdsourced answers. The proposed method contains two integral components: kl-means and Pattern-based Truth Discovery (PTD). With the aid of standard tasks whose ground truth are known, kl-means directly employs a co-clustering reliability model on the correctness matrix to learn the patterns of sources and tasks simultaneously. Then PTD further uses captured patterns to help infer the ground truth. Since these two components do not have any additional requirements for tasks, like the *task-specific* description or *non-specific* answers, our approach is compatible with all tasks. To the best of our knowledge, this is the first work to propose a general fine-grained method with consideration of the patterns of task clusters and source clusters. One important feature of our approach is the utilization of co-clustering reliability model and standard tasks, which allows us to learn the cluster patterns directly from the correctness matrix. Therefore, our method can better demonstrate the relationship between tasks and sources and produce more accurate inference.

In summary, three main contributions of this work are as follows:

(1) To the best of our knowledge, we are the first to recognize the compatibility issues of state-of-art fine-grained truth discovery methods and propose a general one to aggregate noisy crowdsourced answers.
(2) We employ a co-clustering reliability model to learn the patterns of task clusters and source clusters simultaneously and use captured patterns to derive a more accurate inference of ground truth.
(3) We empirically show that the proposed method outperforms existing general truth discovery methods.

2   Problem Formulation

In this section, we first introduce some basic notations used in this paper, then give a formal definition of our problem.

The inputs of our method are m standard tasks $T = \{t_i\}_1^m$, m' target tasks $T' = \{t_i\}_{m+1}^{m+m'}$, n sources $U = \{u_j\}_1^n$, answers matrix $A = \{a_{ij}\}_{i=1,j=1}^{m+m',n}$ and correctness matrix $R = \{r_{ij}\}_{i=1,j=1}^{m,n}$, where standard tasks are those whose ground truth we already know, the target tasks are those whose ground truth we want to find out.

Definition 1. *All tasks are single-choice questions. For task t_i, it has B candidate answers $c_i = \{c_{i1}, c_{i2} \cdots c_{iB}\}$, and its ground truth is denoted by $c_i^* \in c_i$.*

Definition 2. *Answer element a_{ij} denotes the answer given by source u_j to task t_i. $a_{ij} = b, 1 \leq b \leq B$ means source u_j select task t_i's candidate answer c_{ib} as his answer, $a_{ij} = 0$ means source u_j has not answered task t_i.*

Definition 3. *Correctness element r_{ij} denotes the correctness of answer given by source u_j to standard task t_i. $r_{ij} = 1$ means source u_j answers task t_i correctly, $r_{ij} = 0$ means his answer is incorrect, while $r_{ij} = -1$ means source u_j has not answered task t_i.*

Our goal is to derive estimation of target tasks' ground truth $E = \{e_i\}_{m+1}^{m+m'}$, k task clusters $I = \{I_p\}_1^k$, l source clusters $\{J_q\}_1^l$ and co-clustering reliability matrix $\{\eta_{pq}\}_{p=1,q=1}^{k,l}$.

Definition 4. *Estimated ground truth e_i for task t_i is the most accurate answer provided by sources.*

Definition 5. *Co-clustering reliability matrix $\eta^{k \times l}$ is referred as the reliability levels of l source clusters on k task clusters.*

Definition 6. *Sources in same source cluster share an identical pattern on task clusters, and a source's reliability is consistent on the same task cluster.*

Based on these definitions, we can formally define our problem as: given m standard tasks $T = \{t_i\}_1^m$, m' target tasks $T' = \{t_i\}_{m+1}^{m+m'}$, n sources $U = \{u_j\}_1^n$, answers matrix $A = \{a_{ij}\}_{i=1,q=1}^{m+m',n}$ and correctness matrix $R = \{r_{ij}\}_{i=1,q=1}^{m,n}$, our goals are learning k task clusters $I = \{I_p\}_1^k$, l source clusters $\{J_q\}_1^l$, co-clustering reliability matrix $\{\eta_{pq}\}_{p=1,q=1}^{k,l}$ and producing estimation of target tasks' true answers $E = \{e_i\}_{m+1}^{m+m'}$.

3 Co-clustering Reliability Model

The basic idea of proposed approach is to build a co-clustering reliability model on the correctness matrix, which is based on the observation that there exist clusters in both sources and tasks. For each source cluster, its members share same reliability level on tasks belonging to the same task cluster, so do the task clusters.

Therefore, we can divide tasks into k clusters, sources into l clusters by the following functions:

$$\rho : \{t_1, t_2 \cdots t_m\} \rightarrow \{1, 2 \cdots k\}$$
$$\gamma : \{u_1, u_2 \cdots u_n\} \rightarrow \{1, 2 \cdots l\}, \tag{1}$$

where $\rho(t_i) = r$ implies that task t_i is in task cluster r, $\gamma(u_j) = s$ implies that source u_j is in source cluster s.

Let $I = \{I_p\}_1^k$ denote the k task clusters and $J = \{J_q\}_1^l$ denote the source clusters. A co-cluster is referred as the submatrix of correctness matrix R determined by a task cluster I_p and a source cluster J_q.

We evaluate the homogeneity of co-cluster by the sum of squared differences between each entry and the mean of this co-cluster which was used by Hartigan et al. [15]. Consider that the correctness matrix could be highly sparse since it is unnecessary for sources to perform all tasks. We compute the residue of the element r_{ij} differently based on whether source u_j has answered task t_i. Let value -1 represent the unfilled items in the correctness matrix. Formally:

$$h_{ij} = \begin{cases} r_{ij} - \eta_{pq}, & \text{if } r_{ij} \neq -1 \\ 0, & \text{otherwise.} \end{cases} \tag{2}$$

When source u_j has answered task t_i, the residue of r_{ij} is $h_{ij} = r_{ij} - \eta_{pq}$, where p is the cluster label of task t_i, q is the cluster label of source u_j, η_{pq} is the mean of submatrix determined by I_p and J_q. But, when source u_j has not answered task t_i, we set h_{ij} to 0, which in fact ignores the loss caused by unfilled elements.

The objective of this co-clustering reliability model is to learn the patterns of both tasks and sources. Hence, the objective function is to minimize the total squared residues of all co-clusters, which is shown as follows.

$$L = \sum_{I_p, J_q} L_{pq} = \sum_{I_p, J_q} \sum_{i \in I_p, j \in J_q} h_{ij}^2, \tag{3}$$

where I_p, J_q enumerates all co-clusters, L_{pq} represents the sum squared residues of co-cluster determined by I_p and J_q.

4   Algorithms

The proposed approach estimates the ground truth by employing two integral components: (1) using kl-means to learn the co-clustering reliability model from sources performance on standard tasks, which contains the patterns of clusters, and (2) infer the ground truth of target tasks with the help of capture patterns.

4.1 kl-means

We propose kl-means, an extended version of the k-means algorithm, to learn co-clustering reliability of tasks and sources. Differing from previous work

---

Algorithm 1. kl-means

INPUT   : Correctness matrix C, Task cluster number k, Source cluster number l

OUTPUT: task clusters $\{I_p\}_{p=1}^k$, worker clusters $\{J_q\}_{q=1}^l$, co-cluster centers $\{\eta_{pq}\}_{p=1,q=1}^{k,l}$

1 **begin** Initialization
2   Randomly generate k task cluster centers $\{\eta_{p*}|\eta_{p*} \in [0,1]^n\}_{p=1}^k$;
3   Assign tasks to their closest task clusters using Equation 4a;
4   Randomly generate l source cluster centers $\{\eta_{*q}|\eta_{*q} \in [0,1]^m\}_{q=1}^l$;
5   Assign sources to their closest source clusters using Equation 4b;
6   Construct task clusters $\{I_p\}_1^k$ and source clusters $\{J_q\}_1^l$ using Equation 6;
7   Estimate co-cluster centers $\{\eta_{pq}\}_{p=1,q=1}^{k,l}$ using Equation 7;
8 **end**
9 **begin** Iteratively Learning
10   **while** *not converge* **do**
11   Update task cluster centers $\{\eta_{p*}\}_{p=1}^k$ using Equation 8a;
12   Update source cluster centers $\{\eta_{*q}\}_{q=1}^l$ using Equation 8b;
13   Update cluster labels of tasks and sources using Eq. 4a and 4b;
14   Update task clusters and source clusters using Equation 6;
15   Update co-cluster centers using Equation 7;
16   **end**
17 **end**

CoClus [11] which ignores the situation when matrix could be sparse, kl-means treats unfilled elements as perfect elements in their corresponding co-clusters, which makes it compatible with sparse matrix.

Like k-means algorithm, kl-means is composed of a random initialization phase and an iteratively learning phase. It first randomly initializes the task cluster labels $\{\rho(t_i)\}_1^m$, source cluster labels $\{\gamma(u_j)\}_1^m$, co-cluster centers $\{\eta_{pq}\}_{p=1,q=1}^{k,l}$. Based on the initialization results, kl-means iteratively performs re-assignments for task/source cluster labels and co-cluster centers. This iteration learning phase will not end until the objective function converges. The detailed kl-means algorithm is showed as follows:

In the initialization phase, kl-means starts with randomly generating tasks cluster centers $\{\eta_{p*}|\eta_{p*} \in [0,1]^n\}_{p=1}^k$ and source cluster centers $\{\eta_{*q}|\eta_{*q} \in [0,1]^m\}_{q=1}^l$. Then kl-means assigns cluster labels to tasks and sources using Eqs. 4a and 4b:

$$\rho(t_i) = \arg\min_p \{d(t_i, \eta_{p*})\} \tag{4a}$$

$$\gamma(u_j) = \arg\min_q \{d(u_j, \eta_{*q})\}, \tag{4b}$$

where $d(t_i, \eta_{p*})$ (Eq. 5a) is referred as the distance between task t_i and the center of task cluster I_p, $d(u_j, \eta_{*q})$ (Eq. 5b) is referred as the distance between source u_j and the center of source cluster J_q.

$$d(t_i, \eta_{p*}) = \sum_{u_j \in U} \{\mathbb{1}\{r_{ij} \neq -1\} \times (r_{ij} - [\eta_{p*}]_j)^2\} \tag{5a}$$

$$d(u_j, \eta_{*q}) = \sum_{t_i \in T} \{\mathbb{1}\{r_{ij} \neq -1\} \times (r_{ij} - [\eta_{*q}]_i)^2\} \tag{5b}$$

Note that, kl-means treats unfilled elements by ignoring them. Equation 5a uses function $\mathbb{1}\{r_{ij} \neq -1\}$, to filter the empty elements in task' answers, which returns 1 when r_{ij} not equals -1, returns 0 otherwise. Then it computes the sum squared differences between the crowdsourced answers of t_i and center I_p as their distance, where $[\eta_{p*}]_j$ denotes the j-th element in array η_{p*}. Equation 5b computes the distance between sources and source cluster centers in a similar way.

After initializing cluster labels, kl-means constructs task clusters and source clusters for the first time:

$$\begin{aligned} I_p &= \{t_i | \rho(t_i) = p, t_i \in T\} \\ J_q &= \{u_j | \gamma(u_j) = q, u_j \in U\} \end{aligned} \tag{6}$$

At the last of initialization phase, kl-means computes the center of co-clusters using Eq. 7.

$$\eta_{pq} = \frac{\sum_{i \in I_p, j \in J_q} (\mathbb{1}\{r_{ij} \neq -1\} \times r_{ij})}{\sum_{i \in I_p, j \in J_q} \mathbb{1}\{r_{ij} \neq -1\}} \tag{7}$$

For now, kl-means succeeds to produce cluster labels for both tasks and sources, along with co-cluster centers, then it applies iteration phases to find a local optimal solution.

In each iteration, kl-means starts by re-computing the centers of task clusters and the centers of source clusters using Eqs. 8a and 8b, then it updates the cluster labels and re-centers the co-clusters to minimize the objective function. For task cluster I_p, Eq. 8a constructs the task cluster center η_{p*} by setting expected reliability level for each dimension. More specifically, for j-th dimension of η_{p*} which represents the expected reliability of source u_j' on task cluster I_p, its value will be set to $\eta_{p\gamma(u_j)}$. In other words, $[\eta_{p*}]_j$ shall be fixed to the mean of the co-cluster determined by I_p and $\gamma(u_j)$.

$$\eta_{p*} = [\eta_{p\gamma(u_1)}, \eta_{p\gamma(u_2)} \cdots \eta_{p\gamma(u_n)}] \tag{8a}$$

$$\eta_{*q} = [\eta_{\rho(t_1)q}, \eta_{\rho(t_2)q} \cdots \eta_{\rho(t_m)q}] \tag{8b}$$

Based on updated cluster centers η_{p*} and η_{*q}, kl-means updates the cluster labels of tasks and sources, and further estimates the co-cluster centers. Above iterative steps are performed until the loss value converges. By Lemmas 1 and 2, we prove that this iterative learning procedure can always converge for that the loss value decreases when kl-means algorithm runs.

Lemma 1. *For any set $X \subset \mathbb{R}^d$, and any $z \in \mathbb{R}^d$. Let $E(X)$ be the mean of set X and let $cost = \sum_{x \in X} \|x - z\|^2$ be the sum of squared residues between set X and value z.*

$$cost(X, E(X)) <= cost(X, z)$$

Proof. By replacing $x-z$ as $x-E(X)+E(X)-z$, we can get following derivation:

$$cost(X, z) = \sum_{x \in X} \|(x - E(X)) + (E(X) - z)\|^2$$

$$= cost(X, E(x)) + |X| \cdot \|E(X) - z\|^2$$

Therefore, $z = E(X)$ is the value which can minimize $cost(X, z)$.

Lemma 2. *During the kl-means algorithm, the objective value monotonically decreases.*

Proof. Let $I^{(t)} = \{I_p^{(t)}\}_{p=1}^k$ be the task clusters, $J^{(t)} = \{J_q\}_{q=1}^l$ be the source clusters, $\eta^{(t)} = \{\eta_{pq}^{(t)}\}_{p=1,q=1}^{k,l}$ be the co-cluster centers at the start of the t-th iteration of kl-means. By using Eqs. 8a and 8b we can get task cluster centers $\{\eta_{p*}^{(t)}\}_{p=1}^k$ and source cluster centers $\{\eta_{*q}^{(t)}\}_{q=1}^l$ at t-th iteration. Let $loss(t_i, I^{(t)}, \eta^{(t)})$ denote the loss caused by task t_i when task clusters are $I^{(t)}$ and co-cluster centers are $I^{(t)}$, $loss(u_j)^{(t)}$ denote the loss caused by source u_j at the beginning of t-th iteration, which can be calculated by following functions:

$$loss(t_i, I, J, \eta) = \sum_{u_j \in U} h_{ij}^2 \tag{9a}$$

$$loss(u_j, I, J, \eta) = \sum_{t_i \in T} h_{ij}^2, \tag{9b}$$

where h_{ij} denote the loss caused by answer r_{ij}. As stated in Eq. 2, h_{ij} can be computed by $\mathbb{1}\{r_{ij} \neq -1\} \times (r_{ij} - \eta_{\rho(t_i)\gamma(u_j)})$. Notice that, $\rho(t_i)$ denotes the cluster label of task t_i when task clusters are I, $\gamma(u_j)$ represents the cluster label of source u_j when the source clusters are J. Then the objective function can be rewritten as:

$$L(I, J, \eta) = \sum_{I,J} \sum_{i \in I, j \in J} h_{ij}^2$$

$$= \sum_{t_i \in T} loss(t_i, I, J, \eta) = \sum_{u_j \in U} loss(u_j, I, J, \eta)$$

The first step of the iteration assigns each task to its closest center, where each re-assignment in fact decreases the loss caused by re-assigned task. Therefore, the loss values caused by t_i in t-th iteration and $(t+1)$-iteration always satisfies: $loss(t_i, I^{(t)}, J, \eta) \geq loss(t_i, I^{(t+1)}, J, \eta), t_i \in T$, which makes:

$$L(I^{(t)}, J^{(t)}, \eta^{(t)}) = \sum_{t_i \in T} loss(t_i, I^{(t)}, J^{(t)}, \eta^{(t)})$$

$$\geq \sum_{t_i \in T} loss(t_i, I^{(t+1)}, J^{(t)}, \eta^{(t)}) = L(I^{(t+1)}, J^{(t)}, \eta^{(t)})$$

By comparing the loss values caused by sources before and after the re-assignments of source labels, we can prove that the objective value before re-assigning source labels is not less than that after re-assignment, which means

$L(I^{(t+1)}, J^{(t)}, \eta^{(t)}) \geq L(I^{(t+1)}, J^{(t+1)}, \eta^{(t)})$. Furthermore, we can prove that the loss values decreases after re-center co-clusters by using Lemma 1. Therefore, $L(I^{(t+1)}, J^{(t+1)}, \eta^{(t)}) \geq L(I^{(t+1)}, J^{(t+1)}, \eta^{(t+1)})$.

Above all, the objective value at the start of t-th iteration is no larger than that at the start of $(t + 1)$-th iteration. Formally, $L(I^{(t)}, J^{(t)}, \eta^{(t)}) \geq L(I^{(t+1)}, J^{(t+1)}, \eta^{(t+1)})$. Therefore, the objective value monotonically decreases during the kl-means algorithm.

4.2   Pattern-Based Truth Discovery

Based on the output of kl-means, we design a pattern-based approach to aggregate noisy crowdsourced answers. The proposed algorithm takes target tasks $\{t_i\}_{i=m+1}^{m+m'}$, answer matrix $\{a_{ij}\}_{i=m+1,j=1}^{m+m',n}$, source clusters $\{J_q\}_{q=1}^{l}$ and co-cluster centers $\{\eta_{pq}\}_{p=1,q=1}^{k,l}$ as inputs, returns estimated ground truth of target tasks $E = \{e_i\}_{m+1}^{m+m'}$ as outputs. The detailed algorithm as follows:

Algorithm 2. Pattern-based Truth Discovery

INPUT   : Target tasks $T' = \{t_i\}_{i=m+1}^{m+m'}$, Answer matrix $\{a_{ij}\}_{i=m+1,j=1}^{m+m',n}$, Task clusters $I = \{I_p\}_{p=1}^{k}$, Source clusters $J = \{J_q\}_{q=1}^{l}$, Co-cluster centers $\{\eta_{pq}\}_{p=1,q=1}^{k,l}$

OUTPUT: Estimation of ground truth $E = \{e_i\}_{m+1}^{m+m'}$

1 **foreach** $t_i \in T'$ **do** infer the ground truth of task t_i
2 | **foreach** $I_p \in I$ **do** assume task t_i belongs to cluster I_p
3 | | **foreach** $c_{ib} \in c_i$ **do** assume c_{ib} is the ground truth of t_i
4 | | | Calculate expected loss when $t_i \in I_p$ and c_{ib} is the ground truth using Equation 10;
5 | | **end**
6 | **end**
7 | Select the best parameter (p, c_{ib}) minimizing the loss value;
8 | Select the assumptive ground truth c_{ib} as the ground truth of t_i;
9 **end**

For each target task t_i, we enumerate every possible combination of its cluster label and ground truth, denoted by p and e^*. Then we compute the expected loss value of t_i for every combination (p, e^*) using Eq. 10, which sums up the expected squared residues between the sources' correctness and their expected reliability.

$$loss^*(t_i, e^*, p, J, \eta) = \sum_{u_j \in U} (h_{ij}^*(e^*, p))^2 \qquad (10)$$

In Eq. 11, we compare the difference between u_j's answer a_{ij} and expected ground truth e^* and compute the residue of them. When a_{ij} is valid, we set the residue to $1 - \eta_{p\gamma(u_j)}$ if a_{ij} and e_i are identical, set it to $0 - \eta_{p\gamma(u_j)}$ if they are not. When a_{ij} is invalid, the residue is set to 0.

$$h_{ij}^*(e^*, p) = \begin{cases} 1 - \eta_{p\gamma(u_j)}, & \text{if } a_{ij} > 0 \text{ and } a_{ij} = e^*; \\ 0 - \eta_{p\gamma(u_j)}, & \text{if } a_{ij} > 0 \text{ and } a_{ij} \neq e^*; \\ 0, & \text{otherwise.} \end{cases} \qquad (11)$$

In Eq. 12, notation c_i denotes the candidate answer set of task t_i. By enumerating every possible cluster label p and candidate answer e^* for task t_i, we can select the best parameter which minimizes the loss value, then return the best candidate answer $e^* \in c_i$ as inferred correct answer.

$$\underset{p, e^* \in c_i}{\arg \min} \; loss^*(t_i, e^*, p, J, \eta) \qquad (12)$$

5   Experiments

In this section, we firstly use three matrices to evaluate the performance of kl-means on recognizing co-clusters, then we use three real datasets to compare our approach with two baselines. In the last part, we explore how the sparsity of datasets influences the performance of PTD and two comparison algorithms by performing them on two synthetic datasets.

5.1   Performance of kl-means

To explore the capacity of kl-means in recognizing co-clusters, we generate three binary matrices: C-Matrix (a chaos matrix), R-Matrix (a row-clustering matrix) and Co-Matrix (a co-clustering matrix), as showed in Table 2.

Table 2. Co-cluster mean values of C-Matrix, R-Matrix and Co-Matrix

(a) M_C: C-Matrix

	C1
R1	0.5

(b) M_R: R-Matrix

	C1
R1	0.25
R2	0.75

(c) M_O: Co-Matrix

	C1	C2
R1	0.25	0.75
R2	0.75	0.25

For C-Matrix, since it contains no clusters, we treat its rows as belonging to a single row group R1 and dividing its columns to a single column cluster C1. By this means, the chaos matrix is transformed to a matrix with exactly 1×1 co-cluster. As shown in Table 2(a), element 0.5 represents the mean of the co-cluster determined by R1 and C1. Therefore, all elements X in this co-cluster is generated by following distribution $X \sim Bernoulli(0.5)$. For row-clustering matrix (Table 2(b)), by dividing its columns to a single cluster, we transform it to a matrix containing 2×1 co-clusters, where the mean values are 0.25 and 0.75. Likewise, the co-clustering matrix (Table 2(c)) contains 2×2 co-clusters.

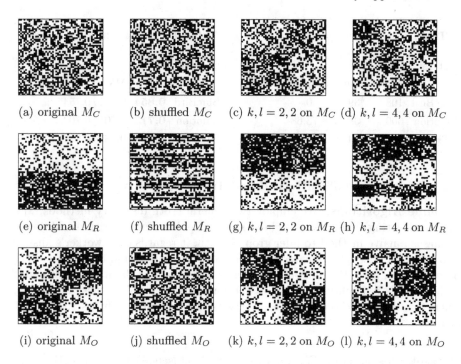

(a) original M_C (b) shuffled M_C (c) $k, l = 2, 2$ on M_C (d) $k, l = 4, 4$ on M_C

(e) original M_R (f) shuffled M_R (g) $k, l = 2, 2$ on M_R (h) $k, l = 4, 4$ on M_R

(i) original M_O (j) shuffled M_O (k) $k, l = 2, 2$ on M_O (l) $k, l = 4, 4$ on M_O

Fig. 1. Performance of kl-means on C-Matrix, R-Matrix and Co-Matrix

Following the above settings, we generate these three matrices composed of 50 rows and 50 columns respectively, and perform kl-means algorithm on them, as shown in Fig. 1. We use Table 1(a) to plot the original where the black points represent elements of 1 and the white points represent elements of 0. We shuffle the rows and columns of C-Matrix and plot the shuffled C-Matrix in Table 1(b). Then we apply kl-means on shuffled C-Matrix with parameters $k = 2, l = 2$ and $k = 4, l = 4$, whose co-clustering results are demonstrated in Fig. 1(c) and (d). As we can see, the co-clusters recognized by kl-means are quite confusing for that the original C-Matrix is a chaos matrix and there are no significant patterns in both rows and columns.

Plot the original R-Matrix with Fig. 1(e) and plot the shuffled R-Matrix with Fig. 1(f). We can easily distinguish the row clusters in the original R-Matrix, but it is not likely to identify them in shuffled one. After applying kl-means on shuffled R-Matrix, we get two co-clustered R-Matrix which are plotted in Fig. 1(g) and (h). In both figures, kl-means significantly recognized the row clusters, even if it falls to an over-fitting situation when trying to identify four different row clusters from R-Matrix. The performance of kl-means on Co-Matrix is shown in Fig. 1(k) and (l), where both row clusters and column clusters are significantly recognized. Above experimental results corroborate with our theoretical analysis that kl-means can effectively recognize co-clusters.

Table 3. Description of datasets

Dataset	Properties		
	Tasks	Sources	Sparsity
BlueBird	108	39	0
Emotion	600	38	73.68
RTE	800	164	93.90

Table 4. Comparison of algorithms

Dataset	Accuracy		
	PTD	MV	WV
BlueBird	**0.854**	0.741	0.754
Emotion	**0.713**	0.701	0.703
RTE	**0.918**	0.887	0.903

5.2 Performance of Pattern-Based Truth Discovery

Baseline Algorithms. We employ two general truth discovery methods, MV (majority voting) and WV (weighted voting), as baseline algorithms. Majority Voting is a naive method to infer ground truth, it regards each worker's answer as a vote and selects the answer with the most votes as ground truth. The other baseline, WV, takes the reliability of each worker into consideration. It estimates each worker's reliability based on his performance on finished tasks and takes. Then this method chooses the answer with the largest weight as the correct answer.

Datasets Description. We employ three real datasets: BlueBird, Emotion, and RTE. As showed in Table 3, BlueBird [12] contains 108 rows and 39 columns. The BlueBird dataset is collected by asking sources whether there is a bird in images, where an image is represented by a row and a source is represented by a column. Emotion [13] is a combination of 6 sub-datasets, respectively judging textual emotions like *anger, disgust, fear, joy, sadness* and *surprise*. The RTE dataset is proposed by Sheshadri et al. [14].

Comparison of Algorithms. Table 4 shows the performance of PTD and baseline algorithms regarding *accuracy* of predicting the ground truth. For BlueBird dataset, the accuracy of PTD improves by 11.3% compared with the worst baseline MV, rising by 10.0% compared with the best baseline WV. For Emotion dataset and RTE dataset, PTD slightly outperforms WV and MV. We see that the performance of PTD is better than that of WV and MV on all three datasets. That is because proposed PTD can learn and utilize the patterns of both task clusters and source clusters. Also, WV outperforms MV on all three datasets for its ability to capture sources' global reliability, which MV fails to do.

Influence of Sparsity Levels. To explore how the sparsity of datasets influences the performance of PTD, we generate two synthetic datasets based on the co-cluster properties extracted from BlueBird and RTE. Table 5(a) and (b) show the co-cluster reliability levels of two real datasets when the parameters of kl-means are $k = 2$ and $l = 4$.

Based on these tables, we generate two synthetic datasets with 600 tasks and 400 workers. For both synthetic datasets, it contains 2 task clusters and 4

Table 5. Co-clustering reliability matrix of two real datasets

Task Cluster	Worker Cluster			
	W1	W2	W3	W4
(a) BlueBird				
T1	0.883	0.896	0.537	0.247
T2	0.663	0.226	0.289	0.844
(b) RTE				
T1	0.956	0.923	0.718	0.251
T2	0.705	0.919	0.838	0.807

worker clusters, making 8 co-clusters. For a specific co-cluster, its elements are generated following distribution $X \sim Bernoulli(p)$, where p is the reliability level of this co-cluster.

By randomly dropping elements of original synthetic datasets, we can change the sparsity in two synthetic datasets. Table 6(a) shows the performance of PTD and two baselines on different sparsity levels. The accuracy of three algorithms decreases when the sparsity increases. From this table, we can see that PTD outperforms WV and MV, and WV beats MV, which accords with previous experimental results on real datasets. We can also observe similar results on synthetic RTE dataset. As shown in Table 6(b), the performance of three algorithms decreases when sparsity increases. Both PTD and WV outperforms MV, but WV can slightly outperform PTD by 0.4% when sparsity is 0.984. One possible reason is that the differences among task/source clusters in BlueBird are larger than those in RTE, making it easier to identify task/source clusters in BlueBird than in RTE. Therefore, the differences in performance between of PTD and WV in RTE are not as obvious as that in BlueBird.

6   Related Work

Some work conducts the crowdsourcing answer aggregation by incorporating the rugged reliability levels of sources [5–7]. David et al. [5] proposed a model in which each source's reliability is estimated, and they used the EM algorithm to infer the ground truth. ZenCrowd [7] also employed EM algorithm to estimate ground truth as well as source reliability. Dong et al. [6] considered copy behavior among sources, and proposed AccuPR to detect dependence for discovering the truth from conflicted information. Despite the differences in their models, they all followed the same principle of assigning larger weights to more reliable sources for estimating the ground truth. However, they ignored the differences among tasks and assumed that each source had same reliability for all tasks. Differently, our method recognizes the differences in individual's rugged reliability for different tasks and estimates sources' fine-grained reliability.

There is also some existing work exploring the sources' fine-grained reliability. FaitCrowd [8] employed topic model on task description and divided

Table 6. Comparison of algorithms on different sparsity levels

Sparsity	Accuracy		
	PTD	MV	WV
(a) Synthetic BlueBird			
0.0	**1.0**	0.663	0.742
0.5	**0.999**	0.675	0.744
0.75	**0.993**	0.689	0.747
0.875	**0.977**	0.705	0.754
0.937	**0.913**	0.706	0.753
0.968	**0.809**	0.683	0.731
0.984	**0.707**	0.648	0.673
(b) Synthetic RTE			
0.0	**1.0**	0.999	**1.0**
0.5	**1.0**	0.999	**1.0**
0.75	**1.0**	0.996	0.999
0.875	**1.0**	0.989	0.998
0.937	**0.998**	0.976	0.991
0.968	**0.981**	0.952	0.971
0.984	0.914	0.867	**0.918**

tasks into clusters, then estimated sources' reliability on each topical-level cluster and inferred the ground truth. Simpson et al. [10] proposed a Bayesian classifier combination method to produce higher accuracy classifications. Moreno et al. [9] considered the existence of user clusters to help conduct more accurate answer aggregation. We roughly classify these approaches into two categories; one tried to utilize the relationship between task clusters and task description, the other modeled sources' fine-grained reliability on different answers. However, they both have additional requirements for tasks, like *task-specific* description and *non-specific* candidate answers, and are only compatible with limited tasks. Distinct from these approaches, our method learns the patterns of task clusters and source clusters directly from correctness matrix and do not obtain additional limitations for tasks, making it a general method.

Truth discovery methods can also be divided into two classes based on the usage of ground truth data. One is those using unsupervised methods; the other uses semi-supervised methods. Approaches like FaitCrowd [8] and ZenCrowd [7] can be divided into the first class, trying to infer the ground truth and source reliability simultaneously. Methods using semi-supervised methods conducts the reliability estimation with the help of ground truth data. Shah et al. [17] employs "gold standard" tasks to help estimate the performance of sources. Yin et al. [16] proposes a semi-supervised approach to identify copy behavior and find correct values with the aid of ground truth data. In our paper, we also employ a semi-supervised approach by using *standard* tasks to help modeling co-clustering reliability.

7   Conclusions

In this paper, we propose a general fine-grained truth discovery approach to aggregate crowdsourced answers. By directly employing a co-clustering reliability model on correctness matrix, the proposed method learns the patterns of both task clusters and source clusters and further uses captured patterns to help produce more precise estimations, making it a general method. To the best of our knowledge, this is the first work to propose a general fine-grained method with consideration of the patterns of task clusters and source clusters. Experimental results show that our method can significantly detect the patterns, and outperform existing general truth discovery methods.

Acknowledgements. This paper was supported by National Natural Science Foundation of China under Grant No. U1301256,61472383, 61472385, 61672369 and 61572342, Natural Science Foundation of Jiangsu Province in China under No. BK20161257, BK20151240 and BK20161258, China Postdoctoral Science Foundation under Grant No. 2015M580470 and 2016M591920.

References

1. Deng, J., Dong, W., Socher, R., Li, L.-J., Li, K., Fei-Fei, L.: Imagenet: a large-scale hierarchical image database. In: IEEE Conference on Computer Vision and Pattern Recognition 2009, CVPR 2009, pp. 248–255. IEEE (2009)
2. Torralba, A., Fergus, R., Freeman, W.T.: 80 million tiny images: a large data set for nonparametric object and scene recognition. IEEE Trans. Pattern Anal. Mach. Intell. **30**, 1958–1970 (2008)
3. Gabriele, P., Jesse, C., Ipeirotis, P.G.: Running experiments on amazon mechanical turk. Judgment Decis. Making **5**, 411–419 (2010)
4. Tim, F., Will, M., Anand, K., Nicholas, K., Justin, M., Mark, D.: Annotating named entities in Twitter data with crowdsourcing. In: Proceedings of the NAACL HLT 2010 Workshop on Creating Speech and Language Data with Amazon's Mechanical Turk, pp. 80–88. Association for Computational Linguistics (2010)
5. Dawid, A.P., Skene, A.M.: Maximum likelihood estimation of observer error-rates using the EM algorithm. Appl. Stat. **28**, 20–28 (1979)
6. Dong, X.L., Berti-Equille, L., Srivastava, D.: Integrating conflicting data: the role of source dependence. Proc. VLDB Endowment **2**, 550–561 (2009)
7. Demartini, G., Difallah, D.E., Cudr-Mauroux, P.: ZenCrowd: leveraging probabilistic reasoning and crowdsourcing techniques for large-scale entity linking. In: Proceedings of the 21st International Conference on World Wide Web, pp. 469–478. ACM (2012)
8. Ma, F., Li, Y., Li, Q., Qiu, M., Gao, J., Zhi, S., Su, L., Zhao, B., Ji, H., Han, J.: Faitcrowd: fine grained truth discovery for crowdsourced data aggregation. In: Proceedings of the 21th ACM SIGKDD International Conference on Knowledge Discovery and Data Mining, pp. 745–754. ACM (2015)
9. Moreno, P.G., Artes-Rodriguez, A., Teh, Y.W., Perez-Cruz, F.: Bayesian nonparametric crowdsourcing. J. Mach. Learn. Res. **16**, 1607–1627 (2015)
10. Simpson, E., Roberts, S.J., Smith, A., Lintott, C.: Bayesian combination of multiple, imperfect classifiers (2011)

11. Cho, H., Dhillon, I.S., Guan, Y., Sra, S.: Minimum sum-squared residue co-clustering of gene expression data. In: Sdm, p. 3. SIAM (2004)

12. Welinder, P., Branson, S., Perona, P., Belongie, S.J.: The multidimensional wisdom of crowds. In: Advances in Neural Information Processing Systems, pp. 2424–2432 (2010)

13. Snow, R., O'Connor, B., Jurafsky, D., Ng, A.Y.: Cheap and fast–but is it good? Evaluating non-expert annotations for natural language tasks. In: Proceedings of the Conference on Empirical Methods in Natural Language Processing, pp. 254–263. Association for Computational Linguistics (2008)

14. Sheshadri, A., Lease, M.: SQUARE: a benchmark for research on computing crowd consensus. In: First AAAI Conference on Human Computation and Crowdsourcing (2013)

15. Hartigan, J.A.: Direct clustering of a data matrix. J. Am. Stat. Assoc. **67**, 123–129 (1972)

16. Yin, X., Tan, W.: Semi-supervised truth discovery. In: Proceedings of the 20th International Conference on World Wide Web, pp. 217–226. ACM (2011)

17. Shah, N.B., Zhou, D., Peres, Y.: Approval voting and incentives in crowdsourcing. In: International Conference on Machine Learning (ICML) (2015)

Learning the Structures of Online Asynchronous Conversations

Jun Chen[1], Chaokun Wang[1(✉)], Heran Lin[1], Weiping Wang[2],
Zhipeng Cai[3], and Jianmin Wang[1]

[1] School of Software, Tsinghua University, Beijing 100084, China
chenjun14@mails.thu.edu.cn, linhr10@gmail.com,
{chaokun,jimwang}@tsinghua.edu.cn
[2] Institute of Information Engineering, CAS, Beijing 100093, China
wangweiping@iie.ac.cn
[3] Department of Computer Science, Georgia State University,
Atlanta, GA 30302, USA
zcai@gsu.edu

Abstract. The online social networks have embraced huge success from the crowds in the last two decades. Now, more and more people get used to chat with friends online via instant messaging applications on personal computers or mobile devices. Since these conversations are sequentially organized, which fails to show the logical relations between messages, they are called asynchronous conversations in previous studies. Unfortunately, the sequential layouts of messages are usually not intuitive to see how the conversation evolves as time elapses. In this paper, we propose to learn the structures of online asynchronous conversations by predicting the "reply-to" relation between messages based on text similarity and latent semantic transferability. A heuristic method is also brought forward to predict the relation, and then recover the conversation structure. We demonstrate the effectiveness of the proposed method through experiments on a real-world web forum comment data set.

Keywords: Asynchronous conversations · Conversation structure · "Reply-to" relation

1 Introduction

With the blooming of Internet in the last two decades, social networks have embraced huge success from Internet users. From the early chatting room on webpage to the later instant messaging application on personal computers till nowadays' chatting APP on mobile devices, more and more people choose to chat with their friends online. The quantity of online conversations generated in a single day is very huge due to the easy access to the Internet world wide, which makes it possible for thousands of millions users to communicate with each other regardless of locations, time zones and devices. Online conversations are free-style where multiple users can be involved and multiple topics can be

© Springer International Publishing AG 2017
S. Candan et al. (Eds.): DASFAA 2017, Part I, LNCS 10177, pp. 19–34, 2017.
DOI: 10.1007/978-3-319-55753-3_2

discussed at the same time. There have been some studies about the analysis of conversations in social networks [14,24]. In Twitter and Weibo[1], people use "@" to engage his/her friends in the conversations [2,4,9,23] where the logical structures of conversations are very clear.

Generally speaking, there will always be a *structure* in each conversation. That is, someone starts a conversation by bringing up a message of a new topic, and each later message in this conversation replies to one or more previous message(s). For example, in the social news and entertainment site Reddit[2], someone first posts a new topic, and another user can comment on the topic as well as on the previous comments by other users. The comments are structured in a *tree* layout on the webpage so that users can understand how a discussion evolves over time.

However, not all online conversations are well-structured. Instead, there are even larger volumes of free-style conversations without clear "reply-to" relations in real-world scenarios like the popular instant group chat in Tencent QQ, WhatsApp, Skype and LINE. When more than two persons are discussing together online, one user of them may reply to a previous message that (s)he is interested in, rather than the last message in the group chat history. These conversations are usually called *asynchronous conversations* [12] where the *temporal order* fails to represent the *logical order* of a message sequence. We attempt to understand the structure of online short-text conversation in this paper by predicting the "reply-to" relation between the messages in it.

The inherent value of this study is to reconstruct the logic of conversations, profile chatters' information and analyze the relations between chatters [28]. It is especially important for the third-party organizations like strategic consultant companies which cannot directly derive the conversation structure by updating the user interface, e.g. add a "reply" button to each of previous messages. Meanwhile, by recovering the conversation structure, we can visualize the conversations with hierarchical layouts like trees or graphs instead of plain message sequences. Besides, reply suggestion [13] and recommendation [3] (e.g. message/chatter recommendation) are other applications of this study.

To learn the conversation structures, we are confronted with the following challenges:

- The asynchrony of messages makes it difficult to figure out the logical relation between messages. There are multiple users engaged in the discussion and multiple topics are discussed at the same time.
- Due to privacy concerns, there is no publicly available conversation data sets before. It means we have to construct the evaluation data set from the scratch.
- Unlike formal articles, the online messages are usually informal, short and context-sensitive. Thus, the traditional natural language processing methods like Latent Dirichlet Allocation (LDA) [1] usually do not work well in dealing with online conversations.

---

[1] http://www.weibo.com.

[2] http://www.reddit.com/.

In this paper, we attempt to address the problem of learning online conversation structures by presenting a domain-independent framework based on text features extracted from the conversation corpus. We summarize the main contributions as follows:

- We studied the problem of asynchronous conversation structure learning based on online short-text messages. A domain-independent method was brought forward to address this problem by employing text similarity feature and latent transferability feature based on message contents.
- We proposed a heuristic method to predict the "reply-to" relations and recover the conversation structure. This method avoids yielding disconnected or cyclic structure. Besides, another graph-based method can be employed to get the optimal tree conversation structure.
- We crawled a new online short-text Chinese conversation corpus and used it to evaluate our method. The experimental results show that our method outperforms the baselines in the prediction accuracy.

The rest of the paper is organized as follows: In Sect. 2, we discuss some related work about the studies of conversations. Then, we formally define the major problem in our study in Sect. 3. In Sect. 4, the proposed method based on text similarity and latent semantic transferability is introduced in detail. We demonstrate the experimental results conducted on the new web forum data set in Sect. 5. Finally, we conclude our study and prospect our future work in Sect. 6.

2   Related Work

As far as we concern, the problem studied in this paper has not been well established before. We discuss the related work on conversation disentanglement and clustering, dialogue act learning and some studies about conversation structures in this section.

2.1   Conversation Disentanglement and Clustering

Similar to the famous cocktail party problem, conversation disentanglement, a.k.a. chat disentanglement, describes the task to isolate the messages belonging to the same topic from a long conversation where multiple users are engaged and multiple topics are discussed [6,7,21,25]. Apparently, this is also a clustering problem. Based on the data like timestamp, mention, cue word and text content, the authors in [6,7] propose a maximum-entropy classifier to judge if two messages are of the same topic. They also propose an algorithm to cluster the messages on a directed weighted graph. Later in [25], the clustering performance is improved by enriching the TF-IDF feature of message m with the TF-IDF features of highly relevant messages which share similar timestamp or username with m. Then, a single-pass clustering algorithm can be used to cluster the messages of a conversation into topics.

2.2 Dialogue Act

As a specialized form of *speech act*, dialogue act [22] studies the role, e.g. *Statement*, *Question*, *Agreement* and *Disagreement*, of messages in a conversation. In [18], the authors propose the Hidden Markov Models (HMM) to study the dialogue acts in a conversation where the words are generated from the act emission distribution or the topic multinomials. In [12], the authors first find that using a graph-based model like the graph partition method [6] to deal with dialogue act annotation does not work well, and then, they use an HMM mixture model and consider the emission of dialogue act as the mixtures of multinomials that generate the words in sentences. The results of dialogue act annotation is improved using their proposed method on the Email and forum data sets.

2.3 Conversation Structure

Unlike the conversation disentanglement and the dialogue act modeling, the ultimate goal in this work is to learn the logical structures of online asynchronous conversation by predicting the "reply-to" relations between messages in a given conversation. The most related work to ours is the thread prediction problem [8,26] where it predicts how each message in a newsgroup style conversation is related to each other. However, it differs from our work in several aspects:

- We are dealing with online conversations where messages are much shorter and more informal than the newsgroup conversations in that work.
- The work in [26] only redefines the TF-IDF features and proposes some time interval constraints to predict the relations between messages without considering the message transferability like what we propose in this paper.

Thus, we are tackling a much more challenging problem here and more features of online conversations are taken into account in the proposed method.

3 Problem Definition

In this paper, we learn the online conversation structures by predicting the "reply-to" relation between messages, through which the directed transition edges can be constructed and then the conversation structure is recovered. In order to focus on online conversation structure learning, we assume that each conversation is only about one topic in this paper, and chat disentanglement could be referred in other work [7].

Definition 1 (Online Short-Text Conversation Corpus). *An online short-text conversation corpus is a set of messages $M = \{m_1, m_2, \ldots, m_{|M|}\}$ from a number of conversations. The message length, i.e. the number of words, of each $m \in M$ is short (e.g. less than 10 words each). The words and phrases in M are usually used in an informal way (e.g. many symbols, abbreviations and Internet words).*

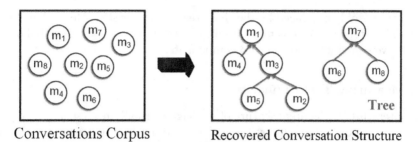

Fig. 1. An illustration of learnt conversation structure. $\mathcal{M}$ here contains two conversations. Each conversation indicates a tree structure.

Definition 2 (Online Short-Text Conversation Structure). *Online short-text conversation structure* $(\mathcal{M}, \prec)$ *is defined by a partial binary operator* $\prec$ *on an online short-text conversation corpus* $\mathcal{M}$*. For* $\forall m_i, m_j \in \mathcal{M}$ *and* $m_i \neq m_j$*, we say* $m_j \prec m_i$ *if and only if: (1)* m_i *and* m_j *are from a same conversation, (2)* m_i *is a reply to* m_j*. Thus,* $(\mathcal{M}, \prec)$ *is namely the "reply-to" structure of a conversation corpus.*

Therefore, our structure learning problem in this paper is to predict the precursor m_j for $\forall m_i \in \mathcal{M}$ based on message content, where m_i and m_j are from a same conversation. This problem is non-trivial due to the asynchrony, informality, and lack of useful cue words in short-text messages.

Figure 1 illustrates an example of the learnt conversation structure. In this example, we assume that each message can at most reply to only one precursor for simplicity, and then the conversation structure is in a tree layout, e.g. web forum conversations. Clearly, the proposed method in this paper can be adapted to deal with DAG structure learning. The study of the directed-acyclic-graph (DAG) structure learning will be our future work.

Figure 1 also shows the difference between our problem and the chat disentanglement problem [7]. The chat disentanglement problem is to cluster messages into different groups (divide the message corpus), but our problem is to predict the "reply-to" relation between messages in a given group (structure learning).

4   Proposed Method

Based on the problem definition, the most basic task of our problem is to identify the "reply-to" relations between messages in a given conversation. We define the "likelihood" that message m_i replies to message m_j as:

$$p_{m_j \prec m_i} = (1 - \gamma)\mathcal{S}(m_i, m_j) + \gamma\mathcal{T}(\mathcal{A}(m_i), \mathcal{A}(m_j)). \tag{1}$$

This likelihood consists of two components: text similarity $\mathcal{S}(m_i, m_j)$ and latent transferability $\mathcal{T}(\mathcal{A}(m_i), \mathcal{A}(m_j))$. In this paper, latent transferability is measured by the latent dialogue act transition, which is proposed to alleviate the sparse text feature problem induced by the short-text characteristics. Here, $\mathcal{A}(m_i)$ represents the latent dialogue act feature of m_i. $\gamma \in [0, 1]$ is a parameter balancing the relative contribution of the two components.

4.1   Measuring Text Similarity

In the literature, the content feature of message is usually represented using the bag-of-words model. We employ the widely-used TF-IDF approach [19] in this study. An alternative is to use Latent Dirichlet Allocation (LDA) [1] to preprocess the corpus. However, we found through experiments that such approach performs poorly in our data set since each message is usually very short (e.g. <10 words) and the phrases are usually informal.

The content feature of message m_i is represented by a W-dimensional column vector $\mathbf{v}_i$ where W is the vocabulary size. The w-th entry of $\mathbf{v}_i$ is the term frequency of the w-th word weighted by the inverse document frequency.

$$v_{i,w} = n_{w,i} \cdot \log \frac{1}{f_w}, \tag{2}$$

where $n_{w,i}$ is the frequency that word w appears in m_i. The document frequency f_w of word w is computed as (Laplace smoothing is applied to avoid division on zero in Eq. (2)):

$$f_w = \frac{n_w + 1}{|\mathcal{M}| + 1}, \tag{3}$$

where n_w is the number of messages which contain word w. Thus, the text similarity between two messages can be measured by their cosine similarity:

$$\mathcal{S}(m_i, m_j) = \frac{\mathbf{v}_i^\top \mathbf{v}_j}{\|\mathbf{v}_i\| \cdot \|\mathbf{v}_j\|}. \tag{4}$$

4.2   Measuring Latent Transferability

"Reply-to" relations are directed. However, the measurement of $\mathcal{S}(m_i, m_j)$ is symmetric. Therefore, we also employ the asymmetric latent transferability between messages based on latent dialogue act features to refine our model. Dialogue acts are high level features of messages. The examples of dialogue acts such as "statement", "question", "answer", or "remark" indicate the roles played by messages in conversations. However, automatic dialogue act classification requires a large amount of user annotation to perform model training [20]. Besides, the performance of these explicit dialogue act classification methods is degraded on the online short-text conversation corpus. Therefore, we propose to use unsupervised learning to get the latent dialogue act feature for each message and use it in the transferability measurement.

TF-DF Feature. Compared with the crucial role that infrequent words play in text mining and information retrieval, the functionality of frequent words is usually ignored in the literature. However, we find that frequent words usually serve as important indicators of the act that each message represents. The benefit to consider frequent words becomes more obvious when the general length of message is short. Therefore, we define the *term-frequency-document-frequency (TF-DF)* feature for each message. The TF-DF of message m_i is an F-dimensional column vector $\mathbf{x}_i$ where $F \ll W$ is the number of the most frequent words in the vocabulary. The reason why only Top-F frequent words rather than all words are used is that we need to reduce the computation cost to learn the dialogue act features without great loss of accuracy. The w-th component of $\mathbf{x}_i$ can be computed as follows:

$$\mathbf{x}_{iw} = n_{w,i} \cdot \frac{1}{1 + e^{-(1 + \ln f_w)}} = n_{w,i} \cdot \frac{f_w}{f_w + e^{-1}}. \tag{5}$$

Note that we use the sigmoid function to rescale the document frequency. The basic idea of TF-DF is that the weights of infrequent words should be less important than those of frequent words which indicate the dialogue act features.

Since the number of existing dialogue acts is much less than the vocabulary size W^3, we need to compress TF-DF feature $\mathbf{x}_i$ into latent dialogue act feature in much lower dimensions. Suppose there are totally K distinct acts to be considered ($K \ll F \ll W$). Let $\mathbf{y}_i \in \mathbb{R}^K$ denote the latent dialogue act feature of message m_i. Then, we need a dialogue act transformation matrix $\mathbf{A} \in \mathbb{R}^{F \times K}$ such that:

$$\mathbf{x}_i = \mathbf{A}\mathbf{y}_i. \tag{6}$$

Suppose $\mathbf{A}$ is already given, we can compute the latent dialogue act feature of each message m_i as:

$$\mathcal{A}(m_i) = \mathbf{y}_i = \mathbf{A}^\dagger \mathbf{x}_i, \tag{7}$$

where $\mathbf{A}^\dagger$ is the pseudo-inverse of $\mathbf{A}$.

Latent Dialogue Act Feature. Now we focus on the estimation of matrix $\mathbf{A}$ from $\mathcal{M}$. Let $\mathbf{X} = (\mathbf{x}_1, \mathbf{x}_2, \ldots, \mathbf{x}_{|\mathcal{M}|})$ and $\mathbf{Y} = (\mathbf{y}_1, \mathbf{y}_2, \ldots, \mathbf{y}_{|\mathcal{M}|})$, then $\mathbf{X} = \mathbf{A}\mathbf{Y}$. Our aim is to estimate $\mathbf{A}$ and $\mathbf{Y}$ given the observations on $\mathbf{X}$. Although this could be considered as a non-negative matrix factorization problem [16,17,27], we choose the independent component analysis (ICA) method [10,11] instead. Because it is more likely that the latent dialogue act feature of each message is separately emitted and mixed from K independent dialogue acts (or latent independent components), but non-negative matrix factorization could not guarantee such independence.

We need to conduct data whitening on $\mathbf{X}$ before performing ICA as discussed in [11]. According to Theorem 1, we firstly make random variable $\mathbf{x} \in \mathbf{X}$ has zero mean by subtracting its expectation (the mean in practice), $\mathbf{x} = \mathbf{x} - \mathrm{E}[\mathbf{x}]$. Then,

---

[3] http://en.wikipedia.org/wiki/Dialog_act.

we perform singular value decomposition (SVD) on $\mathbf{X}$, $\mathbf{X} = \mathbf{U\Sigma V}^\top$, where $\mathbf{U} \in \mathbb{R}^{F \times F}$, $\mathbf{\Sigma} \in \mathbb{R}^{F \times |\mathcal{M}|}$ and $\mathbf{V}^\top \in \mathbb{R}^{|\mathcal{M}| \times |\mathcal{M}|}$.

Theorem 1. *For* $\mathbf{X} = (\mathbf{x}_1, \mathbf{x}_2, \ldots, \mathbf{x}_{|\mathcal{M}|})$, *if random variable* $\mathbf{x} \in \mathbf{X}$ *has zero mean, i.e.* $\mathrm{E}[\mathbf{x}] = \mathbf{0}$, *and* $\mathbf{X}$ *has singular value decomposition* $\mathbf{X} = \mathbf{U\Sigma V}^\top$, *then let* $\mathbf{z} = \left(\frac{1}{\sqrt{|\mathcal{M}|}}\mathbf{\Sigma}\right)^{-1}\mathbf{U}^\top\mathbf{x}$, *random variable* $\mathbf{z}$ *will be whitened.*

Proof. Since $\mathrm{E}[\mathbf{x}] = \mathbf{0}$, then

$$\mathrm{E}[\mathbf{z}] = \sqrt{|\mathcal{M}|}\mathbf{\Sigma}^{-1}\mathbf{U}^\top\mathrm{E}[\mathbf{x}] = \mathbf{0}. \tag{8}$$

We also have:

$$\mathrm{E}[\mathbf{xx}^\top] \approx \frac{1}{|\mathcal{M}|}\mathbf{XX}^\top \tag{9}$$

$$= \frac{1}{|\mathcal{M}|}(\mathbf{U\Sigma V}^\top)(\mathbf{U\Sigma V}^\top)^\top \tag{10}$$

$$= \frac{1}{|\mathcal{M}|}\mathbf{U\Sigma}^2\mathbf{U}^\top. \tag{11}$$

Then, we can prove that:

$$\mathrm{E}[\mathbf{zz}^\top] = \mathrm{E}\left[\left(\frac{1}{\sqrt{|\mathcal{M}|}}\mathbf{\Sigma}\right)^{-1}\mathbf{U}^\top\mathbf{xx}^\top\mathbf{U}\left(\frac{1}{\sqrt{|\mathcal{M}|}}\mathbf{\Sigma}\right)^{-1}\right] \tag{12}$$

$$= |\mathcal{M}|\mathbf{\Sigma}^{-1}\mathbf{U}^\top\mathrm{E}[\mathbf{xx}^\top]\mathbf{U\Sigma}^{-1} \tag{13}$$

$$= |\mathcal{M}|\mathbf{\Sigma}^{-1}\mathbf{U}^\top\frac{1}{|\mathcal{M}|}\mathbf{U\Sigma}^2\mathbf{U}^\top\mathbf{U\Sigma}^{-1} \tag{14}$$

$$= \mathbf{\Sigma}^{-1}\mathbf{U}^\top\mathbf{U\Sigma}^2\mathbf{U}^\top\mathbf{U\Sigma}^{-1} \tag{15}$$

$$= \mathbf{\Sigma}^{-1}\mathbf{\Sigma}^2\mathbf{\Sigma}^{-1} \tag{16}$$

$$= \mathbf{I}. \tag{17}$$

Thus, random variable $\mathbf{z}$ has zero mean and unit variance. That means $\mathbf{z}$ is whitened.

Algorithm 1 shows the procedure to estimate the matrix $\mathbf{A}$ and compute the latent dialogue act features. After performing SVD on $\mathbf{X}$, we compress TF-DF features into much lower K-dimensional space by preserving the K largest singular values and get an approximation matrix $\tilde{\mathbf{X}} \in \mathbb{R}^{K \times |\mathcal{M}|}$ of $\mathbf{X}$ (Line 3–4). Then $\tilde{\mathbf{x}}$ is whitened by transforming to random variable $\tilde{\mathbf{z}}$ based on Theorem 1 (Line 5–7). Since $\tilde{\mathbf{Z}}$ is whitened now, we can perform ICA on it, and let $\tilde{\mathbf{Z}} = \mathbf{AY}$. From the ICA point of view, $\tilde{\mathbf{Z}}$ is a linear mixture of some statistically independent signals. In this paper, we employ the FastICA algorithm [11][4] to get the unmixing matrix $\mathbf{W}$ (Line 8), from which we can get the inverse act transformation matrix $\mathbf{A}^\dagger$ and the dialogue act features $\mathbf{Y}$ of messages in the conversation

---

[4] see the Python library: http://scikit-learn.org/.

Algorithm 1. Latent Dialogue Act Feature Estimation

Input:
    TF-DF features $\mathbf{X} = \{\mathbf{x}_1, ..., \mathbf{x}_{|\mathcal{M}|}\} \in \mathbb{R}^{F \times |\mathcal{M}|}$,
    dimensions of dialogue act features K.

Output:
    inverse act transformation matrix: $\mathbf{A}^\dagger \in \mathbb{R}^{K \times F}$,
    latent dialogue act features: $\mathbf{Y} \in \mathbb{R}^{K \times |\mathcal{M}|}$

1: $\mathbf{X} \leftarrow \mathbf{X} - \{E[\mathbf{x}], ..., E[\mathbf{x}]\}$
2: $\mathbf{U}, \mathbf{\Sigma}, \mathbf{V} \leftarrow \text{SingularValueDecomposition}(\mathbf{X})$
3: $\widetilde{\mathbf{U}}, \widetilde{\mathbf{\Sigma}}, \widetilde{\mathbf{V}} \leftarrow \text{DimentionReduction}(\mathbf{U}, \mathbf{\Sigma}, \mathbf{V}, K)$
4: $\widetilde{\mathbf{X}} \leftarrow \widetilde{\mathbf{U}}\widetilde{\mathbf{\Sigma}}\widetilde{\mathbf{V}}$
5: **for all** $\widetilde{\mathbf{x}}_i \in \widetilde{\mathbf{X}}$ **do**
6:     $\widetilde{\mathbf{z}}_i = \left(\frac{1}{\sqrt{|\mathcal{M}|}}\widetilde{\mathbf{\Sigma}}\right)^{-1} \widetilde{\mathbf{U}}^\top \widetilde{\mathbf{x}}_i,$
7: **end for**
8: $\mathbf{W} \leftarrow \text{FastICA}(\widetilde{\mathbf{Z}})$ /*$\widetilde{\mathbf{Z}} = \{\widetilde{\mathbf{z}}_1, ..., \widetilde{\mathbf{z}}_{|\mathcal{M}|}\}$*/
9: $\mathbf{A}^\dagger \leftarrow \mathbf{W}\left(\frac{1}{\sqrt{|\mathcal{M}|}}\widetilde{\mathbf{\Sigma}}\right)^{-1} \widetilde{\mathbf{U}}^\top$
10: $\mathbf{Y} \leftarrow \mathbf{A}^\dagger \mathbf{X}$
11: **return** $\mathbf{A}^\dagger, \mathbf{Y}$

---

corpus. The result above assumes that the components of the random vector $\mathbf{y}$ is independent. This is usually not the case in reality, especially in our setting where $\mathbf{y}$ encodes the strength of different latent acts. However, previous applications of ICA show that this technique can still gain insights into the data set even if the independence assumption is violated.

Latent Transferability Measurement. We define latent transferability (likelihood of m_i replies to m_j) as below:

$$\mathcal{T}(\mathcal{A}(m_i), \mathcal{A}(m_j)) = \mathcal{T}(\mathbf{y}_i, \mathbf{y}_j) = \hat{\mathbf{y}}_i^\top \mathbf{B} \hat{\mathbf{y}}_j, \tag{18}$$

where $\hat{\mathbf{y}}_i = \frac{abs(\mathbf{y}_i)}{\|abs(\mathbf{y}_i)\|_1}$, $abs(\mathbf{y}_i)$ is the absolute value of $\mathbf{y}_i$. Please note that $\mathcal{T}$ is asymmetric. For a list of messages $m_{p_1}, m_{p_2}, \ldots, m_{p_N}$, suppose the messages they reply to are $m_{q_1}, m_{q_2}, \ldots, m_{q_N}$, respectively. Let

$$\mathbf{Y}_p = \left(\hat{\mathbf{y}}_{p_1}, \hat{\mathbf{y}}_{p_2}, \ldots, \hat{\mathbf{y}}_{p_N}\right), \tag{19}$$

$$\mathbf{Y}_q = \left(\hat{\mathbf{y}}_{q_1}, \hat{\mathbf{y}}_{q_2}, \ldots, \hat{\mathbf{y}}_{q_N}\right). \tag{20}$$

To learn the optimal transition matrix $\mathbf{B}$, we minimize the square error between $\mathbf{Y}_p^\top \mathbf{B}$ and $\mathbf{Y}_q$, as follows:

$$\widehat{\mathbf{B}} = \underset{\mathbf{B} \in \mathbb{R}^{K \times K}}{\arg\min} \|\mathbf{Y}_p^\top \mathbf{B} - \mathbf{Y}_q\|^2. \tag{21}$$

This problem can be solved by employing the non-negative least square algorithm [15]. Thus, we can estimate the final likelihood $p_{m_j \prec m_i}$ that message m_i

replies to m_j using Eq. (1). Please note that although **B** is inferred based on the training data set, it models the transition likelihood between the latent dialogue act features, and thus, it is also generalized to the unseen message pairs in prediction.

4.3   Conversation Structure Recovery

We consider the tree structure recovery problem in this paper, and leave the DAG structure recovery problem in our future work. Actually, the difference between the tree structure recovery and the DAG structure recovery lies in the possibility that each meassage can or cannot reply to more than one previous message. A simple strategy based on this work is to predefine a threshold η of likelihood $p_{m_j \prec m_i}$ to determine the precursor(s) of each message so that n is a reply to m $\forall m, p_{m \prec n} \geq \eta$. For tree structure recovery, since each non-root node has only one parent node, we can predict that for each non-root message $m_i \in \mathcal{M}$, the one that m_i replies to should maximize the "likelihood" $p_{m_j \prec m_i}$ where m_i and m_j are from the same conversation and $m_i \neq m_j$. This strategy is straightforward and simple to implement. Unfortunately, it is also flawed since it may generate an unexpected disconnected or cyclic structure. Figure 2 illustrates a failure example using this strategy. According to the "likelihood" table in Fig. 2, message m_4 and m_5 mutually reply to each other, which makes the conversation structure disconnected and generates a cyclic sub-structure. However, the expected structure is a single rooted tree as shown in Fig. 2(c). The reason for this failure is because this strategy ignores the constraint on the global conversation structure itself, i.e. the structure connectivity and the acyclic property.

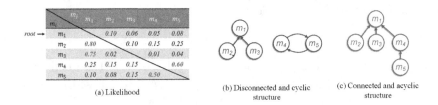

(a) Likelihood   (b) Disconnected and cyclic structure   (c) Connected and acyclic structure

Fig. 2. An example of failure of simple prediction method.

To tackle this problem, we propose a heuristic method for fast computation. Alternatively, we can also use a less-efficient graph-based method to get the optimal results. For the heuristic method, we initialize two sets: **D** as empty set, **M** as the set containing all messages in a given conversation. We iteratively move one message from **M** to **D** until **M** becomes empty. Each time, we move $m \in \mathbf{M}$ so that:

$$\left(\operatorname*{argmax}_{m_i \in \mathbf{M} \cup \mathbf{D}} p_{m_i \prec m} \right) \in \mathbf{D}. \tag{22}$$

Algorithm 2. Conversation Structure Recovery

Input: "reply-to" likelihood table $p_{\prec}$, message set $\mathbf{M}$, total message set $\mathbf{N}$.
Output: conversation structure $\mathcal{G}$.
1: Initialize $\mathcal{G} \leftarrow \emptyset, \mathbf{D} \leftarrow \emptyset$.
2: Identify conversation root message $r \in \mathbf{M}$.
3: $\mathbf{D} \leftarrow \mathbf{D} \cup \{r\}, \mathbf{M} \leftarrow \mathbf{M} \setminus \{r\}$.
4: **while** $\mathbf{M} \neq \emptyset$ **do**
5:    **while** $m, n^* \leftarrow \mathrm{NextToMove}(p_{\prec}, \mathbf{N}, \mathbf{M}, \mathbf{D}) \neq \mathrm{NULL}$ **do**
6:        $\mathbf{D} \leftarrow \mathbf{D} \cup \{m\}, \mathbf{M} \leftarrow \mathbf{M} \setminus \{m\}$
7:        $\mathcal{G} \leftarrow \mathcal{G} \cup \{n^* \prec m\}$
8:    **end while**
9:    **if** $\mathbf{M} \neq \emptyset$ **then**
10:        $n^* \prec m^* \leftarrow \mathrm{argmax}_{m \in \mathbf{M}} \left(\max_{n \in \mathbf{D}} p_{n \prec m} \right)$
11:        $\mathbf{D} \leftarrow \mathbf{D} \cup \{m^*\}, \mathbf{M} \leftarrow \mathbf{M} \setminus \{m^*\}$
12:        $\mathcal{G} \leftarrow \mathcal{G} \cup \{n^* \prec m^*\}$
13:    **end if**
14: **end while**
15: **return** $\mathcal{G}$

Algorithm 3. NextToMove

Input: "reply-to" likelihood table $p_{\prec}$, total message set $\mathbf{N}$, unvisited message set $\mathbf{M}$, visited message set $\mathbf{D}$.
Output: next movable candidate m and its precursor n^*.
1: **for** $m \in \mathbf{M}$ **do**
2:    $n^* \leftarrow \mathrm{argmax}_{n \in \mathbf{N}} p_{n \prec m}$
3:    **if** $n^* \in \mathbf{D}$ **then**
4:        **return** m, n^*
5:    **end if**
6: **end for**
7: **return** NULL

It means that for any message in $\mathbf{M}$, the maximum "reply-to" likelihood should be associated with a message in $\mathbf{D}$. If such an m cannot be found in $\mathbf{M}$, we move the following message,

$$\mathrm{argmax}_{m \in \mathbf{M}} \left(\max_{m_i \in \mathbf{D}} p_{m_i \prec m} \right). \tag{23}$$

After each move, we create a "reply-to" relation from m to $\mathrm{argmax}_{m_i \in \mathbf{D}} p_{m_i \prec m}$. It is apparent that the heuristic method generates a connected and acyclic tree structure. The pseudo code of the heuristic method is show in Algorithm 2. The root message is always chosen as the topic itself in our experiments on the web forum data set.

The heuristic method is fast but also sub-optimal. To get the optimal tree structure, we can consider the messages as nodes in a directed weighted graph and the likelihood $p_{m_i \prec m_j}$ as edge weights. Then, the optimal tree structure can be obtained by applying the Edmond's algorithm [5] to find the maximum spanning arborescence.

Auxiliary Filters. The proposed method is solely based on message contents. However, we can further improve it by employing auxiliary filters.

Time Filter: It is obvious that each message can only reply to the earlier posted message(s). If time information or posted order of messages is available, we can apply this filter in the recovery process.

User Filter: Generally, a chatter does not reply to himself in online conversations. This filter removes the candidates of self-replies in the recovery process, but it works if user identity of posted messages is known.

Both filters are applied in later experiments.

5   Experiments

In this section, we first introduce the new data set we collected. Then, the experimental results on this new data set are demonstrated and discussed.

5.1   Data Set

We investigated Douban Group[5], a popular Chinese web forum. In Douban Groups, users can publish topics for discussion. When someone replies to a comment c under a conversation, the content of c is automatically quoted by the new comment. This makes it possible to reconstruct conversations by tracing the quoting relations among comments. This is how we obtain the ground-truth of "reply-to" relations in our experiments. Please note that we choose web forum chats for evaluation since the ground-truth can be obtained, but we aim to solve the conversation structure learning problem for those unstructured chats, e.g. online group chat. We crawled 10,425 conversations on Douban Group in August, 2013, containing 137,980 messages in total. Each conversation in this data set has a tree structure ground-truth since one user posts a comment (a.k.a. reply) by quoting only one previous comment or the topic. After performing Chinese word cut, each message has about 12 words on average, which is very short.

5.2   Results and Discussion

In the experiments, the method using the conventional text similarity (TF-IDF) only is denoted by "TEXT". The two methods in [26] which redefines the TF-IDF feature and makes some constraints on the time interval between two potentially related messages are denoted by "FIXED" and "TIMED", respectively. The former reduces the number of candidate messages by setting a fixed time intervals, while the latter decreases the importance of candidate messages as the time interval increases. The method only based on latent act transferability is named as ACT. The proposed method is denoted by TACT. We use -H and -E to denote methods using the heuristic and the Edmond's algorithms as the structure recovery strategies, respectively.

---

[5] http://www.douban.com/group.

In the experiment, we randomly choose 80% conversations from our Douban Group data set as the training set (e.g. learn the matrices **A** and **B** in the proposed method), and leave the rest as the testing set. The numbers we reported are the averages after running experiments for 5 times. We use "reply-to" relation prediction accuracy as the major measurement. The accuracy is computed as:

$$\frac{\#\text{correctly predicted reply-to relations}}{\#\text{total reply-to relations}}.$$

Only one precursor is predicted for each message, i.e. Top-1 prediction.

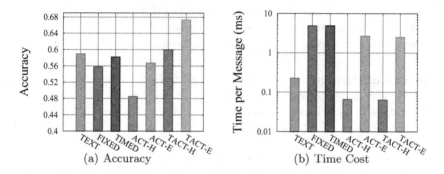

(a) Accuracy   (b) Time Cost

Fig. 3. Comparisons of the accuracy and the efficiency performance.

Predict "Reply-To" Relations. Figure 3(a) shows the results of accuracy performance in the experiments. According to the results, we can see: (1) The proposed methods (TACT-H and TACT-E) generally have the best accuracy performance compared with other baselines. The best accuracy is achieved by TACT-E method at around 67.5%. Considering that the accuracy is obtained on Top-1 precursor prediction, the proposed method is very effective on this data set; (2) Obviously, Edmond's algorithm is always better in accuracy performance than the heuristic method; (3) Both "FIXED" and "TIMED" perform slightly worse than "TEXT", which may result from the redefinition of its TF-IDF features that changes the important signals in representing the short and informal messages; besides, the interval constraint in "FIXED" may also leave the real precursor messages out of consideration and lead to poor performance.

As for the efficiency evaluation, we use the average time cost to predict a single "reply-to" relation as the metric, i.e.

$$\frac{\text{Time to recover a conversation}}{\#\text{ messages in a conversation}}.$$

The results are shown in Fig. 3(b). Apparently, the heuristic method is much efficient than the Edmond's algorithm as well as the other baselines. Both of FIXED and TIMED work very slow in the experiments, while TEXT has the medium efficiency.

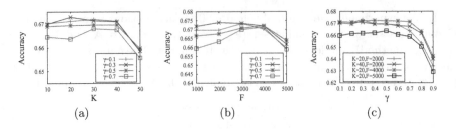

Fig. 4. Accuracy performance under different settings of parameters.

The comparisons show that TACT-H and TACT-E are both effective and efficient in the structure recovery problem. Meanwhile, TACT-H and TACT-E have the advantages of efficiency and effectiveness over the baselines, respectively.

Sensitivity of Parameters. We also analyzed the sensitivity of parameters in TACT, i.e. the dimensions of latent dialogue act feature K, the dimensions of TF-DF feature F, as well as the balancing parameter γ. If not explicitly specified, the default settings of parameters are $K = 20$, $F = 2000$ and $\gamma = 0.5$ in the experiments.

Figure 4(a) shows the performance with different K values. A larger value of K indicates a larger number of latent dialogue acts to consider, but also a larger cost to learn the transition matrix $\mathbf{B}$ and a higher probability to incorporate redundant latent dialogue acts. From the results, we can see that the optimal setting of K value should be around 20.

Figure 4(b) illustrates the performance by changing the dimensions of TF-DF features, i.e., the number of frequent words. The value of F determines the size of matrix $\mathbf{X}$, which means a larger value of F leads to a larger cost to factorize $\mathbf{X}$ with SVD. According to the results, we can see that $F = 2000$ is a good choice in our experiment.

Lastly, the performance with different γ values is shown in Fig. 4(c). It is obvious from the results that the accuracy is very similar when $\gamma \leq 0.7$, and there is an accuracy drop when γ gets closer to 1.0. But the overall performance of the proposed method is stable in the experiments.

6   Conclusion

We investigate the problem of recovering the structure of online short-text conversations. A novel framework combining text similarity and latent semantic transferability between messages is brought forward, and a heuristic method as well as a graph-based one are also presented to recover the conversation structure. The evaluation on the new data set we collected shows the effectiveness and the efficiency of the proposed method. In the future, we are considering to incorporate more linguistic features like syntactic feature and word embeddings in the framework to get more accurate in exploring the relations between messages.

Acknowledgments. This work was supported in part by the National Natural Science Foundation of China (No. 61373023, No. 61133002, No. 61502116), the China National Arts Fund (No. 20164129), and the National Science Foundation (NSF) under grant No. CNS-1252292.

References

1. Blei, D.M., Ng, A.Y., Jordan, M.I.: Latent dirichlet allocation. J. Mach. Learn. Res. **3**, 993–1022 (2003)
2. Boyd, D., Golder, S., Lotan, G.: Tweet, tweet, retweet: conversational aspects of retweeting on twitter. In: Proceedings of the 43rd Hawaii International Conference on System Sciences, pp. 1–10. IEEE (2010)
3. Chen, J., Wang, C., Wang, J.: A personalized interest-forgetting markov model for recommendations. In: AAAI, pp. 16–22 (2015)
4. Cook, J., Kenthapadi, K., Mishra, N.: Group chats on twitter. In: WWW, pp. 225–236 (2013)
5. Edmonds, J.: Optimum branchings. J. Res. Natl. Bur. Stand. B. Math. Math. Phys. **71B**(4), 233–240 (1967)
6. Elsner, M., Charniak, E.: You talking to me? a corpus and algorithm for conversation disentanglement. In: ACL, pp. 834–842 (2008)
7. Elsner, M., Charniak, E.: Disentangling chat. Comput. Linguist. **36**(3), 389–409 (2010)
8. Gandhi, S., Jones, A.R., Nesbitt, P.A., Seacat, L.A.: Instant conversation in a thread of an online discussion forum, November 2015. http://www.freepatentsonline.com/9177284.html
9. Honey, C., Herring, S.C.: Beyond microblogging: conversation and collaboration via twitter. In: Proceedings of the 42nd Hawaii International Conference on System Sciences, pp. 1–10. IEEE (2009)
10. Hyvärinen, A.: Fast and robust fixed-point algorithms for independent component analysis. IEEE Trans. Neural Netw. **10**(3), 626–634 (1999)
11. Hyvärinen, A., Oja, E.: Independent component analysis: algorithms and applications. Neural Netw. **13**(4), 411–430 (2000)
12. Joty, S., Carenini, G., Lin, C.Y.: Unsupervised modeling of dialog acts in asynchronous conversations. In: IJCAI, pp. 1807–1813 (2011)
13. Kannan, A., Kurach, K., Ravi, S., Kaufmann, T., Tomkins, A., Miklos, B., Corrado, G., Lukacs, L., Ganea, M., Young, P., Ramavajjala, V.: Smart reply: automated response suggestion for email. In: KDD, pp. 955–964 (2016)
14. Kumar, R., Mahdian, M., McGlohon, M.: Dynamics of conversations. In: KDD, pp. 553–561 (2010)
15. Lawson, C.L., Hanson, R.J.: Solving Least Squares Problems, vol. 161. Prentice-Hall, Englewood Cliffs (1974)
16. Lee, D.D., Seung, H.S.: Algorithms for non-negative matrix factorization. NIPS **13**, 556–562 (2000)
17. Lin, C.J.: Projected gradient methods for nonnegative matrix factorization. Neural Comput. **19**(10), 2756–2779 (2007)
18. Ritter, A., Cherry, C., Dolan, B.: Unsupervised modeling of twitter conversations. In: NAACL, pp. 172–180 (2010)
19. Salton, G., Buckley, C.: Term-weighting approaches in automatic text retrieval. Inf. Process. Manage. **24**(5), 513–523 (1988)

20. Serafin, R., Eugenio, B.D.: FLSA: extending latent semantic analysis with features for dialogue act classification. In: ACL (2004). No. 692

21. Shen, D., Yang, Q., Sun, J.T., Chen, Z.: Thread detection in dynamic text message streams. In: SIGIR, pp. 35–42 (2006)

22. Stolcke, A., Ries, K., Coccaro, N., Shriberg, E., Bates, R., Jurafsky, D., Taylor, P., Martin, R., Ess-Dykema, C.V., Meteer, M.: Dialogue act modeling for automatic tagging and recognition of conversational speech. Comput. Linguist. **26**(3), 339–373 (2000)

23. Uthus, D.C., Aha, D.W.: The Ubuntu chat corpus for multiparticipant chat analysis. In: Proceedings of the AAAI Spring Symposium (2013)

24. Wang, C., Ye, M., Huberman, B.A.: From user comments to on-line conversations. In: KDD, pp. 244–252 (2012)

25. Wang, L., Oard, D.W.: Context-based message expansion for disentanglement of interleaved text conversations. In: NAACL, pp. 200–208 (2009)

26. Wang, Y.C., Joshi, M., Cohen, W.W., Rosé, C.P.: Recovering implicit thread structure in newsgroup style conversations. In: Proceedings of the 2nd International Conference on Weblogs and Social Media (2008)

27. Wang, Y.X., Zhang, Y.J.: Nonnegative matrix factorization: a comprehensive review. TKDE **25**(6), 1336–1353 (2013)

28. Zhang, J., Wang, C., Wang, J., Yu, J.X., Chen, J., Wang, C.: Inferring directions of undirected social ties. TKDE **28**(12), 3276–3292 (2016)

A Question Routing Technique Using Deep Neural Network for Communities of Question Answering

Amr Azzam[1](✉), Neamat Tazi[1], and Ahmad Hossny[2]

[1] Faculty of Computers and Infomation, Cairo University, Giza, Egypt
{a.tarek,n.eltazi}@Fci-cu.edu.eg
[2] School of Mathematical Sciences, University of Adelaide, Adelaide, Australia
ahmad.hossny@adelaide.edu.au

Abstract. Online Communities for Question Answering (CQA) such as Quora and Stack Overflow face the challenge of providing high quality answers to the questions asked by their users. Although CQA frameworks receive new questions in a linear rate, the rate of the unanswered questions increases in an exponential way. This variation eventually compromise effectiveness of the CQA frameworks as knowledge sharing platforms. The main cause for this challenge is the improper routing of questions to the potential answerers, field experts or interested users. The proposed technique *QR-DSSM* uses *deep semantic similarity model (DSSM)* to extract semantic similarity features using deep neural networks. The extracted semantic features are used to rank the profiles of the answerers by their relevance the routed question. QR-DSSM maps the asked questions and the profiles of the users into a latent semantic space where the relevance is measured using cosine similarity between the two; questions and users' profiles. QR-DSSM achieved MRR score of 0.1737. QR-DSSM outperformed the baseline models such as query likelihood language model (QLLM), Latent Dirichlet Allocation (LDA), SVM classification technique and RankingSVM learning to rank technique.

Keywords: Question routing · Community question answering · Deep learning · Semantic modeling

1 Introduction

CQA websites such as *Yahoo! Answers, Quora and Stack Exchange* became key sources for knowledge sharing and information retrieval for many of Internet users. CQA face the question routing challenge that reduces the participation of users within the community. The increasing number of posted questions makes it harder to the answerer to find the appropriate questions that match their expertise. Meanwhile, the askers may get low-quality answers from non-experts, due to the poor routing of the question. Moreover, the average waiting time to get an answer by a qualified answerer increases significantly.

© Springer International Publishing AG 2017
S. Candan et al. (Eds.): DASFAA 2017, Part I, LNCS 10177, pp. 35–49, 2017.
DOI: 10.1007/978-3-319-55753-3_3

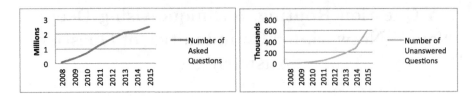

Fig. 1. Stack overflow website statistics

Table 1. The statistics of "stackoverflow.com" for [2008–2014] vs. 2015

	Stack overflow (2008–2014)	Stack overflow (Only 2015)
Total number of questions	8,247,436	2,483,062
Questions without accepted answer	4,697,357 (57%)	1,422,824 (57%)
Unanswered questions	1,230,606 (14%)	596,104 (24%)

CQA need question routing systems to route the questions to their appropriate qualified answerers [10,13,17,23] and consequently minimizing the lead time between posting a question and receiving a high-quality answer. This will encourage both askers to use the CQA and answerers to share their knowledge.

Stack Overflow is an online question answering community for computer programming topics. The best answerer for a given question is selected by either the asker or by opening the question for voting and the best answerer is the one with the most votes. Stack Overflow statistics show a significant increase in the number of posted questions, as well as the number of unanswered questions as shown in Table 1. Figure 1 shows the linear growth of the posted questions over time as well as the exponential growth of unanswered questions respectively.

In this paper, we propose a question routing technique named *QR-DSSM* that uses textual features to predict the semantic similarity between the posted questions and the profiles of their answerers using deep neural networks (DNN). The proposed technique is compared with the standard information retrieval approaches including QLLM [12] and LDA [2]. Moreover, it was compared to routing techniques based on SVM classifier [3] and the learning to rank technique, RankingSVM [8], proposed in [10]. QR-DSSM achieved statistically significant improvement over the aforementioned techniques in all metrics.

QR-DSSM consists of the following three main steps as shown in Fig. 2:

1. Data preparation step: to create a profile for each answerer in the community using his previously answered questions.
2. Learning step: to use the deep structured semantic model (DSSM) [9] is used to capture the semantic similarity between the profiles of the answerers and the posted question.

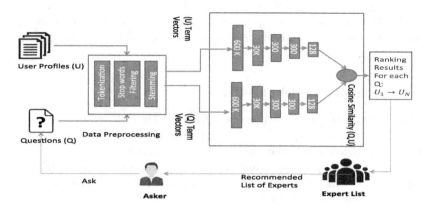

Fig. 2. Proposed question routing technique stages

3. Decision-making step: The question routing technique routes the question to a ranked list of suitable answerers. The ranking of the answerers is built on DSSM which evaluates the similarity between the question and the profiles of the answerers using the cosine similarity.

The rest of the paper is organized as follows. Section 2 provides an overview of the existing question routing approaches. The proposed technique is explained in Sect. 3. In Sect. 4, experiments design and evaluation methodology of the proposed techniques are introduced. Section 5 discusses the results and demonstrates the results of comparing our technique with existing ones. Finally, Sect. 6 presents the conclusion and future research directions.

2   Related Work

There are two main approaches to map the posted question to the potential answerers. The first approach is the profile-based approach [10,17] where the textual representation of the answerers is based on the set of previously answered questions in the CQA. The second approach is the document-based approach which is a two-step process. The first is to find the similarity between the newly posted question and the previously answered questions in the community [10]. While the second is to rank the answerers of the previously answered questions based on the relevance of their answers to the newly posted question.

    Many techniques in question routing systems apply hybrid approaches which combine information retrieval (IR) approaches with non-textual features to find potential answerers, where IR approaches are used to capture the content relevance between the posted questions and the history of the answerers, while the non-textual features are statistical measures that capture certain quality metrics in the community [10,12,13]; which can be user-specific metrics, e.g. social importance, reputation and authority or question-related metrics, e.g. question title length and question body length [10].

In the following subsections, we review several related works that apply hybrid approaches from and this review is categorized by the IR approaches used.

2.1   Vector Space Model (TF-IDF)

In [13] the authors proposed a framework that finds the right experts to a specific question or a category of questions. The framework builds a hybrid approach to create users' profiles. The framework creates users' profiles based on vector space model to process the past questions answered by the users. The users' profiles are converted to term vectors using TF-IDF approach [19]. Non-textual features were extracted and combined with the term vectors. Features such as the users' reputation in the community, link analysis algorithms such as HITS [11] and Page Rank [15] were used to represent the users' authority in the community. Moreover, other features were added such as the number of votes and the time factor that represents the importance of answered questions decays over time.

2.2   Language Modeling

Authors of [1] proposed two general probabilistic language models for finding experts of a given topic. The two models estimate the expertise of each candidate expert as the probability of answering a query q given an expert e as shown in Eq. (1). The first model follows the profile based approach while the second model is a document based one.

$$P(e \mid q) = \frac{P(q \mid e) \cdot P(e)}{P(q)} \tag{1}$$

Work in [12] proposed a question routing framework that uses a query likelihood language model (QLLM) to estimate the expertise of each candidate user. They assumed that the expertise of a candidate users is the conditional likelihood of generating the newly posted question from the users' profiles. Besides the estimated probability of expertise of users, the authors considered the quality of the previous answers of each user and the availability of such users inside the community to answer questions.

In [23], the authors proposed a question routing framework that is based on language modeling techniques. While the authors in [16] proposed question recommendation technique based on the probabilistic latent semantic analysis model (PLSA) that is used to capture the hidden topics of each user in the set of the previously asked questions.

2.3   Topic Modeling

Topic modeling techniques discover hidden topics inside documents collections. Topic modeling proved its ability to find semantic similarity between documents even in lexically different ones. These techniques are applied in different studies like in [17] to capture the similarity between posted questions and users' profiles. Authors of [17] have introduced a question routing framework based on a Segmented Topic Model (STM). The best-answered questions by each user were combined to create the users' profiles. The authors compared STM with multiple information retrieval approaches such as TF-IDF, Language modeling and LDA where STM outperformed the others.

2.4   Learning to Rank

Learning to rank techniques use supervised learning to construct ranking models for information retrieval tasks. These techniques produce a ranking model that is capable of computing the relevance between queries and documents.

Authors of [10] introduced a framework for question routing which combines multiple textual similarity features such as query likelihood language model (QLLM) [12] and Latent Dirichlet Allocation (LDA) [2]. In addition, it includes statistical features such as the length of the question, the number of best answers the user provided, the number of questions the user asked and the number of answers user provided to the community. They [10] estimated the candidate answerers' relevance to the questions based on SVM classifier [3] and a pairwise learning to rank technique called RankingSVM [8].

Authors of [20] formulated the question routing task as a ranking problem. First, LDA is applied to the questions to learn the topic distribution. Then, a 3-order tensor model is constructed to extract semantic relations among the three main entities of the community (asker, answerer, and question). The potential answerers are ranked by maximizing the multi-class area under the ROC curve. Furthermore in [22], Expert finding framework has been developed in based on learning semantic representation of question and users by adopting a random-walk learning with recurrent neural networks to rank CQA users. The main focus of [22] is to model online social relation between the CQA users through heterogeneous network integration.

2.5   Deep Learning Techniques

Deep learning is a class of machine learning techniques that learn multiple levels of representations for the data using artificial neural network through multiple non-linear information processing layers. Deep learning has many applications in speech recognition, computer vision and natural language processing [4].

Deep Learning provides promising results due to its ability to automatically extract such semantic features from the raw text without feature engineering, whereas The other traditional machine learning techniques are limited to optimization functions that use features extracted by humans [4].

It has also been recently utilized in IR tasks where it uses a deep neural network to extract the semantic features from both the query and the documents then ranks those documents based on the relevance to the query [6].

In addition, Deep learning models show great potential in learning the content representation for the content based recommendation tasks such as expert finding [22], music recommendation [14] and cross domain recommendation system [5].

There are two lines of research for improving semantic modeling using deep learning in IR tasks. The first is generative semantic models. Authors of [18] proposed a semantic hashing approach based on two-stage learning for document retrieval; the first is the pre-training stage and the second is the fine-tuning. Another model is proposed by [7] using the deep belief nets where the first layer is a word-count vector and the final layer is a binary code representation of that document and the remaining layers form a Bayesian network.

The semantic hashing approaches do not provide a significant improvement over traditional information retrieval approaches that either depends on lexical matching such as TF-IDF or probabilistic topic models such as latent Dirichlet allocation model (LDA) [2]. These approaches are based on an unsupervised learning method that is optimized to minimize the reconstruction error. In addition, they are generative because the learning optimizes their parameters for document reconstruction while IR approaches usually have a different objective which is obtaining the relevant resources for the given query.

The second line of researches uses click-through data for semantic matching between queries and documents. Click-through data is used to train models built for documents ranking task [6]. A Bi-Lingual Topic Model (BLTM) and a Linear Discriminative Projection Models (DPM) were proposed in [6]. BLTM optimization function objective is to maximize the log-likelihood that is not effectively complied with the ranked-based evaluation metrics. On the other hand, DPM suffers huge computation costs that hinder the training process with large vocabulary size.

DSSM [9] combines the advantages of the both lines of researches; it uses Deep Neural Network (DNN) architecture for semantic matching between the query and the documents. It is also discriminatively trained to optimize DNN parameters in order to maximize the conditional likelihood of the clicked documents given the queries. In addition, DSSM uses DNN to perform a non-linear projection to map the query and the documents into a semantic space where the similarity can be calculated even if the query and the document are lexically different. On the other hand, the second line of research uses linear projection for the same task. DSSM has been recently applied in several tasks such as web search [9], recommendation [7] and media search [21].

3   QR-DSSM for User Profile Ranking in Question Routing System

The proposed model tuned the DSSM to enhance the retrieval of the profiles of the users and rank them based on semantic relevance to the posted questions rather than keyword/term relevance. DSSM is a deep structured semantic model introduced in [9], the main idea of DSSM is to use DNN to map the sparse raw text of the queries and the documents into dense features where the query and the documents can be represented in the same latent semantic space with low dimension feature vectors. The documents are ranked using cosine similarity between the extracted feature vectors.

In the following subsections, we explain building the semantic features of our question routing model architecture (QR-DSSM).

3.1 QR-DSSM Ranking Model Architecture

The QR-DSSM has adapted the DSSM fully connected deep neural network as illustrated in Fig. 3. Here, we define the formal annotation used in the network architecture explanation.

qx: Question input term vectors ux: Users' profiles input term vectors
qwh: qx after word hashing layer embedding
uwh: ux after word hashing layer embedding
h: Hidden layers
y: The output layer
Q: Question
U: User profile
W_i: The $i th$ weight matrix
b_i: The $i th$ bias term
n: The number of nodes in each layer of the network

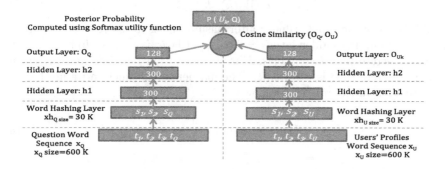

Fig. 3. DNN structure used in question routing technique (QR-DSSM).

3.2 QR-DSSM Ranking Model Input Layer

The text representing questions Q and users' profiles U used in the training phase pass through a data cleansing stage, that is described in detail in Sect. 4, by removing HTML tags, code snippets and stopping words. Questions Q and users' profiles U are then tokenized and converted to a sequence of words that generates two input term vectors qx and ux. The extracted input term vectors are usually sparse which decrease the ability of the neural network to learn and increase the computation cost for neural network training.

To reduce the dimensionality of the generated term vectors, we applied the word hashing embedding method introduced in [9] on qx and ux, respectively.

Word hashing is an effective words representation where each word is decomposed into a vector of fragments. Each fragment consists of three consecutive letters (tri-gram). For instance, the word "Hope" is represented as "#Hope#" where a delimiter is added in the beginning and the end of the word leading to the set of sub-words #Ho, Hop, ope, pe#.

Using this compact representation, our model could accommodate boundless variance of words and convert them into a reasonable number of sub-words. In our work, word hashing secures scalability to our model by reducing the total number of unique words from 600 K to tri-grams of only 30 K.

3.3   QR-DSSM Ranking Model Hidden Layers

The qwh and uwh are the output vectors of word hashing that contain trigrams with their frequency. These vectors are passed to two separate networks with the same architecture but each network has its own parameters set W, b. The architecture of the two DNNs is a fully connected network topology. It consists of two hidden layers with n nodes each where n = 300 nodes. Equation (2) represents QR-DSSM mathematical formulation.

$$h_1(k) = W_1(k) \cdot x$$

$$h_i(k) = F\left(W_i(k) \cdot h_{i-1}(k) + b_i\right) \tag{2}$$

Where h_1 is the input value x after hashing layer in discrete time k. The i value goes from first hidden layer $i = 2$ to ouput layer $i = m$, $W_i(k)$ is the weight value in discrete time k, b is bias, F is the activation function and the output value in discrete time is $y = h_m(k)$. The mathematical function we use here as the activation function is $Tanh$ shown in Eq. (3).

$$F(x) = Tanh(x) = \frac{1 - e^{-2x}}{1 - e^{-2x}} \tag{3}$$

3.4   QR-DSSM Ranking Model Output Layer

The output layer consists of n nodes where $n = 128$. The cosine similarity between the posted question Q and the users' profiles U is based on the output layer vectors of both networks y_Q and y_U shown in Eq. (3).

$$cosineSim(y_Q, y_Q) = \frac{y_Q^T \cdot y_U}{||y_Q|| \cdot ||y_U||} \tag{4}$$

Softmax function applied on the output layer of the network to convert the semantic relevance score between the posted question Q and the users' profiles U into a posterior probability of user profile given the question as shown in Eq. (4). It enforces an important constraint on the network outputs, where the probability should lie between 0 and 1 and the summation of the probabilities of all 128 output nodes is 1.

$$Pr(U \mid Q) = \frac{e^{cosineSim(y_Q, y_U)}}{\sum_{k=1}^{K} e^{cosineSim(y_Q, y_{U_k})}} \tag{5}$$

Ideally, K is the total number of answerers to be ranked. Meanwhile, the huge number of answerers and their answers reduce the performance significantly because of multiple DNN training iterations. So, we approximated K to be the list of the actual answerers of the question together with three randomly selected non-answerers based on NCE sampling strategy.

Log loss function is used to measure the accuracy of the users' profiles ranking given the posted questions. The model produces a probability for each user profile and the log loss function is the cross-entropy between the predicted outputs and the true user profile who actually provided an answer to the question.

The loss function is used to learn the weights by maximizing the product of the predicted probabilities, where these weights are the set of parameters (W_i, b_i) that the model learn. Similarly, we minimize the loss function which is the negative log of the prediction probability as shown in Eq. (6).

$$Pr(U \mid Q) = L(W_i, b_i) = -log \left(\prod_{k=1}^{k} Pr(U_k \mid Q) \right) \tag{6}$$

where U_k is the set of users who actually provided an answer to the question. The loss function in Eq. (6) does not require to be normalized because every training example is a pair of a question and an only one gold answerer at a time and if the question has multiple answerers we created a new separate training example for every answerer of the question.

3.5 Question Routing DNN Model Parameters

The parameters used for building QR-DSSM are explained as follows:

- SGD: A stochastic gradient descent (SGD) is used for optimizing the model objective function.
- BATCHSIZE: The training samples size used in each iteration is 1024.
- ITERATIONSCOUNT: The number of iterations of DNN is 100 iterations.
- LEARNINGRATE: The learning rate used to update the weights and the bias of the DNN is = 0.02

4 Experimental Design

4.1 Dataset Extraction and Pre-processing

We constructed our dataset from a Stack Overflow snapshot according to rules of data extraction introduced in [10]. Our dataset covers 13 months from Jan

2009 to Feb 2010. It has 92,407 CQA sessions, divided into training data (Jan 2009 to Dec 2009) and test data (Jan 2010 to Feb 2010) as shown in Table 2. The extracted questions should have at least one of the tags used in [10] and in [17] to ensure our questions share same characteristics.

We performed several data pre-processing steps including HTML tags removal, tokenization, stop words filtration, and stemming; yielding a clean word vector for each question. Questions are constructed by combining the question title, body and tags. To examine the effect of removing code blocks from the questions, we produced two datasets; one with the other with code blocks removed.

We constructed 3 user sets U_N based on the number of answers N provided by users in each set where $N = 10, 15$ *and* 20 following the experiment design of [10]. Every question in the training set Q_{Trn} and the test set Q_{Tst} has a best answerer and at least one more answerer. The asker, the best answerer and the other answerers all belong to U_N.

We label each question in the training set with its answerers who belong to U_N. In CQAs, many answerers could provide a correct answer for the question not only the best answerer. Table 2 shows the number of the training examples for each U_N which represents the number of pairs generated from pairing each question in Q_{Trn} with the question answerers belonging to U_N.

In the case of $N = 10$ there are 5,759 users who provided an answer for at least 10 questions in the training set. The asker, the best answerer and at least one other answerer for each question in the 16,021 training questions belong to the 5,759 users. For each of the 1,150 test questions, the test questions are routed to the 5,761 users.

Table 2. User set, training set and test set numbers

# of questions answered by user N	# of users U	# of training questions Q_{Trn}	# of training examples	# of test questions Q_{Tst}
10	5,759	16,021	54,218	1,151
15	3,970	11,177	36,238	746
20	2,977	8,371	26,354	517

4.2   Evaluation Methodology

We compare the performance of our proposed technique with two baselines; Query Language Likelihood Model (QLLM) [12] and Latent Dirichlet Allocation (LDA) [2]. We also compare it to SVM classifier [3], the learning to rank technique, RankingSVM introduced in [8] and results summarized in Table 3.

For evaluation purpose, We adopted several statistical metrics such as Mean Reciprocal Rank (MRR), Mean Average Precision (MAP) and Precision at K (P@K); where MRR shows the average rank of the first appearance of an actual answerer for a given test set question, MAP is the arithmetic mean of the average precision score for each test set question which gives an overall retrieval quality score, P@K computes the answerers' profiles precision above a cut-off point K; in our experiments, we use p@5 and p@10.

QLLM is the linear combination of two equally weighted probabilities. The first is the probability of generating the asked question based on the language model of the user's answered questions, while the second is the probability of generating the asked question using both user's answered and asked questions language model. LDA [17], on the other hand, is used to discover topic distribution over the users' profiles. It estimates users' expertise based on the latent topic in their profiles and thus provides semantic mapping between received questions and user profiles. In our experiments, we use Gibbs-LDA with K = 100 hidden topics and we set LDA hyper-parameters to Alpha = 0.1 and Beta = 0.5.

In addition, SVM [3,10] is applied in this experiment as a supervised learning model to classify user profiles into one of two classes; relevant or non-relevant to the posted question. The training examples are created based on the assumption that the asker of the question is a negative example(non-relevant) and the answerer is a positive example (relevant). Moreover, RankingSVM [8,10] is a pairwise learning to rank method based on SVM to produce a ranking model for information retrieval tasks. The predicted score is used for ranking users' profiles based on their relevance to the asked question. The statistical features mentioned in [10] are incorporated with textual features extracted from QLLM and LDA and then these features are provided to both SVM and RankingSVM.

The evaluation ground truth of the proposed technique is the list of answerers in U_N who actually provided an answer to the test question.

Table 3. Comparing the results of the different techniques used for question routing using three datasets (N = 10, 15, 20) where N is the number of questions answered by a single user who is selected in the training and test data. F8 means 8 features and F7 means 7 features where LDA is excluded

	Metrics	QLLM	LDA	STM	SVM (F7)	SVM (F8)	Rank-SVM (F7)	Rank-SVM (F8)	QR-DSSM with <code> elements	QR-DSSM without <code> elements
N=10	MAP	0.0245	0.0386	0.0398	0.0363	0.0364	0.0422	0.0439	0.0514	0.0578
	MRR	0.0504	0.082	0.0949	0.0847	0.085	0.0958	0.0992	0.1198	0.1294
	P@5	0.0143	0.023	0.0298	0.0245	0.024	0.0289	0.0304	0.0361	0.0405
	P@10	0.0124	0.0188	0.0258	0.0227	0.0225	0.0233	0.0252	0.0266	0.2947
N=15	MAP	0.0297	0.0439	0.0460	0.0443	0.0444	0.0508	0.0537	0.0559	0.0662
	MRR	0.0569	0.0895	0.1037	0.098	0.0992	0.1103	0.1156	0.1275	0.1481
	P@5	0.0158	0.0268	0.0322	0.0287	0.0295	0.0316	0.0362	0.0380	0.0447
	P@10	0.016	0.0214	0.0267	0.0261	0.0253	0.0259	0.0272	0.0285	0.0331
N=20	MAP	0.0335	0.0493	0.0557	0.0527	0.0524	0.0587	0.061	0.0749	0.0779
	MRR	0.0618	0.0967	0.1182	0.1149	0.1152	0.1253	0.1299	0.1649	0.1737
	P@5	0.017	0.0279	0.0341	0.0321	0.0313	0.0344	0.0379	0.0487	0.0518
	P@10	0.0166	0.0207	0.0261	0.0286	0.0277	0.0286	0.0298	0.0353	0.0367

5 Discussion and Results Analysis

QR-DSSM significantly improved question routing and achieved better performance than the benchmark techniques. Removing codes elements has proven to reduce the noise in the questions and to increase the performance of QR-DSSM as shown in Fig. 4. The measures show an enhancement that varies between 4% and 16% percent when code snippets were removed. From the training efficiency prespective, Table 4 shows the feature extraction and training time for each QR frameworks included in our experiment that are measured in an environment Intel Core i5 CPU (2.6 GHz) with AMD viga card and 8 GB RAM. Although the training times of SVM and QR-DSSM are less than QR-DSSM training time, both SVM and Rank-SVM depend on QLLM and LDA as input features which increase the overall training time. Moreover, QR-DSSM will be much more efficient in an environment with GPUs.

Applying a minimum for the number of answers provided by user profiles to be included in the training data $X = 10, 15, 20$ affects the performance of all the evaluated techniques. The higher X yields less number of candidate answers and thus higher the ability to route the newly posted questions.

5.1 QR-DSSM Vs Baseline Models

LDA and QR-DSSM construct semantic clusters that group co-occurred terms in the same context into the same cluster. Both consider questions and profiles that might contain lexically different terms but still with high semantic similarity.

Although LDA and QR-DSSM share the same goal, LDA uses unsupervised training approach with an objective function that is not trained to distinguish between relevant and the non-relevant users' profiles to the posted question but it is optimized for reconstructing documents into a set of topics with the words associated with each topic. On the other hand, QR-DSSM adopts supervised learning which, unlike LDA, benefits from having labeled training data. The output of the final layer in QR-DSSM is evaluated using cosine similarity which is popularly used for text similarity.

As shown in Table 3, QR-DSSM significantly outperformed LDA in all metrics presenting better ranking results for expert users. For instance, the mean average precision of QR-DSSM is almost 1.5 times higher than LDA. Moreover, QR-DSSM proved its superiority over the second baseline model, QLLM, in semantically modeling the users' profiles. In Table 3, the best MRR achieved by QLLM indicates that the test set questions should be routed to at least 18 users on average to get an answer while QR-DSSM requires only 6 users on average.

Table 4. The performance of QR frameworks on our dataset (Time, hours)

QLLM	LDA	STM	SVM	Rank-SVM	QR-DSSM
8	10	120	23	21	22

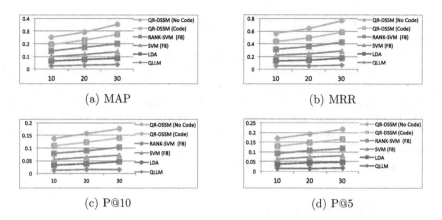

(a) MAP (b) MRR

(c) P@10 (d) P@5

Fig. 4. Comparing question routing results with benchmarks according to the metrics: MAP, P@5, P@10 and MRR

5.2 QR-DSSM Vs SVM Variations

SVM variations use the same combination of textual features and statistical features in order to capture semantic similarity between the posted questions and users' profiles. However, RankingSVM achieves better results than SVM classifier in question routing task. SVM is a pointwise learning to rank approach that reduces the ranking of the users' profiles into a classification problem. On the other hand, RankingSVM is a pairwise approach which considers the partial order of expertise among the answerers which is a more representative formulation to answerers ranking problem.

QR-DSSM also performed much better than SVM variations using semantic features extracted from the raw text. As shown in Table 3. For instance, The highest MRR in this experiment is 0.1737 achieved by QR-DSSM with code removal. While the highest MRR values achieved by SVM and RankingSVM are 0.1152 and 0.1299, respectively. Following the previously mentioned MRR values, RankingSVM should route the test question on average to the top 8 answerers to get an answer which is better than SVM which requires on average 9 answerers. On the other hand, QR-DSSM routes the questions to 6 users on average in order to find a qualified answerer.

Table 3 shows that SVM and RankingSVM surpassed the baseline models in learning a ranking model to the candidate answerers for the new questions. However, SVM variations have not exceeded QR-DSSM for the following reasons:

First, QR-DSSM uses multiple nonlinear transformations to learn the closest curve that represents the relation between the question and its answerers. The network non-linearly combines the outputs of the activation functions of the different layers in order to build a higher level of abstraction for the query and users' profiles from the raw data. Conversely, linear kernel functions are used in SVM and RankingSVM implementations to train the model for ranking answerers relative to the newly posted questions.

Second, QR-DSSM uses a (DNN) jointly performs and optimizes feature extraction and learning process in one step using the word hashing layer. On the contrary, SVM-based methods [10] separate the feature extraction and engineering process from the learning process. Both SVM and RankingSVM, in the feature extraction step, depend on hand-crafted statistical features besides textual features extracted from language models and unsupervised topic modeling (LDA). Then, these features are fed to either SVM classifier or SVM ranker.

Finally, QR-DSSM deep architecture is trained using the questions and users' profiles dataset to optimize the parameters of the question routing technique for one defined target which is users' profiles ranking.

6 Conclusion and Future Work

In this paper, we proposed a technique (QR-DSSM) for question routing in community question answering (CQA) based on a deep learning technique called Deep Semantic Similarity Model (DSSM). The question routing technique estimates users' expertise through capturing the similarity between the newly posted question and users' profiles built from their answered-questions history. The posted questions and the users' profiles pairs are mapped to a low-dimensional semantic space where the relevance between the user profile and the asked question is computed using cosine similarity. The questions are routed to the top ranked users' profiles based on the similarity scores.

Extensive experiments have been conducted on a real-world dataset extracted from Stack Overflow. The results showed that our proposed deep learning technique outperformed the traditional language modeling technique (QLLM), topic modeling technique (LDA), classification technique (SVM) and learning to rank technique (RankingSVM) in all used metrics. Results conclude that a question will get answered if it is routed to the top 6 users. Future work includes studying the effect of combining statistical features such as the percentage of best answers, number of answered questions and the number of votes as well as the effect of reward systems on the performance of QR-DSSM.

References

1. Balog, K., Azzopardi, L., de Rijke, M.: A language modeling framework for expert finding. Inf. Process. Manage. **45**(1), 1–19 (2009)
2. Blei, D.M., Ng, A.Y., Jordan, M.I.: Latent dirichlet allocation. J. Mach. Learn. Res. **3**, 993–1022 (2003)
3. Chang, C.C., Lin, C.J.: LIBSVM: a library for support vector machines. ACM Trans. Intell. Syst. Technol. **2**(3), 27:1–27:27 (2001)
4. Deng, L., Yu, D.: Deep learning: methods and applications. Technical report MSR-TR-2014-21
5. Elkahky, A.M., Song, Y., He, X.: A multi-view deep learning approach for cross domain user modeling in recommendation systems. In: Proceedings of the 24th International Conference on World Wide Web, WWW 2015, pp. 278–288 (2015)

6. Gao, J., He, X., Nie, J.Y.: Clickthrough-based translation models for web search: from word models to phrase models. In: CIKM
7. Gao, J., Pantel, P., Gamon, M., He, X., Deng, L.: Modeling interestingness with deep neural networks. In: Proceedings of the 2014 Conference on Empirical Methods in Natural Language Processing (EMNLP), pp. 2–13 (2014)
8. Herbrich, R.: Learning Kernel Classifiers: Theory and Algorithms
9. Huang, P.S., He, X., Gao, J., Deng, L., Acero, A., Heck, L.: Learning deep structured semantic models for web search using clickthrough data. In: Proceedings of the 22nd ACM International Conference on Information & Knowledge Management, CIKM 2013, pp. 2333–2338 (2013)
10. Ji, Z., Wang, B.: Learning to rank for question routing in community question answering. In: Proceedings of the 22nd ACM International Conference on Information & Knowledge Management, CIKM 2013, pp. 2363–2368 (2013)
11. Kleinberg, J.M.: Authoritative sources in a hyperlinked environment. J. ACM **46**(5), 604–632 (1999)
12. Li, B., King, I.: Routing questions to appropriate answerers in community question answering services. In: Proceedings of the 19th ACM International Conference on Information and Knowledge Management, CIKM 2010, pp. 1585–1588 (2010)
13. Liu, D.R., Chen, Y.H., Kao, W.C., Wang, H.W.: Integrating expert profile, reputation and link analysis for expert finding in question-answering websites. Inf. Process. Manage. **49**(1), 312–329 (2013)
14. van den Oord, A., Dieleman, S., Schrauwen, B.: Deep content-based music recommendation. In: Proceedings of the 26th International Conference on Neural Information Processing Systems, NIPS 2013, pp. 2643–2651 (2013)
15. Page, L., Brin, S., Motwani, R., Winograd, T.: The pagerank citation ranking: Bringing order to the web. Technical report 1999-66, Stanford InfoLab, previous number = SIDL-WP-1999-0120
16. Qu, M., Qiu, G., He, X., Zhang, C., Wu, H., Bu, J., Chen, C.: Probabilistic question recommendation for question answering communities. In: Proceedings of the 18th International Conference on World Wide Web, WWW 2009, pp. 1229–1230 (2009)
17. Riahi, F., Zolaktaf, Z., Shafiei, M., Milios, E.: Finding expert users in community question answering. In: Proceedings of the 21st International Conference on World Wide Web, WWW 2012 Companion, pp. 791–798 (2012)
18. Salakhutdinov, R., Hinton, G.: Semantic hashing. Int. J. Approx. Reasoning **50**(7), 969–978 (2009)
19. Salton, G., Buckley, C.: Term-weighting approaches in automatic text retrieval. Inf. Process. Manage. **24**(5), 513–523 (1988)
20. Yan, Z., Zhou, J.: Optimal answerer ranking for new questions in community question answering. Inf. Process. Manage. **51**(1), 163–178 (2015)
21. Ye, X., Li, J., Qi, Z., He, X.: Enhancing retrieval and ranking performance for media search engine by deep learning. In: 2016 49th Hawaii International Conference on System Sciences (HICSS), pp. 1174–1180 (2016)
22. Zhao, Z., Yang, Q., Cai, D., He, X., Zhuang, Y.: Expert finding for community-based question answering via ranking metric network learning. In: IJCAI
23. Zhou, Y., Cong, G., Cui, B., Jensen, C.S., Yao, J.: Routing questions to the right users in online communities. In: Proceedings of the 2009 IEEE International Conference on Data Engineering, ICDE 2009, pp. 700–711 (2009)

Category-Level Transfer Learning
from Knowledge Base to Microblog Stream
for Accurate Event Detection

Weijing Huang, Tengjiao Wang, Wei Chen[✉], and Yazhou Wang

Key Laboratory of High Confidence Software Technologies (Ministry of Education),
EECS, Peking University, Beijing, China
huangwaleking@gmail.com, {tjwang,pekingchenwei}@pku.edu.cn,
pkuwangyz@gmail.com

Abstract. Many Web applications need the accurate event detection technique on microblog stream. But the accuracy of existing methods is still challenged by microblog's short length and high noise. We develop a novel category-level transfer learning method TRANSDETECTOR to deal with the task. TRANSDETECTOR bases on two facts, that microblog is short but can be enriched by knowledge base semantically with transfer learning; and events can be detected more accurately on microblogs with richer semantics. The following contributions are made in TRANSDETECTOR. (1) We propose a structure-guided category-level topics extraction method, which exploits the knowledge base's hierarchical structure to extract categories' highly correlated topics. (2) We develop a probabilistic model CTrans-LDA for category-level transfer learning, which utilizes the word co-occurrences and transfers the knowledge base's category-level topics into microblogs. (3) Events are detected accurately on category-level word time series, due to richer semantics and less noise. (4) Experiment verifies the quality of category-level topics extracted from knowledge base, and the further study on the benchmark *Edinburgh twitter corpus* validates the effectiveness of our proposed transfer learning method for event detection. TRANSDETECTOR achieves high accuracy, promoting the precision by 9% without sacrificing the recall rate.

Keywords: Event detection · Microblog stream · Transfer learning · Knowledge base

1 Introduction

Many Web applications need the accurate event detection technique on microblog stream, such as public opinion analysis [1], public security [2], and disaster response [3], etc. Although event detection has been a research topic

This research is supported by the Natural Science Foundation of China (Grant No. 61572043), and the National Key Research and Development Program (Grant No. 2016YFB1000704).

© Springer International Publishing AG 2017
S. Candan et al. (Eds.): DASFAA 2017, Part I, LNCS 10177, pp. 50–67, 2017.
DOI: 10.1007/978-3-319-55753-3_4

for a long while [4], event detection in microblog stream is still challenging [5]. According to [6], the characteristics of microblog, which are fast changing, high noise, and short length, raise the challenge.

Knowledge base can be a good supplementary for event detection on microblog stream to address these challenges. Different from the not-well-organized microblog stream, knowledge base (e.g. Wikipedia) is constructed elaborately and contains rich information. For example, the microblog message "Possible Ft. Hood Attack Thwarted (2011-07-28)" is short, but still comprehensible because the words "*Ft. Hood*" is included in the wiki page "*Ft. Hood*", and belongs to the category "*Military*" at a higher conceptual level. By regarding these wiki information and the word *attack* (also highly related to *Military*), the model easily understands the example tweet is about something related to *Military*. In other words, knowledge base enriches the linkages between words and concepts, and provides more comprehensive context for microblogs. Since the transfer learning [7] aims at utilizing the extra information stored in the source dataset to benefit the target dataset, it provides a feasible way to enhance the event detection in microblog stream. Taking Fig. 1 below as an example, the fluctuation of time series of raw word *hood* is not so obvious to reflect the event happened to the military base *Ft. Hood*. After transferring the knowledge about "*Military*" into the microblog stream, the time series of word "*hood*" related to "*Military*" is extracted, which is more vivid to detect the event happened on July 28th, 2011.

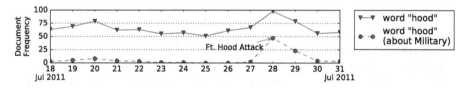

Fig. 1. The comparison of the time series between the raw word *hood* and the *Military* related word *hood*, computed on the *Edinburgh twitter corpus* [8]. The rise of document frequency on July 28th, 2011 is corresponding to the event mentioned in https://en.wikipedia.org/wiki/Fort_Hood#2011_attack_plot.

The benefit of enriching the semantics for micorblogs is attractive, but it's non-trivial to transfer the knowledge stored in the knowledge base into microblog stream directly. The existing RDF model [9] lacks an efficient quick mechanism to transfer the knowledge, since it's mainly designed for managing knowledge as tuples on graph. And the query on large graph is also very expensive [10], which is not suitable for the scene of quickly and accurately detecting events. In existing methods, what meets the demand most is Twevent [11], but it's limited in only treating Wikipedia as a looking-up table and may drop some events incorrectly.

In our paper, to balance the performance and the cost of leveraging knowledge base for event detection, we develop a novel category-level transfer learning method, namely TRANSDETECTOR. It consists of three phases: extracting category-level topics in knowledge base, conducting transfer learning to get category-level topics in microblog stream, and detecting events from category-level word time series, as illustrated in Fig. 2(a). We explain the main idea of the three parts in TRANSDETECTOR, and leave more technique details in Sect. 3. (1) In the **extracting** phase, we propose a structural-guided category-level topics extraction method on the knowledge base. It bases on the following facts. The knowledge base has the three fold hierarchical structure, consisting of the taxonomy graph (*class* → *subclass*), the category-page bipartite graph (*class* → *instance*), and the page-content map (*instance* → *content*), as illustrated in Fig. 2(b). In terms of concept level, the latter part is finer than the former. And the last page-content map goes into the detail at the word level. By considering these three parts together, we can extract a class or category's highly related words, and restore them in *category-level topics in knowledge base*. (2) In the **conducting transfer learning** phase, we propose a novel probabilistic model CTrans-LDA for transferring the knowledge. CTrans-LDA works in the bayesian transfer learning way like [12], by utilizing the extracted *category-level topics in knowledge base* as the informative priors to bridge two data domains. CTrans-LDA labels whether a word in a microblog message should link to a category in knowledge base, or just label it as no-category-related word. Applying CTrans-LDA on more microblogs, it gets the *category-level topics in microblog stream* and *category-level word time series*. (3) In the **detecting** phase, since the words in microblogs have been enriched semantically and the meaningless words are labeled as no-category-related, the event detection on *category-level word time series* is much more accurate than other methods which don't conduct the transfer learning. And the experiment on the *Edinburgh twitter corpus* demonstrates the effectiveness of our proposed TRANSDETECTOR.

To sum up, the contribution of this paper is mainly in four aspects. (1) We propose a structure-guided category-level topics extraction method, exploiting the knowledge base's hierarchical structure to extract categories' highly correlated topics. (2) We develop a probabilistic model CTrans-LDA for category-level transfer learning, which utilizes the word co-occurrences and transfers the knowledge base's category-level topics into microblogs. (3) Events are detected accurately on category-level word time series, due to richer semantics and less noise. (4) Experiment verifies the quality of category-level topics extracted from knowledge base, and the further study on the benchmark *Edinburgh twitter corpus* validates the effectiveness of our proposed transfer learning method for event detection.

2   Related Works

Event Detection. Based on how much data are used, the methods are mainly in two groups, *without extra information* and *leveraging extra information*.

Most existing methods are implemented *without extra information*. They are mainly carried out by clustering articles [13–15], analyzing word frequencies

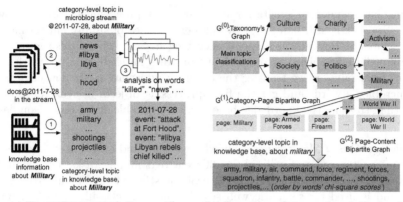

(a) TRANSDETECTOR's Process Flow

(b) Extracting Category-Level Topics in Knowledge Base

Fig. 2. (a) TRANSDETECTOR's process flow, taking *Military* related events in microblogs as an example. TRANSDETECTOR includes three phases: extracting category-level topics in knowledge base, conducting transfer learning to get category-level topics in microblog stream, and detecting events from category-level word time series; (b) Illustration on how to extract category-level topics in knowledge base via its three fold hierarchical structure, taking *Military* as an example.

[16,17], or finding bursty topics [18,19] etc. (1) By clustering articles, UMass [13], LSH [14], k-Term-FSD [15] and other similar methods model the occurred events as clusters of articles. The decision is based on whether the dissimilarity between the incoming article and existed event clusters is over the user-specified threshold. (2) By analyzing word frequencies, EDCoW [17] exploits wavelet transformation, which converts signal from time domain to time-scale domain, to detect the change point of word signal. However they treat the word as the most basic unit in analysis, without regarding polysemy words. (3) By finding bursty topics via topic modelling, such as TimeUserLDA [18], EUTB [20], and BurstyBTM [19]. TimeUserLDA and EUTB distinguish user's long term interests and short term bursty events, and BurstyBTM utilizes the burstiness of biterms as prior knowledge. This kind method needs to set the appropriate number of topics, and detects the "large" events but ignore the "small" ones. To summarize, due to microblog's short length and high noise, it's not easy to set the directly applicable parameter for existing methods to achieve high precision and high recall simultaneously.

The second group methods detect events by *leveraging extra information*. [21] incorporates user's and the followers' profiles to help to detect the events of public concern. However it's not easy to get this kind contextual information. [22] compares two time series generated by event-related tweets and the corresponding Wikipedia article's page views, and further filter out spurious events. Twevent [11] divides the tweet into segments according to the Microsoft Web N-Gram service and Wikipedia, then detect the bursty segments and cluster these segments into candidate events for necessary post processings. But as shown in [19], Twevent is still hampered by the performance of simply clustering the bursty segments. Beyond the scope of microblog stream, [23] utilizes the concurrent wikipedia edit spikes for event detection. In a nutshell, existing works are different from ours, as we transfer the category information of knowledge base into microblog's words and get the finer processing objects in event detection, rather than treating the knowledge base as a lookup table or a comparison base.

Knowledge Base. Many general knowledge bases and customized knowledge bases are constructed and utilized for different text mining tasks. For example, Probase [24] constructs a large general probabilistic *IsA* taxomomy from web-pages, and is used for semantic web search, and text classification [25]. Such kind efforts are also made by DBPedia [26], Yago [27], and Freebase [28]. These works manages knowledge as tuples on graph. However the query on large graph is very expensive [10], which is not very suitable for the scene of quickly and accurately detecting events. The customized knowledge bases such as EVIN [29], Event Registry [30], and Story-base [31], are designed for managing events. These knowledge bases are mainly built on news articles. EVIN maps the existing event-related news articles into semantic classes. Event Registry collects the articles by the API of News Feed Service [32], then detects events, and provides the structural information of the events, such as the related Wikipedia article, timestamp, and location etc. Storybase introduces Wikipedia current events[1] as the resources for constructing event-and-storyline knowledge base on news articles, which are provided by GDELT project [33]. As the news articles may lag behind the microblogs when the emergency of events, it's not enough to detect events from microblog stream by directly applying these customized knowledge bases.

Transfer Learning. Although the training examples in the target domain is very sparse, transfer learning [7] can utilize the domain-independent knowledge as the bridge to fill the information gap between source domain and target domain [34] to get more "examples" for training. By bayesian transfer learning, [12] uses the distribution in source domain as the prior knowledge for text classification. And inspired by this idea, we propose the *category-level topic* as the bridge for transferring knowledge.

3   TRANSDETECTOR

In this section, we present a novel category-level transfer learning method for event detection, namely TRANSDETECTOR. Illustrated in Fig. 2(a), TRANSDETECTOR consists of three parts: (1) extracting category-level topics in knowledge base, (2) conducting transfer learning to get category-level topics in microblog stream, and (3) detecting events from category-level word time series. The following definitions are used by our method TRANSDETECTOR. Definition 1 is used in

---

[1] https://en.wikipedia.org/wiki/Portal:Current_events.

the **extracting** phase, Definitions 2 and 3 are used in the **conducting transfer learning** phase, and Definitions 4 to 6 are used in the **detecting** phase. Because Definitions 2 to 6 are very easy to understand, what needs to be specified are Definition 1. *Category-level topics in knowledge base* is a set of tuples extracted from knowledge base, which weighs the importance of given words to the specific category by the chi-square score, defined in Definition 1 formally. Taking the *Military* category in Fig. 2(b) as an example, the category-level topic of *military* contain the words *army, military,* and *shootings* etc., which are highly related to the category *Military*.

Definition 1 (Category-Level Topics in Knowledge Base). Given category c, the category level topic in knowledge base is defined by a set of tuples, in which the first element is the word $w_i^{(c)}$ related to the category c, and the second element is the chi-square score $chi(c, w_i^{(c)})$ under the category c. And we denote c's category-level topic in knowledge base as $\boldsymbol{h}_c = \{< w_i^{(c)}, chi(c, w_i^{(c)}) >\}_{i=1}^{N_c}$.

Definition 2 (Category-Level Topics in Microblog Stream). Given category c, the category level topic in microblog stream at time t is defined by a set of tuples, and denoted as $\boldsymbol{r}_{c,t} = \{< w_i^{(c)}, n(c, t, w_i^{(c)}) >\}_{i=1}^{N_c}$, in which word $w_i^{(c)}$ is related to the category c, and $n(c, t, w_i^{(c)})$ is its document frequency in the time window t.

Definition 3 (Category-Level Word Time Series). Given category c and word w, the category-level word time series is a list of document frequencies extracted from $\{\boldsymbol{r}_{c,t}\}_{t=1}^{T}$, and denoted as $\{n(c, t, w_i^{(c)})\}_{t=1}^{T}$.

Definition 4 (The Set of Events' Candidate Words). The set of events' candidate words $\mathcal{B}_{c,t}$ are groups of the bursty words in $\boldsymbol{r}_{c,t}$.

Definition 5 (Event Phrase). Event phrase $\mathcal{C}_{c,t,i}$ is the i-th combination of words which occurred in the set of events' candidate words $\mathcal{B}_{c,t}$, and represents the i-th event happened in time t under the category c.

Definition 6 (Event Related Microblogs). Event related microblogs $\mathcal{D}_{c,t,i}$ are articles relating to the i-th event in time t of category c, and correspond to the event phase $\mathcal{C}_{c,t,i}$.

Section 3.1 explains the details of how to extract *category-level topics in knowledge base* by using the three fold hierarchical structure. Section 3.2 shows how to conduct transfer learning to get *category-level word time series* in microblog stream. And Sect. 3.3 interprets the high accuracy event detection based on the processed *category-level word time series*.

3.1   Extracting Category-Level Topics in Knowledge Base

In this part, we discuss in detail how to extract the *category-level topics in knowledge base* on the given categories. The knowledge base such as Wikipedia has the structure of classes, subclasses, instances, and the edges between them. This structure usually can be represented as triples in RDF graph [9], which is adopted to build DBPedia [26] and YAGO [27] from Wikipedia. But the the query on the large graph is usually very expensive [10]. To make a trade-off between cost and performance, we use the lightweight data structure *topics* to represent the knowledge at category(class) level. And the knowledge base's threefold structure $G^{(0)}$, $G^{(1)}$, $G^{(2)}$ benefits the extraction of *category-level topics*.

Taxonomy Graph $G^{(0)}$. The directed edges in $G^{(0)}$ represent the *class→subclass* relations in the knowledge base. Taking Wikipedia for example (Fig. 2(b)), the node *Main topic classifications*[2] has the subclass *Society*, further contains the subclass *Politics*, which is the ancestor of the subclass *Military*. As $G^{(0)}$ is not a Directed Acyclic Graph originally [35], we remove the cycles according to nodes' PageRank-HITS [36] score. Specifically, the edges *class→subclass* are preserved only when the node *class* has the higher PageRank-HITS score than the node *subclass*, which is shown in the line 2 of Algorithm 1. After removing cycles, the taxonomy structure on the knowledge base is better represented by the directed acyclic graph $G^{(0)'}$. As shown in line 3 of Algorithm 1, by visiting the category *Military* in the DAG $G^{(0)'}$, the breadth-first traverse can reach its successor sub-category nodes such as *Firearms*, *The World Wars*, and *World War II*, etc.

Category-Page Bipartite Graph $G^{(1)}$. The directed edges in $G^{(1)}$ represent the *class→instance* relations in the knowledge base. In Wikipedia, by considering $G^{(0)'}$ and $G^{(1)}$ together as shown in line 5 of Algorithm 1, we can get all the pages related to the given category.

Page-Content Map $G^{(2)}$. For a specific Wikipedia dumps version, the edges *page →content* in $G^{(2)}$ define a one-to-one mapping. There are a bulk of information stored in the wiki text content. In order to extract the key words in wiki page's content, which distinguish it from other pages, we use the chi-square statistics [37,38] to measure the importance of each word for the specific page.

For a given category in $G^{(0)'}$, chi-square statistics also can evaluate the importance of each word appeared in text contents. For example, the chi-square statistic of the term *shooting* under the category *military* is 2888.7 under chi-square test with 1 degree of freedom. That means the term *shooting* is highly related to the category *military*. Finally, we can get the *category-level topic*, which are composed by the category related words and their corresponding importance.

Considering the full Wikipedia's contents, the word's score in the *category-level topic* evaluates the importance of word to the concerned category accurately. For example, in *Military*'s *category-level topic*, the top words are *army, military, air, command, force*, and *regiment*, etc. The document which contains these words is related to the category *Military* with high probability.

---

[2] https://en.wikipedia.org/wiki/Category:Main_topic_classifications.

Algorithm 1. Extraction of Category-Level Topics in Knowledge Base

---

Input: Taxonomy's Graph $G^{(0)}$, Category-Page Bipartite Graph $G^{(1)}$, Page-Content
  Bipartite Graph $G^{(2)}$, topic related category node c
Output: c's category-level topic in knowledge base $\boldsymbol{h}_c$

1   $Pages(c) \leftarrow \varnothing, \boldsymbol{h}_c \leftarrow \varnothing$
2   DAG $G^{(0)'} \leftarrow$ Remove Cycles of $G^{(0)}$ by nodes' HITS-PageRank scores.
3   $SuccessorNodes(c) \leftarrow$ Breadth-first-traverse($G^{(0)'}, c$)
4   **for** $node \in SuccessorNodes(c)$ **do**
5 | $Pages(c) \leftarrow Pages(c) \cup G^{(1)}.neighbours(node)$
6   Word frequency table $n(c,.) \leftarrow$ do word count on the text contents of $Pages(c)$
7   Word frequency table $n(All,.) \leftarrow$ do word count on the text contents of all pages in $G^{(2)}$.
8   **for** $word\ w$ in $WordFrequencyTable(All).keys()$ **do**
9 | $chi(c, w) \leftarrow w$'s chi-square statistics on $WordFrequencyTable(c)$ and
          $WordFrequencyTable(All)$.
10 | $h_{c,w} \leftarrow chi(c, w)$
11 **return** $\boldsymbol{h}_c$

3.2 Conducting Transfer Learning

In this subsection, we describe how the proposed probabilistic model CTrans-LDA transfers the category-level topics into microblogs.

There are two facts inspiring CTrans-LDA. (1) The topics in the document may contain category-level topics. As an example, the tweet "Libyan rebel chief gunned down in Benghazi (2011-07-28)" contains the topic that is similar to the *Military*'s *category-level topic in knowledge base*. (2) The *category-level topics in microblog stream* can reuse the information stored in the corresponding *category-level topics in knowledge base*. The word *libyan* in the aforementioned example tweet ranks much higher in the *Military* and the *Middle East* category-level topics than the other categories' topics. After considering the relatedness between the remaining context and the *category-level topics in knowledge base*, the learned topics of the example tweet include *Military* and *Middle East*. The generative process of CTrans-LDA can be described as follows.

1. Draw corpus prior distribution $\boldsymbol{m} \sim Dir(\alpha \boldsymbol{u})$, where $\boldsymbol{u}$ is the uniform distribution.
2. For each topic $k \in \{1, \cdots, K\}$,
   (a) word distribution on the topic $\boldsymbol{\phi_k} \sim Dir(\boldsymbol{\beta} + \boldsymbol{\tau_k})$.
3. For each document index $d \in \{1, \cdots, D\}$,
   (a) topic distribution on the document $\theta_d \sim Dir(\boldsymbol{m})$,
   (b) for each word index $n \in \{1, \cdots, N_d\}$,
     i. word's topic assignment $z_{dn} \sim Multinomial(\theta_d)$,
     ii. word $w_{dn} \sim Multinomial(\phi_{z_{dn}})$.

In the above generative process, the line 2 is the key point to distinguish CTrans-LDA from LDA [39], where $\boldsymbol{\tau_k}$ is defined by Eq. (1). K_{KB} is the number of pre-defined categories used for transferring, and S_k is the set of words appeared in the k-th category-level topic. As [40] mentioned that the asymmetric prior distribution can significantly improve the quality of topic modelling, $\boldsymbol{\phi_k} \sim Dir(\boldsymbol{\beta} + \boldsymbol{\tau_k})$ incorporates *category-level topics in knowledge base* into the asymmetric prior of the word distribution on topic. The effect of τ_k is obvious, e.g., $\tau_{Military,army}/\tau_{Military,basketball} = 203$ leads to that topic *Military* prefers to contain the word *army* other than *basketball*. The parameter λ controls how much the knowledge is used for transferring, which can be chosen by cross-validated grid-search.

$$\tau_{kv} = \begin{cases} \lambda \dfrac{h_{kv}}{\sum_{v \in S_k} h_{kv}}, v \in S_k \text{ and } k \leq K_{KB} \\ 0, v \notin S_k \text{ or } k > K_{KB} \end{cases} \tag{1}$$

To solve CTrans-LDA, the gibbs sampling is adopted to determine the hidden variable z_{dn} and the model parameter ϕ_k. In the initialization phase of Gibbs sampling for CTrans-LDA, the hidden variable z_{dn} is initialized to topic k with probability $\hat{q}_{k|v}$ as Eq. (2). For the word v that belongs to any category-level topic, $\hat{q}_{k|v}$ is proportional to its importance in the category-level topic τ_{kv} as Eq. 2(a). For the new word v in text stream, $\hat{q}_{k|v}$ is set uniformly $1/(K - K_{KB})$ on the other topics as Eq. 2(b, c). The initialization makes sure that the learned topics are aligned to the pre-defined category-level topic.

$$\hat{q}_{k|v} = \begin{cases} \dfrac{\tau_{kv}}{\sum_{k=1}^{K} \tau_{kv}}, \sum_k \tau_{kv} > 0 & (a) \\ 0, \sum_k \tau_{kv} = 0 \text{ and } k \leq K_{KB} & (b) \\ 1/(K - K_{KB}), \sum_k \tau_{kv} = 0 \text{ and } k > K_{KB} & (c) \end{cases} \tag{2}$$

The sampling process uses the conditional probability $p(z_{dn} = k|.) \propto (n_{dk}^{(d)} + \alpha m_k)(n_{kv}^{(w)} + \tau_{kv} + \beta)/(n_{k,.}^{(w)} + \tau_{k,.} + V\beta)$, where $n_{dk}^{(d)}$ is the number of words in document d assigned to topic k, and $n_{kv}^{(w)}$ is the times of word v assigned to topic k. We set $\alpha = 0.1$ and $\beta = 0.005$ according to the suggestion by [41].

After Gibbs sampling on discrete time windows, CTrans-LDA learned all hidden topics of words in microblog stream. And for a specific word type w and its category assignment c in time window t, its document frequency is counted as $n(c, t, w^{(c)})$, which is the element of category-level word time series. Event detection on category-level word time series is discussed in the following subsection.

3.3 Detecting Events from Category-Level Word Time Series

After transfer learning, it's much easier to detect the event from category-level word time series, which is shown in Fig. 1. We take the detection in three sub-phases: (1) detecting events' candidate words; (2) generating event phrases; and (3) retrieving event related microblogs. Note that these sub-phases are also available for the common text stream, but they perform better when words are enriched by category concepts of knowledge base.

Detecting events' candidate words. For a word w and a category c, on its time series $\{n(c, t, w^{(c)})\}_{t=1}^{T}$, many bursty detection methods can be applied to check if the word w is bursty in category c by given time t. In our paper, we assume the document frequency of $w^{(c)}$ follows a poisson distribution, which is also used in [18]. And the document frequency of nonbursting $w^{(c)}$ has the poisson distribution with the parameter $\lambda = \mu_0$; while the bursting one with the parameter $4\mu_0$ means the document frequency $n(c, t, w^{(c)})$ is much higher when bursting. We empirically set μ_0 as the moving average $\mu_0^{(t)} = \frac{1}{T} \sum_{\tau=t-T}^{t-1} n(c, \tau, w^{(c)})$, and set T=10. Hence, at time t, bursty or not is determined by comparing the probabilities of above poisson distributions. After bursty detection on the time series, we add each time t's bursty word $w^{(c)}$ into the set of event's candidate words $\mathcal{B}_{c,t}$.

Generating event phrases. In the set of event's candidate words $\mathcal{B}_{c,t}$, some words appear together and should be grouped into event phrases. For example, there are six bursty words in $\mathcal{B}_{military,2011-07-28}$ belonging to two event phrases "*ft, hood, attack*" and "*libyan, rebel, gunned*" in the time window 2011-07-28 to represent the events happened to the US military establishment and the Libyan rebel respectively.

To group the bursty words together, we construct the directed weighted graph $\mathcal{G}_{c,t} = (\mathcal{B}_{c,t}, \mathcal{E}_{c,t}, \mathcal{W}_{c,t})$, where $\mathcal{E}_{c,t} = \{(a,b)|a \in \mathcal{B}_{c,t}, b \in \mathcal{B}_{c,t}, PMI(a,b) > 0\}$, and $\mathcal{W}_{c,t}$ gives the PMI scores on the edges. The graph $\mathcal{G}_{c,t}$ means the words in $\mathcal{B}_{c,t}$ are connected if and only if their PMI score in the given time window's microblogs is over 0. Given the graph $\mathcal{G}_{c,t}$, the spectral clustering [42] is utilized for exploring the best partition of words. To get the optimum cluster number, we use the graph density as the criteria. The graph density is the ratio of the number of edges to that of complete graph (the graph with all possible edges). We search the best cluster number from 1 to $|\mathcal{B}_{c,t}|$, and stop when all the resulting subgraphs satisfy the criteria that the density is over the threshold, which is set to 0.6 empirically. In this way, the generated event phrase combines the co-occurred bursty words and excludes the unrelated.

Retrieving event related microblogs. To better understand the event, we retrieve the event microblogs $\mathcal{D}_{c,t,i}$ by using the event phrase $\mathcal{C}_{c,t,i}$. Generally, according to the number of bursting words in the event phrase $\mathcal{C}_{c,t,i}$, there are two situations to be addressed. (1) $|\mathcal{C}_{c,t,i}| = 1$, we directly add the microblog in time window t, which contains the bursting category-word $w^{(c)}$, into the set $\mathcal{D}_{c,t,i}$. (2) $|\mathcal{C}_{c,t,i}| \geq 2$, it's not necessary that all the bursting words are included in the event related microblog. For example, the tweet "*Soldier wanted to attack Fort Hood troops*" contains the bursting words *attack* and *hood*, not all the event phrase "*ft, hood, attack*". To tackle this problem, we consider the microblog, that contains any pair of category-words in the event phrase, as the event related microblog. Finally, TRANSDETECTOR gets the detected events $\{(\mathcal{C}_{c,t,i}, \mathcal{D}_{c,t,i})\}_{i=1}^{|\mathcal{C}_{c,t}|}$, containing the event phrase and the corresponding event related microblogs for the given category and the time window.

4 Experiments

In this subsection, we demonstrate the effectiveness of TRANSDETECTOR, by evaluating the *category-level topics in knowledge base, category-level topics in microblog stream* learned by transfer learning, and the effect of event detection.

Knowledge Base. We construct the taxonomy graph $G^{(0)}$, the category-page bipartite graph $G^{(1)}$ from the latest dump of category links[3] and the page-content map $G^{(2)}$ from Wikipedia pages[4]. We set $K_{KB} = 50$, which means 50 categories are selected manually to cover the topics of Wikipedia and the target corpus as widely as possible. There are two kinds of categories considered. The mid-high categories in the taxonomy graph $G^{(0)'}$, which are representative, are likely to be selected, such as *Aviation, Military*, and *Middle East*, etc. And the mid categories, which reflect the main interests of the target corpus, are also taken into consideration, such as *American Football, Basketball*, and *Baseball*, etc.

Microblog Stream Dataset. We conduct the empirical analysis on a text stream benchmark *Edinburgh twitter corpus* which is constructed by [8] and widely used by previous event detection researches [15,43]. Due to the developer policy of Twitter, [8] only redistributes tweets' IDs[5]. We collected the tweets' contents according to the IDs with the help of Twitter API. Though we cannot get the whole dataset due to the limit of Twitter API, after necessary pre-processing, our rebuilt dataset still contains 36,627,434 tweets, which also spans from 2011/06/30 to 2011/09/15. More details of the original dataset are described in [44].

4.1 Evaluation on Categroy-Level Topics in Knowledge Base

The category-level topics are initialized on the pre-defined categories from the knowledge base as Algorithm 1.

Evaluation Metrics. We compare the topic coherence of category-level topics with the topics learned from Wikipedia by LightLDA [45] in terms of NPMI [46]. Different from traditional experiments that only compute the topic coherence of top words, we want to check whether it can hold for more words. Due to the limit of NPMI computing module[6], which computes the coherence of up to 10 words each time, we compute NPMI on the combination of top five words with each next five words as Table 1.

Results. Observe that even for the combination of 96th to 100th words with the top five words in *Aviation*'s category-level topics in knowledge base, $NPMI = 0.131$ shows that the topic coherence still holds well without drifting. More generally, Fig. 3 illustrates that the category-level topics are much more

---

[3] https://dumps.wikimedia.org/enwiki/latest/enwiki-latest-categorylinks.sql.gz.

[4] https://dumps.wikimedia.org/enwiki/latest/enwiki-latest-pages-articles.xml.bz2.

[5] http://demeter.inf.ed.ac.uk/cross/docs/fsd_corpus.tar.gz.

[6] https://github.com/AKSW/Palmetto.

Table 1. The comparison on the topic coherence (NPMI) between our method and LightLDA [45], taking *Aviation* as an example. (GID is short for Group Id. * means each group contains the five top words. NPMI is computed on a group of ten words. ~ stands for the top five words.)

Category-Level Topics extracted from Wikipedia by TRANSDETECTOR				Topics Learned from Wikipedia by LightLDA[47]			
GID	#words*	words	NPMI	GID	#words*	words	NPMI
-	1-5	aircraft air airport flight airline	-	-	1-5	engine aircraft car air power	-
0	1-5, 6-10	~, airlines aviation flying pilot squadron	0.113	0	1-5, 6-10	~, design flight model production speed	0.112
1	1-5, 11-15	~, flights pilots raf airways fighter	0.155	1	1-5, 11-15	~, system vehicle cars engines mm	0.062
2	1-5, 16-20	~, boeing runway force crashed flew	0.092	2	1-5, 16-20	~, fuel vehicles designed models type	0.072
3	1-5, 21-25	~, airfield landing passengers plane aerial	0.179	3	1-5, 21-25	~, version front produced rear electric	0.035
4	1-5, 26-30	~, bomber radar wing bombers crash	0.137	4	1-5, 26-30	~, space control motor standard development	0.085
5	1-5, 31-35	~, airbus airports operations jet helicopter	0.189	5	1-5, 31-35	~, film range light using available	-0.002
6	1-5, 36-40	~, squadrons base flown havilland crew	0.088	6	1-5, 36-40	~, wing powered wheel weight launch	0.087
7	1-5, 41-45	~, combat luftwaffe aerodrome carrier fokker	0.159	7	1-5, 41-45	~, developed low test ford cylinder	0.007
8	1-5, 46-50	~, planes fly engine takeoff fleet	0.186	8	1-5, 46-50	~, equipment side pilot hp aviation	0.091
9	1-5, 51-55	~, fuselage helicopters aviator naval aero	0.157	9	1-5, 51-55	~, systems us sold body drive	-0.051
10	1-5, 56-60	~, glider command training balloon faa	0.166	10	1-5, 56-60	~, gear introduced class safety seat	0.069
...	...	...		...	...	...	...
18	1-5, 96-100	~, scheduled carriers military curtiss biplane	0.131	18	1-5, 96-100	~, transmission special replaced limited different	0.059
19	1-5, 101-105	~, accident engines iaf albatross rcaf	0.068	19	1-5, 101-105	~, features machine nuclear even unit	0.011

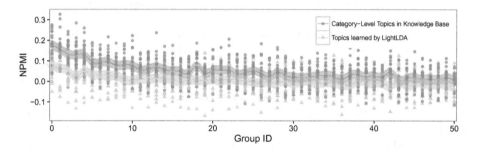

Fig. 3. More topics are compared at the NPMI metrics between our method and LightLDA

stable than the topics learned by LightLDA. Taking the group 10 (the combination of 56th to 60th words with the top five words) as an example, the median one of category-level topic performs better than those learned by LightLDA. And more topics are compared shown in Fig. 3, proving that our method performs better at extracting key features for categories.

For CTrans-LDA, K is set to be 200, which means CTrans-LDA learned 50 category-level topics and 150 other topics. After cross-validated grid-search, λ is set to be 12.8. The other parameters are set as $\alpha = 0.1$, $\beta = 0.005$. We run CTrans-LDA window by window on text stream, and learn specific categories' topics. The result in Table 2 demonstrates the effectiveness of transferring category-level topics into the microblog stream, and finding more new words in the stream which is not stored in the knowledge base.

4.2 Effects of Event Detection

Baseline Methods. We compare our proposed method against the following methods, Twevent [11], BurstyBTM [19], LSH [14], EDCoW [17], and

Table 2. Category-Level Topics extracted from knowledge base and the corresponding topics on microblog stream learned from CTrans-LDA, taking the categories *Aviation, Health, Middle East, Military,* and *Mobile Phones* as examples. The words in **bold italic** font are newly learned on the microblog stream by the transfer learning, which semantic meanings are verified consistent with the categories; while the words in normal font play the role as the bridge in transfer learning, and appear in the both category-level topics in two domains.

| Aviation | | Health | | Middle East | | Military | | Mobile Phones | |
Knowledge Base	Microblog Stream	Knowledge Base	Microblog Stream	Knowledge Base	Microblog Stream	Knowledge Base	Microblog Stream	Knowledge Base	Microblog Stream
aircraft	air	health	weight	al	***#syria***	army	killed	android	iphone
air	plane	patients	loss	israel	***#bahrain***	military	news	mobile	apple
airport	flight	medical	diet	iran	people	air	***#libya***	nokia	android
flight	time	disease	health	arab	israel	command	libya	ios	app
airline	airlines	treatment	cancer	israeli	police	force	rebels	phone	ipad
airlines	news	hospital	lose	egypt	***#libya***	regiment	people	samsung	samsung
aviation	boat	patient	fat	egyptian	#egypt	forces	police	game	mobile
flying	airport	clinical	tips	ibn	news	squadron	war	app	blackberry
pilot	force	symptoms	treatment	jerusalem	***#israel***	infantry	libyan	iphone	tablet
squadron	fly	cancer	body	syria	world	battle	attack	htc	apps

TimeUserLDA [18], which are mentioned in the Sect. 2. We implement these competing methods based on the open source community versions, e.g. EDCoW[7], or the authors' releases, e.g. BurstyBTM[8]. The above methods are set according to the descriptions in their papers. More precisely, (1) for LSH, 13 bits per hash table, 20 hash tables are set, and top 500 clusters with high entropy are selected as the event candidates; (2) for Twevent, the number of candidate bursty segments is set to be the square root of the window size, the newsworthiness threshold is set to be 4, and 375 candidate events are detected; (3) for EDCoW, the parameter γ is set to be 40, and 349 bursty "phrases" are found for evaluation; (4) for TimeUserLDA, the topic number is set to be 500, and the most 100 bursty topics are selected as candidate events; (5) for BurstyBTM, the topic number is set to be 200, which is also the number of bursty topics in the model. The information about the number of events to be evaluated is listed in Table 3, where TransDetector detects 457 events after filtering out too niche events containing less than 20 tweets.

Benchmarks and Evaluation Metrics. The evaluation is conducted on two benchmarks. The first benchmark on *Edinburgh twitter corpus* contains 27 manually labeled events [43][9], which all exist in our rebuilt dataset on the *Edinburgh twitter*'s IDs. These labeled events focus on the events that are both mentioned in twitter and newswire, e.g. *"Oslo Attacks"* and *"US Increasing Debt Ceiling"*, but still miss many important events such as *"Hurricane Irene"*, *"Al-Qaida's No. 2 Leader Being Killed"*, and popular events such as *"Harry Potter and the Deathly Hallows (Part 2)"*. To include these important events and enlarge the ground truth of realistic events pool, we build the second benchmark carefully.

---

[7] https://github.com/Falitokiniaina/EDCoW.

[8] https://github.com/xiaohuiyan/BurstyBTM.

[9] http://demeter.inf.ed.ac.uk/cross/docs/Newswire_Events.tar.gz.

We manually evaluate the candidate events detected by LSH, TRANSDETEC-TOR, Twevent, EDCoW, BurstyBTM, and TimeUserLDA, with the help of the *Wikipedia Current Event Portal*[10] and a local search engine built on Lucene. The labeling process generates the Benchmark2, and contains 395 events. We use precision and recall to evaluate each method on both benchmarks, and utilize the DERate (Duplicate Event Rate) metric [11] to measure the readability of detected events. The smaller the metric DERate, the less duplicate events to be filtered out in the application.

Results. In general, our method is better than the existing methods in terms of the precision and the recall, only sacrificing in the DERate slightly, as shown in Table 3. This is because an event could be grouped into multiple categories (e.g. the event *"S&P downgrade US credit rating"*, related to the politics category and the financial category simultaneously), but is not a problem as the method TRASNDETECTOR has already achieved the high precision and recall.

Comparing to TRASNDETECTOR, the existing methods are suffered from having to choose between precision and recall, but not both. Taking Twevent as an example, which also utilizes the knowledge base for promoting the event detection, it has three parameters to trade off, in which the newsworthiness is the most impact one. In our experiment, when setting newsworthiness to be 4, Twevent achieves its best performance in terms of F value; lower or higher of value sacrifices the recall or precision. The other methods also meet the same problem of tuning parameters, such as the topic number for TimeUserLDA and BurstyBTM, and the distance threshold for LSH.

Table 3. Overall Performance on Event Detection

Method	Number of events to be evaluated	Recall@ Bench-mark1	Precision@ Bench-mark2	Recall@ Bench-mark2	F@ Benchmark2	DERate[a] (Duplicate Event Rate)@ Benchmark2
LSH	500	0.704	0.788	0.651	0.713	0.348
TimeUserLDA	100	0.370	0.790	0.177	0.289	0.114
Twevent	375	0.741	0.808	0.658	0.725	0.142
EDCoW	349	0.556	0.748	0.511	0.607	0.226
BurstyBTM	200	0.667	0.825	0.384	0.497	**0.079**
TRANSDETECTOR	457	**0.889**	**0.912**	**0.876**	**0.894**	0.170

[a] DERate = (the number of duplicate events)/(the total number of detected realistic events)[11]

In order to understand why the existing methods cannot perform well, we dive into the results on the Benchmark1, and show the relation between the recall and the event size (number of microblogs related to the events) in Fig. 4.

---

[10] https://en.wikipedia.org/wiki/Portal:Current_events.

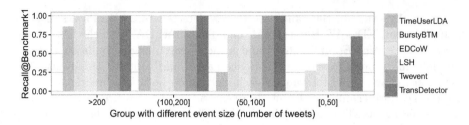

Fig. 4. The relation between the recall and the event size

Table 4. Events about *military* detected by systems between 2011-07-22 and 2011-07-28

Date	Event key words	Representative event tweet	Number of event tweet	Methods[a]					
				L	TU	TW	E	B	TD
7/22/11	Norway, Oslo, attacks, bombing	Terror Attacks Devastate Norway: A bomb ripped through government offices in Oslo and a gunman... http://dlvr.it/cLbk8	557	✓	✓	✓	✓	✓	✓
7/23/11	Gunman, rink	Gunman Kills Self, 5 Others at Texas Roller Rink http://dlvr.it/cLcTH	43	-	-	✓	✓	-	✓
7/26/11	Kandahar, mayor, suicide, attack	TELEGRAPH]: Kandahar mayor killed by Afghan suicide bomber: The mayor of Kandahar, the biggest city in south_	47	✓	-	✓	✓	-	✓
7/28/11	Ft., Hood, attack	Possible Ft. Hood Attack Thwarted http://t.co/BSJ33hk	52	-	-	-	-	-	✓
7/28/11	Libyan, rebel, gunned	Libyan rebel chief gunned down in Benghazi http://sns.mx/prfvy1	44	-	-	-	-	-	✓

[a] L = LSH [14], TU = TimeUserLDA [18], TW = Twevent [11], E = EDCoW [17], B = BurstyBTM [19], TD = TransDetector.

The 27 events are divided into 4 groups according to its size. The first group contains 5 very popular events, having more than 200 microblogs, such as *"S&P downgrade US credit rating"* with 656 microblogs, *"Atlantis shuttle lands"* with 595 microblogs, etc. The remaining groups' sizes are listed in the x-axis of Fig. 4.

The fourth group contains 11 not-so-popular events, each containing less than 50 microblogs, such as *"War criminal Goran hadzic arrested"* with only 27 microblogs reported. In our experimental settings, the methods all perform well on the very popular events, but perform worse on the not-so-popular events. For the not-so-popular events, the underlying bursty pattern is not so obvious that leads many methods to fail to catch the pattern. We further find out that the number of microblogs in an event follows a power-law distribution, which similar phenomenon is also reported in [11]. The widespread existence of such not-so-popular events challenge almost all the existing event detection methods. More result in Table 4 shows the examples of military related events detected by TRANSDETECTOR on the real dataset.

5   Conclusions and Future Work

Knowledge base is constructed elaborately and contains rich information, which can benefit the not-well-organized text stream. In our proposed TRANSDE-TECTOR method, by using category-level topic in knowledge base as the prior knowledge, we transfer abundant knowledge from knowledge base into microblog stream, which enriches the semantics of microblogs and further enhances the accuracy of microblogs event detection. As a part of our future work, we will explore the effects of transfer learning from knowledge base to text stream for more tasks, such as text classification and key words extraction, especially for short texts.

Acknowledgments. Thanks to Dr. Jian Tang and the anonymous reviewers for giving valuable suggestions on this work.

References

1. Thelwall, M., Buckley, K., Paltoglou, G.: Sentiment in twitter events. J. Am. Soc. Inf. Sci. Technol. **62**(2), 406–418 (2011)
2. Li, R., Lei, K.H., Khadiwala, R., Chang, K.C-C.: TEDAS: a Twitter-based event detection and analysis system. In: ICDE (2012)
3. Yin, J., Karimi, S., Robinson, B., Cameron, M.A.: ESA: emergency situation awareness via microbloggers. In: CIKM (2012)
4. Allan, J., Carbonell, J.G., Doddington, G., Yamron, J., Yang, Y.: Topic detection and tracking pilot study final report (1998)
5. Atefeh, F., Khreich, W.: A survey of techniques for event detection in twitter. Comput. Intell. **31**(1), 132–164 (2015)
6. Huang, J., Peng, M., Wang, H., Cao, J., Gao, W., Zhang, X.: A probabilistic method for emerging topic tracking in microblog stream. In: World Wide Web (2016)
7. Pan, S.J., Yang, Q.: A survey on transfer learning. TKDE **22**(10), 1345–1359 (2010)
8. Petrović, S., Osborne, M., Lavrenko, V.: Using paraphrases for improving first story detection in news and twitter. In: NAACL-HLT (2012)
9. Klyne, G., Carroll, J.J.: Resource description framework (RDF): concepts and abstract syntax (2006)

10. Huang, J., Abadi, D.J., Ren, K.: Scalable SPARQL querying of large RDF graphs. In: VLDB (2011)
11. Li, C., Sun, A., Datta, A.: Twevent: segment-based event detection from tweets. In: CIKM (2012)
12. Dai, W., Xue, G.-R., Yang, Q., Yu, Y.: Transferring naive bayes classifiers for text classification. In: AAAI (2007)
13. Allan, J., Lavrenko, V., Malin, D., Swan, R.: Detections, bounds, timelines: UMass and TDT-3. In: Proceedings of Topic Detection and Tracking Workshopp (2000)
14. Petrovic, S., Osborne, M., Lavrenko, V.: Streaming first story detection with application to twitter. In: HLT-NAACL (2010)
15. Wurzer, D., Lavrenko, V., Osborne, M.: Twitter-scale new event detection via K-term hashing. In: EMNLP (2015)
16. Mathioudakis, M., Koudas, N.: TwitterMonitor: trend detection over the twitter stream. In: SIGMOD (2010)
17. Weng, J., Yao, Y., Leonardi, E., Lee, F.: Event detection in twitter. In: ICWSM (2011)
18. Diao, Q., Jiang, J., Zhu, F., Lim, E.-P.: Finding bursty topics from microblogs. In: ACL (2012)
19. Yan, X., Guo, J., Lan, Y., Xu, J., Cheng, X.: A probabilistic model for bursty topic discovery in microblogs. In: AAAI (2015)
20. Yin, H., Cui, B., Lu, H., Huang, Y.: A unified model for stable and temporal topic detection from social media data. In: ICDE (2013)
21. Huang, W., Chen, W., Zhang, L., Wang, T.: An efficient online event detection method for microblogs via user modeling. In: Li, F., Shim, K., Zheng, K., Liu, G. (eds.) APWeb 2016. LNCS, vol. 9931, pp. 329–341. Springer, Heidelberg (2016). doi:10.1007/978-3-319-45814-4_27
22. Osborne, M., Petrovic, S., McCreadie, R., Macdonald, C., Ounis, I.: Bieber no more: first story detection using twitter and wikipedia. In: SIGIR Workshop on Time-aware Information Access (2012)
23. Steiner, T., Van Hooland, S., Summers, E.: MJ no more: using concurrent wikipedia edit spikes with social network plausibility checks for breaking news detection. In: WWW (2013)
24. Wu, W., Li, H., Wang, H., Zhu, K.Q.: Probase: a probabilistic taxonomy for text understanding. In: SIGMOD (2012)
25. Wang, F., Wang, Z., Li, Z., Wen, J.-R.: Concept-based short text classification and ranking. In: CIKM (2014)
26. Auer, S., Bizer, C., Kobilarov, G., Lehmann, J., Cyganiak, R., Ives, Z.: DBpedia: a nucleus for a web of open data. In: Aberer, K., et al. (eds.) ASWC/ISWC - 2007. LNCS, vol. 4825, pp. 722–735. Springer, Heidelberg (2007). doi:10.1007/978-3-540-76298-0_52
27. Suchanek, F.M., Kasneci, G., Weikum, G.: YAGO: a core of semantic knowledge unifying wordnet and wikipedia. In: WWW (2007)
28. Bollacker, K., Evans, C., Paritosh, P., Sturge, T., Taylor, J.: Freebase: a collaboratively created graph database for structuring human knowledge. In: SIGMOD (2008)
29. Kuzey, E., Weikum, G.: EVIN: building a knowledge base of events. In: WWW (2014)
30. Leban, G., Fortuna, B., Brank, J., Grobelnik, M.: Event registry: learning about world events from news. In: WWW (2014)
31. Wu, Z., Liang, C., Giles, C.L.: Storybase: towards building a knowledge base for news events. In: ACL-IJCNLP (2015)

32. Trampuš, M., Novak, B.: Internals of an aggregated web news feed. In: Proceedings of 15th Multiconference on Information Society, pp. 431–434 (2012)
33. Leetaru, K., Schrodt, P.A.: GDELT: global data on events, location, and tone, 1979–2012. In: ISA Annual Convention, vol. 2. Citeseer (2013)
34. Xiang, E.W., Cao, B., Hu, D.H., Yang, Q.: Bridging domains using world wide knowledge for transfer learning. TKDE **22**, 770–783 (2010)
35. Faralli, S., Stilo, G., Velardi, P.: Large scale homophily analysis in twitter using a twixonomy. In: IJCAI (2015)
36. Yan, R., Song, Y., Li, C.-T., Zhang, M., Hu, X.: Opportunities or risks to reduce labor in crowdsourcing translation? Characterizing cost versus quality via a pagerank-HITS hybrid model. In: IJCAI (2015)
37. Yang, Y., Pedersen, J.O.: A comparative study on feature selection in text categorization. In: ICML (1997)
38. Liu, Y., Loh, H.T., Sun, A.: Imbalanced text classification: a term weighting approach. Expert Syst. Appl. **36**(1), 690–701 (2009)
39. Blei, D.M., Ng, A.Y., Jordan, M.I.: Latent dirichlet allocation. JMLR **3**, 993–1022 (2003)
40. Wallach, H.M.: Structured topic models for language. Ph.D. thesis, University of Cambridge (2008)
41. Tang, J., Meng, Z., Nguyen, X., Mei, Q., Zhang, M.: Understanding the limiting factors of topic modeling via posterior contraction analysis. In: ICML (2014)
42. Von Luxburg, U.: A tutorial on spectral clustering. Stat. Comput. **17**(4), 395–416 (2007)
43. Petrović, S., Osborne, M., McCreadie, R., Macdonald, C., Ounis, I.: Can twitter replace newswire for breaking news? In: ICWSM (2013)
44. Petrović, S., Osborne, M., Lavrenko, V.: The edinburgh twitter corpus. In: NAACL-HLT (2010)
45. Yuan, J., Gao, F., Ho, Q., Dai, W., Wei, J., Zheng, X., Xing, E.P., Liu, T.-Y., Ma, W.-Y.: LightLDA: big topic models on modest computer clusters. In: WWW (2015)
46. Röder, M., Both, A., Hinneburg, A.: Exploring the space of topic coherence measures. In: WSDM (2015)

Indexing and Distributed Systems

AngleCut: A Ring-Based Hashing Scheme for Distributed Metadata Management

Jiaxi Liu[1], Renxuan Wang[1], Xiaofeng Gao[1(✉)], Xiaochun Yang[2],
and Guihai Chen[1]

[1] Shanghai Key Laboratory of Scalable Computing and Systems,
Department of Computer Science and Engineering,
Shanghai Jiao Tong University, Shanghai, China
{liujiaxi,penny0619}@sjtu.edu.cn, {gao-xf,gchen}@cs.sjtu.edu.cn
[2] School of Computer Science and Engineering, Northeastern University,
Shenyang, China
yangxc@mail.neu.edu.cn

Abstract. Today's file systems are required to store PB-scale or even EB-scale data across thousands of servers. Under this scenario, distributed metadata management schemes, which store metadata on a group of metadata servers (MDS's), are used to alleviate the workload of a single server. However, they present a significant challenge as the group of MDS's should maintain a high level of metadata locality and load balancing, which are practically contradictory to each other. In this paper we propose a novel and specially designed hashing scheme called AngleCut to partition metadata namespace tree and serve large-scale distributed storage systems. AngleCut first uses a locality preserving hashing (LPH) function to project the namespace tree into linear keyspace, i.e., multiple Chord-like rings. Then we design a history-based allocation strategy to adjust the workload of MDS's dynamically. The metadata cache mechanism is also adopted in AngleCut to improve the query efficiency. In general, our scheme preserves the metadata locality essentially as well as maintaining high load balancing between MDS's. The theoretical proof and extensive experiments on trace data exhibit the superiority of Angle-Cut over the previous literature.

Keywords: Metadata management · Locality preserving hashing · Distributed storage system · Namespace tree

1 Introduction

Nowadays, based on the improved storage techniques and the increased network bandwidth, file systems are able to support EB-scale datasets with a group of storage servers, meeting the needs of data-intensive applications. In file systems,

This work has been supported in part by the program of International S&T Cooperation (2016YFE0100300), the China 973 Project (2014CB3-40303), and National Natural Science Foundation of China (Nos. 61672353, 61472252 and 61322208).

S. Candan et al. (Eds.): DASFAA 2017, Part I, LNCS 10177, pp. 71–86, 2017.
DOI: 10.1007/978-3-319-55753-3_5

metadata plays a crucial role. Metadata is the data describing not only the organization of the data files, but also the structure of the whole system. It usually includes directory contents, file block pointers, file attributes, etc. [8,16]. Although the size of metadata is typically 0.1% to 1% of the size of actual data [6], it can still reach up to being PB-scale in an EB-scale file system. More importantly, around 50% to 60% access of the files is related to metadata [7]. Therefore, it is meaningful to develop an efficient metadata management scheme to improve the performance of the whole file system.

Traditionally, metadata is stored in a single metadata server (MDS) or in several MDS's by a shared-disk architecture. However, such schemes fail to satisfy the rapid growth of file systems as a single server has a maximum capability of handling intensive access. Therefore, a distributed metadata management is urgently required for current large-scale storage systems. In distributed metadata management, maintaining the hierarchy locality of metadata namespace tree and the load balancing between MDS's are the most essential parts. However, this is a significant challenge as they are actually contradictory to each other.

For example, in Fig. 1, the basic goal is to allocate the nodes of a metadata namespace tree to 3 MDS's. According to the POSIX-style permission checking and interaction standard, accessing a metadata node requires the access of all its ancestor nodes (prefix inodes) up to the root to check the R/W permission or other properties. Thus, the access to "/home/a/g.pdf" requires accessing 4 nodes: "/", "/home", "/home/a", and "/home/a/g.pdf" in order. One kind of partition scheme in Fig. 1(a), e.g. subtree partition [19], is POSIX-compliant and keeps the metadata locality well since all the 4 nodes are stored in the same MDS. However, it may face serious accessing unbalance problem since MDS #1 with a popular namespace subtree could be overloaded under burst access. The other scheme as shown in Fig. 1(b), e.g., hash-based mapping [1,15], deals well with load balancing, but it may have metadata locality issue. E.g, it needs 3 "switches" between MDS's when we access "h1.tex".

In general, hash-based mapping [1,15], static subtree partition [10,19] and dynamic subtree partition [9], along with other novel designs like Mantle [3] and DROP [17] are major ways to separate metadata into different servers. However, most literature simply designed an architecture and some algorithms but failed to

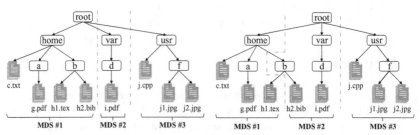

(a) Scheme with good metadata locality   (b) Scheme with good load balancing

Fig. 1. Two typical schemes of partitioning metadata to MDS's

give a uniform definition or description on the distributed metadata management problem, let alone a trade-off research among metadata hierarchy locality and load balancing among MDS's brought from the distributed manner.

In this paper, we first give concise definitions of both locality and load balancing degree to interpret the measurement of the system. Then, we present a novel scheme called AngleCut to partition metadata namespace tree and serve large-scale distributed storage systems. Our scheme first uses a novel locality preserving hashing (LPH) function to make a ring-projection and angle-assignment on the namespace tree and adopts multiple Chord-like rings as the keyspace (also called *identifier* space). The LPH function preserves the relative location of nodes from metadata namespace tree to a linear keyspace, which essentially keeps the metadata locality. Then, we design a novel history-based allocation strategy to allocate metadata uniformly to MDS's and adjust the workload dynamically. For each MDS, it adopts a sampling method based on a random walk [2] on metadata nodes to estimate *cumulative distribution function* (CDF) of the access frequency and then dynamically adjusts the workload. The strategy maintains good load balancing between MDS's and also keeps the metadata locality, which is compatible with the LPH design. Moreover, the cache mechanism is also adopted in AngleCut in order to improve the metadata query efficiency.

We evaluate the performance of AngleCut from both theoretical and practical manner to validate its efficiency. First, we theoretically prove the well-definedness on LPH and present a space complexity analysis of the hashing key. Then we implement AngleCut on 3 trace data called **MSN**, **DevTool** and **LiveMap** and illuminate various comparisons with static subtree partition [19], dynamic subtree partition [9], hash-based mapping [11] and DROP [17]. The comparison results exhibit the superiority of AngleCut over the previous literature.

The rest of the paper is organized as follows. We introduce the related work in Sect. 2 and give some definitions in Sect. 3. Then in Sect. 4 we present the AnlgeCut scheme. The performance of AngleCut is evaluated both theoretically and practically in Sects. 5 and 6. Finally, we conclude this paper in Sect. 7.

2   Related Work

Existing work on metadata management can be classified into *centralized* strategy and *distributed* strategy. Previous literature like GFS [4] adopted centralized manner, while the distributed strategy is the research orientation nowadays.

Hash-Based Mapping. Hash-Based Mapping usually maps a file's pathname or other identifiers into some hash keys and distributes the metadata to MDS's by the projection from keys of metadata to keys of MDS's. zFS [11] and CalvinFS [15] adopt this idea. Usually they efficiently balance the metadata workload among MDS's, but the hierarchical directory structure used to support directory operations for POSIX-compliant purpose is severely broken.

Subtree Partition. Subtree partition includes static and dynamic scenarios. The conventional technique is to partition the global namespace into subtrees and each MDS is in charge of one or some of them. Some distributed file systems such as CODA [12] and PVFS2 [19] follow this approach. It provides better locality and greater MDS independence than hash-based mapping, while the workload may not be evenly partitioned among MDS's, let alone handling with growth or contraction of individual subtrees over time.

Dynamic subtree partition is an optimization of static manner. File systems like GIGA+ [9] use this scheme. The key idea is that a directory hierarchy subtree can be subdivided and delegated to different MDS's and the metadata workload will be dynamically redistributed along with the variations of workloads. This approach requires an accurate load measurement method and all servers need to periodically exchange the load information.

Other Schemes. Among other state-of-the-art schemes of metadata management, Dynamic Dir-Grain [16] observes that the partition granularity of static subtree partition and dynamic subtree partition may be too large. It proposes a triple ⟨D,D,F⟩ to determine the maximum granularity. Mantle [13] introduces a programmable metadata load balancer for the Ceph file system. DROP [17] is an efficient and scalable distributed scheme to serve EB-scale file systems. It exploits an LPH to distribute metadata among MDS's and adopts HDLB strategy to quickly adjust the metadata distribution. However, DROP requires much extra space since it preserves many bytes to store an augmented hash key. Also, it does not provide the analysis on its locality and balancing features.

Distributed Hash Table. In peer-to-peer (P2P) systems, a distributed hash table (DHT) is a class of a decentralized distributed data structure that provides a lookup service. Chord [14] is one of the most classic DHT protocol design which adopts a variant of consistent hashing to assign keys to Chord nodes. It applies an m-bit identifier ring as the hash key space, which improves the scalability and enhances the performance of the system.

The LPH techniques are usually adopted in DHT [20] and some other areas [18] since they preserve the neighborhood structure of data set, i.e., ensure that close keys are assigned to close objects. In general, our AngleCut scheme is inspired by several designs in DHT. It is essentially an LPH and adopts a variant of *identifier* ring in Chord as the keyspace.

3   Problem Formulation

In this section, we present concise definitions of the distributed metadata management problem in a formal way. We first give a clear mathematical definition of the metadata locality, load balancing degree between MDS's and migration cost to interpret the measurement of the system. Next, we present the goal of our optimization problem.

Locality. Given a typical metadata namespace tree with N metadata items, let $T = \{n_i \mid 0 \leq i \leq N\}$ denote the set of all metadata nodes in which n_0 is the root. We define *Jumps* J_i of a metadata node n_i as the number of "switches" (i.e. the hop distance) between MDS's during the recursive search from top to bottom on the tree. Considering that different metadata nodes have different popularity and contribute to the total *Jumps* distinctively, we use access frequency f_i to describe the popularity of n_i. Notably, the access of each metadata node may come from the dedicate access to itself or the process of accessing its descendants. Now, we give the definition of the *locality* to describe the aggregation degree of the metadata nodes stored in several MDS's.

Definition 1. *Define Loc as the global locality value of the whole system, which can be deduced from:*

$$Loc = 1 \Big/ \sum_{i=1}^{N} J_i \cdot f_i \tag{1}$$

Good *locality* means that "close" metadata nodes are more likely to be assigned to a same MDS, which reduces the response delay and improves the efficiency of the distributed system (The definition of "close" will be discussed in Sect. 5.1). In our algebraic expression, large *Loc* means the *locality* is good. *Loc* is $+\infty$ under single server scenario intuitively.

Load Balancing Degree. Earlier studies like [6] have proposed several load balancing algorithms, targeting static, small-scale and/or homogeneous environments. We now provide a concise review of them. Assume there are M MDS's, let $L_i, 1 \le i \le M$ be the current load, i.e. the sum of access frequencies, of the ith MDS and C_i be the maximum capacity. Define the global ideal load factor μ as $\mu = \frac{\sum_{i=1}^{M} L_i}{\sum_{i=1}^{M} C_i}$, then we can compute the ideal load I_i of the ith MDS with $I_i = \mu C_i$. Correspondingly, the remaining capacity is $R_i = L_i - I_i$. If $R_i > 0$, it means that the ith MDS is light. Otherwise, it is heavily loaded.

Considering the load of each MDS as *sample* in *Statistics*, the *load balancing degree* of the whole system is defined like the inverse of the *sample variance* of the load of all MDS's, which is consistent with the literal meaning of "balance".

Definition 2. *The* load balancing degree *of all MDS's is defined as*

$$Bal = 1 \Big/ \frac{1}{M-1} \sum_{i=1}^{M} \left(\frac{L_i}{C_i} - \mu \right)^2 = \frac{M-1}{\sum_{i=1}^{M} (\frac{L_i}{C_i} - \mu)^2} \tag{2}$$

Good *load balancing degree* means that every MDS's current load is close to its ideal load, i.e., *Bal* is large.

Migration Cost. As the access frequency of the system varies with time, we may need to migrate metadata between MDS's to keep a balanced workload. In this paper, we simply define Mig_k as the migration cost of the k-th MDS involved in a dynamic adjustment without considering the topological structure of the MDS cluster.

Problem Statement. To keep the metadata locality and the load balancing of the whole system, our goal is to improve the *locality* and the *load balancing* *degree* while partitioning the metadata tree. Since excessive migration may cause huge update cost, the *migration cost* is also worth to be considered. Thus, we define the problem as an optimization problem. We set 2 coefficients λ and μ on *locality* and *load balancing degree* respectively and a constraint σ on *migration cost*. Then, the goal is to maximize a linear combination of *Loc* and *Bal*.

$$\max \qquad \lambda Loc + \mu Bal \qquad (3)$$

$$s.t. \qquad \bigcup_{i=0}^{N} n_i = T, \qquad (4)$$

$$Mig_k \leqslant \sigma, \forall 1 \leq k \leq M \qquad (5)$$

Equation (4) means MDS's should contain all nodes while Eq. (5) shows the constraint on *migration cost*.

4   The AngleCut Scheme

In this section, we present a detailed description of our AngleCut scheme, in which the LPH design and the dynamic allocation strategy help maintain a high level $\lambda Loc + \mu Bal$. We first present the hashing design of AngleCut in Sects. 4.1 and 4.2. In Sect. 4.3, we introduce the history-based allocation strategy which adjusts the workload of MDS's dynamically. Finally, we give a discussion on the metadata cache management and metadata query mechanism in Sect. 4.4.

4.1   The Hashing Design

As mentioned above, AngleCut uses a novel LPH function to hash the metadata namespace tree onto a linear keyspace, i.e., it projects the metadata nodes to multiple Chord-like rings by assigning each node with an angle value. Fig. 2(a) illustrates an example of projection in AngleCut from a simple metadata tree to a single ring. The tree is composed of 4 layers and we define node 0 as the 0-th layer. For nodes except the root, they are projected to the ring in a left-hand depth-first-search order, in which "invalid" means the nodes may have been deleted. Notably, node C and D are projected to the same location as they have the same angle. In Fig. 2(b), we present more information on the angle of the corresponding nodes in Fig. 2(a), e.g., the angle of both node C and D is 120^δ.

In our hashing design, the hash key of a metadata node n_i is represented by two key components: *tag* and *ang*. The $n_i.tag$ is used to store some identification information of n_i while the $n_i.ang$ is used to store its angle. We now give a detailed interpretation of the design.

tag: The *tag* is a 1 byte key including effective bit, the order number *ord* of the ring and some reserved bits. The effective bit indicates whether the metadata node is valid, e.g., the node is invalid after being deleted. The order number *ord* is used to show which ring the metadata node locates on. At present we take the single ring scenario as an example, which means all metadata nodes have $ord = 0$. The multiple rings design is relevant to the update and maintenance of the system, which will be introduced in Sect. 4.2.

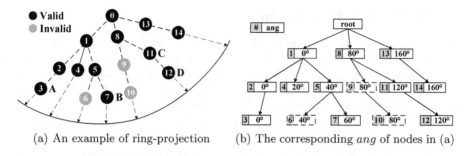

(a) An example of ring-projection   (b) The corresponding *ang* of nodes in (a)

Fig. 2. An example illustrating the hashing design on calculating the *ang*

ang: *ang* is the kernel component of the hash key, which is closely related to the *locality*. Given a metadata tree and a ring, the *ang* of one metadata node is the angle of the projective node on the ring, ranging in $[0°, 360°)$. To calculate the *ang*, we first give a definition of *coordinate* to represent the "location" of node n_i in a metadata tree.

Definition 3. *Given a metadata tree, the coordinate of a non-root node n_i is $(c_1, c_2, \ldots, c_{d_i})$. The dimension d_i of the coordinate is the depth of n_i from the root and $c_j, 1 \leq j \leq d_i$ represents the number of left siblings of n_i on the j-th layer in counter-clockwise direction.*

For example shown in Fig. 2, the *coordinates* of node A, B, C and D are $(0, 0, 0)$, $(0, 2, 1)$, $(1, 1)$ and $(1, 1, 0)$ respectively.

With the *coordinate* of metadata node n_i, we are ready to calculate its angle $n_i.ang$. Firstly, all nodes on the 1-st layer share the $360°$ angle uniformly. We denote the *ang* difference between any two adjacent nodes on the 1-st layer by θ. For example in Fig. 2(b), we have $n_8.ang - n_1.ang = \theta = 80°$. Secondly, for children nodes on the 2-nd layer, all nodes on the rightmost location such as node 4 and 11 own an *ang* increment of $\theta/2$ while the other nodes share the $\theta/2$ angle range uniformly since they are treated equally, e.g., $n_{11}.ang = 120°$ and $n_4.ang = 20°$. Similarly, the *ang* of the rightmost node in the 3-rd layer increases by $\theta/4$ and its left siblings share the $\theta/4$ angle range uniformly. We continue this process until all nodes are assigned with the *ang* key. The angle-assignment is executed inductively and the details are shown in Algorithm 1.

The angle-assignment of our hashing design can keep the *locality* of the namespace tree. Note that from the 2-rd layer to bottom, the largest *ang* increments of rightmost children nodes are $\theta/2, \theta/4, \theta/8 \ldots$ respectively. Now, for a node n_0 on the k-th layer with $ang = \delta$, the *ang* of its any child node δ_c satisfies $\delta_c < \varphi + \theta/2^k + \theta/2^{k+1} + \theta/2^{k+2} + \cdots < \varphi + \theta/2^{k-1}, k \geq 1$. It means any child node's *ang* will not go beyond the *ang* of the adjacent right sibling of n_0. This is how the AngleCut scheme keeps the *locality*. We will give a formal and detailed proof in Sect. 5.1.

4.2   The Multiple Rings Design

The aforementioned hashing design illustrates the angle-assignment of a given metadata namespace tree, which deals well with the static scenario with only read, write requests. However, when meeting dynamic adjustment transactions such as insert request on a metadata node, there may exist no reserved angle on the corresponding layer, i.e., the angle for the new node may be occupied by the nodes on the existing rings. To deal with this, we propose to use multiple rings as hash keyspace, i.e., use an identifier ord which is stored in tag to denote which ring the metadata node locates on. Consider insert request on a new metadata node n_i, AngleCut checks the existing rings to make sure whether there is a reserved ang. If not, AngleCut would create a new ring to place it, i.e., set a new value to the ord of n_i. The ang are computed still in the same way, which means the $coordinate$ and $n_i.ang$ are the same as some nodes in other rings. The ord is the only identifier to distinguish them.

When meeting delete request, AngleCut needs to set the requested metadata node and its descendant nodes which may be distributed on the multiple rings invalid. As shown in Fig. 2(a), when deleting node 9, both node 9 and node 10 are set invalid.

Algorithm 1. Cycle-Projection

Input: A namespace Tree
Output: The ang keyspace
1 **for** $i \in [0, N]$ **do**
2 $\quad$ set n_i valid;
3 $\quad$ **if** $i > 0$ **then**
4 $\quad\quad$ Get the $coordinate$ of n_i;
5 $\quad$ $C_{n_i} = $ # of children nodes of n_i;
6 $\theta = 360/C_{n_0}$;
7 **for** $i \in [1, N]$ **do**
8 $\quad$ $n_i.ang = $
$\quad\quad \sum\limits_{j=1}^{d_i} \dfrac{c_j\theta}{(C_{n_i.ancestor}-1)2^{j-1}}$;
9 add n_i to the rings;

Algorithm 2. Boundaries Adjustment

Input: Metadata accessing patterns
Output: New boundaries ϕ_i
1 **foreach** MDS **do**
2 $\quad$ sample recent accessing patterns;
3 estimate temp. CDF;
4 calculate ψ_1 to ψ_{M-1} by temp. CDF;
5 calculate ϕ_1 to ϕ_{M-1} by opt. CDF;
6 **for** $i \in [1, M-1]$ **do**
7 $\quad$ $\phi_i = \alpha \cdot \phi_i + (1-\alpha) \cdot \psi_i$;
8 unite temp. CDF into opt. CDF;
9 update α;
10 **return** $\phi_1, \dots, \phi_{M-1}$;

4.3   Metadata Allocation

Since we have obtained multiple rings as the keyspace, the next step is to allocate metadata nodes uniformly to the M MDS's, which is an NP-hard problem reduced from the 0-1 Knapsack problem [17]. We now propose a novel history-based allocation strategy to allocate metadata uniformly to MDS's and adjust the workload dynamically.

In oder to allocate the metadata nodes, an essential step is to get the distribution of metadata access frequency as to ang in the keyspace, while it is impractical to leverage all metadata nodes to simulate the *cumulative distribution function* (CDF) since the quantity is considerably huge. Thus, AngleCut adopts a sampling strategy based on a random walk [2] on metadata nodes to estimate a temporal CDF of the access frequency with respect to ang. Notably, the temporal CDF can only indicate the accessing patterns in the recent past. With this strategy, we are able to divide the angle range $[0, 360°)$ to M arcs in an approximately uniform manner and derive $M-1$ temporary angle range boundaries $\psi_1, \ldots, \psi_{M-1}$.

Compared with some traditional methods [14] which divide the keyspace to more subspace, the allocation is straightforward but reasonable due to one remarkable thing that any partitioning of the keyspace may break the hierarchical locality of metadata. So it is optimal by letting each MDS be responsible for a continuous arc. Figure 3 illustrates the metadata allocation strategy in which the projective nodes of metadata located on the rings are allocated to 3 MDS's. It also includes the angle-assignment which is consistent with Fig. 2 and the multiple rings design mentioned in Sect. 4.2.

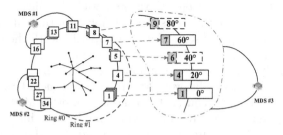

Fig. 3. An illustration for metadata allocation in AngleCut

As the access frequencies of nodes change over time, the workload of MDS's may become unbalanced. Once over a period of time, the MDS's need to renew the sampling and update the CDF estimation of access frequency. In our strategy, each MDS also keeps a history record of optimal CDF, which indicates the optimal accessing patterns in each period of accessing history. Similarly, we calculate the optimal angle range boundaries $\phi_1, \ldots, \phi_{M-1}$ based on it. To avoid the abrupt change of boundaries and match the subsequent accessing patterns, in each update process we set the optimal boundaries ϕ_i as a weighted sum of previous optimal boundaries and temporary boundaries by assigning weight $\alpha \in (0, 1)$, i.e., $\phi_i = \alpha \cdot \phi_i + (1 - \alpha) \cdot \psi_i$, $\forall i \in [1, M-1]$. The optimal boundaries adjustment algorithm is presented in Algorithm 2.

There are many factors like similarities between recent accessing patterns that affect α. Though it is better to determine α by a Deterministic Finite Automaton (DFA), we keep it as a simple parameter to avoid redundant records and calculation of the weight between several factors. In this way, α increases monotonically since the optimal CDF includes more and more information about the accessing patterns of the system. However, it should have an upper bound since the temporal CDF can indicate the future accessing trend, i.e., it should always get a certain weight. The upper bound is not easy to be theoretically determined while we prefer to adjust it in experiments.

For each update of the optimal boundaries, each MDS needs to update its hash tables which store the hash keys allocated from the keyspace to it.

The corresponding metadata nodes will be migrated as well. We use experiments in Sect. 6 to show that AngleCut averts a large-scale metadata migration.

4.4   Cache Management and Metadata Query

POSIX-style standard requires many metadata operations to perform pathname traversal and permission checking across each ancestor node [15]. This accessing pattern is not well balanced across MDS's because metadata nodes near the top of the namespace tree are accessed much more frequently than those at the lower part of the tree. The other disadvantage is that it may bring repetitive *Jumps* between MDS's and reduce the *Loc*. To improve read and write efficiency of metadata, cache technology is adopted in AngleCut. Cache module is placed on each MDS and saves hot or being-written metadata which are not allocated to this MDS, especially the top parts of the namespace tree. Metadata cache is also organized by a mapping table, which is compatible with the hashing design of AngleCut. Each newly cached metadata is denoted by its *tag* and *ang* key components. The key is added to the mapping table then.

With cache mechanism, the working process of a metadata query is: one MDS receives an read/write requests from end user or other MDS and checks the ancestor nodes (prefix inodes) from its own cache. If cache hits, it reads/writes cache directly, which can assure the returned metadata is the latest. Otherwise, to read requests the MDS checks the hash tables and forwards the requests to the corresponding MDS, while write requests are recorded in cache until they are updated to MDS at the same time. The cache design greatly decreases network communication load and improves the query efficiency.

5   Theoretical Analysis

5.1   Well-Definedness on Locality Preserving Hashing

Locality preserving hashing (LPH) is a kind of hash function satisfying the property of preserving or keeping the *locality* as well as reducing the dimensionality of the input data. For three points A, B and C in a high dimensional metric space, an LPH function f needs to satisfy the condition: $|A - B| < |B - C| \Rightarrow |f(A) - f(B)| < |f(B) - f(C)|$.

Consider the previous example shown in Fig. 2(a). Intuitively it seems that C and D have shorter "distance" than B and D while the *ang* difference of C and D is smaller too. However, this has no theoretical basis. We now present a proof of well-definedness of AngleCut on LPH.

Theorem 1. *The hashing to ang in AngleCut is an LPH which maps the namespace tree space to linear space.*

Proof. First, we propose a definition of *distance* between any two nodes in a typical metadata namespace tree.

Definition 4. *Consider two non-root metadata nodes n_1 and n_2 with respective coordinate denoted by $(x_1, x_2, \ldots, x_i)$ and $(y_1, y_2, \ldots, y_j)$, where i, j are the respective dimensions of coordinate. The distance between n_1 and n_2 is defined as follows:*

$$f_c(n_1, n_2) = \sum_{k=1}^{\min\{i,j\}} w_k \cdot |x_k - y_k| \qquad (6)$$

The setting of weight function w_k is because each coordinate component should have different weight since it represents "location" on different layers of the namespace tree. Typically, we define that the weight function is $w_k = 1/2^k$ here. In Fig. 2(a), we have $f_c(B, D) = 7/8$ while $f_c(C, D) = 0$.

As to the definition of *distance* in *ang* keyspace, for nodes n_1, n_2, we define their distance $f_a(n_1, n_2)$ is just the difference of their *ang*. e.g., $f_a(B, D) = 60°$ and $f_a(C, D) = 0°$. Finally, we get the deduction:

$$f_c(C, D) < f_c(B, D) \Rightarrow f_a(C, D) < f_a(B, D) \qquad (7)$$

Without loss of generality, we choose nodes B, C and D as examples in the proof. In fact, for any two nodes in the namespace tree, it is not hard to see the hashing to *ang* does preserve the locality.

5.2 Space Complexity of AngleCut

Theorem 2. *For a metadata namespace tree with N non-root nodes, the total space to store hashing key is $17N$ bytes on overage.*

Proof. As mentioned in Sect. 4.1, first, each node uses one byte for *tag* to store the effective bit, *ord* and so on. Second, each node needs a *double* value to record its *ang*, adding up to $8N$ bytes for the whole system. Besides, to find the MDS's where the child nodes are stored, each node should also record the angle of its children nodes. Let $D_i = \{d_i^1, d_i^2, \ldots, d_i^{|D_i|}\}$ denote the children nodes set of node n_i, the space for storing D_i is $8D_i$ bytes then. However, as each node has exactly one parent node, the space for storing all children nodes is $8N$ bytes in total. To summarize, the total space to store hashing key is $17N$ bytes on overage.

For DROP [17], it preserves $48N$ bytes in total for the same namespace tree in order to append an augmented hash key to each node. It indicates that AngleCut outperforms DROP a lot as to space complexity.

6   Performance Evaluation

To validate our scheme, we implement AngleCut in C++ and evaluate it based on our distributed storage framework [5]. We create several virtual machines on a server with eight-core 4.0 GHz Intel i7-6700K CPU and 64GB memory. Each virtual machine has 1 GB memory and 20 GB storage. Experimental scale of MDS's ranges from 5 to 30, with an increment of 5. We also implement and make comparisons with subtree partition [19], dynamic subtree partition [9], hash-based mapping [11] and DROP [17]. Some new schemes like CalvinFS [15] are not included since we can not find enough technique details on metadata management part. The datasets we use are three real-world traces[1] called **MSN**, **DevTool** and **LiveMap**. The detailed information is shown in Table 1.

Table 1. The data traces used in experiments

Trace source	Data size	Records count	Duration for trace collection
MSN storage file server	5.77 GB	26,657,063	6 h
Development tools release	5.90 GB	17,952,090	24 h
Live maps back end	15.1 GB	44,755,552	24 h

Experiments are conducted in the following ways. For each scale of MDS's and specific dataset, we evaluate the namespace *locality*, the *load balancing degree* and the *migration cost* of AngleCut. We interact with the running MDS's with one master MDS in a centralized manner, mainly in charge of: (1) the distribution of query tasks. (2) the generation of *ang* hash tables. For each experiment, the master MDS continuously distributes query tasks randomly over all MDS's. After each 5% of all query records in the traces, every MDS renews a sampling on metadata nodes stored in it to estimate a temporal CDF of access frequency and sends it to the master. The master updates the hash table and broadcasts it to all MDS's. The update process is executed 20 times in total.

Metadata Locality. In this experiment, we use the *Loc* of static subtree partition scheme as a baseline as it is the ideal scenario for keeping high metadata locality. In Fig. 4, we present the *Loc* ratio from the other mechanisms mentioned above to static subtree partition on three traces. The results of varying M (the number of MDS's) show that AngleCut has great performance on *locality*, which is twice as good as DROP and three times better than the hash-based mapping for all the three traces. It has almost the same performance as dynamic subtree partition, especially when the number of MDS's is small. This is because when we have less MDS's, dynamic subtree partition distributes many small subtrees over MDS's, so the tree is more separated, while AngleCut cuts the tree into less parts. However, as the scale of MDS's increases, AngleCut breaks more edges of the tree, which adds to more "switches" between MDS's and the locality becomes worse then. In contrast, the small granularity in Dynamic subtree partition shows its advantage.

Load Distribution. The adjustment of load balancing between MDS's is a dynamic process. In our experiment, the workload rebalance and boundary adjustment are executed 20 times in total. Similarly, we set the *Bal* of hash-based mapping as the baseline and present the ratio of *Bal* to it after the last rebalance in Fig. 4. The results show that the static subtree partition performs the

---

[1] Data source: http://iotta.snia.org/traces/158.

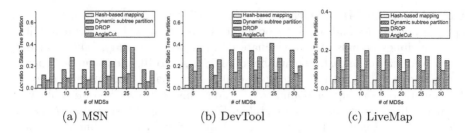

Fig. 4. The locality performance under different schemes

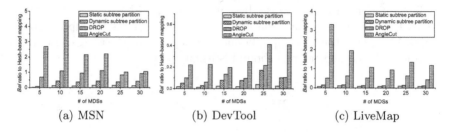

Fig. 5. The load balancing performance under different schemes

worst while dynamic subtree partition is slightly better, since it adopts dynamic adjustment to make up for load imbalance. Both AngleCut and DROP have much better load balancing performance and AngleCut performs twice as good as DROP in most cases, sometimes even much more better (Fig. 5).

The results also show that the load balancing performance of hash-based mapping on **DevTool** trace is remarkably distinctive. It is different from the previous *locality* case in which the static subtree partition always has the best *locality*. The reason is that, in hashed-based mapping, all metadata items are just evenly assigned to each MDS without taking access frequencies into consideration. For example, in **MSN** and **LiveMap**, queries concentrate on one or two subtrees and the corresponding MDS's are heavily loaded. Thus, AngleCut can even perform better than hashed-based mapping on these two traces.

The Migration Cost. With the insert or delete of metadata nodes and the join or leaving of MDS's, AngleCut needs to migrate metadata between MDS's for maintaining a good *load balancing degree*.

Note that two factors may affect the migration cost Mig_k of the kth MDS. One is the scale of the trace, the other is the number of MDS's since more MDS's sharing the migration may decrease each Mig_k in general. Thus, we use the Mig which denotes the means of migration proportion, i.e., $\frac{\sum Mig_k}{\# \ of \ MDS's}$, to measure the performance. The result in Fig. 6 shows that less than 2% metadata items are migrated on average. The cost is quite low under a maintenance of high Bal. The Mig of **MSN** trace is relatively lower than others in that its accesses are quite evenly distributed over different subtrees.

Upper Bound of α. As mentioned in Sect. 4.3, α controls the proportion of optimal boundaries and temporal boundaries. We conduct the experiment with different upper bounds of α to evaluate its influence on *locality, load balancing degree* and *migration cost*. The number of MDS is set to 10, and the upper bound varies from 0.2 to 1, with an increment of 0.2. Note that setting upper bound to 1 means there is no upper bound to α. Under this scenario, the *ang* table will gradually become more and more stable. Figures 7, 8 and 9 show how *locality, load balancing degree* and *migration cost* change with respect to different upper bounds of α. The results show that neither *locality* nor *load balancing degree* has absolute positive correlation or negative correlation with the upper bound of α. Thus, it is hard to decide the best upper bound. Moreover, in the **MSN** and **DevTool** trace, the migration cost decreases rapidly as the upper bound increases. In general, the best upper bound of α depends on the accessing pattern of a data set, but a higher upper bound is usually better.

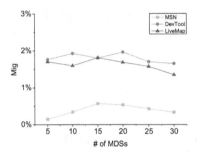

Fig. 6. Average migration cost

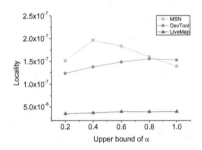

Fig. 7. The effect of α on locality

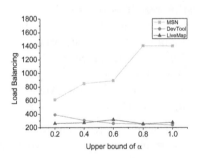

Fig. 8. Effect on load balancing

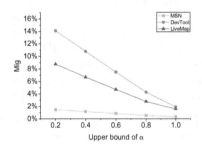

Fig. 9. Effect on migration cost

7 Conclusion

In this paper, we proposed a novel AngleCut scheme to partition metadata namespace tree and serve large scale storage systems. AngleCut first uses a locality preserving hashing function to project the namespace tree into multiple

Chord-like rings. Then, we design a novel history-based allocation strategy to allocate metadata uniformly to MDS's and adjust the workload dynamically. The metadata cache mechanism which decreases network communication load and improves the query efficiency is also adopted in AngleCut. AngleCut preserves the metadata *locality* essentially as well as maintaining a high *load balancing degree* between MDS's. The theoretical proof and extensive experiments on three trace data exhibit the superiority of AngleCut over the previous literature.

References

1. Brandt, S.A., Miller, E.L., Long, D.D., Xue, L.: Efficient metadata management in large distributed storage systems. In: MSST, pp. 290–298 (2003)
2. Fonseca, P., Rodrigues, R., Gupta, A., Liskov, B.: Full-information lookups for peer-to-peer overlays. TPDS **20**(9), 1339–1351 (2009)
3. Fu, Y., Xiao, N., Zhou, E.: A novel dynamic metadata management scheme for large distributed storage systems. In: HPCC, pp. 987–992 (2008)
4. Ghemawat, S., Gobioff, H., Leung, S.T.: The Google file system. In: SOSP, vol. 37, pp. 29–43 (2003)
5. Hong, Y., Tang, Q., Gao, X., Yao, B., Chen, G., Tang, S.: Efficient R-tree based indexing scheme for server-centric cloud storage system. TKDE **28**(6), 1503–1517 (2016)
6. Miller, E.L., Greenan, K., Leung, A., Long, D., Wildani, A.: Reliable and efficient metadata storage and indexing using NVRAM (2008), http://dcslab.hanyang.ac.kr/nvramos08/EthanMiller.pdf
7. Ousterhout, J.K., Da Costa, H., Harrison, D., Kunze, J.A., Kupfer, M., Thompson, J.G.: A trace-driven analysis of the UNIX 4.2 BSD file system. In: SOSP, vol. 19 (1985)
8. Park, J.-H., Kim, B.-K., Lee, Y.-H., Lee, M.-W., Jung, M.-O., Kang, J.-H.: XQuery-based TV-anytime metadata management. In: Zhou, L., Ooi, B.C., Meng, X. (eds.) DASFAA 2005. LNCS, vol. 3453, pp. 151–162. Springer, Heidelberg (2005). doi:10.1007/11408079_15
9. Patil, S., Gibson, G.A.: Scale and concurrency of giga+: file system directories with millions of files. In: FAST, vol. 11 (2011)
10. Pawlowski, B., Juszczak, C., Staubach, P., Smith, C., Lebel, D., Hitz, D.: NFS version 3: design and implementation. In: USENIX ATC, pp. 137–152 (1994)
11. Rodeh, O., Teperman, A.: zFS-a scalable distributed file system using object disks. In: MSST, pp. 207–218 (2003)
12. Satyanarayanan, M., Kistler, J.J., Kumar, P., Okasaki, M.E., Siegel, E.H., Steere, D.C.: CODA: a highly available file system for a distributed workstation environment. TC **39**(4), 447–459 (1990)
13. Sevilla, M.A., Watkins, N., Maltzahn, C., Nassi, I., Brandt, S.A., Weil, S.A., Farnum, G., Fineberg, S.: Mantle: a programmable metadata load balancer for the ceph file system. In: SC, pp. 21:1–21:12 (2015)
14. Stoica, I., Morris, R., Liben-Nowell, D., Karger, D.R., Kaashoek, M.F., Dabek, F., Balakrishnan, H.: Chord: a scalable peer-to-peer lookup protocol for internet applications. TON **11**(1), 17–32 (2003)
15. Thomson, A., Abadi, D.J.: Calvinfs: consistent wan replication and scalable metadata management for distributed file systems. In: FAST, pp. 1–14 (2015)

16. Xiong, J., Hu, Y., Li, G., Tang, R., Fan, Z.: Metadata distribution and consistency techniques for large-scale cluster file systems. TPDS **22**(5), 803–816 (2011)
17. Xu, Q., Arumugam, R.V., Yong, K.L., Mahadevan, S.: Efficient and scalable metadata management in EB-scale file systems. TPDS **25**(11), 2840–2850 (2014)
18. Xue, M., Papadimitriou, P., Raïssi, C., Kalnis, P., Pung, H.K.: Distributed privacy preserving data collection. In: Yu, J.X., Kim, M.H., Unland, R. (eds.) DASFAA 2011. LNCS, vol. 6587, pp. 93–107. Springer, Heidelberg (2011). doi:10.1007/978-3-642-20149-3_9
19. Yu, W., Liang, S., Panda, D.K.: High performance support of parallel virtual file system (pvfs2) over quadrics. In: ICS, pp. 323–331 (2005)
20. Zhang, X., Shou, L., Tan, K.-L., Chen, G.: iDISQUE: tuning high-dimensional similarity queries in DHT networks. In: Kitagawa, H., Ishikawa, Y., Li, Q., Watanabe, C. (eds.) DASFAA 2010. LNCS, vol. 5981, pp. 19–33. Springer, Heidelberg (2010). doi:10.1007/978-3-642-12026-8_4

An Efficient Bulk Loading Approach of Secondary Index in Distributed Log-Structured Data Stores

Yanchao Zhu[1], Zhao Zhang[1,2($\boxtimes$)], Peng Cai[1], Weining Qian[1],
and Aoying Zhou[1]

[1] School of Data Science and Engineering,
East China Normal University, Shanghai, China
yczhu@stu.ecnu.edu.cn, {zhzhang,pcai,wnqian,ayzhou}@sei.ecnu.edu.cn
[2] School of Computer Science and Software Engineering,
East China Normal University, Shanghai, China

Abstract. How to improve reading performance of Log-Structured-Merge (LSM)-tree gains much attention recently. Meanwhile, constructing secondary index for LSM data stores is a popular solution. And bulk loading of secondary index is inevitable when a new application is developed on an existing LSM data stores. However, to the best of our knowledge there are few studies on research of bulk loading of secondary index in distributed LSM-tree. In this paper, we study the performance improvement of bulk loading of secondary index in *distributed* LSM-tree data stores. We propose an efficient bulk loading approach of secondary index in Log-Structured Data Stores. Firstly, we design secondary index structure based on distributed LSM-tree to guarantee the scalability and consistency of secondary index. Secondly, we propose an efficient framework to handle bulk loading of secondary index in a distributed environment, which can provide a good load balancing for query processing by using *equal-depth histogram* to capture data distribution. Analysis of theoretical and experimental results on standard benchmark illustrate the efficacy of the proposed methods in a distributed environment.

Keywords: Secondary index · Distributed bulk loading · Load balancing

1 Introduction

Recently, NoSQL databases are becoming more and more popular to support scale-out applications, such as BigTable [11], LevelDB [3] and Hbase [1], etc. Most NoSQL storage engines are implemented by *distributed LSM-tree* [13], which is a tree-like data structure that has high performance in write-intensive workloads.

For distributed LSM-trees, fast read operations are challenging because the data is partitioned by the primary key. Each partition is distributed to different

© Springer International Publishing AG 2017
S. Candan et al. (Eds.): DASFAA 2017, Part I, LNCS 10177, pp. 87–102, 2017.
DOI: 10.1007/978-3-319-55753-3_6

nodes in a distributed environment. Furthermore, in order to get consistent query results, the corresponding relation tuples will be merged from disk stores and in-memory stores [13]. The system must scan the whole data set distributed to different nodes in order to answer queries on non-primary key attributes.

Building a secondary index is a popular solution to decrease response time of the queries on non-primary key attributes. For an empty LSM-tree data store, several methods have been proposed to build secondary indices efficiently. For example, Tan et al. [14] introduce the *Diff-index* structure to support index maintenance schemes. However, constructing a secondary index on an existing distributed LSM data store is very inefficient if traditional insertion operator is adopted. Unfortunately, bulk loading of a secondary index is inevitable when a new application is developed on a given LSM-tree data store.

In this study, we address the issues of constructing secondary index for a given LSM-tree data set, i.e. bulk loading of the secondary index on distributed LSM-tree data stores. The major difficulties include (1) the data is distributed to several nodes; and (2) the data store in the memory and in the disk at the same time. If each node maintains its own local index, high selective queries will be inefficient.

To construct the secondary index on an existing LSM-tree data store, the key task is to do efficient global sorting on search keys of the secondary index for bulk loading. An approach has three stages, including local index construction, index partition division and global index construction. Firstly, each data node builds its own local index and sends statistic information of search keys to the coordinator. Secondly, the coordinator generates multiple uniform ranges on search keys based on statistic information from data nodes, and sends the ranges to corresponding data nodes. Finally, data nodes receive ranges from the coordinator to create index partitions by shuffling data. Particularly, we propose *equal-depth histograms* to capture data distributions on search keys in Stage 1.

We have made the following contributions in this paper.

- We design a global sorting approach for bulk loading of secondary index on distributed LSM-tree.
- We employ *sampling equal-depth histogram* to capture data distribution.
- We integrate the bulk loading algorithm into our distributed database CEDAR [2] and measure the system performance using a standard benchmark. Experimental results show the efficacy of the proposed methods.

The rest of paper is organized as follows: Sect. 2 introduces background knowledge related to our work. Section 3 gives an overview of index construction. Discussion of details of bulk loading is in Sect. 4. We then provide an evaluation of the approach in Sect. 5. Finally, we describe related work in Sect. 6, and conclude our paper in Sect. 7.

2   Background

We begin with an overview of the LSM-tree. An implementation of LSM-tree is shown in Sect. 2.1. And the index structure based on the LSM-tree in distributed systems is introduced in Sect. 2.2.

2.1   System Model

The LSM-tree is a data structure that is optimized for frequent updates. It comprises a tree-like in-memory store and several tree-like immutable disk stores. Writing into LSM-tree equals to an insertion into the in-memory store. When the size of the in-memory store reaches a threshold, its content will be flushed to the disk. Multiple disk stores will be generated after several data merge processes. Thus read operations need to merge disk stores with the in-memory store to get consistent data, which greatly affects the performance of query. In order to improve the performance of query processing and to save disk space, multiple small disk stores can be merged into a large disk store periodically.

A typical implementation of a distributed storage system employing the LSM-tree is shown in Fig. 1. A table in the system is divided into partitions by a continuous primary key. Each partition is denoted by a range of primary keys $[start\_key, end\_key)$. Partitions are stored on the SSD on data nodes in a replicated way. The system meta data is stored on a node called the coordinator. Modifications of the database are stored in the memory of a node called the transaction node. When the size of in-memory store reaches a threshold, a merge process is launched. A new in-memory store will accept modifications and the old in-memory store will be merged with disk stores on data nodes, applying changes of data to the disk. After the merge process, new partitions are in service and old partitions are deleted. Read operations will first ask the coordinator for data location and then merge the corresponding data from disk stores and the in-memory store for response.

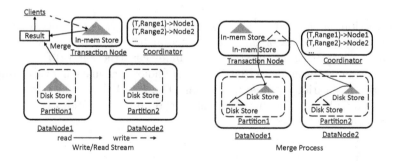

Fig. 1. Implementition of LSM-tree

2.2 Index Structure

A common structure of an index is an index table [15], which is partitioned and stored on a cluster of nodes as an ordinary one. A record in the index table is combination of an index column (*search_key*) and a primary key column of the data table, like as (*search_key, primary_key*). Figure 2 shows an example of the index. The primary key of the item table is column "Item_Id". If we create an index on the column "Sale", the schema of index table is shown in Fig. 2. A query is executed by accessing the index table to get the primary key of the data table, and then getting results from the data table according to the aforementioned primary key. Certainly, we also permit users to build the covering index, which is an index that contains all, and possibly more, the columns that you need for your query.

We organize the secondary index as the ordinary data table for three reasons. Firstly, modifications of indices are not subject to the CAP theorem [10] since modifications of the data and the index table are on the same node, which is important to a distributed write-intensive system. Secondly, bulk loading of indices can be done without blocking transactions since modifications of indices and data tables are stored in the in-memory store. Thirdly, the management of index data can reuse components for the data table, such as load balancing, scalability and high availability, etc.

Item_Id	Sale	Stock
3014	480	150
3015	320	180
3016	180	190

Table Item

Sale	Item_Id
180	3016
320	3015
480	2014

Table Index_Sale

Index on Sale

Fig. 2. Index structure

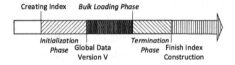

Fig. 3. Index construction process

3 Overview of Index Construction

In this section, a general process of secondary index construction will be described. The procedure of index construction has three phases: (1) Initialization phase, preparing for the start of the index construction. (2) Bulk loading phase, creating uniform index partition. (3) Index termination phase, implementing the replication of index partitions. The Fig. 3 shows the more details on above three phases. The more explanations on process of index construction will be described as follows.

3.1 Initialization Phase

When receiving the command of creating index, the system enters the initialization phase. In this phase, the system waits for a time point when the system reaches a global consistent data version V. After the time point, modifications of indices will be maintained in the in-memory store. Since modifications of indices and data tables are maintained in the in-memory store, constructing indices on data version V will guarantee the consistency of indices and data tables without blocking transactions. As is mentioned in Sect. 2, after data merge processing, data in the old in-memory store is merged with disk stores, modifications of system are maintained in the new in-memory store. Hence, the completion of the merge processing means that initialization phase finished.

3.2   Bulk Loading Phase

The system will complete the construction of index partitions in this phase. Furthermore, the construction of the index is divided into three stages: local index construction, balanced index partition division and global index construction. In the first stage, each data node builds its own local index and sends statistic information of search key to the coordinator. In the second stage, the coordinator generates multiple uniform ranges on search keys based on statistic information from data nodes, and sends the ranges to corresponding data nodes. In the third stage, data nodes receive ranges from the coordinator and create index partitions by shuffling data. Algorithm 1 is the main program. Algorithm 1 calls local index construction routine (Algorithm 2), index partition division routine (Algorithm 3) and global index construction (Algorithm 4) routine in the proper order.

Local index construction. The first key task of bulk loading is efficient global sorting on search keys of the secondary index. On the one hand, construction of local index can reduce communication overhead when global sorting on search keys is executed. On the other hand, the statistic information denoted by a equi-depth histogram collected on the local index can make partition division more uniform, which can guarantee good balancing for query processing.

Balanced index partition division. If the system wants to get uniform index divisions, the coordinator must understand the distributions of search keys on the whole data set and the number of index partitions. So, statistic information received from data nodes is summarized at the coordinator node. And the the number of index partitions is computed. Finally, the coordinator uniformly divides the index ranges according to the number of index partitions and sends the ranges to corresponding data nodes.

Global index construction. The data nodes are executors of global index construction. The data node communicates with other data nodes according to index ranges received from the coordinator to create index partitions in the global search key order.

3.3   Index Termination

After the bulk loading phase, the coordinator will schedule the task for replication of the index for high availability of the index. The replication mechanism of index table is the same to the original table since an index is organized as an ordinary table. Index construction is completely finished after index replication. And then the index is available for query.

4 Bulk Loading Process

The bulk loading phase has three stages, including local index construction, index partition division and global index construction. Firstly, each data node builds its own local index and sends statistic information of search keys to the coordinator. Secondly, the coordinator generates multiple uniform ranges on search keys based on statistic information from data nodes, and sends the ranges to corresponding data nodes. Finally, data nodes receive ranges from the coordinator create index partitions by shuffling data. Note that Algorithm 1 is the main program. Algorithm 1 calls local index construction routine (Algorithm 2), index partition division routine (Algorithm 3) and global index construction (Algorithm 4) routine in proper order.

Algorithm 1 illustrates the framework of the bulk loading approach in distributed LSM-tree data stores. We assume we will construct a secondary index on column search_key of table T. And partitions P of table T are distributed to multiple data nodes. So, all data nodes contained $p \in P$ run the function Run-LocalIndex(p, interval, serach_key, storing) (Algorithm 2) to build local index and report the histogram H containing search_key distribution information to the coordinator (at line 6). Afterwards, the coordinator runs RunPartitionDivision(H,N) (Algorithm 3) to get balanced index data partitions based on histogram H and the number of partitions of index data (at line 9), and assigns partition ranges to appropriate data nodes (at line 10). Finally, the data nodes receive range information from the coordinator and run RunGlobalIndex(L) (Algorithm 4) to build the global index based on index partition range list L (at line 12).

4.1 Local Index Construction

Local index construction starts after the initialization phase. Algorithm 2 illustrates the process of local index construction. Note that Algorithm 2 must run at data nodes. Each data node scans the original data table partitions located in itself in order to construct index entries (at lines 6–14). If we need to build the covering index, i.e. storing columns are not NULL, an index entry needs to contain storing columns besides the search key and primary key of the original table (at line 8). Otherwise, an index entry only contains the search key and primary key (at line 11). And an equi-depth histogram H_i is adopted to capture data distributions on search keys, where each bucket is defined by an interval which is left-closed and right-open, i.e. $[start\_key, end\_key)$ (at lines 15–26). And *interval* represents the depth of the bucket h. The size of depth of bucket can be adjusted. Smaller bucket depth means more accurate information of data distributions, but it also needs more space to store more statics information. Finally, the local index records are written to the disk (at line 25).

Algorithm 1. Index Bulk Loading

---

1 Let P denote partition set of table T which needs to construct index;
2 Let h_i and H denote a bucket and an array of equi-depth histogram
   respectively;
3 Let R denote an array of range /* construct local index   */
4 **foreach** *partition $p \in P$ in all data nodes* **do**
5      $h_i \leftarrow$ RunLocalIndex(p, *depth*) ;
6      $H.add(h_i)$;
7 **end**
8 report H to Coordinator;
   /* divide index partition range   */
9 $R \leftarrow$ RunPartitionDivision(H, N) ;
10 Coordinator sends the partition range $R[i]$ to all data node with respect to $R[i]$;
   /* construct index partition   */
11 **foreach** *datanode correlated to $R[i]$* **do**
12   *index_partition* $\leftarrow$ RunGlobalIndex($R[i]$);
13 **end**

Example 1. To help explain this process, we refer to Fig. 4 as our running example. Table Item has 16 records and contains three partitions, i.e., partition1, partition2 and partition3 on DataNode1, DataNode2. Local index construction in Fig. 4 describes the process local index construction. Each record in a partition is mapped to an index record and index records are sorted by the primary key of the index table. In our example, the primary key of the index table is $(Sale, Item\_Id)$. After local index construction, three *local_indexes* *local_index*1, *local_index*2, *local_index*3 are constructed. The equal-depth histogram of partition1 is shown in Fig. 5. The depth of bucket is 2. Thus three buckets are sampled. See $< (101, 3001), (201, 3002) >$, $< (201, 3002), (400, 3004) >$ and $< (400, 3004), (600, 3006) >$ in Fig. 5.

4.2   Balanced Index Partition Division

After local index construction, the coordinator will receive equal-depth histograms of all *local_index* and then divide the index data into multiple uniform partitions. It is very important to decide the number of partitions and the division strategy of the index data. Subsequently, more details will be explored.

     Generally, the larger size of partition means less efficient for high selective queries, while the smaller size of partition means more space for meta data in a distributed database. Thus, we have to set an appropriate size of partition for the index data. Considering a table $T(col_1, col_2...col_n)$ with an index I on col_j. Let $size(T)$ and M denote the size of T and the maximal limitation of partition size of T respectively. In fact, the partition number P of T can be defined by Eq. (1). Similarly, the partition number of the index data on table T denoted by P' can be defined by Eq. (2). The relationship between the number

---

Algorithm 2. RunLocalIndex

---

1 <u>**Function**</u> RunLocalIndex(*P, search_key, storing*) /* local index
  construction   */
2   Let r and r' denote a record of data table and index table respectively ;
3   Let P' denote the set of index records sorted by search_key ;
4   Let H and $h[j]$ denote a equi-depth histogram and a bucket respectively ;
5   int $i, j \leftarrow 0$; $P' \leftarrow \emptyset$; $Flag \leftarrow TRUE$;
6   **for** *each* $r \in P$ **do**
7   **if** *storing is NULL* **then**
8             $r' \leftarrow (r.search\_key, r.primary\_key, storing)$;
9   **end**
10   **else**
11            $r' \leftarrow (r.search\_key, r.primary\_key)$;
12   **end**
13        $P' \leftarrow P'$ inserted r' based on the order of search-key ;
14   **end**
15   **while** $i <> |P'|$ **do**
16   i++ ;
17   **if** *Flag* **then**
18            $h[j].start\_key \leftarrow r'[i-1].search\_key$; $Flag \leftarrow FALSE$;
19   **end**
20   **if** i *mod interval* $== 0$ **then**
21            $h[j].end\_key \leftarrow r'[i].search\_key$;
22            $Flag \leftarrow TRUE$; j++;
23            $H \leftarrow H \cup h[j]$;
24   **end**
25   write r' to local index partition on disk ;
26   **end**
27   **return** H
28 **end**

---

of index data partitions and the number of original data table partitions can be defined by Eq. (3). Furthermore, the index partition number can be calculated by Eq. (4) for fixed-length storage systems. However, we have to capture the size of index records during local index construction to calculate the partition number of the index by Eq. (2) for varying-length storage systems. In practice, we set the number of partitions of the index data same to the number of partitions of the original table data in order to reduce query response time.

$$P = \frac{size(T)}{M} \tag{1}$$

$$P' = \frac{size(I)}{M} \tag{2}$$

$$P' = P \times \left(\frac{size(I)}{size(T)} \right) \tag{3}$$

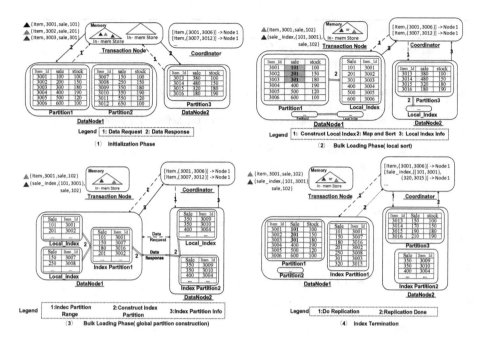

Fig. 4. Index bulk loading

$$P' = P \times \left(\frac{size(search\_key, primary\_key)}{size(col_1, col_2 ... col_n)} \right) \tag{4}$$

Algorithm 3 illustrates the procedure where the coordinator divides the index data into multiple uniform partitions. So, Algorithm 3 must run at the coordinator node. First, the coordinator puts together a whole histogram H' with the histogram h_i received from each individual data node based on the order of bucket (at line 5). Afterwards, The coordinator will divide the H' into $\lceil (|H'| - 1)/N \rceil$ ranges, where $|H'|$ represents the number of buckets and N is the number of index data partitions (at line 6). Note that each index data partition is defined by an interval [start_key,end_key) which is left-closed and right-open. Furthermore, the range is continuous, such as $(MIN, index\_key_1]$, $(index\_key_1, index\_key_2]...(index\_key_p, MAX)$. After the division, the coordinator will allocate ranges to data nodes and data nodes will construct corresponding partitions, which are described in the main program (Algorithm 1).

The key point of Algorithm 3 is to guarantee each data node maintains almost same size of the index, aiming at providing a good load balancing for query processing. The coordinator adopts two strategies to assign index ranges to data nodes: (1) **overlap priority**, it means that the coordinator assigns the index range to the data node which has maximal overlaps with the range. (2) **Load priority**, it means that the coordinator assigns the index range to the data node which has the lightest load. The overlap priority can reduce communication overhead, and the load priority can provide a quicker response for queries.

Algorithm 3. RunPartitionDivision

---

1 **<u>Function</u>** RunPartitionDivision(H, N)
2   Let H' denote a equi-depth histogram which bucket is sorted by its start_key ;
3   Let R denote an array of range defined by [start_key, end_key) ;
4   Let $h[i]$ be the i-th bucket of H' ;
5   Load all buckets of H based on the order of bucket.startkey into H' ;
6     $n \leftarrow \lceil (|H'| - 1)/N \rceil$;
7     $R \leftarrow \emptyset$; $i, j \leftarrow 0$; $Flag \leftarrow FALSE$;
8     $R[1].start\_key \leftarrow h[1].start\_key$;
9   **while** $i <> |H'|$ **do**
10   i++ ;
11   **if** $Flag$ **then**
12 | $R[j].start\_key \leftarrow h[i-1].end\_key$; $Flag \leftarrow FALSE$;
13   **end**
14   **if** $i \bmod n == 0 || i == |H'|$ **then**
15 | $R[j].end\_key \leftarrow h[i].end\_key$;
16 | $Flag \leftarrow TRUE$; j++ ;
17   **end**
18   **end**
19   **return** R;
20 **end**

The end users can adopt different policies according to application scenarios. Certainly, the communication overhead cannot be totally avoided even when the overlap priority is adopted. Fortunately, communication overhead is generated in the offline stage. It has no effect on the response of query processing.

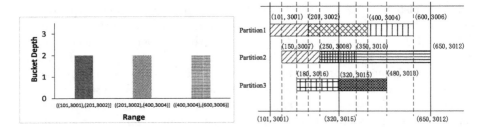

Fig. 5. Equal-depth histogram   **Fig. 6.** Partition division

Example 2. Take the example in Fig. 4. After local index construction, the coordinator receives three histograms and will divide ranges of index partitions. The process of division is shown in Fig. 6. Since the partition number of the data table is 3. For fixed-length storages, the partition number of the index is 2. Thus, ranges of partitions are: (MIN,(320,3015)] and ((320,3015),MAX). The first partition includes 7 records and the second partition includes 9 records.

4.3   Global Index Construction

Global index construction starts after the division of index partitions. During global index construction, a data node may receive two types of information. One is the control information from the coordinator node. Another is the data information from other data nodes. The former is the index partition construction command with a list of index partition ranges, which means the data node will maintain the index data corresponding to ranges received from the coordinator. The latter is index records loaded from other data nodes in the given range. Certainly, the index records are sorted by the same search key. After getting all the index records in the given range, the sorted index data will be flushed to the disk serving as an index table partition.

Algorithm 4 illustrates the procedure of global index construction. It is worth noting that Algorithm 4 must run at all data nodes received construction index messages from the coordinator. Firstly, the algorithm prepares the data for each range according to a range list L received from the coordinator (at line 4, line 8). Meanwhile, it may receive requests from other data nodes for index records and it will response to the these requests. After getting all the records of P, it will sort these records and then flush them to disk as an index partition P (at line 5, line 9). In practice, the procedure of constructing index partitions can be implemented by multi-thread technique, i.e. each range can be assigned a thread.

Algorithm 4. Global Index Construction

---

1 **Function** RunGloalIndex(L)
2   Let P denote a index partition ;
3   **for** *each $l_i \in L$* **do**
4   **if** *the current data node contains all index data in range l_i* **then**
5   Flush index data to disk to construct index partition P ;
6   **end**
7   **else**
8   shuffle and sort index data which is not located in current nodes with other data nodes correlated to l_i ;
9   Flush index data to disk to construct index partition P ;
10   **end**
11   **end**
12 **end**

Example 3. Take the example of global index construction in Fig. 4. After index partition division, DataNode1 receives a partition range $(MIN, (320, 3015)]$ and DataNode2 receives a partition range $((320, 3015), MAX]$. After getting all the records of a range, index records are flushed to the disk serving as an index partition, such as index partition1 and index partition2 in this example.

5 Experimental Evaluation

5.1 Experimental Setup

For experimental analysis, we integrate the approach with our system CEDAR based on OCEANBASE 0.4.2 [4], which is a scalable open source RDBMS developed by Alibaba. It consists of four modules: master server (Coordinator), update server (TransactionNode), baseline data server (DataNode) and data merge server (MergeServer). Now we describe our experimental setup and give a brief overview of the database employment we have used for evaluation.

Cluster platform: We run the experiments on a cluster of 9 machines for most of our experiments except scalability. Each machine is equipped with an Intel(R) Xeon(R) CPU E5-2620@2.00 GHz (a total of 12 physical cores), 96 GB RAM and 3TB Raid5 while running CentOS version 6.5. All machines are connected by a gigabit Ethernet switch.

Database deployment: The database is configured with four MergeServe and four DataNode and each of them is deployed on a single machine in the cluster. The cluster is also configured with a Coordinator and a TransactionNode and they are deployed on a same machine. For better performance, we cache the index and the data table in memory. We choose the load priority policy for guarantee load balancing of index.

Benchmark: Sysbench [8] is a popular open source benchmark to test open source DBMSs. We extend Sysbench [8] by adding an item table in which each row has a unique *item_id* as the primary key and 3 columns. Among them, *item_price* is the column to index. The rest 2 columns are *item_desc* fed with 100 Byte long random byte and *item_title* fed with 92 Byte long random byte, Altogether each row is approximately of 200 Byte in size. We change the workload by varying client threads. For tests of scalability, we vary DataNode from 1 to 4. We test the performance of the index with a query with a predicate on the item price attribute: *Q1: SELECT item_id, item_desc FROM item WHERE item_price = "?"*. By adjusting the size of return size, we can define the selectivity of the test queries.

5.2 Index Construction

We first evaluate the efficiency of index construction. We generate 50 million to 125 million and load data to the system. We compare our approach with a common approach to construct index table denoted as "Read-Insert" where the index table is constructed by first reading from the data table and then inserting into the index table. We launch 100 threads for the "Read-Insert" approach and 10 threads for our bulk loading approach on each DataNode. As is shown in Fig. 7, our approach constructs index much faster than the common approach, for there is less network overhead and no random disk I/O in our approach.

We then test the index partition strategy. We vary the distribution of data on the index key and test three types of distribution: Uniform distribution, Gaussian

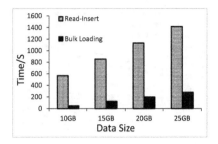

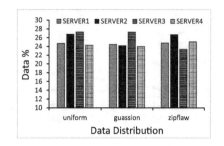

Fig. 7. Index construction time   **Fig. 8.** Partition distribution

distribution, Zipf distribution (Zipf factor = 2). We generate 50 million records for each distribution and create indices on *item_price* for them respectively. After index construction, data distributions of indices are shown in Fig. 8. As we can see, benefiting from balanced index partition strategy, the index data on each node relatively equals to each other.

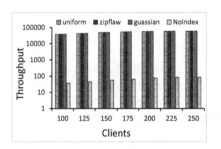
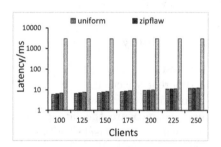

Fig. 9. Query throughput   **Fig. 10.** Query latency

5.3 Query Performance

In this test, we first study the query performance of indices with different data distribution properties. Since the index key has different distribution properties, the result size of test queries may differ, however, since the query distribution is uniform, average result size for queries is relatively the same under different data distribution properties. Figures 9 and 10 demonstrate that query throughput and latency of the system under different data distribution properties are almost same, for our approach captures the different properties of data and guarantees the load balancing of index.

We then test the latency of queries with different selectivities, using concurrent client threads from 50 to 275. We vary selectivity from 0.0001% (50 rows in result) to 0.001% (500 rows in result). As is shown in Figs. 11 and 12, for queries with higher selectivities, the system need less time for query processing

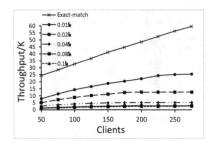

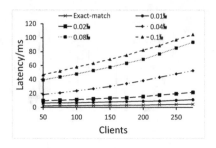

Fig. 11. Query throughput (selectivity)   **Fig. 12.** Query latency (selectivity)

and network communication, thus the system has better latency and throughput with higher selectivity. Figure 11 also shows that the latency of queries with high selectivity is less affected by the workload than that of queries with low selectivity. Queries with lower selectivity requires more system resources such as CPU and Network than queries with high selectivity. As the workload increases, more queries will compete for CPU. Thus, the throughput keeps unchanged and the latency of queries increases.

5.4   Scalability

In this test, we test the scalability of index construction and query performance with different selectivities. We vary the size of data node from 1 to 4. For index construction, we measure the construction time for index. For query performance, we measure the latency and throughput of exact match queries and queries with different selectivities.

As we can see in Fig. 13, time for index construction reduces with the increment of the number of data nodes, this is due to the fact that we allocate index partitions to data nodes uniformly. However, time for index construction does not reduce linearly because adding nodes may cause more network communication.

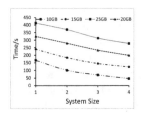   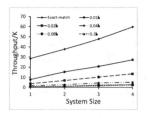   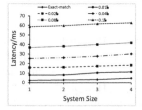

Fig. 13. Index construction   **Fig. 14.** Query throughput   **Fig. 15.** Query latency

As can be seen from Figs. 14 and 15, the system scales well with nearly flat query latency when the size of data node increases due to the fact that the index is organized as a normal table and can take advantage of the load balancing of the system. In this test, the workload submitted to the system is proportional to the data node size. As we can see, by adding data node, more workload can be handled. The latency for queries with a specific selectivity remains essentially unchanged with different number of data nodes.

6 Related Work

There are two types of secondary index models on distributed data stores: local index models and global index models. In local index models, the index is built on each node which indexes local data. Global indices support high selectivity queries better. Huawei's Hindex [6] realizes the local index on the LSM-based storage system Hbase [1], however, load balancing of index is not supported. Diff-index (Differentiated Index) [14] realizes the global index on Hbase which supports index creation on an empty table.

There have been several existing bulk loading approaches of secondary indices in distributed log-structed data stores. Phoenix [5] use map-reduce [12] to construct distributed index, modifications of index are maintained in memory. However, when the data set is huge, it needs to create index in an asynchronous way by an external tool. AsterixDB [9] presents a bulk loading approach for converting existing index structures to LSM-based index structures. The approach is for AsterixDB and employs multiple data structures realizing bulk loading of indices, which means it is hard to do load balancing of indices. Another approach is by external storages such as solr [7]. Indices are stored in an external system. Thus, the construction of indices needs to read from the database and insert into the external system.

7 Conclusion

We introduce a new bulk loading approach of secondary index in distributed log-structed data stores. The approach supports efficient bulk loading of index by taking rational use of resources in a cluster. Whats more, by using equal-depth histogram, load balancing of index is guaranteed. We perform an extensive evaluation of our approach on a LSM-based distributed system. The results from our experiments show that our approach take rational use of cluster for bulk loading of index and guarantee the load balancing of index. In addition, since an index is organized as a normal table and is integrated in the load balancing of the system, the index is scalable and adding nodes to the system will improve the performance of the index.

Acknowledgements. This work is partially supported by National High-tech R&D Program (863 Program) under grant number 2015AA015307, National Science Foundation of China under grant numbers 61402180, 61432006 and 61672232, Natural Science Foundation of Shanghai under grant numbers 14ZR1412600, and Guangxi Key Laboratory of Trusted Software (kx201602). The corresponding author is Zhao Zhang.

References

1. Apache HBase website. http://hbase.apache.org/
2. CDEAR website. https://github.com/daseECNU/Cedar/
3. LevelDB website. http://leveldb.org/
4. OceanBase website. https://github.com/alibaba/oceanbase/
5. PHOENIX website. http://phoenix.apache.org/
6. Secondary Index for HBase. https://github.com/Huawei-Hadoop/hindex
7. SOLR website. http://lucene.apache.org/solr/
8. Sysbench website. http://dev.mysql.com/downloads/benchmarks.html
9. Alsubaiee, S., Asterixdb, A., et al.: A scalable, open source bdms. Proc. VLDB Endowment **7**(14), 1905–1916 (2014)
10. Brewer, E.: Pushing the cap: strategies for consistency and availability. Computer **45**(2), 23–29 (2012)
11. Chang, F., Dean, J., Bigtable, G., et al.: A distributed storage system for structured data. ACM Trans. Comput. Syst. (TOCS) **26**(2), 4 (2008)
12. Dean, J., Ghemawat, S.: Mapreduce: simplified data processing on large clusters. In: Conference on Symposium on Opearting Systems Design & Implementation, pp. 107–113 (2004)
13. ONeil, P., Cheng, E., Gawlick, D., O'Neil, E.: The log-structured merge-tree (lsm-tree). Acta Informatica **33**(4), 351–385 (1996)
14. Tan, W., Tata, S., Tang, Y., Fong, L.L.: Diff-index: differentiated index in distributed log-structured data stores. In: EDBT, pp. 700–711 (2014)
15. Zou, Y., Liu, J., Wang, S., Zha, L., Xu, Z.: CCIndex: a complemental clustering index on distributed ordered tables for multi-dimensional range queries. In: Ding, C., Shao, Z., Zheng, R. (eds.) NPC 2010. LNCS, vol. 6289, pp. 247–261. Springer, Heidelberg (2010). doi:10.1007/978-3-642-15672-4_22

Performance Comparison of Distributed Processing of Large Volume of Data on Top of Xen and Docker-Based Virtual Clusters

Haejin Chung and Yunmook Nah[✉]

Department of Computer Science and Engineering, Graduate School,
Dankook University, 152 Jukjeon-ro,
Suji-gu, Yongin-si, Gyeonggi-do 16890, Korea
haejini.chung@gmail.com, ymnah@dankook.ac.kr

Abstract. Recently, with the advent of cloud computing, it becomes essential to run distributed computing tasks, such as Hadoop MapReduce tasks, on top of virtual computing nodes instead of physical computing nodes. But, distributed big data processing on top of virtual machines usually causes unbalanced use of physical resources, such as memory, disk I/O and network resources, thus resulting in severe performance problems. In this paper, we show how virtualization methods affect distributed processing of very large volume of data, by comparing Hadoop MapReduce processing performance on top of Xen-based virtual clusters versus Docker-based virtual clusters. In our experiments, we compare the performance of two different virtual clusters by changing virtualization methods, block sizes and node numbers. Our results show that, in terms of the distributed big data processing performance, Docker-based virtual cluster is usually faster than Xen-based virtual cluster, but there exist some cases where Xen is faster than Docker according to the parameters, such as block size and virtual node numbers.

1 Introduction

The need to manage massive volume of continuously produced information is ever increasing rapidly in many future applications, such as wearable computing, wireless sensor networks and IoT-enabled applications. Distributed computing platforms, such as Google MapReduce and Hadoop MapReduce, have been emerged to support efficient handling of very large volume of data [1, 2]. Distributed processing tasks, such as MapReduce jobs, are usually executed on top of physical computing clusters to avoid performance degradation. Also, due to trends like cloud computing, virtualization technologies are gaining increasing importance, allowing efficient use and sharing of physical computing resources [3]. There have been research works to compare the performance between different hypervisors, such as Xen hypervisor and KVM (Kernel-based Virtual Machine) driver [4, 5] and to compare container-based virtualization versus full virtualization [6–8].

Recently, with the wide spread use of cloud computing, such as Amazon Web Service, MS Azure and IBM Bluemix, it becomes essential to run distributed computing

© Springer International Publishing AG 2017
S. Candan et al. (Eds.): DASFAA 2017, Part I, LNCS 10177, pp. 103–113, 2017.
DOI: 10.1007/978-3-319-55753-3_7

tasks on top of virtual computing nodes instead of physical computing nodes. But, distributed big data processing on top of virtual machines usually causes unbalanced use of physical resources, such as memory, disk I/O and network resources, thus resulting in severe performance problems. The performance of distributed processing of large volume of data on top of virtualized clusters is also affected by the characteristics of running applications [9, 10].

In this paper, we show how virtualization methods affect distributed processing of very large volume of data, by comparing Hadoop MapReduce processing performance on top of Xen-based virtual clusters versus Docker-based virtual clusters. In our experiments, we compare the performance of two different virtual clusters by changing virtualization methods, block sizes and node numbers. In general, Docker is known as faster virtualization method than Xen. Our results show that, in terms of the distributed big data processing performance, Docker-based virtual cluster is usually faster than Xen-based virtual cluster, but there exist some cases where Xen is faster than Docker according to the parameters, such as block size and node numbers.

The remainder of this paper is organized as follows. Section 2 explains related work and Sect. 3 provides an overview of various virtualization methods. In Sect. 4, we show the impact of virtualization methods on the performance of distributed processing of large volume of data. Section 5 concludes this paper.

2 Related Work

Virtualization technologies are gaining increasing importance due to trends like cloud computing and there have been research works to compare the performance between different hypervisors [4–8]. There have been research works to compare the performance between different hypervisors, such as Xen and KVM [4, 5]. In [4], the overall performance, performance isolation, and scalability of virtual machines running on Xen and KVM are studied. KVM had substantial problems with guests crashing, beginning with 4 guests. KVM had better performance isolation than Xen, but Xen's isolation properties were also quite good. The overall performance results were mixed, with Xen outperforming KVM on a kernel compile test and KVM outperforming Xen on I/O-intensive tests.

Researches to compare container-based virtualization versus hypervisor-based virtualization are usually focused on virtualization itself [6–8]. In the study for the performance comparison between KVM and container [8], they use a suite of workloads that stress CPU, memory, storage, and networking resources. Their results show that containers result in equal or better performance than VMs in almost all cases. Both VMs and containers require tuning to support I/O-intensive applications.

Distributed processing tasks, such as Hadoop MapReduce jobs, are usually executed on top of physical computing clusters, but, with the wide spread use of virtualization technologies, it becomes essential to run distributed computing tasks on top of virtual computing nodes instead of physical computing nodes. But, in distributed big data processing on top of virtual clusters, physical resources, such as memory, disk I/O and network, can be bottleneck, because multiple virtual machines can access such physical resources simultaneously, as shown in Fig. 1. Especially, disk I/O is considered as one

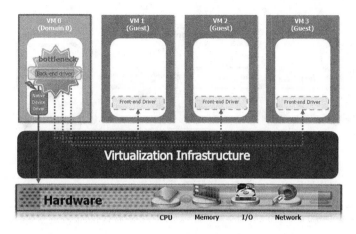

Fig. 1. I/O bottleneck with multiple virtual machines

of the critical reasons of performance degradation for distributed big data processing on virtual clusters. The performance of distributed processing of large volume of data on top of virtualized clusters can be also largely affected by the characteristics of applications [9, 10, 13, 14].

The main purpose of our research team is to achieve 10 times faster disk I/O in such virtualized big data processing environments by combining virtualization technologies and parallel database technologies. In our previous research [10], we applied Hadoop-based distributed big data processing on the Xen-based virtualized cluster environment, with special focus on more than 200 configuration variables of Hadoop to improve the overall performance of distributed processing. We performed various experiments to find the optimal configuration variables and analyzed the impact of the size of input data set and the number of data nodes on the optimal number of Reduce tasks.

3   Virtualization Methods

Virtualization is the process of presenting computing resources in ways that users and applications can easily get value from them. Virtualization provides a logical rather than a physical view of data, computing power, storage capacity, and other resources. A hypervisor is runtime executive software in charge of logically replicating the physical platform. Within each logical replica virtual machine, a guest OS may run independently of other guest OS running in other virtual machines. Such replication is achieved by partitioning and/or virtualizing platform resources [11, 12].

There have been lots of researches to compare performance among various virtualization platforms. Figure 2 shows the performance-influencing factors grouped under virtualization type, VMM architecture, resource management configuration and workload profile [5]. We slightly modified this figure and represented the major factors considered in this paper in yellow box.

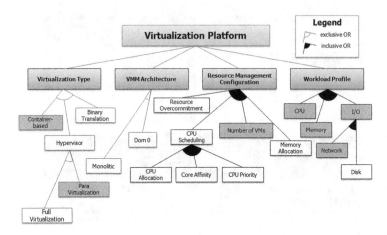

Fig. 2. Performance-influencing factors of virtualization platform (adapted from [5]) (Color figure online)

Virtualization using hypervisors can be classified as full virtualization type and para-virtualization type. A full virtualization technique is used to implement a certain kind of VM environment, namely, one that provides complete simulation of the underlying hardware. The result is a system in which all software (including OS) capable of execution on the raw hardware can be run in the virtual machine. When para-virtualization is used, the logically replicated hardware differs slightly from the underlying physical hardware. Instead of fully supporting the physical hardware ISA (instruction set architecture), para-virtualization offers a special API that can be used only by modifying the guest OS. KVM is the most famous virtualization software among full virtualization types and Xen is the most popular virtualization software among para-virtualization types.

Docker provides a way to run applications securely isolated in a container, packaged with all its dependencies and libraries. Docker containers are lightweight than para-virtualization type hypervisor and run without the extra load of a hypervisor. Docker runs applications without guest OS and using only binaries and libraries. But, whenever a container is created, it instantiates new file system thus destructs data stored in the container. Therefore, Docker is currently not suitable for applications which require persistent data store.

4 Performance Comparison

We present a performance comparison of MapReduce processing in the two virtualized cluster environments, one is built by using the Xen hypervisor and another is built by using container-based Docker. Table 1 shows the major specification of our experimental environments.

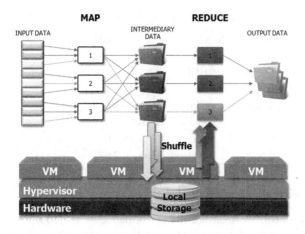

Fig. 3. MapReduce processing on top of Xen-based virtual cluster

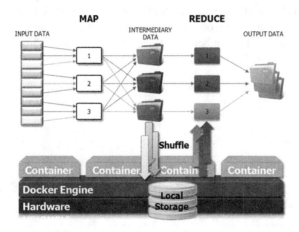

Fig. 4. MapReduce processing on top of Docker-based virtual cluster

In all experiments in this section, we built virtual clusters on one physical node. The ratio of physical server to virtual servers for one physical node is up to 1:7 (usually 1:5) in commercial data centers, so in our experiment, we generated 4 to 6 virtual machines or containers for one physical server.

Figure 3 shows Hadoop MapReduce processing system architecture on top of Xen-based virtual clusters. Virtual machines are generated by the Xen hypervisor and Hadoop tasks are run on these virtual computing nodes. Figure 4 shows Hadoop MapReduce processing system architecture on top of container-based virtual clusters using Docker. Multiple containers are generated by the Docker engine and Hadoop tasks are run on these virtual computing nodes.

Table 1. Experimental platform

Physical node	CPU	Intel 2.93 GHz 6core * 2
	Memory	36 GB
	HDD	1.8 TB
	OS	CentOS 6.6
Virtual node	CPU	Vcpu * 2
	Memory	2 GB
	OS	CentOS 6.6
	Hadoop	Hadoop 2.6

4.1   Effect of Virtualization Methods

In our first experiment, we measure the performance of Hadoop processing without data replication, to show the pure impact of hypervisor on the performance of distributed processing of big data. For experiments, we used TeraGen and TeraSort, which are provided as standard benchmarks by Hadoop. We randomly generated 10 GB of data by using TeraGen and executed TeraSort 5 times. We created 4 virtual nodes using Xen and Docker. Among them, one node is used as name node and other three nodes are used as data nodes. The number of replica is set as 1. Figure 5(a) shows TeraGen processing time and Fig. 5(b) shows TeraSort processing time.

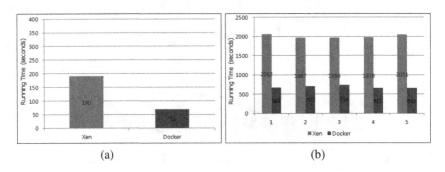

(a)   (b)

Fig. 5. TeraGen and TeraSort processing time

TeraGen execution time is 190 s for Xen and 70 s for Docker. Docker is 2.71 times faster than Xen. TeraSort execution time for Xen is 2,004 s, while Docker requires 687 s, which is 2.92 times faster than Xen. This performance gap results from the difference of resource allocation policies of each virtualization methods. Docker enables resource sharing by virtualizing host operating system and it allocates minimum resources to each application, thus maximizing utilization of host resources. However, Xen hypervisor allocates computing resources to each virtual machine, which limits the degree of resource sharing compared to Docker, thus resulting in performance degradation.

4.2   Effect of Block Sizes

In our second experiment, we examine the effect of block sizes on the performance of Hadoop processing. We created 6 virtual nodes using Xen and Docker. Among them, one node is used as name node and other five nodes are used as data nodes. The number of replica is set as 3. We repeated 5 times of reading and writing of 10 GB data, for block sizes 32 MB, 64 MB, 128 MB. For experiments, we used TestDFSIO, a standard Hadoop benchmark, which allows to measure data throughput for read and write operations on HDFS. Figure 6 shows read performance and Fig. 7 shows write performance.

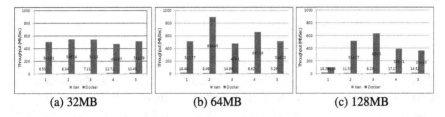

| (a) 32MB | (b) 64MB | (c) 128MB |

Fig. 6.   Throughput of read operations

As similarly reported from other experiments related with virtualization perfor-mance, in terms of throughput of data read operations, the performance of Docker-based virtual cluster is much better than the performance of Xen-based virtual cluster. Docker shows 29.33–59.1 times better throughput than Xen.

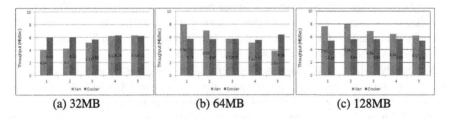

| (a) 32MB | (b) 64MB | (c) 128MB |

Fig. 7.   Throughput of write operations

As shown in Fig. 7, for the write operations, the average data throughputs for Xen and Docker are similar with each other. That is, the read operation efficiency of Docker is much better than Xen, but the write operation efficiency is similar with each other. The main reason lies in the data replication. To keep consistency for replicated data, the traffic between name node and data nodes and between containers increases, thus resulting in performance degradation in both cases.

But, for the block size 128 MB, the write performance of Xen-based virtual cluster is always better than Docker-based virtual cluster. In this case, Docker shows 0.79 times slower performance than Xen. Also, for the block size 64 MB, there exist some cases where Xen is better than Docker. For the case of Xen, the increase of block size seems to be related with decrease of block number, decrease of metadata and decrease of name node overhead. This phenomenon shows that the block size can affect the overall performance of distributed processing, regardless of virtualization methods. Therefore, it seems necessary to select carefully the appropriate block size according to the application purposes.

Table 2. Best data throughput for read and write operations

		Block size (MB)	Throughput (MB/sec)
Read	Xen	128	13.55
	Docker	64	610.47
Write	Xen	128	7.01
	Docker	32	6.04

Table 2 summarizes the best average data throughput. For the applications having more write operations than read operations, Xen-based cluster with block size 128 MB is better than Docker-based cluster. For the applications having more read operations than write operations, Docker-based cluster with block size 64 MB is preferable than Xen-based cluster.

4.3   Effect of Node Numbers

In our third experiment, we examine the effect of number of virtual nodes on the performance of Hadoop big data processing. We compared the case of 4 virtual nodes having 3 data nodes with the case of 6 virtual nodes having 5 data nodes. The number of replica is fixed as 1, to eliminate the additional workload for data replication. We randomly generated 10 GB of data by using TeraGen and executed TeraSort 5 times. Figure 8 shows the TeraGen processing time.

Noticeably, when we increase the number of data node from 3 to 5, the execution time of TeraGen for Xen-based virtual cluster decreases from 190 s to 144 s, while the execution time of TeraGen for Docker-based virtual cluster steeply increases from 70 s to 223 s. With the increase of the number of data nodes from 3 to 5, Xen-based cluster requires less time (only 35% of time), while Docker-based cluster requires more time (2.85 times longer than 3 data node case). For the 3 data node case, Xen is 0.41 times slower than Docker, but, for the 5 data node case, Xen becomes 1.54 times faster than Docker.

The main reason of this phenomenon lies in the difference of nature of two virtualization methods. Xen hypervisor is based on para-virtualization, which allocates computing resources to each virtual machine, while Docker is based on light-weight container, which virtualizes host operating system for resource sharing. The increase of data nodes results in the increase of the number of virtual machines for Xen-based cluster and the increase of the number of containers for Docker-based cluster. For the Xen-based cluster, the increase of virtual machines, even all they run on one physical

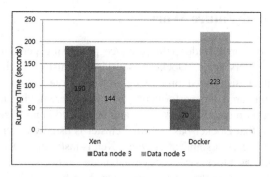

Fig. 8. TeraGen processing time

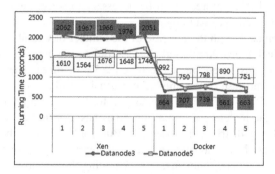

Fig. 9. TeraSort processing time

computing node, has a positive effect of improving the performance of the distributed processing. But, for the Docker-based cluster, the increase of containers, all run on one physical computing node, has a negative effect of increasing the resource management overhead in time-sharing environments, thus degrading the performance of distributed processing.

Figure 9 shows the TeraSort processing time. The average execution time of TeraSort for Xen-based virtual cluster is 2,004 s, and with the increase of data nodes, it decreases to 1,648 s. The average execution time of TeraSort for Docker-based virtual cluster is 687 s, and with the increase of data nodes, it increases to 836 s. With the increase of the number of data nodes from 3 to 5, Xen-based cluster requires less time (82% of time), while Docker-based cluster requires more time (1.22 times longer than 3 data node case). For the 3 data node case, Xen is 0.34 times slower than Docker, and, for the 5 data node case, Xen is 0.51 times slower than Docker. For the TeraSort case, Xen is always slower than Docker, but the speed gap decreases with the increase of virtual data node numbers.

5   Conclusion and Future Work

In this paper, we compared Hadoop MapReduce processing performance on top of Xen-based virtual clusters versus Docker-based virtual clusters to show how virtualization methods affect distributed processing of very large volume of data. In our experiments, we compared the performance of two different virtual clusters, while changing virtualization methods, block sizes and virtual node numbers.

Our results show that Docker-based virtual cluster is usually faster than Xen-based virtual cluster. But there exist some cases where Xen is faster than Docker according to the parameters, such as block size and virtual node numbers. For the applications having more write operations than read operations, Xen-based cluster with block size 128 MB and 64 MB is better than Docker-based cluster. This result shows that the block size can affect the overall performance of distributed processing, regardless of virtualization methods. Therefore, it will be necessary to select the appropriate block size according to application requirements.

For the TeraGen case, with the increase of the number of virtual data nodes from 3 to 5 (on one physical node), Xen-based cluster requires less time (only 35% of time), while Docker-based cluster requires more time (285% of time). The increase of data nodes results in the increase of the number of virtual machines for Xen-based cluster and the increase of the number of containers for Docker-based cluster. For the Xen-based cluster, the increase of virtual machines has a positive effect of improving the performance of the distributed processing. But, for the Docker-based cluster, the increase of containers has a negative effect of increasing the resource management overhead in time-sharing environments, thus degrading the performance of distributed processing.

This means that we cannot conclude that Docker is always faster than Xen, from the perspective of the distributed processing of very large volume of data on top of virtualized clusters.

We are currently considering hybrid virtualization environment, simultaneously utilizing Xen and Docker on different computing nodes, according to the characteristics of applications. Additional experiments using TPC-H and TPC-C benchmark will be helpful to find out more performance-influencing factors for massive data processing on top of virtualized cluster environments.

Acknowledgements. This research was supported by the MSIP (Ministry of Science, ICT and Future Planning), Korea, under the ITRC (Information Technology Research Center) support program (IITP-2016-R0992-16-1012) supervised by the IITP (Institute for Information & communications Technology Promotion). This work was also supported by Institute for Information & communications Technology Promotion (IITP) grant funded by the Korea government (MSIP) (No. R0190-16-1085, Over 1,000 Cores Scale Out Clustered Database Platform Development).

References

1. Ghemawat, S., Gobioff, H., Leung, S.-T.: The google file system. In: Proceeding of the SOSP (2003)
2. Dean, J., Ghemawat, S.: MapReduce: simplified data processing on large clusters. CACM **51**(1), 107–113 (2008). ACM
3. Barham, P., Dragovic, B., Fraser, K., Hand, S., Harris, T., Ho, A., Neugebauery, R., Pratt, I., Warfield, A.: Xen and the art of virtualization. In: Proceedings of the SOSP, pp. 164–177. ACM (2003)
4. Deshane, T., Shepherd, S., Matthews, Z.N., Ben-Yehuda, M., Shah, A., Rao, B.: Quantitative comparison of Xen and KVM. In: Xen Summit, pp. 1–3 (2008)
5. Huber, N., von Quast, M., Hauck, M., Kounev, S.: Evaluating and modeling virtualization performance overhead for cloud environments. In: Proceedings of the CLOSER, pp. 563–573 (2011)
6. Lina, Q., Qia, Z., Wua, J., Dongb, Y., Guana, H.: Optimizing virtual machines using hybrid virtualization. J. Syst. Softw. **85**(11), 2593–2603 (2012). Elsevier
7. Morabito, R., Kjällman, J., Komu, M.: Hypervisors vs. lightweight virtualization: a performance comparison. In: Proceedings of the International Conference on Cloud Engineering (IC2E), pp. 386–393. IEEE (2015)
8. Felter, W., Ferreira, A., Rajamony, R., Rubio, J.: An updated performance comparison of virtual machines and linux containers. In: Proceedings of the 2015 International Symposium on Performance Analysis of Systems and Software (ISPASS 2015). IEEE (2015)
9. Xavier, M.G., Neves, M.V., de Rose, C.A.F.: A performance comparison of container-based virtualization systems for MapReduce clusters. In: Proceedings of the 22nd Euromicro International Conference on Parallel, Distributed and Network-Based processing (PDP), pp. 299–306. IEEE (2014)
10. Kim, Y., Chung, H., Choi, W., Kim, J., Choi, J.: Effects of reduce task number on performance of I/O-intensive MapReduce applications in virtualization environment. J. KIISE: Comput. Pract. Lett. **19**(7), 403–407 (2013). KIISE
11. SCOPE Alliance: Virtualization: State of the Art (2008)
12. Heiser, G.: The role of virtualization in embedded systems. In: Proceedings of the 1st Workshop on Isolation and Integration in Embedded Systems, pp. 11–16. ACM (2008)
13. Zaharia, M., Konwinski, A., Joseph, A.D., Katz, R.H., Stoica, I.: Improving MapReduce performance in heterogeneous environments. OSDI **8**(4), 277–298 (2008)
14. Ibrahim, S., Jin, H., Lu, L., Qi, L., Wu, S., Shi, X.: CLOUDLET: towards MapReduce implementation on virtual machines. In: Proceedings of the 18th ACM International Symposium on High Performance Distributed Computing, pp. 65–66. ACM (2009)

An Adaptive Data Partitioning Scheme for Accelerating Exploratory Spark SQL Queries

Chenghao Guo[1,2], Zhigang Wu[1,2], Zhenying He[1,2], and X. Sean Wang[1,2(✉)]

[1] School of Computer Science, Fudan University, Shanghai, China
{chguo15,wuzg16,zhenying,xywangcs}@fudan.edu.cn
[2] Shanghai Key Laboratory of Data Science, Shanghai, China

Abstract. For data analysis, it's useful to explore the data set with a sequence of queries, frequently using the results from the previous queries to shape the next queries. Thus, data used in the previous queries are often reused, at least in part, in the next queries. This fact may be used to accelerate queries with data partitioning, a widely used technique that enables skipping the irrelevant data for better I/O performance. For getting effective partitions which are likely to cover the query workload in the future, we propose an adaptive partitioning scheme, combining the data-driven metrics and user-driven metrics to guide the data partitioning as well as a heuristic model using the metric plugin system to support different exploratory patterns. For partition storage and management, we propose an effective partition index structure for quickly searching for appropriate partitions to answer queries. The system is quite helpful in improving the performance of exploratory queries.

1 Introduction

Users often do not have a clear query intent when exploring unfamiliar datasets [11]. When working with a multi-dimensional dataset, people often have no idea about what predicates should be used to filter the dataset at first. This could occur because the user is unfamiliar with the dataset or the query specification can only be derived by prior data exploration [10]. For instance, a user wants to find some interesting topic among a journal dataset. At first, he even has no idea about what there is or what is interesting for him. After an overview of the data, he finds that most of the journals are published between the years 1990 and 1920. Then he begins to dig into the journal articles among these years and finally find the journal Science is just the one he likes. So he extracts all the Science articles during these years. After a short period, a few other users are exploring the dataset. Coincidentally, some users among them have the same interest as the first user and repeat the process like the first one did. For the scenario above, if storing the data from 'journaltitle=Science' or 'years between 1900 and 1920' as additional partitions, we can accelerate the exploratory process for the following users and the whole runtime can be reduced.

The work is supported by the NSFC (No.61370080, No.61170007) and the Shanghai Innovation Action Project (Grant No.16DZ1100200).

S. Candan et al. (Eds.): DASFAA 2017, Part I, LNCS 10177, pp. 114–128, 2017.
DOI: 10.1007/978-3-319-55753-3_8

For a common Spark SQL [6] with a SELECT-FROM-WHERE clause to extract data, the Spark SQL engine will read the whole data from HDFS and apply the where predicates to filter. As the random access is not supported well in most distributed file systems such as HDFS, the I/O cost of reading the whole dataset and then filtering with where clauses is quite high. Rather than thoroughly scanning the whole dataset, if we can correctly answer the query by using a subset of the dataset, the I/O cost will be much lower.

Data partitioning is widely used for improving the performance of distributed systems. Partitioning approaches based on static workloads are not suitable for the exploratory queries as workloads may vary over time and users gain more information from previous queries. Our work is based on the assumption that users' intentions are relatively stable over a period of time and previous queries reflect a hint which can guide the query process afterwards.

We implement our APEQ (Adaptive Partitioning Scheme for Exploratory Queries) system upon Spark SQL. The implementation is mainly about how to perform data partitioning and organize the partitions for query optimization. In summary, the contributions of this paper are:

1. We propose a metric plugin library and a learning model for generating partitions. We present several data-driven plugins according to the dataset features and a user-driven plugin based on query workload.
2. For partition management and indexing, we design a multi-dimensional index structure suitable for partitions with various numbers of dimensions.
3. Our system is helpful in improving the overall I/O performance for exploratory queries especially when the original dataset is quite large, and can also be used to supplement the original Hive [1] partition which is weak in reacting to exploratory queries.

The rest of this paper is organized as follows: Sect. 2 introduces related works. In Sect. 3, we describe the system architecture. Sections 4 and 5 present the partition learning strategy with metric plugins and the partition index structure. In Sect. 6, we give a performance evaluation of the APEQ system. Finally we discuss our work and conclude this paper in Sects. 7 and 8 respectively.

2 Related Work

Many related works about the exploratory query optimization and data partitioning techniques have been proposed in the past.

SnapToQuery [10] provides visual and interactive feedback for exploratory queries. The paper inspires our work as it proposes a data-driven method to influence users' query behaviour. The paper introduces the concept of SnapTo-Query, a feedback mechanism for guiding users to their intended queries during exploratory query specification. The target of the paper is to help users snap to the correct regions to sponsor exploratory queries accurately.

SEEDB [17] proposes a visualization recommendation engine to facilitate fast visual analysis. It adopts a deviation based metric for visualization utility.

It is quite useful for a set of problems especially when users are interested in the deviation of the data. The deviation based idea is quite enlightening for our metric plugins.

AQWA [5] proposes an adaptive partitioning strategy for spatial data. The paper uses an optimized K-d trees for the problem and a cost model to determine whether to divide a new partition. However, the model issued by the paper is quite limited for high-dimensional data. Splitting query workspace into cells is not efficient for high-dimensional data as the number of cells will be quite large.

Cracking [9] is a technique designed for adapting the placement of data and indexes in main-memory column-stores. Stochastic Cracking [8] introduces random physical reorganization steps for efficient incremental index building. The paper introduces various stochastic cracking algorithms which combine a data-driven approach with each query as a hint on adaptive partition, while we design a set of data-driven plugins which focus on the interesting parts of data together with a query workload plugin to generate a partition learning model. Furthermore, as the ways of random access and index are quite different between main-memory column-stores and distributed systems, the cracking algorithms which are efficient in main-memory column-stores may cause high repartitioning cost in distributed environments.

Adaptive data skipping techniques use scans with skipping methods for fast data filtering. Adaptive Zonemaps [15] is a data skipping technique in main-memory systems, which logically segments data into contiguous zones and keeps metadata for each zone. While in ditributed systems, the Fine-grained Partioning technique [16] based on feature vectors reorganizes the data tuples into blocks, with a goal of partitioning the data into fine-grained and balance-sized blocks so that queries can maximally skip the blocks. The idea proposed by our paper is also a data skipping technique without maintaining the full index and is slightly more flexible than the feature-based partitioning in multi-dimensional situations.

The implementation of our system adopts partial ideas from Hive's existing partitioner. In Hive's implementation of partitioning, each partition is stored as a sub-directory within the table's directory on HDFS. When the table is queried, only the required partitions of the table are queried, thereby reducing the I/O and time. The main shortcoming is that Hive's partitions are hierarchical and need to be decided by users. Users who have no idea about the dataset can not partition the dataset efficiently. CPS [13] proposes a condition-based partitioning scheme to implement optimization strategy on Hive. The paper takes advantage of the concept of logical partition. For a given table, the data belonging to the same logical partition is placed into one directory on HDFS while the files are physically distributed onto multiple nodes in a cluster. Thus, organizing the partition in the same logical directory is helpful for balance. Our APEQ adopts the idea and creates every partition by using a new Hive table with the raw data stored in the same logical directory for maximizing the balance and flexibility.

3   System Overview

The overall architecture of our system is presented in Fig. 1. Every query issued by users first goes to the query optimization engine. The engine will extract the dimensions and predicates from the query and then search the partition index for any partition which can provide the data source needed by the query. Upon finding a qualified partition, the optimization engine will replace the original table with the new partition and then use Spark SQL to execute the query. We define the query process happened in the past N hours or recent K queries as a query period and the interval between query periods as idle status. When the system is idle, the partition learning engine will work by using the metric plugins to generate partitions with feedback from the last query period. The generated partitions will be placed to HDFS using Spark SQL if not existing and the index will be inserted into the partition index tree.

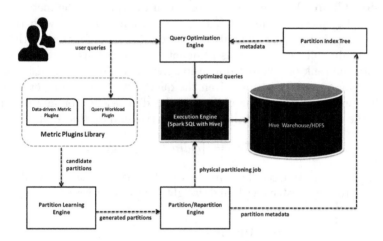

Fig. 1. APEQ architecture

Data-driven Metric Plugin. We design several partition strategy plugins according to the properties inheriting from the dataset. The partitions generated by data-driven metric plugins are created at the initial phase before the queries are executed. They are static and physically stored.

User-driven Metric Plugin. Currently, the only user-driven metric plugin is the query workload metric. We implement a heuristic algorithm based on clustering to generate the partitions dynamically.

Partition Index Tree. The partitions are stored in HDFS using independent Hive tables while the indexes are stored using the index tree. We design a multi-dimensional partition index tree as the Sect. 5.1 explains.

Query Optimization Engine. The query optimization engine will search in partition indexes to determine whether the data source can be replaced with partitions. APEQ can answer the query by accessing only the relevant data and reduce the input size if qualified partitions are found.

Execution Engine. The Spark supports reading data from HDFS and the Spark SQL can use the Spark engine to replace the MapReduce engine in Hive for a better performance.

4   Partition Learning Strategy

In this section, we propose our partition learning strategy. The data-driven metric plugins accept a random sample of datasets as input and generate initial partitions. The user-driven metric plugin is based on the query workloads. In the end, we propose a naive partition learning strategy by combining the data-driven plugins with the query workload plugin.

Dimension Chain. We use a dimension chain to represent the metadata of a partition and a query. For instance, P is a partition generated with the filter clause $20 \leq$ L_quantity ≤ 40 and $0.04 \leq$ L_discount ≤ 0.07, for this partition the dimension chain will represent it as [L_quantity[20, 40], L_discount[0.04, 0.07]]. To estimate the rank of different dimensions, we assign every dimension with a weight. During one query period, given the query set $\mathcal{Q}$ containing every query q and the dimension set $\mathcal{A}$ containing all the dimensions, for any dimension $dim \in \mathcal{A}$, the weight of dim can be decided by the sum of dim in all queries.

$$weight(dim) = |\{q|dim \in q, q \in \mathcal{Q}\}| \qquad (1)$$

The dimension rank reflects the importance of every dimension. Analysts can focus on the top dimensions with high weight for exploratory tasks.

4.1   Data-Driven Metric Plugins

Initially, our system will forecast and produce several partitions according to the data-driven properties inheriting from the dataset. APEQ has implemented several metric plugins based on deviation and clustering. Partitions generated by data-driven metric plugins are static and often created at the initial phase. As exploring a high-dimensional dataset is quite challenging, generally we choose a few featured dimensions to apply data-driven metrics. If there are existing query logs, we can focus on the top dimensions from the dimension rank list. We have defined the following metric plugins in our APEQ system.

Metric-Clustering. We define the Metric-Clustering plugin as follows

$$Partitions = \{cluster|size(cluster(norm(dim_1), norm(dim_2), ..) \geq T\} \qquad (2)$$

The Metric-Clustering will provide several clusters whose size is above the threshold T. The function $norm()$ represents for normalization, dim means dimension and $cluster$ will get several clusters from these dimensions. For a metric plugin that needs clustering, the high dimension is quite complicated. Using dimension chains, we investigate the dimension from beginning to end. Every time we iteratively scan two dimensions with clustering to estimate the distribution pattern of the data and finally generate the candidate partitions. In order to fit a wide range of queries, data-driven plugins usually generate partitions with low dimension and big size.

Metric-Deviation. Deviation based metric is widely used in spotting interesting subset. We propose a metric for generating one-dimensional partition.

$$Partitions = \{boundary | \Delta count(boundary(groupby(dimension))) \geq T\} \quad (3)$$

The Metric-Deviation will spot the boundaries where the results of aggregate function have vast drop-down near the border. The function $count()$ can also be replaced with other aggregate functions. The histogram is quite useful for spotting partitions with deviation near boundaries. Every time we select one dimension to generate partitions using metric-deviation plugin. We often only take the top dimensions into consideration. Usually the data-driven partitions are large in size and can answer many queries, while the user-driven metric is more powerful in generating more accurate partitions with smaller size.

Metric-Experiences. Previous experiences may exist if users have explored the dataset. The previous partitions and pattern can contribute to the current query task. Users or data analysts can also define new data-driven metric plugins according to different datasets.

The partitions generated from data-driven metrics are decided by the feature of the dataset and the metric plugins we choose, which can predict more query workloads in the future even the existing query workloads have not arrived yet.

4.2   Query Workload Metric Plugin

Currently, the only user-driven metric plugin is the query workload metric. Algorithm 1 shows the query workload learning method. The algorithm will generate candidate partitions using query workloads.

We adopt an unsupervised and heuristic method to generate the representative query workloads in every dimension. With the DBSCAN [7] clustering method, the representative workloads would obey the locality principle around the users' query workloads to some degree.

In our algorithm, we will first find the representative query workloads in one dimension and recursively do it in the others. We use a DBSCAN clustering method to estimate the approximate range of query workloads in one dimension, and then we further dig into every clustering region. For every clustering region we have ever got, we will find the overlapping region with the most overlap and extend it to cover the boundaries as a new partition. For other dimensions followed in the dimension chain, we filter the query workloads which can fit the predicates of previous partition and recursively get the representative workloads in other dimensions. In this way, we can generate new partitions with all the previous representative workloads step by step.

Algorithm 1. User-driven Metric: Query Workload Learning

Input: Query logs: queries
Output: Query workload partitions
 1: Result ← null
 2: dimensionlists←getDimensionRank(queries)
 3: **for** i=0 → len(dimensionlists) **do**
 4:     predicateslist←queries.filterPredicatesBy(dimensionlists[i],previous_workload[i])
 5:     clusters←DBSCANClustering(workloads.center)
 6:     **for** cluster in clusters **do**
 7:         overlap ← findMostOverlap(cluster.workloads)
 8:         exworkload ← extendWorkload(overlap)
 9:         Partitions[i].add(exworkload)
10:     **end for**
11:     previous_workload[i+1] ← Partitions[i]
12:     Result.add(CombinePartitionChain(Partitions[i]))
13: **end for**
14: **return** Result

4.3 Partition Learning Framework

The APEQ partition learning framework consists of multi plugins. We build the model using different weights of the plugins. The weights of the metric plugins mainly depend on the partitions' feedback.

When a partition is created, whether it is good or not depends on the hit rate and the selection rate. A better partition prefers to be smaller but cover more queries. We define the feedback of a single partition P as follows:

$$Feedback(P) = \frac{\sum_{i=1}^{N_{hit}} \frac{sizeof(workload_i)}{sizeof(P)}}{N_{queries}} \tag{4}$$

$N_{queries}$ means the number of queries since the partition was created. N_{hit} means the times when the partition P is hit by queries. $workload_i$ means the workload of the $query_i$. For simplicity, we use the count(*) to represent the size of the partition and workload.

We use the Algorithm 2 to figure out the best model during a query period. When faced with multi plugins and partitions, we assign weight to each plugin for an exploratory task and we generate each weight of the metric plugin using the total feedback of all the partitions generated by the plugin. A naive method is to select different numbers of partitions from the plugins according to their weights. With the RandomSelect function, everytime we need a total number of N partitions, we will retrieve $N * w_i$ partitions from the plugin X_i and unite the partitions from metric plugins to get the final partitions.

Additionally, there is an extra metric plugin in the algorithm, which is chosen from the existing metric plugins randomly. As the weights of plugins may decrease to zero if no partitions from them are hit, APEQ will randomly give a chance to a plugin with an additional weight in case a plugin discarded by previous users is useful for others.

Algorithm 2. Partition Learning Strategy

Input: Metric plugins with partitions
Output: Final partitions
  1: Select data-driven metric plugins:$X_1,X_2.\ .\ .\ .\ .$
  2: Select user-driven metric plugin: Query Workload Metric
  3: Select the learning algorithm: Naive Linear
  4: Select feedback model, evaluate the total feedback of every metric plugin from last
     query period: $f_1, f_2...f_n$
  5: Randomly select the metric X_e from the existing metric X_i with extra weight w_e
  6: $w_i \leftarrow (1 - w_e)*(\frac{f_i}{\sum_{i=1}^{n} f_i})$
  7: Compute the parameters and run the learning
  8: Select the Partition Retrieving Strategy: $F = RandomSelect(X, w)$
  9: Partitions= $F(X_1, w_1) \cup F(X_2, w_2) \cup .\ .\ .\ F(X_n, w_n) \cup F(X_e, w_e)$
 10: **return** Partitions

---

5   Partition Organization

The learning algorithm will generate the metadata of a partition and then the partition is created using Spark SQL. Instead of using the native Hive partition, we use a new Hive table to store the physical data of every partition in a directory and maintain the metadata in Hive meta database. Rather than modifying the physical data in HDFS, we organize the metadata of partitions with indexes to support the replacement of the table occurred in the user queries. The metadata can fit in memory as the number of partitions will not be too large due to our partition strategy.

5.1   Multi-dimensional Partition Interval Tree

The Multi-dimensional Partition Interval Tree (MPIT) is the index structure of our partition storage as the Fig. 2 shows. MPIT is organized using the dimension chains. The system will build an interval tree on every dimension of all the working partitions. An interval tree [14] is a tree data structure to hold intervals. Every node of interval tree stores following information.

1. i: An interval which is represented as a pair [low, high].
2. max: Maximum high value in subtree rooted with this node.

   The interval tree stores an interval and max value in subtree rooted with the node. For the insertion of the nodes we create a recursive method that runs through the tree until it finds the correct point in which the node has to be inserted. And it also updates the additional information about the maximum value of the subtree when needed. Figure 2 shows an example of a multi-dimensional partition interval tree. The structure consists of the metadata of all the partitions stored in dimension chain with partition id. For each dimension existing in these partitions, we build a one-dimensional interval tree by extracting the workloads from the corresponding dimension.

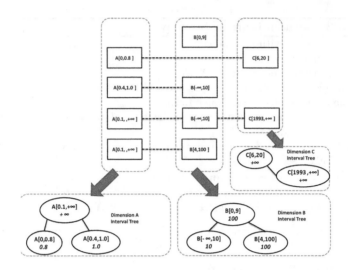

Fig. 2. Example of multi-dimensional partition interval tree

When inserting a partition into index, the partition will be represented as a dimension chain and every chain node will be distributed to the corresponding dimension tree. For every node, the system will scan the matched interval tree for a best place like the original one does.

5.2   Partition Search Engine

The partition search engine is responsible for replacing the original data source in the query with the partition if the data source filtered by the where clauses can be fully provided by the partition. We use the following Algorithm 3 to optimize the query process by quickly searching the qualified partitions in the index tree.

When receiving a new query, APEQ will extract its dimensions with predicates into query dimension chains. For every dimension, we use the interval tree search algorithm to find the partitions that can cover the predicates. For partition qualified at one dimension, we push it into stack and check the qualification in other dimensions if the partitions have more dimensions. Every dimension of the partition must match the query predicates. We implement the multi-dimensional partition search using a Depth-First-Search algorithm with stack. After we get a bucket of qualified partitions, we choose the smallest partition among the results for the best performance.

5.3   Partition Reorganization

When a new partition is going to be created physically, we search the predicates of the partition in the MPIT index. If there is a partition which can fully cover the new partition, we can create the new partition incrementally. To estimate

---

Algorithm 3. Multi-dimensional Partition Interval Tree Search

---

Input: MPIT T, query
Output: Qualified partition
1: **function** INTERVALSEARCH($root, i, res$)
2:    **if** Cover($root.node, i$) **then**
3:        $res.add(root.node)$
4:    **end if**
5:    **if** $root.left \neq null$ **and** $root.left.max \geq i.low$ **then**
6:        **return** IntervalSearch($root.left, i, res$);
7:    **end if**
8:    **return** IntervalSearch($root.right, i, res$);
9: **end function**
10:
11: **function** MULTI-DIMENSION SEARCH
12:    Select any $dimension \in query$
13:    IntervalSearch($T.dimension, query.dimension, res$)
14:    **if** $res.Partition$ has no other dimensions **then**
15:        Partitions.add($res.Partition$)
16:    **else**
17:        DFS other dimension $d\_next \in res.Partition$
18:        **if** res not in IntervalSearch($T.d\_next, query.d\_next, res\_next$) **then**
19:            discard $res.Partition$
20:        **end if**
21:        DFS until all other dimensions matched: Partitions.add($res.Partition$)
22:    **end if**
23:    **return** Minimum(Partitions)
24: **end function**

---

the time of creating a partition using Spark SQL, the most important factor is the I/O cost. A partition can be placed into HDFS physically only when the decrease in the cost of queries is greater than the partitioning cost.

Incremental Partitioning. A partition can also be derived from the existing partitions. For instance, if we need to generate partition with the where clauses $x_1 \leq A \leq y_1, x_2 \leq B \leq y_2$ and we already have a partition P_1 with the where predicates $x_1 \leq A \leq y_1$. We can easily generate the new partition using the Spark SQL "*select * from P_1 where $x_2 \leq B \leq y_2$*". The incremental partition can greatly reduce the I/O cost of creating a new partition.

Repartitioning Cost Estimation. Before a new partition is going to be created in HDFS, we consider the benefit time. For instance, if an existing partition is small enough and the new partition derived from the existing one will not reduce the size of the partition greatly, there is no need to split the partition. If a potential partition P can provide the data source used by N_{hit} queries in the past query period, the APEQ system will create it only if the following benefit model can be fitted.

$$\sum_{q \in hit\_queries} (T(q_{withoutP}) - T(q_{withP})) \geq T(P_{creation}) \qquad (5)$$

As $T(q_{withP})$ is unknown before P is really created, the APEQ uses a naive method for estimation. By ignoring the calculation time of queries, the system simply represents the I/O cost of a query q using the empirical equation $T(q) = w_{read} * sizeof(input) + w_{write} * sizeof(output)$ and uses count(*) to estimate size. Thus, the benefit equation can be transformed to the following form,

$$N_{hit}*w_{read}*(count(\mathcal{D})-count(P)) \geq (w_{read}*count(\mathcal{D})+w_{write}*count(P)) \quad (6)$$

$$\Rightarrow \frac{count(P)}{count(\mathcal{D})} \geq \frac{(N_{hit}-1)*w_{read}}{w_{write}+N_{hit}*w_{read}} \quad (7)$$

The item $\mathcal{D}$ is the original data source which can generate the new partition P, it can also be an existing partition for incremental partitioning. How expensive writes are compared to reads depends on different systems. In our system the average cost of writing is about 3.76 times the cost of reading. The benefit equation itself is empirical.

Partition Deletion. Partition Deletion operation is implemented in the repartitioning engine for the purpose of limiting the number and size of existing partitions. Using lazy strategy, the deletion operation will be performed when the feedback of partition is lower than the threshold during several query periods, or the number/size of the overall partitions reaches the limit.

6   Experiments

We evaluate our APEQ system over various occasions. The rest of this section is organized as follows. We first describe our experimental setup. Then we evaluate the performance on TPC-H as well as two real datasets respectively and analyse the query execution process.

6.1   Experimental Setup

We conduct our experiments on a Spark cluster with one master node and 9 workers. Each machine has a 2.1 GHz Intel Xeon processor, 16 GB RAM, running 64-bit Ubuntu server 14.04 with the software Hive 1.2.1, Spark 2.0.0 and Hadoop 2.7.2. We use dbgen in TPC-H [4] to generate the synthetic dataset. We choose the table lineitem and the scale factor is 100 with 74.11 GB in size. For query logs, we use the queries $q1, q4, q6, q10, q14, q15, q18, q19$ which are related to the table lineitem as the base templates. Then we use the template queries to build the synthetic exploratory queries. For the purpose of evaluating the I/O performance, we use the count(*) as the aggregation function to avoid additional calculation and output. The total number of queries executed is 120.

For the real case study, we evaluate APEQ on two datasets. The first dataset we used is called EJC (Early Journal Content) Data Bundle [2] including a few journal articles published from 1665 to 1920. The dataset is originally in XML format with 10 dimensions, such as *title*, *year*, *journaltitle* and *content*. The dataset is skewed in the *year* and *journaltitle* dimension, with 87.2% of articles published from 1871 to 1922 in the *year* dimension and 20.4% of articles published on the top three journals in the *journaltitle* dimension. The original size is 6.16 GB and we expand to 98.54 GB to make it relative large for our experimental environment and the number of queries issued by 5 users is 84. For the second real case study, we use the dataset and the query logs both from the public SDSS (Sloan Digital Sky Survery) [3] website. The SDSS dataset contains data from the astronomy domain. Our experiment uses the view PhotoPrimary from the BestDr8 database, with related query logs collected from the Sqllog of SDSS on March 2011. The PhotoPrimary describes primary photo objects with 454 dimensions. The size we used is 103.05 GB, which is a small fraction extracted by the casjobs tool. The number of queries is 169 and we modify several queries to fit in Spark SQL form.

6.2 Synthetic Data Evaluation

We define every 30 queries as a query period in the TPC-H experiment. Upon the ending of a query period, the APEQ system will do the partition generating work and create partitions using Spark SQL. Then the system will proceed to another query period with new partitions provided for answering new queries. The execution time of every query in different occasions are shown in Fig. 3(a), (b) and (c).

The queries without partitions get a steady performance in the execution time as all the data is fully scanned. For the Hive static partitions, we apply partitioning in the two most occurred dimensions named *l_discount* and *l_receiptdate*. When the queries are hit by these partitions, the execution time will decrease. For the APEQ partitioning, during the first period, there are only partitions generated by the data-driven metric plugins. During the second and the third period, more partitions are provided by the query workload plugin. And in the last period, most queries are hit by the partitions and the execution time is greatly decreased. The worst case is that queries are not exploratory and the workloads in the following query periods have no relation with the preceding ones. In this way, the query workload metric cannot generate appropriate partitions and APEQ can only depend on data-driven metrics.

The total time used is shown in Fig. 3(d). We evaluate the running time regardless of the initialization time of static partitions. So the time cost of Hive static partitioning and data-driven partitioning in APEQ will be ignored as they do not influence the execution time. Partitions generated from the query workload plugin during the query periods will cost overhead time in APEQ. The entire queries take 2.91 h (with 0.35 h of overhead) to run on APEQ, compared to 3.67 h without partitions (26% improvement) and 3.33 h with Hive static partitions (14% improvement).

6.3 Real Case Study

For the first real case study, we invited 5 users to explore the EJC dataset with APEQ running. The users have no basic idea about what the dataset is.

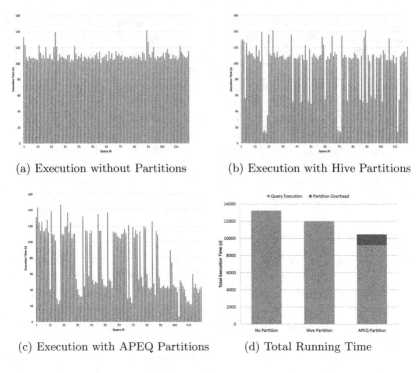

(a) Execution without Partitions   (b) Execution with Hive Partitions

(c) Execution with APEQ Partitions   (d) Total Running Time

Fig. 3. TPC-H queries execution time

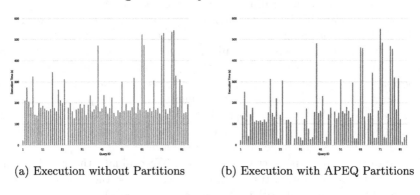

(a) Execution without Partitions   (b) Execution with APEQ Partitions

Fig. 4. Early journal content case study

We provide APEQ for them to issue queries and run their queries on Spark SQL without partitions for comparison. There are various range queries related with different dimensions as well as aggregate functions for analysis. The execution time of every query in different occasions is shown in Fig. 4. The final runtime of execution with APEQ partitions is nearly 4.20 h (with 0.60 h of partition overhead), compared to 5.06 h without partitions (20% improvement).

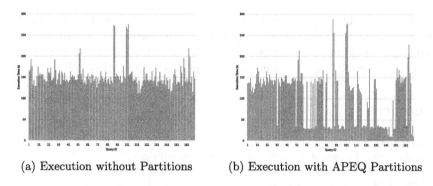

(a) Execution without Partitions   (b) Execution with APEQ Partitions

Fig. 5. SDSS case study

For the second real case study using the SDSS dataset, we define every 5 days as a query period according to the query distribution. As the Fig. 5 shows, queries are greatly optimized from query id 62 to 112. However, in the last query period after query id 152, few queries are optimized as the workloads seem to change rapidly. The final runtime of execution with APEQ partitions is 6.05 h (with 0.95 h of partition overhead), compared to 7.20 h without partitions (19% improvement).

7   Discussion

Replication and overlapping is quite common among the APEQ partitions. If the dataset itself is continuously changing, our system may be faced with high cost as all the related partitions need to be updated. This is our weakness compared with cell-based method like AQWA [5] as it stores only one copy. Whether a partitioning technique is efficient depends on the dataset and the task. For the data warehouse which is not frequently updated, APEQ is suitable for the exploratory task with the principle "overview first then detail" [12].

For improving the learning model, rather than naive method, supervised learning can be used when there are enough queries for training. The queries happening in the future can be regarded as a ground truth to evaluate the model. This is our extensive work to be done in the future.

Comparing with the Hive static partitioning, our work can also act as a materialize view optimization engine to support the Hive partitioning in response to dynamic workloads. Our work is based on Spark SQL with Hive and contributes to an adaptive partitioning scheme for Hive on exploratory queries. It has been used in our internal project and proved to be efficient.

8   Conclusion

In the paper, we proposed an adaptive data partitioning framework consisting of data-driven and user-driven strategies. We designed our APEQ system based on Spark SQL. The implementation was mainly about how to perform data partitioning and organize partitions for query optimization. Our work can speed up the data exploration and analysis process. The results of the experiments showed considerable improvements in the overall execution time.

References

1. Apache Hive. https://hive.apache.org/
2. Early Journal Content Data Bundle. http://dfr.jstor.org/
3. Sloan Digital Sky Surver (SkyServer). http://cas.sdss.org/dr8/en/
4. TPC-H, Benchmark Specification. http://www.tpc.org/tpch/
5. Aly, A.M., Mahmood, A.R., Hassan, M.S., Aref, W.G., Ouzzani, M., Elmeleegy, H., Qadah, T.: Aqwa: adaptive query workload aware partitioning of big spatial data. Proc. VLDB Endowment **8**(13), 2062–2073 (2015)
6. Armbrust, M., Xin, R.S., Lian, C., Huai, Y., Liu, D., Bradley, J.K., Meng, X., Kaftan, T., Franklin, M.J., Ghodsi, A., et al.: Spark SQL: Relational data processing in spark. In: Proceedings of the 2015 ACM SIGMOD International Conference on Management of Data, pp. 1383–1394. ACM, New York (2015)
7. Ester, M., Kriegel, H.P., Sander, J., Xu, X., et al.: A density-based algorithm for discovering clusters in large spatial databases with noise (1996)
8. Halim, F., Idreos, S., Karras, P., Yap, R.H.: Stochastic database cracking: towards robust adaptive indexing in main-memory column-stores. Proc. VLDB Endowment **5**(6), 502–513 (2012)
9. Idreos, S., Kersten, M.L., Manegold, S., et al.: Database cracking. In: CIDR, vol. 3, pp. 1–8 (2007)
10. Jiang, L., Nandi, A.: Snaptoquery: providing interactive feedback during exploratory query specification. Proc. VLDB Endowment **8**(11), 1250–1261 (2015)
11. Nandi, A., Jagadish, H.: Guided interaction: rethinking the query-result paradigm. Proc. VLDB Endowment **4**(12), 1466–1469 (2011)
12. Paurat, D., Garnett, R., Gärtner, T.: Interactive exploration of larger pattern collections: a case study on a cocktail dataset. In: Workshop on Interactive Data Exploration and Analytics. IDEA (2014)
13. Peng, S., Gu, J., Wang, X.S., Rao, W., Yang, M., Cao, Y.: Cost-based optimization of logical partitions for a query workload in a hadoop data warehouse. In: Chen, L., Jia, Y., Sellis, T., Liu, G. (eds.) APWeb 2014. LNCS, vol. 8709, pp. 559–567. Springer, Cham (2014). doi:10.1007/978-3-319-11116-2_52
14. Preparata, F.P., Shamos, M.: Computational Geometry: An Introduction. Springer, New York (2012)
15. Qin, W., Idreos, S.: Adaptive data skipping in main-memory systems. In: ACM SIGMOD International Conference on Management of Data (2016)
16. Sun, L., Franklin, M.J., Krishnan, S., Xin, R.S.: Fine-grained partitioning for aggressive data skipping. In: Proceedings of the 2014 ACM SIGMOD International Conference on Management of Data, pp. 1115–1126. ACM, New York (2014)
17. Vartak, M., Rahman, S., Madden, S., Parameswaran, A., Polyzotis, N.: SeeDB: efficient data-driven visualization recommendations to support visual analytics. Proc. VLDB Endowment **8**(13), 2182–2193 (2015)

Network Embedding

Semi-Supervised Network Embedding

Chaozhuo Li[1]($\boxtimes$), Zhoujun Li[1], Senzhang Wang[2,3], Yang Yang[1],
Xiaoming Zhang[1], and Jianshe Zhou[4]

[1] State Key Lab of Software Development Environment,
Beihang University, Beijing, China
{lichaozhuo,lizj,yangyang}@buaa.edu.cn

[2] Nanjing University of Aeronautics and Astronautics, Nanjing, China
szwang@nuaa.edu.cn

[3] Collaborative Innovation Center of Novel Software Technology and
Industrialization, Nanjing, China

[4] Capital Normal University, Beijing 100048, People's Republic of China
yolixs@buaa.edu.cn

Abstract. Network embedding aims to learn a distributed representation vector for each node in a network, which is fundamental to support many data mining and machine learning tasks such as node classification, link prediction, and social recommendation. Current popular network embedding methods normally first transform the network into a set of node sequences, and then input them into an unsupervised feature learning model to generate a distributed representation vector for each node as the output. The first limitation of existing methods is that the node orders in node sequences are ignored. As a result some topological structure information encoded in the node orders cannot be effectively captured by such order-insensitive embedding methods. Second, given a particular machine learning task, some annotation data can be available. Existing network embedding methods are unsupervised and are not effective to incorporate the annotation data to learn better representation vectors. In this paper, we propose an order sensitive semi-supervised framework for network embedding. Specifically, we first propose an novel order sensitive network embedding method: StructuredNE to integrate node order information into the embedding process in an unsupervised manner. Then based on the annotation data, we further propose an semi-supervised framework SemNE to modify the representation vectors learned by StructuredNE to make them better fit the annotation data. We thoroughly evaluate our framework through three data mining tasks (multi-label classification, network reconstruction and link prediction) on three datasets. Experimental results show the effectiveness of the proposed framework.

1 Introduction

Curse of dimensionality is a fundamental problem that makes many learning tasks difficult. This issue is particularly obvious when one wants to model the joint distribution among many discrete random variables, such as the words in

© Springer International Publishing AG 2017
S. Candan et al. (Eds.): DASFAA 2017, Part I, LNCS 10177, pp. 131–147, 2017.
DOI: 10.1007/978-3-319-55753-3_9

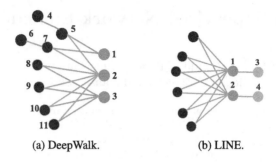

(a) DeepWalk.   (b) LINE.

Fig. 1. Two illustrations of DeepWalk and LINE.

a sentence and the nodes in a sparse network. To address this issue, recently a large body of works [1,14,17,21,25] focused on investigating how to represent the node in a sparse network as a dense, continuous, and low-dimensional vector. Such a distributed vector is defined as the representation vector of each node and the learning of the representation vectors can be also referred to *network embedding* [17]. As a more promising way to represent nodes in the network, network embedding is the foundation for many machine learning and data mining tasks, such as node classification [3,5], link prediction [11], network reconstruction [21,23] and recommendation [24].

Recently, some proven neural network techniques are introduced to perform network embedding [7,9,14,21]. Neural network based methods are easier to generalize and can usually achieve more promising embedding performance, and thus have attracted rising research interests recently. Although several network embedding methods are proposed as mentioned above, it is still very difficult for them to learn promising representation vectors which can both effectively preserve the network structure information and fit a particular data mining task. In general, there are two major limitations for existing methods. Firstly, existing methods [14,25] directly apply order insensitive embedding methods to generate node representations, while the node order information in the node sequences is largely ignored. DeepWalk assumes that two nodes with similar contextual nodes in the node sequences tend to be similar, and the contextual nodes of a center node are defined as a fixed-size window of its previous nodes and after nodes in the sequences. As shown in Fig. 1(a), node 1 and 2 share common contextual nodes {4,5,6,7}, while node 2 and 3 share common contextual nodes {8,9,10,11}. DeepWalk considers that the similarity between nodes 1 and 2 should be close to the similarity between nodes 2 and 3, since the nodes in each pair share equal number of common contextual nodes. However, according to literature [6], the influence of the direct neighbor nodes on a specific node can be significantly larger than those that are farther from it. Hence in Fig. 1(a), the similarity between nodes 2 and 3 should be larger than that between nodes 2 and 1 because the common contextual nodes of nodes 2 and 3 are all their direct neighbors. LINE [17] and SDNE [21] both utilize first-order and second-order proximity to conduct network embedding. As shown in Fig. 1(b), according to the first-order proximity, nodes 1 and 3 are similar to each other because they are directly connected; according to second-order proximity, nodes 1 and 2 are similar to each other because they share many common neighbors. However, these two proximities only preserve limited global network structure information. As shown in Fig. 1(b), nodes 3 and 4 are not directly connected and share no neighbors, but they are also very likely to be similar due to the fact that their only neighbor nodes 1 and 2 are similar.

The second limitation is that existing network embedding methods are unsupervised. Given a particular machine learning task, a small number of manual annotation data might be available [4]. Previous unsupervised network embedding methods are not effective to integrate the annotation data into the embedding process. In such a case, directly utilizing a general purposed network embedding method and ignoring these labeled samples may not achieve desirable performance for such a particular task.

In this paper, we study the order sensitive semi-supervised network embedding problem. The studied problem is difficult to address due to the following three challenges. Firstly, the objective function of previous embedding methods cannot effectively capture the node order information, and thus a new order-sensitive embedding method is necessary. Secondly, due to the expensive cost of manual annotation, in many scenarios there are only a few labeled nodes available. Consider the scarcity of labeled data, it is challenging to propagate the label information from the labeled nodes to the vast unlabeled ones. Finally, since the network topology and the annotation data potentially encode different types of information, combining them is expected to give a better performance. However, it is not obvious how best to do this in general.

To address the above challenges, we propose a two-stage semi-supervised network embedding framework. In the first stage, we learn an initial representation vector for each node in an unsupervised way. To effectively capture the node order information, we propose StructuredNE to encode the node order information into network embedding by exploiting an novel order-sensitive Skip-Gram algorithm. In the second stage, to incorporate a small number of labeled samples, we further propose a semi-supervised network embedding framework SemNE to modify the initial node representations learned from the first stage. Based on the observation that center node and its contextual nodes are likely to share the same label, we try to modify the representation vectors of the contextual nodes according to the label of the center node.

To summarize, we make the following primary contributions:

- We show that the order information in a sampled node sequence can affect the performance of network embedding, then we propose a novel model to incorporate the node order information into the embedding process.
- We propose a novel neural network based semi-supervised learning schema to integrate a small number of labeled data into network embedding process.
- We extensively evaluate our approach through multi-label classification, network reconstruction and link prediction tasks on three datasets. Experimental results show the effectiveness of the proposed embedding framework.

The rest of this paper is organized as follows. Section 2 formally defines the problem of semi-supervised network embedding. Section 3 introduces the proposed semi-supervised embedding framework in details. Section 4 presents the experimental results. Section 5 summarizes the related works. Finally we conclude this work in Sect. 6.

2   Problem Definition

A network G is defined as $G = (V, E)$, where V represents nodes in the network and E contains the edges. Existing unsupervised network embedding methods only utilize the network G. Different from previous works, in our setting we also have some labeled data, and we aim to fully utilize them to learn better node embeddings. The annotation matrix $Y \in \mathbb{R}^{|V_t| \times |\mathcal{Y}|}$ contains partial label information, in which $\mathcal{Y}$ is the set of label categories, and node set V_t contains a small number of labeled nodes. We formally define the problem of semi-supervised network embedding as below:

Definition 1. (Semi-supervised Network Embedding) *Given a network $G = (V, E)$ and the label matrix Y, the problem of* semi-supervised network embedding *aims to learn such a matrix $X \in \mathbb{R}^{|V| \times d}$, where d is the number of latent dimensions with $d \ll |V|$, and each row vector X_i of X is the representation vector of node i. We want to make the node similarity calculated by their representation vectors preserves both the network structure and manual annotation information.*

3   Semi-supervised Network Embedding Framework

In this section, we present the details of the semi-supervised network embedding framework. Firstly we introduce the frame diagram of the proposed embedding framework. Then DeepWalk, the basis of our work, is briefly introduced. After that, we present two critical components of the proposed framework: the order sensitive unsupervised embedding model StructuredNE and the semi-supervised embedding model SemNE.

3.1   Framework

As shown in Fig. 2, our network embedding framework contains two stages. In the first stage, the network is transformed into a set of node sequences using random walk. Based on these sequences, the order sensitive embedding model StructuredNE is introduced to perform network embedding. The learned representation vectors from the first stage preserve the network topology and node order information. In the second stage, based on the annotation data, we further propose an semi-supervised network embedding framework SemNE to modify the initial representations learned from the first stage. In order to transform the annotation information from labeled nodes to unlabeled ones, we generate a set of training units and their labels. Different colors of the training units represent

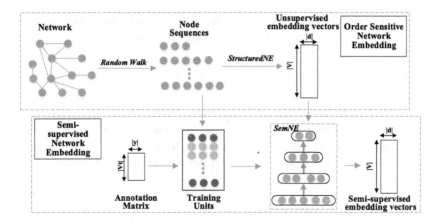

Fig. 2. Framework of the semi-supervised network embedding model.

different labels. A labeled center node together with its contextual nodes in a window are treated as a training unit, and we consider the label of the training unit to be the same as the center node. We propose a neural network model to predict the labels of the training units, and with error back propagation algorithm, the original node representation vectors are modified to match the labels of training units.

3.2   DeepWalk

In this subsection, we briefly present DeepWalk [14], the basis of our work. Deep-Walk introduced Skip-Gram algorithm [12,13] into the study of social network to learn the representation vector for each node. Firstly, DeepWalk generates a set of node sequences by random walk. Given a node sequence $T = \{n_1, n_2, ..., n_{|T|}\}$, nodes $n \in \{n_{t-w}, ..., n_{t+w}\} \backslash n_t$ are regarded as the contextual nodes of the center node n_t, where w is the window size. The objective function of the Skip-Gram model is to maximize the likelihood of the prediction of contextual nodes given the center node. More formally, given a node sequence $T = \{n_1, n_2, ..., n_{|T|}\}$, Skip-Gram aims to maximize the following objective function:

$$L = \frac{1}{T} \sum_{t=1}^{T} \sum_{\substack{-w \leq j \leq w \\ j \neq 0}} \log p(n_{t+j}|n_t) \tag{1}$$

In order to obtain the output probability $p(n_{t+j}|n_t)$, the Skip-Gram model estimates a matrix $O \in \mathcal{R}^{|V| \times d}$, which maps the representation vector $X_{n_t} \in \mathcal{R}^{1 \times d}$ of node n_t, into a $|V|$-dimensional vector o_{n_t}:

$$o_{n_t} = X_{n_t} \cdot O^T \tag{2}$$

where d is the embedding dimension. Then, the probability of the node n_{t+j} given the node n_t can be calculated by the following softmax objective function:

$$p(n_{t+j}|n_t) = \frac{e^{O_{n_t}(n_{t+j})}}{\sum_{n \in V} e^{O_{n_t}(n)}} \tag{3}$$

In the network embedding process, the representation vector X_{n_t} is modified to maximize L. After the learning process, we can obtain a matrix $X \in \mathbb{R}^{|V| \times d}$ which contains the representation vectors of all nodes.

Figure 3(a) shows the framework of Skip-Gram model. Skip-Gram model uses each center node as an input to a log-linear classifier with continuous projection layer. The projection layer convert each input node to the corresponding representation vector, and the log-linear classifier predict the contextual nodes given the center node. From the objective function of Skip-Gram model, we can clearly see that the node order information is ignored. In Formula 3, the probability $P(n_{t+j}|n_t)$ is independent from the node order index j, which may leads to suboptimal results.

3.3   Order Sensitive Unsupervised Network Embedding

In this subsection we propose an order sensitive network embedding method StructuredNE, which introduces the structured Skip-Gram model [22] for order sensitive network embedding. By introducing changes that make the embedding process aware of the relative positioning of contextual nodes, our goal is to improve the learned representation vectors while maintaining the simplicity and efficiency.

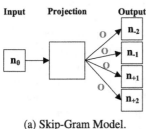

(a) Skip-Gram Model.

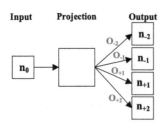

Fig. 3. Skip-Gram and Structured Skip-Gram Model.

Algorithm 1: StructuredNE

Input:
1      $G(V, E)$;
2   window size w;
3   embedding size d;
4   walks per node r;
5   walk length t;
Output:
6      $X \in \mathbb{R}^{|V| \times d}$;
7   $\mathcal{P} = [\,]$;
8   **for** $i = 0; i < r; i{+}{+}$ **do**
9   │ $\mathcal{O} = \text{Shuffle}(V)$;
10 │   **for** v_i *in* $\mathcal{O}$ **do**
11 │   │ $path = \text{RandomWalk}(G, v_i, t)$;
12 │   │ $\mathcal{P}.\text{append}(path)$;
13 │   **end**
14 **end**
15 $X = \text{StructuredSkipGram}(\mathcal{P}, w, d)$;
16 **return** X;

Figure 3(b) shows the framework of the structured Skip-Gram model. Different from the original Skip-Gram model, structured Skip-Gram model defines a set of $w \times 2$ output predictors $O_{-w}, ..., O_{-1}, O_1, O_w$, with each $O_i \in \mathcal{R}^{|V| \times d}$. w is the the window size of contextual nodes. Each of the output matrices is dedicated to predicting the output for a specific relative position to the center node. When making a prediction $p(n_j | n_i)$, we select the appropriate output matrix O_{j-i} to project the network embeddings to the output vector. More formally, given a node sequence $T = \{n_1, n_2, ..., n_{|T|}\}$ in $\mathcal{P}$, the structured Skip-Gram model aims to maximize:

$$L = \frac{1}{T} \sum_{t=1}^{T} \sum_{\substack{-w \leq j \leq w \\ j \neq 0}} \log p(n_{t+j} | n_t) \tag{4}$$

in which

$$p(n_{t+j} | n_t) = \frac{e^{o_{j,n_t}(n_{t+j})}}{\sum_{n \in V} e^{o_{j,n_t}(n)}} \tag{5}$$

o_{n_t} is a $|V|$-dimensional vector calculated by:

$$o_{j,n_t} = X_{n_t} \cdot O_j^T \tag{6}$$

One can see that the objective function of structured Skip-Gram model is affected by the relative positions between the center node and its contextual nodes. Hence, order information in the node sequences is integrated into the embedding process.

Algorithm 1 shows the main steps of the proposed StructuredNE. Lines 7 to 14 reveal the generation process of node sequences. In an iteration from line 9 to 13, after shuffling all nodes in V, the random walk generator function $RandomWalk(G, v_i, t)$ generates a single node sequence. In network G, starting from node v_i, a node sequence of length t is generated by random walk. This kind of iteration will perform r times to obtain enough sequences. All node sequences are appended into list $\mathcal{P}$. Then, the structured Skip-Gram model is introduced to learn the node representations with dimension d, and w represents the window size of contextual nodes.

Note that, the proposed order-sensitive embedding model does not increase the time cost. The number of calculations that must be performed for the forward and backward passes in the learning process of StructuredNE remains the same to the DeepWalk, as we are simply switching the output layer O for each different relative node positions. The number of parameters in matrix O is linearly increased by a factor of $w \times 2$.

3.4   Semi-supervised Network Embedding

In this subsection we propose an semi-supervised network embedding framework SemNE to incorporate the annotation data. In NLP area, many kinds of neural

Table 1. Generation of training units

(a) Example of node sequence and categories

Node sequence	$\{n_1, n_2, n_3, n_4,$ $n_5, n_6, n_7, n_8, n_9\}$
Categories	n_3 : category 1
	n_5 : category 2
	n_7 : category 3

(b) Training units extracted from above data with window size as 2

Training unit	Label
$\{n_1, n_2, n_3, n_4, n_5\}$	category 1
$\{n_3, n_4, n_5, n_6, n_7\}$	category 2
$\{n_5, n_6, n_7, n_8, n_9\}$	category 3

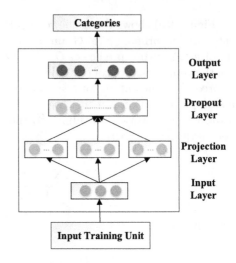

Fig. 4. Framework of SemNE.

networks have been proposed to process the corresponding tasks, such as the log-linear networks [12] and deeper feed-forward neural networks [8,16]. Motivated by these works, we modify the embedding vectors obtained from unsupervised methods by an neural network classification model to obtain the semi-supervised network embedding results.

The insight of SemNE is to utilize the manual annotations to tune the initial representation vectors by a neural network learning framework. The initial representation vectors of nodes are learned by the unsupervised embedding methods. In order to transform the annotation information from labeled nodes to unlabeled ones, we propose to utilize the training unit as the input of SemNE. Assume node n_i is a labeled node in V_t, given a node sequence contains n_i, the combination of n_i and its contextual nodes in a fixed window size w are treated as a training unit, and the label of the training unit is the same as the center node n_i. Table 1 shows an example of training units. Table 1(a) contains a node sequence and some labeled nodes, and Table 1(b) shows training units and their labels generated from the data of Table 1(a). We try to transfer annotation information from the center node to its contextual nodes in the same unit, and jointly learn their semi-supervised embeddings. As we sample a large number of node sequences for learning, thus statistically the unlabeled nodes in a training unit have a higher probability to be assigned the same label to many its neighbors sharing the same label, while a lower probability to be assigned other labels otherwise.

Given a training unit u, our neural network model tries to learn its distribution on different classes $P(\mathcal{Y}_i|u)$. As shown in Fig. 4, SemNE includes following layers:

- **Input Layer**: A training unit is submitted into the input layer. Given the window size w as 2, assume the input training unit is $\{n_{-2}, n_{-1}, n_0, n_1, n_2\}$, where $\{n_{-2}, n_{-1}, n_1, n_2\}$ are the contextual nodes of n_0. The label of the training unit is as same as the label of n_0. Input layer maps nodes in the input training unit to their corresponding indexes. The output vector of this layer is a w-dimensional vector containing the indexes of nodes in the input training unit.
- **Projection Layer**: This layer maps the input w-dimensional vector to a $(w \cdot d)$-dimensional vector. According to the input node indexes, project layer looks up their corresponding representation vectors in the embedding matrix X, which is initialized by unsupervised network embedding methods. After that, these representation vectors are concatenated as the output vector of project layer.
- **Dropout Layer**: In order to avoid overfitting, we add a dropout layer to the network to make use of its regularization effect. During training, the dropout layer copies the input to the output but randomly sets some of the entries to zero with a probability, which is set to 0.5.
- **Output Layer**: This layer linearly maps the input vector $I \in \mathbb{R}^{1 \times (w \cdot d)}$ into a $|\mathcal{Y}|$-dimensional vector $\mathcal{O}$:

$$\mathcal{O} = W \cdot I^T + b; \tag{7}$$

where $W \in \mathbb{R}^{|\mathcal{Y}| \times (w \cdot d)}$ and $b \in \mathbb{R}^{|\mathcal{Y}| \times 1}$ are parameters to be learned. $\mathcal{Y}$ is the set of categories. Then softmax function is introduced as the activation function:

$$P(\mathcal{Y}_i | u) = \frac{e^{\mathcal{O}_i}}{\sum_{0 \le k \le |\mathcal{Y}| - 1} e^{\mathcal{O}_k}} \tag{8}$$

The output of this layer can be interpreted as a conditional probability over categories given the input unit.

The entire formulation of our framework is:

$$\mathcal{O} = W \cdot DropOut(Concat_{0 \le i \le w-1}(X_{u_i})) + b;$$
$$P(\mathcal{Y}_i | u) = \frac{e^{\mathcal{O}_i}}{\sum_{0 \le k \le |\mathcal{Y}| - 1} e^{\mathcal{O}_k}} \tag{9}$$

where u is the input training unit and w denotes the window size. $X \in \mathbb{R}^{|V| \times d}$ contains the pre-trained node representations. $Concat$ is the concatenation function used in project layer. $DropOut$ denotes the dropout layer. W and b are parameters in the output layer. We utilize SGD and error back propagation to minimize negative log-likelihood cost function for each training example (u, l). The loss function is defined as

$$Loss(u, l) = -\log\left(\frac{e^{\mathcal{O}_l}}{\sum_{0 \le k \le |\mathcal{Y}| - 1} e^{\mathcal{O}_k}}\right) = -\mathcal{O}_l + \log\left(\sum_{0 \le k \le |\mathcal{Y}| - 1} e^{\mathcal{O}_k}\right) \tag{10}$$

During the training process, parameters in the neural network are modified to minimize the loss function. The embedding matrix X, which is used as parameters in the project layer, is also modified to adapt the annotation data. After the training process, the modified matrix X contains the semi-supervised representation vector for each node.

4   Experiments

In this section, firstly we introduce the datasets and comparison methods used in our experiments. Then we evaluate the proposed embedding methods through three data mining tasks on three social network datasets. Also we discuss the experimental results and investigate the sensitivity across core parameters.

4.1   Experiment Setup

In this section, we introduce the setup of our experiments, including datasets, baseline methods and the parameter setup.

Datasets. We use the following network datasets for evaluation:

- **BlogCatalog**: In the BlogCatalog website, bloggers submit blogs and specify the categories of these blogs. Blogger can follow other bloggers and form a social network. Nodes are the bloggers and labels represent their interest categories.
- **Flickr**: Flickr is a popular website where users can upload photos. Besides, there are various groups that are interested in some specific subjects. Users can join several groups they are interested in. In this data, nodes are users and labels represent interest groups users join in. We select top-10 most popular interest groups as categories of this dataset.
- **YouTube**: YouTube is a popular video sharing website. Users can upload videos and follow other similar interest users, which forms the social network. In this data, nodes are users and labels represent the interest groups of the users.

Table 2 displays the detailed statistics of the three datasets used in our experiments.

Table 2. Statistics of the datasets

Data	BlogCatalog	Flickr	YouTube
Categories	39	10	47
Nodes	10, 312	80, 513	1, 138, 499
Links	333, 983	5, 899, 882	2, 990, 443

Embedding Methods. We compare the proposed embedding methods with state-of-the-art methods as follows, including both matrix factorization based methods and neural network based methods.

- **TruncatedSVD**: TruncatedSVD [20] performs linear dimensionality reduction through truncated singular value decomposition (SVD).
- **LINE**: LINE [17] is a state-of-the-art network embedding method. As shown in [17], we concatenate the first-order and second-order embedding results as the final representation vector.
- **DeepWalk**: DeepWalk [14] introduces the Skip-Gram embedding method into social representation learning.
- **StructuredNE**: StructuredNE is the proposed order sensitive embedding method.

To further evaluate the performance of various semi-supervised embedding methods, we conduct experiments with the following methods.

- **DeepWalk + SemNE(p)**: The embedding results of DeepWalk are treated as the initial representation vectors of nodes. Given p percents nodes as known labeled information, the initial vectors are modified by using SemNE model to generate the final representation vectors.
- **StructuredNE + SemNE(p)**: Similar to DeepWalk + SemNE(p), StructuredNE first learns the initial representation vector for each node, and then SemNE is utilized to generate the final representation vectors.

Parameter Setup. Following most network embedding methods [14,17], we set the dimension of representation vector to 128. In LINE method, the parameters are set as follows: negative = 5 and samples = 100 million. In DeepWalk methods, parameters are set as window size w = 5, walks per node r = 80 and walk length t = 40. In semi-supervised embedding model SemNE, the size of training unit w = 5 and the percentage of labeled node set V_t used in learning process p = 0.05.

4.2 Network Reconstruction

In this subsection, we represent the network reconstruction performance of different methods. Wang et al. [21] propose network reconstruction task as a basic evaluation of different embedding methods. A good network embedding method should ensures that the learned embedding vectors are able to reconstruct the original network.

Assume the node representation vectors has been generated by network embedding methods. Given a specific node in the network, we can calculate the similarity scores of representation vectors between it and other nodes, and sort other nodes on descending order of the similarity scores. The existing links in the original network are known and can serve as the ground-truth. Neighbor nodes of the specific node are regarded as positive samples and nonadjacent nodes are negative samples. We choose $pre@k$ to evaluate the number of positive samples appeared in top k similar nodes, which is defined as:

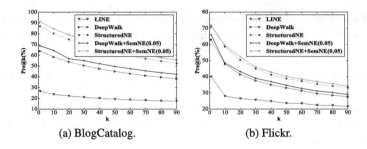

(a) BlogCatalog.   (b) Flickr.

Fig. 5. Network reconstruction performance of *pre@k*.

$pre@k = \frac{|\{j|i,j \in V, index(j)k, E_{ij}=1\}|}{k}$, in which $index(j)$ is the ranked index of the j-th node and $E_{ij} = 1$ refers to node i and j are directly connected.

We conduct experiments with k from 10 to 90 and Fig. 5 presents the corresponding results. From the experimental results, one can tell that our proposed methods consistently outperform all the other baselines on two datasets. Among all unsupervised embedding methods(Line, DeepWalk, StructuredNE), StructuredNE achieves the best performance and beats DeepWalk by 20% in BlogCatalog and 10% in Flickr, which proves the order information is helpful to preserve the network structure information. LINE performs worst on this task because it directly concatenate first-order and second-order embedding vectors as the final vector. The direct linked relationships between nodes are ignored by the second-order embedding vectors, which will introduce noise into the similarity calculation. Furthermore, after introduced a few annotation data by SemNE model, the performance is further improved.

4.3   Multi-Label Classification

In this subsection, we provide and analyze the multi-label classification performance of different embedding methods. We design the following classification methods.

- **Embedding Based Classification:** This kind of classification methods utilize representation vectors generated by various network embedding methods in Sect. 4.1. The representation vector of each node is treated as its feature vector, and then we use a one-vs-rest logistic regression model to return the most likely labels.
- **SpectralClustering** [19]: SpectralClustering conducts a low-dimension embedding of the affinity matrix between samples, followed by a k-means algorithm.
- **EdgeCluster** [18]: This method uses k-means clustering to cluster the adjacency matrix of G, which can handle the huge graphs.

In the embedding based classification methods, the one-vs-rest logistic regression classifier is implemented using scikit-learn[1] tool. A portion (T_r) of the

---

[1] http://scikit-learn.org/stable/.

labeled nodes are randomly picked as the training samples, and the rest nodes are treated as the test samples. We repeat this process 10 times, and report the average performance in terms of both macro-F1 and micro-F1. In EdgeCluster and SpectralClustering methods, the parameters are set the same as reported in the literatures. Figure 6 shows the results of classification on the three datasets. When the percentage of training data is set to 0.2, the improvement over the baseline methods (DeepWalk, LINE) is statistically significant (sign test, p-value < 0.05).

Experimental Results and Analysis. For the BlogCatalog dataset, we conduct experiments with the training data ratio increasing from 10% to 90%. From the experimental results shown in Fig. 6(a), one can see that among the unsupervised embedding methods, by incorporating the node order information, StructuredNE achieves better performance than other baselines. One can also see that, by incorporating a small number of annotation data into the proposed SemNE model, the performance of semi-supervised methods increases nearly 5% compared to the corresponding unsupervised methods. In Flickr dataset, our proposed order sensitive embedding method StructuredNE performs better than other unsupervised methods. The proposed semi-supervised method SemNE further improves the classification performance. Overall, StructuredNE + SemNE achieves the best performance on this data set and beats the best baseline method DeepWalk by nearly 5%. YouTube dataset is much larger than the other two datasets. Some baseline methods such as TruncatedSVD and SpectralClustering are not applicable due to the huge size of the data. We increase the training data ratio (T_r) from 1% to 9%. Among unsupervised embedding methods, our proposed order sensitive embedding method StructuredNE outperforms DeepWalk. By utilizing the labeled data, SemNE model can further improve the micro-F1 and macro-F1 scores.

From the results on the three datasets, we can obtain the following conclusions. (1) The proposed order sensitive embedding method StructuredNE outperforms the order insensitive embedding method DeepWalk; (2) Compared to the unsupervised network embedding methods, the proposed semi-supervised embedding method SemNE does improve the performance, and the initial representation vectors in SemNE can affect the classification performance.

Parameter Sensitivity. As the percentage p of the available labels can remarkably affect the performance of SemNE, here we conduct experiments to study the effect of different p. We evaluate the classification performance on BlogCatalog and Flickr datasets with various p. As shown in Fig. 7, we can see that on both datasets, the micro-F1 and macro-F1 scores consistently increase with the increase of p. The results also show that the initial representation vectors can affect the classification performance of SemNE, because StructuredNE + SemNE consistently outperforms DeepWalk + SemNE.

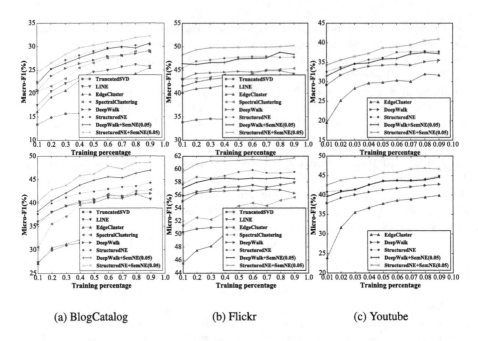

(a) BlogCatalog   (b) Flickr   (c) Youtube

Fig. 6. Classification performance (Micro-F1 and Macro-F1) on three datasets.

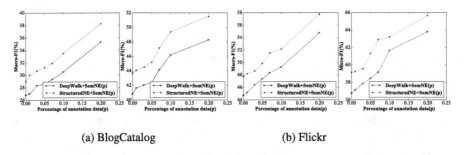

(a) BlogCatalog   (b) Flickr

Fig. 7. The classification performance w.r.t the percentage p of labeled nodes.

4.4   Link Prediction

In this task, we wish to predict whether a pair of nodes should have an edge connecting them. Link prediction is useful in many domains, such as in social networks, it helps us predict friend relationships between users. Different from the network reconstruction, this task predicts the future links instead of reconstructing the existing links.

Firstly we remove a portion of existing links from the input network. According to the remaining subnetwork, we generate the representation vector for each node using different embedding methods. Node pairs in the removed edges are considered as the positive samples. Besides, we randomly sample an equal

Table 3. Link prediction performance (AU score)

Method	BlogCatalog	Flickr
LINE	0.693	0.526
DeepWalk	0.742	0.628
Common neighbors	0.654	0.502
StructuredNE	0.796	0.661
StructuredNE + SemNE(0.05)	**0.821**	**0.683**

number of node pairs which are not directly connected as the negative samples. The similarity score between two nodes in a sampled node pair are calculated according to their representation vectors. We choose the Area Under Curve (AUC) score to measure the consistency between labels and the similarity scores. In addition to the embedding based methods, we also choose an popular and proven method *Common Neighbor* [10].

Table 3 shows the experimental results of link prediction. After removed 40% edges from the input network, our proposed embedding method StructuredNE + SemNE outperforms other baseline methods by 8% in BlogCatalog and 6% in Flickr dataset, which prove our proposed embedding method is still effective in sparse networks.

5   Related Work

To address the network embedding problem, conventional matrix factorization based approaches first construct the affinity matrix from the input network, and then generate the representation vectors during the decomposition process of the affinity matrix. Locally linear embedding [15] generate the embedding vectors which preserve distances within local neighborhoods. Spectral Embedding [2] finds the non-linear embeddings of the input affinity matrix using a spectral decomposition of the Laplacian graph. However, previous methods rely on the factorization of the affinity matrix, the complexity of which is too high to apply on a large network.

Recently the neural network techniques are introduced to learn the representation vectors of nodes, which can counter the limitations of matrix factorization based approaches. DeepWalk [14] introduces an order-irrelevance word embedding algorithm (Skip-Gram) to perform network embedding. Tang et al. [17] propose two types of proximities to preserve the global and local topology information, and they further propose a general embedding model LINE to utilize these proximities. Wang et al. [21] propose a deep non-linear embedding model to capture the first-order and second-order proximities in the network. Node2vec [9] first defines the notion of network neighborhood, and then proposed an improved random walk procedure, which efficiently explores diverse neighborhoods. However, previous related works are unsupervised learning methods which only considers the network structure information. Different from previous works, given a few annotation data, our work aims to propose an semi-supervised learning framework which is able to incorporate these label information for better learning the embedding vectors.

6   Conclusion

This paper firstly proposes an order sensitive unsupervised embedding model StructuredNE. Then we propose an novel semi-supervised network embedding framework SemNE, which is a neural network based model to integrate the annotation information into network embedding process. Experimental results of multi-label classification, link prediction and network reconstruction tasks on three datasets demonstrate the effectiveness of the proposed methods.

Acknowledgement. This work was supported by Beijing Advanced Innovation Center for Imaging Technology (No.BAICIT-2016001), the National Natural Science Foundation of China (Grand Nos. 61370126, 61672081, 61602237, U1636211, U1636210), National High Technology Research and Development Program of China (No.2015AA016004), the Fund of the State Key Laboratory of Software Development Environment (No.SKLSDE-2015ZX-16).

References

1. Ahmed, A., Shervashidze, N., Narayanamurthy, S., Josifovski, V., Smola, A.J.: Distributed large-scale natural graph factorization. In: WWW, pp. 37–48 (2013)
2. Belkin, M., Niyogi, P.: Laplacian eigenmaps for dimensionality reduction and data representation. Neural Comput. **15**(6), 1373–1396 (2003)
3. Bhagat, S., Cormode, G., Muthukrishnan, S.: Node classification in social networks. In: Social Network Data Analytics, pp. 115–148. Springer, US (2011)
4. Cao, J., Wang, S., Qiao, F., Wang, H., Wang, F., Yu, P.S.: User-guided large attributed graph clustering with multiple sparse annotations. In: Bailey, J., Khan, L., Washio, T., Dobbie, G., Huang, J.Z., Wang, R. (eds.) PAKDD 2016. LNCS (LNAI), vol. 9651, pp. 127–138. Springer, Cham (2016). doi:10.1007/978-3-319-31753-3_11
5. Cao, S., Lu, W., Grarep, Q.: Learning graph representations with global structural information. In: CIKM, pp. 891–900. ACM, New York (2015)
6. Cha, M., Haddadi, H., Benevenuto, F., Gummadi, P.K.: Measuring user influence in twitter: the million follower fallacy. In: ICWSM 2010 (2010)
7. Chang, S., Han, W., Tang, J., Qi, G.-J., Aggarwal, C.C., Huang, T.S.: Heterogeneous network embedding via deep architectures. In: KDD, pp. 119–128. ACM, New York (2015)
8. Collobert, R., Weston, J., Bottou, L., Karlen, M., Kavukcuoglu, K., Kuksa, P.: Natural language processing (almost) from scratch. JMLR **12**, 2493–2537 (2011)
9. Grover, A., Leskovec, J.: node2vec: Scalable feature learning for networks. In: KDD (2016)
10. Liben-Nowell, D., Kleinberg, J.: The link prediction problem for social networks. In: CIKM, pp. 1019–1031 (2003)
11. Liben-Nowell, D., Kleinberg, J.: The link-prediction problem for social networks. J. Am. Soc. Inf. Sci. Technol. **58**(7), 1019–1031 (2007)

12. Mikolov, T., Chen, K., Corrado, G., Dean, J.: Efficient estimation of word representations in vector space. arXiv preprint arXiv:1301.3781 (2013)
13. Morin, F., Bengio, Y.: Hierarchical probabilistic neural network language model. In: Aistats, vol. 5, pp. 246–252. Citeseer (2005)
14. Perozzi, B., Al-Rfou, R., Skiena, S.: Deepwalk: Online learning of social representations. In: KDD, pp. 701–710. ACM, New York (2014)
15. Roweis, S.T., Saul, L.K.: Nonlinear dimensionality reduction by locally linear embedding. Science **290**(5500), 2323–2326 (2000)
16. Taghipour, K., Ng, H.T.: Semi-supervised word sense disambiguation using word embeddings in general and specific domains. In: NAACL, pp. 314–323 (2015)
17. Tang, J., Qu, M., Wang, M., Zhang, M., Yan, J., Mei, Q.: Line: Large-scale information network embedding. In: WWW, pp. 1067–1077 (2015)
18. Tang, L., Liu, H.: Leveraging social media networks for classification. In: CIKM, pp. 1107–1116 (2009)
19. Tang, L., Liu, H.: Leveraging social media networks for classification. Data Min. Knowl. Disc. **23**(3), 447–478 (2011)
20. Tropp, A., Halko, N., Martinsson, P.: Finding structure with randomness: Stochastic algorithms for constructing approximate matrix decompositions. Technical report (2009)
21. Wang, D., Cui, P., Zhu, W.: Structural deep network embedding. In: KDD (2016)
22. Wang, L., Dyer, C., Black, A., Trancoso, I.: Two/too simple adaptations of word2vec for syntax problems. In: NAACL, pp. 1299–1304 (2015)
23. Wang, S., Hu, X., Yu, P.S., Li, Z.: Mmrate: inferring multi-aspect diffusion networks with multi-pattern cascades. In: ACM SIGKDD International Conference on Knowledge Discovery and Data Mining, pp. 1246–1255 (2014)
24. Yan, S., Xu, D., Zhang, B., Zhang, H.-J., Yang, Q., Lin, S.: Graph embedding and extensions: a general framework for dimensionality reduction. IEEE Trans. Pattern Anal. Mach. Intell. **29**(1), 40–51 (2007)
25. Yang, C., Liu, Z., Zhao, D., Sun, M., Chang, E.Y.: Network representation learning with rich text information. In: IJCAI, pp. 2111–2117 (2015)

CirE: Circular Embeddings
of Knowledge Graphs

Zhijuan Du, Zehui Hao, Xiaofeng Meng$^{(\boxtimes)}$, and Qiuyue Wang

School of Information, Renmin University of China, Beijing, China
{2237succeed,jane0331,xfmeng,qiuyuew}@ruc.edu.cn

Abstract. The embedding representation technology provides convenience for machine learning on knowledge graphs (KG), which encodes entities and relations into continuous vector spaces and then constructs $\langle entity, relation, entity \rangle$ triples. However, KG embedding models are sensitive to infrequent and uncertain objects. Furthermore, there is a contradiction between learning ability and learning cost. To this end, we propose circular embeddings (CirE) to learn representations of entire KG, which can accurately model various objects, save storage space, speed up calculation, and is easy to train and scalable to very large datasets. We have the following contributions: (1) We improve the accuracy of learning various objects by combining holographic projection and dynamic learning. (2) We reduce parameters and storage by adopting the circulant matrix as the projection matrix from the entity space to the relation space. (3) We reduce training time through adaptive parameters update algorithm which dynamically changes learning time for various objects. (4) We speed up the computation and enhance scalability by fast Fourier transform (FFT). Extensive experiments show that CirE outperforms state-of-the-art baselines in link prediction and entity classification, justifying the efficiency and the scalability of CirE.

Keywords: Knowledge graph · Circular embedding · Circulant matrix · Holographic projection · Dynamic learning · FFT

1 Introduction

Entities are the basic unit of human knowledge and are linked by relations. They are very important to learn relational knowledge representation. KG is a collection of multi-relational knowledge. It is mathematically represented as a multi-graph that linked with facts $\langle head\ entity,\ relation,\ tail\ entity \rangle$, abbreviated as $\langle h,\ r,\ t \rangle$, where h and t indicate nodes and are linked by the edge r.

This research was partially supported by the National Key Research and Development Program of China (No. 2016YFB1000603, 2016YFB1000602); the grants from the Natural Science Foundation of China (No. 61532010, 61379050, 91646203, 61532016); Specialized Research Fund for the Doctoral Program of Higher Education (No. 20130004130001), and the Fundamental Research Funds for the Central Universities, the Research Funds of Renmin University (No. 11XNL010).

© Springer International Publishing AG 2017
S. Candan et al. (Eds.): DASFAA 2017, Part I, LNCS 10177, pp. 148–162, 2017.
DOI: 10.1007/978-3-319-55753-3_10

They are widely utilized in nature language processing applications, such as question answering, web search, knowledge inference, fusion and completion, etc. However, applications of KG are suffering from data sparsity and computational inefficiency with the increasing size of KG. Thus, the embedding representation technology for KG is born and become a hot trend, including three branches: translation-based models, compositional models, SE/SME and neural network models.

Although these models have strong capability, they still face two challenges: the **sensitive issue** and the **contradiction issue**.

Firstly, about 48.9% 1-to-1 and N-to-1 triplets are over 85% accuracy ($Hits@10^1$), and 28.3% 1-to-N triplets are less than 60% [1,2]. This mainly results from uncertainty objects in KG, as shown in Fig. 1(a). For examples, *US President* can be *Herbert Bush* or *Walker Bush*. In addition, the occurring frequency of objects (entities or relations) is not balanced in KG, as shown in Fig. 1(b). We define objects with occurring less than 10 times as infrequent objects, and more than 50 times as frequent objects. Infrequent objects are highly informative and discriminative, but usually suffer an overfitting problem when training. Frequent ones some times are underfitting. Improper fitting losses information and slows down convergence. Thereby, embedded models are sensitive for uncertainty objects, infrequent objects and frequent objects.

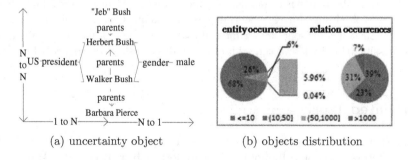

(a) uncertainty object   (b) objects distribution

Fig. 1. Entities and relationships distribution statistics

Secondly, existing embedding models cannot balance learning cost, computational cost and fitting accuracy. As shown in Table 1, TransE [1] has the shortest training time because there is only the vector addition and subtraction in triple fitting. TransR [3] has better accuracy since entities and relations lie in different semantic spaces. But its matrix-vector multiplication needs longer training time. RESCAL [4,5] represents fact via the tensor product and captures rich interactions. HOLE [6] outperforms others in fitting accuracy $Hits@10$). But they require a higher dimensions and larger storage.

In this paper, we propose an embedding model called CirE to overcome above two problems. Our contributions are as follows:

---

[1] *Hit@10*:proportion of correct entities in top-10 ranked entities.

Table 1. Example of contradiction issue

FB15K	TransE	TransR	RESCAL	HOLE
Fitting operation	v	m	Tensor product	Dot product
$Hits$@10	47.1	68.7	44.1	70.3
Training times	5 min	3 h	8 h	2 h
Dimensions	50d	50d	150d	200d
Semantic space	Same	Different	Same	Same

v:vector add or subtract; m:matrix-vector multiplication

- **accuracy:** We use the holographic projection and dynamic learning to improve the accuracy of triple fitting.
- **cost:** We use the circulant matrix to project **h** and **t** to **r**, which reduces model parameters and the storage space.
- **convergence:** We use adaptive parameters update algorithm dynamically adjusting learning time on various objects to prevent improper fitting, which accelerates convergence and reduces training time.
- **scalability:** For high-dimensional data, we employ FFT to speed up the computation and further reduce storage space, which enhances scalability.

The rest of the paper is organized as follows. Section 2 describes related work in learning embeddings for KG. Section 3 expounds our CirE model and analyzes its ability. Section 4 presents experimental results on the link prediction task, the entity classification task and the scalability analysis task. Section 5 concludes the paper.

2   Related Work

The mathematical notations are as follows: a triple is denoted by (h, r, t). Its column vectors is denoted by bold lower case letters $\mathbf{h}, \mathbf{r}, \mathbf{t}, \mathbf{e} = \{\mathbf{h}, \mathbf{t}\}$; matrices by bold upper case letters, such as $\mathbf{M}$; projection vector is denoted by subscript p; score function is represented by $f_r(\mathbf{h}, \mathbf{t})$. As mentioned in Sect. 1, the related work is introduced through the following 3 branches.

2.1   Translation-Based Models

The first branch is the translation-based model. Our CirE also belongs to this branch, which is inspired by translation invariant of semantic and syntactic relations. This means $\mathbf{h} + \mathbf{r}$ is supposed to be close to $\mathbf{t}$ for a triplet (h, r, t). In the score function $f_r(\mathbf{h}, \mathbf{t}) = \|\mathbf{h} + \mathbf{r} - \mathbf{t}\|_{\ell1, \ell2}$, $\mathbf{r}$, $\mathbf{h}$ and $\mathbf{t}$ can be within the same semantic space or from separate semantic spaces. TransE [1] is a innovative work, which models $\mathbf{r}$ and $\mathbf{e}$ within the same semantic space. Here, $\mathbf{e}_i = \mathbf{e}_j, \forall i \neq j$ when r link many h or t. Hence, it is unsuitable for N-to-1, 1-to-N and N-to-N relations. To break the limitation, TransH characterizes the entity embedding as

$\mathbf{e}_r = \mathbf{e} - \mathbf{w}_r^{\mathrm{T}}\mathbf{e}\mathbf{w}_r$ to enable an entity to have distinct representations and play different roles when involved in different relations [7].

Later, as a milestone, TransR [3] inspires series of models putting $\mathbf{r}$ and $\mathbf{e}$ in different semantic spaces, such as TransD [8] and TranSparse [2] which set the projection matrix $\mathbf{M}_r$ for each $\mathbf{r}$ to project $\mathbf{e}$ from the entity space to the relation space, e.g. $\mathbf{e}_p = \mathbf{M}_r\mathbf{e}$. The score function is $f_r(\mathbf{h},\mathbf{t}) = \|\mathbf{h}_p + \mathbf{r} - \mathbf{t}_p\|_{\ell 1,\ell 2}$. $\mathbf{M}_r$ is a common matrix and is shared with $\mathbf{h}$ and $\mathbf{t}$ in TransR [3]. TransD [8] believes that $\mathbf{h}$ and $\mathbf{t}$ should be distinguished, and $\mathbf{M}_r$ should be related to both entities and relations. Thus, $\mathbf{M}_r$ is replaced by $\mathbf{M}_{re} = r_p e_p + I^{n \times m}$. TranSparse [2] replaces $\mathbf{M}_r$ with sparse matrix $\mathbf{M}_r^e(\theta_r^e)$. The sparse degree θ_r^e is determined by the number of entities linked to $\mathbf{r}$.

Although translation-based models outperform others, their convergence and calculation are slow. TranSparse [2] only updates nonzero elements, which only reduces the complexity of updating. Moreover, sparse matrix-vector multiplication has a inherent bottleneck of frequently memory access. However, CirE can accelerate convergence via dynamic learning time for different objects. FFT speeds up the computation and further reduces storage to enhance scalability.

2.2 Compositional Models

The second branch is the compositional model that fits triplet based on tensor product, such as LFM [9,10], RESCAL [4,5], DistMult [11], which lets $f_r(\mathbf{h},\mathbf{t}) = \mathbf{h}^{\mathrm{T}}\mathbf{M}_r\mathbf{t}$. LFM only optimizes the nonzero elements. RESCAL optimizes entire $\mathbf{M}_r$, but brings a lot of parameters. Thus, DistMult [11] uses diagonal of $\mathbf{M}_r$ to reduce parameters. But this approach can only model symmetric relations.

HOLE [6] uses dot product rather than tensor product. It employs circular correlation between $\mathbf{h}$ and $\mathbf{t}$ to represent the entity pairs, denoted as $[\mathbf{h} * \mathbf{t}]_k = \sum_{i=0}^{d-1} \mathbf{h}_i \mathbf{t}_{(i+k) \bmod d}$, $f_r(\mathbf{h},\mathbf{t}) = \sigma(\mathbf{r}^{\mathrm{T}}(\mathbf{h} * \mathbf{t}))$, $\sigma(x) = 1/(1+e^{-x})$. It has an advantage on non-commutative relation and equivalence relation ($\mathbf{h}$ similar to $\mathbf{t}$), and speeds up calculation by FFT.

The basic ideal of HOLE [6] is the circular correlation between two entities while CirE focuses on cross-correlation between an entity and a relation. Thus, HOLE is limited to relation types. Yet, CirE can avoid this.

2.3 SE/SME and Neural Network Models

The third branch is the earliest. SE [12] transforms the entity space with the head-specific matrix $\mathbf{M}_{rh}$ and tail-specific matrix $\mathbf{M}_{rt}$ by $f_r(h,t) = \|M_{rh}h - M_{rt}t\|_1^1$, but it cannot capture relations between entities [13]. SME [14] can handle correlations between entities and relations by $f_r(\mathbf{h},\mathbf{t}) = (\mathbf{M}_1\mathbf{h} \odot \mathbf{M}_2\mathbf{r} + \mathbf{b}_1)^{\mathrm{T}}(\mathbf{M}_3\mathbf{h} \odot \mathbf{M}_4\mathbf{r} + \mathbf{b}_2)$, $\odot = +, \otimes$. In addition, there are some neural network models, such as SLM [13] and NTN [13]. SE [12] cannot be used to accurately depict the semantic relations between entities and relations. SLM [13] uses the nonlinear operation of single-layer neural network to solve this problem and meanwhile reduces the parameters of SE. But it only provides a relatively weak link between entities and relations. The NTN [13] model defines a score function combining SLM [13] and LFM [9,10], $f_r (\mathbf{h}, \mathbf{t}) = \mathbf{u}_r^{\mathrm{T}} g (\mathbf{h} \mathbf{M}_r \mathbf{t} + \mathbf{M}_{r,1} \mathbf{h} + \mathbf{M}_{r,2} \mathbf{t} + \mathbf{b}_r)$, where $\mathbf{u}_r^{\mathrm{T}}$ is the relation-specific linear layer, g is the *tanh* function, $\mathbf{M}_r (\mathbf{M}_r \in \mathbb{R}^{d \times d \times k})$ is a 3-way tensor. Therefore, this kind of models need much triplets for training and does not suit for sparse KG [1].

3   Methodology

CirE is a translation-based model. More particularly, entities and relations are from separate semantic spaces. They are converted by the projection matrix, which usually uses common matrix or sparse matrix. But our CirE uses circulant matrix. Circulant matrix [15] is a special kind of Toeplitz matrix and is widely used in dimensionality reduction, binary embedding and so on. Each row vector rotates one element to the right relative of the preceding row vector. However, projection matrix aims at linear space transformation. Therefore, the elements in the circulant matrix can be rotated to the right or to the left. Here, We extend the circulant matrix, as shown in Definition 1.

Definition 1 (circulant matrix). As shown in Fig. 2, each row in circulant matrix rotates one element to the right or left relative to the preceding row vector. Thus, it can be divided into left and right circulant matrix, respectively denoted as $\mathbf{A}^L = CircL(\mathbf{a})$ and $\mathbf{A}^R = CircR(\mathbf{a})$, where a is the first row vector, called circulant vector.

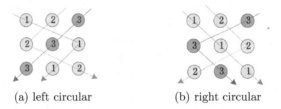

(a) left circular   (b) right circular

Fig. 2. The schematic of circulant matrix

The elements of left and right circulant matrix are denoted by Eqs.(1) and (2) respectively.

$$\mathbf{a}_{ij}^L = \mathbf{a}_{(i-1)((k-j) \bmod n)}^L, k = 1, \ldots, n, \mathbf{a}_{ij} \in A^L \tag{1}$$

$$\mathbf{a}_{ij}^R = \mathbf{a}_{(i-1)((k+j) \bmod n)}^R, k = 1, \ldots, n, \mathbf{a}_{ij} \in A^R \tag{2}$$

The relation between the left and the right circulant matrix is shown in Eq. (3)

$$\begin{cases} \mathbf{a}_{1j}^L = \mathbf{a}_{1j}^R, i = 0 \\ \\ \mathbf{a}_{ij}^L = \mathbf{a}_{i(-j \bmod n)}^R, i > 0 \end{cases} \tag{3}$$

3.1 CirE

In CirE, for each triple (h, r, t), we set entities as $e = \{h, t\}$ and its embeddings vector as $e = \{\mathbf{h}, \mathbf{t}\} \in \mathbb{R}^n$ and relation embedding as $\mathbf{r} \in \mathbb{R}^m$. For each $\mathbf{r}$, we project $\mathbf{e}$ from $\mathbb{R}^n$ to $\mathbb{R}^m$ via projection matrix $\mathbf{A}_e \in \mathbb{R}^{n \times m}$ or $\mathbf{A}_e \in \mathbb{R}^{n \times n}$ $(m > n)$. $\mathbf{A}_e$ is the circulant matrix. The **circulant vector** is denoted as $\mathbf{a}^e = (\mathbf{a}_i^e, i = 1, \ldots, n)$, $\mathbf{a}^{le} = (\mathbf{a}_i^{le}, i = 1, \ldots, m - n + 1)$. The projected vector of $\mathbf{e}$ is $\mathbf{e}^r$. Here, let $\mathbf{A}_e$ be a right circulant matrix. Left circulant matrix is similar. The k_{th} element of $\mathbf{e}^r$ can be obtained by Eqs. (4), (5) and (6), and the score functions is defined as Eq. (7).

$$if \quad n = m, \mathbf{e}_k^r = \sum_{i=0}^{i=n-1} \mathbf{e}_i \mathbf{a}_{(i+k) \bmod n}^e, k \in [0, n-1] \tag{4}$$

$$if \quad n > m, \mathbf{e}_{k1}^r = \mathbf{e}_k^r, k_1 = k \in [0, m-1] \tag{5}$$

$$else \quad \begin{cases} \mathbf{a}^e = [0, \mathbf{a}^{le}], \mathbf{a}^{le} \in \mathbb{R}^{(m-n+1)} \\ \\ \mathbf{e}_k^r = \sum_{i=0}^{i=n-1} \mathbf{e}_i \mathbf{a}_{(i+k) \bmod n}^e, k \in [0, n-1] \end{cases} \tag{6}$$

$$f_r(\mathbf{h}, \mathbf{t}) = ||\mathbf{h}^r + \mathbf{r} - \mathbf{t}^r||_{\ell1, \ell2}^2 \tag{7}$$

$$||\mathbf{o}||_{\ell1, \ell2} \le 1, \mathbf{o} = \mathbf{h}, \mathbf{r}, \mathbf{t}, \mathbf{e}^r$$

From the Eq. (4), we can see $\mathbf{e}^r$ is a cross-correlation [15] between $\mathbf{e}$ and $\mathbf{a}$. Similarly, it is a circular convolution [16] when $\mathbf{A}_e$ is a left circulant matrix. We called this projection a holographic projection. In Eq. (7), $||\mathbf{o}||_{\ell1, \ell2} \le 1$ is an enforce constraints on the norms of the embedding $\mathbf{h}, \mathbf{r}, \mathbf{t}$ and $\mathbf{e}^r$.

Furthermore, in Eq. (4), the calculation process of the projected vector $\mathbf{e}^r$ is shown in Fig. 3(b), which is a inverse process of Fig. 3(a). In Fig. 3(a), the projected vectors $\mathbf{e}^r$ can be interpreted as a circular convolution between $\mathbf{e}$ and $\mathbf{a}$. In signal processing, $\mathbf{e}$ and $\mathbf{e}^r$ are input and output signal, $\mathbf{a}$ is an activation function [17]. According to the theory of structural calculations [18], circulant matrix $\mathbf{A}$ is internal constraints between each component in $\mathbf{e}$, and $\mathbf{e}^r$ is external conditions. From holographic memories [19], $\mathbf{e}^r$ can preserve similar properties as $\mathbf{e}$. $\mathbf{a}$ is a cue. Thus, $\mathbf{a}$ and $\mathbf{e}$ are solved by the correlation of the cue with memory trace in learning phase. From the aspect of linear algebra, $\mathbf{e}^r = \mathbf{A}\mathbf{e}$ is a circular linear equation, the solution of which is stable [20].

Similarly in Fig. 3(b), $\mathbf{e}^r$ is cross-correlation between $\mathbf{e}$ and $\mathbf{a}$, which equal to circular convolution when $\mathbf{a}_i^R = \mathbf{a}_{(-i \bmod n)}^L$ [20]. Thus, we can employ FFT [21] to speed up the computation. Then, the runtime complexity is quasilinear (loglinear) to d, as $\mathbf{e}^r$ can be computed via Eq. (8).

$$\mathbf{e}_r^k = F^{-1}\left(\overline{F(\mathbf{e}_i)} \odot \overline{F(\mathbf{a}_i^e)}\right), \mathbf{e} = \mathbf{h}, \mathbf{t} \tag{8}$$

where F and F^{-1} denote the FFT [20] and its inverse, $\odot$ denotes the Hadamard (entrywise) product. The computational complexity of FFT is $O(d \log d)$.

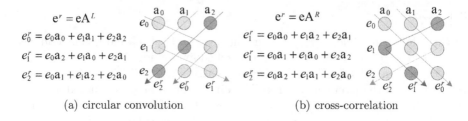

(a) circular convolution   (b) cross-correlation

Fig. 3. The solution of the holographic projection

3.2   Training Method and Algorithm Implementation

At last, Eq. (7) is transformed into a minimization problem with constraints $\|\mathbf{o}\|_{\ell 1, \ell 2} \leq 1, \mathbf{o} = \mathbf{r}, \mathbf{h}, \mathbf{t}$. The objective function $\mathcal{L}$ and score function $f_r(\mathbf{h}, \mathbf{t})$ are shown in Eqs.(9) and (10).

$$\mathcal{L} = \sum_{(h,r,t) \in \Delta} \sum_{(h',r,t') \in \Delta'} \max\left(0, f_r(\mathbf{h}, \mathbf{t}) + \gamma - f_r(\mathbf{h}', \mathbf{t}')\right) \tag{9}$$

$$\Delta' = \left\{ \left(h', r, t\right) | h' \in \mathcal{E} \cup \left(h, r, t'\right) | t' \in \mathcal{E} \right\} \tag{10}$$

where γ is the margin, Δ and Δ' are the set of correct and incorrect triples respectively. Δ' is the negative sampling set of Δ. Then minimize $(\mathcal{L})$ can be solved by using Stochastic Gradient Descent (SGD) [22], as also proposed in most models [1–7,9–14]. SGD use a global learning rate η to updating all parameters, regardless of their characteristic. But according to analyze in Sect. 1, we can see frequent objects need longer time to learn and infrequent objects need shorter time to learn. We adopt Adadelta [23] which dynamically adapts over time to make small gradients have larger learning rates, while large gradients have smaller learning rates. Firstly, Adadelta restricts the window of past gradients that are accumulated to be with in a fixed size w, and then implements this accumulation as an exponentially decaying average of the squared gradients. Assume at time, t this running average is $E\left[g^2\right]_t$ as shown in Eq. (11).

$$E\left[g^2\right]_t = \rho E\left[g^2\right]_{t-1} + (1 - \rho) g_t^2 \tag{11}$$

where ρ is a decay constant. Since it requires the square root of this quantity when updating parameters, this effectively becomes the RMS (as shown in Eq. (12)) of previous squared gradients up to time.

$$\text{RMS}[g]_t = \sqrt{E[g^2]_t + \varepsilon} \tag{12}$$

where ε is a constant, the resulting parameter update is shown in Eq. (13).

$$\Delta x_t = -\frac{\eta}{\text{RMS}[g]_t} g_t \tag{13}$$

Therefore, the algorithm of Op-TransE is shown in Algorithm 1.

Algorithm 1. Learning CirE

Input: Training set Δ and Δ', entity and relation set $\mathcal{E}$ and $\mathcal{R}$, margin γ, embeddings dim. m, n.
Output: h, r, t
1: initializer $\leftarrow$ uniform$(-6/\sqrt{k}, 6/\sqrt{k})$ for each $\mathbf{r} \in \mathcal{R}$ //or the result of TransE
2: $\mathbf{r} \leftarrow \mathbf{r}/||\mathbf{r}||$ for each $\mathbf{r} \in \mathcal{R}$
3: $\mathbf{e} \leftarrow$ uniform$(-6/\sqrt{k}, 6/\sqrt{k})$ for each $\mathbf{e} \in \mathcal{E}$
4: Let each projection matrix be $\mathbf{A}_e = Circ\_R(1, 0, \ldots, 0)$ //also $Circ\_L$
5: loop
6:    $\mathbf{e} \leftarrow \mathbf{e}/||\mathbf{e}||$ for each $\mathbf{e} \in \mathcal{E}$
7:    $\Delta_{batch} \leftarrow sample(\Delta, b)$ //sample a minibatch of size b
8:    $T_{batch} \leftarrow \emptyset$ // initialize the set of pairs of triplets
9:    for $(h, r, t) \in \Delta_{batch}$ do
10:      $(h', r, t') \leftarrow sample(\Delta'_{(h', r, t')})$//sample a corrupted triplet
11:      $T_{batch} \leftarrow T_{batch} \cup \{(h, r, t), (h', r, t')\}$
12:   end for
13:   update embeddings and projection matrices
14:   for $\ell \in$ entities or relations in T_{batch} // normalized vector
15:      if $||\ell||^2_{\ell 1, \ell 2} > 1$ then
16:         $\ell \leftarrow \ell \big/ ||\ell||^2_{\ell 1, \ell 2}$ //constrains: $||\mathbf{e}||_{\ell 1, \ell 2} \leq 1, ||\mathbf{r}||_{\ell 1, \ell 2} \leq 1, ||\mathbf{e}^r||_{\ell 1, \ell 2} \leq 1, \mathbf{e} = \mathbf{h}, \mathbf{t}$
17:      end if
18:   end for
19:   for $t \in [1, T]$ do //update parameters, loop over # of updates
20:      $g_t \leftarrow \sum_{((h,r,t),(h',r,t')) \in T_{batch}} \nabla[\gamma + f_r(\mathbf{h}, \mathbf{t}) - f_r(\mathbf{h}', \mathbf{t}')]_+$ // compute gradient
21:      $E[g^2]_t \leftarrow \rho E[g^2]_{t-1} + (1 - \rho) g_t^2$ // accumulate gradient
22:      $\Delta x_t = -\frac{\eta}{\text{RMS}[g]_t} g_t$ // compute update
23:      $E[\Delta x^2]_t = \rho E[\Delta x^2]_{t-1} + (1 - \rho) \Delta x_t^2$ //accumulate update
24:      $x_t \leftarrow x_{t-1} + \Delta x_t$ // update parameters
25:   end for
26: end loop

3.3 Complexity Comparisons

The scalability of an algorithm lies not only in high accuracy, but also low time and space complexity. We compare CirE with other models as shown in Table 2.

In Table 2, complexity is measured by the number of parameters, time and memory of multiplication operations in an epoch. An epoch is a single pass over all triples. n_e, n_r and n_{tr} represent the number of entities, relations and triplets in a KG respectively. n_k is the number of hidden nodes of a neural network and n_s is the number of slice of a tensor. d represent $d_e = d_r$. d_e and d_r represent the dimension of entity and relation embedding space respectively. $\theta(0 \ll \theta \ll 1)$ denotes the average sparse degree of all transfer matrices. We can see that the number of parameters in CirE is same as TransH. The time of multiplication operations is similar to TranSpare, and between TransD and TransR, which shows the high efficiency of our approach. In addition, the higher dimension is, the more obvious our advantage is.

4 Experiments and Analysis

Our approach, CirE, is evaluated by the tasks of link prediction, triplet classification efficiency and scalability on two subsets of WN11 [13] and WN18 [15] from WordNet and two subsets of FB15k [15] and FB13 [13] from Freebase. The statistics of the 4 datasets are given in Table 3.

Table 2. Complexities comparison

Model	Time complexity	Memory complexity	Opt.
SE [12]	$O(2d^2 n_{tr})$	$O(n_e\ d + 2n_r d^2)$	SGD
SME [14]	$O(4dn_k n_s n_{tr})$	$O(n_e\ d + n_r d + (4dn_s + 1)n_k)$	SGD
SLM [14]	$O((2dn_k + n_k)n_{tr})$	$O(n_e d + 2n_r n_k(1 + d))$	SGD
NTN [13]	$O(((d^2 + d)n_s + (2d + 1)n_k)n_{tr})$	$O(n_e d + n_r n_s(d^2 \times 2d + 2))$	LBFGS
LFM [10]	$O((d^2 + d)n_{tr})$	$O(n_e d + n_r d^2)$	SGD
RESCAL [4]	$O(pqd_e (1 + q) + q^2*(3d_e + q + pq))$	$O(n_e d + n_r d^2)$	SGD
DistMult [11]	$O(pd_e (1 + q) + q(3d_e + q + pq))$	$O(n_e d + n_r d^2)$	AdaGrad
HOLE [6]	$O(pqd_e (1 + \log q) + q \log q(3d_e + q + p \log q))$	$O((n_e + n_r)d)$	SGD
TransE [1]	$O(dn_{tr})$	$O((n_e + n_r)d)$	SGD
TransH [7]	$O(2dn_{tr})$	$O(n_e d_e + 2n_r d_r)$	SGD
TransR [3]	$O(2d_e d_r n_{tr})$	$O(n_e d_e + (d_e + 1)n_r d_r)$	SGD
TransD [8]	$O(2d_r n_{tr})$	$O(2(n_e d_e + n_r d_r))$	AdaDelta
TranSparse [2]	$O\left(\lambda(1 - \theta)\ d_e d_r n_{tr}\right)(0 \ll \theta \ll 1)$	$O(n_e d_e + \lambda(1 - \theta)* (d_e + 1)n_r d_r)$	SGD
CirE	$O(d_e \log d_e n_{tr})$	$O(n_e d_e + 2n_r d_r)$	AdaDelta

$n_s = 1$: linear; $\lambda = 2$: separate, $\lambda = 1$: share

Table 3. Statistics of the datasets used in this paper

Dataset	WN11	WN18	FB13	FB15k
#Rel	11	18	13	1,345
#Ent	38,696	40,493	75,043	14,951
#Train	112,581	141,442	316,232	483,142
#Valid	2,609	5,000	5908	50,000
#Test	10,544	5,000	23,733	59,071

For evaluation, we use the same metrics as TransE [1]: (1) average rank of correct entities *MeanRank* and (2) proportion of correct entities in top-10 ranked entities(*Hits@*10). Firstly, we replace the head or tail entity for each test triplet with all entities in the KG, and rank them in descending order of score function $f_r(\mathbf{h}, \mathbf{t})$. Secondly, we filter out the corrupted triples which have appeared in KG. We report *MeanRank* and *Hits@10* with two settings: the original one is named *Raw*, while the newer one is *Filter*. We compare with translation-based

Table 4. Evaluation results on link prediction

Datasets	WN18				FB15K			
Metric	MeanRank		Hits@10%		Mean Rank		Hits@10%	
	Raw	Filter	Raw	Filter	Raw	Filter	Raw	Filter
TransE [1]	263	251	75.4	89.2	243	125	34.9	47.1
TransH(u/b) [7]	318/401	303/388	75.4/73.0	86.7/82.3	211/212	84/ 87	42.5/45.7	58.5/64.4
TransR(u/b) [3]	232/238	219/225	78.3/79.8	91.7/92.0	226/198	78/77	43.8/48.2	65.5/68.7
CTransR(u/b [3]	243/231	230/218	78.9/79.4	92.3/92.3	233/199	82/75	44.0/48.4	66.3/70.2
TransD(u/b) [8]	242/224	229/212	79.2/79.6	92.5/92.2	211/194	67/91	49.4/53.4	74.2/77.3
TranSparse(t1) [2]	248/237	236/224	79.7/80.4	93.5/93.6	226 /194	95/88	48.8/53.4	73.4/77.7
TranSparse(t2) [2]	242/233	229/221	79.8/80.5	93.7/93.9	231/191	101/86	48.9/53.5	73.5 /78.3
TranSparse(t3) [2]	235 /224	223/221	79.0/79.8	92.3/92.8	211/187	**63/82**	50.1/53.3	77.9/79.5
TranSparse(t4) [2]	233/223	**221/211**	79.6/80.1	93.4/93.2	216/190	66/ 82	50.3/**53.7**	78.4/79.9
TranSparse(ave)	239.5/229.3	227.3/219.3	79.5/80.2	93.2/93.4	221/190.5	81.3/84.5	49.5/53.5	75.8/78.9
CirE(u/b)	**228/221**	**220/213**	**81.25/82.1**	**94.2/94.6**	**203/163**	68/85	**52.4/54.1**	**80.3/80.5**

u/b: unif/bern; t1: share, S, unif/bern; t2: share, US, unif/bern; t3: separate, S, unif/bern; t4: separate, US, unif/bern; ave: (t1 + t2 + t3 + t4)/4

models. Since the datasets are the same, we directly report the results of several baselines from TranSparse [2].

To be fair, CirE is limited to $d <= 100$, our projection matrix is square. We select the margin $\gamma = \{0.1, 0.5, 1, 2, 5, 10\}$, dimension $n = m = \{20, 50, 100\}$ and the mini-batch size $B = \{100, 200, 480, 1440\}$, the dissimilarity measure $d = \{\ell_1, \ell_2\}$.

4.1 Link Prediction

The role of link prediction is to predict the missing h or t for a given fact (h, r, t) [1,2]. We evaluate our CirE on WN18 and FB15K, the results are shown in Table 4. In addition, to compare *Hits*@10 of different kinds of relations, we evaluate our CirE on FB15K according to TransE [1], the detailed results as shown in Table 5.

As expected, in Table 4, the filtered setting provides lower *MeanRank* and higher *Hits@10*. Compared to TranSparse(t4), CirE improves 0.6% on *Hits@10* on FB15K, and outperforms all the compared methods. On WN18, CirE improves 1.4% and 2.4% on *Hits@10* compared to TranSparse and TransD respectively, and improves 12.1% compared to TransR. Therefore, we can conclude that the entity vectors projected from the entity space to the relation space are more reasonable in CirE than TransH, TransR, TransD and TranSparse. And the performance achieved by CirE is significant through the combination of holographic projection and dynamic learning. On average, CirE improves 8.31, 14.56 on *MeanRank* and 1.46%, 2.42% on *Hits@10* than TranSparse on WN18 and FB15K respectively while TranSparse improves $-2.06, -3.56, 0.71\%, 0.84\%$ than TransD.

Table 5. Experimental results on FB15k by mapping properties of relations (%)

Tasks	Prediction Head(Hits@10)				Prediction Tail(Hits@10)			
Types	1-to-1	1-to-N	N-to-1	N-to-N	1-to-1	1-to-N	N-to-1	N-to-N
TransE [1]	43.7	65.7	18.2	47.2	43.7	19.7	66.7	50.0
TransH(u/b) [7]	66.7/66.8	81.7/87.6	30.2/28.7	57.4/ 64.5	63.7/ 65.5	30.1/39.8	83.2/83.3	60.8/67.2
TransR(u/b) [3]	76.9/78.8	77.9/89.2	38.1/34.1	66.9/69.2	76.2/79.2	38.4/37.4	76.2/90.4	69.1/72.1
CTransR(u/b) [3]	78.6/81.5	77.8/89.0	36.4/34.7	68.0/71.2	77.4/80.8	37.8/38.6	78.0/90.1	70.3/73.8
TransD(u/b) [8]	80.7/86.1	85.8/95.5	47.1/39.8	75.6/78.5	80.0/85.4	54.5/50.6	80.7/94.4	77.9/81.2
TranSparse(t1) [2]	83.2/87.5	86.4/95.4	50.3/44.1	73.9/78.7	84.8/87.6	57.7/55.6	83.3/93.9	75.3/80.6
TranSparse(t2) [2]	83.4/87.1	86.7/95.8	49.8/44.2	73.4/79.1	84.8/87.2	57.3/55.5	78.2/94.1	76.4/81.7
TranSparse(t3) [2]	82.3/86.8	85.2/95.5	51.3/44.3	79.6/80.9	82.3/86.6	59.8/56.6	84.9/94.4	82.1/83.3
TranSparse(t4) [2]	83.2/87.1	85.2/95.8	51.8/44.4	80.3/81.2	82.6/87.5	60.0/57.0	85.5/94.5	82.5/83.7
TranSparse(ave)	83.0/87.1	85.9/95.8	50.8/44.3	76.8/80.0	83.6/87.2	58.7/56.2	83.0/94.2	79.1/782.3
CirE(u/b)	**84.8/87.8**	**85.5/96.1**	**54.6/50.2**	**82.0/83.0**	**85.5/88.0**	**62.3/60.0**	**93.8/94.5**	**84.0/84.3**

l/b: linear/bilinear; u/b: unif/bern; t1: share, S, unif/bern; t2: share, US, unif/bern; t3: separate, S, unif/bern; t4: separate, US, unif/bern; ave: (t1 + t2 + t3 + t4)/4

Table 5 shows CirE outperforms other methods on FB15k. For all types of relations, CirE is higher than 50% on *Hits@10*. Especially, compared to TranSparse(t4), CirE improves 4.4% and 1.2% on *Hits@10* for N-to-1 and N-to-N relations. On average, CirE improves 1.28%, 4.30%, 5.21%, 3.78% than TranSparse on 1-to-1, 1-to-N, N-to-1, N-to-N respectively. And CirE improve more significantly than other methods. This shows that multiple relation semantics can be accurately represented in CirE. Although the improvement of CirE is slightly lower in the simple relationship, for the complex relationship, the improvement is significant. There are mainly two reasons: (1) each dimension of the vector obtained by the holographic projection measures some correlation between the entity vector and the relation vector, which is more suitable for projection; (2) the accuracy of infrequent objects is further improved by dynamic parameters learning which learns frequent objects with longer times and learns infrequent objects with shorter times.

4.2 Triplet Classification

Triplet classification is a binary classification, which aims to judge whether a given triplet (h, r, t) is correct or not. We use 3 datasets WN11, FB13 and FB15K to evaluate CirE. The test sets of WN11, FB13 is provided by TransE. It contains positive and negative triplets. But the test set of FB15K only contains correct triples, so we make up some negative triples used by TranSparse for it. The compared results are described in Table 6 and Fig. 4.

Table 6 shows that CirE almost outperforms all the compared models. It obtains the best accuracy of 87.3% on WN11. It is near to the best accuracy of 89.1% of TransD on FB13 and 88.5% of TranSparse on FB15K, but higher than that of other models significantly. Figure 4 shows that CirE improves the performance of TransE and TranSparse both on simple and complex relations. Moreover, the classification accuracies of different relations are more than 70%. That is why we use joint holographic projection and adaptive gradient update to

Table 6. Triples classification accuracies (%)

Data sets	WN11	FB13	FB15
TransE(u/b) [1]	75.9/75.9	70.9/81.5	77.3/79.8
TransH(u/b) [7]	77.7/78.8	76.5/83.5	74.2/79.9
TransR(u/b) [3]	85.5/85.9	74.7/82.5	81.1/82.1
CTransR(bern) [3]	85.7	-	84.3
TransD(u/b) [8]	85.6/86.4	85.9/**89.1**	86.4/88.0
TranSparse(t1) [2]	86.2/86.3	85.5/87.8	85.7/87.9
TranSparse(t2) [2]	86.3/86.3	85.3/87.7	86.2/88.1
TranSparse(t3) [2]	86.2/86.4	86.7/88.2	87.1/88.3
TranSparse(t4) [2]	86.8/86.8	86.5/87.5	87.4/**88.5**
CirE(u/b)	**87.2/87.3**	**87.4/88.6**	**88.1/88.4**

u/b: unif/bern; t1: share, S, u/b; t2: share, US, u/b; t3: separate, S, u/b n; t4: separate, US, u/b

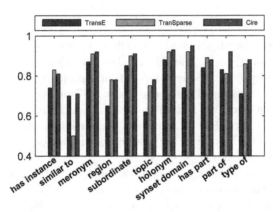

Fig. 4. Classification accuracies of different relations on WN11.

improve the accuracy of learning infrequent objects but not affect the frequent objects. Therefore, We believe that our model CirE can deal with the multi-relational data very well.

4.3 Scalability Evaluation

Another advantage of CirE is easy to train. We perform experiments on WN18 and FB15K to evaluate the scalability of CirE. Which is tested by number of iterations and time of each iteration on the one hand. The results are described in Table 7.

From Table 7, we can see CirE converges after 300 epochs on both WN18 and FB15K. The convergence is fast compared to 1000 epochs for other methods. The reason is that joint holographic projection and adaptive parameter update avoid

Table 7. Train time of embedding models

Method	d	Epoch	Time/Epoch	
			WN18	FB15K
TransE	50	1000	5 s	5.4 s
TransH	50	500	7 s	12 s
TransR	50	500	1.4 min	3.08 min
TransD	50	1000	12 s	20 s
TranSparse	50	1000	0.87 min	—
TranSparse	100	1000	3.83 min	6.4 min
CirE	50	300	1.17 min	—
	100	300	4.82 min	8.1 min
	128	300	3.94 min	6.8 min
	256	300	8.96 min	13 min
	512	300	13.2 min	17.9 min

underfitting of frequent objects and overfitting of infrequent objects. Moreover, the training time of TransE, TransH and TransD are about 5, 7 and 12 s per epoch on WN18 respectively. The computation complexity of TransR is higher, which takes about 1.4 min. TranSparse costs 0.87 min per epoch. CirE costs 3.94 min per epoch with 128d close to 3.83 min of TranSparse with 100d on WN18. Because we use FFT to accelerate computing starting from the 128d. The convergence of CirE is far below TranSparse. It can be seen that our model is easy to train and expand.

On the other hand, we design query answer experiments with different dimensions to test runtime. The experiments demonstrate that CirE performs inference as efficiently as embedding methods when d = {50, 100, 128, 256, 512}. Figure 5 depicts the number of answers that CirE offers per second.

In Fig. 5, we use FFT to accelerate computing starting from 128d. We can see, the query answer per second rates presents exponential reduction with the

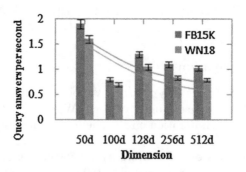

Fig. 5. Query answers per second rates for different dimensions.

exponential growth of dimension without FFT, such as at 50d and 100d. However, if we use FFT to accelerate computing on 128d, 256d, 512d, the query answer per second rates presents logarithmic reduction rather than exponential reduction with the exponential growth of dimension. And the query answer per second rates with FFT is higher than without FFT, such as on 128d higher than 100d. We believe that the higher dimension, the more advantageous for CirE. Therefore, CirE is easy to expand.

5   Conclusion and Future Work

In this paper we propose CirE, a translation based embedding model for KGs, which is based on the holographic projection with cross-correlation between entities and relations, and via adaptive parameters update avoid underfitting for frequent objects and overfitting for infrequent objects to accelerate convergence and reduce training time. An attractive property of CirE is that it balances learning ability and learning cost, and is easy to train. Experiments show that CirE provides state-of-the-art performance on a variety of benchmark datasets. However, Our experiments are on binary relations, and have not tested on higherarity yet. In addition, KG is simply seen as symbol triples in these embedding models, which focuses on the explicit semantic of the knowledge, but ignores the implied semantics. The explicit semantic is reflected by triple fitting, but implicit semantic is omitted from contextual information, which is supposed to be most critical in natural language understanding. To this end, we will further exploit CirE in higher-arity relations or via semantic-aided with background knowledge in future.

References

1. Bordes, A., Usunier, N., Garcia-Duran, A, et al.: Translating embeddings for modeling multi-relational data. In: Advances in Neural Information Processing Systems, pp. 2787–2795. MIT Press Massachusetts (2013)
2. Ji, G., Liu, K., He, S., et al.: Knowledge graph completion with adaptive sparse transfer matrix. In: Proceedings of the Thirtieth AAAI Conference on Artificial Intelligence, pp. 985–991. AAAI Press, MenloPark (2016)
3. Lin, Y., Liu, Z., Sun, M., et al.: Learning entity and relation embeddings for knowledge graph completion. In: Proceedings of the Twenty-Ninth AAAI Conference on Artificial Intelligence, pp. 2181–2187. AAAI Press, MenloPark (2015)
4. Nickel, M., Tresp, V., Kriegel, H.P.: A three-way model for collective learning on multi-relational data. In: Proceedings of the 28th International Conference on Machine Learning, pp. 809–816. ACM Press, New York (2011)
5. Nickel, M., Tresp, V., Kriegel, H.: Factorizing YAGO: scalable machine learning for linked data. In: Proceedings of the 21st World Wide Web Conference, pp. 271–280. ACM, New York (2012)
6. Nickel, M., Rosasco, L., Poggio, T.: Holographic embeddings of knowledge graphs. In: Proceedings of the Thirteenth AAAI Conference on Artificial Intelligence, pp. 1955–1961. AAAI Press, MenloPark (2016)

7. Wang, Z., Zhang, J., Feng, J., et al.: Knowledge graph embedding by translating on hyperplanes. In: Proceedings of the Twenty-Eighth AAAI Conference on Artificial Intelligence, pp. 1112–1119. AAAI Press, MenloPark (2014)

8. Ji, G., He, S., Xu, L., et al.: Knowledge graph embedding via dynamic mapping matrix. In: Proceedings of the 53rd Annual Meeting of the Association for Computational Linguistics and the 7th International Joint Conference on Natural Language Processing of the Asian Federation of Natural Language Processing, pp. 687–696. MIT Press, Massachusetts (2015)

9. Sutskever, I., Tenenbaum, J.B., Salakhutdinov, R.: Modelling relational data using Bayesian clustered tensor factorization. In: Proceedings of Advances in Neural Information Processing Systems 22: 23rd Annual Conference on Neural Information Processing Systems, pp. 1821–1828. MIT Press, Massachusetts (2009)

10. Jenatton, R., Roux, N.L., Bordes, A., et al.: A latent factor model for highly multi-relational data. In: Proceedings of Advances in Neural Information Processing Systems 25: 26th Annual Conference on Neural Information Processing Systems, pp. 3167–3175. MIT Press, Massachusetts (2012)

11. Yang, B., Yih, W., He, X., Entities, E., et al.: Relations for learning, inference in knowledge bases. In: Proceedings of ICLR (2015). Engelmore, R., Morgan, A. (eds.): Blackboard Systems. Addison-Wesley, Reading (1986)

12. Bordes, A., Weston, J., Collobert, R., et al.: Learning structured embeddings of knowledge bases. In: Proceedings of the the Twenty-Fifth AAAI Conference on Artificial Intelligence, pp. 301–306. AAAI Press, MenloPark (2011)

13. Socher, R., Chen, D., Manning, C.D., et al.: Reasoning with neural tensor networks for knowledge base completion, In: Proceedings of Advances in Neural Information Processing Systems 26: 27th Annual Conference on Neural Information Processing Systems, pp. 926–934. MIT Press, Massachusetts (2013)

14. Erbas, C., Tanik, M.M., Nair, V.S.S.: A circulant matrix based approach to storage schemes for parallel memory systems. In: Proceedings of the Fifth IEEE Symposium on Parallel and Distributed Processing, pp. 92–99. IEEE Press, Piscataway (1993)

15. Plate, T.A.: Holographic reduced representations. IEEE Trans. Neural Netw. **6**(3), 623–641 (1995)

16. Zhang, J., Fu, N., Peng, X.: Compressive circulant matrix based analog to information conversion. IEEE Sig. Process. Lett. **21**(4), 428–431 (2014)

17. Gentner, D.: Structure-mapping: a theoretical framework for analogy. Cogn. Sci. **7**(2), 155–170 (1983)

18. Gabor, D.: Associative holographic memories. IBM J. Res. Dev. **13**(2), 156–159 (1969)

19. Schnemann, P.H.: Some algebraic relations between involutions, convolutions, and correlations, with applications to holographic memories. Biol. Cybern. **56**(5–6), 367–374 (1987)

20. Angel, E.S.: Fast Fourier transform and convolution algorithm. Proc. IEEE **70**(5), 527–527 (1982)

21. Duchi, J., Hazan, E., Singer, Y.: Adaptive subgradient methods for online learning, stochastic optimization. J. Mach. Learn. Res. **12**, 2121–2159 (2011)

22. Zeiler, M.D.: ADADELTA: an adaptive learning rate method. arXiv preprint arXiv:1212.5701 (2012)

23. Bordes, A., Glorot, X., Weston, J., et al.: A semantic matching energy function for learning with multi-relational data. Mach. Learn. **94**(2), 233–259 (2014)

PPNE: Property Preserving Network Embedding

Chaozhuo Li[1]([✉]), Senzhang Wang[2,3], Dejian Yang[1], Zhoujun Li[1], Yang Yang[1],
Xiaoming Zhang[1], and Jianshe Zhou[4]

[1] State Key Lab of Software Development Environment, Beihang University,
Beijing, China
{lichaozhuo,ydj1994,lizj,yangyang,yolixs}@buaa.edu.cn
[2] Nanjing University of Aeronautics and Astronautics, Nanjing, China
szwang@nuaa.edu.cn
[3] Collaborative Innovation Center of Novel Software Technology and
Industrialization, Nanjing, China
[4] Capital Normal University, Beijing 100048, People's Republic of China

Abstract. Network embedding aims at learning a distributed representation vector for each node in a network, which has been increasingly recognized as an important task in the network analysis area. Most existing embedding methods focus on encoding the topology information into the representation vectors. In reality, nodes in the network may contain rich properties, which could potentially contribute to learn better representations. In this paper, we study the novel problem of property preserving network embedding and propose a general model PPNE to effectively incorporate the rich types of node properties. We formulate the learning process of representation vectors as a joint optimization problem, where the topology-derived and property-derived objective functions are optimized jointly with shared parameters. By solving this joint optimization problem with an efficient stochastic gradient descent algorithm, we can obtain representation vectors incorporating both network topology and node property information. We extensively evaluate our framework through two data mining tasks on five datasets. Experimental results show the superior performance of PPNE.

1 Introduction

Networks are ubiquitous in our daily lives and many real-life applications focus on mining information from the networks. A fundamental problem in network mining is how to learn the desirable network representations [4,22]. To address this problem, *network embedding* is presented to learn the distributed representations of nodes in the network. The main idea of network embedding is to find a dense, continuous, and low-dimensional vector for each node as its distributed representation. Representing nodes into the distributed vectors can form up a potentially powerful basis to generate high-quality node features for many data mining and machine learning tasks, such as node classification [6], link prediction [11] and recommendation [24,27].

© Springer International Publishing AG 2017
S. Candan et al. (Eds.): DASFAA 2017, Part I, LNCS 10177, pp. 163–179, 2017.
DOI: 10.1007/978-3-319-55753-3_11

Most related works investigate the topology information for network embedding, such as DeepWalk [16], LINE [19], Node2Vec [6] and SDNE [22]. The basic assumption of these topology-driven embedding methods is that nodes with similar topology context should be distributed closely in the learned low dimensional representation space. However, in many real scenarios, it is insufficient to learn desirable node representations by purely relying on the network topology structure. For example in the social networks, it is possible that two users share very similar interests but they are not connected and share no common friends, thus their similarity on interests cannot be effectively captured by the topology based network embedding methods. In such a case, other types of information should be incorporated as the complementary content to help us learn better representations.

Usually, nodes in the network may be associated with a set of properties, such as the profiles of each user in a social network and the metadata of each paper in a citation network. Node property information is also important to measure the similarity between nodes, but are largely ignored by previous network embedding methods. For example in a social network, if two users share some common tags or tweet topics, they are very likely to be similar even if they are topologically far away from each other. Since the node properties potentially encode different types of information from the network topology, integrating them into the embedding process is expected to achieve a better performance. As the first attempt, TADW [25] incorporates the text features of nodes into network embedding process under a framework of matrix factorization. However, there are two limitations of TADW. Firstly, the very time and memory consuming matrix factorization process of TADW makes it infeasible to scale up to large networks. Secondly, TADW only considers the texts associated to each node, and it is difficult to apply TADW to handle the node properties of rich types in general.

In this paper, we propose a general network embedding framework which can effectively encode both the topology information of the network and the rich properties of the nodes. This task is difficult to address due to the following challenges. Firstly, nodes in a network may contain several types of properties, and in different networks, the types of node properties are different. It is non-trivial to model various types of properties into an unified format and utilize these property information. Secondly, although combining network topology and node properties into the embedding process is expected to achieve better performance, it is not obvious how best to do this under a general framework. There are sophisticated interactions between network topology and node properties, and it is difficult to integrate node properties into the existing topology-derived models. For example, DeepWalk and Node2Vec cannot easily handle additional information during the random walk process in a network.

To address the above challenges, we propose a general and flexible property preserving network embedding model PPNE. We formulate the learning process of property preserving network embedding as a joint optimization problem, where the topology-derived and property-derived objective functions are optimized jointly. Specifically, we propose a negative sampling based objective function to capture the topology information, which aims to maximize the likelihood of the prediction of the center node given a specific contextual node.

Besides, we extract a set of constraints according to the property similarity between each pair of nodes, and a property-derived objective function is proposed to restrict the learned representation vectors to satisfy the extracted constraints. Finally, we utilize the stochastic gradient descent (SGD) algorithm to solve this joint optimization problem.

To summarize, we make the following contributions:

- In this paper we propose and study the novel problem of property preserving network embedding, and propose a general embedding framework to effectively incorporate both network topological information and node property information into the network embedding process.
- To utilize the property similarity information, we propose two ways of extracting constraints from the property similarity matrix. A carefully designed objective function with such constraints is also proposed.
- We extensively evaluate our approach through multi-class classification and link prediction tasks on five datasets. Experimental results show the superior performance of PPNE over state-of-the-art embedding methods.

The rest of this paper is organized as follows. Section 2 summarizes the related works. Section 3 formally defines the problem of property preserving network embedding. Section 4 introduces the proposed model PPNE in details. Section 5 presents the experimental results. Finally we conclude this work in Sect. 6.

2   Related Work

Network embedding aims to learn a distributed representation vector for each node in a network, which essentially is an unsupervised feature learning process. In general, network embedding is related to the problem of *graph embedding* or *dimensionality reduction*. Most existing network embedding methods can be categorized into two broad categories: matrix factorization based and neural network based methods.

Matrix factorization based methods first express the input network with a affinity matrix in which the entries represent the relationships between nodes, and then embed the affinity matrix into a low dimensional space using matrix factorization techniques. Locally linear embedding [17] seeks a lower-dimensional projection of the input affinity matrix which preserves distances within local neighborhoods. Spectral Embedding [2] is one method to calculate the non-linear embeddings. It finds a low dimensional representation of the input data using a spectral decomposition of the graph Laplacian. Sparse random projection [10] reduces the dimensionality of data by projecting the original input space using a sparse random matrix. However, matrix factorization based methods rely on the decomposition of the affinity matrix, which is too expensive to scale efficiently to large real-world networks. Besides, the manually predefined node similarity measurements are needed to construct the affinity matrix, which can significantly affect the quality of learned representation vectors.

Recently neural network based models are introduced to solve the network embedding problem. As the first attempt, DeepWalk [16] introduces an word embedding algorithm (Skip-Gram) [14] to learn the representation vectors of nodes. Tang et al. propose LINE [19], which optimizes a carefully designed objective function that preserves both the local and global structure. Wang et al. propose SDNE [22], a deep embedding model to capture the highly non-linear network structure and preserve the global and local structures. SDNE exploits the first-order and second-order proximities to preserve the network structure. Node2Vec [6] learns a mapping of nodes to a low-dimensional space of features that maximizes the likelihood of preserving distances between network neighborhoods of nodes. TADW [25] incorporates the text features of nodes into network embedding process under a framework of matrix factorization. Compared to the matrix factorization based methods, neural network based methods are easier to generalize and own strong representation ability. However, most previous works only take the network topology information into consideration. Different from the previous typology-only works, our work aims to propose a general property preserving network embedding model which integrate the rich types of node properties in the network into the embedding process.

Finally, there is a body of works focus on the problem of node classification [8,20,21,26] or link prediction [11,13]. However, the objective of our work is totally different from these works. We aim to learn better representation vectors for nodes, while the node classification or link prediction tasks are only utilized to evaluate the quality of the embedding results.

3   Problem Definition

In this section, we formally define the studied problem. The input network G is defined as $G = (V, T, P)$, where V represents nodes in the network. The topology matrix $T \in \mathbb{R}^{|V| \times |V|}$ is the adjacency matrix of the network. P is the property similarity matrix with each entry $P_{i,j} \in [0, 1]$ denoting the property similarity score between node i and j. Here we formally define the problem of property preserving network embedding:

Definition 1. *(Property Preserving Network Embedding): Given a network G = (V, T, P), the problem of property preserving network embedding aims to learn a matrix $X \in \mathbb{R}^{|V| \times d}$, where d is the number of latent dimensions with $d \ll |V|$. Each row vector X_i in X is the embedding vector of node i. The objective of property preserving network embedding is to make the learned representation vectors explicitly preserve both the network topology and node property information.*

4   Property Preserving Network Embedding

In this section, we present the details of the proposed property preserving network embedding model PPNE. Firstly we briefly introduce the framework of PPNE. Then the topology-derived objective function and property-derived objective function are introduced separately. After that we present the joint optimization process of the above two objective functions. Finally we discuss several practical issues of the proposed model.

4.1   Framework

Figure 1 shows the framework of PPNE. One can see that each node in the network is associated with a set of properties. The first step of PPNE is to construct two matrices from the input network: the topology matrix and the property similarity matrix. Topology matrix is a 0–1 adjacency matrix which represents the connections among nodes. The property similarity matrix contains the property similarities between each pair of nodes, which is calculated by a predefined similarity measurement. Given a particular machine learning or data mining task, users can flexibly choose or design the property similarity measurement. For example, in order to serve a geographic mining task, the geography related properties (address, geographic tag) should be more important than other properties. How to calculate the node property similarities is not the focus of this paper, as it varies to different networks and applications. We assume the property similarity matrix has been given by domain experts. For the topology matrix, we utilize the random walk algorithm to generate a set of node sequences. A topology-derived objective function is proposed to capture topology information preserved in the node sequences. For the property similarity matrix, we extract a set of constraints, which ensures the embedding vectors of nodes with similar properties should be distributed closely in the learned representation space. Here we define two kinds of constraints: inequality constraints and numeric constraints. We also propose a property-derived objective function for each type of the constraints. Finally, the topology-derived and property-derived objective functions are jointly optimized sharing same parameters.

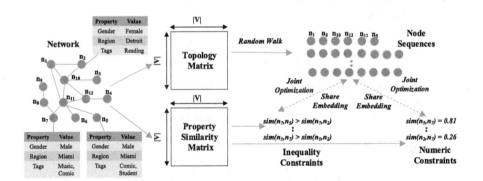

Fig. 1. Framework of PPNE.

4.2   Topology-Derived Objective Function

Following the idea of DeepWalk [16], we assume that nodes with similar topology context tend to be similar. With such an assumption, we aim to maximize the likelihood of the prediction of the center node given a specific contextual node. The contextual nodes of a center node are defined as a fixed-size window w of its previous nodes and after nodes in the node sequences generated by random walk. We propose an novel negative sampling based optimization objective function in which the node representation vectors are considered as the parameters. In the optimization procedure, the representation vectors are updated and the finally learned vectors preserve the topology information.

Firstly, the network topology matrix T is converted into a set of node sequences $\mathcal{C}$ by random walk. In a random walk iteration, for each node n in V, we generate a node sequence of length t which starts from node n. This iteration will repeat r times to generate enough sequences.

Based on the node sequences $\mathcal{C}$, we try to solve the following objective function:

$$maximize \quad D_T = \prod_{n \in \mathcal{C}} \prod_{z \in context(n)} \prod_{u \in \{\{n\} \cup NEG(n)\}} p(u|z) \tag{1}$$

Given a sampled center node n and its contextual nodes $context(n)$, $NEG(n)$ is the set of negative samples of the center node n with a predefined size ns. Nodes far away from the center node have a larger chance to be picked as the negative samples. $p(u|z)$ defines the probability of the center node u given the contextual node z. Given the contextual node z, we aim to maximize the probability of the positive samples $u \in \{n\}$, while minimize the probability of the negative samples $u \in NEG(n)$.

For each node n in V, we design two corresponding vectors: the embedding vector and the parameter vector. The embedding vector v_n is the representation of node n when it is treated as the contextual node, while the parameter vector θ_n is the representation of n when it is treated as the center node. In our model, $p(u|z)$ is defined as

$$p(u|z) = \begin{cases} \sigma(v_z^T \theta_u), & L^n(u) = 1 \\ 1 - \sigma(v_z^T \theta_u), & L^n(u) = 0 \end{cases}$$

in which σ is the sigmoid function:

$$\sigma(v_z^T \theta_u) = \frac{1}{1 + e^{-(v_z^T \theta_u)}}$$

$L^n(u)$ is an indicator function:

$$L^n(u) = \begin{cases} 1, & u \in \{n\} \\ 0, & u \in NEG(n) \end{cases}$$

Moreover, $p(u|z)$ can be represented as

$$p(u|z) = [\sigma(v_z^T \theta_u)]^{L^n(u)} \cdot [1 - \sigma(v_z^T \theta_u)]^{1-L^n(u)}$$

Hence, the objective function can be rewritten as follows:

$$maximize \quad D_T = \prod_{n \in \mathcal{C}} \prod_{z \in context(n)} \prod_{u \in \{\{n\} \cup NEG(n)\}}$$
$$[\sigma(v_z^T \theta_u)]^{L^n(u)} \cdot [1 - \sigma(v_z^T \theta_u)]^{1-L^n(u)} \tag{2}$$

By maximizing the likelihood of prediction of positive samples and minimizing the likelihood of negative samples, the proposed objective function encodes the topology information into the representation vectors of nodes.

Compare to DeepWalk, a popular topology-derived method, the proposed model is faster and more effective. Firstly, the objective function of DeepWalk is defined as

$$maximize \quad D = \prod_{n \in C} \prod_{z \in context(n)} p(z|n)$$

The hierarchical softmax method is introduced to design the probability $p(z|n)$, which reduces the computational complexity of calculating $p(z|n)$ from $O(|V|)$ to $O(\log_2 |V|)$. Hence the computational complexity of DeepWalk is $O(|C| \cdot 2w \cdot \log_2 |V|)$, in which w is the window size of the contextual nodes. In the objective function (2), the computational complexity is further reduced to $O(|C| \cdot 2w \cdot (ns + 1))$, in which ns is a constant number irrelevant to the size of network. Thus the model training time is significantly reduced. Secondly, by choosing the negative samples according to their distances from the center node, the rich global topology information is integrated into our model. This strategy ensures our model not only considers the local information of the contextual nodes, but also can encode the information of the nodes which have farther topological distance from the center node.

4.3   Property-Derived Objective Function

In this subsection we present the details of the property-derived objective function. In natural language processing area, SWE [12] and RC-NET [23] incorporate the semantic knowledges into the word embedding process. Inspired by the above works, we propose two ways to extract constraints from the property similarity matrix P. Based on these constraints, we introduce the property-derived objective functions.

Related works incorporate the original property features of single type into the embedding process [25]. Different from such works, in our approach the property matrix $P \in \mathbb{R}^{|V| \times |V|}$ contains the property similarity scores between each pair of nodes, which is calculated by a predefined similarity measurement. According to the requirements of targeting data mining tasks and the types of networks, users can flexibly design an appropriate property similarity measurement to cast the input network into the property similarity matrix. This strategy guarantees the proposed embedding framework can be applied on various kinds of networks and serves different types of data mining tasks.

Inequality Constraints. The first way we proposed to utilize the property similarity matrix P is to extract a set of inequalities from matrix P as the constraints. For each node n in V, according to the matrix P, we construct its most similar node set pos_n and most dissimilar node set neg_n. pos_n contains top k similar nodes of n and neg_n contains top k dissimilar nodes, where $k \ll |V|/2$. With pos_n and neg_n, we can obtain the following inequalities for node n:

$$P_{np} > P_{nq} \quad p \in pos_n, q \in neg_n$$

in which P_{np} is the property similarity score between node n and p. The final representation vectors of the corresponding nodes should satisfy the following constraint:

$$sim(v_n, v_p) > sim(v_n, v_q) \quad p \in pos_n \ q \in neg_n$$

in which v_n is the embedding vector of node n. $sim(v_n, v_p)$ is the cosine similarity between the embedding vectors of node n and p. After constructing such constraints for all the nodes, we can obtain a set of constraints $\mathcal{S}$, which contains triples in the form of $\{(i,j,k), \ sim(v_i, v_j) > sim(v_i, v_k)\}$.

Based on the constraint set $\mathcal{S}$ and the node sequences $\mathcal{C}$, we propose the following objective function to force the embedding vectors to satisfy the extracted constraints:

$$minimize \ \ D_I = \sum_{n \in \mathcal{C}} \sum_{(i,j,k) \in \mathcal{S}} I_{i,j,k}(n) \cdot f(i,j,k) \tag{3}$$

where $I_{i,j,k}(n)$ is an indicator function:

$$I_{i,j,k}(n) = \begin{cases} 1, & i = n \ or \ j = n \ or \ k = n \\ 0, & else \end{cases}$$

The function $f(.)$ is a normalization hinge loss function:

$$f(i,j,k) = max(0, sim(v_i, v_k) - sim(v_i, v_j))$$

The objective function D_I ensures that the similarity score between embedding vectors of two nodes with similar properties should be no less than the nodes with dissimilar properties. For a sampled node in the sequences, we select the inequality constraints associated with this node and judge whether the current embedding vectors of the corresponding nodes satisfy these constraints. If the constraints are satisfied, these embedding vectors will remain unchanged, otherwise they will be updated towards the direction of satisfying these constraints.

Numeric Constraints. The second way of utilizing the matrix P is to consider the property similarity scores as the numeric constraints for refining the network embedding process. The motivation is that the learned representation vectors of two nodes should be distributed closer to each other if they have similar properties, namely their property similarity score in P is high.

Similar to the generation process of the inequality constraints, according to the matrix P, for each node n in $|V|$ we select its top k similar and dissimilar node sets as the pos_n and neg_n. Nodes in the two sets own strong discrimination ability over node n. Based on the property similarity matrix P and node sequences $\mathcal{C}$, we propose the following objective function:

$$minimize \ \ D_N = \sum_{n \in \mathcal{C}} \sum_{i \in \{pos_n \cup neg_n\}} P_{ni} d(v_n, v_i) \tag{4}$$

in which $d(v_i, v_j)$ is the Euler distance function to measure the distance between the embedding vectors of node i and j: $d(v_i, v_j) = \sqrt{(v_i - v_j)^T (v_i - v_j)}$.

One can see that the optimization process of the objective function D_N is affected by the property similarity score P_{ni}, and the distances between nodes with similar properties decrease faster than that between dissimilar ones. Hence, D_N can ensure that nodes with similar properties are distributed closer in the learned embedding space.

4.4 Joint Optimization

In this subsection, we show the joint optimization process of the topology-derived and property-derived objective functions. Here we propose two types of PPNE model: PPNE$_{ineq}$ and PPNE$_{num}$. PPNE$_{ineq}$ aims to jointly optimize the topology-derived objective function D_T and the inequality constraint based objective function D_I. PPNE$_{num}$ aims to jointly optimize D_T and the property-derived function D_N with numeric constraints. We utilize the SGD algorithm to solve the optimization problems. Firstly we introduce the optimization process of D_T, D_I and D_N separately, and the derivative results are utilized to jointly update the embedding vectors.

To maximize the objective function D_T in the Formula (2), we try to maximize the following log-likelihood function:

$$\log D_T = \sum_{n \in C} \sum_{z \in context(n)} \sum_{u \in \{\{n\} \cup NEG(n)\}}$$
$$\{L^n(u) \cdot \log[\sigma(v_z^T \theta_u)] + [1 - L^n(u)] \cdot \log[1 - \sigma(v_z^T \theta_u)]\}$$

Given a sample of $(n, context(n))$ in C, with sampled z and u, we set

$$\mathcal{L} = L^n(u) \cdot \log[\sigma(v_z^T \theta_u)] + [1 - L^n(u)] \cdot \log[1 - \sigma(v_z^T \theta_u)] \tag{5}$$

Firstly we calculate the following partial derivative:

$$\frac{\partial \mathcal{L}}{\partial \theta_u} = [L^n(u) - \sigma(v_z^T \theta_u)] \cdot v_z$$

Thus θ_u can be updated by

$$\theta_u = \theta_u + \eta[L^n(u) - \sigma(v_z^T \theta_u)] \cdot v_z \tag{6}$$

where parameter η is the learning rate. In Formula (5), θ_u and v_z are symmetric, so v_z can be updated as

$$v_z = v_z + \eta \sum_{u \in \{\{n\} \cup NEG(n)\}} [L^n(u) - \sigma(v_z^T \theta_u)] \cdot \theta_u \tag{7}$$

Then we show the optimization process of D_I in Formula (3). For convenience, we use s_{np} to represent $sim(v_n, v_p)$. Given a sampled $(n, context(n))$ in $\mathcal{C}$, firstly we calculate the partial derivative of D_I:

$$\frac{\partial D_I}{\partial v_n} = \sum_{(i,j,k) \in S} f' \cdot (\delta_{ik}(n) \frac{\partial s_{ik}}{\partial v_n} - \delta_{ij}(n) \frac{\partial s_{ij}}{\partial v_n}) \qquad (8)$$

in which $\delta_{ij}(n)$ are defined as

$$\delta_{ij}(n) = \begin{cases} 1, & i = n \ or \ j = n \\ 0, & others \end{cases}$$

and

$$f' = \begin{cases} 1, & s_{ij} < s_{ik} \\ 0, & s_{ij} \geq s_{ik} \end{cases}$$

The partial derivatives in Formula (8) can be easily calculated. For example, given a constraint contains n and assume $i = n$, we can get

$$\frac{\partial s_{ij}}{\partial v_n} = \frac{\partial s_{nj}}{\partial v_n} = -\frac{s_{nj} v_n}{|v_n|^2} + \frac{v_j}{|v_n||v_j|}$$

For the sample $(n, context(n))$, the embedding vectors of node n will be updated as follows:

$$v_n = v_n - \beta \cdot \eta \sum_{(i,j,k) \in S} \frac{\partial D_I}{\partial v_n} \qquad (9)$$

in which β is the balance parameter to control the weight of the node properties in the embedding process.

Here we present the optimization process of D_N in Formula (4). Given a sampled $(n, context(n))$, we calculate the partial derivative of D_N:

$$\frac{\partial D_N}{\partial v_n} = \sum_{i \in \{pos_n \cup neg_n\}} P_{ni} \cdot (v_n - v_i)[(v_n - v_i)^T (v_n - v_i)]^{-\frac{1}{2}}$$

The embedding vector of node n will be updated as follows:

$$v_n = v_n - \beta \cdot \eta \sum_{i \in \{pos_n \cup neg_n\}} \frac{\partial D_N}{\partial v_n} \qquad (10)$$

Finally, Algorithm 1 shows the joint optimization process of PPNE$_{\text{ineq}}$ model. For PPNE$_{\text{num}}$ model, we only need to modify line 14 to update embedding vector following Formula (10).

Algorithm 1. PPNE$_{\text{ineq}}$

Input:
1   Network $G(V, T, P)$;
2   Embedding size d;
Output:
3   Embedding matrix $X \in \mathbb{R}^{|V| \times d}$;
4 **for** *node n in V* **do**
5   $\quad$ initialize embedding vector $v_n \in \mathbb{R}^{1 \times d}$;
6   $\quad$ initialize parameter vector $\theta_n \in \mathbb{R}^{1 \times d}$;
7 **end**
8 node sequences $\mathcal{C} = \text{RandomWalk}()$;
9 **for** $(n, context(n))$ *in* $\mathcal{C}$ **do**
10 $\quad$ #$Topology - derived\ Objective\ Function$;
11 $\quad$ update embedding vectors following Formula (7) ;
12 $\quad$ update parameter vectors following Formula (6) ;
13 $\quad$ #$Property - derived\ Objective\ Function$;
14 $\quad$ update embedding vector following Formula (9) ;
15 **end**
16 **for** $i = 0; i < |V|; i + +$ **do**
17 $\quad$ $X_i = v_i$
18 **end**
19 **return** X

4.5   Discussion

We discuss several issues of the PPNE model in detail.

Choice of the Measurements. We choose the cosine similarity measurement to measure the similarity between embedding vectors, and the Euler distance to measure the distance between embedding vectors. The proposed model is still effective with these simple and popular measurements, which can better show the generality of our model.

Property Similarity Matrix. The property similarity matrix of the input network is the basis of the proposed model. For a large network, the calculation process of the pairwise similarities between nodes seems to be time consuming. However there are several effective strategies which can significantly reduce the time cost. Firstly, we can precompute the norm of each vector and store it using a lookup table. Secondly, this process can be implemented easily in parallel. These improvements has been implemented in a popular machine learning toolkit: scikit-learn[1]. With scikit-learn, it takes only several hours to construct the similarity matrix for the largest network in our experiments.

---

[1] http://scikit-learn.org/stable/.

5   Experiments

In this section, firstly we introduce the datasets and baseline methods used in this work. Then we thoroughly evaluate our proposed methods through two classic data mining tasks on four paper citation networks and one social network. Finally we analyze the quantitative experimental results and investigate the sensitivity across parameters.

5.1   Experiment Setup

Datasets. In order to thoroughly evaluate the proposed methods, we conduct experiments on four paper citation networks and one social network with different scale of nodes. Table 1 shows the detailed information of the five datasets. The four paper citation networks are Citeseer[2], Cora (see Footnote 2), PubMed (see Footnote 2) [18] and DBLP[3] [9]. In the paper citation networks, nodes refer to papers and links refer to the citation relationships among papers. Papers are classified into several categories according to the belonged domains. In Citeseer, Wiki and PubMed networks, each paper has abstract as its property, and in DBLP dataset, each paper has properties like title, authors, publication venue and abstract. Google+ (see Footnote 3) is a social network in which nodes refer to users and links represent friend relationships among users. Each user has gender, job title, university and workplace as his properties. The institution of each user is considered as his category. We select top 6 popular institutions as the final categories. For the networks with single type of node property (Citeseer, Cora and PubMed), the property similarity matrix contains the pairwise cosine similarity scores between nodes. For the networks with richer and more complex node properties (DBLP and Google+), we calculate the cosine similarity between nodes over each type of properties separately, and then weighted linearly combine them. The weights are tuned in a few random sampled instances.

Embedding Methods. We compare PPNE with the following baseline methods:

- **DeepWalk:** DeepWalk [16] is a topology-only network embedding method, which introduces the Skip-Gram algorithm to learn the node representation vectors.
- **LINE:** LINE [19] is a popular topology-only network embedding method, which considers the first-order and second-order proximities information.
- **Property Features:** In this method nodes are represented by the property features.
- **Naive Combination:** We simply concatenate the vectors from both Property Features and DeepWalk as the final representation vectors.
- **TADW:** TADW [25] incorporates the text features of each node into the embedding process under a framework of matrix factorization.

---

[2] http://linqs.cs.umd.edu/projects/projects/lbc/index.html.
[3] https://snap.stanford.edu/data/index.html.

- **PPNE$_{ineq}$** : PPNE$_{ineq}$ is the PPNE model with the inequality constrains.
- **PPNE$_{num}$** : PPNE$_{num}$ is the proposed PPNE model with the numeric constrains.

Table 1. Statistics of the datasets

Data	Nodes	Links	Categories
Citeseer	3,312	4,732	6
Wiki	2,405	17,981	11
PubMed	19,717	44,338	3
DBLP	244,021	4,354,534	9
Google+	107,614	13,673,453	6

Parameter Setup. For all datasets, the dimension of the learned representation vector is set to $d = 160$. In DeepWalk method, parameters are set as window size $w = 10$, walks per node $r = 80$ and walk length $t = 40$. In LINE method, the parameters are set as follows: $negative = 5$ and $samples = 10$ million. In Property Features method, we reduce the dimension of node property features to 160 via SVD [7] algorithm. In TADW method the parameters are set to the same as given in the original paper. In PPNE method, the number of negative samplings $ns = 5$, the balance parameter $\beta = 0.3$, learning rate $\eta = 0.1$, walks per node $r = 80$, walk length $t = 40$.

5.2   Multi-class Classification

We utilize the representation vectors generated by various network embedding methods to perform multi-class node classification task. The representation vector of each node is treated as its feature vector, and then we use a linear support vector machine model [3] to return the most likely category. The classification model is implemented using scikit-learn. For each dataset, a portion (T_r) of the labeled nodes are randomly picked as the training data, and the rest of nodes are the test data. We repeat this process 10 times, and report the average performance in terms of classification accuracy.

Table 2 shows the classification performance on four datasets. Here "-" means TADW can not handle large networks due to its very time and memory consuming process of matrix factorization. From Table 2, one can see PPNE consistently outperforms other baseline methods. For Citeseer dataset, PPNE$_{ineq}$ achieves the best performance and beat the best baseline TADW by 5%. In Wiki dataset, PPNE$_{ineq}$ beat baselines by 3%. PPNE$_{ineq}$ improves the classification performance by 5% on DBLP dataset and PPNE$_{num}$ beat the best baseline by nearly 6% on Google+ dataset. Besides, the improvement over TADW is statistically significant (sign test, p-value < 0.05) on Citeseer and Wiki datasets. PPNE$_{ineq}$ extracts the inequalities between nodes from the property similarity

Table 2. Classification performance (Accuracy) on four datasets.

Methods	Citeseer			Wiki			DBLP			Google+		
T_r	10%	20%	30%	10%	20%	30%	1%	2%	3%	1%	2%	3%
DeepWalk	47.8	53.5	56.5	56.9	61.8	64.0	63.0	64.8	65.7	55.9	56.9	57.4
LINE	41.2	45.8	49.5	57.6	59.4	63.2	61.2	62.8	63.8	53.7	54.6	55.1
Property features	53.4	55.8	58.4	58.1	63.3	65.4	68.6	69.8	71.2	59.8	60.9	61.2
Naive combination	54.1	56.5	60.5	64.4	69.3	72.3	69.4	71.2	72.9	61.8	62.7	64.2
TADW	55.9	58.5	61.8	71.0	74.9	77.3	-	-	-	-	-	-
$PPNE_{ineq}$	**60.4**	**63.2**	**66.1**	**74.5**	**77.7**	**80.0**	**76.2**	**77.8**	**79.2**	67.2	69.1	70.3
$PPNE_{num}$	58.5	62.7	65.5	71.4	75.0	76.7	75.5	76.9	78.7	**68.7**	**70.9**	**71.8**

matrix, which is a robust method to represent the similarity information. As a more delicate method, the optimization process of $PPNE_{num}$ is affected by the numeric similarity scores, which essentially introduces the degree of inequalities between nodes. If these degrees match the label information, such as in Google+, $PPNE_{num}$ performs better than $PPNE_{ineq}$, otherwise it may introduce noise into $PPNE_{num}$ as shown on Citeseer, Wiki and DBLP datasets.

The experimental results demonstrate the effectiveness of the proposed embedding methods. By incorporating the network topology and node property information into a unified embedding framework, the quality of the learned representation vectors are improved. Meanwhile, PPNE performs consistently better when training data is small.

Parameter Sensitivity Analysis. PPNE has two major parameters: dimension d and the balance parameter β. We fix the training proportion to 30% and test the classification accuracies with different d and β. We let β varies from 0.1 to 0.9 and d varies from 10 to 500. Figure 2 shows the classification performances with different β and d on Citeseer and Wiki datasets. With the increase of β, the classification accuracy first increases and then decreases. When we increase the dimension d, the classification accuracy first increases and then keeps stable. It shows that, PPNE achieves the best performance when β varies within a reasonable range and d is larger than a specific threshold.

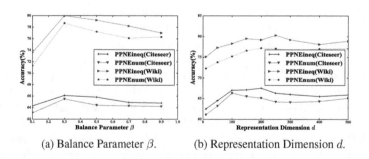

(a) Balance Parameter β.   (b) Representation Dimension d.

Fig. 2. The classification performance w.r.t the balance parameter β and dimension d.

5.3   Link Prediction

Given a snapshot of the current network, the link prediction task refers to predicting the edges that will be added in the future time [1,13]. Link prediction can show the predictability of different network embedding methods. To process the link prediction task, a portion of existing links (50%) are removed from the input network. Based on the residual network, node representation vectors are learned by different embedding methods. Node pairs in the removed edges are considered as the positive samples. We also randomly sample the same number of node pairs that are not connected as the negative samples. Positive and negative samples form a balanced data set. Given a node pair in the samples, the cosine similarity score is calculated according to their representation vectors. Area Under Curve (AUC) [5] is used to evaluate the consistency between the labels and the similarity scores of the samples. We also choose *Common Neighbors* as a baseline method because it has been proved as an effective method [11,15].

Table 3. Link prediction performance (AUC score)

Method	Citeseer	PubMed
LINE	0.725	0.751
DeepWalk	0.743	0.78
Common neighbors	0.691	0.714
TADW	0.757	0.792
$PPNE_{ineq}$	**0.791**	**0.846**
$PPNE_{num}$	0.783	0.812

Table 3 shows the experimental results. One can see that PPNE outperforms other embedding methods. Compare to TADW, $PPNE_{ineq}$ improves the AUC score by 4% in Citeseer and 5% in PubMed, which demonstrates the effectiveness of PPNE in learning good node representation vectors for the task of link prediction.

6   Conclusion

This paper proposes a general network embedding model PPNE to incorporate both network topology information and node property information. We formulate the learning of property preserving network embedding as a joint optimization problem. Firstly we propose the topology-derived objective function and property-derived objective function, and then the above objective functions are optimized jointly sharing the same parameters. Experimental results on the multi-class classification and link prediction tasks over five datasets demonstrate the effectiveness of PPNE.

Acknowledgement. This work was supported by Beijing Advanced Innovation Center for Imaging Technology (No. BAICIT-2016001), the National Natural Science Foundation of China (Grand Nos. 61370126, 61672081, 61602237, U1636211, U1636210), National High Technology Research and Development Program of China (No. 2015AA016004), the Fund of the State Key Laboratory of Software Development Environment (No. SKLSDE-2015ZX-16).

References

1. Al Hasan, M., Zaki, M.J.: A survey of link prediction in social networks. In: Aggarwal, C.C. (ed.) Social Network Data Analytics, pp. 243–275. Springer, New York (2011)
2. Belkin, M., Niyogi, P.: Laplacian eigenmaps for dimensionality reduction and data representation. Neural Comput. **15**(6), 1373–1396 (2003)
3. Chang, C.-C., Lin, C.-J.: Libsvm: a library for support vector machines. ACM Trans. Intell. Syst. Technol. (TIST) **2**(3), 27 (2011)
4. Chang, S., Han, W., Tang, J., Qi, G.J., Aggarwal, C.C., Huang, T.S.: Heterogeneous network embedding via deep architectures. In: The ACM SIGKDD International Conference, pp. 119–128 (2015)
5. Fawcett, T.: An introduction to roc analysis. Pattern Recogn. Lett. **27**(8), 861–874 (2006)
6. Grover, A., Leskovec, J.: node2vec: scalable feature learning for networks. In: The ACM SIGKDD International Conference (2016)
7. Halko, N., Martinsson, P.G., Tropp, J.A.: Finding structure with randomness: probabilistic algorithms for constructing approximate matrix decompositions. SIAM Rev. **53**(2), 217–288 (2010)
8. Kipf, T.N., Welling, M.: Semi-supervised classification with graph convolutional networks (2016)
9. Leskovec, J., Krevl, A.: SNAP datasets: Stanford large network dataset collection, June 2014. http://snap.stanford.edu/data
10. Li, P., Hastie, T.J., Church, K.W.: Very sparse random projections. In: The ACM SIGKDD International Conference, pp. 287–296 (2006)
11. Liben-Nowell, D., Kleinberg, J.: The link-prediction problem for social networks. J. Assoc. Inf. Sci. Technol. **54**(7), 1345–1347 (2007)
12. Liu, Q., Jiang, H., Wei, S., Ling, Z.-H., Hu, Y.: Learning semantic word embeddings based on ordinal knowledge constraints. In: ACL, pp. 1501–1511 (2015)
13. Menon, A.K., Elkan, C.: Link prediction via matrix factorization. In: Gunopulos, D., Hofmann, T., Malerba, D., Vazirgiannis, M. (eds.) ECML PKDD 2011. LNCS, vol. 6912, pp. 437–452. Springer, Heidelberg (2011). doi:10.1007/978-3-642-23783-6_28
14. Mikolov, T., Sutskever, I., Chen, K., Corrado, G.S., Dean, J.: Distributed representations of words and phrases and their compositionality. In: Advances in Neural Information Processing Systems, pp. 3111–3119 (2013)
15. Newman, M.E.: Clustering and preferential attachment in growing networks. Phys. Rev. E **64**(2), 025102 (2001)
16. Perozzi, B., Al-Rfou, R., Skiena, S.: Deepwalk: online learning of social representations. In: The ACM SIGKDD International Conference, pp. 701–710. ACM (2014)
17. Roweis, S.T., Saul, L.K.: Nonlinear dimensionality reduction by locally linear embedding. Science **290**(5500), 2323–2326 (2000)

18. Sen, P., Namata, G., Bilgic, M., Getoor, L., Galligher, B., Eliassi-Rad, T.: Collective classification in network data. AI Mag. **29**(3), 93 (2008)
19. Tang, J., Qu, M., Wang, M., Zhang, M., Yan, J., Mei, Q.: Line: large-scale information network embedding. In: International Conference on World Wide Web, pp. 1067–1077 (2015)
20. Tang, L., Liu, H.: Leveraging social media networks for classification. In: CIKM, pp. 1107–1116 (2009)
21. Tang, L., Liu, H.: Leveraging social media networks for classification. Data Min. Knowl. Discov. **23**(3), 447–478 (2011)
22. Wang, D., Cui, P., Zhu, W.: Structural deep network embedding. In: The ACM SIGKDD International Conference (2016)
23. Xu, C., Bai, Y., Bian, J., Gao, B., Wang, G., Liu, X., Liu, T.-Y.: Rc-net: a general framework for incorporating knowledge into word representations. In: CIKM, pp. 1219–1228 (2014)
24. Yan, S., Xu, D., Zhang, B., Zhang, H.-J., Yang, Q., Lin, S.: Graph embedding and extensions: a general framework for dimensionality reduction. IEEE Trans. Pattern Anal. Mach. Intell. **29**(1), 40–51 (2007)
25. Yang, C., Liu, Z., Zhao, D., Sun, M., Chang, E.Y.: Network representation learning with rich text information. In: International Conference on Artificial Intelligence, pp. 2111–2117 (2015)
26. Yang, Z., Cohen, W.W., Salakhutdinov, R.: Revisiting semi-supervised learning with graph embeddings. In: ICML (2016)
27. Zhang, H., Li, Z., Chen, Y., Zhang, X., Wang, S.: Exploit latent dirichlet allocation for one-class collaborative filtering. In: CIKM, pp. 1991–1994 (2014)

HINE: Heterogeneous Information Network Embedding

Yuxin Chen[1(✉)] and Chenguang Wang[2]

[1] Key Laboratory of High Confidence Software Technologies (Ministry of Education),
EECS, Peking University, Beijing, China
chen.yuxin@pku.edu.cn
[2] IBM Research Almaden, San Jose, CA, USA
chenguang.wang@ibm.com

Abstract. Network embedding has shown its effectiveness in embedding homogeneous networks. Compared with homogeneous networks, heterogeneous information networks (HINs) contain semantic information from multi-typed entities and relations, and are shown to be a more effective model for real world data. The existing network embedding methods fail to explicitly capture the semantics in HINs. In this paper, we propose an HIN embedding model (HINE), which consists of local and global semantic embedding. Local semantic embedding aims to incorporate entity type information via embedding the local structures and types of the entities in a supervised way. Global semantic embedding leverages multi-hop relation types among entities to propagate the global semantics via a Markov Random Field (MRF) to impact the embedding vectors. By doing so, *HINE* is capable to capture both local and global semantic information in the embedding vectors. Experimental results show that HINE significantly outperforms state-of-the-art methods.

Keywords: Heterogeneous information network · Network embedding · Semantic embedding

1 Introduction

Network embedding has recently been proposed as a new representation of networks. The representation consists of low-dimensional vectors carrying the most important information about the network. It thus benefits lots of network-based applications, such as visualization [18], node classification [3], as well as link predication [15] and web search [32]. The common factor shared by various network embedding approaches (e.g., DeepWalk [23], LINE [27] and Node2vec [11]) is: *the network structure embedding.*

We are grateful to Tengjiao Wang for invaluable guidance, support and contribution in regard to this research and resulting paper. This research is supported by the Natural Science Foundation of China (Grant No. 61572043), and the National Key Research and Development Program (Grant No. 2016YFB1000704).

S. Candan et al. (Eds.): DASFAA 2017, Part I, LNCS 10177, pp. 180–195, 2017.
DOI: 10.1007/978-3-319-55753-3_12

The existing network embedding approaches are mainly focusing on leveraging structural information to embed homogeneous networks. Compared to homogeneous networks, heterogeneous information networks (HINs) have been demonstrated as a more efficient way to model real world data for many applications, such as similarity search [26,34,38], classification [35,43], clustering [33] et al. The reason is that HINs are graphs consisting of multi-typed entities and relations. The various type information carries rich semantics about networks other than the basic structural information. It is thus of great need to study HIN embedding.

It is non-trivial to apply the existing homogeneous network embedding methods to HINs, due to the following two reasons.

Incorrect embedding results. Only considering structural information in HIN embedding will not only lose the semantics provided by HINs, but also lead to incorrect embedding vectors. For example, two entities "New York City" and "The New York Times" will probably have dissimilar embedding vectors by only considering the structural information, since the near neighbors (i.e., local structure) of two entities are different. However HINs could provide relation type *publishedIn* (as global information) between the two entities, thus the embedding vectors of two entities should be similar.

Lack of user-guided semantics. HIN based approaches often require user-guided semantics [20]. For example, in similarity search [42], users are often asked to provide the example entities which are similar to the target entity. However the low-dimensional vectors generated by the existing embedding methods are distributed representations, thus lack of semantic interpretation. We expect the HIN embedding vectors could still preserve the semantics, to facilitate various HIN based applications. Therefore, we are considering an HIN embedding approach to incorporate the HIN semantics in the embedding model and preserve the semantics in the embedding vectors.

In this paper, we propose an HIN embedding (*HINE*) model to embed an HIN into a low-dimensional semantic vector space. In particular, HINE contains two embedding mechanisms: (1) *local semantic embedding* aims to incorporate entity types in HINs via embedding the local structures and types of entities in a supervised way; and (2) *global semantic embedding* leverages multi-hop relation types among entities to propagate the global semantics of similar entities via a Markov Random Field (MRF) [24] to impact the HIN embedding. Then we carefully design a generative model to encode both local semantics and global semantics. By doing so, HINE is capable to capture both local and global semantic information in the embedding vectors. Notice that each dimension of the embedding vectors is a distribution over entities, thus is able to preserve the user-guided semantics. We demonstrate the effectiveness of HINE over existing state-of-the-art techniques on several multi-label classification tasks in two real world networks. The experimental results show that the HINE is able to leverage semantics for better network embedding while preserving the semantics in the resultant embedding vectors.

The main contributions of this paper can be highlighted as below:

- We study the problem of HIN embedding, which is important and has broad applications (e.g., node classification).
- We propose HINE model to embed HINs into low-dimensional semantic vector spaces by consuming both local and global semantic information in HINs.
- We conduct various multi-label network classification tasks on two HIN datasets. The results show that HINE provides significant improvements over state-of-the-art methods with even less training data.

2 Problem Definition

In this section, we first formally introduce HIN, then define the problem of heterogeneous information network embedding (HINE).

Definition 1. *A **heterogeneous information network** (HIN) is a graph $\mathcal{G} = (\mathcal{V}, \mathcal{E}, \rho, \psi)$, where $\mathcal{V}$ denotes the node (or entity) set, and $\mathcal{E} \subseteq \mathcal{V} \times \mathcal{V}$ denotes the set of edges (or relations) connecting the nodes in $\mathcal{V}$, with entity type mapping function $\rho: \mathcal{V} \rightarrow \mathcal{Y}$ and relation type mapping function $\psi: \mathcal{E} \rightarrow \mathcal{R}$. $\mathcal{Y}$ denotes the set of node types, and $\mathcal{R}$ denotes the set of edge types. The number of entity types $|\mathcal{Y}| > 1$ or the number of relation types $|\mathcal{R}| > 1$.*

Definition 2 *Heterogeneous information network embedding.* *Given a network $\mathcal{G} = (\mathcal{V}, \mathcal{E}, \rho, \psi)$, the heterogeneous information network embedding aims at incorporating semantic information in $\mathcal{G}$ to map the entities into a low-dimensional space $\mathbb{R}^d$, where $d << |\mathcal{V}|$. The embedding vectors preserve the semantics in $\mathcal{G}$.*

3 HINE: HIN Embedding

To enable embedding semantics for HINs, we propose HINE model to embed both local and global semantic information in HINs into low-dimensional vectors. To incorporate local semantics, we design a local semantic embedding layer to embed the local structure of each entity as well as its type information in the embedding vectors. To incorporate global semantics, we design a global semantic embedding layer to propagate multi-hop relation type information via an MRF to impact the embedding vectors.

3.1 Model Description

The graphical model representation of HINE is shown in Fig. 1, which has global semantic embedding layer and local semantic embedding layer. Let $\boldsymbol{\theta}_i$ be the embedding vector of entity v_i ($v_i \in \mathcal{V}$) on HIN $\mathcal{G}$, which is a K dimensional multivariate random variable. Let $\boldsymbol{\theta}$ be $\{\boldsymbol{\theta}_1, ..., \boldsymbol{\theta}_N\}$, where $N = |\mathcal{V}|$. In global semantic embedding, we construct a Markov Random Field (MRF), referred as G, over the embedding vectors $\boldsymbol{\theta}$ to describe the dependency relationships

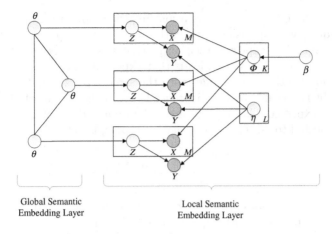

Global Semantic
Embedding Layer

Local Semantic
Embedding Layer

Fig. 1. Model description of HINE. HINE includes global and local semantic embedding layers.

among local semantic embedding, following the topology structure of the HIN. Local semantic embedding consists of generative models for each entity. We assume that each entity can be represented by its local structure in local semantic embedding. Let x_i be the local structure for entity v_i, and z_i be the embedding vectors of its local structure, while y_i is the type of v_i and $y_i \in \mathcal{Y}$. Local semantic embedding is used to embed the local structure x_i for v_i, under the supervision of y_i. The joint embedding probability of both global and local semantic embedding is defined as:

$$p(X, Y, \theta, Z | \beta, \mathcal{G}, \eta) = p(\theta | \mathcal{G}) p(X, Y, Z | \theta, \beta, \eta)$$
$$= p(\theta | \mathcal{G}) \int \left(\prod_{i=1}^{N} p(x_i, y_i, z_i | \theta_i, \phi, \eta) \right) p(\phi | \beta) \, d\phi, \tag{1}$$

where it can be decomposed into global semantic embedding $p(\theta | \mathcal{G})$ and local semantic embedding $p(X, Y, Z | \theta, \beta, \eta)$. Once θ on HIN $\mathcal{G}$ are given in global semantic embedding, the local semantic embedding of entities is conditional independent with each other.

Global Semantic Embedding Layer. By defining an MRF on HIN $\mathcal{G}$, we give the definition of the global semantic embedding $p(\theta | \mathcal{G})$. Inspired by [25] which modeling the document relationships with MRF, we use Markov Random Field [24], a graphical way to represent cyclic dependencies, to model the dependency relationships between entities and propagate global structural and semantic information. Since the links between entities are multi-typed in the HIN, different types of relations may have a broad range of frequencies and weights. Thus we construct the MRF on the HIN, by normalizing multi-typed relation frequencies and weights.

Motivated by community modularity [10] which measures the density comparison between the actual subgraph and random subgraph with the same degree distribution, we build multi-typed relation frequency normalization which measures the frequency comparison between the actual multi-typed relation and the expected relation. The expected relation is what would be expected if the link was randomly placed. The basic idea is that expected relation is viewed as the average relation for those pair nodes with the same type, so the frequency normalized weight is revealed by the difference between the actual relation and the corresponding expected relation. Expected relation w_{ij}^e, the probability of having entity v_i connected to entity v_j with relation type r, is defined as:

$$w_{ij}^e = \frac{\sum_{k \in N_r(v_i)}^{out} w_{ik} \sum_{k \in N_r(v_j)}^{in} w_{kj}}{W_r}, \qquad (2)$$

where $N_r(v_i)$ are neighbor entities connected v_i with type r, and W_r is the sum of weights of all relations with type r on HIN $\mathcal{G}$, while $r \in \mathcal{R}$. The frequency normalized weight w_{ij}^f is defined as:

$$w_{ij}^f = \frac{1}{W_r} \left(w_{ij} - w_{ij}^e \right). \qquad (3)$$

Then we use Min-Max Normalization [2] to normalize multi-typed relation weights for each relation type. Let W' be the result of W, after normalizing multi-typed relation frequencies and weights.

MRF G is constructed following the topological structure of the HIN with normalized weights W' which considering multi-typed relations. Now we introduce the definition of MRF over $\boldsymbol{\theta}$. Since we assume modeling the entity's $\boldsymbol{\theta}$ by using its neighbors' $\boldsymbol{\theta}$, our MRF satisfy local Markov property. Thus the joint density function can be factorized over the cliques of $\mathcal{G}$:

$$p(\boldsymbol{\theta}|\mathcal{G}) = \frac{1}{Z} \prod_{c \in \mathcal{C}} V_c(\boldsymbol{\theta}_c), \qquad (4)$$

where $\mathcal{C}$ is the set of cliques of $\mathcal{G}$, and $Z = \sum_{\boldsymbol{\theta}} \prod_{c \in \mathcal{C}} V_c(\boldsymbol{\theta}_c)$ is the partition function. Since $\boldsymbol{\theta}_i$ is affected by its neighbors $\boldsymbol{\theta}_{N(i)}$, the global semantic embedding $p(\boldsymbol{\theta}|\mathcal{G})$ is defined as:

$$p(\boldsymbol{\theta}|\mathcal{G}) = \frac{1}{Z} \prod_{i=1}^{N} p(\boldsymbol{\theta}_i|\boldsymbol{\theta}_{N(i)}), \qquad (5)$$

where $p(\boldsymbol{\theta}_i|\boldsymbol{\theta}_{N(i)})$ is a Dirichlet distribution as following:

$$p(\boldsymbol{\theta}_i|\boldsymbol{\theta}_{N(i)}) \sim Dir \left(\sum_{j \in N(i)} w'_{ij} \boldsymbol{\theta}_j \right). \qquad (6)$$

Local Semantic Embedding Layer. Now we define the probability of local semantic embedding $p(\boldsymbol{X}, \boldsymbol{Y}, \boldsymbol{Z}|\boldsymbol{\theta}, \boldsymbol{\beta}, \boldsymbol{\eta})$ of the joint embedding in Eq.(1). Since there are multi-typed entities in the HIN, motivated by supervised LDA [19] which uses documents' values or labels to supervise topics, we use types of the entities to supervise their local semantic embedding. We assume that each entity in local semantic embedding can be represented by its local structure which is defined as surrounding nodes, such as neighbors, and the corresponding normalized weights from W'. Then $\boldsymbol{x}_i$ is the local structure for entity v_i, which consists of surrounding nodes $x_{i,1}, ..., x_{i,m}, ..., x_{i,M_i}$ with weights $w'_{i1}, ..., w'_{im}, ..., w'_{iM_i}$. Let U be the node set of the HIN, which is used as the set of tokens for local semantic embedding layer. By generating all surrounding nodes for each entity, we produce the local embedding vectors for entities. To generate each surrounding node $x_{i,m}$, we first draw a surrounding node vector $z_{i,m}$ which is a multinomial distribution $Mult(\boldsymbol{\theta}_i)$, then choose a token u from U following multinomial distribution $Mult(\boldsymbol{\phi}_{z_{i,m}})$, where ϕ are sampled from $Dir(\boldsymbol{\beta})$. Let L be the number of types of all entities. For entity v_i, we use entity type y_i to supervise its local semantic embedding, by drawing entity type y_i which sampled from multinomial distribution $Mult(\frac{exp(\boldsymbol{\eta}_{y_i}\bar{\boldsymbol{z}_i}^T)}{\sum_{l=1}^{L} exp(\boldsymbol{\eta}_l\bar{z}_i^T)})$, where $\bar{z}_i := \frac{1}{\sum_{m=1}^{M_i} w'_{im}} \sum_{m=1}^{M_i} w'_{im} z_{i,m}$. Then the probability of local semantic embedding $p(\boldsymbol{X}, \boldsymbol{Y}, \boldsymbol{Z}|\boldsymbol{\theta}, \boldsymbol{\beta}, \boldsymbol{\eta})$ is defined as:

$$
\begin{aligned}
&p(\boldsymbol{X}, \boldsymbol{Y}, \boldsymbol{Z}|\boldsymbol{\theta}, \boldsymbol{\beta}, \boldsymbol{\eta}) \\
&= \int \left(\prod_{i=1}^{N} p(\boldsymbol{x}_i, z_i|\boldsymbol{\theta}_i, \boldsymbol{\phi})p(y_i|z_i, \boldsymbol{\eta}) \right) p(\boldsymbol{\phi}|\boldsymbol{\beta})\, d\boldsymbol{\phi} \\
&= \int \left\{ \prod_{i=1}^{N} \left[\prod_{m=1}^{M_i} \left(p(z_{i,m}|\boldsymbol{\theta}_i)p(x_{i,m}|z_{i,m}, \boldsymbol{\phi}_{z_{i,m}}) \right)^{w'_{im}} \right] p(y_i|z_i, \boldsymbol{\eta}) \right\} \prod_{k=1}^{K} p(\boldsymbol{\phi}_k|\boldsymbol{\beta})\, d\boldsymbol{\phi}.
\end{aligned}
\tag{7}
$$

3.2 Model Inference

The key inference problem of HINE is to compute the posterior $p(\boldsymbol{\theta}, \boldsymbol{Z}|\boldsymbol{X}, \boldsymbol{Y}, \mathcal{G})$ of latent variables $\boldsymbol{\theta}$ and $\boldsymbol{Z}$ with observed data X, Y on HIN $\mathcal{G}$. HINE is an undirected MRF coupled with a directed graphic, which makes the posterior inference tough. Since exact inference is generally intractable, we use Gibbs sampling method to perform approximate inference.

Since $p(\boldsymbol{\theta}_i|\boldsymbol{\theta}_{N(i)})$ is a Dirichlet distribution and $p(z_i|\boldsymbol{\theta}_i)$ is a Multinomial distribution, the posterior distribution of $\boldsymbol{\theta}_i$ is a Dirichlet distribution. Then each $\boldsymbol{\theta}_i$ is updated as:

$$
\begin{aligned}
p(\boldsymbol{\theta}_i|\boldsymbol{\theta}_{\neg(i)}, \boldsymbol{Z}, \boldsymbol{X}, \boldsymbol{Y}, \boldsymbol{\beta}, \boldsymbol{\eta}) &\propto p(\boldsymbol{\theta}_i|\boldsymbol{\theta}_{N(i)}, z_i) \\
&= Dir(\boldsymbol{\theta}_i|\boldsymbol{n}_i + \sum_{j\in N(i)} w'_{ij}\boldsymbol{\theta}_j),
\end{aligned}
\tag{8}
$$

where $\boldsymbol{n}_i = (n_{i,1}, ..., n_{i,k}, ..., n_{i,K})$ and $n_{i,k}$ is the weighted sum of tokens in entity v_i on k^{th} dimension. Once $\boldsymbol{\theta}$ are given, the local embedding of all entities is conditional independent with each other. Then every $z_{i,m}$ will be updated in turn as:

$$
\begin{aligned}
&p(z_{i,m}|z_{\neg(i,m)}, \boldsymbol{X}, \boldsymbol{Y}, \boldsymbol{\theta}, \boldsymbol{\beta}, \boldsymbol{\eta}) \\
&\propto \frac{p(\boldsymbol{Z}, \boldsymbol{X}, \boldsymbol{Y}|\boldsymbol{\theta}, \boldsymbol{\beta}, \boldsymbol{\eta})}{p(z_{\neg(i,m)}, X_{\neg(i,m)}, Y_{\neg z_{(i,m)}}|\boldsymbol{\theta}, \boldsymbol{\beta}, \boldsymbol{\eta})} \\
&= \theta_{i,z_{(i,m)}} \frac{n_{z_{(i,m)},x_{(i,m)}}^{\neg(i,m)} + \beta_{x_{(i,m)}}}{\sum_{u=1}^{U} (n_{z_{(i,m)},u}^{\neg(i,m)} + \beta_u)} \frac{exp[\boldsymbol{\eta}_{y_i}(\bar{z}_i - \bar{z}_i^{\neg(i,m)})^T]}{\sum_{l=1}^{L} exp[\boldsymbol{\eta}_l(\bar{z}_i - \bar{z}_i^{\neg(i,m)})^T]},
\end{aligned}
\tag{9}
$$

where $n_{z_{(i,m)},x_{(i,m)}}^{\neg(i,m)}$ is the weighted sum of tokens $x_{(i,m)}$ which are assigned to $z_{(i,m)}$ except for m^{th} token of ith entity, and $\bar{z}_i^{\neg(i,m)} = \frac{1}{\sum_{j=1}^{M_i} w'_{ij} - w'_{im}} (\sum_{j=1}^{m-1} w'_{ij} z_{i,j} + \sum_{j=m+1}^{M_i} w'_{ij} z_{i,j})$.

After sampling all entities, we update each η_l through MLE, where $l \in L$. Since the maximum of likelihood function cannot be solved analytically, we use gradient descent as following:

$$\eta_l := \eta_l - \lambda \left\{ -\frac{1}{N} \sum_{i=1}^{N} \left[\bar{z}_i \left(\mathbf{1}\{y_i = l\} - \frac{exp(\eta_{y_i} \bar{z}_i^T)}{\sum_{l=1}^{L} exp(\eta_l \bar{z}_i^T)} \right) \right] \right\}, \quad (10)$$

where $\mathbf{1}\{\}$ is the indicator function and λ is the learning rate. The outer loop will be terminated, once all the parameters Z, θ, η are equilibrium.

4 Experiments

4.1 Data and Evaluation Measures

We use the following two representative HIN datasets to evaluate HINE.

- **DBLP** [14]: is the network used most frequently in the study of HINs. It has four node types: Paper, Author, Conference, Term, and four edge types: authorOf, publishedIn, containsTerm, and cites.
- **PubMed:** is the bibliographic network for medicine area, which has the same node and edge types with DBLP.

To promote the comparison between HINE and the comparable methods, we use the same task, multi-label classification, as in [11,23,27]. In research bibliography networks, "research domain" information is critical for many applications. Thus, the aims of our multi-label classification tasks are to classify researchers' fields. We exploit the domain information crawled from Microsoft Academic Search to derive the gold standard. After mapping conferences' and authors' names, about 2 K authors and 1 K conferences are matched. For paper nodes, we use their conferences' domains to be their labels. Since there is no ground truth for terms' domains, we only evaluate three former type nodes in tasks. The statistics of two datasets are represented in Table 1.

Table 1. Statistics of two datasets

Datasets	#(Author)	#(Paper)	#(Conference)	#(Term)	$|V|$	$|E|$	$|y|$
DBLP	885	5,952	921	8,811	16,569	129,186	24
PubMed	530	580	152	4,594	5,856	22,268	23

We use the same metrics (Micro-F1 and Macro-F1) as in [11,23,27] to evaluate the multi-label classification performance for network embedding. Besides, we choose example-based metric Exact Match [31] to show exact match performance. Given a multi-label dataset involving N instances and J category labels, let D be the $(N \times J)$ matrix whose each row is a vector of an instance's ground true labels. P denotes a $(N \times J)$ matrix whose each row is a vector of an instance's predicted labels. We use the following metrics to evaluate multi-label task performance. For those metrics, the bigger the value, the better the performance.

- **Micro-F1** [31]: evaluates both micro average of Precision [31] and Recall [31]. It would be more affected by the performance of the categories with more instances.

$$Micro - F1 = \frac{2 \sum_{j=1}^{J} \sum_{i=1}^{N} D_{i,j} P_{i,j}}{\sum_{j=1}^{J} \sum_{i=1}^{N} D_{i,j} + \sum_{j=1}^{J} \sum_{i=1}^{N} P_{i,j}}. \tag{11}$$

- **Macro-F1** [31]: computes both Precision and Recall on each type of label separately, then evaluates the average of them. It would be more affected by the performance of the categories with fewer instances.

$$Macro - F1 = \frac{1}{J} \sum_{j=1}^{J} \frac{2 \sum_{i=1}^{N} D_{i,j} P_{i,j}}{\sum_{i=1}^{N} D_{i,j} + \sum_{i=1}^{N} P_{i,j}}. \tag{12}$$

- **Exact Match** [31]: is a very rigorous evaluation measure due to requiring the predicted label set to be an exact match of the true label set.

$$ExactMatch = \frac{1}{N} \sum_{i=1}^{N} 1\{P_i = D_i\}, \tag{13}$$

where $1\{\}$ is the indicator function.

4.2   Compared Methods

We use the following eight methods as the comparable methods. The first four are the latest representative homogeneous network embedding methods. Since knowledge graphs consist of entity-relation types, they can be regarded as one typical type of heterogeneous information networks. We incorporate the comparison with recently typical knowledge graph embedding methods to show the robustness of the proposed embedding model.

- **DeepWalk** [23]: is a network representation method which converts the graph structure to linear sequences though fixed length random walks and learns the sequences with skip-gram.
- **LINE** [27]: is a network representation algorithm that maintains the first and second order proximity between the vertexes.
- **GraRep** [5]: is a network representation method that captures k-step (with $k > 2$) proximity information, called global structure, between the vertexes.
- **Node2vec** [11]: is a semi-supervised network representation method that preserves flexible neighborhood information for vertexes.

- **TransE** [4]: is a typical neural-based knowledge base representation learning method which embeds both entities and relations into a low-dimensional space, by treating the relations as translation operations between head and tail entities.
- **TransH** [40]: models relations using hyperplanes and translation vectors, which enables entities having different representations in different relationships.
- **TransR** [17]: embeds entities and relations into separate spaces and builds translations between entities which projected to the corresponding relation space.
- **PTransE** [16]: encodes multiple-step relation paths to learn knowledge base representation, which includes PTransE-ADD, PTransE-MUL, and PTransE-RNN. Since the performance of three models in our tasks is similar, we use PTransE-RNN to represent PTransE.

4.3   Effectiveness Analysis

To compare our method with baselines properly, we use the similar experimental procedure as in [11,23,27]. Different percentages of the vertexes are randomly sampled for training, and the rest are used as the test data for evaluation. We report average performance of Exact Match, Macro-F1 and Micro-F1 over ten different runs. For all models, the multi-label classification problems are decomposed into multiple binary classifications. We use logistic regression implemented by LibLinear [9] for the binary classification. For Node2vec, we search $p, q \in \{0.5, 1, 2, 4\}$. We set p as 1 and q as 4, which makes Node2vec achieving the best performance in tasks generally. We present results for GraRep with $k = 4$, which is enough for DBLP and PubMed.

Table 2 shows the results of training ratio from 1% to 9% for all models with 300 dimensions on DBLP dataset. Numbers in parenthesis represent the percentage improvement, comparing with the highest score of baselines in the column. HINE performs significantly better than all the other methods. As results, with only 4% of the entities used for training, HINE outperforms all the baselines when they are given 9% of the entities. Among all the baselines, knowledge base representation methods, including TransE and its extensions, perform much worse than homogeneous network embedding methods (DeepWalk, LINE, GraRep and Node2vec). That is because the types of relations used in knowledge base representation are very fine-grained, which make models easy to overfit on HINs. Besides, they also ignore the weights of the relations. Although homogeneous network embedding methods achieve better performance among the baselines, the Macro-F1 of HINE achieves 20.42%–72.96% improvement and the Exact Match and Micro-F1 of HINE achieve around 30% increase. It is not surprising because the multi-typed entities and relations encode semantic insights for heterogeneous information network representation learning. This experiment also demonstrates the advantage of joint structural and semantic information for HIN embedding.

Table 2. Results of multi-label classification on DBLP (Numbers in parenthesis represent the percentage improvement, comparing with the highest score of baselines in the column.)

Metric	Algorithm	1%	2%	3%	4%	5%	6%	7%	8%	9%
Exact Match	DeepWalk	6.36	8.57	12.97	13.94	14.01	13.68	15.26	15.38	15.19
	GraRep	11.92	15.97	17.15	17.93	18.19	22.8	22.06	23.85	22.16
	LINE	6.13	8.86	9.98	11.79	13.73	16.33	15.17	18.07	19.45
	Node2vec	8.62	11.12	13.23	13.84	17.21	18.71	21.54	20.91	21.94
	PTransE	4.37	3.96	3.06	2.32	1.81	2.22	2.6	2.15	2.65
	TransE	5.11	2.09	2.61	2.75	2.76	2.73	2.78	3.5	3.04
	TransH	3.36	3.52	3.11	2.88	2.62	2.96	4.05	2.76	3.23
	TransR	4.45	3.29	2.92	3.79	3.28	2.56	3.24	3.5	3.52
	HINE	**14.27**	**17.49**	**21.7**	**26.07**	**30.17**	**29.22**	**32.28**	**34.16**	**34.47**
		(19.71%)	(9.51%)	(26.53%)	(45.40%)	(65.86%)	(28.16%)	(46.33%)	(43.23%)	(55.55%)
Micro-F1	DeepWalk	16.82	16.67	18.63	17.94	19	17.57	17.39	18.56	18.45
	GraRep	17.09	25.71	26.35	30.09	30.31	35.86	33.36	37.5	36.09
	LINE	12.18	16.43	18.13	21.92	23.85	27.85	27.24	31.29	32.84
	Node2vec	14.4	18.44	22.49	25.29	29.05	31.31	35.81	34.1	37.91
	PTransE	11.48	12.29	10.67	9.96	8.22	8.63	9.66	8.77	8.87
	TransE	13.14	10.72	10.5	9.01	10.46	9.28	9.76	10.07	10.13
	TransH	12	12.43	10.64	10.62	10.78	9.86	10.86	9.39	9.84
	TransR	13.22	12.42	11.46	11.96	10.87	9.84	10.11	10.52	10.53
	HINE	**22.63**	**27.69**	**33.99**	**38.46**	**43.31**	**42.25**	**45.3**	**47.42**	**48.64**
		(32.41%)	(7.7%)	(28.99%)	(27.81%)	(42.89%)	(17.81%)	(26.50%)	(26.45%)	(28.3%)
Macro-F1	DeepWalk	5.27	7.39	9.72	10.4	11.06	11.52	11.79	12.26	12.47
	GraRep	5.76	11.21	10.79	14.06	16.31	19	17.83	20.62	19.67
	LINE	5.88	7.92	9.29	12.34	12.27	14.62	16.79	18.74	19.63
	Node2vec	4.82	9.19	11.78	12.75	14.47	17.28	19.34	22.03	23.15
	PTransE	3.88	4.39	4.76	4.74	4.19	4.02	4.76	4.65	4.2
	TransE	4.42	4.67	4.68	4.16	4.76	4.33	4.27	4.33	5.02
	TransH	4.28	5.15	4.58	5.3	5.13	5.15	4.61	4.37	4.07
	TransR	4.52	5.13	5.2	5.03	5.18	5.45	5.35	5.01	4.98
	HINE	**9.25**	**13.5**	**20.1**	**23.49**	**28.21**	**26.9**	**30.48**	**33.16**	**35.49**
		(57.31%)	(20.42%)	(70.62%)	(67.06%)	(72.96%)	(41.57%)	(57.60%)	(50.52%)	(53.30%)

Table 3 presents the results of varying the training ratio from 10% to 90% on PubMed dataset. Since PubMed network is sparser than DBLP, we use 400 dimensions to present the results. The performance of HINE is significantly better than all the baselines, which is consistent with the previous experiment. Comparing to all the baselines, the Exact Match of HINE achieves 24.72%–39.37% improvement and the Micro-F1 and Macro-F1 of HINE achieve around 15–30% increase. Besides, with only 40% of the entities used for training, the performance of all the metrics for HINE exceeds all the baselines even when they have been given 90% of the entities. That is HINE can beat all the baselines with 50% less training data. Comparing to the previous experiment, the performance of knowledge base representation methods remain worse with more training data, while the other methods including HINE achieve significant increase generally. This indicates that the types of relations used in knowledge base representation make models easy to overfit on HINs.

Table 3. Results of multi-label classification on PubMed (Numbers in parenthesis represent the percentage improvement, comparing with the highest score of baselines in the column.)

Metric	Algorithm	10%	20%	30%	40%	50%	60%	70%	80%	90%
Exact Match	DeepWalk	27.62	34.09	37.46	40.94	43.98	45.26	48.29	50.08	49.28
	GraRep	29.55	38.94	43.12	48.63	51.63	53.73	55.15	54.26	53.06
	LINE	26.51	36.92	42.29	47.89	51.23	55.88	57.68	58.59	58.94
	Node2vec	30.45	38.48	42.59	45.96	46.47	48.13	52.88	51.13	54.84
	PTransE	2.98	3.96	6.91	11.12	12.3	15.1	17.64	20.03	21.54
	TransE	3.84	4.96	6.36	9.22	11.62	13.92	16.85	18.81	20.89
	TransH	4.41	5.39	8.07	10.74	15.05	16.58	19.26	21.07	25.26
	TransR	4.21	4.48	7.73	10.36	13.51	15.03	16.98	18.58	22.58
	HINE	**41.76**	**49.91**	**60.1**	**64.54**	**69.03**	**70.34**	**71.94**	**73.55**	**78.2**
		(37.14%)	(28.17%)	(39.37%)	(32.71%)	(33.70%)	(25.87%)	(24.72%)	(25.53%)	(32.67%)
Micro-F1	DeepWalk	40.96	43.23	46.82	49.27	52.31	54.57	55.68	57.85	60.75
	GraRep	48.79	60.3	64.72	67.97	71.21	73.42	75.38	75.25	74.88
	LINE	43.52	57.04	62.82	67.75	70.91	73.91	76.48	77.49	77.57
	Node2vec	49.78	60.83	64.62	68.83	69.16	70.13	73.6	75.97	75.52
	PTransE	6.36	8.65	14.84	22.77	25.69	29.44	33.39	36.12	38.76
	TransE	8.18	10.86	13.97	19.92	24.28	29.06	33.46	35.92	39.45
	TransH	9.73	12.15	16.13	21.28	29.18	30.79	35.02	37.29	42.64
	TransR	9.37	10.19	16.14	22.08	26.94	29.82	33.05	34.73	41.4
	HINE	**61.09**	**68.71**	**76.1**	**79.59**	**83.03**	**83.88**	**84.54**	**86.92**	**89.16**
		(22.71%)	(12.95%)	(17.58%)	(15.63%)	(16.59%)	(13.48%)	(10.53%)	(12.16%)	(14.94%)
Macro-F1	DeepWalk	25.2	29.37	32.04	33.92	36.22	37.35	38.6	39.32	39.96
	GraRep	27.26	37.15	45.9	48.68	52.04	53.73	53.89	57.46	39.79
	LINE	24.2	40.65	45.38	54.1	55.32	54.33	60.53	55.78	52.16
	Node2vec	29.21	47.03	53.02	59.83	56.09	56.16	58.32	62.04	57.77
	PTransE	2.44	3.3	5.89	9.95	11.36	12.81	14.49	16.88	17.1
	TransE	3.27	4.45	5.76	8.86	10.94	12.48	14.27	16.18	17.21
	TransH	3.75	5.07	6.98	9.13	12.71	13.9	15.59	16.39	18
	TransR	3.8	4.45	7.01	9.77	11.71	13.2	14.68	15.23	18.15
	HINE	**42.17**	**55.1**	**63.7**	**64.12**	**67.9**	**71.27**	**68.53**	**66.54**	**60.32**
		(44.36%)	(17.15%)	(20.14%)	(7.17%)	(21.05%)	(26.90%)	(13.21%)	(7.25%)	(4.41%)

These experiments indicate that properly using multi-typed entities and relations to embedding HINs is critical.

4.4 HINE Parameter Study

To evaluate how the number of dimensions affects the performance, we test the changes in performance of HINE on multi-label classifications task on PubMed dataset. Figure 2 shows the performance of the HINE model with different dimensions and training rates. In Fig. 2, increasing the number of dimensions improves performance. Then improvements tend to be gentle once the numbers of dimensions reach around 300. It is not surprising since HINE captures network structural and semantic information in a top-down manner. The smaller the number of dimensions is, more generalized information is captured. With more dimensions, more detailed information is added into the embedding vectors. Once the global and local information is enough for the current task, the performance tends

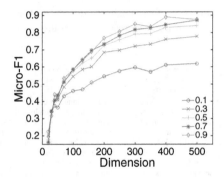

Fig. 2. Performance over dimensions on PubMed

to increase slightly. Besides, results show that the optimal number of dimension which is determined by Elbow criteria grows with training rates. This is mainly because the larger number of dimensions brings more information, which increases the performance with more labeled data. This experiment suggests that HINE captures more and more structural and semantic information (starting from generalized ones to specific ones), with the growing number of dimensions.

4.5   Case Study of HINE Vectors

To provide the readers more insights about the semantics of embedding vectors, Table 4 empirically shows part of dimensions and two entities' vectors on DBLP dataset. Numbers in bold represent the values of top 3 highest dimensions for two vectors. Since each dimension is a distribution on all entities in the network, the last row of results shows the top 15 entities and their weights from corresponding distributions for those dimensions, where the letter before @ is the abbreviation of the entity's type. For example, "a" is the abbreviation of node type *Author*. We can see that dimension #59 and #130 are mainly about information retrieval, while #121 focuses on XML data search and #41 is more concern of feedback and safety information retrieval. Entity "SIGIR" is mainly distributed on dimension #130 and #59 which are highly related to it. Comparing to "SIGIR", the distribution of entity "search" on dimensions is more gentle. It is not surprising since "search" is used on a much broader scale. By using the distributions of entities to represent dimensions, the embedding vectors preserve semantics, which will significantly improve the understanding of the HIN embedding.

5   Related Work

Network embedding technology has been widely studied in these years. The classical methods, belonging to graph embedding, embed graph matrix into a low dimensional space, such as linear methods based on SVD [28,29], IsoMap

Table 4. Demonstration for part of dimensions and the vectors of entity "SIGIR" and "search" on DBLP (Numbers in bold represent the values of top 3 highest dimensions for two vectors.)

	Dimension #41	Dimension #59	Dimension #121	Dimension #130	...
SIGIR	**0.089481**	**0.173554**	0.009164	**0.301765**	...
search	0.003949	**0.117275**	**0.122546**	**0.142530**	...
	p@Fine-grained relevance feedback for XML retrieval:0.209451, p@Warping-Based Offline Signature Recognition:0.171994, a@Suneel Suresh:0.097583, v@IEEE Transactions on Information Forensics and Security:0 .096428, t@feedback:0.079638, t@relev:0.043808, t@structur:0.036836, t@xml:0.030606, t@signatur:0.018158, t@offlin:0.017672, t@grain:0.016550, t@fine:0.016550, t@retriev:0.014943, t@recognit:0.0 14522, t@ir:0.008294	v@SIGIR Forum:0.123274, p@Hierarchical Fuzzy Intelligent Controller for Gymnastic Bar Actions:0.103294, p@Report on INEX 2008:0.096149, p@The first joint international workshop on entity-oriented and semantic search (JIWES):0.076257, p@Temporal index sharding for space-time efficiency in archive search:0.071717, p@A novel hybrid index structure for efficient text retrieval:0.069568, p@Index maintenance for time-travel text search:0.066497, p@Report on INEX 2010:0.059012, p@Report on INEX 2009:0.058984, v@JACIII:0.053933, t@report:0.044425, t@entiti:0.007204, t@joint:0.006976, t@fuzzi:0.006553, t@search:0.004659	p@Exploiting Structure, Annotation, and Ontological Knowledge for Automatic Classification of XML Data:0.122698, p@Intelligent Search on XML Data, Applications, Languages, Models, Implementations, and Benchmarks:0.119706, v@Intelligent Search on XML Data:0.118028, p@Classification and Focused Crawling for Semistructured Data:0.102976, p@Ontology-Enabled XML Search:0.097273, v@WebDB:0.041657, a@Dominique A. Winne:0.041657, t@focus:0.031450, t@xml:0.030943, t@ontolog:0.028016, t@data:0.024637, t@classif:0.023185, t@search:0.022258, t@enabl:0.018745, t@crawl:0.015353	v@SIGIR:0.215321, p@Efficient and self-tuning incremental query expansion for top-k query processing:0.159374, p@Making SENSE: socially enhanced search and exploration:0.148130, p@Efficient top-k querying over social-tagging networks:0.131191, t@tag:0.031723, t@search:0.026990, t@user:0.021907, t@recommend:0.019541, t@work:0.013616, t@item:0.012864, t@sens:0.009380, t@make:0.009368, t@enhanc:0.009309, t@effici:0.008464, t@content:0.007294	...

[30], MDS [8], and graph factorization [1]. Due to their high complexity, various neural network embedding methods are proposed. DeepWalk [23] converts the network structure to linear sequences though fixed length random walks and learns the sequences with skip-gram. LINE [27] maintains the first and second order proximity between the nodes, while GraRep [5] and HOPE [21] consider high-order proximities. DNGR [6] and SDNE [39] adopt deep neural network to capture graph structural information. TriDNR [22] and TADW [41] learn network representation with text information. Node2vec [11] proposes a semi-supervise algorithm to learn network representation flexibly. We note that these methods focus on homogeneous networks. Besides, HNE [7] aims at embedding networks consisting of various data sources of nodes (such as text, image, and video). All the above methods discard the semantic information carried by the multi-typed entities and relations during the embedding. Thus they can not be adapt to HINs.

Since knowledge graphs consist of billions of entity-relation types, they can be regarded as one typical type of heterogeneous information networks [36,37]. TransE [4] is a typical neural-based knowledge base representation method which embeds both entities and relations into a low-dimensional space, by treating the relations as translation operations between head and tail entities. There are various methods proposed to expand TransE, such as TransH [40], TransR [17], PTransE [16], TransD [12], TranSparse [13], and so on. However, the types of relations in TransE and its extensions are very fine-grained, which makes models easy to overfit on HINs. In contrast, by properly incorporating the HIN semantics in the embedding model and preserving the semantics in the embedding vectors, HINE can learn the embedding for HINs.

6   Conclusion

We propose HINE, a novel model for learning semantic representations of entities for HINs. Our method incorporates the local and global HIN semantics in the embedding model and preserves the semantics in the embedding vectors. Each dimension of our embedding vectors is a distribution of semantic entities, which will significantly improve the understanding of the HIN embedding and be very useful for later follow-up HIN studies. Extensive experiments over existing state-of-the-art methods exhibit the effectiveness of our method on various real world HINs.

References

1. Ahmed, A., Shervashidze, N., Narayanamurthy, S., Josifovski, V., Smola, A.J.: Distributed large-scale natural graph factorization. In: WWW, pp. 37–48 (2013)
2. Al Shalabi, L., Shaaban, Z., Kasasbeh, B.: Data mining: a preprocessing engine. J. Comput. Sci. 2(9), 735–739 (2006)
3. Bhagat, S., Cormode, G., Muthukrishnan, S.: Node classification in social networks. In: Social Network Data Analytics, pp. 115–148. Springer, US (2011)
4. Bordes, A., Usunier, N., Garcia-Duran, A., Weston, J., Yakhnenko, O.: Translating embeddings for modeling multi-relational data. In: NIPS, pp. 2787–2795 (2013)
5. Cao, S., Lu, W., Xu, Q.: Grarep: Learning graph representations with global structural information. In: CIKM, pp. 891–900 (2015)
6. Cao, S., Lu, W., Xu, Q.: Deep neural networks for learning graph representations. In: AAAI, pp. 1145–1152 (2016)
7. Chang, S., Han, W., Tang, J., Qi, G.J., Aggarwal, C.C., Huang, T.S.: Heterogeneous network embedding via deep architectures. In: KDD, pp. 119–128 (2015)
8. Cox, T.F., Cox, M.A.: Multidimensional Scaling. CRC Press, Boca Raton (2000)
9. Fan, R.E., Chang, K.W., Hsieh, C.J., Wang, X.R., Lin, C.J.: Liblinear: a library for large linear classification. JMLR 9, 1871–1874 (2008)
10. Fortunato, S.: Community detection in graphs. Phys. Rep. 486, 75–174 (2010)
11. Grover, A., Leskovec, J.: node2vec: Scalable feature learning for networks. In: KDD (2016)
12. Ji, G., He, S., Xu, L., Liu, K., Zhao, J.: Knowledge graph embedding via dynamic mapping matrix. In: ACL, pp. 687–696 (2015)
13. Ji, G., Liu, K., He, S., Zhao, J.: Knowledge graph completion with adaptive sparse transfer matrix. In: AAAI, pp. 985–991 (2016)
14. Ley, M.: DBLP: some lessons learned. VLDB 2, 1493–1500 (2009)
15. Liben-Nowell, D., Kleinberg, J.: The link-prediction problem for social networks. J. Am. Soc. Inf. Sci. Technol. 58, 1019–1031 (2007)

16. Lin, Y., Liu, Z., Luan, H., Sun, M., Rao, S., Liu, S.: Modeling relation paths for representation learning of knowledge bases. In: EMNLP, pp. 705–714 (2015)
17. Lin, Y., Liu, Z., Sun, M., Liu, Y., Zhu, X.: Learning entity and relation embeddings for knowledge graph completion. In: AAAI, pp. 2181–2187 (2015)
18. van der Maaten, L., Hinton, G.: Visualizing data using t-SNE. JMLR **9**, 2579–2605 (2008)
19. Mcauliffe, J.D., Blei, D.M.: Supervised topic models. In: Advances in Neural Information Processing Systems, pp. 121–128 (2008)
20. Meng, C., Cheng, R., Maniu, S., Senellart, P., Zhang, W.: Discovering meta-paths in large heterogeneous information networks. In: WWW, pp. 754–764 (2015)
21. Ou, M., Cui, P., Pei, J., Zhu, W.: Asymmetric transitivity preserving graph embedding. In: KDD (2016)
22. Pan, S., Wu, J., Zhu, X., Zhang, C., Wang, Y.: Tri-party deep network representation. In: IJCAI, pp. 1895–1901 (2016)
23. Perozzi, B., Al-Rfou, R., Skiena, S.: Deepwalk: online learning of social representations. In: KDD, pp. 701–710 (2014)
24. Rue, H., Held, L.: Gaussian Markov Random Fields: Theory and Applications. CRC Press, Boca Raton (2005)
25. Sun, Y., Han, J., Gao, J., Yu, Y.: iTopicModel: information network-integrated topic modeling. In: 2009 Ninth IEEE International Conference on Data Mining, pp. 493–502 (2009)
26. Sun, Y., Han, J., Yan, X., Yu, P.S., Wu, T.: Pathsim: meta path-based top-k similarity search in heterogeneous information networks. In: VLDB, 992–1003 (2011)
27. Tang, J., Qu, M., Wang, M., Zhang, M., Yan, J., Mei, Q.: Line: large-scale information network embedding. In: WWW, pp. 1067–1077 (2015)
28. Tang, L., Liu, H.: Scalable learning of collective behavior based on sparse social dimensions. In: CIKM, pp. 1107–1116 (2009)
29. Tang, L., Liu, H.: Leveraging social media networks for classification. In: DMKD, pp. 447–478 (2011)
30. Tenenbaum, J.B., De Silva, V., Langford, J.C.: A global geometric framework for nonlinear dimensionality reduction. Science **290**, 2319–2323 (2000)
31. Tsoumakas, G., Katakis, I., Vlahavas, I.: Mining multi-label data. In: Data Mining and Knowledge Discovery Handbook, pp. 667–685 (2009)
32. Wang, C., Duan, N., Zhou, M., Zhang, M.: Paraphrasing adaptation for web search ranking. In: ACL, pp. 41–46 (2013)
33. Wang, C., Song, Y., El-Kishky, A., Roth, D., Zhang, M., Han, J.: Incorporating world knowledge to document clustering via heterogeneous information networks. In: KDD, pp. 1215–1224 (2015)
34. Wang, C., Song, Y., Li, H., Zhang, M., Han, J.: Knowsim: a document similarity measure on structured heterogeneous information networks. In: ICDM, pp. 1015–1020 (2015)
35. Wang, C., Song, Y., Li, H., Zhang, M., Han, J.: Text classification with heterogeneous information network kernels. In: AAAI, pp. 2130–2136 (2016)
36. Wang, C., Song, Y., Roth, D., Wang, C., Han, J., Ji, H., Zhang, M.: Constrained information-theoretic tripartite graph clustering to identify semantically similar relations. In: IJCAI, pp. 3882–3889 (2015)
37. Wang, C., Song, Y., Roth, D., Zhang, M., Han, J.: World knowledge as indirect supervision for document clustering. TKDD **11**(2), 13:1–13:36 (2016)
38. Wang, C., Sun, Y., Song, Y., Han, J., Song, Y., Wang, L., Zhang, M.: Relsim: relation similarity search in schema-rich heterogeneous information networks. In: SDM (2016)

39. Wang, D., Cui, P., Zhu, W.: Structural deep network embedding. In: KDD (2016)
40. Wang, Z., Zhang, J., Feng, J., Chen, Z.: Knowledge graph embedding by translating on hyperplanes. In: AAAI, pp. 1112–1119 (2014)
41. Yang, C., Liu, Z., Zhao, D., Sun, M., Chang, E.Y.: Network representation learning with rich text information. In: IJCAI, pp. 2111–2117 (2015)
42. Yu, X., Sun, Y., Norick, B., Mao, T., Han, J.: User guided entity similarity search using meta-path selection in heterogeneous information networks. In: CIKM, pp. 2025–2029 (2012)
43. Zhou, Y., Liu, L.: Activity-edge centric multi-label classification for mining heterogeneous information networks. In: KDD, pp. 1276–1285 (2014)

Trajectory and Time Series Data Processing

DT-KST: Distributed Top-k Similarity Query on Big Trajectory Streams

Zhigang Zhang[1], Yilin Wang[1], Jiali Mao[1], Shaojie Qiao[2], Cheqing Jin[1(✉)], and Aoying Zhou[1]

[1] School of Data Science and Engineering,
East China Normal University, Shanghai, China
{zgzhang,ylwang,jlmao1231}@stu.ecnu.edu.cn,
{cqjin,ayzhou}@sei.ecnu.edu.cn
[2] College of Information Security Engineering,
Chengdu University of Information Technology, Chengdu, China
sjqiao@cuit.edu.cn

Abstract. During the past decade, with the widespread use of smartphones and other mobile devices, big trajectory data are generated and stored in a distributed way. In this work, we focus on the distributed top-k similarity query over big trajectory streams. Processing such a distributed query is challenging due to the limited network bandwidth. To overcome this challenge, we propose a communication-saving algorithm DT-KST (Distributed Top-K Similar Trajectories). DT-KST utilizes the multi-resolution property of Haar wavelet, and devises a level-increasing communication strategy to tighten the similarity bounds. Then, a local pruning strategy is imported to reduce the amount of data returned from distributed nodes. Theoretical analysis and extensive experiments on a real dataset show that DT-KST outperforms the state-of-the-art approach in terms of communication cost.

Keywords: Top-k similarity query · Trajectory stream · Communication cost

1 Introduction

Recently, the explosive development of positioning techniques leads to the widespread of various location-acquisition devices. These devices, monitoring the motions of vehicles, people, animals, and goods, are producing massive and high-speed distributed trajectory streams. Analyzing this kind of stream data enables the understanding and forecasting of moving behaviors, and brings out novel applications and services.

In this paper, we are aiming at processing such a distributed query: "given a reference trajectory Q, compare it against a crowd of trajectory streams stored on spatially distributed nodes, and find the top-k similar ones". Q can be an actual trajectory of a moving object or a virtual movement describing a desired moving pattern. This kind of query is practical in many scenarios. For example,

© Springer International Publishing AG 2017
S. Candan et al. (Eds.): DASFAA 2017, Part I, LNCS 10177, pp. 199–214, 2017.
DOI: 10.1007/978-3-319-55753-3_13

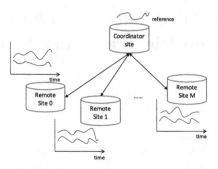

Fig. 1. Distributed processing model

video cameras are set up in many roads to capture the moving traces of vehicles continuously. The transport department is interested in finding trajectories similar to a given driving pattern such as "waving" or "swerving" to detect potential drunk drivers [2]. Another typical scenario is that, when a city manager plans to add or remove a bus route, he needs to know whether a specific route is taken by at least k passengers during 8:00–9:00. To deal with this issue, he may ask a crowd of organizations such as the bus, taxi companies and even smart phone owners to get trajectories similar to the given route [14]. In the above cases, it is inefficient to gather all distributed trajectory streams into a central site in advance, due to the limited network resources and privacy issues. Hence, centralized methods cannot be applied directly [3,8,10,11].

We firstly abstract the distributed model as a network which consists of a coordinator site and M remote sites (shown in Fig. 1). Each remote site maintains some local trajectory streams, and only communicates with the coordinator site. Query references are submitted to the coordinator to get the query results. In a naive solution, the coordinator directly transmits the query reference Q to all remote sites. Then, each remote site computes the similarity between Q and its local streams to report the k closest ones to the coordinator site. Finally, the coordinator site determines the final top-k results after receiving the candidates from all remote sites. Despite the simplicity, the communication overhead of this method is huge, because the coordinator site needs to send Q to all remote sites. The overhead becomes unacceptable when a lot of remote sites exist. To reduce the communication cost, multi-resolution based techniques that decompose the original data into different resolutions have been proposed [4,13]. In these works, the original data are decomposed into different resolutions, and data in a coarser resolution provide a rough outline of the original data, while those in a finer resolution disclose more details. LEEWAVE exploits the multi-resolution property of Haar wavelet that can tighten the similarity bounds gradually after iterations [13]. [4] utilizes multi-resolution property of "bounding envelope". The core idea of LEEWAVE is to compute Euclidean distance based similarity bounds for each candidate trajectory, while [4] computes a DTW distance based lower bound. Compared with [4], LEEWAVE is specially designed for processing stream data which matches our problem. However, it suffers the following problems: (i) It aims at processing one-dimensional time series, while trajectories are essentially multi-dimensional. (ii) It collects data from remote sites to prune candidates in the coordinator, which requires much communication cost when the number of candidate trajectories or that of remote sites is large.

In this work, we show that the Haar wavelet can be used to compress trajectory data, and we can compute a similarity bound for the compressed data. Then, we propose an iterative algorithm, called DT-KST, to process the distributed top-k similarity query. DT-KST prunes the candidate trajectories in a level-increasing manner and gradually tightens the similarity bound for candidates. In comparison with LEEWAVE-CL — an improved version of LEEWAVE by adopting our tighter lower bound, DT-KST outperforms it in two aspects: (i) Only the local top-k upper bounds are required to be sent to the coordinator, while two parameters of each candidate are required in LEEWAVE-CL. (ii) In each iteration, DT-KST only sends the global k-th smallest upper bound to remote sites, while LEEWAVE-CL needs to send a list containing IDs of all candidates. The main contributions are summarized as follows:

- We show that the Haar wavelet based technology can compress the trajectory streams.
- We propose a new algorithm DT-KST to process top-k similarity query over distributed trajectory streams. DT-KST sends the coefficients of the query reference one level at a time in a top-down manner, and prunes the candidates progressively. In comparison of LEEWAVE-CL which collects information from remote sites and prunes results in the coordinator site, DT-KST prunes candidates in the remote sites directly.
- We give theoretical analysis and extensive experimental results to show that DT-KST can save more communication cost than LEEWAVE-CL.

The rest of the paper is organized as follows. Section 2 discusses the related work. In Sect. 3, we define the problem formally and show that normalized Haar wavelet can be used to compress trajectory data. Furthermore, Sect. 4 proposes DT-KST algorithm and analyzes the performance in theory. Section 5 shows the experimental results. Finally, a brief conclusion is given in Sect. 6.

2   Related Work

In this section, we review recent works related to ours, including distributed top-k query on data streams and distributed trajectory similarity query.

2.1   Distributed Top-k Query on Data Streams

There exist some works on reducing the communication cost for distributed streaming top-k query. [9] proposes two schemes similar to the naive idea in Sect. 1, called CP and PRP. However, both of them suffer from heavy communication overhead due to the necessity to send the query reference to all remote

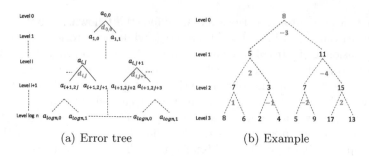

(a) Error tree   (b) Example

Fig. 2. Haar wavelet transform

sites. To overcome this weakness, LEEWAVE, a level-wise approach, can tighten the similarity bound gradually by leveraging the multi-resolution property of Haar wavelet [13]. It only sends a fraction of the query reference to remote sites iteratively to approach the final result. A more compact lower bound is proposed in [5]. These two works use the Euclidean distance as the similarity metric. [4] investigates the same problem using DTW distance. As all of them aim to process one-dimensional time series, they cannot be adopted to process multi-dimensional trajectories directly.

2.2 Distributed Trajectory Similarity Query

[6,12] study how to deal with similarity join by utilizing the MapReduce framework, with the goal to reduce the data transmission between map and reduce tasks, and they are optimized for reducing the communication cost among data nodes. But in our distributed environment, remote sites only communicate with the coordinator site. In [15], each trajectory is divided into a few subsequences and these subsequences are stored in different remote sites. This work does not take communication into consideration. Communication cost is considered in Smart Trace [1] and Smart Trace[+] [14]. However, they only consider a special case where each remote cite (represented by a smartphone) only contains one trajectory. In contrary, we consider a more general scenario where each trajectory is fully stored in a remote site, and each site can maintain multiple trajectories. So, the aforementioned techniques [1,14,15] cannot be applied to solve our problem.

3 Preliminary

3.1 Problem Statement

We assume the time span is divided into a series of basic "time-bins" (i.e., 10 s), and each object will generate a location per time-bin. A trajectory point p_i is the location of a moving object at time-bin t_i. For each object, a trajectory stream is an infinite sequence of trajectory points, $\{p_0, p_1, \cdots\}$, generated at time-bins $\{t_0, t_1, \cdots\}$. To process infinite stream data, we only maintain data in recent n time-bins in cache. The query reference $\mathcal{Q}$ and a candidate trajectory are represented as: $\mathcal{Q} = \{q_0, q_1, \cdots, q_{n-1}\}$, $\mathcal{C} = \{c_0, c_1, \cdots, c_{n-1}\}$ respectively. Each trajectory point is in d-dimensional space, i.e., $q_i, c_i \in R^d$. The Euclidean Distance between a pair of trajectory points, and that between a pair of trajectories are computed as follows:

$$ED(q_i, c_i) = ||q_i - c_i|| = \sqrt{||q_i||^2 + ||c_i||^2 - 2q_i \cdot c_i} \qquad (1)$$

$$ED(\mathcal{Q}, \mathcal{C}) = \sqrt{\sum_{i=0}^{n-1} ED(q_i, c_i)^2} \qquad (2)$$

Here, $||q_i||$ (or $||c_i||$) denotes the L^2-norm value of vector q_i (or $||c_i||$), and $q_i \cdot c_i$ denotes the dot/inner product of two vectors q_i and c_i. For simplicity, we also use Squared Euclidean Distance (SED) to measure the similarity, i.e., $SED(q_i, c_i) = ED(q_i, c_i)^2$ and $SED(\mathcal{Q}, \mathcal{C}) = ED(\mathcal{Q}, \mathcal{C})^2$. We formalize our query below.

Definition 1 (Distributed Top-k Similarity Query, DTSQ($\mathcal{Q}, TS$)). *Given a query reference $\mathcal{Q}$ and a set of all trajectory streams TS which are maintained in all remote sites, this query returns a result set S such that (i) $|S| = k$, $S \subseteq TS$, and (ii) $\forall \mathcal{C} \in S$, $\mathcal{C}' \in TS - S$, $SED(\mathcal{Q}, \mathcal{C}) \le SED(\mathcal{Q}, \mathcal{C}')$.*

3.2 Review of Haar Wavelet

Haar wavelet is effective to compress time series [5,13]. Its transforming procedure can be regarded as a series of averaging and differencing operations at different resolutions. Working in a bottom-up manner, this procedure won't stop until the average for the whole time series is obtained. This progress can be described as an error tree, as shown in Fig. 2(a). For simplicity, an arbitrary non-leaf node has two entries, a_i^j and d_i^j, where the former is the pair-wise average of its children and the latter is the corresponding difference. The leaf nodes contain the original time series. In Fig. 2(b), we show an example of transforming a one-dimensional time series $\{8, 6, 2, 4, 5, 9, 17, 13\}$. The first pair-wise averages in the time series are $\{\frac{8+6}{2} = 7, \frac{2+4}{2} = 3, \frac{5+9}{2} = 7, \frac{17+13}{2} = 15\}$, and the corresponding differences are $\{\frac{8-6}{2} = 1, \frac{2-4}{2} = -1, \frac{5-9}{2} = -2, \frac{17-13}{2} = 2\}$. Next, based on $\{7, 3, 7, 15\}$, we obtain new averages $\{5, 11\}$ and new differences $\{2, -4\}$. Finally, the overall average $\{8\}$ and the corresponding difference $\{-3\}$ are obtained. Thus, the wavelet coefficients are $\{8, -3, 2, -4, 1, -1, -2, 2\}$, containing the overall average and all the differences. Note that the non-normalized factor, $\frac{1}{2}$, is used to compute the averages and differences in our example. In fact, $\frac{1}{\sqrt{2}}$, which is called the normalized factor, is adopted in this paper.

4 DT-KST Algorithm

4.1 Haar Wavelet for Trajectory

Trajectory is a special kind of time series in which each position is a data point in d-dimensional space. In this section, we show that the normalized Haar wavelet can be extended to decompose trajectory data. For two trajectories $\mathcal{Q}$ and $\mathcal{C}$ of the same length n (n is a power of 2), the depth of their error trees are $L+1$, where $L = \log_2 n$, and the Haar wavelet coefficients for them are $H(\mathcal{Q}) = \{a_{0,0}^{\mathcal{Q}}, d_{0,0}^{\mathcal{Q}}, d_{1,0}^{\mathcal{Q}}, \cdots, d_{L-1,n/2-1}^{\mathcal{Q}}\}$ and $H(\mathcal{C}) = \{a_{0,0}^{\mathcal{C}}, d_{0,0}^{\mathcal{C}}, d_{1,0}^{\mathcal{C}}, \cdots, d_{L-1,n/2-1}^{\mathcal{C}}\}$, where $a_{0,0}^{\mathcal{Q}}$ and $a_{0,0}^{\mathcal{C}}$ are the overall average of $H(\mathcal{Q})$ and $H(\mathcal{C})$ respectively, $d_{i,j}^{\mathcal{Q}}$ and $d_{i,j}^{\mathcal{C}}$ are the differences, and $a_{i,j}^{\mathcal{Q}}, a_{i,j}^{\mathcal{C}}, d_{i,j}^{\mathcal{Q}}, d_{i,j}^{\mathcal{C}} \in R^d$.

According to the normalized Haar wavelet transform procedure, the pair-wise average and difference for two successive nodes of $\mathcal{Q}$'s error tree are computed as: $a_{i,j}^{\mathcal{Q}} = \frac{a_{i+1,2j}^{\mathcal{Q}} + a_{i+1,2j+1}^{\mathcal{Q}}}{\sqrt{2}}$, $d_{i,j}^{\mathcal{Q}} = \frac{a_{i+1,2j}^{\mathcal{Q}} - a_{i+1,2j+1}^{\mathcal{Q}}}{\sqrt{2}}$. Two adjacent averages in the $(i+1)$-th level of $\mathcal{Q}$'s error tree who share the same farther node are in the form of $\{a_{i+1,2j}^{\mathcal{Q}}, a_{i+1,2j+1}^{\mathcal{Q}}\}$. Then, the sum of pair-wise distance is:

$$SED(a_{i+1,2j}^{\mathcal{Q}}, a_{i+1,2j}^{\mathcal{C}}) + SED(a_{i+1,2j+1}^{\mathcal{Q}}, a_{i+1,2j+1}^{\mathcal{C}})$$

$$= SED(\frac{a_{i,j}^{\mathcal{Q}} + d_{i,j}^{\mathcal{Q}}}{\sqrt{2}}, \frac{a_{i,j}^{\mathcal{C}} + d_{i,j}^{\mathcal{C}}}{\sqrt{2}}) + SED(\frac{a_{i,j}^{\mathcal{Q}} - d_{i,j}^{\mathcal{Q}}}{\sqrt{2}}, \frac{a_{i,j}^{\mathcal{C}} - d_{i,j}^{\mathcal{C}}}{\sqrt{2}})$$

$$= ||\frac{a_{i,j}^{\mathcal{Q}} + d_{i,j}^{\mathcal{Q}}}{\sqrt{2}}||^2 + ||\frac{a_{i,j}^{\mathcal{C}} + d_{i,j}^{\mathcal{C}}}{\sqrt{2}}||^2 - 2\frac{a_{i,j}^{\mathcal{Q}} + d_{i,j}^{\mathcal{Q}}}{\sqrt{2}} \cdot \frac{a_{i,j}^{\mathcal{C}} + d_{i,j}^{\mathcal{C}}}{\sqrt{2}} + ||\frac{a_{i,j}^{\mathcal{Q}} - d_{i,j}^{\mathcal{Q}}}{\sqrt{2}}||^2$$

$$+ ||\frac{a_{i,j}^{\mathcal{C}} - d_{i,j}^{\mathcal{C}}}{\sqrt{2}}||^2 - 2\frac{a_{i,j}^{\mathcal{Q}} - d_{i,j}^{\mathcal{Q}}}{\sqrt{2}} \cdot \frac{a_{i,j}^{\mathcal{C}} - d_{i,j}^{\mathcal{C}}}{\sqrt{2}} = SED(a_{i,j}^{\mathcal{Q}}, a_{i,j}^{\mathcal{C}}) + SED(d_{i,j}^{\mathcal{Q}}, d_{i,j}^{\mathcal{C}})$$

Thus, we get the following theorem:

Theorem 1. *For two trajectories $\mathcal{Q}$ and $\mathcal{C}$, $SED(\mathcal{Q},\mathcal{C}) = SED(H(\mathcal{Q}), H(\mathcal{C}))$.*

Proof 1. *Let $S_i(\mathcal{Q},\mathcal{C})$ denote the sum of distances between averages at level i of $\mathcal{Q}$ and $\mathcal{C}$'s error trees, and $SED_i(\mathcal{Q},\mathcal{C})$ denote the sum of distances between differences. That is:*

$$S_i(\mathcal{Q},\mathcal{C}) = \sum_{j=0}^{2^i - 1} SED(a_{i,j}^{\mathcal{Q}}, a_{i,j}^{\mathcal{C}}) \quad SED_i(\mathcal{Q},\mathcal{C}) = \sum_{j=0}^{2^i - 1} SED(d_{i,j}^{\mathcal{Q}}, d_{i,j}^{\mathcal{C}})$$

Now, we have:

$$S_{i+1}(\mathcal{Q},\mathcal{C}) = \sum_{j=0}^{2^{i+1} - 1} SED(a_{i+1,j}^{\mathcal{Q}}, a_{i+1,j}^{\mathcal{C}})$$

$$= \sum_{j=0}^{2^i - 1} SED(a_{i,j}^{\mathcal{Q}}, a_{i,j}^{\mathcal{C}}) + \sum_{j=0}^{2^i - 1} SED(d_{i,j}^{\mathcal{Q}}, d_{i,j}^{\mathcal{C}})$$

$$= S_i(\mathcal{Q},\mathcal{C}) + SED_i(\mathcal{Q},\mathcal{C})$$

For the bottom level of the error trees, we have:

$$S_L(\mathcal{Q},\mathcal{C}) = S_0(\mathcal{Q},\mathcal{C}) + \sum_{i=0}^{L-1} SED_i(\mathcal{Q},\mathcal{C}) = SED(H(\mathcal{Q}), H(\mathcal{C}))$$

As the bottom level of the error tree maintains the original data, we have $SED(\mathcal{Q},\mathcal{C}) = S_L(\mathcal{Q},\mathcal{C})$. Finally, $SED(\mathcal{Q},\mathcal{C}) = SED(H(\mathcal{Q}), H(\mathcal{C}))$ holds.

According to [13], if the length of the reference is not a power of 2, then it can be divided into a few subsequences, each with a length equal to a power of 2. In this way, the overall similarity can be computed by summing the distances. So, Haar wavelet can be used to decompose trajectories of any length.

4.2   Level-Increasing Bounds

The core idea of DT-KST is that the coordinator iteratively sends coefficients in a top-down manner and only one level is dispatched at a time. Remote sites gradually prune the candidates with more and more coefficients of $\mathcal{Q}$. To better illustrate our idea, we give the following definition:

Definition 2. *For two trajectories $\mathcal{Q}$ and $\mathcal{C}$, the accumulated distance from level 0 to level l is computed as:* $accSED_l(\mathcal{Q},\mathcal{C}) = S_0(\mathcal{Q},\mathcal{C}) + \sum_{i=0}^{l} SED_i(\mathcal{Q},\mathcal{C})$.

Then, we have the following equation:

$$
\begin{aligned}
SED(\mathcal{Q},\mathcal{C}) &= accSED_{L-1}(\mathcal{Q},\mathcal{C}) \\
&= accSED_l(\mathcal{Q},\mathcal{C}) + \sum_{i=l+1}^{L-1} SED_i(\mathcal{Q},\mathcal{C})
\end{aligned}
\tag{3}
$$

According to Eq. 3, if only coefficients at the first few levels are received by remote sites, each site cannot determine how much those not-yet-seen coefficients at lower levels will contribute to the whole distance. So, we expand the latter part of Eq. 3 and get the following formula:

$$
\sum_{i=l+1}^{L-1} SED_i(\mathcal{Q},B) = \sum_{i=l+1}^{L-1} \sum_{j=0}^{2^i-1} (||d_{i,j}^{\mathcal{Q}}||^2 + ||d_{i,j}^{\mathcal{C}}||^2 - 2d_{i,j}^{\mathcal{Q}} \cdot d_{i,j}^{\mathcal{C}})
\tag{4}
$$

In Eq. 4, $\sum_{i=l+1}^{L-1} \sum_{j=0}^{2^i-1} ||d_{i,j}^{\mathcal{Q}}||^2$ is computed by $SSQ - \sum_{i=0}^{l} \sum_{j=0}^{2^i-1} ||d_{i,j}^{\mathcal{Q}}||^2$ where SSQ is the sum of squared coefficients of $H(\mathcal{Q})$ and can be computed in advance by the coordinator, $\sum_{i=0}^{l} \sum_{j=0}^{2^i-1} ||d_{i,j}^{\mathcal{Q}}||^2$ can be computed according to the known coefficients received by remote sites. Similarly, the second part in Eq. 4 is computed in remote sites. The main challenge is how to compute the value of the third part of Eq. 4 without knowing coefficients below level l. In DT-KST, instead of computing the third part exactly, we compute a compact bound for it using the Cauchy-Schwarz inequality. Then, we have the following bound:

$$
-2\sqrt{\sum_{i=l+1}^{L-1} \sum_{j=0}^{2^i-1} ||d_{i,j}^{\mathcal{Q}}||^2} \cdot \sqrt{\sum_{i=l+1}^{L-1} \sum_{j=0}^{2^i-1} ||d_{i,j}^{\mathcal{C}}||^2}
$$

$$
\leq -2\sum_{i=l+1}^{L-1} \sum_{j=0}^{2^i-1} ||d_{i,j}^{\mathcal{Q}}|| \cdot ||d_{i,j}^{\mathcal{C}}|| \leq \sum_{i=l+1}^{L-1} \sum_{j=0}^{2^i-1} -2d_{i,j}^{\mathcal{Q}} \cdot d_{i,j}^{\mathcal{C}} \leq 2\sum_{i=l+1}^{L-1} \sum_{j=0}^{2^i-1} ||d_{i,j}^{\mathcal{Q}}|| \cdot ||d_{i,j}^{\mathcal{C}}||
$$

$$
\leq 2\sqrt{\sum_{i=l+1}^{L-1} \sum_{j=0}^{2^i-1} ||d_{i,j}^{\mathcal{Q}}||^2} \cdot \sqrt{\sum_{i=l+1}^{L-1} \sum_{j=0}^{2^i-1} ||d_{i,j}^{\mathcal{C}}||^2}
\tag{5}
$$

In combination of Eqs. 3, 4 and Ineq. 5, we get the following similarity bound:

$$\sum_{i=l+1}^{L-1}\sum_{j=0}^{2^i-1}(||d_{i,j}^{\mathcal{Q}}||^2 + ||d_{i,j}^{\mathcal{C}}||^2) - 2\sqrt{\sum_{i=l+1}^{L-1}\sum_{j=0}^{2^i-1}||d_{i,j}^{\mathcal{Q}}||^2} \cdot \sqrt{\sum_{i=l+1}^{L-1}\sum_{j=0}^{2^i-1}||d_{i,j}^{\mathcal{C}}||^2}$$
$$+ accSED_l(\mathcal{Q},\mathcal{C}) \le SED(\mathcal{Q},\mathcal{C}) \le accSED_l(\mathcal{Q},\mathcal{C})$$

$$+ \sum_{i=l+1}^{L-1}\sum_{j=0}^{2^i-1}(||d_{i,j}^{\mathcal{Q}}||^2 + ||d_{i,j}^{\mathcal{C}}||^2) + 2\sqrt{\sum_{i=l+1}^{L-1}\sum_{j=0}^{2^i-1}||d_{i,j}^{\mathcal{Q}}||^2} \cdot \sqrt{\sum_{i=l+1}^{L-1}\sum_{j=0}^{2^i-1}||d_{i,j}^{\mathcal{C}}||^2} \quad (6)$$

In Ineq. 6, we maintain both lower and upper bounds of the similarity in a level-increasing manner. Note that the same upper bound is used by LEEWAVE and DT-KST. But LEEWAVE directly uses $accSED_l(\mathcal{Q},\mathcal{C})$ as the lower bound. Obviously, our lower bound is larger than $accSED_l(\mathcal{Q},\mathcal{C})$. The correctness of DT-KST is based on the assumption that both bounds become progressively tighter. We give the proof of this assumption below.

Theorem 2 (Lower Bound Theorem). *The lower bound of a similarity range is non-decreasing when we move from level l to level $l + 1$.*

Proof 2. *Let Lb_l denote the lower bound of the similarity on level l, then:*

$$Lb_{l+1} - Lb_l = SED_{l+1}(\mathcal{Q},\mathcal{C}) - \sum_{j=0}^{2^{l+1}-1}(||d_{l+1,j}^{\mathcal{Q}}||^2 + ||d_{l+1,j}^{\mathcal{C}}||^2)$$

$$+2\left(\sqrt{\sum_{i=l+1}^{L-1}\sum_{j=0}^{2^i-1}||d_{i,j}^{\mathcal{Q}}||^2} \cdot \sqrt{\sum_{i=l+1}^{L-1}\sum_{j=0}^{2^i-1}||d_{i,j}^{\mathcal{C}}||^2}\right.$$

$$\left. -\sqrt{\sum_{i=l+2}^{L-1}\sum_{j=0}^{2^i-1}||d_{i,j}^{\mathcal{Q}}||^2} \cdot \sqrt{\sum_{i=l+2}^{L-1}\sum_{j=0}^{2^i-1}||d_{i,j}^{\mathcal{C}}||^2}\right) \quad (7)$$

To show that the lower bound will not decrease as level goes down, we need to prove that $Lb_{l+1} - Lb_l \ge 0$. In Eq. 7, $SED_{l+1}(\mathcal{Q},\mathcal{C})$ can be substituted with $\sum_{j=0}^{2^{l+1}-1}(||d_{l+1,j}^{\mathcal{Q}}||^2 + ||d_{l+1,j}^{\mathcal{C}}||^2 - 2d_{l+1,j}^{\mathcal{Q}} \cdot d_{l+1,j}^{\mathcal{C}})$. Then our problem is transformed into proving the following inequation holds:

$$\sum_{j=0}^{2^{l+1}-1} d_{l+1,j}^{\mathcal{Q}} \cdot d_{l+1,j}^{\mathcal{C}} \le \sqrt{\sum_{i=l+1}^{L-1}\sum_{j=0}^{2^i-1}||d_{i,j}^{\mathcal{Q}}||^2} \cdot \sqrt{\sum_{i=l+1}^{L-1}\sum_{j=0}^{2^i-1}||d_{i,j}^{\mathcal{C}}||^2}$$

$$-\sqrt{\sum_{i=l+2}^{L-1}\sum_{j=0}^{2^i-1}||d_{i,j}^{\mathcal{Q}}||^2} \cdot \sqrt{\sum_{i=l+2}^{L-1}\sum_{j=0}^{2^i-1}||d_{i,j}^{\mathcal{C}}||^2} \quad (8)$$

In Ineq. 5, we have $\sum_{j=0}^{2^{l+1}-1} d_{l+1,j}^{\mathcal{Q}} \cdot d_{l+1,j}^{\mathcal{C}} \le \sqrt{\sum_{j=0}^{2^{l+1}-1}||d_{l+1,j}^{\mathcal{Q}}||^2} \cdot \sqrt{\sum_{j=0}^{2^{l+1}-1}||d_{l+1,j}^{\mathcal{C}}||^2}$.

So, our target changes to prove that $\sqrt{\sum_{j=0}^{2^{l+1}-1}||d_{l+1,j}^{\mathcal{Q}}||^2} \cdot \sqrt{\sum_{j=0}^{2^{l+1}-1}||d_{l+1,j}^{\mathcal{C}}||^2}$

is less than the right part of Ineq. 8. For ease of expression, we set $x = \sum_{j=0}^{2^{l+1}-1} ||\boldsymbol{d}_{l+1,j}^{\mathcal{Q}}||^2$, $y = \sum_{j=0}^{2^{l+1}-1} ||\boldsymbol{d}_{l+1,j}^{\mathcal{C}}||^2$, $\alpha = \sum_{i=l+2}^{L-1} \sum_{j=0}^{2^l-1} ||\boldsymbol{d}_{i,j}^{\mathcal{Q}}||^2$, $\beta = \sum_{i=l+2}^{L-1} \sum_{j=0}^{2^l-1} ||\boldsymbol{d}_{i,j}^{\mathcal{C}}||^2$. Ineq. 8 is transformed to:

$$\sqrt{x \cdot y} + \sqrt{\alpha \cdot \beta} \le \sqrt{(\alpha + x) \cdot (\beta + y)} \tag{9}$$

We square both sides of Ineq. 9 and get the following inequation:

$$2\sqrt{x \cdot y \cdot \alpha \cdot \beta} \le \alpha \cdot y + \beta \cdot x \tag{10}$$

Ineq. 10 holds according to the arithmetic-geometric mean inequality, so does Ineq. 8.

Theorem 3 (Upper Bound Theorem). *The upper bound of a similarity range is non-increasing when we move from level l to level $l + 1$.*

The proof of Theorem 3 is omitted because it is similar to that of Theorem 2.

4.3 Implementation of DT-KST Algorithm

Algorithm 1 shows the level-increasing pruning work in the coordinator site. Initially, coordinator site transforms the reference $\mathcal{Q}$ using the normalized Haar wavelet and sends the sum of squared coefficients SSQ to all remote sites (line 1–2). Then, it runs an iterative pruning procedure (line 3–11) until the *Done* flag (initialized with *false*) is *true*. During the i-th iteration, it sends the level i coefficients to the candidate sites (line 4). Here, a candidate trajectory implies that it still has a chance to be the result. Similarly, candidate sites refer to the remote sites that contain at least one candidate trajectory. It receives local upper bounds from candidate sites, and sends the global k-th smallest one to candidate sites for pruning (line 5–6). After pruning procedure, the coordinator site recomputes the total number of candidate trajectories. If there are still more than k candidates, the iteration continues. Otherwise, DT-KST terminates the iteration and informs the candidate sites to stop running (line 9). Finally, DT-KST receives the overall top-k trajectories from remote sites and returns the result (line 12).

Algorithm 2 details the work in the remote site. Initially, the remote site extracts the coefficients for each local trajectory, and maintains the upper and lower bounds for each trajectory. The remote site stores such information in set S_r (line 2). When receiving the level i coefficients of the query reference, the remote site updates the similarity bounds for each candidate according to Ineq. 6 and sends k smallest upper bounds to the coordinator (line 6–8). After receiving the global k-th smallest upper bound, it prunes candidates and sends the number of candidates to the coordinator (line 11–12). Two cases will lead to the termination of iteration in Algorithm 2: (i) No candidate is left after pruning (line 13–15); (ii) Remote site receives the finish signal (line 4). Finally, the remote site will send the final result to the coordinator before stopping.

Algorithm 1. DT-KST in coordinator site

Input: reference trajectory $\mathcal{Q}$, k;
Output: the k most similar trajectories to $\mathcal{Q}$;
  1: Extract coefficients of $\mathcal{Q}$ and get $H(\mathcal{Q})$, $SSQ = \sum_{i=0}^{n-1} ||H(\mathcal{Q})_i||^2$;
  2: Send SSQ to all remote sites;
  3: **for** $i = 0$; !**Done**; $i + +$ **do**
  4:    Send the level i coefficients of $H(\mathcal{Q})$ to all candidate sites;
  5:    **if** Receive local upper bounds from all candidate sites **then**
  6:       Sort these bounds and send the k-th smallest one, $gkub$, to candidate sites;
  7:    **else**
  8:       /* Receive $|S_r|$ from all candidate remote sites; */
  9:       If the total number of candidate trajectories is equal to k, send the $finish$
         signal to candidate sites and set **Done** to $true$;
10:    **end if**
11: **end for**
12: return the final k trajectories;

4.4 Performance Analysis

We compare DT-KST with LEEWAVE-CL (an improved version of LEEWAVE with our tighter lower bound) in terms of communication cost and time complexity. We use M to denote the number of remote sites, N to denote the number of trajectories, and n to denote the length of trajectories. Moreover, we use $|C_i|$ to refer to the number of candidate trajectories after the i-th iteration, and $|CS_i|$ to denote the number of remote sites that contain candidates.

The iterative processing strategy of LEEWAVE-CL inherits from LEEWAVE. In each iteration, it computes two summary parameters: $\sqrt{\sum_{i=l+1}^{L-1} \sum_{j=0}^{2^i-1} ||d_{i,j}^{\mathcal{C}}||^2}$ and $accSED_l(\mathcal{Q}, \mathcal{C})$ at the remote sites for each candidate. Then, the coordinator receives the two parameters to generate a tighter bound for each candidate. Finally, it prunes candidates according to the updated similarity bounds.

Time complexity: The running time of DT-KST and LEEWAVE-CL consist of two parts: time for Haar wavelet transforming which is $O(N \cdot n)$ and the iterative pruning time. Since both of them compute in iterative way and tighten the similarity bounds in each iteration, the pruning time complexity is $O(N \cdot \log_2 n)$ for the two algorithms (the sorting time is omitted as k is usually small). As $\log_2 n \ll n$, the whole running time is dominated by the transforming time. In conclusion, the overall time complexity is $O(N \cdot n)$ for both of them.

Communication cost of DT-KST: In the 0-th iteration, the coordinator sends the level 0 efficients to all M remote sites and receives at most N upper bounds from these sites, which requires a bandwidth cost of $O(d \cdot (M + N))$. In the i-th iteration ($i \geq 1$), the main cost lies in that the coordinator sends the i-th level coefficients to candidate sites, then receives at most $|C_{i-1}|$ upper bounds, which requires a communication cost of $O(d \cdot (2^i \cdot |CS_{i-1}| + |C_{i-1}|))$ bytes. So, after λ iterations, the total communication cost upper bound is $O(d \cdot (M + N + \sum_{i=1}^{\lambda-1}(2^i \cdot |CS_{i-1}| + |C_{i-1}|)))$.

Algorithm 2. DT-KST in remote site r

Input: a set of trajectory TS_r;

 1: Extract coefficients for each trajectory;
 2: Create a set S_r, which maintains a triple $< ID, ub, lb >$ for each trajectory;
 3: Receive SSQ from the coordinator;
 4: **while** The received value is not the $finish$ signal **do**
 5:    **if** Receive the level i coefficients of $H(Q)$ **then**
 6:       Update the similarity bounds for trajectories in S_r;
 7:       Sort triples in S_r according to the upper bounds;
 8:       Send k smallest upper bounds to coordinator;
 9:    **else**
10:       /* Receive $gkub$; */
11:       Remove the triples whose lower bounds are greater than $gkub$ from S_r;
12:       Send the number of triples in S_r to the coordinator;
13:       **if** $|S_r| = 0$ **then**
14:          break;
15:       **end if**
16:    **end if**
17: **end while**
18: Send the contents of trajectories whose IDs are in S_r;

Communication cost of LEEWAVE-CL: The communication cost in the 0-th iteration is $O(d \cdot (M + 2 \cdot N))$, because the coordinator site sends the level 0 coefficients to each site and receives the two summary parameters that are computed for each trajectory stream. In the i-th iteration ($i \geq 1$), there are three steps. Firstly, the coordinator transmits the i-th level coefficients and the IDs of candidates to candidate remote sites, which leads to a communication cost of $O(|CS_{i-1}| \cdot (d \cdot 2^i + |C_{i-1}|))$. Then, each remote site sends two summary parameters for each candidate trajectory which costs $O(d \cdot |C_{i-1}|)$. Finally, the coordinator computes the global candidates C_i for next iteration. Hence the communication cost in the i-th iteration is $O(|CS_{i-1}| \cdot (d \cdot 2^i + |C_{i-1}|) + d \cdot |C_{i-1}|)$. After λ iterations, the accumulated communication cost reaches $O(d \cdot (M + 2N + \sum_{i=1}^{\lambda-1} 2^i \cdot |CS_{i-1}| + \sum_{i=1}^{\lambda-1} |C_{i-1}|) + \sum_{i=1}^{\lambda-1} |C_{i-1}| \cdot |CS_{i-1}|)$.

Discussion: We summarize our key findings regarding to the communication comparison with DT-KST and LEEWAVE-CL. Firstly, in the 0-th iteration, the communication cost of DT-KST is smaller than that of LEEWAVE-CL. Secondly, in other iterations, both of them are same in $|C_i|$ and $|CS_i|$, because they use the same bound to prune candidates. However, their communication strategies are different. In LEEWAVE-CL, each candidate sends two summary values to the coordinator firstly. Then, the coordinator sends the list of pruned results to each remote site. However, DT-KST computes the bounds for candidate trajectories in the candidate remote sites, and only sends local k smallest upper bounds to the coordinator. So, the amount of data received in the coordinator is reduced. Moreover, LEEWAVE-CL sends the list of candidate trajectories to all candidate remote sites. DT-KST only needs to send the

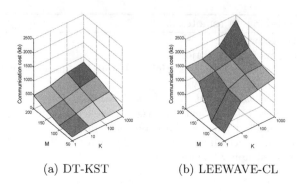

(a) DT-KST   (b) LEEWAVE-CL

Fig. 3. Communication cost in respect of k and M

global k-th upper bound to candidate sites. So, the amount of data sent by the coordinator is also reduced in DT-KST. In summary, DT-KST saves at least $O(d \cdot (N + \sum_{i=1}^{\lambda-1} |C_{i-1}|) + \sum_{i=1}^{\lambda-1} |C_{i-1}| \cdot |CS_{i-1}|)$ communication cost than LEEWAVE-CL.

5   Experiments

5.1   Experimental Setup

We evaluate the performance of DT-KST in this section. We compare DT-KST with LEEWAVE-CL algorithm. Both algorithms are implemented in Java, and all experiments are conducted on a Server with an 8 core Intel E5335 2.0 GHz processor and 16 GB memory.

We use a real world trajectory dataset of Beijing taxis [7] for experiment. Each trajectory point contains position (longitude and latitude), speed and other information of the taxi at the given timestamp. We select the trajectories from 1 to 7 October, 8:00 to 10:00 am and choose 10,000 longest ones for experiment. We align the length of trajectories to 10,000 using the dead-reckoning technique [16].

5.2   Results

We first examine the impact of k and M on bandwidth consumption when the length of all trajectories is set to 1,024. The value of k varies from 1 to 1,000, and M varies from 50 to 200. The results are shown in Fig. 3. For both algorithms, a larger k or M indicates the increment of communication cost. Because when k grows, both the number of candidate remote sites and that of candidate trajectories increase in each iteration. Similarly, when M increases, there are more candidate sites in each iteration. Finally, we can conclude that in comparison with LEEWAVE-CL, DT-KST is more communication-efficient in respect of k and M.

The communication cost of DT-KST and LEEWAVE-CL is determined by two factors: sending and receiving data by the coordinator site. Figure 4 evaluates

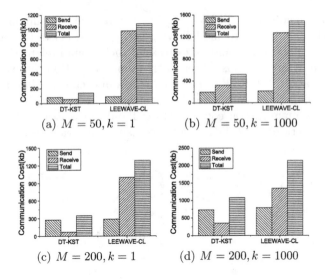

(a) $M = 50, k = 1$ (b) $M = 50, k = 1000$

(c) $M = 200, k = 1$ (d) $M = 200, k = 1000$

Fig. 4. Communication cost comparison

the performance from these two aspects. Figure 4 shows that the difference in the amount of data sent by the coordinator site is small. But, the difference in the amount of received data is obviously large. In LEEWAVE-CL, the amount of received data is much larger than that of sent data. As DT-KST only receives the upper bound of a subset of whole candidate trajectories, the amount of received data is largely reduced. Especially, when k is small (in Fig. 4(a) and (c)), the amount of data received in DT-KST is 10% less than that of LEEWAVE-CL.

We next evaluate the impacts of n and k on bandwidth consumption when $M = 200$, as shown in Fig. 5. In this experiment, k varies from 1 to 1,000, and n varies from 1,024 to 8,192. The bandwidth consumption of both DT-KST and LEEWAVE-CL increase steadily as the length of query reference increases. This is due to that more iterations are required and more relevant coefficients

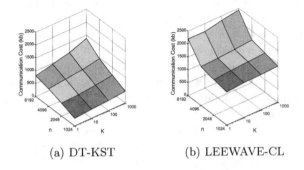

(a) DT-KST   (b) LEEWAVE-CL

Fig. 5. Communication cost in respect of k and n, $M = 200$

are sent to the remote sites. LEEWAVE-CL requires more communication cost than DT-KST, mainly due to the following two reasons: (1) In the initial step, LEEWAVE-CL requires all candidates to send their bound-related information to the coordinator site which is rather costly (about 469 KB in this experiment); (2) In each iteration, DT-KST prunes candidates in each candidate remote site instead of sending them back directly. Meanwhile, as n increases from $1,024$ to $8,192$, the communication cost in DT-KST only increases to 2 times. This confirms the superiority of DT-KST for processing long trajectories.

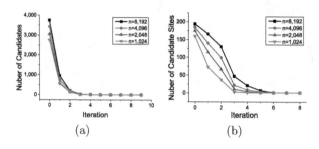

Fig. 6. Pruning performance in each iteration

We then evaluate the pruning effect of DT-KST in respect of the following metrics: the number of candidates and that of candidate sites, which have been used in [13,14]. Figure 6 shows the result when n varies from 1,024 to 8,192. At the first few iterations, the number of candidate trajectories decreases exponentially (in Fig. 6(a)) and the number of candidate sites also decreases fast at the first few iterations (in Fig. 6(b)). That's because with the increment of iterations, the coefficients of the query reference obtained by remote sites also increase exponentially. More coefficients lead to tighter similarity bounds for pruning candidates. The number of candidates and that of candidate sites keep steady after the first few iterations because the top-k results have been found (after the 6-th iteration). In combination of communication cost analysis in Sect. 4.4 and Fig. 6, longer query references usually mean more communication cost.

Finally, we evaluate the time efficiency of DT-KST. Figure 7(a) shows the running time in terms of n, k and N when $M = 200$. We find that the running time increases linearly with the increment of n and N, which validates that the time complexity is proportional to n and N. That's due to the following reasons: (i) The Haar wavelet decomposition time is linear to n and N; (ii) The Haar wavelet transforming time dominates the total running time. So, although with the increase of k, the iterative pruning time increases accordingly. The overall running time is not significantly affected. Next, we evaluate the running time in respect of N and M when $k = 1, n = 8,192$. The result in Fig. 7(b) shows that for a given query, when data are distributed in more sites, the time efficiency is improved.

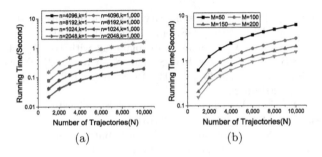

Fig. 7. Time efficiency

6   Conclusion

In this paper, we firstly show that Haar wavelet can be used to compress multidimensional time series, and we can compute a similarity bound from the compressed data. Then, we present DT-KST — a communication cost-saving approach to process distributed top-k similarity query over trajectory streams. To be specific, DT-KST distributes the relevant wavelet coefficients of query reference to remote sites in a level-increasing fashion. Starting from the top level and moving down by one level at a time, DT-KST tightens the similarity bounds for candidates and prunes results accordingly. Theoretical analysis and extensive experiment results show that DT-KST saves more bandwidth than state-of-the-art algorithms.

Acknowledgement. Our research is supported by the National Key Research and Development Program of China (2016YFB1000905), NSFC (61370101, 61532021, U1501252, U1401256 and 61402180), Shanghai Knowledge Service Platform Project (No. ZF1213).

References

1. Costa, C., Laoudias, C., Zeinalipour-Yazti, D., Gunopulos, D.: SmartTrace: finding similar trajectories in smartphone networks without disclosing the traces. In: Proceedings of the 27th ICDE, pp. 1288–1291 (2011)
2. Dai, J., Teng, J., Bai, X., Shen, Z., Xuan, D.: Mobile phone based drunk driving detection. In: Proceedings of the 2010 International Conference on Pervasive Computing Technologies for Healthcare, pp. 1–8. IEEE (2010)
3. Ding, H., Trajcevski, G., Scheuermann, P.: Efficient similarity join of large sets of moving object trajectories. In: Proceedings of the 15th TIME, pp. 79–87. IEEE (2008)
4. Hsu, C.C., Kung, P.H., Yeh, M.Y., Lin, S.D., Gibbons, P.B.: Bandwidth-efficient distributed k-nearest-neighbor search with dynamic time warping. In: Proceedings of the 2015 IEEE ICBD, pp. 551–560. IEEE (2015)
5. Kashyap, S., Karras, P.: Scalable kNN search on vertically stored time series. In: 2011 Proceedings of the 17th ACM SIGKDD, pp. 1334–1342 (2011)

6. Kim, Y., Shim, K.: Parallel top-k similarity join algorithms using MapReduce. In: Proceedings of the IEEE 28th ICDE, pp. 510–521. IEEE (2012)
7. Liu, H., Jin, C., Zhou, A.: Popular route planning with travel cost estimation. In: Navathe, S.B., Wu, W., Shekhar, S., Du, X., Wang, X.S., Xiong, H. (eds.) DASFAA 2016. LNCS, vol. 9643, pp. 403–418. Springer, Heidelberg (2016). doi:10. 1007/978-3-319-32049-6_25
8. Ma, C., Lu, H., Shou, L., Chen, G.: KSQ: top-k similarity query on uncertain trajectories. TKDE **25**(9), 2049–2062 (2013)
9. Papadopoulos, A.N., Manolopoulos, Y.: Distributed processing of similarity queries. Distrib. Parallel Databases **9**(1), 67–92 (2001)
10. Sacharidis, D., Skoutas, D., Skoumas, G.: Continuous monitoring of nearest trajectories. In: Proceedings of the 22nd ACM SIGSPATIAL, pp. 361–370. ACM (2014)
11. Skoumas, G., Skoutas, D., Vlachaki, A.: Efficient identification and approximation of k-nearest moving neighbors. In: Proceedings of the 21st ACM SIGSPATIAL, pp. 264–273. ACM (2013)
12. Vernica, R., Carey, M.J., Li, C.: Efficient parallel set-similarity joins using MapReduce. In: Proceedings of the 16th ACM SIGMOD, pp. 495–506. ACM (2010)
13. Yeh, M.Y., Wu, K.L., Yu, P.S., Chen, M.S.: LeeWave: level-wise distribution of wavelet coefficients for processing kNN queries over distributed streams. PVLDB **1**(1), 586–597 (2008)
14. Zeinalipour-Yazti, D., Laoudias, C., Costa, C., Vlachos, M., Andreou, M.I., Gunopulos, D.: Crowdsourced trace similarity with smartphones. TKDE **25**(6), 1240–1253 (2013)
15. Zeinalipour-Yazti, D., Lin, S., Gunopulos, D.: Distributed spatio-temporal similarity search. In: Proceedings of the 2006 CIKM, pp. 14–23 (2006)
16. Zheng, Y., Zhou, X.: Computing with Spatial Trajectories. Springer, New York (2011)

A Distributed Multi-level Composite Index for KNN Processing on Long Time Series

Xiaqing Wang, Zicheng Fang, Peng Wang$^{(\boxtimes)}$, Ruiyuan Zhu, and Wei Wang

School of Computer Science, Fudan University, Shanghai, China
{xiaqingwang15,zcfang16,pengwang5,ryzhu14,weiwang1}@fudan.edu.cn

Abstract. Recently, sensor-based applications have emerged and collected plenty of long time series. Traditional whole matching similarity search can only query full length time series. However, for long time series, similarity search on arbitrary time windows is more attractive and important. In this paper, we address the problem of window-based KNN search of time series data on HBase. Based on PAA approximation, we propose a composite index structure comprising Horizontal Segment Tree and Vertical Inverted Table. VI-Table is capable to prune time series by data summary in high levels, while HS-Tree leverages data summary in low levels to reduce access of the raw time series data. Both VI-Table and HS-Tree can be built parallel and incrementally. Our experiment results show the effectiveness and robustness of the proposed approach.

1 Introduction

Time series data has important applications in numerous domains, such as network analysis, image processing, financial data analysis, and sensor network monitoring. As the applications of Internet-of-things explodes, massive time series have been generated and collected.

Increasing applications choose HBase [1], a popular cloud key-value store, as their storage engine of the collected time series data. The most attractive trait of HBase is that it provides both high write throughput and realtime rowkey-based read capacity simultaneously. Recent years, both open-source and commercial time series databases choose HBase as the storage platform, like *OpenTSDB* [4].

Not only the storage capacity of huge data volume, these systems also support simple aggregations, such as SUM, AVG, MAX and so forth. However, how to process the complex mining tasks on cloud storage engine has not been addressed yet. In the last decade, extensive works have been done to mine time series data, such as similarity and correlation analysis [6], subsequence matching [7,15,16], motif discovery [14], and shapelet mining [13]. So far, how to consolidate these abundant works to the distributed platform is urgent.

The work was supported by the Ministry of Science and Technology of China, National Key Research and Development Program under No. 2016YFB1000700, National Key Basic Research Program of China under No. 2015CB358800 and NSFC (61672163, U1509213).

S. Candan et al. (Eds.): DASFAA 2017, Part I, LNCS 10177, pp. 215–230, 2017.
DOI: 10.1007/978-3-319-55753-3_14

In this paper, we focus on a fundamental time series mining problem, KNN similarity search, which is defined as follows. Given a set S of time series, a query time series Q and a distance function, KNN search retrieves K number of time series in S which are most similar to Q. This definition works on short or middle length time series, such as dozens of hundred length. However, due to frequency increasing of sensors and rapid progress in distributed storage systems, many up-to-date applications may produce a large number of extremely long time series, multiple million or even billion for instance. Considering the smart grid as an example, if a smart meter collects meter data once per minute, it would collect more than 1.5 million meter data per three years. In Zhejiang Province of China, currently 20 million smart meters have been deployed [11], which suggests 20 million of million-length time series are collected. In this case, instead of the whole time span (3 years), users are more interested in similarity search in an arbitrary time window to analyze the user behavior of a certain period, such as "finding K households whose electricity consumption habits *in the first quarter of 2015* are most similar to a certain household".

We call this type of query as *window-based KNN search*, or top-$K(r, l, Q)$, for time window $[r, r+l-1]$ and length-l query time series Q. In order to address this problem, we faced three key challenges. First, the existing approaches working on single-machine are not fit in this scenario. Most existing approaches first transform the time series into lower-dimension space, and then build tree-like index, such as R-tree. Yet building a tree-like index in the distributed cluster is not easy. Moreover, similarity search processing needs to traverse the tree paths sequentially, which cannot utilize the parallel processing of the cluster. Second, it is common that applications of generating time series data will continuously append new data points, or even new time series. Continuously updating the tree-like index based on the new arrival data on key-value store can slow down the speed of data ingestion. Third, for the window-based query, the number of possible windows is quadratic to the length of time series n. We can build multi-resolution indices, yet maintaining a large amount of indices is inevitable. Additionally, for an arbitrary query window, we have to combine multiple indices to process the query, which further amplifies the above challenges.

Here we study the problem of processing window-based KNN search on HBase. Different from the work discussing how to deal with traditional time series dataset, our goal is to query tens of thousands of time series with length 10^8 on any arbitrary windows. We first introduce a pyramid-shaped segment-level framework. For each time series, we abstract its data summary information according to the framework. In higher levels summary information would save more space yet with more information loss; in lower levels it would raise accuracy at the expense of extra space. Based on that, we propose a composite index structure comprising Horizontal Segment Tree and Vertical Inverted Table. VI-Table is capable to prune time series by data summary in high levels, while HS-Tree leverages data summary in low levels to verify the rest of time series with fewer raw time series readings from the disk. Their common advantage is to exploit the *Scan* operation of HBase, which differs from the stand-alone environment with page as the unit of index node organization. Another advantage is that our approach is fit for incremental construction, unlike B+-tree or R-tree, which require dynamic adjustment.

The rest of the paper is organized as follows. In Sect. 2 some essential definitions are presented. Then we describe the structure and building process of the proposed index in Sect. 3, and illustrate query processing based on the index in Sect. 4. Some improvements for index building and query processing are introduced in Sect. 5. In Sect. 6, we evaluate the index structure with a series of experiments. We list related work in Sect. 7 and conclude for the paper in Sect. 8.

2 Preliminaries

2.1 Basic Concepts and Problem Statement

Definition 1 (Time Series & Subsequence). *A time series is a sequence of values ordered by their arrival timestamps. For the simplification of description, we omit the timestamps and abstract a time series of length n as an n-dimensional vector, $S = (s_1, s_2, \cdots, s_k, \cdots, s_n)$, where each s_k is the k-th value in the original sequence. A subsequence $S[b:e]$ is a sequence of continuous values in time series S, where b and e are the offset of the first value and that of the last value in S respectively.*

Due to the popularity of Euclidean distance, we choose it as the metric distance. Now we give the definition of *query window*:

Definition 2 (Query Window). *A query window $w(r, l)$ is an interval of offset, by which we can obtain a subsequence $S[r : r + l - 1]$ from time series S of at least $r + l - 1$ length. We limit l by $L_{min} \leq l \leq n$, where n is the length of time series S, and L_{min} is the minimal length of query time series supported by our index structure.*

Given two time series, $S_i = (s_1^i, s_2^i, \cdots, s_n^i)$ and $S_j = (s_1^j, s_2^j, \cdots, s_n^j)$, with query window $w(r, l)$, we can obtain two subsequences $S_i[r : r + l - 1]$ and $S_j[r : r + l - 1]$ respectively. The distance of the two subsequences is:

$$Dist(S_i[r : r + l - 1], S_j[r : r + l - 1]) = \sqrt{\sum_{k=r}^{r+l-1} (s_k^i - s_k^j)^2}. \tag{1}$$

Now we give the definition of the problem we intend to solve:

Definition 3 (K-Nearest Neighbors of Query Time Series). *Let S be a set of m time series of length n: $S = \{S_1, S_2, \cdots, S_m\}$. Given an arbitrary positive integer $K(1 \leq K \leq m)$ as well as a query time series Q with the query window $w(r, l)$, find a set $\mathcal{R}$ of K time series in S to satisfy the following constraint: For any $S_i \in S(S_i \notin \mathcal{R})$ and any $S_j \in \mathcal{R}$:*

$$Dist(S_i[r : r + l - 1], Q) \geq Dist(S_j[r : r + l - 1], Q). \tag{2}$$

2.2   Utilization of KVStores

In order to support an index structure atop of the disk in an efficient and scalable way, we adopt HBase, a modern cloud Key-Value Store, to manage the storage for data of both original long time series and the index structure. We exploits some traits of HBase to obtain better locality and scalability. All records stored in HBase are organized as key-value pairs, and are maintained in lexicographic order. We leverage it to assemble relative rows together in order to get a better IO performance. HBase provides some simple but convenient operations for users to manipulate data stored in it. We use *Put* to write rows into HBase when building index, and use *Scan* to batched read substantial successive rows, which is much more efficient than a series of *Get*.

3   Index Building

To give an effective and efficient solution to the problem, we propose a composite index structure comprising Horizontal Segment Tree and Vertical Inverted Table, which both are based on a pyramid-shaped segment-level framework.

3.1   Pyramid-Shaped Segment-Level Framework

To tackle the problem, we adopt a well-known data summary method, Piecewise Aggregate Approximation (abbreviated as PAA), as the foundation of the index structure. We would show how to apply this method in each segment on each level within a pyramid-shaped segment-level framework. Before that, we give the definition of *segment* and *level*:

Definition 4 (Segment & Level). *Let S be a time series of length n, $S = (s_1, s_2, \cdots, s_n)$. Without loss of generality, we assume $n = 2^H$ for a non-negative integer H. For each integer $h(1 \leq h \leq H)$, we construct a separation point series P, denoted by $P = (p_0, p_1, \cdots, p_i, \cdots, p_{\frac{n}{2^h}})$, where $p_i = 2^h \cdot i$. Thus we can divide the whole time series into $\frac{n}{2^h}$ disjoint pieces of length 2^h as: $[p_0 + 1, p_1], [p_1 + 1, p_2], \cdots, [p_{i-1} + 1, p_i], \cdots, [p_{\frac{n}{2^h}-1} + 1, p_{\frac{n}{2^h}}]$. We define each of these pieces as a segment on level h, and denote the t-th segment on level h as $SG_{h,t}$.*

Given a time series S, for each segment $SG_{h,t}$, we calculate the mean value of all values in S within the segment, denoted by $\mu_{h,t}$. We regard all these mean values as the data summary for S. All these mean values $\mu_{h,t}$ in segments are actually organized in a multi-level style, and the higher the level, the less the segments on the level. Thus we call it a pyramid-shaped segment-level framework.

Now we treat mean values of S in continuous segments on a certain level h as an approximation time series, denoted by $A_h(S)$. Clearly, for $S = (s_1, s_2, \cdots, s_n)$, we have $A_h(S) = (\mu_{h,1}, \mu_{h,2}, \cdots, \mu_{h,\frac{n}{2^h}})$. According to the proof in [10], we have a lower bound for the distance of approximation time series of S_i and S_j in different levels ($1 \leq h_2 < h_1 \leq H$):

$$2^{h_1-h_2} Dist^2(A_{h_1}(S_i), A_{h_1}(S_j)) \leq Dist^2(A_{h_2}(S_i), A_{h_2}(S_j)). \qquad (3)$$

We can get tighter bounds from data summary in lower levels than that in higher levels. In particular, if we extend the concept of *level* downward and treat the original time series S as $A_0(S)$, we have:

$$2^h Dist^2(A_h(S_i), A_h(S_j)) \leq Dist^2(S_i, S_j). \qquad (4)$$

The lower bound above also holds for subsequences and can be extended to situations of *n-norm distance*.

3.2   Structure of the Index

The proposed index structure consists of two parts, Vertical Inverted Table and Horizontal Segment Tree (VI-Table and HS-Tree for short). Despite of their differences in the form of organizing and presenting data, both of them keep the information about mean values of time series in the pyramid-shaped segment-level framework. Among all levels in the framework, generally, we build VI-Table on high levels, and HS-Tree on low levels.

Structure of VI-Table. VI-Table is an inverted index containing several groups of records. Each of these groups is built from a perspective of all time series in S but within just one segment on a certain level. Records among different groups share a similar structure. We depict that in Fig. 1.

Before we describe the structure of VI-Table, we give the definition of *bucket*:

Definition 5. *Without loss of generality, we assume that the domain of values of time series in S is $[d_{min}, d_{max})$, denoted by D. We select an appropriate width W to divide D into several equal-width ranges as $[d_{min}, d_{min} + W), [d_{min} + W, d_{min} + 2W), \cdots, [d_{max} - W, d_{max})$. We define each of these ranges as a bucket, and use B_k to represent k-th bucket $[d_{min} + (k-1)W, d_{min} + kW)$.*

Based on the definition above, we get a series of buckets: $B_1, B_2, \cdots, B_{\frac{d_{max}-d_{min}}{W}}$. Given a segment $SG_{h,t}$, for each time series S in S, we calculate its corresponding mean value $\mu_{h,t}$ in $SG_{h,t}$. Then we map $\mu_{h,t}$ to one of the buckets previously divided from the domain. Specifically, if $d_{min} + (k-1)W \leq \mu_{h,t} < d_{min} + kW$, then $\mu_{h,t}$ should be mapped into bucket B_k. When we have mapped mean values of all time series in S into buckets, for a segment $SG_{h,t}$, we combine each bucket and IDs of time series whose mean values $\mu_{h,t}$ exist in the bucket, and store this information into an HBase table. For each $SG_{h,t}$ and B_k that is non-empty within $SG_{h,t}$, we store its information as a single row. We take $h|t|k$ as the rowkey, and the list of all IDs of time series whose mean values $\mu_{h,t}$ exist in B_k as the value, where $|$ represents string concatenation operation.

Structure of HS-Tree. HS-Tree is a tree-like structure built on time series in $\mathcal{S}$. Contrary to VI-Table, HS-Tree is built from a perspective of all segments on all levels but for just one time series in $\mathcal{S}$. We depict that in Fig. 2.

Given a time series S_i in $\mathcal{S}$, for each segment $SG_{h,t}$, we calculate the mean value $\mu_{h,t}$ of S_i in $SG_{h,t}$. Then we store the ID of the time series S_i, the information of the segment and its mean value $\mu_{h,t}$ together into another HBase table. We take $i|h|t$ as the rowkey, and $\mu_{h,t}$ as the value.

Although there is no explicit pointer between the nodes (if we treat information in a certain segment as a *node*) on adjacent levels, we could easily calculate the start offset value of the segment in lower level with that of the higher level, and further get the rowkey of the corresponding row in HBase table for HS-Tree. Therefore, it's sufficient to call this structure a *tree*.

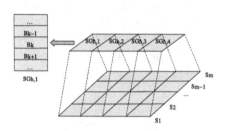

Fig. 1. Basic Structure of VI-Table   **Fig. 2.** Basic Structure of HS-Tree

3.3   Building Process

Before we start index building, we need to select an appropriate level h_C as the lowest level of VI-Table. We only build HS-Tree on levels lower than h_C and VI-Table on levels equal or higher than h_C. If h_C is too large, the granularity of VI-Table is relatively coarse, which may influence the pruning performance of VI-Table. If h_C is too small, space overhead of the index may substantially increase. In this paper, we define h_C as the highest level making the following constraint hold: length of $SG_{h_C,t}$ is less than $\frac{1}{\omega}$ of the shortest query window length L_{min}. According to our experiments, we select 16 as the default value for ω. We would show that VI-Table on levels equal or higher than h_C is sufficient to work in Sect. 6 and explain that HS-Tree on levels higher than h_C has little chance to be used in Sect. 4.

Building HS-Tree. We build HS-Tree through one pass for every time series in $\mathcal{S}$. For each level h on which we calculate mean values, we accumulate the sum of values in the current segment, say the t-th segment, $SG_{h,t}$. Then we obtain the mean value when scanning to the offset $2^h \cdot t$, the end of the segment $SG_{h,t}$.

To reduce the space overhead, we optimize the storage strategy by compressing certain number of successive mean values into a single row. We choose a segment on level h_C as the basic *storage unit*: we encode all successive mean values in such a storage unit into one row. We also put all rows into a *Put-List* to batch-import them into HBase later, which can significantly improve the efficiency of building process.

Building VI-Table. Since we only build VI-Table on levels equal or higher than h_C, it would be inefficient to calculate mean values from original time series. We could get mean values on the highest level of HS-Tree from what we have stored in HBase table in last building step.

In the t-th segment on level h, $SG_{h,t}$, we calculate the mean values for all time series in S. Then we map these mean values into corresponding buckets. After that, we recompose information of buckets to rows and batch-import them into the HBase table.

4 Query Processing

The algorithm to process the query is consist of four steps: preprocess the query (Sect. 4.1); generate initial candidates using VI-Table (Sect. 4.2); prune unnecessary time series by VI-Table (Sect. 4.3); verify time series in undetermined set by HS-Tree (Sect. 4.4). Now we introduce them in turn.

4.1 Preprocess of the Query

We preprocess the query Q similarly as building HS-Tree. We calculate mean values for each segment on each level of the query time series, and store them in a hash map. Assume that we have built VI-Table from level h_C to level H. Given Q with query window $w(r, l)$, we calculate the mean value $\mu^q_{h,t}$ of Q in the segment $SG_{h,t}$ for $1 \le h \le H$ and $t^h_b \le t \le t^h_e$, where t^h_b and t^h_e refer to the first and the last segments on level h completely covered by $w(r, l)$ respectively. We indicate the mean value of the t-th segment of level h by $\mu^q_{h,t}$.

4.2 Candidates Generation

The second step is to select K time series from S as the initial top-K results. We intend to select high-quality candidates to alleviate the computation of the next steps. The selected K time series are defined as a *candidate set*, denoted by C. Moreover, we calculate the real distances between Q and all time series in C, and maintain the largest one as the *threshold* of C, denoted by ϵ. In the next two steps, we update C by replacing time series in it with those have smaller distances. Apparently, if a time series S holds $Dist(S(r, l), Q) > \epsilon$, it cannot be in KNN results of Q, because there exist at least K time series in C with smaller distances to Q.

We generate C based on h_C, the lowest level of VI-Table, because it keeps the finest granularity information. The rationale behind is that we try to find

K time series which are the most similar to Q on the granularity of h_C. In order to do that, we maintain a length-m list of counters, denoted as $CL = \{C_1, C_2, \cdots, C_m\}$, with all C_i are initialized to 0. For each $\mu^q_{h_C,t}$ ($t^{hc}_b \leq t \leq t^{hc}_e$), we find a bucket in which $\mu^q_{h_C,t}$ falls, denoted as $B_{M(t)}$. Then we deal with $B_{M(t)}$ sequentially. If S_i is in $B_{M(t)}$, we increase C_i by 1. After this process, we sort CL in a descending order, and select top-K ones as $\mathcal{C}$. Moreover, for these time series, we fetch the series from HBase, and compute the real distances between them and Q. The largest distance will be considered as the *threshold* of current candidate set $\mathcal{C}$, denoted as ϵ.

In some cases, we might find non-zero counters in CL are less than K after this process. In such case, we recursively *expand* buckets from which we count IDs, until we get enough non-zero counters.

4.3   Filter Phase

Given $\mathcal{C}$ and ϵ, if a time series whose distance to Q exceeds ϵ, then it cannot belong to the final result set $\mathcal{R}$, because there already exist K time series with smaller distances. In this phase, we try to filter some time series via ϵ. It is sufficient to filter time series based on the lower bound of distance without computing the real distance. That is, for time series S, if we find that the lower bound of $Dist(S[r, r+l-1], Q)$ already exceeds ϵ, we can prune S safely.

We maintain a length-m list of distance accumulations, denoted as $DL = (D_1, D_2, \cdots, D_m)$, for all S_i in $\mathcal{S}$, and initiate all D_i to 0. We select an appropriate level of VI-Table, say $h_F(h_C \leq h_F \leq H)$, and define the *distance increment* $DI_{k,t}$ for the bucket B_k of level h_F and segment $SG_{h_F,t}$. Suppose the mean value $\mu^q_{h_F,t}$ of Q in $SG_{h_F,t}$ is mapped into bucket B_k, i.e., $M(t) = k$, we define

$$DI_k = \begin{cases} 2^{h_F}(d_{min} + (k-1)W - \mu^q_{h_F,t})^2 & d_{min} + (k-1)W > \mu^q_{h_F,t} \\ 2^{h_F}(d_{min} + kW - \mu^q_{h_F,t})^2 & d_{min} + kW < \mu^q_{h_F,t} \\ 0 & \text{else} \end{cases} \tag{5}$$

For each segment $SG_{h_F,t}$ of Q and all buckets B_k that $DI_{k,t}$ is no more than ϵ^2, we obtain information from VI-Table via a single Scan operation. We calculate the *distance increment* $DI_{k,t}$ for each of these buckets. For time series S_i ($S_i \notin \mathcal{C}$) whose ID is in the ID list of VI-Table index with rowkey $h|t|k$, we increase the corresponding distance accumulation D_i by DI_k. Considering that D_i is a lower bound of $Dist(S_i, Q)$, based on Eq. 4, we have

$$D_i \leq 2^h Dist^2(A_h(S_i[r : r+l-1]), A_h(Q)) \leq Dist^2(S_i, Q) \tag{6}$$

Thus, once the distance accumulation D_i exceeds ϵ^2, we can make sure that S_i cannot be in the final result set $\mathcal{R}$. Therefore S_i can be pruned without further processing. When we have applied the procedure above for all segments $SG_{h_F,t}$, the remaining time series is defined as an *undetermined set* and denoted as $\mathcal{U}$, which is the input of the verification phase.

4.4   Verification Phase

In this phase, we intend to verify every time series in undetermined set $\mathcal{U}$. Fetching series from HBase and computing the distance for each time series in $\mathcal{U}$ are too costly. So we use HS-Tree to speed up this process. For each time series S_i in $\mathcal{U}$, we recursively verify it by each level of HS-Tree from top to bottom. Using a lower level to verify S_i can get a tighter lower bound for $Dist(S_i[r:r+l-1], Q)$. When we find that the lower bound of S_i exceeds ϵ in a certain level h_T, we could terminate the iteration for S_i immediately.

Specifically, for level $h(h \leq h_C)$, we have mean values $\mu_{h,t}^q$ in all segments $SG_{h,t}$ of the query time series Q. These mean values form a subsequence $A_h(Q)$. For each S_i in $\mathcal{U}$, we can get a subsequence of $A_h(S_i)[t_b : t_e]$ in the same segments. We calculate the lower bound of $Dist(S_i[r:r+l-1], Q)$:

$$2^h Dist^2(A_h(S_i)[t_b : t_e], A_h(Q)) \leq Dist^2(S_i[r:r+l-1], Q) \qquad (7)$$

If the lower bound exceeds ϵ^2, we would be sure that S_i does not belong to $\mathcal{R}$, and terminate the verification of S_i. Otherwise we perform the same procedure on lower levels until the lowest level. If the lowest level is reached, we have to read raw series of S_i from the disk and calculate the real distance $Dist(S_i[r : r+l-1], Q)$ to verify it. If $Dist(S_i[r+l-1], Q)$ does not exceed ϵ, we replace the time series in $\mathcal{C}$ whose distance to Q equals ϵ with S_i, and update the threshold ϵ with the current K-th largest distance. After verification of all S_i in $\mathcal{U}$, the candidate set $\mathcal{C}$ is exactly the result set $\mathcal{R}$ we search for.

According to our experiments, the major time overhead in this phase is for reading raw time series from the disk when updating $\mathcal{C}$ and ϵ is necessary. Fortunately, for the case of KNN, the expectation of updating times $E(m)$ is no more than $O(Klog(m))$. However, due to space limitation, we omit the proof here.

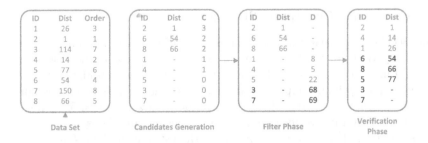

Fig. 3. Example for query processing

We give a concrete example for the whole procedure of query processing, and depict it in Fig. 3. To find KNN results for $\mathcal{S}$ with 8 time series in it, assume $K = 3$, after we have preprocessed Q, we use CL to count for all S_i and sort it. We take the top-3 time series, S_2, S_6 and S_8 as candidates. We calculate real distances for them, and treat the largest one, 66, as the current threshold.

Then in filter phase, we use DL to maintain distance accumulations for all S_i, and prune those whose D_i exceeds ϵ^2, i.e., S_3 and S_7. Finally, we verify the rest time series while updating C and ϵ if necessary, and get the results: S_2, S_4 and S_1.

Correctness Analysis. We analyze the correctness for each phase. In candidates generation, we intend to find K candidates which can provide a smaller ϵ, thus improving the efficiency of the next two phases. In fact, we can still get correct results even using random candidates. The reason is that no matter what the initial ϵ is, the time series pruned in next two phases via ϵ cannot belong to $\mathcal{R}$.

In filter phase, we use distance accumulations D_i to determine whether or not to eliminate time series S_i. Considering that D_i is actually a sum of lower bound of distances for different segments, thus D_i itself is also a lower bound of the distance. Our algorithm eliminates S_i if and only if D_i exceeds ϵ^2. Then according to Eq. 6, we can derive that the real distance exceeds ϵ as well. Therefore S_i can be pruned safely.

In verification phase, the result of processing any time series in $\mathcal{U}$ belongs to one of three cases. First, during verification of certain level, if the lower bound exceeds ϵ, we prune it safely. Second, we fetch the raw series and find that the real distance exceeds ϵ, then we still prune it. Third, if the real distance is less than ϵ, we add it to C and update ϵ. In summary, any pruned time series cannot belong to the true result set $\mathcal{R}$.

5   Improvements

On the basis of the index structure described detailedly in Sects. 3 and 4, we further propose some significant optimizations.

5.1   Candidates Generation Strategy

From Sect. 4, it is clear that candidates generation is the key phase in query processing. With an ideal and tight threshold ϵ, we could prune a substantial part of $\mathcal{U}$ in filter phase and reduce updating times of the candidate set C and ϵ. However, in our primary implementation, the threshold ϵ is not tight enough due to the inaccurate candidates. Thus we introduce a new approach for candidates generation, with a pre-verification step to get better candidates.

Aiming to find KNN of Q, we maintain the list of counters CL as described in Sect. 4.2. However, we need to guarantee that there are at least $g \cdot K$ (g is a parameter we would investigate in Sect. 6) counters in CL larger than 0, and select the largest $g \cdot K$ counters from sorted CL instead of K. We calculate the real distances for the top-K of them to get a temporary ϵ and take these K time series as the current C. Then we use HS-Tree to apply verification for the rest of $g \cdot K$ time series in descending order of their counts in the same manner as Sect. 4.4. During this procedure, C and ϵ might be updated. After that we would get a better C with a much tighter threshold ϵ.

We use this strategy because we believe that with more counts larger than 0 we can get more stable candidates. Besides, time series whose counts are not within top-K but relatively large are more likely to be similar to Q. We apply pre-verification for them to get a tighter ϵ beforehand. As shown in Sect. 6, our candidates generation strategy works well.

5.2   Heuristic Building for HS-Tree

In the proposed index structure, it is obvious that index in lower levels consumes much more space. Therefore, we intend to optimize structure and storage of HS-Tree which contains the lower levels part in our design.

To apply a heuristic building for HS-Tree, we need to build VI-Table ahead, thus changing the building order of VI-Table and HS-Tree. This may increase a little building time but can reduce space overhead remarkably.

In the building process of HS-Tree, we process queries from a query set, which is produced by real query simulation. We record the average processing time per query T_h for each level h. Assume level h_1 and h_2 are two adjacent levels $(1 \leq h_2 < h_1 < h_C)$. When we have built HS-Tree for level h_1, we compare T_{h_1} and T_{h_2}. If $T_{h_1} \leq T_{h_2}$, we could infer that HS-Tree on level h_1 has better performance than that on level h_2. This is because HS-Tree on level h_1 has fewer mean values to read from the disk and take into calculation, though there may be several time series that HS-Tree on level h_2 can verify but that on h_1 cannot. We keep HS-Tree on level h_1 and drop it on h_2, then apply a similar process on higher levels, until we find that the lower level, say h_T, has better performance than the higher level. In such case, we retain HS-Tree on h_T as the only HS-Tree structure and merely use it to apply verification.

In Sect. 6 we would show that this heuristic building can significantly reduce space overhead with only a little penalty on query processing time.

5.3   Iterative Filter by VI-Table

Another issue necessary for discussion is how to select an appropriate level h_F for filter phase. We introduce a parameter ω in the original implementation. This simple decision may not work well in all cases, since the data set size of S and the window size l of $w(r, l)$ differ a lot. Thus we decide to apply iterative filter in filter phase with VI-Table in multiple levels.

We apply all steps of filter phase in the same manner as Sect. 4.3, except processing on multiple levels rather than on a single level h_F. We start iterative filter from the top level of VI-Table H. For each level h in VI-Table, we get a subset S_h of S with time series not pruned in it, and P_h, where $P_h = \frac{|S_{h+1}| - |S_h|}{|S|}$. When we find that $P_h \geq \phi$ and $\frac{P_h - P_{h+1}}{P_{h+1}} \leq \psi$, we terminate filter phase and take the current S_h as $\mathcal{U}$. According to our experiments, we select 30% and 0 as the default value of ϕ and ψ respectively.

6   Experiments

We set up the experiment environment on a cluster of 8 nodes, each one of which is a server with a single 8-core CPU, 64 GB memory and 5.2 TB disk space. We deploy Hadoop 2.7.2 and HBase 1.1.5 on all servers.

In order to evaluate the proposed index structure in datasets large enough, and with time series long enough, we use synthetic datasets we produced for all experiments. We make data of all time series distributed in a relatively uniform way by controlling the variation in a certain range to simulate the real time series. The datasets include: $1k$ time series of length 2^{24}, $10k$ time series of length 2^{22} and $100k$ time series of length 2^{20}. We also produce query time series sets with different window sizes for our experiments. These query time series are actually pieces of time series sampled from datasets. The query sets include: query time series of length 2^{16}, length 2^{17} and length 2^{18}. All the experiment results are the averages of the results from 50 independent experiments.

6.1   Performance Evaluations

In performance evaluations, there are 3 methods compared with the proposed index on query processing. In Figs. 4 and 5, *Naive* refers to the naive approach: we calculate the distance of each time series S in the dataset $\mathcal{S}$ to Q, and pick the top-K as the result. *HST only* refers to the method that only HS-Tree is used, in other words, we take $\mathcal{S}$ as the undetermined set $\mathcal{U}$ with an empty candidate set $\mathcal{C}$, and directly start verification phase for all series in $\mathcal{S}$. *VIT only* refers to the method that only VI-Table is used, so after filter phase, we verify each time series S in $\mathcal{U}$ by calculating the real distance with raw time series read from the disk. *HST+VIT* refers to the approach based on the proposed index.

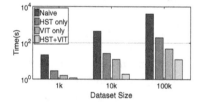
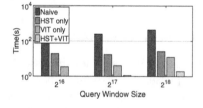

Fig. 4. Time vs. $|\mathcal{S}|$ **Fig. 5.** Time vs. $|\mathcal{Q}|$

In Fig. 4, we process a set of queries with length from 2^{16} to 2^{18} in 3 kinds of datasets: $1k$, $10k$ and $100k$. In Fig. 5, we choose $10k$ as the dataset size and process queries in different window sizes: 2^{16}, 2^{17} and 2^{18}. In all experiments above, we use 5 as the default value for K. It is obvious that our approach significantly outperforms others despite the size of datasets and the length of query windows. Besides, we also find that both HS-Tree and VI-Table are useful for the problem and reduce some query processing time, but either of them is insufficient to provide an efficient approach. Therefore, our composite index structure consisting both HS-Tree and VI-Table is necessary.

Next in Fig. 6, we depict the effect of the value K to our approach. This experiment is performed on $1k$ dataset. We see that our approach shows a tendency of degeneration as K increases. This is because that for a very large K, we need to produce a large $\mathcal{C}$, and calculate the real distance for each time series in $\mathcal{C}$. Thus it is inevitable to read a large number of raw time series from the disk with little optimizing space. However, we consider that in practical situations K is not likely to be a very large number, and it is still a quite effective approach as K does not exceed 50.

Table 1. Statistic for space costs of different implementations

Index	$1k$	$10k$	$100k$
Data	27.430	65.584	171.235
HST+VIT (all levels)	15.000	37.560	94.432
HST-only (all levels)	14.988	37.531	94.432
VIT-only	0.012	0.029	0.074
HST-only (one level)	0.041	0.093	0.206
HST+VIT (one level)	0.053	0.122	0.280

6.2 Scalability Analyses

We show the space costs for the proposed index in Table 1, in which *Data* refers to the space cost for original time series data as a baseline. *All Levels* and *One Level* refer to the proposed index structure without heuristic building for HS-Tree and that with heuristic building for HS-Tree respectively. Specifically, the index structure without heuristic building occupies 56% space of the original time series. However, using heuristic building for HS-Tree we have mentioned in Sect. 5.2, the space cost is reduced to only a negligible size. This is because HS-Tree on the lower levels accounts for a big proportion for total space costs. Thus our optimization for this part of index structure makes an impressive effect.

Table 2. Statistic for query phases in different datasets

Dataset size	$1k$		$10k$		$100k$	
Phase	Time	$|\mathcal{U}|$	Time	$|\mathcal{U}|$	Time	$|\mathcal{U}|$
Candidates generation	316	9980	363	9980	425	9980
Filter phase	76	98	168	91	113	145
Verification phase	370	19	657	20	1002	19

We also give detailed time expense of each phase in Table 2. For the small dataset, candidates generation is a considerably costly phase, because we are likely to process more expansions for its low density of time series. As the dataset grows, verification phase becomes the most costly phase, because there are more time series need to verify.

6.3   Parameter Selection

Figure 7 shows the effect of parameter g in candidates generation strategy mentioned in Sect. 5.1. *Candidates Position* refers to the maximum percentage of candidates in similarity order. We find that we cannot get much more benefits as g grows when $g \geq 3$. So we select 3 as the default value for g in our experiments.

Figure 8 shows the average response time for a query using a certain level of HS-Tree only, compared to the approach using all levels. Level 5 and Level 6 are quite close, we choose level 6 since it has lower space overhead for storage.

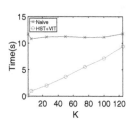

Fig. 6. Time vs. K

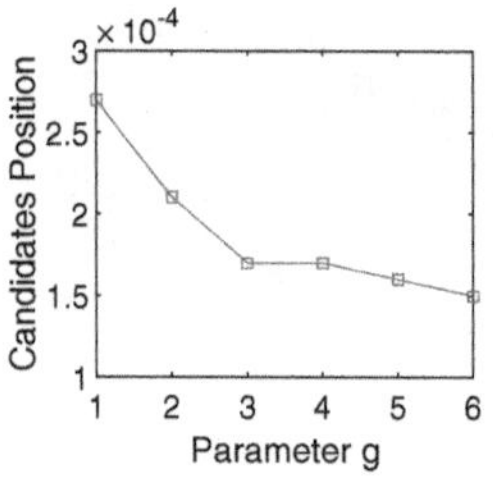

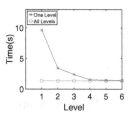

Fig. 7. Quality of C vs. g

Fig. 8. Time vs. HS-Tree level

7   Related Work

Similarity Search on Time Series. Many studies on similarity search over time-series databases have been conducted in the past decade. They can be grouped into two categories, whole matching and subsequence matching. For the former [17], sequences in query sets and databases have equal length. The state-of-the-art approach is to transform each time series to a vector of lower dimensions by DFT, DWT, PAA, etc. Then R-tree is built based on the vectors to support the query. For the latter, the query sequences are potentially shorter than sequences in databases. The goal is to find certain subsequences that are similar to the query sequence, such as FRM [7] and general match [12]. Window-based similarity search is different from both of them, because the search space is the set of aligned subsequences of all time series in the database, which can be considered as the combination of whole matching and subsequence matching.

Multi-window Processing. The second category of the related work is search and aggregation on an arbitrary time window for a set of aligned time series. [8] deals with *aggregate top-k query* on time series data. [9] focuses on discovering the longest subsequence in the database whose correlation coefficient with the query exceeds the threshold. These works share the same input and the flexibility of any length windows. However, both of them are center-based approaches.

Secondary Index on Key-Value Stores. HBase is an efficient system to support key-based data query. But querying on values is time-consuming due to the lack of secondary index on values. In order to solve this problem, some works are devoted to implementing the secondary index for Key-Value Stores. ITHBase [3] is the pioneer work to implement secondary index on HBase, but it is only prototype and incompatible to the latest version due to the lack of maintenance. Afterwards, hindex [2] and Phoenix [5] are proposed. They implemented the secondary index through Coprocessor framework, which is better-designed and light-weighted. But these secondary indices are mainly reverse indices, which are not suitable for time series analysis.

8 Conclusion

In this paper, we propose a novel composite index for KNN processing on time series. We design and build this index structure in a pyramid-shaped segment-level framework. We use VI-Table to generate candidates and prune unnecessary time series, and use HS-Tree to verify the remaining time series. We exploit the efficiency of HBase to achieve high performance in a distributed environment. We also give a detailed evaluation based on experiments to show this approach works well and outperforms other methods by an order of magnitude.

References

1. Apache HBase. http://hbase.apache.org
2. hindex. https://github.com/Huawei-Hadoop/hindex
3. ITHBase. https://github.com/hbase-trx/hbase-transactional-tableindexed
4. OpenTSDB. http://opentsdb.net
5. Phoenix. http://phoenix.apache.org
6. Camerra, A., Palpanas, T., Shieh, J., Keogh, E.: iSAX2.0: indexing and mining one billion time series. In: ICDM 2010 (2010)
7. Faloutsos, C., Ranganathan, M., Manolopoulos, Y.: Fast subsequence matching in time-series databases. In: SIGMOD 1994 (1994)
8. Jestes, J., Phillips, J.M., Li, F., Tang, M.: Ranking large temporal data. In: VLDB 2012 (2012)
9. Li, Y., Hou, U.-L., Yiu, M.-L., Gong, Z.: Discovering longest-lasting correlation in sequence database. In: VLDB 2013 (2013)
10. Lian, X., Chen, L., Yu, J.X., Wang, G., Yu, G.: Similarity match over high speed time-series streams. In: ICDE 2007 (2007)

11. Liu, Y., Songlin, H., Rabl, T., Liu, W., Jacobsen, H.-A., Kaifeng, W., Chen, J., Li, J.: DGFIndex for smart grid: enhancing hive with a cost-effective multidimensional range index. In: VLDB 2014 (2014)

12. Moon, Y.-S., Whang, K.-Y., Han, W.-S.: General match: a subsequence matching method in time-series databases based on generalized windows. In: SIGMOD 2002 (2002)

13. Mueen, A., Keogh, E.J., Young, N.: Logical-shapelets: an expressive primitive for time series classification. In: KDD 2011 (2011)

14. Mueen, A., Keogh, E.J., Zhu, Q., Cash, S., Westover, M.B.: Exact discovery of time series motifs. In: SDM 2009 (2009)

15. Papapetrou, P., Athitsos, V., Potamias, M., Kollios, G., Gunopulos, D.: Embedding-based subsequence matching in time-series databases. In: TODS 2011 (2011)

16. Athitsos, V., Papapetrou, P., Potamias, M., Kollios, G., Gunopulos, D.: Approximate embedding-based subsequence matching of time series. In: SIGMOD 2008 (2008)

17. Wang, Y., Wang, P., Pei, J., Wang, W., Huang, S.: A data-adaptive and dynamic segmentation index for whole matching on time series (2013)

Outlier Trajectory Detection: A Trajectory Analytics Based Approach

Zhongjian Lv[1], Jiajie Xu[1(✉)], Pengpeng Zhao[1], Guanfeng Liu[1], Lei Zhao[1], and Xiaofang Zhou[2,1]

[1] School of Computer Science and Technology,
Soochow University, Suzhou, China
20165227001@stu.suda.edu.cn, {xujj,ppzhao,gfliu,zhaol}@suda.edu.cn
[2] School of ITEE, The University of Queensland, Brisbane, Australia
zxf@itee.uq.edu.au

Abstract. Trajectories obtained from GPS-enabled devices give us great opportunities to mine out hidden knowledge about the urban mobility, traffic dynamics and human behaviors. In this paper, we aim to understand historical trajectory data for discovering outlier trajectories of taxis. An outlier trajectory is a trajectory grossly different from others, meaning there are few or even no trajectories following a similar route in a dataset. To identify outlier trajectories, we first present a prefix tree based algorithm called PTS, which traverses the search space on-the-fly to calculate the number of trajectories following similar routes for outlier detection. Then we propose two trajectory clustering based approaches PBOTD and DBOTD to cluster trajectories and extract representative routes in different ways. Outlier detection is carried out on the representatives directly, and the accuracy can be guaranteed by some proven error bounds. The evaluation of the proposed methods on a real dataset of taxi trajectories verifies the high efficiency and accuracy of the DBOTD algorithm.

1 Introduction

With recent improvements in satellites and GPS-enabled devices, it is possible to accumulate a great amount of trajectory data which provides us much information about behaviors or certain variation. The vast deposits of information contained in such a big dataset can help us to understand animal migration pattern, urban traffic situation, human driving behavior and so on. So far a number of amusing applications are being developed, including routes planning [2,18], pattern mining [5,12], trajectory-based inference and prediction [8,11], moving together mining [7,20], etc.

In addition to these mentioned above, outlier trajectory detection is being paid increasingly attention in recent years for moving object surveillance. The pioneer studies [9,17,22] mainly focus on the trajectories in free-space which means that the objects tend to move freely and their movement are not restricted by a given road network. These approaches can achieve a good result but are

© Springer International Publishing AG 2017
S. Candan et al. (Eds.): DASFAA 2017, Part I, LNCS 10177, pp. 231–246, 2017.
DOI: 10.1007/978-3-319-55753-3_15

not for those trajectories under road network constraints. Hence, it is necessary to study these trajectories limited by road network. More recently, increasing attentions [1,13,15,23,24] are paid to detect outlier moving object trajectories while considering the topology of the underlying road network. These studies help us to address some needs in our daily life, such as to identify if a taxi driver deliberately takes unnecessary detours to overcharge passengers who are not familiar to the area. Similarly, it is highly useful in monitoring the vehicles that carry noxious chemicals or other dangerous cargo. It is no doubt that outlier trajectory detection have become more important in the Internet era.

In effect, we hold that a precise definition with strong physical meaning is necessary for us to solve the problem and the result would be more accurate and convincing under the fine-grained definition. Unfortunately, previous methods seem to lack such a definition or the definition is not reasonable enough. For example, TPRO [24] gives an outlier score for given trajectory by comparing it with top-k popular routes. Apparently, it has a weak physical meaning because a normal trajectory other than the outliers may be different from the popular routes too. It means that some normal trajectories would be easily regarded as outliers. According to [6], an outlier is a data object that is grossly inconsistent with the remaining data. In this paper, we also hold that an outlier trajectory should be grossly different from the other trajectories. That is to say, if a taxi driver takes a little detour when encountering a traffic jam or does not choose a very popular route, the generated trajectory should not be regarded as an abnormal because there are still existing certain number of trajectories that are similar to it. So it is concluded that a trajectory can be seen as an outlier if there are almost no similar trajectories in dataset. It is our definition for outlier trajectory.

Although having the appropriate definition, detecting driving outliers is still a uphill work. There are mainly two challenges in our work as follows: on the one hand, there might be many different normal driving paths from the origin to the destination and it is difficult to describe all those trajectories covering these paths. So we need to cast about for a proper method to directly or indirectly portray those trajectories about their driving routes and quantity; on the other hand, when we do outlier trajectory detection based on our structure, the performance should be efficient and the result should be accurate. [15] obtained the number of relevant trajectories by comparing the given trajectory with each trajectory in dateset. There is no doubt that the accuracy is high but the efficiency will be in a dilemma when facing with a large dataset. So we should determine quickly and accurately whether a given trajectory is an outlier.

In this paper, we propose two methods to model our trajectories. Firstly, we develop a baseline algorithm called Prefix Tree Searching (PTS). In this algorithm, we generate a prefix tree by searching routes similar to a given test trajectory on the road network. Then we can get the proportion of those similar trajectories in all trajectories by this tree for outlier detection. It is time consuming although the method can give us an accurate result. Aiming to improve the performance, we present our second method based on clustering, this is because it can put similar trajectories together effectively. And then we can quickly judge whether a given trajectory is an outlier by comparing representative trajectories chosen from cluster results with the given one. Based on such idea, we present Prototype Based Outlier Trajectory Detection (PBOTD) based on the idea of k-medoids [16] as our first cluster-based algorithm and choose medoids as representative routes. But the method cannot achieve a high accurate because of the problem of the selection of k and local optimum. For overcoming these drawbacks, we present another algorithm called Density Based Outlier Trajectory Detection (DBOTD), where we refer to DBSCAN [4] to cluster and choose core routes as our representatives. The algorithm can quickly and accurately do outlier detection because the approach much more fit our definition. In summary, the main contributions of this paper include:

- To detect outliers more accurately, we map trajectories into bit-map and present PTS algorithm to calculate the proportion of similar quantity in total quantity.
- Taking efficiency into consideration, we provide a PBOTD and a better DBOTD algorithm to cluster trajectories and extract representative routes which can summarize those trajectories well and by using these representatives we can quickly do outlier detection.
- We demonstrate, by using various real dataset, that DBOTD can effectively and accurately do outlier detection compared with other algorithms.

2   Related Work

In this section, we introduce some related works which can be categorized into two groups. The first group focuses on analyzing or exploiting trajectories with research issues other than outlier detection. For example, Krumm et al. [8] and Liao et al. [11] predicted the destination and inferred transportation routines respectively. Jeung et al. [7] aimed at discovering a group of moving objects that have traveled together for some time. Tang et al. [20] presented a strong online mining method for the streaming trajectories. Chen et al. [2] defined popular routes which embody the common law of driving and designed an efficient search strategy on the given road network for popular routes. In [10], Lee et al. developed TRACLUS to cluster sub-trajectories for discovering the most common mode of movement in free-space. Other includes trajectory preprocessing like calibrating [19] and trajectory management as well as query [3,21].

The second group includes outlier trajectory detection. Lee et al. [9] put forward a group-and-detect framework and develop an algorithm called TRACLUS to detect outlier sub-trajectories in free-space. Pnueli et al. [17] proposed a new framework named ROAM to represent trajectories in a feature space oriented on the discrete fragments and developed a general-purpose, rule-space classifier to detect abnormal traces. Zhang et al. [23] proposed an isolation-based method called IBAT. The key idea is to randomly pick a point from the test trajectory and remove other trajectories which do not contain this point. If the test trajectory is an outlier, this process will end very soon. Zhu et al. [24] took the travel time into consideration and proposed a novel algorithm called TPRO to discover those outlier trajectories during some periods which may be normal during other periods. Different from the previous ones' ideas, we hold that an outlier trajectory is grossly inconsistent with the trajectories in the dataset, which means that there are almost no trajectories following similar routes. Masciari et al. [15] keeps a similar definition but they do outlier detection by comparing the test one with each trajectory in the dataset which is unacceptable when the dataset is large enough. In contrast, we propose different methods including searching similar routes on the given road network and clustering trajectories which pass from a certain origin and a certain destination (od-pair).

3   Problem Definition

This part presents some definitions and gives a formal statement of the problem this paper focuses on.

Definition 1 (Road Network). A road network is modeled as a directed graph $G = (V, E)$, where each vertex $v_i \in V$ denotes an intersection or end of a road, and each edge $e_i^j \in E$ denotes a road segment from vertex v_i to v_j (the direction is $v_i \rightarrow v_j$).

Given a road network G, a route is a path that a moving object passes, and it is denoted as $r = (v_{c_1}, ..., v_{c_n})$, where $1 \le c_i \le |V|$. Trajectory data can be derived if we tackle how object moves on a route by sampling spatial and temporal information.

Definition 2 (Trajectory). A raw trajectory $t_{raw} = (p_1, p_2, ..., p_n)$ is a sequence of geo-tagged sampling points, where p_i denotes a geographic coordinate. After map-matching, we can derive the mapped trajectory $t_{mapped} = (v_{c_1}, v_{c_2}, ..., v_{c_m})$. For simplicity, we will drop the "mapped" qualifier and use t to replace t_{mapped}. We use $t_{i->j}$ to denote the sub-trajectory $(v_{c_i}, ..., v_{c_j})$ of t, where $1 \le i < j \le m$.

We use $Count(r)$ to denote how many trajectories or sub-trajectories cover the route r totally. A route may not be covered by existing trajectory, it means that the route's count is zero, namely $Count(r) = 0$.

Definition 3 (Route Distance Function). Route Distance Function $\varphi(r, r')$ is a formula which gives a difference score between routes which have the same origin and destination. It is noted that the distance function can also give score between a route and a trajectory. If we have a route $r = (v_{c_1}, ..., v_{c_n})$ and another route $r' = (v_{c_1'}, ..., v_{c_m'})$, where $c_1 = c_1'$ and $c_n = c_m'$, then their score can be denoted as

$$\varphi(r, r') = Min \begin{cases} \varphi(Rest(r), r') + add\_cost(e_{c_{n-1}}^{c_n}) \\ \varphi(r, Rest(r')) + add\_cost(e_{c_{m-1}'}^{c_m'}) \\ \varphi(Rest(r), Rest(r')) \\ \quad + replace\_cost(e_{c_{n-1}}^{c_n}, e_{c_{m-1}'}^{c_m'}) \end{cases} \quad (1)$$

where $Rest(r) = (v_{c_1}, ..., v_{c_{n-1}})$ is prefix of r after removing the last edge $e^{c_n}_{c_{n-1}}$, $add\_cost(e) = 1$ and $replace\_cost(e^{c_n}_{c_{n-1}}, e^{c'_m}_{c'_{m-1}}) = 0$ if $c_{n-1} = c'_{m-1}$ and $c_n = c'_m$, otherwise 2 which is as twice as big than add_cost because it takes both add and delete into consideration.

If $Rest(r) = (v_{c_1})$, then we have $\varphi(Rest(r), r') = \sum_{e \in r'} add\_cost(e)$. This also applies to the situation where $Rest(r')$ only contains starting vertex. If both $Rest(r')$ and $Rest(r)$ have one vertex, their difference score is 0.

Definition 4 (Similar Route). A route r is said to be an $\alpha - similar$ route with respect to another route r' if $\varphi(r, r') < \alpha$, where α is a given threshold.

Note that the route distance function in Definition 3 and similar route here are also applicable to a route and a trajectory because a trajectory is a route after map-matching.

Definition 5 (Outlier Trajectory). Given a trajectory t which passes from v_{c_1} to v_{c_m} and an outlier score threshold θ, we can calculate the outlier score

$$S(t) = \frac{\sum_{r \in SR(t)} Count(r)}{\sum_{r \in AR(t)} Count(r)}, \tag{2}$$

where $SR(t) = \{r \mid r \text{ is } \alpha - similar \text{ to } t\}$, denotes all $\alpha - similar$ routes respect to t and $AR(t) = \{r \mid r.c'_1 = c_1 \text{ and } r.c'_n = c_m\}$, denotes all routes which can go from t's starting vertex to t's ending vertex. If $S(t) < \theta$, then we say the trajectory is a $\theta - outlier$ trajectory.

Problem: Given a history trajectory set T, a trajectory t, a similar threshold α and an outlier score threshold θ, we need to judge whether the trajectory t is a $\theta - outlier$.

4   Road Network Search Based Approach

In this section, we firstly introduce our baseline algorithm in which we search similar routes about the given trajectory on the road network. We start from origin and traverse the road network graph to arrive at destination. The process will generate a prefix tree and so the algorithm is also called Prefix Tree Searching algorithm. Then we can do aggregation with the prefix tree for accurately detecting outliers.

4.1   Prefix Tree Based Search

Before introducing searching algorithm, we need to solve a problem on how we obtain the quantity of trajectories covering a route. For solving this, we can use a bitmap for each vertex. The bitmap records whether each trajectory goes through the vertex or not and we can get the number of the trajectories going through the vertex by it. Then we can get the quantity of trajectories covering a route by doing and-operation on the bitmaps of all vertices of this route.

Now we can introduce our prefix tree searching algorithm. Given a trajectory t, we start from the origin of t and adopt breadth-traversal to search on the road network until arriving at the destination of t. After this process, we would generate a structure like a prefix tree. However, the search space is huge and it is necessary for us to find out how to prune when searching. We mainly consider the following three aspects: firstly, if Best Match Distance between a partial route and given trajectory reaches or exceeds our threshold α, we can stop searching on this path; secondly, if there are no trajectories covering the path, namely the bitmap after and-operation is 0, we can stop too; thirdly, if the vertex is the same as another vertex in this path, we also prune the branch because a normal route usually does not contain same vertices.

It is easy to understand the second and third points. And for the first point, we firstly define the Best Match Distance function and give an example.

Definition 7 (Best Match Distance Function). Given a partial route $pr = (v_{c_1}, ..., v_{c_k})$ and a trajectory $t = (v_{c'_1}, ..., v_{c'_n})$, where $v_{c_1} = v_{c'_1}$, we can calculate their minimum route distance which is called Best Match Distance. And the Best Match Distance function $\psi_{bm}(pr, t)$ is defined as

$$\psi_{bm}(pr, t) = \varphi(part(pr), part(t)) + \sum_{e \in rest(pr)} add\_cost(e) \qquad (3)$$

where $part(pr) = (v_{c_1}, ..., v_{separation})$, $part(t) = (v_{c'_1}, ..., v_{separation})$ and $rest(pr) = (v_{separation}, ..., v_{c_k})$. Here, a separation point $v_{separation}$ refers to the last coincidence vertex of the given trajectory and the partial route. Note that the rest part on the partial route may be empty and we do not need to calculate add_cost in this situation. If we arrive at destination, it is evident that $\psi_{bm} = \varphi$.

Example 1. Supposed that we have the complete trajectory $t = (v_1, v_3, v_6, v_7)$ and partial-route $pr = (v_1, v_2, v_4)$ in the left of Fig. 1. The separation point is v_1. So we can know that $part(t) = (v_1)$, $part(pr) = (v_1)$ and $rest(pr) = (v_1, v_2, v_4)$. Then the Best Match Distance can be calculated by Eq. 3 which equals 2.

Then we have the following lemma to prove the first point:

Lemma 1. *Given a trajectory t and a partial route pr_{+i}, where $i \geq 0$, we can get that $\psi_{bm}(pr_{+i}, t) \leq \psi_{bm}(pr_{+(i+1)}, t)$. Here, we use $pr_+(i + 1)$ to denote a partial route that we pass through another vertex based on pr_{+i}.*

Proof. Supposed that $pr_{+i} = (v_{c_1}, ..., v_{c_i})$ and $pr_{+(i+1)} = (v_{c_1}, ..., v_{c_i}, v_{c_{i+1}})$, on the one hand, if $v_{c_{i+1}}$ is a vertex that does not exist in trajectory t, then $\psi_{bm}(pr_{+(i+1)}, t) = \psi_{bm}(pr_{+i}, t) + add\_cost(e_{c_i}^{c_{i+1}}) = \psi_{bm}(pr_{+i}, t) + 1 \geq \psi_{bm}(pr_{+i}, t)$; if $v_{c_{i+1}}$ exists in t, it is evident that $\psi_{bm}(pr_{+(i+1)}, t) = \varphi(pr_{+(i+1)}, part(t)) \geq \psi_{bm}(pr_{+i}, t)$ because $replace\_cost$ in $\varphi(pr_{+(i+1)}, part(t))$ not only contains add_cost but also takes deleting edges into consideration. So Lemma 1 can be proven.

Lemma 2. *Given a partial-route pr and a trajectory t, if $\psi_{bm}(pr, t) \geq \alpha$, then any route r covering pr must not be $\alpha - similar$ with t.*

Proof. We know that $pr = pr_{+0}$ and $\exists\ m \geq 1$, let $r = pr_{+m}$. According to Lemma 1 and our condition, we know that $\alpha \leq \psi_{bm}(pr, t) = \psi_{bm}(pr_{+0}, t) \leq \psi_{bm}(pr_{+1}, t) \leq ... \leq \psi_{bm}(pr_{+m}, t) = \varphi(r, t)$, so r is not $\alpha - similar$ with t by Definition 4.

Example 2. For brevity, we just show a subgraph of our road network in the left of Fig. 1 in which we can start from v_1 and go to v_7 and there are three trajectories being mapped into vertices' bitmaps. We can find that v_8 is another vertex that a moving object can go to from v_1. The vertex's bitmap is 000 which means that the three trajectories do not pass through this vertex.

If we want to test a trajectory $t = (v_1, v_3, v_6, v_7)$ and the similar threshold α is set to 2, we can get a prefix tree (in the right of Fig. 1) when searching on the road network. We firstly do and-operation on the bitmaps of origin v_1 and destination v_7. Then the result is taken as the bitmap of root node which ensures our search marching to the destination. Now we can start from the origin and search all routes which are similar to our given trajectory. We can find that we stop searching at v_8 because the bitmap after doing and-operation is 000. v_4 is also a stop node because we can get $\psi_{bm} = 2$ and this score has reached the preset limit. The path $(v1, v3, v6, v7)$ is a valid path and so we get a similar route about our test trajectory.

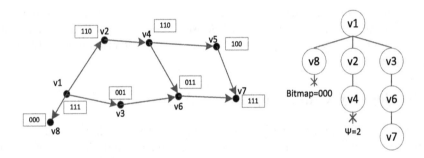

Fig. 1. A SubGraph of Road Network and A Prefix Tree of Searching

4.2   Outlier Detection

After searching, we can get paths result set and each path represents a similar route and contains a bitmap after doing and-operation. Intuitively, we can compute $Count(r)$ by bitmap. However, it has the following faults:

1. A trajectory which covers similar route's reverse vertices will also be contained in this bitmap, which is easy to understand.
2. Some trajectories will be contained in another bitmap which also results in the wrong number of $Count(r)$ as shown in Observation 1.

Observation 1. In Fig. 2, we can get two routes: $r_1 = (v_1, v_2, v_3)$ and $r_2 = (v_1, v_3)$. If we use bitmap to calculate $Count(r)$, then $Count(r_1) = 2$ and $Count(r_2) = 3$. However, there are only three trajectories which proves that our result is wrong because there are two trajectories accidentally being contained in the bitmap of r_2.

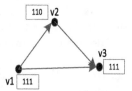

Fig. 2. An Observation of Computing Count(r)

For solving it, we need to do verification. After searching, we achieve all similar routes from prefix tree. For each route r, we can get a last bitmap resulting from doing and-operation on all bitmaps of vertices in this route and the last bitmap contains all possible trajectories. Then we compare these trajectories with this route to compute $Count(r)$. By doing so, we can do outlier detection by Eq. 2.

Though baseline algorithm can provide an accurate result, it is not efficient in two aspects: searching and verifying. So we are required to vigorously pursue other methods to describe trajectories and do outlier detection efficiently under certain precision.

5   Trajectory Clustering Based Approach

To solve the efficiency problem discussed above, this section introduces our approaches based on clustering trajectories. Clustering analysis is a division of data into groups according to given rules for similarity and it is a common means in data mining which can help us describe similar things. After putting similar trajectories into together, it is necessary to extract routes from each group that can represent clusters at a high level. Then we can do outlier detection by comparing these representatives with the test trajectory. Compared with the road network searching method, it is no doubt that the kind of method can be faster and the accuracy will not be lost too much if we choose representative routes properly.

Based on these ideas, we propose two cluster-based outlier trajectory detection algorithms: PBOTD based on prototype-based clustering and DBOTD based on density-based clustering. To make result of trajectory clustering meaningful, it is necessary to divide trajectories and sub-trajectories into different groups according to od-pairs in advance. The bitmap structure mentioned in Sect. 4.1 can help us solve the problem. For improving performance of clustering, those trajectories covering a certain route can be as one object to cluster.

We can build a prefix tree to get these routes and their quantity. Now we can cluster them and save necessary cluster's representative routes. At last, we get related routes and do outlier detection when a test trajectory comes.

5.1 Prototype Based Outlier Trajectory Detection

Prototype-based clustering algorithms, such as k-means [14], are frequently used to find the structure of a dataset by grouping all objects into clusters based on their similarities. These algorithms initialize a prototype and then update it by some strategies to achieve the final cluster results. Among them, k-means is a basic method. However, the calculation of the mean for trajectories is very difficult and it is easily affected by abnormal trajectories because of poor quality of dataset. Then we decide to refer to another algorithm called k-medoids. The algorithm divides data into k clusters and each cluster has a medoid which we can use to represent the cluster. It is a great advantage that we only save k routes for each od-pair and compare k times when doing outlier detection.

Algorithm Description. After deriving all routes subject to an od-pair, we use k-medoids method to cluster the routes and find representative routes. At first, we should select initial k medoids. By default we can use Route Distance Function to calculate distance, however, it is more appropriate to take the quantity of trajectories covering these routes into consideration. And the weighted route distance function used in this algorithm is given belows:

Definition 8 (Weighted Route Distance Function). Given two routes r and r', in order to measure their weighted distance (taking quantity into account), we define the distance function as follows:

$$\varphi_q(r, r') = \frac{\varphi(r, r')}{Count(r) + Count(r')} \tag{4}$$

where $Count(r)$ and $Count(r')$ are quantity of r and r' respectively. It is obvious that the larger the quantity is, the smaller the distance should be.

Then we can calculate an initial score for each route r_j by the following formula:

$$s_j = \sum_{i=1}^{n} \frac{\varphi_q(r_i, r_j)}{\sum_{l=1}^{n} \varphi_q(r_i, r_l)} \tag{5}$$

And we select k routes having the first k smallest scores as initial medoids, which tends to select k most middle routes. Secondly, we obtain initial cluster result by assigning each route to the nearest medoid and calculate the sum of distances from all routes to their medoids. At last, we need to choose a new route which has the minimal sum of distances from it to other routes in this cluster to update current medoid for each cluster and assign again. We will not stop these steps until the sum of distances from all routes to their medoids does not change. Note that the distance we mention here is calculated by Eq. 4.

After clustering, we need to choose routes to represent clusters which are used to detect outliers. In this algorithm, we choose the medoid as our representative route for each cluster. Then we can get a representative routes set *representative_routes*.

Outlier Detection. Given a test trajectory t, the starting and ending vertices of t need be used to get related representative routes set *representative_routes*. The main issue is how to make use of these routes from trajectory clustering for outlier detection. Then we need to calculate a Min Core Distance which means the minimum distance between representative routes and a trajectory. We have the following definition:

Definition 9 (Min Core Distance Function). Given a trajectory t, we can get its relevant *reference_routes*, and the Min Core Distance Function $\rho(t)$ is defined as

$$\rho_{mc}(t) = min(\varphi(r_1, t), ..., \varphi(r_n, t)) \tag{6}$$

where $r_1, ..., r_n \in reference\_routes$.

If $\rho_{mc}(t) < \alpha$, where α is similarity threshold given, we can say that the trajectory can appear in the cluster. And a trajectory is an outlier if it does not appear in the clusters. It means that if $\rho_{mc}(t) \geq \alpha$, then the test trajectory t is an outlier. However, we cannot guarantee by theory that the result satisfies our definition and we also cannot give out the corresponding error bound.

5.2   Density Based Outlier Trajectory Detection

As mentioned above, PBOTD cannot provide guarantee of the high accuracy of the result. To address this issue, we provide another algorithm called DBOTD. In this algorithm, we refer to the idea of density-based clustering algorithms which expands result of clustering on the connectivity between the objects from the angle of their density. Correspondingly, we cluster trajectories according to the quantity of them nearby a certain route and it is extremely consistent with our outlier definition.

Algorithm Description. DBSCAN [4] is a classical algorithm in density-based cluster methods. It can discover clusters of arbitrary shape and filter out outliers. Here we apply this algorithm to cluster trajectory data. After deriving all clusters, we extract representative routes which can generalize the whole in each cluster. Now we can do outlier trajectory detection by these routes.

For portraying the compactness degree of objects, DBSCAN defines core object and connectivity on the help of neighborhood parameters $(\epsilon, MinPts)$. Similarly, we need to define a core route as follows:

Definition 10 (Core Route). Given a radius d and a MinPtsRate γ, if a route r satisfies $|N(r,d)| \geq \gamma \times |N(group\ of\ r)|$, where $N(group\ of\ r) = AR(r)$ denotes all routes which go from r's starting vertex to r's ending vertex, $N(r,d) = \{r'|\varphi(r,r') < d\ and\ r' \in N(group\ of\ r)\}$ and $|N| = \sum_{r \in N} Count(r)$, then we can say r is a core route.

According to our definitions, we should set $\gamma = \theta$ and $d = \alpha$. By doing so, we discover from Definition 10 that a core route must be a non-outlier route because the proportion of those trajectories around the core route in all trajectories reaches or exceeds our outlier threshold θ.

In this algorithm, we also expand clusters on the basis of connectivity and cores. We first choose a core route cr and then $N(cr,d)$ should also be included in this cluster. Then if there are new core routes being included in this cluster, these new core routes will be expanded too according to prior step. After these steps, we can get a certain number of clusters.

Now we need to choose representative routes to describe clusters comprehensively. On the one hand, if we record only one representative route, it is prone to generate error; on the other hand, it wastes storage space to store all core routes. For solving it, when we expand clusters from a core route, the core route will be chosen if existing routes not be clustered around it. It means that a core route will not be chosen if it can be deduced by other core routes.

Outlier Detection. After clustering, we also need to judge whether a trajectory is outlier. Namely, given a trajectory t, if Min Core Distance $\rho_{mc}(t) \geq \alpha$, it is clear that the trajectory is an outlier. We have the following lemma:

Lemma 3. *Given a trajectory t, if $\rho_{mc}(t) \geq \alpha$ wrt. $\gamma = \theta$ and $d = \alpha$, then the trajectory is an outlier.*

Proof. We know that $\rho_{mc}(t)$ denotes the minimum route distance between t and representative routes. And these routes describe clusters comprehensively. So $\rho_{mc}(t) \geq \alpha$ shows that the trajectory is not be included by any clusters. Then the trajectory must not be a core route. It means that $|N(t,d)| < \theta \times |N(group\ of\ t)|$, namely $\sum_{r \in SR(t)} Count(r) < \theta \times \sum_{r \in AR(t)} Count(r)$. So it is an outlier.

One problem an acute reader would immediately notice is that border routes in the cluster are also outliers because they also do not satisfy our definition. However, there is an error bound: we can be sure that these border routes are not-outliers within the scope of the $2d$ and we have the following lemma:

Lemma 4. *Given a trajectory t, if $\rho_{mc}(t) < \alpha$ wrt. $\gamma = \theta$ and $d = \alpha$, it must not be an outlier with the scope of the $2d$.*

Proof. From the condition $\rho_{mc}(t) < \alpha$, we know that there is at least a core route cr making $\varphi(cr,t) < \alpha$. So $N(t,2\alpha)$ must contains $N(cr,\alpha)$, where $N(t,2\alpha) = \{r'|\varphi(t,r') < 2\alpha\}$ and $N(cr,\alpha) = \{r'|\varphi(cr,r') < \alpha\}$. We know that $|N(cr,\alpha)| \geq \theta \times |N(group\ of\ cr)|$, then $|N(t,2\alpha)| \geq \theta \times |N(group\ of\ t)|$, where $N(group\ of\ cr) = N(group\ of\ t)$, which proves obviously our lemma.

Compared with the PBOTD algorithms, this algorithm can guarantee giving out a theoretic error bound. It can extract routes which represent clusters better and do outlier detection efficiently.

6   Experimental Study

This section presents our experiment of study. We introduce our experiment settings and evaluation criteria in the first subsection. Then we present experiment results and analysis in the second subsection.

6.1   Experiment Settings

The experiment is taken under a real-world dataset which contains 5,660,692 trajectories in Beijing. We picked up about 5,300 trajectories from the dataset as testing set and asked volunteers to manually label whether each trajectory is outlier or not. The rest trajectories are used as the training dataset. And the Beijing road network data in our experiment contains 253,180 vertices and 557,134 edges.

Table 1. Default values of parameters

Parameter	Default value	Description
θ	3%	Threshold of outlier score
α	25	Threshold of similar routes
k	6	K clusters in PBOTD
D	5000 K	The number of training trajectories

We compare the average time cost of detection and the accuracy in the next subsection. The default values for parameters are given in Table 1. In the experiments, we vary one parameter and keep the others constant to investigate the effect of this parameter. All algorithms are implemented in Java and run on a server with two 6-core Intel(R) Xeon(R) CPUs (2.6GHz) and 256GB memory.

6.2   Performance Evaluation

In this part, we vary the values of parameters in table 1 to compare our three algorithms and investigate the effect of each parameter. After this, we compare our best DBOTD algorithm with TPRO and IBAT algorithms which are proposed in [24] and [2] respectively.

Effect of θ. In the first part of experiments, we study the effect of θ which mainly affects our PTS and DBOTD algorithm. As shown in Fig. 3(a), our PTS algorithm achieves high accuracy compared with other algorithms and DBOTD algorithm is also good which pales slightly beside PTS. Within our expectations, PBOTD works worst. In the aspect of average time costing, as shown in Fig. 3(b), PTS consumes too much time which we cannot accept while PBOTD and DBOTD can detect quickly within tens of milliseconds. Taking into account of two aspects, it holds that DBOTD works best. We can also find out that the result is stable, especially in DBOTD, when θ varies from 1% to 5%.

Effect of k. We study the effect of different k which only PBOTD has. From Fig. 4(a), we can find out that the accuracy of PBOTD is relatively poor when k is small and it will increase as k grows until reaching the limit which is not satisfactory. On the other hand, the average time of detecting a trajectory in PBOTD almost linearly increases with k increasing, which is shown in Fig. 4(b) Although PBOTD just saves k routes for each od-pair, the problem of low accuracy will make us choose other algorithms.

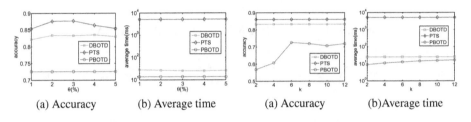

| (a) Accuracy | (b) Average time | (a) Accuracy | (b)Average time |

Fig. 3. Effect of θ **Fig. 4.** Effect of k

Effect of α. We investigate the performance of these algorithms when the threshold of similar route α is varying. Figure 5 shows the results of our experiment. With the increase of α, PTS will incur significantly time cost while cluster-based algorithms will not because they have saved corresponding information which is used to detect outliers. On the other hand, DBOTD and PTS always outperform PBOTD in accuracy when α varies from 15 to 35. It is also shown that PTS and DBOTD have a stable and good performance in an interval of 20 and 30, which is what we expect. Taking both accuracy and time cost into consideration, DBOTD is a great algorithm.

Effect of D. In order to evaluate the effect of D, we sample the dataset to generate training dataset with different number of trajectories varying from 1000 K to 5000K, and report the average time cost and the accuracy in Fig. 6 There is a marginal increase in the regard of accuracy with the training data size growing. Meanwhile, PTS and DBOTD algorithms are always better than PBOTD in accuracy no matter how D changes. In the aspect of time cost, PTS will increasingly spend time because bitmaps' size and search space will expand as training dataset size grows. In general DBOTD performs well when D varies from 1000 K to 5000K.

DBOTD vs. TPRO and IBAT. This paragraph gives a comparison among DBOTD, TPRO and IBAT. All of the three algorithms are tested in their best

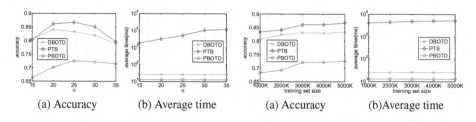

(a) Accuracy   (b) Average time   (a) Accuracy   (b)Average time

Fig. 5. Effect of α **Fig. 6.** Effect of D

parameters, which are listed in table 2. It is noted that we remove the attribute of time in TPRO's route distance function and such a move will not destroy the model of TPRO. And $S_{threshold}$ denotes the best threshold of outlier score in our experiment.

Table 2. Parameter setting of DBOTD, TRRO and IBAT

Algorithm	DBOTD	TPRO	IBAT
Parameters	$\alpha = 20, \theta = 3\%$	$k = 5, S_{threshold} = 115$	$m = 256, \psi = 100, S_{threshold} = 0.55$

Figure 7(a) shows the accuracy of DBOTD, TPRO and IBAT in different training data sizes. Firstly, we can see that the results of DBOTD and IBAT are almost the same in the aspect of accuracy and they both outperform TPRO greatly. It is easy to understand because the routes that TPRO chooses may be similar and then TPRO will ignore other normal routes which results in low accuracy. While DBOTD chooses routes as completely as possible and so it has a high accuracy. Secondly, it shows the difference of the average time cost in Fig. 7(b). We can know that DBOTD is slightly slower than TPRO but faster that IBAT. For all concerned, DBOTD is a great algorithm.

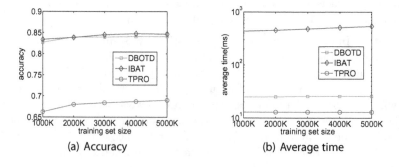

(a) Accuracy   (b) Average time

Fig. 7. Accuracy and average time cost of DBOTD, TPRO and IBAT

It can be concluded from the above experimental results that DBOTD can achieve a relatively high accuracy within short time in the algorithms that we proposed. And compared with others' algorithms, DBOTD also works best when we take both accuracy and time cost into consideration.

7   Conclusion and Future Work

In this paper, we aim at understanding historical trajectory data for discovering outlier routes of taxis. We believe that an outlier route is a route grossly different from others, meaning there are few trajectories or even no trajectory following a similar route in a dataset. Based on this idea, we firstly propose an accurate algorithm called PTS but it consumes too much time. Then we propose two trajectory clustering-based approaches PBOTD and DBOTD to cluster and extract representative routes for detecting outliers directly. Among all the three algorithms, DBOTD performs the best with a relatively high accuracy and low time cost which has been demonstrated by extensive experiment results on real datasets.

In the future, we plan to enhance our algorithm in two aspects. Firstly, we adopt transfer probability model and propose a better algorithm to optimize PTS. Secondly, we try to seek out a better cluster-based algorithm to improve accuracy.

Acknowledgement. This work was partially supported by Chinese NSFC project under grant numbers 61402312, 61232006, 61472263, 61572335, 61532018, 61502324, and Australia Research Council discovery projects under grant number DP170101172.

References

1. Chen, C., Zhang, D., Castro, P.S., Li, N., Sun, L., Li, S., Wang, Z.: iBoat: isolation-based online anomalous trajectory detection. IEEE Trans. Intell. Transp. Syst. **14**, 806–18 (2013)
2. Chen, Z., Shen, H.T., Zhou, X.: Discovering popular routes from trajectories. In: ICDE, pp. 900–911 (2011)
3. Ding, Z., Jiajie, X., Yang, Q.: Seaclouddm: a database cluster framework for managing and querying massive heterogeneous sensor sampling data. J. Supercomput. **66**, 1260–84 (2013)
4. Ester, M., Kriegel, H.-P., Sander, J., Xiaowei, X., et al.: A density-based algorithm for discovering clusters in large spatial databases with noise. In: KDD, pp. 226–231 (1994)
5. Gonzalez, H., Han, J., Li, X., Myslinska, M., Sondag, J.P.: Adaptive fastest path computation on a road network: a traffic mining approach. In: VLDB, pp. 794–805 (2007)
6. Han, J., Pei, J., Kamber, M.: Data Mining: Concepts and Techniques. Elsevier (2011)
7. Jeung, H., Yiu, M.L., Zhou, X., Jensen, C.S., Shen, H.T.: Discovery of convoys in trajectory databases. In: PVLDB, pp. 1068–1080 (2008)

8. Krumm, J., Horvitz, E.: Predestination: inferring destinations from partial trajectories. In: Dourish, P., Friday, A. (eds.) UbiComp 2006. LNCS, vol. 4206, pp. 243–60. Springer, Heidelberg (2006). doi:10.1007/11853565_15

9. Lee, J.-G., Han, J., Li, X.: Trajectory outlier detection: a partition-and-detect framework. In: ICDE, pp. 140–149 (2008)

10. Lee, J.-G., Han, J., Whang, K.-Y.: Trajectory clustering: a partition-and-group framework. In: SIGMOD, pp. 593–604 (2007)

11. Liao, L., Patterson, D.J., Fox, D., Kautz, H.: Learning, inferring transportation routines. Artif. Intell. **171**, 311–1 (2007)

12. Liu, L., Andris, C., Ratti, C.: Uncovering cabdrivers behavior patterns from their digital traces. In: Computers, Environment and Urban Systems, pp. 541–548 (2010)

13. Liu, S., Ni, L.M., Krishnan, R.: Fraud detection from taxis' driving behaviors. IEEE Trans. Veh. Technol. **63**, 464–72 (2014)

14. Lloyd, S.: Least squares quantization in PCM. IEEE Trans. Inf. Theory **28**, 129–136 (1982)

15. Masciari, E.: Trajectory outlier detection using an analytical approach. In: ICTAI, pp. 377–384 (2011)

16. Park, H.-S., Jun, C.-H.: A Simple and fast algorithm for k-meds clustering. Expert Syst. Appl. **36**, 3336–3341 (2009)

17. Pnueli, A.: Roam: Rule-and motif-based anomaly detection in massive moving object data sets. In: SDM, pp. 273–284 (2007)

18. Sacharidis, D., Patroumpas, K., Terrovitis, M., Kantere, V., Potamias, M., Mouratidis, K., Sellis, T.: On-line discovery of hot motion paths. In: EDBT, pp. 392–403 (2008)

19. Han, S., Zheng, K., Huang, J., Wang, H., Zhou, X.: Calibrating trajectory data for spatio-temporal similarity analysis. VLDB J. **24**, 93–116 (2015)

20. Lu-An Tang, Y., Zheng, J.Y., Han, J., Leung, A., Hung, C.-C., Peng, W.-C.: On discovery of traveling companions from streaming trajectories. In: ICDE, pp. 186–197 (2012)

21. Wang, H., Zheng, K., Jiajie, X., Zheng, B., Zhou, X., Sadiq, S.: Sharkdb: an in-memory column-oriented trajectory storage. In: CIKM, pp. 1409–1418 (2014)

22. Yanwei, Y., Cao, L., Rundensteiner, E.A., Wang, Q.: Detecting moving object outliers in massive-scale trajectory streams. In: KDD, pp. 422–431 (2014)

23. Zhang, D., Li, N., Zhou, Z.-H., Chen, C., Sun, L., Li, S.: iBAT: detecting anomalous taxi trajectories from gps traces. In: UbiComp, pp. 99–108 (2011)

24. Zhu, J., Jiang, W., Liu, A., Liu, G., Zhao, L.: Time-dependent popular routes based trajectory outlier detection. In: Wang, J., Cellary, W., Wang, D., Wang, H., Chen, S.-C., Li, T., Zhang, Y. (eds.) WISE 2015. LNCS, vol. 9418, pp. 16–30. Springer, Heidelberg (2015). doi:10.1007/978-3-319-26190-4_2

Clustering Time Series Utilizing a Dimension Hierarchical Decomposition Approach

Qiuhong Li[1], Peng Wang[1(✉)], Yang Wang[1], Wei Wang[1], Yimin Liu[2],
Jiaye Wu[1], and Danyang Dou[1]

[1] School of Computer Science, Fudan University, Shanghai, China
{qhli09,pengwang5,081024004,weiwang1,wujy16,dydou16}@fudan.edu.cn
[2] Third Affiliated Hospital of Second Military Medical University, Shanghai, China
liuyiminzsh@aliyun.com

Abstract. Time series clustering has attracted amount of attention recently. However, clustering massive time series faces the challenge of the huge computation cost. To reduce the computation cost, we propose a novel Dimension Hierarchical Decomposition (DHD for short) method to represent time series and a corresponding tree structure, denoted as DHDTree, to reorganize the time series collections to achieve the best separation effect. The main idea of DHDTree is to adapt k-d tree for time series by utilizing the DHD representation. When splitting, we select the most separable splitting strategy according to a predefined cost model. A fundamental feature of DHDTree is that it overcomes dimension curse by leveraging dimension compositions instead of selecting only one dimension when splitting, aiming to acquire the maximal separation effect. We illustrate that DHDTree obtains both the balance and the locality properties, which are important factors for the efficiency of time series organization for clustering. By the support of DHDTree, we improve clustering in two aspects. First, the DHD representation decreases the computation cost between time series dramatically. Secondly, we acquire the centers benefiting from the reorganization of the time series using our proposed DHDTree structure. Both the synthetic and real data sets verify the effectiveness and efficiency of the proposed method.

1 Introduction

The increasing instrumentation of physical and computing processes has given us unprecedented capabilities to collect massive volumes of data. Applications for data center management, environmental monitoring, financial engineering, scientific experiments, and mobile asset tracking produce massive time series streams (or, signals) from various (physical and virtual) sensors. The dramatic cost decreasing and volume increasing of storage devices makes it possible to

The work was supported by the Ministry of Science and Technology of China, National Key Research and Development Program under No.2016YFB1000700, National Key Basic Research Program of China under No.2015CB358800 and NSFC (61672163, U1509213).

© Springer International Publishing AG 2017
S. Candan et al. (Eds.): DASFAA 2017, Part I, LNCS 10177, pp. 247–261, 2017.
DOI: 10.1007/978-3-319-55753-3_16

archive data for a long period of time and efficiently support various statistical and data mining queries on historic data. Clustering such time series becomes more and more popular in different areas [16,17].

Recall some popular clustering methods. K-means algorithm has maintained its popularity for large-scale data sets clustering. The k-means clustering problem is one of the oldest and most important questions in all of computational geometry. Given an integer k and a set of n data points in R^d, the goal is to choose k centers so as to minimize ϕ, the total squared distance between each point and its closest center. A high-quality initialized centers are important for both accuracy and efficiency of k-means. K-means++ [5] improves the quality of initial centers by selecting the points far away from the already selected centers. However, they need to make k passes over the whole data set sequentially to find k centers, which limits their applicability to massive data. Scalable k-means++ [6] overcomes the sequential nature by selecting more than one centers at an iteration. However, it still faces huge computation when updating the weights for all points.

The large volume of time series poses two main challenges for the center-based clustering methods. The first is how to alleviate the affect of dimension curse when selects initial k centers for time series data; the second is how to decrease the huge computation cost while comparing with the centers. In this paper, we solve these problems by proposing a dimensional hierarchical decomposition tree (DHDtree, for short) as the underlying data organization scheme, aiming to provide a quick partitioning method for massive time series data, which groups similar time series together. We propose a dimension composition tree (DCTree, for short) to provide candidate dimension compositions for the node splitting when constructs a DHDtree. We notice that a coarse partitioning may use less info than the refined partitioning and define a series of dimension compositions with different granularity in a DCTree. By this way, we adapt k-d tree for time series by utilizing a novel dimension composition method. The main difference between DHDTree and k-d tree is that k-d tree selects only one dimension for the node splitting, while DHDTree selects a dimension composition and defines a cost model to measure the different choices of dimension compositions. Considering the fact that on top of the tree structure, we need less info to separate the time series than the lower part. That is, an overall synopsis may separate the time series for a coarse granularity. While walking down the tree, more detailed synopses are needed for the separation. Here we use the mean values as the synopsis for different compositions of dimensions.

Notice that for massive time series, it is expensive to assign time series to partitions by calculating the distances between the time series and the center. To decrease the comparison cost, we define the DHD distance based on DHD representations instead of the original time series.

The contributions are listed below:

1. We propose a dimension hierarchical decomposition representation for time series.

2. We propose a dimension hierarchical decomposition tree structure for time series collections.
3. We propose a clustering method based on the DHD representation and the DHDTree structure.

The remainder of the paper is structured as follows. Section 2 introduces the preliminary and background. Section 3 presents DHD tree. Section 4 presents DHD-Clustering algorithm. Section 5 presents the related work. Section 6 reports the experiment results and Sect. 7 concludes the paper.

2   Preliminary Knowledge and Background

2.1   K-d Tree

K-d tree [7] is a recursive space partitioning tree. The canonical method of k-d tree construction is the following:

1. As one moves down the tree, one cycles through the axes used to select the splitting planes. (For example, the root would have an x-aligned plane, the root's children would both have y-aligned planes, the root's grandchildren would all have z-aligned planes, the next level would have an x-aligned plane, and so on.)
2. Points are inserted by selecting the median of the points being put into the subtree, with respect to their coordinates in the axis being used to create the splitting plane. (Note the assumption that we feed the entire set of points into the algorithm up-front.)

This construction method leads to a balanced k-d tree, in which each leaf node is about the same distance from the root.

2.2 K-means++ and K-means ||

Let $\mathcal{X} = \{x_1, x_2, ..., x_n\}$ be a set of observations in the d-dimensional Euclidean space. Let $\|x_i - x_j\|$ denote the Euclidean distance between x_i and x_j. Let $\mathcal{C} = \{c_1, c_2, ..., c_k\}$ be the k centers. We denote the cost of $\mathcal{X}$ with respect to $\mathcal{C}$ as

$$\phi_{\mathcal{X}}(\mathcal{C}) = \sum_{x \in \mathcal{X}} d^2(x, \mathcal{C}) = \sum_{x \in \mathcal{X}} \min_{1 \leq i \leq k} \|x - c_i\|^2 \tag{1}$$

The k-means++ is proposed by Arthur and Vassilvitskii [5], which focuses on improving the quality of the initial centers. The main idea is to choose the centers one by one in a controlled fashion, where the current set of chosen centers will stochastically bias the choice of the next center. The details of k-means++ are presented in Algorithm 1. The sampling probability for a point is decided by the distance between the point and the center set (Line 3). The distances are considered as weights for sampling. After an iteration, the weights should be

---

Algorithm 1. k-means++ initialization

Require: $\mathcal{X}$: the set of points, k: number of centers;
1: $C \leftarrow$ sample a point uniformly at random from $\mathcal{X}$
2: **while** $|C| \leq k$ **do**
3:     Sample $x \in \mathcal{X}$ with probability $\frac{d^2(x,c)}{\phi_{\mathcal{X}}(C)}$
4:     $C \leftarrow C \bigcup x$
5: **end while**

---

changed because of the new centers added into the center set. The advantage of k-means++ is that the initialization step itself obtains an $(8 \log k)$-approximation to the optimization solution in expectation. However, its inherent sequential nature makes it unsuitable for massive data set.

The second variant of k-means is k-means $\|$, which is the underlying algorithm for the scalable k-means++ in [6]. The MapReduce implementation of k-means $\|$ is given in [6], which is called the scalable k-means++. The details of k-means $\|$ are presented in Algorithm 2. The k-means $\|$ improves the parallelism of k-means++ by selecting l centers at one iteration. The k-means $\|$ picks an initial center and computes the initial cost of the clustering. It then proceeds in $\log \psi$ iterations. In each iteration, given the current set of centers C, it samples each x with probability $\frac{l \cdot d^2(x,c)}{\phi_{\mathcal{X}}(C)}$ and obtains l new centers. The sampled centers are then added to C, the quantity $\phi_{\mathcal{X}}(C)$ updated, and the iteration continued.

Algorithm 2. k-means $\|$ initialization

Require: $\mathcal{X}$: the set of points, l: the number of centers sampled for a time, k: number of centers;
1: $C \leftarrow$ sample a point uniformly at random from $\mathcal{X}$
2: $\psi \leftarrow \phi_{\mathcal{X}} C$
3: **for** $o(\log \psi)$ **do**
4:     Sample C' each point $x \in \mathcal{X}$ with probability $\frac{l \cdot d^2(x,C)}{\phi_{\mathcal{X}}(C)}$
5:     $C \leftarrow C \bigcup C'$
6: **end for**
7: For $x \in C$, set w_x to be the number of points in $\mathcal{X}$ closer to x than any other point in C
8: Recluster the weighted points in C into k

---

3   Dimension Hierarchical Decomposition Tree

In this section, we present a dimension hierarchical decomposition tree index on time series.

3.1   Definitions

A *time series* $x = (x_1, \ldots, x_n)$ is a sequence of values. Without loss of generality, in this paper we assume that every time series has a value at every time instant $t = 1, 2, \ldots, n$. We denote by $|x| = n$ the *length* of the time series x.

We know that a time series is made up of continuous instants. To dimension reduction purpose, we define $dc(i, j)$ as the continuous dimensions in a time series from i to j. For example, $dc(7, 16)$ represents a dimension composition, which means the dimensions from 7 to 16. We define a synopsis for a dimension composition as the mean value of the included instants in the dimension composition. That is, for a given time series $x = (x_1, \ldots, x_n)$ and a dimension composition $dc(i, j)$, the mean of x on dc is defined as $\mu_{dc}(x) = \frac{(x_i + \ldots + x_j)}{sizeof(dc)}$. Here i and j are the first and last dimensions of the dimension composition. $sizeof(dc)$ means the count of the dimensions in dc (Fig. 1).

3.2   DHD Representation

We propose a novel representation of time series, denoted as DHD representation, by using a dimension hierarchical decomposition. We first introduce the dimension composition tree (DCTree, for short), which is illustrated in Fig. 3. The possible dimension compositions are organized in a binary tree called DCTree. Each node of a DCTree represents a dimension composition. The dimensions in root D_1 includes all the dimensions from 1 to n, denoted as $dc(1, n)$. Its children D_2 and D_3 include half of the dimensions in D_1, which are denoted as $dc(1, \lfloor \frac{n}{2} \rfloor)$ and $dc(\lfloor \frac{n}{2} \rfloor + 1, n)$ respectively. By this way, we can acquire more refined dimension composition by splitting nodes D_2 and D_3. The dimension composition level h is specified by the depth of the DCTree in Fig. 3.

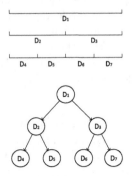

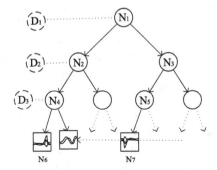

Fig. 1. DCTree   **Fig. 2.** DHDTree

The DHD representation of a time series x, denoted as $DHD(x)$ is defined as the following:

Definition 1. (DHD representation of a time series)

$$DHD_h(x) = \{\mu_{D_1}(x), \ldots, \mu_{D_{2^h - 1}}(x)\} \tag{2}$$

where h is the depth of the DCTree for the given dimension n, $\mu_{D_i} = \frac{\Sigma_{t \in D_i} x_t}{sizeof(D_i)}$,
$1 \leq i \leq 2^h - 1$.

For two time series ts_1 and ts_2, the corresponding DHD representations are

$$DHD_h(ts_1) = \{\mu_{D_1}(ts_1), ..., \mu_{D_{2^h-1}}(ts_1)\} \tag{3}$$

$$DHD_h(ts_2) = \{\mu_{D_1}(ts_2), ..., \mu_{D_{2^h-1}}(ts_2)\} \tag{4}$$

We define the DHD distance as:

$$dist_{dhd}(ts_1, ts_2) = D(DHD_h(ts_1), DHD_h(ts_2)) \tag{5}$$

Here D is the Euclidean distance.

3.3   DHDTree

DHDTree is an extension of k-d tree on time series. The main difference between DHDTree and k-d tree is that k-d tree selects only one dimension for the node splitting. While DHDTree selects a dimension composition. As illustrated in Fig. 2, DHDTree organizes the time series into a hierarchy. A DHDTree is built on the DHD representations of a collection of time series, instead of on the original data set. The node capacity ψ is specified as the maximum number of time series in a node. DHDTree partitions the data in a top-down way. At first, all time series are assigned to the root node N_1. If the number of the time series in N_1 exceeds the node capacity ψ, N_1 is split into two children N_2 and N_3. DHDTree uses a novel splitting strategy by choosing a dimension composition from the corresponding DCTree. In Fig. 2, a circle represents an internal node, and a rectangle represents a leaf node. The dotted circle represents a dimension composition chosen for the node splitting from the corresponding DCTree. The details of the node splitting will be introduced in Sect. 3.5. There are two types of nodes: internal nodes and leaf nodes. Each node contains the following information.

1. The number C of the time series indexed in the subtree rooted at this node.
2. The identifiers of the time series indexed in the subtree rooted at this node.
3. The dimension composition dc for splitting if it is non-leaf node, which will be discussed in detail in Sect. 3.5.
4. Two pointers pointing to children nodes (only internal nodes).

3.4   DHDTree Construction

Given a set of time series $\mathcal{X}$, each of length n, and the node capacity is set to ψ, the construction of the DHDTree on $\mathcal{X}$ is depicted in Algorithm 3.

Algorithm 3. DHDTree Building

1: **Input:** a set of time series $\mathcal{X}$, node capacity ψ;
2: Calculate the height of DCTree h by $h = \left\lceil \lg \frac{sizeof(\mathcal{X})}{\psi} \right\rceil + 1$;
3: Construct the DCTree with the level d based on n
4: Calculate the DHD representation for each time series
5: Assign all time series to the root N_{root}
6: **if** $sizeof(\mathcal{X}) < \psi$ **then**
7:     return;
8: **end if**
9: Initialize N_{root} with $\mathcal{X}$
10: $D_{best} = $ BestSplitDimensionComposition();
11: Create two children nodes A and B for N_{root};
12: Sort the time series using the *quickSort* algorithm, time series x is represented by $\mu_{D_{best}}(x)$;
13: Calculate the median of the time series in N_{root} as the splitting value, denoted as *split*;
14: **for** each time series x in N_{root} **do**
15:     **if** $\mu_{D_{best}}(x) < split$ **then**
16:         Route x to the left child node A;
17:     **else**
18:         Route x to the right child node B;
19:     **end if**
20: **end for**
21: Build DHDTree for node A recursively;
22: Build DHDTree for node B recursively;

3.5   Node Splitting Strategies

DHDTree is designed to partition the time series collections utilizing a dimension hierarchical decomposition method and the clustering based on it can benefit from the data partitioning. We need to find a proper dimension composition for splitting a node to guarantee the best separation effect. In this subsection, we first demonstrate how to find the candidate splitting dimension compositions, and then present a measure to acquire the best dimension composition for splitting.

To alleviate the effect of the number of dimensions, we assign weights to all the nodes in the DCTree according to the level the node belongs to. The closer node from the root has a bigger weight value. At first, only D_1 can be selected for the splitting for the root node. The children nodes of the root marked D_1 as their parent splitting strategy (Here splitting strategy denotes a dimension composition presented by a node in Fig. 3). Recall that we use a DCTree to represent the dimension composition hierarchy. We propose a node splitting method based on the DCTree. We use a top-down method to select the dimension composition. As Fig. 3(a) shows, at first, we select the root D_1 in the DCTree as the splitting dimension composition for the root N_1. We use the *quickSort* algorithm to sort the time series according to the values of μ_{D_1} and calculate

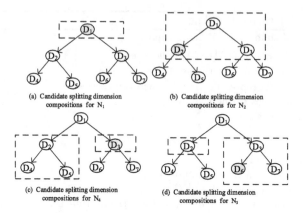

Fig. 3. Node Splitting

the median value as the splitting value to split the time series in N_1. The time series with μ_{D_1} greater than the splitting value are routed to the left child and the others are routed to the right child. Thus the time series in N_1 are routed into N_2 and N_3. For an internal node, the dimension composition of its parent, denoted as D_{parent}, and the children of D_{parent} are selected as the candidate dimension compositions. At the same time, the sibling of D_{parent} is added into the candidate lists. By this way, as Fig. 3(b) and (c) show, the dotted rectangles include the candidate dimension compositions. The blue ones are selected as the splitting dimension compositions. We choose the best splitting dc from the candidate dcs according to the separation degree (SD) defined below:

$$SD = \Sigma_{dhd \in N}(w_i * \sigma_{dc_i}(dhd)) \qquad (6)$$

$$w_i = \frac{sizeof(dc_i)}{sizeof(D_1)} \qquad (7)$$

Here, w_i is the weight of the dimension composition dc_i, which is defined as the ratio of the number of dimensions in dc_i against the number of all the dimensions in D_1; dhd is a DHD representation in Node N; $\sigma_{dc_i}(dhd)$ is the deviation of the dhd on the dc_i. There are five main steps when we split a DHDTree node into two children nodes:

1. Calculate the available candidate dimension compositions DC.
2. Calculate the separation degree SD of each candidate.
3. Find the $dc \in DC$ to maximize the SD.
4. Sort the time series in the node according to the dhd values of the dc and find the median dhd value as the splitting value.
5. Split the node into two parts by the splitting value.

3.6   Storage Analysis

For a data set with m time series, the height of the dimension composition tree (DCTree) is decided by $h = \left\lceil \lg \frac{m}{\psi} \right\rceil + 1$, here ψ is the node capacity, which denotes the maximum number of time series in a node. By adopting DHD representation, a time series is reduced to $2^h - 1$ dimensions. If storing a double needs 8 bytes, we need $8 \cdot m \cdot (2^h - 1)$ bytes to store the DHD representations. Concerning of the DHDTree, a DHDTree node needs to record the information of time series contained in the node, the splitting information and the tree information. We use a pointer to denote the time series and two position pointers to denote the beginning position and the ending position. Thus three 8 bytes pointer can record the time series information. We use an array of Integers to record the available dimension compositions and an integer to record the splitting dimension composition. Furthermore, we need two pointers to record the left child and the right child. Because we do not store the DHD representations and the real time series in DHDTree, the storage of the DHDTree is small compared with the original time series.

4   DHD-Clustering

In this section, we use DHDTree to cluster time series.

4.1   DHD-Clustering Overview

Similar with traditional k-means algorithm, DHD-Clustering is a clustering method based on centers. Given the set of time series $\mathcal{X}$, the number of clusters k, we define DHD-Clustering as the following.

1. Build DHDTree on $\mathcal{X}$, for each leaf node N_i of DHDTree, get the center c_{N_i}.
2. Acquire the initial k centers by clustering the centers of the DHDTree leaves.
3. For each x in $\mathcal{X}$, assign it to the cluster with the nearest center.
4. For each cluster, compute the new center.
5. Repeat 3 and 4 until ϕ keeps unchanged (optional).

The DHD-Clustering algorithm has two advantages. First, because DHDTree is a balanced tree, the seeding from the centers of the DHDTree leaves benefits from the real data distribution. Secondly, to reduce the computation cost during the iteration phase, we benefit from using DHD distance as the measure to acquire the nearest center for each time series in $\mathcal{X}$.

4.2   Seeding Phase

The seeding phase is to select the initial centers for center-based clustering methods, which is important for both the quality of clustering and the run-time. K-means++ [6] exploits the fact that a good clustering is relatively spread out. Based on it, we propose a seeding method by selecting the medians of the DHDTree nodes to serve as the candidate centers. Then we re-cluster the candidate centers to get the final k centers. Because DHDTree uses the dimension composition with the maximum sum of deviation to split a node, we can achieve the best separation effect by using the centers of the DHDTree leaves as the candidates. From the candidates, we use k-means++ to select k centers.

4.3 Iteration Phase

In the iteration phase of center-based clustering methods, all time series need to be compared with the centers and the nearest center is found for each time series. However, it is expensive to calculate the real Euclidean distance between time series and the centers if we have a big center number k. In the paper, we decrease the computation cost during the iteration phase of the DHD-clustering algorithm by adopting DHD distance as the measure to compare the distances. Notice that we calculate ϕ using the Euclidean distance instead of DHD distance. When DHD-Clustering converges, we can employ k-means to get a better result by utilization the final centers of DHD-Clustering as the input.

5 Related Work

Clustering problems have attracted interests of study for the past many years by data management and data mining researchers. The k-means algorithm keeps popular for its simplicity. Despite its popularity, k-means suffers several major shortcomings such as the need of specified k value and proneness of the local minima. There are many variants of naive k-means algorithm. Ordonez and Omiecinski [20] studied disk-based implementation of k-means, taking into account the requirements of a relational DBMS. The X-means [21] extends k-means with efficient estimation of the number of clusters. Joshua Zhexue Huang [14] proposes a k-means type clustering algorithm that can automatically calculate variable weights. Alsabti et al. [4] proposed an algorithm based on the data structure of the k-d tree and used a pruning function on the candidate centroid of a cluster. The k-d tree fulfills the space partitioning and a partition is treated a unit for processing. The processing in batch can reduce the computation substantially.

The k-means algorithm has also been considered in a parallel environment. Dhillon and Modha [11] considered k-means in the message-passing model, focusing on the speed up and scalability issues in this model. MapReduce [10] as a popular massive-scale parallel data analysis model gains more and more attention and a lot of enthusiasm in parallel computing communities. Hadoop [1] is a famous open-source implementation of MapReduce model. There are many applications on top of Hadoop. Mahout [19] is a famous Apache project which serves as a scalable machine learning libraries, including the k-means implementation on Hadoop. Robson L.F.Cordeiro [9] proposed a method to cluster multi-dimensional datasets with MapReduce. Yingyi Bu proposes HaLoop [8], which is a modified version of Hadoop, and gives the implementation of k-means algorithm on it. Ene et al. [13] considered the k-median problem in MapReduce and gave a constant-round algorithm that achieves a constant approximation.

D. Arthur and S. Vassilvitskii propose k-means++ [5], which can improve the initialization procedure. Scalable k-means++ [6] is proposed by Bahman Bahmani and Benjamin Moseley, which can cluster massive data efficiently.

Time series clustering has attracted a significant amount of effort in the last two decades. Jessica Lin. proposed I-kmeans [18], which is an iterative clustering of time series by leveraging off the the multi-resolution property of wavelets. Thanawin Rakthanmanon [22] proposed a meaningful clustering of subsequences from a time series stream based on a Minimum Description Length (MDL) framework. Except using Euclidean and DTW as the similarity measure, Jesin Zakaria. [15] proposed a clustering method using some local patterns, called shapelets as the similarity measure. However, the discovery of shapelets is computationally expensive Josif Grabocka. Liudmila Ulanova. [23] proposed a scalable clustering method using U-shapelets. Rui Ding. [12] proposed yarding, which uses L_1 norm as similarity measure and the multi-density approach as the clustering method.

6   Evaluation

In this section, we report experimental results to verify the efficiency and the scalability of our proposed DHD-Clustering.

6.1   Experimental Settings

Data Sets and Platform. We use 85 UCR real data sets [2], two UCI data sets and a synthetic data set for the evaluation.

- Stock: Stock data set is from the UCR archive, which contains 93 stock streams, each one of 3000 length.
- TAO: TAO data set is from the Tropical Atmosphere Ocean project [3], which includes 12218 streams, each one of 963 length.
- Synthetic: We generate 2000000 time series with length 512. The generation method is depicted below. We use Synthetic-50 to denote the first 500000 time series, Synthetic-100 to denote the first 1000000 time series and Synthetic-200 to denote the whole Synthetic data set.

The synthetic data set is a combination of four types of time series as follows.

- Random walk times series. The start point is picked randomly from range $[-5, 5]$ and the step length is chosen randomly in range $[0, 2]$;
- One-segment Gaussian time series. The values in the whole time series are picked from a Gaussian Distribution with mean value and standard deviation randomly selected in ranges $[-5, 5]$ and $[0, 2]$, respectively;
- Multi-segment Gaussian time series. Such a time series is concatenated by multiple one-segment Gaussian time series. The number of segments is randomly set between 3 to 10.

– A mixed sine time series. Each time series is a mixture of several sine waves whose period is randomly set in range $[2, 10]$, amplitude is randomly set in range $[2, 10]$, and mean value randomly chosen in range $[-5, 5]$.

To generate a time series, the synthetic data generator first randomly chooses a type, and then picks the corresponding parameters randomly to generate the time series.

The experimental platform includes a machine with 64 Cores (1.7 GHz) CPU and 64 GB of memory.

Baselines and Metrics. We use I-kmeans [18], k-means++ and k-means $\|$ as the baseline methods.

We use two metrics to evaluate clustering accuracy. For UCR datasets, we use "Rand Index" to evaluate clustering accuracy over the fused training and test sets of each datasets. The reason is that UCR data sets have the class labels. The "Rand Index" metric is related to the classification accuracy and is defined as $R = \frac{TP+TN}{TP+TN+FP+FN}$, where TP denotes the number of time series pairs that belong to the same class and are assigned to the same cluster, TN denotes the number of time series pairs that belong to different classes and are assigned to different clusters, FP denotes the number of time series pairs that belong to different classes but are assigned to the same cluster, and FN denotes the number of time series pairs that belong to the same class but are assigned to different clusters.

For other datasets without class labels, we use ϕ defined in Eq. 1 to evaluate clustering accuracy. The less ϕ is, the better is clustering quality.

6.2   Performance of Clustering

First, we compare our approach on 85 UCR datasets with I-kmeans [18], k-means++ and k-means $\|$. The statistical results are represented in Table 1. In Table 1, Columns ">", "l=", and "<" denote the number of datasets over which

Table 1. Comparison against baselines

Algorithm	>	<	=	Rand index	Runtime (s)
I-kmeans (level=4)	29	54	2	0.679	24.4
I-kmeans (level=5)	30	51	4	0.669	25.2
k-means++	35	47	3	0.677	29.44
k-means $\|$	33	47	5	0.676	29.36
DHD-Clustering (h=8)	-	-	-	0.684	20.47
DHD-Clustering (h=6)	35	40	10	0.681	15.47
DHD-Clustering (h=4)	31	44	10	0.682	13.49

Table 2. Experimental results of clustering on synthetic dataset

	Average ϕ			Minimum ϕ			Converge time T		
k	k-means++	k-means ||	DHD	k-means++	k-means ||	DHD	k-means++	k-means ||	DHD
100	11356	11683	11234	11321	11519	11143	336.9	339.2	77.6
200	10920	10926	10813	10871	10898	10734	728.2	538.4	165.2
500	10432	10689	10424	10401	10518	10311	2751.5	1610.2	315.4

Table 3. Experimental results of clustering on original stock dataset

	Average ϕ			Minimum ϕ			Converge time T		
k	k-means++	k-means ||	DHD	k-means++	k-means ||	DHD	k-means++	k-means ||	DHD
5	9251	9207	9200	9070	9070	9044	0.11	0.09	0.06
10	7736	7726	7686	7508	7709	7400	0.11	0.12	0.08
15	6721	6570	6500	6595	6381	6244	0.12	0.13	0.09

Table 4. Experimental results of clustering on original TAO dataset

	Average ϕ			Minimum ϕ			Converge time T		
k	k-means++	k-means ||	DHD	k-means++	k-means ||	DHD	k-means++	k-means ||	DHD
100	9259	9287	9184	10494	9230	9164	81.1	171.3	23.1
200	9115	9107	9023	9109	9089	9020	245.4	393.3	29.4
500	8914	9127	8837	8898	8935	8833	490.2	417.2	69.5

an algorithm is better, equal, or worse, respectively, in comparison to DHD-Clustering (The default parameter h is set to 8). "Rand Index" denotes the average accuracy achieved in the 85 datasets.

The previous evaluation focuses on the comparisons for the whole UCR datasets. To evaluate the clustering quality and time efficiency for single dataset, we use two UCI datasets (Stocks and Tao) compared with k-means++ and k-means \|\|. Both the running time and the quality of the clustering are reported in Table 2, 3 and 4. Because UCI datasets have not the class label, we use average ϕ to measure the clustering quality. We see that DHD-Clustering has the best clustering quality as well as the best time efficiency.

6.3 Varying Node Capacity ψ and Height of DCTree h

In the paper, we use the medians of the leaf nodes serving as the candidates for the clustering seeding. A small capacity ψ leads to a big number of leaves. Smaller capacity leads to a better clustering quality. However, the clustering efficiency decreases because we need to acquire the centers for more candidates. We use TAO data set to evaluate the effect of the capacity parameter ψ. The results are illustrated in Table 5. We vary the capacity ψ and the decomposition level h and record the quality ϕ and running time. From Table 5, we conclude that the bigger the height of DCTree h, the better quality we can acquire. The reason is that the number of the reduced dimensions is defined as $2^h - 1$. The bigger h is, the more accurate we compute the distances. It is easy to understand that it takes more time when h increases.

Table 5. Varying parameter ψ and h

ψ	$h=2$	$h=4$	$h=6$	$h=8$
10	5315/1.5	5062/1.6	5035/3.0	5033/43.3
20	5325/1.5	5052/6.5	5034/2.9	5023/45.6
50	5324/1.4	5072/1.4	5052/3.5	5023/15.4
100	5326/1.1	5063/1.4	5047/3.8	5039/36.1
200	5342/1.0	5149/1.3	5108/2.9	5119/10.1

6.4 Scalability

To test the scalability of DHD-Clustering, we use three synthetic data sets with the sizes of 0.5 million, 1 million and 2 million respectively, each of length 512. We vary the k number with 100, 200 and 500. The experimental results are shown in Fig. 4, with the increase of the number of records, the time increases linearly. As Fig. 4 shows, DHD-Clustering has good scalability.

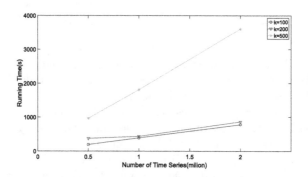

Fig. 4. Scalability of DHD-Clustering

7 Conclusion

We propose a novel Dimension Hierarchical Decomposition (DHD for short) method to represent time series. Based on DHD representation, we propose DHDTree to reorganize the time series collections to achieve the best separation effect. Furthermore, we propose a DHD-Clustering method to improve both the seeding and the iteration of center-based clustering method by utilizing the DHDTree to partition the time series collections and achieve the best separation effect. Both the real and synthetic datasets verify the efficiency of the DHD-Clustering algorithm.

References

1. http://hadoop.apache.org/
2. http://www.cs.ucr.edu/~eamonn/time_series_data/
3. http://www.pmel.noaa.gov/tao/data_deliv/
4. AlSabti, K., Ranka, S., Singh, V.: An efficient space-partitioning based algorithm for the K-means clustering. In: Zhong, N., Zhou, L. (eds.) PAKDD 1999. LNCS (LNAI), vol. 1574, pp. 355–360. Springer, Heidelberg (1999). doi:10.1007/3-540-48912-6_47
5. Arthur, D., Vassilvitskii, S.: k-means++: the advantages of careful seeding. In: Proceedings of the Eighteenth Annual ACM-SIAM Symposium on Discrete Algorithms, SODA, New Orleans, Louisiana, USA, January 7–9, pp. 1027–1035 (2007)
6. Bahmani, B., Moseley, B., Vattani, A., Kumar, R., Vassilvitskii, S.: Scalable k-means++. PVLDB 5(7), 622–633 (2012)
7. Bentley, J.L.: K-d trees for semidynamic point sets, pp. 187–197 (1990)
8. Yingyi, B., Howe, B., Balazinska, M., Ernst, M.: Haloop: efficient iterative data processing on large clusters. PVLDB 3(1), 285–296 (2010)
9. Cordeiro, R.L.F., Traina Jr. C., Traina, A.J.M., López, J., Kang, U., Faloutsos, C.: Clustering very large multi-dimensional datasets with mapreduce. In: KDD, pp. 690–698 (2011)
10. Dean, J., Ghemawat, S.: Mapreduce: simplified data processing on large clusters. In: OSDI (2004)
11. Dhillon, I.S., Modha, D.S.: A data-clustering algorithm on distributed memory multiprocessors. In: Large-Scale Parallel Data Mining, pp. 245–260 (1999)
12. Ding, R., Wang, Q., Dang, Y., Qiang, F., Zhang, H., Zhang, D.: YADING: fast clustering of large-scale time series data. PVLDB 8(5), 473–484 (2015)
13. Ene, A., Im, S., Moseley, B.: Fast clustering using mapreduce. In: KDD, pp. 681–689 (2011)
14. Huang, J.Z., Ng, M.K., Rong, H., Li, Z.: Automated variable weighting in k-means type clustering. IEEE Trans. Pattern Anal. Mach. Intell. 27(5), 657–668 (2005)
15. Koperski, K., Han, J.: Discovery of spatial association rules in geographic information databases. In: Egenhofer, M.J., Herring, J.R. (eds.) SSD 1995. LNCS, vol. 951, pp. 47–66. Springer, Heidelberg (1995). doi:10.1007/3-540-60159-7_4
16. Kriegel, H.-P., Kröger, P., Zimek, A.: Clustering high-dimensional data: a survey on subspace clustering, pattern-based clustering, and correlation clustering. TKDD 3(1), 1–58 (2009)
17. Kuo, H.-C., Lee, T.-L., Huang, J.-P.: Cluster analysis on time series gene expression data. IJBIDM 5(1), 56–76 (2010)
18. Lin, J., Vlachos, M., Keogh, E., Gunopulos, D.: Iterative incremental clustering of time series. In: Bertino, E., Christodoulakis, S., Plexousakis, D., Christophides, V., Koubarakis, M., Böhm, K., Ferrari, E. (eds.) EDBT 2004. LNCS, vol. 2992, pp. 106–122. Springer, Heidelberg (2004). doi:10.1007/978-3-540-24741-8_8
19. Mondal, A., Lifu, Y., Kitsuregawa, M.: P2PR-tree: an R-tree-based spatial index for peer-to-peer environments. In: EDBT Workshops, pp. 516–525 (2004)
20. Ordonez, C., Omiecinski, E.: Efficient disk-based k-means clustering for relational databases. IEEE Trans. Knowl. Data Eng. 16(8), 909–921 (2004)
21. Pelleg, D., Moore, A.W.: X-means: Extending k-means with efficient estimation of the number of clusters. In: ICML, pp. 727–734 (2000)
22. Rakthanmanon, T., Keogh, E.J., Lonardi, S., Evans, S.: Mdl-based time series clustering. Knowl. Inf. Syst. 33(2), 371–399 (2012)
23. Ulanova, L., Begum, N., Keogh, E.J.: Scalable clustering of time series with u-shapelets. In: Proceedings of the SIAM International Conference on Data Mining, Vancouver, BC, Canada, April 30–May 2, pp. 900–908 (2015)

Data Mining

Fast Extended One-Versus-Rest Multi-label SVM Classification Algorithm Based on Approximate Extreme Points

Zhongwei Sun, Zhongwen Guo$^{(\boxtimes)}$, Xupeng Wang, Jing Liù, and Shiyong Liu

Department of Computer Science and Technology, Ocean University of China,
Qingdao, China
guozhw@ouc.edu.cn

Abstract. In large-scale multi-label classification framework, applications of non-linear kernel extended one-versus-rest multi-label support vector machine (OVR-ESVM) classification algorithm are severely restricted by excessive training time. To deal with this problem, we improve the OVR-ESVM classification algorithm and propose fast OVR-ESVM classification algorithm based on approximate extreme points (AEML-ESVM). The AEML-ESVM classification algorithm integrates the advantages of OVR-ESVM classification algorithm and binary approximate extreme points support vector machine (AESVM) classification algorithm. In other words, it can not only shorten the training time greatly, but also reflect label correlation of individual instance explicitly. Meanwhile, its classification performance is similar to that of the OVR-ESVM classification algorithm. Experiment results on three public data sets show that AEML-ESVM classification algorithm can substantially reduce training time and its classification performance is comparable with that of the OVR-ESVM classification algorithm. It also outperforms existing fast multi-label SVM classification algorithms in both training time and classification performance.

Keywords: Support vector machine · Multi-label classification · Approximate extreme points · Label correlation · Non-linear kernel

1 Introduction

In the field of machine learning and data mining, multi-label classification is an important research topic. Compared with the traditional classification problem, in the multi-label classification problem, each individual example is also represented by one instance. However, every instance is associated with more than one label, thus these labels are no longer mutually exclusive [1]. In recent years, multi-label classification has been applied successfully in many fields, such as text classification [2], video and image semantic annotation [3,4], medical prediction [5], music emotion classification [6] and so on. Until now, researchers have put forward many multi-label classification methods, such as methods based on support

© Springer International Publishing AG 2017
S. Candan et al. (Eds.): DASFAA 2017, Part I, LNCS 10177, pp. 265–278, 2017.
DOI: 10.1007/978-3-319-55753-3_17

vector machine, neutral network, decision tree, nearest neighbor and so on [7]. Simultaneously, label correlation has also been utilized in many multi-label classification algorithms to improve the classification performance. Furthermore, label correlation has been utilized in three levels, that is, individual instance, partial instances and different labels [8]. But in the real world, multi-label classification methods are implemented on large-scale data sets which makes many existing multi-classification algorithms fail to work well, especially in algorithms based on SVM [9]. The main reason of this phenomenon is that many algorithms are restricted by excessive training time.

In the multi-label classification, OVR-ESVM [8] classification algorithm has been widely used. However, due to excessive training time, it is restricted from being used on large-scale data sets. In the real world, many data sets are nonlinear. To get better classification performance, OVR-ESVM classification algorithm adopts non-linear kernels and this further restricts it from being used on large-scale data sets.

To solve the problem that OVR-ESVM encounters when it is used on the large-scale data sets, we combine OVR-ESVM with binary AESVM [10] classification algorithm to construct a new one-versus-rest multi-label SVM classification algorithm, that is, AEML-ESVM classification algorithm. It can reduce training time greatly while exploiting the label correlation of individual instance directly. Experiment results on three public data sets demonstrate that compared with existing multi-label classification algorithms: OVR-ESVM, RCP (Ranking by pairwise comparison)-SVM [11] and ML-CVM (Multi-label, Core vector machine) [12], our algorithm spends the least training time but achieves a similar classification performance compared with OVR-ESVM classification algorithm on the five evaluation metrics. In addition, its classification performance also outperforms RCP-SVM and ML-CVM classification algorithms on the five evaluation metrics.

The rest of this paper is organized as follows. Firstly, some related works are introduced in Sect. 2. Secondly, the new AEML-ESVM classification algorithm is proposed in Sect. 3. After that, the analysis of the experiment results is presented in Sect. 4. Finally, the paper is summarized in Sect. 5.

2   Related Work

In the past few decades, many multi-label classification methods have been proposed and used widely. These algorithms can be categorized into two classes, one class is based on problem transformation strategy and the other is based on algorithm adaptation strategy [13].

Problem transformation strategy is to solve the multi-label classification problem by transforming it into a few single-label classification problems. So multi-label classification algorithms based on problem transformation strategy implement multi-label classification by combining problem transformation trick with existing binary classification methods. Problem transformation trick mainly includes OVR (One-Versus-Rest), OVO (One-Versus-One), OBO (One-By-One), LP (Label Powerset) and so on [8]. Among them, OVR, OVO and OBO indirectly reflect label correlation of individual instance by reusing training instances and LP directly reflects label correlation of partial instances by considering possible label combinations. Common-used binary classification methods mainly include SVM, CVM, neural network, decision tree, nearest neighbor and so on [7].

OVR transformation trick transforms multi-label data set with q labels into q binary data subsets. It is worth noting that the i-th binary data subset consists of the positive instances that are labeled with the i-th label and negative ones that are labeled with other labels. In [14], it presents three problems of OVR transformation trick. Firstly, the correlations and interdependences among labels are neglected as OVR transformation trick presumes that the labels are independent. Secondly, OVR transformation trick may lead to data imbalance. Lastly, if the number of labels is too large, the data imbalance problem will deteriorate and the number of classifiers will increase. Despite of these problems, OVR transformation trick is simple and practical, and the data set is reversible. Firstly, it can use any existing binary classifier as base classifier to implement multi-label classification. Secondly, its complexity is linearly with respect to the number of labels. Thirdly, it can be easily parallelized. Finally, its main advantage is that it can optimize several loss functions [15]. The popular OVR transformation trick is adopted to implement multi-label classification in this paper. In addition, the OVR transformation trick's shortcoming of neglecting label correlation is overcome in our proposed algorithm.

Algorithm adaptation strategy solves the multi-label classification problem by modifying existing classification algorithms. Algorithms based on algorithm adaptation strategy include multi-label classification algorithm of C4.5 type [16], Rank-SVM classification algorithm [17], Rank-CVM classification algorithm [18], BP-MLL classification algorithm [19] and so on. Algorithm of C4.5 type can directly reflect label correlation of partial instances by considering possible label combinations which is similar to the LP transformation trick. Rank-SVM, Rank-CVM and BP-MLL all reflect label correlation of individual instance by using an approximate expression of ranking loss. They are mainly used on small-scale data sets.

To sum up, although many multi-label classification algorithms are widely used and they have utilized the label correlation as far as possible, they are restricted from being used on large-scale data sets to some extent, especially in algorithms based on SVM. The proposed AEML-ESVM classification algorithm in this paper can solve this problem efficiently. It can not only shorten the training time greatly but also achieve the comparable classification performance to that of the OVR-ESVM. Furthermore, it also further improves the algorithm's classification performance by exploiting label correlation of individual instance explicitly.

3   Fast OVR-ESVM Classification Algorithm Based on Approximate Extreme Points

In this section, we will first introduce binary AESVM. Secondly, we will improve OVR-ESVM with AESVM and put forward the extended one-versus-rest multi-label support vector machine based on approximate extreme points (AEML-ESVM). Then we will introduce the design and implementation of AEML-ESVM classification algorithm. Finally, we will analyze the time and space complexity.

3.1   Binary Approximate Extreme Points SVM

Assume that there is a binary classification training data set X which contains N instances, that is, $X = \{x_i : x_i \in R^D, i = 1, 2, ..., N\}$ and its corresponding label set is $Y = \{y_i : y_i \in [-1, 1], i = 1, 2, ..., N\}$.

Manu Nandan [10] et al. propose a binary approximate extreme points support vector machine (AESVM) classification algorithm. It firstly selects representative set from the training data set. The representative set is $X^* = \{x_t : x_t \in R^D, t = 1, 2, ..., M\}$ and its corresponding label set is $Y^* = \{y_t : y_t \in [-1, 1], t = 1, 2, ..., M\}$, $M \ll N$. Then the SVM classification algorithm is trained on the representative set. Its primal optimization problem is transformed into the following unconstrained optimization problem.

$$\min_{w,b} F(w, b) = \frac{1}{2}\|w\|^2 + C \sum_{t=1}^{M} \beta_t l(w, b, \Phi(x_t)), \tag{1}$$

Here $l(w, b, \Phi(x_t)) = max\{0, 1 - y_t(w^T\Phi(x_t) + b)\}, \forall x_t \in X^*$, $\Phi : R^D \longrightarrow H, b \in R$ and $w \in H$, H a Hilbert space.

It should be noted that $l(w, b, \Phi(x_t))$ is the hinge loss of the training instance x_t, $\|w\|^2$ is used to reflect the model's complexity, C is used to balance the model's complexity and the sum of losses of the training data set [9], β_t is related to the approximate extreme points method.

To get the representative set from the training data set, Manu Nandan et. al put forward the DeriveRS algorithm which is based on the approximate extreme points technology and its time complexity is linear. And they demonstrate that binary AESVM classification algorithm can enhance the training speed greatly and get the comparable classification performance to that of the traditional SVM classification algorithm. But the existing binary AESVM cannot be used directly in the multi-label classification field.

3.2   The Extended One-Versus-Rest Multi-label SVM Based on Approximate Extreme Points

Assume that there is a multi-label training data set S which contains N instances and $S = \{x_i : x_i \in R^D, i = 1, 2, ..., N\}$. Its corresponding label set is $L = \{Y_i \subseteq Q, i = 1, 2, ..., N\}$. Among them, $Q = \{1, 2, ..., k\}$ is the label set and k indicates the number of labels. $\overline{Y_i} = Q \backslash Y_i$ is the label complement of the i-th training instance. For convenience, we define a binary label vector $\boldsymbol{y_i} = [y_{i1}, y_{i2}, ..., y_{ik}]^T$ for training instance $\boldsymbol{x_i}$. And if $k \in Y_i$, then $y_{ik} = +1$, otherwise $y_{ik} = -1$.

Xu Jianhua [8] et al. propose the extended one-versus-rest multi-label support vector machine (OVR-ESVM) classification algorithm in which a new approximate ranking loss is defined and the formula is as follows.

$Approximate\ ranking\ loss =$

$$\frac{1}{2N} \sum_{i=1}^{N} [\frac{1}{|Y_i||\overline{Y_i}|} \sum_{(m,n)\in(Y_i \times \overline{Y_i})} (l(\boldsymbol{w_m}, b_m, \Phi(\boldsymbol{x_i})) + l(\boldsymbol{w_n}, b_n, \Phi(\boldsymbol{x_i})))], \tag{2}$$

Here $l(\boldsymbol{w_m}, b_m, \Phi(\boldsymbol{x_i})) = max\{0, 1 - y_{im}(\boldsymbol{w_m^T}\Phi(\boldsymbol{x_i}) + b_m)\}, l(\boldsymbol{w_n}, b_n, \Phi(\boldsymbol{x_i})) = max\{0, 1 - y_{in}(\boldsymbol{w_n^T}\Phi(\boldsymbol{x_i}) + b_n)\}, \forall \boldsymbol{x_i} \in S, \Phi : R^D \longrightarrow H, b_m \in R, b_n \in R, \boldsymbol{w_m} \in H, \boldsymbol{w_n} \in H, H$ a Hilbert space.

Approximate ranking loss is actually the upper bound of ranking loss. Xu Jianhua et al. construct the OVR-ESVM by using approximate ranking loss as the empirical loss term and the optimization formula of the OVR-ESVM is as follows.

$$min\frac{1}{2} \sum_{j=1}^{k} \boldsymbol{w_j^T}\boldsymbol{w_j} + $$
$$kC \sum_{i=1}^{N} [\frac{1}{|Y_i|} \sum_{m\in Y_i} l(\boldsymbol{w_m}, b_m, \Phi(\boldsymbol{x_i})) + \frac{1}{|\overline{Y_i}|} \sum_{n\in\overline{Y_i}} l(\boldsymbol{w_n}, b_n, \Phi(\boldsymbol{x_i}))] \tag{3}$$

Here $l(\boldsymbol{w_j}, b_j, \Phi(\boldsymbol{x_i})) = max\{0, 1 - y_{ij}(\boldsymbol{w_j^T}\Phi(\boldsymbol{x_i}) + b_j)\}, \forall \boldsymbol{x_i} \in S, \Phi : R^D \longrightarrow H, b_j \in R, \boldsymbol{w_j} \in H, j \in Q, H$ a Hilbert space.

OVR-ESVM introduces approximate ranking loss as the empirical loss term and thus it can exploit the label correlation of individual instance directly. However, its use on large-scale data set is severely restricted by excessive training time.

So we combine the advantages of AESVM and OVR-ESVM to propose the AEML-ESVM classification algorithm. Firstly, it uses the approximate extreme points technique to obtain representative sets from the multi-label training data set. Then the OVR-ESVM classification algorithm is trained on the representative sets. Based on formula (2), we propose new approximate ranking loss as an empirical loss term. Its formula is as follows.

$New\ approximate\ ranking\ loss =$

$$\frac{1}{2M} \sum_{t=1}^{M} [\frac{1}{|Y_t||\overline{Y_t}|} \sum_{(m,n)\in(Y_t \times \overline{Y_t})} (l(\boldsymbol{w_m}, b_m, \Phi(\boldsymbol{x_t})) + l(\boldsymbol{w_n}, b_n, \Phi(\boldsymbol{x_t})))], \tag{4}$$

Here $l(\boldsymbol{w_m}, b_m, \Phi(\boldsymbol{x_t})) = max\{0, 1 - y_{tm}(\boldsymbol{w_m^T}\Phi(\boldsymbol{x_t}) + b_m)\}, l(\boldsymbol{w_n}, b_n, \Phi(\boldsymbol{x_t})) = max\{0, 1 - y_{tn}(\boldsymbol{w_n^T}\Phi(\boldsymbol{x_t}) + b_n)\}, \forall \boldsymbol{x_t} \in S^*, \Phi : R^D \longrightarrow H, b_m \in R, b_n \in R, \boldsymbol{w_m} \in H, \boldsymbol{w_n} \in H, H$ a Hilbert space.

It should be noted that S^* represents the representative set of the training data set S and M is the size of the representative set S^*. New approximate ranking loss is also actually the upper bound of ranking loss. We construct the AEML-ESVM by using new approximate ranking loss as the empirical loss term and the optimization formula of the AEML-ESVM is as follows.

$$
min\frac{1}{2}\sum_{j=1}^{k}\boldsymbol{w}_j^T\boldsymbol{w}_j+
$$

$$
kC\sum_{t=1}^{M}\beta_t[\frac{1}{|Y_t|}\sum_{m\in Y_t}l(\boldsymbol{w}_m,b_m,\Phi(\boldsymbol{x}_t))+\frac{1}{|\overline{Y_t}|}\sum_{n\in\overline{Y_t}}l(\boldsymbol{w}_n,b_n,\Phi(\boldsymbol{x}_t))]
$$

(5)

Here $l(\boldsymbol{w}_j,b_j,\Phi(\boldsymbol{x}_t))=max\{0,1-y_{tj}(\boldsymbol{w}_j^T\Phi(\boldsymbol{x}_t)+b_j)\},\forall\boldsymbol{x}_t\in S^*,\Phi:R^D\longrightarrow H,b_j\in R,\boldsymbol{w}_j\in H,j\in Q,H$ a Hilbert space.

After analyzing the optimization problem of AEML-ESVM, we can get a conclusion that this optimization problem is decoupled among different labels. It means that it can be transformed into k sub-problems by using the OVR problem transformation strategy and each sub-problem can be solved by using improved binary SVM. More specifically, it firstly transforms the multi-label training data set S into k independent binary training data subsets using OVR problem transformation strategy, i.e., $S_j(j\in Q)$. Then it can obtain the representative set for each binary training data subset by using the approximate extreme points technique and $S_j^*(j\in Q)$ represents the representative set of the j-th binary training data subset S_j. Finally, the improved binary SVM is trained on each representative set. We construct the following improved binary SVM formula to solve the sub-problem.

$$
min\frac{1}{2}\boldsymbol{w}_j^T\boldsymbol{w}_j+
$$

$$
kC\sum_{t=1}^{|S_j^*|}\beta_t[\frac{1}{|Y_t|}\sum_{j\in Y_t}l(\boldsymbol{w}_j,b_j,\Phi(\boldsymbol{x}_t))+\frac{1}{|\overline{Y_t}|}\sum_{j\in\overline{Y_t}}l(\boldsymbol{w}_j,b_j,\Phi(\boldsymbol{x}_t))]
$$

(6)

Here $l(\boldsymbol{w}_j,b_j,\Phi(\boldsymbol{x}_t))=max\{0,1-y_{tj}(\boldsymbol{w}_j^T\Phi(\boldsymbol{x}_t)+b_j)\},\forall\boldsymbol{x}_t\in S_j^*,\Phi:R^D\longrightarrow H,b_j\in R,\boldsymbol{w}_j\in H,j\in Q,H$ a Hilbert space.

It should be noted that $|S_j^*|$ is the size of the representative set S_j^*. The AEML-ESVM method adopts the following decision function, the formula is as follows.

$$
Y=\{j,s.t.f_j(\boldsymbol{x})\geq 0,j\in Q\}
$$

(7)

Avoiding to get a null relevant label set, the following rule is used [20].

$$
Y=\{j,s.t.\max f_j(\boldsymbol{x}),f_j(\boldsymbol{x})<0,j\in Q\},
$$

(8)

In a word, the AEML-ESVM classification algorithm integrates the advantages of AESVM and OVR-ESVM classification algorithm. It can substantially shorten the training time as well as obtain the comparable classification performance to that of the OVR-ESVM classification algorithm. Additionally, it also utilizes the label correlation of individual instance directly.

3.3   AEML-ESVM Algorithm

The proposed AEML-ESVM classification algorithm firstly calculates the number of relevant label for each instance, and then adopts OVR problem transformation strategy to transform the multi-label data set into k binary data subsets. Then ModDeriveRS algorithm is proposed on the basis of the rewritten and slightly modified DeriveRS algorithm. It can be effectively used to get the representative set from each of the binary data subsets. Then the number of the relevant labels of each training instance in the representative sets is found in the original multi-label data set. After that, different values are set to the empirical loss term according to whether the training instance is positive instance or not. Finally, modified algorithm of the most popular LIBSVM [21] is used to deal with the representative sets and k binary classifiers are obtained. Formulas (7) and (8) are used to integrate the results of the k binary classifiers to realize fast multi-label classification. The reason why LIBSVM is used is that it can implement the SMO algorithm excellently [22]. Basing on the introduction of the binary AESVM and OVR-ESVM in the above two subsections, it can be expected that AEML-ESVM classification algorithm with non-linear kernel can be effectively applied to large-scale multi-label data sets. In other words, it can not only reduce the training time greatly while exploiting the label correlation of individual instance directly. In addition, it can also achieve similar classification performance with that of the OVR-ESVM classification algorithm.

The pseudocode of AEML-ESVM algorithm is shown in Algorithm 1. In Algorithm 1, $|Y_i|$ represents the number of labels of the i-th training instance, S_j^* represents the representative set of the j-th training subset S_j, $|S_j^*|$ represents the size of the representative set S_j^*, β_j is a constant related with the approximate extreme points method, f_j represents the prediction model for the j-th label trained on the representative set S_j^*, $f_j(\boldsymbol{x})$ represents the prediction value of the j-th prediction model for the instance $\boldsymbol{x}$.

3.4   Complexity Analysis of AEML-ESVM Algorithm

The time complexity of AEML-ESVM in the calculation of representative set is $O(km)$. SMO algorithm is used in the optimization of SVM [22]. So the training time complexity of AEML-ESVM is between $O(km)$ and $O(km^{2.2})$. Its space complexity is $O(km)$. The symbol k represents the number of labels and the symbol m is the size of the training data set. We can conclude that AEML-ESVM implements fast training mainly by reducing the size of training data set.

4   Experiments

In this section, we will compare AEML-ESVM with three existing multi-label classification algorithms. Before presenting our experimental results, we will introduce briefly these three multi-label classification methods, three large-scale multi-label data sets and five common-used multi-label classification evaluation metrics.

$Y = AEML - ESVM(S, \boldsymbol{x}, P, V, k, c)$

Input : S *The multi-label training data set*
          $\{(\boldsymbol{x}_1, Y_1), (\boldsymbol{x}_2, Y_2), \cdots, (\boldsymbol{x}_N, Y_N)\}$,

      $\boldsymbol{x}$ *The test instance $\boldsymbol{x} \in X$,*

      P *Maximum size of subset after first level of*
          segregation.

      V *Maximum size of subset after second level*
          of segregation.

      k *The number of labels.*

      c *constant of positive real number.*

Output: Y The prediction label set of testing instance $\boldsymbol{x}(Y \subseteq \{1, 2, \cdots, k\})$;

begin

for *each training instance* $(\boldsymbol{x}_i, Y_i)(i = 1, 2, \cdots, N)$ **do**
  | Calculate the number of relevant label for the training instance $|Y_i|$;

end

For multi-label training data set $S = \{(\boldsymbol{x}_1, Y_1), (\boldsymbol{x}_2, Y_2), \cdots, (\boldsymbol{x}_N, Y_N)\}$,
transform the multi-label training data set S into k independent binary training
data subsets using OVR problem transformation strategy, i.e., $S_1, S_2, \cdots, S_k$.

for *each binary training data subset* $S_j(j = 1, 2, \cdots, k)$ **do**
  | Compute its representative set using
    $[S_j^*, \beta_j] = ModDeriveRS(S_j, P, V)$;
    for *each training instance in the representative set*
        $(\boldsymbol{x}_m, y_m) \in S_j^*, (m = 1, 2, \cdots, |S_j^*|)$ **do**
      | Find out the training instance in its corresponding multi-label training
        set S and obtain the number of its relevant labels $|Y_i|$;
      if $y_m = +1$ **then**
        | $C[m] = \frac{ck}{|Y_i|}$;
      else
        | $C[m] = \frac{ck}{k-|Y_i|}$;
      end
    end
    Train an improved SVM classifier using $f_j = SVMTrain(S_j^*, \beta_j, C)$,
    based on non-linear kernel, using the formulation (5) and (6).

end

Obtain the prediction label set of the testing instance $\boldsymbol{x}$,

if *all* $f_j(\boldsymbol{x}) < 0$ **then**
  | $Y = \{j, s.t. \max f_j(\boldsymbol{x}), j = 1, 2, \cdots, k\}$, using the formulation (8).

else
  | $Y = \{j, s.t. f_j(\boldsymbol{x}) \geq 0, j = 1, 2, \cdots, k\}$, using the formulation (7).

end

return Y;

Algorithm 1. AEML-ESVM

4.1   Three Existing Multi-label Classification Methods

Three existing multi-label classification algorithms, OVR-ESVM, ML-CVM and
RCP-SVM are chosen to be compared with the proposed classification algorithm
AEML-ESVM. OVR-ESVM method combines OVR problem transformation

strategy with OVR-ESVM to implement multi-label classification [8] detailed in Sect. 3.2. ML-CVM method combines OVR problem transformation strategy with CVM to implement multi-label classification [12]. RCP-SVM combines OVO problem transformation strategy with binary SVM algorithm to implement multi-label classification [11]. OVR-ESVM and RCP-SVM are all realized with the modified popular SVM software package called LIBSVM. ML-CVM algorithm [12] can be implemented through modifying the CVM algorithm.

4.2 Three Large-Scale Multi-label Data Sets

In this experiment, three public large-scale multi-label data sets are chosen and they can be downloaded from the LIBSVM website [23]. The size and attributes of them are shown in Table 1.

Table 1. A description of data sets

Data set	Training instances	Testing instances	Features	Labels
Mediamill (exp1)	30993	12914	120	101
Siam-competition2007	21519	7077	30438	22
Rcvlv2 (topics;full sets)	23149	12000	47236	103

4.3 Performance Evaluation Metrics

The evaluation metrics of multi-label classification are more complicated compared with that of single-label classification [1,24,25]. We choose five popular evaluation metrics, that is, hamming loss, one-error, coverage, ranking loss and average-precision.

(1) Hamming loss: it is used to evaluate how many times, on average, an instance-label pair is misclassified, i.e., a relevant label is not in the prediction result or an irrelevant is predicted. The smaller the value of it, the better the classifier's performance.

(2) One-error: it is used to evaluate how many times the top-ranked label is not in the set of possible labels. The smaller the value of it, the better the classifier's performance.

(3) Coverage: it is used to evaluate how many steps are required, on average, to move along the ranked label list so as to get all the relevant labels of an example. The smaller the value of it, the better the classifier's performance.

(4) Ranking loss: it is used to evaluate the average of pairs of labels that are misordered for the instance. The smaller the value of it, the better the classifier's performance.

(5) Average-precision: it is used to evaluate the average fraction of relevant labels ranked higher than a particular label. The greater the value of it, the better the classifier's performance.

Briefly, the greater the value of average-precision, the better the multi-label classifier's performance. The smaller the value of the other four metrics, the better the multi-label classifier's performance.

4.4 Experiment Setting and Result Analysis

In the experiment, radial basis function, $K(x, y) = exp(-\gamma \|x - y\|_2^2)$ is adopted by the proposed AEML-ESVM and other three multi-label classification algorithms. Among them, γ represents the ratio factor of the kernel. $\| \cdot \|_2$ represents the Euclidean distance. In the AEML-ESVM algorithm, three parameters P, V and ε are set to get optimum representative set. At the same time, the following parameters are needed to be set for the four multi-label classification algorithms, i.e., allowing termination condition parameter e and the loss function parameter C. To obtain a fair and reasonable experiment result, same values are set for parameters of the four multi-label classification algorithms in the same data set, and different values are set for the parameters of different data sets. The experiment is conducted on a desktop PC with memory of 8 GB, processor of i5-4690 and main frequency of 3.5 GHz. The specific parameter settings and experiment results for specific data sets are analyzed in the following.

For data set of mediamill(exp1), to get optimum representative set, the specific parameter settings are as follows, P = 200, V = 150 and ε = 0.105. And the parameters of e and C of the four multi-label classification algorithms are set as e = 1.95e−5 and C = 1.0 respectively. The experiment results of the four different multi-label classification algorithms on this data set are shown in Tables 2 and 3.

Data in Table 2 shows that compared with the OVR-ESVM classification algorithm, AEML-ESVM algorithm drops by 0.3801 in terms of coverage, 0.1% in ranking loss and 0.42% in average-precision, increases by 0.02% in hamming loss and 0.07% in one-error. These fully demonstrate that AEML-ESVM can obtain similar classification performance with that of OVR-ESVM and better

Table 2. Experiment results on data set of mediamll(exp1)

Evaluation metrics	AEML-ESVM	OVR-ESVM	ML-CVM	RCP-SVM
Coverage	**20.7694**	21.1495	**21.0280**	34.0844
Ranking loss	**0.0574**	**0.0584**	0.0722	0.1402
Hamming loss	**0.0314**	**0.0312**	0.0550	0.0512
One-error	**0.1317**	**0.1310**	0.8690	0.5732
Average-precision	**0.7177**	**0.7219**	0.4311	0.3366

Table 3. Training time on data set of mediamill(exp1)

Time	AEML-ESVM	OVR-ESVM	ML-CVM	RCP-SVM
Training time	**1492.9**	15425.5	1920.5	6401.6

Table 4. Experiment results on data set of rcv1v2

Evaluation metrics	AEML-ESVM	OVR-ESVM	ML-CVM	RCP-SVM
Coverage	**5.2555**	**4.7665**	17.6724	5.7908
Ranking loss	**0.0115**	**0.0091**	0.0787	0.0135
Hamming loss	**0.0142**	**0.0104**	0.0608	0.0161
One-error	**0.0470**	**0.0365**	0.9203	0.0489
Average-precision	**0.8874**	**0.9212**	0.3264	0.8727

Table 5. Training time on data set of rcv1v2

Time	AEML-ESVM	OVR-ESVM	ML-CVM	RCP-SVM
Training time	**1916.3**	16164.7	2290.2	8137.9

classification performance than that of ML-CVM and RCP-SVM. Meanwhile, from Table 3, we can find out that the training time of AEML-ESVM is the shortest and it is 1/10.33 of that of OVR-ESVM, 1/1.29 of ML-CVM and 1/4.29 of RCP-SVM.

For data set of rcv1v2, to get optimum representative set, the specific parameter settings are as follows, $P = 200$, $V = 180$ and $\varepsilon = 0.925$. And the parameters of e and C of the four multi-label classification algorithms are set as $e = 9.5e-5$ and $C = 1.0$ respectively. The experiment results of the four different multi-label classification algorithms on this data set are shown in Tables 4 and 5.

Data in Table 4 shows that compared with the OVR-ESVM classification algorithm, AEML-ESVM algorithm increases by 0.489 in terms of coverage, 0.24% in ranking loss, 0.38% in hamming loss and 1.05% in one-error, drops by 3.38% in average-precision. These fully demonstrate that AEML-ESVM can obtain similar classification performance with that of OVR-ESVM and better classification performance than that of ML-CVM and RCP-SVM. Meanwhile, from Table 5, we can find out that the training time of AEML-ESVM is the shortest and it is 1/8.44 of that of OVR-ESVM, 1/1.20 of ML-CVM and 1/4.25 of RCP-SVM.

For data set of siam-competition2007, to get optimum representative set, the specific parameter settings are as follows, $P = 120$, $V = 90$ and $\varepsilon = 0.85$. And the parameters of e and C of the four multi-label classification algorithms are set as $e = 9.5e-5$ and $C = 1.0$ respectively. The experiment results of the four different multi-label classification algorithms on this data set are shown in Tables 6 and 7.

Data in Table 6 shows that compared with the OVR-ESVM classification algorithm, AEML-ESVM algorithm increases by 0.2491 in terms of coverage, 0.9% in ranking loss, 0.22% in hamming loss and 4.7% in one-error, drops by 3.05% in average-precision. These fully demonstrate that AEML-ESVM can obtain similar classification performance with that of OVR-ESVM and outperform ML-CVM and RCP-SVM greatly. Meanwhile, from Table 7, we can find

Table 6. Experiment results on data set of siam-competition2007

Evaluation metrics	AEML-ESVM	OVR-ESVM	ML-CVM	RCP-SVM
Coverage	2.6568	**2.4077**	4.5531	**2.6027**
Ranking loss	**0.0591**	**0.0501**	0.1423	0.0606
Hamming loss	**0.0592**	**0.0570**	0.1437	0.0687
One-error	**0.2854**	**0.2384**	0.8038	0.2942
Average-precision	**0.7843**	**0.8148**	0.4763	0.7707

Table 7. Training time on data set of siam-competition2007

Time	AEML-ESVM	OVR-ESVM	ML-CVM	RCP-SVM
Training time	**1115.7**	12114.6	1244.7	2840.5

out that the training time of AEML-ESVM is the shortest and it is $1/10.86$ of that of OVR-ESVM, $1/1.12$ of ML-CVM and $1/2.55$ of RCP-SVM.

In conclusion, experiment results on three public data sets show that the proposed AEML-ESVM classification algorithm gets similar classification performance with that of OVR-ESVM and better classification performance than that of the RCP-SVM and ML-CVM algorithms on the five common-used evaluation metrics. At the same time, the training time of AEML-ESVM is the shortest and this will greatly improve the applicability of AEML-ESVM classification algorithm on large-scale multi-label data sets.

5   Conclusion

To overcome the excessive training time problem that OVR-ESVM encounters when it is used on large-scale multi-label data sets, the AEML-ESVM classification algorithm is proposed. It can not only shorten the training time greatly but also exploit the label correlation of individual instance directly to improve the classification performance. In addition, it also achieves similar classification performance compared with that of OVR-ESVM. Experiment results on three public large-scale multi-label data sets demonstrate that its classification performance is comparable with that of the OVR-ESVM and better than that of RCP-SVM and ML-CVM. In the future work, we are going to improve the algorithm by using the label correlation between partial instances and different labels.

Acknowledgments. This work is supported by the National Natural Science Foundation of China (NSFC) under the grant number 61170258, 61103196, 61379127, 61379128, 61572448, by the National Key R&D Program of China under the grant number 2016YFC1401900 and by the Shandong Provincial Natural Science Foundation of China under the grant number ZR2014JL043.

References

1. Tsoumakas, G., Katakis, I., Vlahavas, I.: Mining multi-label data. In: Maimon, O., Rokach, L. (eds.) Data Mining and Knowledge Discovery Handbook, pp. 667–685. Springer, Heidelberg (2009)
2. Elghazel, H., Aussem, A., Gharroudi, O., Saadaoui, W.: Ensemble multi-label text categorization based on rotation forest and latent semantic indexing. Expert Syst. Appl. **57**(C), 1–11 (2016)
3. Hou, S., Zhou, S., Chen, L., Feng, Y., Awudu, K.: Multi-label learning with label relevance in advertising video. Neurocomputing **171**(C), 932–948 (2016)
4. Jing, X.Y., Wu, F., Li, Z., Hu, R., Zhang, D.: Multi-label dictionary learning for image annotation. IEEE Trans. Image Process. **25**(6), 2712–2715 (2016)
5. Zufferey, D., Hofer, T., Hennebert, J., Schumacher, M., Ingold, R., Bromuri, S.: Performance comparison of multi-label learning algorithms on clinical data for chronic diseases. Comput. Biol. Med. **65**(C), 34–43 (2015)
6. Liu, Y., Liu, Y., Wang, C., Wang, X.: What strikes the strings of your heart?-multi-label dimensionality reduction for music emotion analysis via brain imaging. IEEE Trans. Autonom. Mental Dev. **7**(3), 176–188 (2015)
7. Gibaja, E., Ventura, S.: A tutorial on multi-label learning. ACM Comput. Surv. **47**(3), 1–38 (2015)
8. Xu, J.: An extended one-versus-rest support vector machine for multi-label classification. Neurocomputing **74**(17), 3114–3124 (2011)
9. Cortes, C., Vapnik, V.: Support-vector networks. Mach. Learn. **20**(3), 273–297 (1995)
10. Nandan, M., Khargonekar, P.P., Talathi, S.S.: Fast SVM training using approximate extreme points. J. Mach. Learn. Res. **15**(1), 59–98 (2014)
11. Fürnkranz, J., Hüllermeier, E., Mencía, E.L., Brinker, K.: Multilabel classification via calibrated label ranking. Mach. Learn. **73**(2), 133–153 (2008)
12. Tsang, I.W., Kwok, J.T., Cheung, P.M.: Core vector machines: fast SVM training on very large data sets. J. Mach. Learn. Res. **6**(1), 363–392 (2005)
13. Wang, S., Wang, J., Wang, Z., Ji, Q.: Enhancing multi-label classification by modeling dependencies among labels. Pattern Recogn. **47**(10), 3405–3413 (2014)
14. Zhou, Z.H., Zhang, M.L., Huang, S.J., Li, Y.F.: Multi-instance multi-label learning. Artif. Intell. **176**(1), 2291–2320 (2012)
15. Luaces, O., Díez, J., Barranquero, J., Coz, J.J., Bahamonde, A.: Binary relevance efficacy for multilabel classification. Prog. Artif. Intell. **1**(4), 303–313 (2012)
16. Clare, A., King, R.D.: Knowledge discovery in multi-label phenotype data. In: Raedt, L., Siebes, A. (eds.) PKDD 2001. LNCS (LNAI), vol. 2168, pp. 42–53. Springer, Heidelberg (2001). doi:10.1007/3-540-44794-6_4
17. Elisseeff, A., Weston, J.: A kernel method for multi-labelled classification. In: Advances in Neural Information Processing Systems, pp. 681–687 (2001)
18. Xu, J.: Fast Multi-label core vector machine. Pattern Recogn. **46**(3), 885–898 (2013)
19. Zhang, M.L., Zhou, Z.H.: Multi-label neural networks with applications to functional genomics and text categorization. IEEE Trans. Knowl. Data Eng. **18**(10), 1338–1351 (2006)
20. Boutell, M.R., Luo, J., Shen, X., et al.: Learning multi-label scene classification. Pattern Recogn. **37**(9), 1757–1771 (2004)
21. Chang, C.C., Lin, C.J.: LIBSVM: a library for support vector machines. ACM Trans. Intell. Syst. Technol. **2**(3), 27 (2011)

22. Platt, J.C.: Fast training of support vector machines using sequential minimal optimization. In: Advances in Kernel Methods, pp. 185–208 (1999)
23. LIBSVM datasets. https://www.csie.ntu.edu.tw/~cjlin/libsvmtools/datasets/
24. Zhang, M.L., Zhou, Z.H.: A review on multi-label learning algorithms. IEEE Trans. Knowl. Data Eng. **26**(8), 1819–1837 (2014)
25. Schapire, R.E., Singer, Y.: BoosTexter: a boosting-based system for text categorization. Mach. Learn. **39**(2), 135–168 (2000)

Efficiently Discovering Most-Specific Mixed Patterns from Large Data Trees

Xiaoying Wu[1] and Dimitri Theodoratos[2(✉)]

[1] State Key Laboratory of Software Engineering, Wuhan University, Wuhan, China
xiaoying.wu@whu.edu.cn
[2] New Jersey Institute of Technology, Newark, USA
dth@njit.edu

Abstract. Discovering informative tree patterns hidden in large datasets is an important research area that has many practical applications. Along the years, research has evolved from mining induced patterns to mining embedded patterns. Mixed patterns allow extracting all the information extracted by embedded or induced patterns but also more detailed information which cannot be extracted by the other two. Unfortunately, the problem of extracting unconstrained mixed patterns from data trees has not been addressed up to now.

In this paper, we address the problem of mining unordered frequent mixed patterns from large trees. We propose a novel approach that non-redundantly extracts most-specific mixed patterns. Our approach utilizes effective pruning techniques to reduce the pattern search space. It exploits efficient homomorphic pattern matching algorithms to compute pattern support incrementally and avoids the costly enumeration of all pattern matchings required by older approaches. An extensive experimental evaluation shows that our approach not only mines mixed patterns from real and synthetic datasets up to several orders of magnitude faster than older state-of-the-art embedded tree mining algorithms applied to large data trees but also scales well empowering the extraction of informative mixed patterns from large datasets for which no previous approaches exist.

1 Introduction

Mining frequent complex patterns from loosely structured data is a central process in the analysis of large datasets like those encountered in big data applications. The most frequently used data structure for such data is the tree. Trees are used for representing data in a plethora of applications.

As tree pattern mining is very important for data analysis, it has been the subject of extensive research [3–5, 10, 11, 13, 14, 16–18]. The tree patterns extracted from the data can be basically characterized by two parameters: (a) the type of edges the tree patterns have, and (b) the type of morphism used to map

---

X. Wu—The research of this author was supported by the National Natural Science Foundation of China under Grant No. 61202035 and 61272110.

S. Candan et al. (Eds.): DASFAA 2017, Part I, LNCS 10177, pp. 279–294, 2017.
DOI: 10.1007/978-3-319-55753-3_18

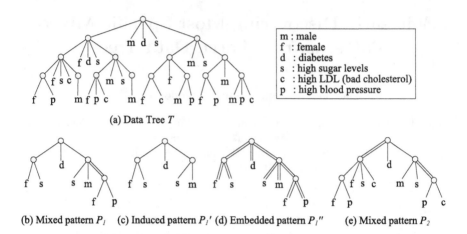

(a) Data Tree T

m : male
f : female
d : diabetes
s : high sugar levels
c : high LDL (bad cholesterol)
p : high blood pressure

(b) Mixed pattern P_1 (c) Induced pattern P_1' (d) Embedded pattern P_1'' (e) Mixed pattern P_2

Fig. 1. Examples of mixed, embedded, and induced patterns from a data tree.

the tree patterns to the data structure. An edge in a tree pattern can be a child edge (in which case it is mapped to an edge in the data tree) or a descendant edge (in which case it is mapped to a path in the data tree). The morphisms considered in the bibliography can be, almost exclusively, of two types: isomorphisms (injections that map a pattern child edge to a data tree edge) [3,5] and embeddings (injections that map a pattern descendant edge to a data tree path and preserve not-ancestor relationships among the pattern nodes) [16,17].

Over the years the extracted tree patterns have evolved from *induced* patterns [3] to *embedded* patterns [10,11,13,16–18]. The induced patterns have only child edges and are mapped to the data tree using isomorphisms. The embedded patterns have only ancestor edges and are mapped to the data tree using embeddings. *Mixed patterns* are patterns that allow child *and* descendant relationships. They generalize both induced and embedded patterns as they allow extracting all the information extracted with embedded or induced patterns but also more detailed information which cannot be extracted neither by induced patterns nor by embedded patterns. Unfortunately, the problem of extracting unconstrained mixed patterns from data trees has not been addressed up to now.

Furthermore, existing tree mining algorithms focus almost exclusively on extracting patterns from a set of small trees. The problem of mining patterns from a large data tree has been neglected. This is a serious omission since a multitude of information rich datasets from different application areas are in the form of a single large tree.

In this paper we address the problem of extracting frequent unordered mixed tree patterns from a large data tree. Mixed patterns are larger and more informative than induced patterns, and they are more specific and therefore more informative than embedded patterns.

Consider the data tee T of Fig. 1(a) which records medical conditions of maternal and paternal ancestors of a male individual. Every recorded person is represented by a node and his/her conditions and gender are represented by its child nodes which are labelled accordingly. This type of information for multiple patients can be recorded by collecting trees for different patients into one tree and hanging them under a new root. Such a tree can grow to be very large in terms of breadth and also depth but for the needs of our example se show a tree for a single patient restricted to four generations. Let the support of a pattern P be defined as the number of images of the root of P under all possible embeddings of P to T. Assuming that the frequency threshold is 2, we can see that the pattern P_1 of Fig. 1(b) is a frequent pattern that can be extracted from T. This is a mixed pattern: it involves child and descendant edges. Child edges are shown in a pattern by a single line between the nodes while descendant edges are shown by a double line edge. The "closest" induced pattern that can be extracted from T is the pattern P_1' of Fig. 1(c), while the "closest" embedded pattern that can be extracted from T is the pattern P_1'' of Fig. 1(d). As one can see, both types of patterns—induced and embedded—can miss information which can be captured by mixed patterns. Figure 1(e) shows another larger mixed frequent pattern[1].

Contribution. The main contributions of the paper can be summarized as follows:

- We introduce a specificity relation on mixed patterns to reduce the search space of mined mixed patterns. We formally define the problem of extracting most-specific frequent mixed patterns from a large data tree (Sect. 2).
- We design rules to prune non-useful candidate patterns which either do not have an embedding to the data tree or do not contribute to the generation of any pattern in the final result. We further develop an efficient method which checks if a pattern is most-specific on-the-fly without comparing with previously generated patterns (Sect. 3).
- We design a novel pattern frequency computation approach which exploits a holistic twig-join algorithm [2] to compute the homomorphisms of a pattern to the data tree, and then filters out homomorphisms which are not embeddings. Our approach involves an incremental method that computes the embeddings of a new candidate pattern based on the embeddings of already computed frequent patterns. Our approach turns out to be much more efficient than computing the embeddings directly as older embedded pattern mining approaches do (Sect. 4).
- We run extensive experiments to evaluate the performance and scalability of our approach on real and synthetic datasets. The experimental results show that: (1) our approach mines most-specific mixed patterns up to several orders of magnitude faster than a state-of-the-art algorithm mining embedded tree patterns when applied to a large data tree; (2) our on-the-fly pattern specificity checking strategy greatly reduces the mixed pattern search space on dense data; and (3) our techniques of early-detecting and pruning non-useful or infrequent candidates are effective in improving pattern mining performance and scalability (Sect. 5).

---

[1] Even though there is documented relationship between diabetes, high LDL and sugar levels and high blood pressure, this specific example dataset is fictitious.

2   Preliminaries and Problem Definition

Trees and inverted lists. We consider rooted labeled unordered (i.e., no pre-defined left-to-right ordering among the children of each node) trees. For every label a in an input data tree T, we construct an inverted list L_a of the data nodes with label a ordered by their pre-order appearance in T. Figure 2(a) and (b) shows a data tree and inverted lists of its labels.

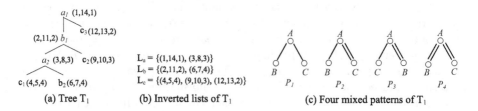

| (a) Tree T_1 | (b) Inverted lists of T_1 | (c) Four mixed patterns of T_1 |

Fig. 2. A tree T_1, its inverted lists and four mixed patterns.

Tree morphisms. Two types of tree patterns have been considered in the literature of tree pattern mining: patterns whose edges represent child relationships (*child edges*) and patterns whose edges represent descendant relationships (*descendant edges*). Different types of morphisms are employed to determine if a tree pattern is included in a tree.

Let lb be a labeling function mapping nodes of patterns to labels. Given a pattern P and a tree T, a *homomorphism* from P to T is a function m mapping nodes of P to nodes of T, such that: (1) for any node $x \in P$, $lb(x) = lb(m(x))$; and (2) for any edge $(x, y) \in P$, if (x, y) is a child edge, $(m(x), m(y))$ is an edge of T, while if (x, y) is a descendant edge, $m(x)$ is an ancestor of $m(y)$ in T.

Previous contributions have constrained the homomorphisms considered for tree mining in different ways. Let P be a pattern P with *child* edges. An *isomorphism* from P to T is an injective function m mapping nodes of P to nodes of T, such that: (1) for any node x in P, $lb(x) = lb(m(x))$; and (2) (x, y) is an edge of P *iff* $(m(x), m(y))$ is an edge of T. Clearly, an isomorphism is a special case of a homomorphism. If isomorphisms are considered, the mined patterns are qualified as *induced* [3].

Another type of constrained homomorphism considered in the pattern mining literature is the embedding: let P be a descendant-only pattern (i.e., a pattern that has only descendant edges). An *embedding* from P to T is an injective function m mapping nodes of P to nodes of T, such that: (1) for any node $x \in P$, $lb(x) = lb(m(x))$; and (2) (x, y) is an edge in P *iff* $m(x)$ is a parent or an ancestor of $m(y)$ in T. Patterns mined using embeddings are called *embedded* patterns [16,17]. Observe that, contrary to a homomorphism, an embedding can map two sibling nodes in P to two nodes on the same path in T or to the same node in T (the latter in case these nodes have the same label). Clearly, an isomorphism on a pattern with descendant edges is also an embedding.

We extend the above embedding definition for mixed tree patterns: let P be a mixed pattern. A *mixed embedding* from P to T is an injective function m mapping nodes of P to nodes of T, such that: (1) for any node $x \in P$, $lb(x) = lb(m(x))$; and (2) (x, y) is a child (resp. descendant) edge in P iff $m(x)$ is a parent (resp. ancestor) of $m(y)$ in T. In the rest of discussions, unless otherwise specified, we refer to mixed embeddings as embeddings.

Support. We identify an *occurrence* of P on T by a tuple indexed by the nodes of P whose values are the images of the corresponding nodes in P under a homomorphism of P to T. The values in a tuple are the positional representations of nodes in T. An *embedded occurrence* of P on T is an occurrence defined by an embedding from P to T.

The set of occurrences of P under all possible homomorphisms of P to T is a relation OC whose schema is the set of nodes of P. If X is a node in P labeled by label a, the *occurrence list of X on T* is a sublist L_X of L_a containing only those nodes that occur in the column for X in OC. Let OC^e be the subset of OC containing all the embedded occurrences from P to T. The *embedded occurrence list of X on T*, denoted L_X^e, is defined similarly to L_X over OC^e instead of OC. Clearly, L_X^e is a sublist of L_X.

We define the *occurrence list set of P on T* as the set OL of the occurrence lists of the nodes of P on T; that is, $OL = \{L_X \mid X \in nodes(P)\}$. Similarly, we define the *embedded occurrence list set of P on T*, OL^e, as the set of the embedded occurrence lists of the nodes of P on T.

As an example, Fig. 3 shows the occurrence relations and lists as well as their embedded versions for the pattern P on the tree T_1 of Fig. 2(a).

The *support* of pattern P on T is defined as the size of the embedded occurrence list of the root R of P on T, L_R^e. A pattern is *frequent* if its support is no less than a user defined threshold *minsup*. We denote by F_k the set of all frequent patterns of size k, also known as *k-patterns*.

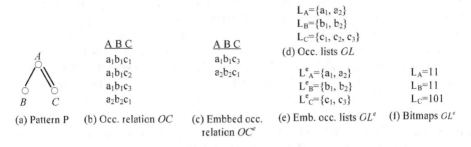

Fig. 3. Occurrence information for pattern P on the tree T_1 of Fig. 2.

Most-specific patterns. When mixed tree patterns are considered, it is possible that there are multiple distinct mixed tree patterns that can be extracted from a data tree, and they have the same embeddings to the data tree. Clearly, it has no benefit to extract all such mixed patterns. We therefore define next most-specific patterns. We first introduce a specificity relation on patterns: A pattern P_1 is *more specific* than a pattern P_2 (and P_2 is *less specific* than P_1) iff there is an embedding from P_2 to P_1 but not from P_1 to P_2. If a pattern P_1 is more specific than a pattern P_2, we write $P_1 \prec P_2$. For instance, the specificity relationships among the four patterns in Fig. 2 are $P_1 \prec P_2 \prec P_4$ and $P_1 \prec P_3 \prec P_4$. Clearly, $\prec$ is a strict partial order. If $P_1 \prec P_2$, P_1 conveys more information on the dataset than P_2.

A frequent pattern P is *most-specific* if there is no other frequent pattern P_1, such that P_1 and P_2 have the same embedding to the input data tree and $P_1 \prec P$. In Fig. 2, patterns P_3 is not most-specific, as it has the same embedding as P_1 to T_1.

Problem statement. Given a large tree T and a minimum support threshold *minsup*, our goal is to mine all the frequent unordered most-specific mixed patterns.

3   Candidate Pattern Generation

Our approach for mining embedded tree patterns from a large tree iterates between the candidate generation phase and the support counting phase. In order to systematically generate candidate mixed tree patterns, we extend the equivalence class-based pattern generation method proposed in [16,17]. A candidate pattern may have multiple alternative isomorphic representations. To minimize the redundant generation of the isomorphic representations of the same pattern, we extend the canonical form for descendant-only tree patterns [16] to mixed patterns. The details are omitted in the interest of space.

Pattern equivalence class. Let P be a pattern. Each node of P is identified by its *depth-first position* in the tree, determined through a depth-first traversal of P, by sequentially assigning numbers to the first visit of the node. The *rightmost leaf* of P, denoted *rml*, is the node with the highest depth-first position. The *immediate prefix* of P is the sub-pattern of P obtained by deleting the *rml* from P. A sub-pattern of P obtained by a series of *rml* deletions is called *prefix* of P.

Given a pattern P of size k-1 ($k \geq 1$), its *equivalence class* is the set of all the patterns of size k that have P as immediate prefix. We denote the equivalence class of P as $[P]$.

We use the notation $P^i_{/,x}$ (resp. $P^i_{//,x}$) to denote a k-pattern in $[P]$ formed by adding a child node labeled by x, using a child (resp. descendant) edge, to the node with position i in P as the rightmost leaf node.

Equivalence class-based pattern expansion. Given an equivalence class $[P]$, we obtain its successor equivalence classes by expanding patterns in $[P]$. The main idea of this expansion is to *join* each pattern $P^i_{u,x} \in [P]$ with any other pattern in $[P]$, including itself (self expansion), to produce the patterns of the equivalence class $[P^i_{u,x}]$. There can be up to three outcomes for each pair of patterns to be joined. Formally, let $P^i_{u,x}$ and $P^j_{v,y}$ denote any two elements in $[P]$. The join operation $P^i_{u,x} \otimes P^j_{v,y}$ is defined as follows:

- if $j = i$, return two patterns $Q^{k-1}_{/,y}$ and $Q^{k-1}_{//,y}$, where $Q = P^i_{u,x}$.
- if $j \leq i$, return the pattern $Q^j_{v,y}$ where $Q = P^i_{u,x}$.

No new candidate is possible if $i < j$. The two outcome patterns $Q^{k-1}_{/,y}$ and $Q^{k-1}_{//,y}$ are called *child* expansions of P^i_x by P^j_y, whereas $Q^j_{v,y}$ is called the *cousin* expansion. By iteratively joining $P^i_{u,x}$ with all elements $P^j_{v,y}$ of $[P]$, we generate all possible k-patterns in class $[P^i_{u,x}]$. We call patterns $P^i_{u,x}$ and $P^j_{v,y}$ the *left-parent* and *right-parent* of a join outcome, respectively. Figure 4 shows an example of child and cousin expansions.

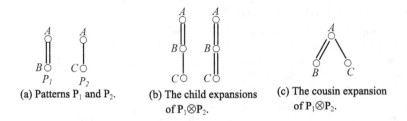

(a) Patterns P_1 and P_2.

(b) The child expansions of $P_1 \otimes P_2$.

(c) The cousin expansion of $P_1 \otimes P_2$.

Fig. 4. An example of pattern expansions.

Non-useful candidates pruning. The candidate set should be non-redundant to the extent possible; ideally, each pattern should be generated as most once. For this, the pattern expansion will only consider canonical patterns for further expansions; that is, it only picks from class $[P]$ those $P^i_{u,x}$ that are in canonical form as the left operand in a join operation. However, even with this requirement, naively applying the above pattern expansion method will still generate a large number of non-useful candidates, which either have empty embedding on the data tree or will not contribute to the generation of any pattern in the final result. We design the following two rules to prune non-useful candidate patterns:

- if $v = /$, prune child pattern expansions $Q^{k-1}_{/,y}$ and $Q^{k-1}_{//,y}$.
- if $v = /$ and the cousin expansion is not canonical, prune the cousin expansion.

Intuitively, the first rule says that, a candidate should be pruned if it is generated by inserting a node between two nodes connected by a child edge in its parent pattern, since it cannot have matches in a data tree. The second rule says that, a candidate should be pruned if it is a cousin expansion outcome, non-canonical, and its rightmost leaf edge is a child edge; the reason is that no patterns can be grown out from it; further, it will not be used to generate child expansions by rule 1. The expansions given in Fig. 4 are examples of non-useful candidates.

Finding most-specific patterns. The pattern generation process may produce candidates which are not most-specific (defined in Sect. 2). We have developed an efficient method to check if a pattern P is most-specific. It does on-the-fly checking without conducting any pattern specificity testing on P. It is realized by identifying *refinable* edges of a pattern. Let $X//Y$ denote the descendant edge from node X to node Y in P. We say edge $X//Y$ is refinable to be the child edge X/Y, if every embedding of P to T maps $X//Y$ to an edge of T.

We can show that a pattern is most-specific if and only if it has no refinable edges. We will describe in the next section how to check refinable edges without even enumerating pattern embeddings. Any non most-specific pattern is discarded right away since any pattern that can be grown from it can be generated from a more specific pattern.

Input: Pattern Q, Q's parents P_1 and P_2, OL^es of P_1 and P_2 on T, and support threshold $minsup$
Output: the embedded occurrence list set OL^e of Q

1. Let S_1 denote the OL^e intersection of P_1 and P_2 on their common prefix;
2. Let S_2 denote the set of emb.occu.lists of the rightmost leaves of P_1 and P_2;
3. Intersect S_2 with OL^e of M_{12} in case Q is a child expansion of $P_1 \otimes P_2$, where M_{12} is the pattern corresponding to the rightmost edge of Q;
4. Let S_{12} denote the union of S_1 and S_2;
5. Compute OL of Q using the twig-join algorithm on S_{12};
6. Compute OL^e of Q from OL by filtering out elements not in any embedding of Q on T.

Fig. 5. Procedure CompEmbOL.

4   A Novel Technique for Support Computation

To compute the support of a pattern P on a data tree T, a majority of existing tree mining algorithms [3–5,10,11,16,17] explicitly compute and store the matches of P (P is a child-only or descendant-only tree pattern) to T. Unfortunately, the problem of finding an unordered embedding of P to T is NP-Complete [6]. On the other hand, finding a homomorphism from P to T can be done in PTIME [9]. Moreover, in recent years, efficient algorithms have been designed for evaluating tree pattern queries over a data tree [1,2].

Inspired by the above observations, we develop a generation-filtering approach which first computes homomorphisms of P to T, and then, it filters out homomorphisms which are not embeddings (remember that the embeddings are special cases of homomorphisms). The outline of our approach, called *CompEmbOL*, is shown in Fig. 5. Its implementation details are described below.

4.1   A Holistic Twig-Join Approach for Computing Pattern Homomorphisms

Holistic twig-join algorithms (e.g., *TwigStack* [2]) are the state of the art algorithms for computing all the homomorphic occurrences of tree-pattern queries on tree data. We briefly outline below such an algorithm to explain how it is integrated in our approach.

The twig-join approach for computing relation OC. A typical holistic twig-join algorithm for evaluating tree patterns works when the input data tree is represented by a set of inverted lists and the tree nodes are encoded (Sect. 2). It is called holistic as it joins multiple inverted lists at a time to avoid generating intermediate join results. It proceeds in two phases.

In the first phase, the algorithm computes *path occurrences* of a given pattern P which are occurrences of the individual root-to-leaf paths of the pattern. In the second phase, it merge-joins the path occurrences to compute the occurrences for P. When all edges in P are descendant relationships, the algorithm can guarantee worst-case performance *linear* on the size of the data tree inverted lists (the input) and the size of relation OC (the output), i.e., it is optimal [2].

Sub-optimality of twig join for mixed patterns. The twig-join method is sub-optimal for evaluating mixed tree patterns, since it may generate *redundant* path occurrences (path occurrences that do not contribute to an occurrence of the pattern) for patterns involving child edges [2]. Clearly, the existence of redundant path occurrences impacts the pattern evaluation performance negatively, since these path occurrences are processed and possibly joined with other path occurrences without producing an occurrence of the pattern. Further, the twig-join method can be expensive for evaluating a large number of candidates, since it needs to fully scan the inverted lists corresponding to the node labels of the candidate pattern.

To address the problems of the twig-join method, we design an incremental method, which evaluates a pattern by leveraging the computation done at patterns evaluated earlier. Further, the incremental method encodes the embeddings of previously computed frequent patterns in inverted lists, a technique which records in polynomial space a potentially exponential number of embeddings.

4.2   An Incremental Method Using Bitmaps

Using a holistic twig-join algorithm, a pattern is evaluated by iterating over the inverted lists corresponding to the labels of the pattern nodes. Let P be a pattern and X be a node in P labeled by a. If there is a sublist, say L_X, of L_a such that P can be evaluated on T using L_X instead of L_a, we say that P can be *evaluated using L_X for X on T*. Since L_X is non-strictly smaller than L_a, both the data access cost and the computation cost can be reduced.

Given a pattern Q, we can easily identify a homomorphism from each parent of Q to Q. The following proposition can be shown.

Proposition 1. *Let X' be a node in a parent Q' of Q and X be the image of X' under a homomorphism from Q' to Q. The occurrence list L_X of X on T, is a sublist of the occurrence list $L_{X'}$ of X' on T.*

Recall that $L_{X'}$ is the inverted list of data tree nodes that participate in the occurrences of Q' to T. By Proposition 1, pattern Q can be evaluated using $L_{X'}$

for X instead of using the corresponding label inverted list. Further, if X is the homomorphic image of the nodes X_1 and X_2 in the left and right parent of Q, respectively, we can evaluated Q using the *intersection*, $L_{X_1} \cap L_{X_2}$, of L_{X_1} and L_{X_2} for X. This is the sublist of L_{X_1} and L_{X_2} comprising the nodes that appear in both L_{X_1} and L_{X_2}.

In general, if a pattern P has a homomorphism to pattern Q, we can use the occurrence list set OL of P to evaluate Q, and we say that the occurrence set of Q can be *computed using OL*. By Proposition 1, we can compute the occurrence set of Q using only the occurrence list sets of its parents. When Q is a cousin expansion, the intersection of its parent occurrence list sets can minimize the occurrence lists used to compute the occurrence set of Q, compared to using the occurrence lists of other patterns having a homomorphism to Q (lines 1–2 in Procedure *CompEmbOL* of Fig. 5). When Q is a child expansion, to obtain the minimality, we use additionally occurrence lists of an edge pattern, corresponding to the rightmost leave edge of Q, since it is the only edge of Q that does not appear in Q's parents (line 3 in *CompEmbOL*).

Therefore, we store with each frequent pattern its occurrence list set. Our method is space efficient since the occurrence lists can encode in linear space an exponential number of occurrences for the pattern [2]. In contrast, the state-of-the-art methods for mining embedded patterns [16,17] have to store information about all the occurrences of each given pattern in T.

We can show that, when the twig-join algorithm evaluates mixed patterns using occurrence list sets of their parent patterns (and of the edge patterns for child expansions), redundant path occurrences are avoided. We omit the details in the interest of space.

Another advantage offered by the incremental method is that, it allows a quick identification of some non-frequent candidates before they are evaluated, by just inspecting the intersection results of occurrence lists of the left and right parents of a candidate. As verified by our experimental results, substantial CPU cost can be saved using this early-detection of infrequent candidate patterns.

Representing Occurrence Lists As Bitmaps. The occurrence list L_X of a pattern node X labeled by a on T can be represented by a bitmap on L_a that has a '1' bit at position i iff L_X comprises the tree node at position i of L_a. Then, the occurrence list set of a pattern is the set of bitmaps of the occurrence lists of its nodes. Examples of bitmaps for pattern occurrence lists can be found in Fig. 3(f).

Clearly, storing the occurrence lists of multiple patterns as bitmaps results in important space savings. Moreover, the intersection of the occurrence lists of pattern nodes can be implemented by a bitwise operation on the corresponding bitmaps. This offers both CPU and I/O cost savings [15].

Generating embedded occurrence list set OL^e. Recall that, in contrast to a homomorphism, an embedding cannot map two siblings of P to two nodes on the same path in the data tree T or to the same node of T in case these nodes have the same label. Since $OL^e \subseteq OL$, to obtain OL^e, we can prune elements from OL that violates the above constraint. For this, we design a procedure which has the following two steps: (1) compute embedded occurrence lists for a chosen set of one-level induced subtrees of P; and (2) filter elements from the resulted occurrence lists of the last step to obtain OL^e of P. We omit the details in the interests of space.

While doing the two-step processing, the procedure checks if every descendant edge $X//Y$ of P is refinable. For this, it suffices to check if every element $y \in L_Y^e$ has an element $x \in L_X^e$, such that x and y maps to an edge in the data tree T; if $X//Y$ is the rightmost edge, no elements in L_X^e are mapped to nodes on a same path in T. All these checking can be easily implemented by the twig-join method.

The above techniques are summarized in our mixed tree pattern mining algorithm called $mixTM$, whose pseudocode and descriptions are omitted due to lack of space.

5   Experimental Evaluation

We implemented our mixed tree pattern mining approach and we conducted experiments to: (a) compare the features of the extracted mixed patterns with those of induced patterns and (descendant edge-only) embedded patterns, and (b) study the performance of our approach in terms of execution time, memory consumption and scalability.

Specifically, we implemented and compared the following algorithms that mine patterns from a single large tree: (1) the algorithm which mines all the frequent most-specific mixed patterns (denoted $mixTM$); (2) the baseline algorithm which mines all the possible frequent mixed tree patterns without checking whether they are most-specific (denoted as $mixTM$-bas). (3) the algorithm which mines all the frequent embedded tree patterns (denoted as $embTM$); and (4) the algorithm which mines all the frequent induced tree patterns (denoted as $indTM$).

To the best of our knowledge, there is no previous algorithm computing mixed patterns from data trees. Therefore, we compared the performance of our algorithm with a state-of-the-art unordered embedded tree mining algorithm $sleuth$ [16]. Algorithm $sleuth$ was designed to mine embedded patterns from a set of small trees. In order to allow the comparison, we adapted $sleuth$ to a large single tree setting by making it to return as support of a pattern the number of root occurrences of this pattern in the data tree.

Our implementation was coded in Java. All the experiments reported here were performed on a workstation having an Intel Xeon CPU 3565 @3.20 GHz processor with 8 GB memory running JVM 1.7.0 in Windows 7 Professional. The Java virtual machine memory size was set to 4 GB.

Datasets. We have ran experiments on three real and benchmark datasets. Due to space limitation, we only present results of our experimental study on one synthetic benchmark tree dataset called $XMark^2$ modeling an auction website.

---

[2] http://xml-benchmark.org.

The dataset is deep and has many regular structural patterns. It includes some recursive elements (which are elements on the same root-to-leaf path with the same label). Its statistics are shown below.

Dataset	Tot. #nodes	#labels	Max/Avg depth	#paths
XMark	180769	24245	13/6.4	138840

Time Performance. We measure the total elapsed time of the five algorithms for producing frequent patterns at different support thresholds. All four algorithms *mixTM*, *mixTM-bas*, *embTM*, and *indTM* employ the same approach described in Sect. 4 for computing pattern support. In contrast, algorithm *sleuth* [16] uses a completely different method.

Figure 6(a) shows the time spent by *sleuth,mixTM*, *mixTM-bas*, *mixTM*, *embTM*, and *indTM* under different support thresholds on the XMark. Notice that a logarithmic scale is used in the Y-axis of the figure. We have the following observations.

The induced pattern miner *indTM* has clearly the best time performance. Both *sleuth* and *mixTM-bas* run slower than others by at least one order of magnitude. Also, their runtime increase rate is much sharper than others as the support level decreases.

The poor performance of *mixTM-bas* can be explained by the substantially larger number of candidates and patterns it has to evaluate and produce, respectively, compared to *mixTM* and the other algorithms at the same support threshold (Fig. 6(c)). This demonstrates the importance of detecting and pruning mixed patterns which are not most-specific.

As *mixTM* produces more frequent patterns than *embTM*, we expect that the former runs slower than *embTM* (this is also confirmed by the experiments). However, we found that the performance advantage of *embTM* over *mixTM* on XMark is only marginal. This can be explained by the fact that, for most of the testing cases, the number of candidates evaluated by *mixTM* is only slightly larger than that evaluated by *embTM*. This shows the effectiveness of the pruning technique used by *mixTM* for non-useful candidates. For instance, when *minsup* = 650 *mixTM* will have to evaluate 95% more candidates without the pruning than *embTM* and this which will negatively affect the performance of *mixTM*.

The poor time performance of *sleuth* [16] results from its way of computing pattern support. To find all the embedded frequent patterns, *sleuth* keeps track of all possible embedded occurrences of a candidate to a data tree, and to perform expensive join operations over these occurrences. Therefore, its performance suffers for mining dense data (that is, data which contain a large number of patterns).

The poor time performance of *sleuth* [16] is the result of its way for computing pattern support. To find all the embedded frequent patterns, *sleuth* keeps track of all possible embedded occurrences of a candidate to a data tree, and

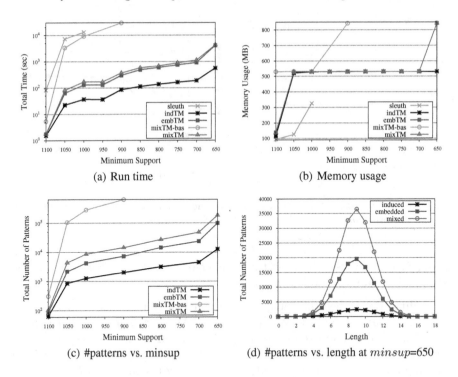

Fig. 6. Performance comparison on XMark

performs expensive join operations over these occurrences. Therefore, its performance suffers for mining dense data (that is, data which contain a large number of patterns).

Memory Usage. Figure 6(b) shows the memory consumption of the five methods under different support thresholds on the XMark. As aforementioned, the four algorithms *mixTM*, *mixTM-bas*, *embTM*, and *indTM* use the same strategy for computing pattern support.

As we can see from Fig. 6(b), for most of the testing cases, the memory consumption of *mixTM*, *embTM*, and *indTM* remains almost constant and has no noticeable differences. This can be explained by the fact that the differences among the sizes of bitmaps stored by them in that range are small. The sudden increase in memory usage for both *mixTM* and *embTM* at support level 650 can be explained by the significantly larger number of candidates computed by the two algorithms at that point. The memory consumption of *mixTM-bas* is larger than that of the other algorithms in all the applicable cases, due to its substantially larger amount of candidates computed and patterns produced.

Algorithm *sleuth* needs to store in memory all the embedded pattern occurrences for candidates under consideration. Nevertheless, the memory consumption of *sleuth* is not large in all cases where it can finish within a reasonable

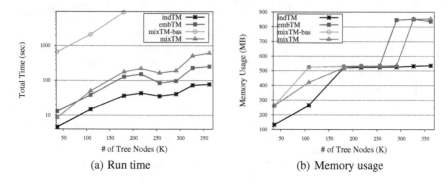

Fig. 7. Scalability comparison on XMark with increasing size for mining closed and maximal patterns ($minsup = 0.5\%$).

amount of time. This is because the number of embedded occurrences at the corresponding support levels is not large.

Scalability. We studied the scalability of the five algorithms, as we increase the size of input data on XMark. We generated 8 XMark trees by setting $factor = 0.01, 0.02, \ldots, 0.08$ and fixed $minsup$ at 0.55%. Even on the smallest fragment, *sleuth* was unable to finish within 12 h. We therefore didn't continue testing it on larger fragments. Figure 7 shows only the scalability results of *mixTM*, *mixTM-bas*, *embTM*, and *indTM*.

As we increase the input data size, the growth of the running time of *mixTM-bas* is much sharper than the other three. It is outperformed by *mixTM* by a factor of up to 77. The memory usage of *mixTM-bas* is also larger than the others. While *embTM* runs slightly faster than *mixTM*, they have compatible memory usage. *indTM* has the best runtime performance and memory usage. The results are consistent with the results tested on a single sized dataset.

Comparison of mined patterns. Figure 6(d) shows the distribution of the frequent induced, embedded, and (most-specific) mixed patterns by length at the support threshold of 650 on XMark. All of them exhibit a symmetric distribution. The longest frequent induced, embedded, and mixed patterns has 16, 17, and 17 nodes, respectively. For all the three types of patterns, the number of patterns having a length of 9 nodes is the largest.

6   Related Work

The problem of mining tree patterns has been studied since the last decade. Our work is related more closely to unordered embedded tree mining. Among the many tree mining algorithms studied in the literatures, only few mine unordered embedded patterns [12,16,18].

TreeFinder [12] is the first unordered embedded tree pattern mining algorithm. It is a two-step algorithm. In the first step, it clusters the input trees by the co-occurrence of labels pairs. In the second step, it computes maximal trees that are common to all the trees of each cluster. A known limitation of *Tree-Finder* is that it tends to miss many frequent patterns and is computationally expensive.

Sleuth [16] extends the ordered embedded pattern mining algorithm *Tree-Miner* [17]. *Sleuth* maintains a list of embedded occurrences with each pattern. It defines also a quadratic join operation over pattern occurrence lists to compute support for candidates. The join operation becomes inefficient when the size of pattern occurrence lists is large. Unlike *Sleuth* and *TreeFinder*, *Frestm* [18] mines *restrictedly* embedded patterns from a set of unordered trees, by enforcing some structure restrictions on pattern nodes.

The work on mining tree patterns in a single large tree/graph setting has so far been very limited. The only known papers are [4,5] which focus on mining tree patterns with only child edges from a single graph, [13,14] which respectively discover embedded tree patterns and homomorphic tree patterns with descendant edges from a single large tree.

The issue of mining mixed patterns has been neglected. A couple of works studied mining some very restricted forms of mixed patterns from a set of small trees [7,8]. *FAT-miner* [8] extracts embedded patterns with frequent common attributes (which can be seen as child edges) from a set of tree structured data with attributes. The algorithm of [7] extracts k-ee patterns from a set of trees. A k-ee pattern consists of set of induced patterns connected by at most k descendant edges for a user specified value k. Both *FAT-miner* and k-ee miner do not allow an arbitrary sequence of child/descendant edges in patterns. To the best of our knowledge, our work is the first one for mining unconstrained mixed patterns in the large data tree setting.

7   Conclusion

In this paper we have addressed the problem of mining most specific mixed patterns from a single large tree for which there are no previous algorithms. Mixed patterns allow the extraction of interesting information about the dataset which cannot be extracted from induced and/or embedded patterns. To formulate this problem in the single large tree setting we have introduced a novel way for definition the support of a pattern. We have designed effective techniques to prune the pattern search space. We also have designed a novel approach for efficiently computing the support of a candidate pattern which combines different techniques from tree databases: (a) holistic twig-join algorithms for efficiently finding the homomorphic occurrences of a pattern to a data tree, (b) materializing tree-pattern queries as bitmaps of inverted lists, and (c) incrementally computing the homomorphic occurrences of a pattern using exclusively the bitmaps of previously computed patterns. An extensive experimental evaluation shows that our approach is feasible and efficient in mining informative patterns beyond induced and embedded patterns from large data trees.

We are currently working on incorporating user-specified constraints to the proposed approach to enable constraint-based mixed pattern mining.

References

1. Baca, R., Krátký, M., Ling, T.W., Lu, J.: Optimal and efficient generalized twig pattern processing: a combination of preorder and postorder filterings. VLDB J. **22**(3), 369–393 (2013)
2. Bruno, N., Koudas, N., Srivastava, D.: Holistic twig joins: optimal XML pattern matching. In: SIGMOD (2002)
3. Chi, Y., Xia, Y., Yang, Y., Muntz, R.R.: Mining closed and maximal frequent subtrees from databases of labeled rooted trees. IEEE Trans. Knowl. Data Eng. **17**(2), 190–202 (2005)
4. Dries, A., Nijssen, S.: Mining patterns in networks using homomorphism. In: SDM (2012)
5. Kibriya, A.M., Ramon, J.: Nearly exact mining of frequent trees in large networks. Data Min. Knowl. Disc. **27**(3), 478–504 (2013)
6. Kilpeläinen, P., Mannila, H.: Ordered and unordered tree inclusion. SIAM J. Comput. **24**(2), 340–356 (1995)
7. Kim, S., Kim, H., Weninger, T., Han, J., Kim, H.D.: Authorship classification: a discriminative syntactic tree mining approach. In: SIGIR, pp. 455–464 (2011)
8. Knijf, J.D.: Fat-miner: mining frequent attribute trees. In: SAC, pp. 417–422 (2007)
9. Miklau, G., Suciu, D.: Containment and equivalence for a fragment of XPath. J. ACM **51**(1), 2–45 (2004)
10. Tan, H., Hadzic, F., Dillon, T.S., Chang, E., Feng, L.: Tree model guided candidate generation for mining frequent subtrees from XML documents. TKDD **2**(2), 9 (2008)
11. Tatikonda, S., Parthasarathy, S., Kurç, T.M.: TRIPS and TIDES: new algorithms for tree mining. In: CIKM (2006)
12. Termier, A., Rousset, M.-C., Sebag, M., TreeFinder: a first step towards XML data mining. In: ICDM (2002)
13. Wu, X., Theodoratos, D.: Leveraging homomorphisms and bitmaps to enable the mining of embedded patterns from large data trees. In: Renz, M., Shahabi, C., Zhou, X., Cheema, M.A. (eds.) DASFAA 2015. LNCS, vol. 9049, pp. 3–20. Springer, Cham (2015). doi:10.1007/978-3-319-18120-2_1
14. Wu, X., Theodoratos, D., Peng, Z.: Efficiently mining homomorphic patterns from large data trees. In: Navathe, S.B., Wu, W., Shekhar, S., Du, X., Wang, X.S., Xiong, H. (eds.) DASFAA 2016. LNCS, vol. 9642, pp. 180–196. Springer, Cham (2016). doi:10.1007/978-3-319-32025-0_12
15. Wu, X., Theodoratos, D., Wang, W.H., Sellis, T.: Optimizing XML queries: bitmapped materialized views vs. indexes. Inf. Syst. **38**(6), 863–884 (2013)
16. Zaki, M.J.: Efficiently mining frequent embedded unordered trees. Fundamenta Informaticae **66**(1–2), 35–52 (2005)
17. Zaki, M.J.: Efficiently mining frequent trees in a forest: algorithms and applications. IEEE Trans. Knowl. Data Eng. **17**(8), 1021–1035 (2005)
18. Zhang, S., Du, Z., Wang, J.T.: New techniques for mining frequent patterns in unordered trees. IEEE Trans. Cybern. **45**(6), 1113–1125 (2015)

Max-Cosine Matching Based Neural Models for Recognizing Textual Entailment

Zhipeng Xie[✉] and Junfeng Hu

Shanghai Key Laboratory of Data Science, School of Computer Science,
Fudan University, Shanghai, China
{xiezp,15210240075}@fudan.edu.cn

Abstract. Recognizing textual entailment is a fundamental task in a variety of text mining or natural language processing applications. This paper proposes a simple neural model for RTE problem. It first matches each word in the hypothesis with its most-similar word in the premise, producing an augmented representation of the hypothesis conditioned on the premise as a sequence of word pairs. The LSTM model is then used to model this augmented sequence, and the final output from the LSTM is fed into a softmax layer to make the prediction. Besides the base model, in order to enhance its performance, we also proposed three techniques: the integration of multiple word-embedding library, bi-way integration, and ensemble based on model averaging. Experimental results on the SNLI dataset have shown that the three techniques are effective in boosting the predicative accuracy and that our method outperforms several state-of-the-state ones.

Keywords: Textual entailment · Recurrent neural networks · LSTM

1 Introduction

In natural language text, there are always different ways to express the same meaning. This surface-level variability of semantic expressions is fundamental in tasks related to natural language processing and text mining. Textual entailment recognition (or RTE in short) is a specific semantic inference approach to model surface-level variability. As formulated by Dagan and Glickman [8], the task of Recognizing Textual Entailment is to decide whether the meaning of a text fragment Y (called the Hypothesis) can be inferred (is inferred) from another text fragment X (called the Premise). Giampiccolo et al. [10] extended the task to include the additional requirement that systems identify when the Hypothesis contradicts the Premise. The semantic inference needs are pervasive in a variety of NLP or text mining applications [7], inclusive of but not limited to, question-answering [13], text summarization [19], and information extraction. Given a pair of premise X and hypothesis Y, the relation between them may be: Entailment (Y can be inferred from X), Contradiction (Y is inferred to contradict X), or Neutral (X and Y are unrelated to each other). Table 1 presents a simple example to illustrate these three relations.

© Springer International Publishing AG 2017
S. Candan et al. (Eds.): DASFAA 2017, Part I, LNCS 10177, pp. 295–308, 2017.
DOI: 10.1007/978-3-319-55753-3_19

Table 1. An illustrative RTE example

Premise	Hypothesis	Relation
If you help the needy, God will reward you	Giving money to a poor man has good consequences	Entailment
	Giving money to a poor man has no consequences	Contradiction
	Giving money to a poor man will make you a better person	Neutral

A lot of research work has been devoted to the RTE problem in the last decade. The mainstream methods for recognizing textual entailment can be roughly divided into two categories:

- The first category attempts to provide a sequence of transformations allowing to derive the hypothesis Y from the premise X, by applying one transformation rule at each step.
- The second category simply thinks of RTE problem as a classification problem, where features (manually defined or automatically constructed) are extracted from the premise-hypothesis pairs.

In transformation-based RTE methods (also called rule-based methods), the underlying idea is to make use of inference rules (or entailment rules) for making transformation. However, the lack of such knowledge has been a major obstacle to improving the performance on RTE problem. The acquisition of entailment rules can be done either by learning algorithms which extract entailment rules from large text corpora, or by methods which extract rules from manually constructed knowledge resources.

Some research works have focused on extraction of entailment rules from manually constructed knowledge resources. WordNet [9] is the most prominent resource to extract entailment rules from. The synonymy and hypernymy relations (called substitutable relations) can be exploited to do direct substitution. To make use of the other non-substitutable relations (such as entailment and cause relations), Szpektor and Dagan [27] populated these non-substitutable relations with argument mapping which are extracted various resource, and thus extended WordNet's inferential relations at the syntactic representation level. FrameNet [2] is another manually constructed lexical knowledge base for entailment rule extraction. Aharon et al. [1] detected the entailment relations implied in FrameNet, and utilized FrameNet's annotated sentences and relations between frames to extract both the entailment relations and their argument mappings.

Although the entailment rules extracted from manually constructed knowledge resources have achieved sufficiently accuracy, their coverage is usually severely limited. A lot of research work has been devoted to learning entailment rules from a given text corpus. The DIRT algorithm proposed by Lin and Pantel [18] was based on the so-called Extended Distributional Hypothesis which states that phrases occurring in similar contexts are similar. An inference rule extracted by DIRT algorithm is actually a paraphrase rule, which is a pair of language patterns that can replace each other in a sentence. In DIRT, the language patterns are chains in dependency trees, with placeholders for nouns at the end of this chain. Different from the Extended Distributional Hypothesis adopted by DIRT, Glickman and Dagan [11] proposed an instance-based approach, which uses linguistic filters to identify paraphrase instances that describe the same fact and then rank the candidate paraphrases based on a probabilistically motivated paraphrase likelihood measure. Sekine [25] extracted the phrase between two named entities as candidate linear pattern, then identified a keyword in each phrase and joined phrases with the same keyword into sets, and finally linked sets which involve the same pairs of individual named entities. The sets or the links can be treated as paraphrases.

Besides paraphrase rules (which can be thought of as a specific case of entailment rules), a more general notion needed for RTE is that of entailment rules [8]. An entailment rule is a directional relation between two language patterns, where the meaning of one can be entailed or inferred from the meaning of the other, but not vice versa. Pekar [22] proposed a three-step method: it first identifies pairs of discourse-related clauses, and then creates patterns by extracting pairs of verbs along with relevant information as to their syntactic behaviour, and finally scores each verb pair in terms of plausibility of entailment. Recently, Szpektor et al. [28] presented a fully unsupervised learning algorithm for Web-based extraction of entailment rules. The algorithm takes as its input a lexical-syntactic template and searches the Web for syntactic templates that participate in an entailment relation with the input template.

Recently, with the availability of large high-quality dataset, especially with the Stanford Natural Language Inference (SNLI) corpus published in 2015 by Bowman et al. [4], there comes an upsurge of end-to-end neural models for RTE, where the fundamental problem is how to model a sentence pair (X, Y). The first and simplest model, proposed by Bowman et al. [4], encodes the premise and the hypothesis with two separate LSTMs, and then feeds the concatenation of their final outputs into a MLP for classification. This model does not take the interaction between the premise and the hypothesis into consideration. Several follow-ups have been proposed to solve this problem by modeling their interaction with a variety of attentive mechanisms [6,24,29]. These models treat sentences as word sequences, but some others adopt more principled choice to work on the tree-structured sentences, by explicitly model the compositionality and the recursive structure of natural language over trees. Such kind of work includes the Stack-augmented Parser-Interpreter Neural Network (SPINN) [5] and Tree-based Convolutional Neural Network (TBCNN) [21]. In Sect. 2, we shall have a look at all these neural models in more detail.

In this paper, we propose a simple neural method, called MaxConsine-LSTM, based on max-cosine matching for natural language inference. It first matches each word in the hypothesis (or the premise) with its most-similar word in the premise (or the hypothesis), and obtains a representation of hypothesis (or the premise conditioned on the premise (or the hypothesis). Then, LSTM is used to model the enhanced representations of hypothesis and premise into dense vectors. And finally, we concatenate the two dense vectors and feed it into a softmax layer to make the final decision about the relation between them. Experimental results have shown that our method achieves better or comparable performance when compared with state-of-the-art methods.

2 Related Work

In this section, we review some neural models that work for recognizing textual entailment.

The first and simplest neural model to RTE was proposed by Bowman et al. [4] in 2015, which uses separate LSTMs [15] to encode the premise and the hypothesis as dense fixed-length vectors and then feeds their concatenation into a multi-layer perceptron (MLP) or other classifiers for classification. It learns the sentence representation of premise and hypothesis independently, and does not take their interaction into consideration.

This first neural model suffer from the fact that the hypothesis and the premise are modeled independently, and thus the information cannot flow between them. To solve this problem, a sequential LSTM model is proposed in [24]. An LSTM reads the premise, and a second LSTM with different parameters reads a delimiter and the hypothesis, but its memory state is initialized with the final cell state of the previous LSTM. In this way, information from the premise can flow to the encoding of hypothesis.

To further facilitate the information flow between the premise and the hypothesis, Rocktäschel et al. [24] applied a neural attention model which can achieve better performance. When the second LSTM processes the hypothesis one word at a time, the first LSTM's output vectors are attended over, generating attention weights over all output vectors of the premise for every word in the hypothesis. The final sentence-pair representation is obtained from the last attention-weighted representation of the premise and the last output vector of the hypothesis. It outperforms Bowman et al. (2015) in that it checks for entailment or contradiction of individual word- and phrase-pairs.

Wang and Jiang [29] used an LSTM to perform word-by-word matching of the hypothesis with the premise. It is expected that the matching results that are critical for the final prediction will be "remembered" by the LSTM while less important matching results will be "forgotten".

The attentive mechanisms used in [24,29] are both between the hypothesis and the premise. It is sometimes helpful to exploit the attentive mechanism within the hypothesis or the premise. The long short-term memory-network (LSTMN) proposed by Cheng et al. [6] induces undirected relations among tokens as an intermediate step of learning representations, which can be thought of as an intra-attention mechanism. It has also been manifested that the intra-attention mechanism can lead to representations of higher quality.

The algorithms described above all deal with sentences as sequences of word vectors, and learn sequence-based recurrent neural networks to map them to sentence vectors. Another more principled choice is to learn the tree-structured recursive networks. Recursive neural networks explicitly model the compositionality and the recursive structure of natural language over tree. Bowman et al. [5] introduced the Stack-augmented Parser-Interpreter Neural Network (or SPINN in short) to combine parsing and interpretation within a single tree-sequence hybrid model by integrating tree-structured sentence interpretation into the linear sequential structure of a shift-reduce parser.

Mou et al. [21] proposed a tree-based convolutional neural network (TBCNN) to capture sentence-level semantics. TBCNN is more robust than sequential convolution in terms of word order distortion introduced by determinators, modifiers, etc. In TBCNN, a pooling layer aggregates information along tree, serving as a way of semantic compositionality. Finally, two sentences' information is combined and fed into a softmax layer for output.

3   Base Method

Recognizing textual entailment is concerned about the relation between two sequences - the premise X and the hypothesis Y. The commonly used encoder-decoder architecture processes the second sequence conditioned on the first one.

In this paper, we establish the connection of the hypothesis to the premise at the word level, where each word in the hypothesis is matched to and paired with its most-similar word in the premise. It leads to a simple base method called $MaxCosineLSTM$, consisting of three steps as illustrated in Fig. 1:

- **Step 1:** (Word Matching) Each word y_t in the hypothesis Y is matched to its most-similar word (denoted as $\gamma(y_t)$ in X, where the similarity between two words is measured as the cosine similarity between their embeddings. Such a match strategy can be thought of as a conditional representation of the hypothesis on the given premise;
- **Step 2:** (Sequence Modeling) For each time step $1 \leq t \leq m$, the concatenation of the word embedding of y_i and that of $\gamma(y_i)$ is fed into an LSTM layer, yield a vector representation of the hypothesis conditional on the premise;
- **Step 3:** (Decision Making) The final output of the LSTM layer is fed into a softmax layer to get the final decision about the relation between X and Y: Entailment, Contradiction, or Neutral.

3.1   Word Matching with Cosine Similarity

To judge whether a hypothesis Y can be inferred from a given premise X, it is of importance to get to know whether each word in Y expresses a similar meaning as one word in the premise X. Distributional Hypothesis proposed by Harris [12] has provided a guiding principle, which states that words appearing in similar contexts tend to have similar meanings. This principle has led to a variety of distributional semantic models (DSM) that use multidimensional vectors as word-sense representation. Latent semantic analysis [17] is a representative

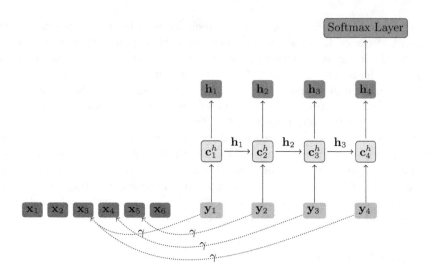

Fig. 1. Architecture of MaxCosine-LSTM

method of this kind, which applies truncated Singular Value Decomposition to a matrix of word-context co-occurrence matrix. Recently, neural network-based methods, such as Skip-Gram [20] and Glove [23], have been proposed to represent words as low-dimensional vectors called word embeddings. Compared with traditional DSM methods, these word embeddings have shown superior performance in similarity measurement between words.

Let $\mathbf{D} \in \mathbb{R}^{d \times |V|}$ be a learned embedding matrix for a finite vocabulary V of $|V|$ words, with d denoting the dimensionality of word embeddings. The i-th column, $(\mathbf{D}(i) \in \mathbb{R}^d$, represents the embedding of the i-th word in the vocabulary V. Given two words x and y in the vocabulary, we can measure their semantic similarity as the most commonly-used cosine similarity between their word embeddings $\mathbf{x} = \mathbf{D}(x)$ and $\mathbf{y} = \mathbf{D}(y)$:

$$sim(x, y) = cosine(\mathbf{x}, \mathbf{y}) = \frac{\langle \mathbf{x}, \mathbf{y} \rangle}{\|\mathbf{x}\| \cdot \|\mathbf{y}\|} \tag{1}$$

Therefore, for each word $y_t (1 \leq t \leq m)$ in the hypothesis $Y = y_1 y_2 ... y_m$, we use $\gamma(y_t)$ to denote the word in premise $X = x_1 x_2 ... x_n$ that is of highest semantic similarity with y_t:

$$\gamma(y_t) = \arg \max_{x_s} sim(x_s, y_t) \tag{2}$$

Such a mapping γ can build the connection from the hypothesis to the premise, at the word level. Each sentence pair (X, Y) can then be represented as a sequence $Z = z_1 z_2 ... z_m$ where $z_t = (y_t, \gamma(y_t)), 1 \leq t \leq m$, denotes the t-th word in hypothesis Y paired with its most-similar word $\gamma(y_t)$ in the premise Y. This pairing process do associate the most relevant words from the hypothesis

Y to the premise X. We use $\mathbf{z}_t \in \mathbb{R}^{2d}$ to denote the concatenation of word embeddings of y_t and $\gamma(y_t)$ for $z_t = (y_t, \gamma(y_t))$:

$$\mathbf{z}_t = \begin{bmatrix} \mathbf{D}(y_t) \\ \mathbf{D}(\gamma(y_t)) \end{bmatrix} \tag{3}$$

Thus, we can get a sequence $\mathbf{Z} = (\mathbf{z}_1, \mathbf{z}_2, \ldots, \mathbf{z}_m)$ of vectors in $\mathbb{R}^{2d}$, which can be thought of as an augmented representation of the hypothesis with reference to the premise.

3.2 Sequence Modeling with LSTM

Next, we would like to transform the sequence $\mathbf{Z}$ into a vector, as the representation of the hypothesis conditioned on the premise. Recurrent neural networks are naturally suited for modeling variable-length sequences, which can recursively compose each $(2d)$-dimensional vector $\mathbf{z}_t$ with its previous memory. Traditional recurrent neural networks often suffer from the problem of vanishing and exploding gradients [3,14], making it hard to train models. In this paper, we adopt the Long Short-Term Memory (LSTM) model [15] which partially solves the problem by using gated activation function.

The LSTM maintains a memory state $\mathbf{c}$ through all the time steps, in order to save the information over long time periods. Concretely, at each time step t, the concatenation of two word embeddings, $\mathbf{z}_t$, is fed into the LSTM as the input. The LSTM updates the memory state from the previous $\mathbf{c}_{t-1}$ to the current $\mathbf{c}_t$, by adding new content that should be memorized and erasing old content that should be forgotten. The LSTM also outputs current content that should be exposed. relying on memory is updated by partially forgetting the previous memory $\mathbf{c}_{t-1}$ and adding a new memory content $\tanh(\mathbf{W}^c\mathbf{H}+\mathbf{b}^c)$. The output gates $\mathbf{o}_t$ modulates the amount of memory content exposure.

$$\mathbf{H} = \begin{bmatrix} \mathbf{z}_t \\ \mathbf{h}_{t-1} \end{bmatrix} \tag{4}$$

$$\mathbf{i}_t = \sigma(\mathbf{W}^i\mathbf{H} + \mathbf{b}^i) \tag{5}$$

$$\mathbf{f}_t = \sigma(\mathbf{W}^f\mathbf{H} + \mathbf{b}^f) \tag{6}$$

$$\mathbf{o}_t = \sigma(\mathbf{W}^o\mathbf{H} + \mathbf{b}^o) \tag{7}$$

$$\mathbf{c}_t = \mathbf{f}_t \odot \mathbf{c}_{t-1} + \mathbf{i}_t \odot \tanh(\mathbf{W}^c\mathbf{H} + \mathbf{b}^c) \tag{8}$$

$$\mathbf{h}_t = \mathbf{o}_t \odot \tanh(\mathbf{c}_t) \tag{9}$$

To prevent overfitting, dropout is applied to regularize the LSTMs. Dropout has shown a great success when working with feed-forward networks [26]. As indicated in [30], our method drops the input and the output of the LSTM layer, with the same dropout rate.

3.3   Decision Making with Softmax Layer

The final output, $\mathbf{h}_m$, generated by the LSTM on the enhanced representation $\mathbf{Z}$ of the hypothesis Y conditioned on the premise X, is fed into a softmax layer which performs the following two steps.

As the first step, $\mathbf{h}_m$ goes through a linear transformation to get a 3-dimensional vector $\mathbf{p}$:

$$\mathbf{p} = \mathbf{W}^s \mathbf{h}_m + \mathbf{b}^s \tag{10}$$

where the weight matrix $\mathbf{W}^s \in \mathbb{R}^{3 \times k}$, and bias vector $\mathbf{b}^s \in \mathbb{R}^3$ are the parameters of the softmax layer.

The $\mathbf{p} = (p_1, p_2, p_3$ is then transformed by a nonlinear softmax function, resulting in a probabilistic prediction $(\hat{t}_1, \hat{t}_2, \hat{t}_3)$ over the three possible labels (Entailment = 1, Contradiction = 2, or Neutral = 3):

$$\hat{t}_i = \frac{\exp p_i}{\sum_{j=1}^3 \exp p_j} \tag{11}$$

During the training phase, the cross-entropy error function is used as the cost function. At test time, the label with the highest probability, $\arg\max_{1 \le i \le 3} p_i$, is output as the predicted label.

4   Improvements over the Base Method

To obtain better predictive performance, three optional techniques are applied on the base method described above:

- Multiple word-embedding libraries, which improves the vector representations of words;
- Biway-LSTM integration, which enhances the representations of relations between text pairs;
- Ensemble based on model averaging, which produces more accurate predictions.

4.1   Mutliple Embeddings

Word2vec and Glove are two popular software for learning word embeddings from text corpus. We use $\mathbf{D}_{w2v}$ and $\mathbf{D}_{glove}$ to denote their induced word embedding libraries, respectively. For each word x, we can represent it as a vector $\mathbf{D}(x)$ as the concatenation of its embedding from $\mathbf{D}_{w2v}$ and that from $\mathbf{D}_{glove}$, that is,

$$\mathbf{D}(x) = [\mathbf{D}_{w2v}(x)\mathbf{D}_{glove}(x)].$$

Or equivalently, a new embedding matrix $\mathbf{D} \in \mathbb{R}^{(2d) \times |V|}$ is constructed by concatenating $\mathbf{D}_{w2v}$ with $\mathbf{D}_{glove}$. The semantic similarity between words from the hypothesis and the premise is calculated with regards to this new embedding matrix. The aim of using this technique is to integrated the potentially complementary information provided by different word embedding libraries. As another possible solution, we can also make use of canonical correlation analysis (CCA) to project the two embedding libraries into a common semantic space, and thus induce a new embedding library. However, we do not make any further exploration, because it is out of the scope of this paper.

4.2 Biway-LSTM Integration

In RTE problem, the premise should also play an important role as the hypothesis. In the previous two subsections, we have described how to model the hypothesis conditioned on the premise. This idea can also be applied the other way round, i.e. to model the premise conditioned on the hypothesis.

To justify this statement, let us consider the following simple example. Let $X^{(1)}$ = "John failed to pass the exam.", $X^{(2)}$ = "John succeeded in passing the exam.", and Y = "John passed the exam." It is clear that $X^{(2)}$ entails Y, but $X^{(1)}$ contradicts Y. However, the enhanced representation of Y conditioned on $X^{(1)}$ is the same as that of Y conditioned on $X^{(2)}$. Therefore, we cannot discriminate these two situations based on only the enhanced representation of Y.

To do a remedy, we extend our base model to the biway architecture illustrated in Fig. 2. Two LSTMs are used to separately model the enhanced representation of the premise and that of the hypothesis. Their final output vectors are then concatenated and fed into a softmax layer to do the final decision.

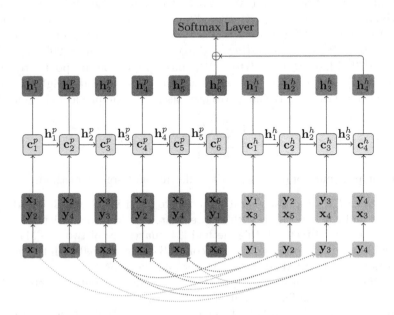

Fig. 2. The biway architecture of MaxCosine-LSTM

4.3 Ensemble by Model Averaging

Combining multiple models generally leads to better performance. The component models are expected to be diverse and accurate, in order to produce an ensemble of high quality. In this paper, all the components are homogeneous, that is, all of them are induced by our base method (optionally enhanced with bi-embedding integration and/or biway integration). The diversity of these components comes from random initialization of the model parameters, with different random seeds. The predictions from these component models are averaged to make the final decision.

5   Experiments

In the experimental part, we evaluate our method on the Stanford Natural Language Inference (SNLI) dataset [4] which consists of about 570 K sentence pairs. After filtering sentence pairs with unknown class labels, we get a train data of 549,367 pairs, a validation data of 9,842 pairs, a test data of 9,824 pairs. This dataset has been commonly used by previous state-of-the-art neural models.

To train our model, we use cross-entropy loss $J(\theta)$ in Eq. (12), the B is the mini-batch size, $t^{(i)}$ is the true label of sample i, $\{t^{(i)} = j\}$ is 1 if $t^{(i)}$ equals j else 0. We use stochastic mini-batch gradient descent with the ADAM optimizer [16], we set ADAM's hyperparameters $\beta_1 = 0.9$ and $beta_2 = 0.999$ and the initial learning rate to 0.001. We use both the pre-trained Glove [23] model **glove.840B.300d** and Word2vec [20] model **GoogleNews-vectors-negative300** to initialize word embeddings. We don't tune the word embeddings when train, OOV words' vectors are set to be the average of their window words' vectors, follow same setting in the Match-LSTM paper [29]. We fix the length of LSTM hidden states k to 300D, and apply various dropout rate on the input layer. We don't apply any regularization to the network weights, and use batch size $B = 128$ when training. We use Lasagne[1] to implement these models.

$$J(\theta) = -\frac{1}{B} \sum_{i=1}^{B} \sum_{j=1}^{3} \mathbf{1}\{t^{(i)} = j\} \log \hat{t}_j \tag{12}$$

We compare our approach with the following state-of-art methods:

- Separate-LSTM: the first neural method proposed in [4], which encodes the premise and the hypothesis with two separate LSTMs independently.
- Sequential-LSTM method: this method [24] makes use of two LSTMs, where an LSTM reads the premise, and a second LSTM initialized with the final cell state of the first LSTM reads a delimiter and the hypothesis.
- Attention-LSTM: the method in [24] that attends over output vectors of the premise only for the final output of the hypothesis.
- Word-by-Word Attention-LSTM: the method in [24] that attends over output vectors of the premise for every word in the hypothesis.

---

[1] http://lasagne.readthedocs.io/en/latest/.

- matchLSTM with word embedding: the method in [29] that performs word-by-word matching of the hypothesis with the premise.

We implemented our versions of Attention-LSTM, Word-by-Word Attention-LSTM, and matchLSTM, because the codes of the original papers are not made publicly available.

Results on the SNLI corpus are summarized in Table 2. Our method MaxCosine-LSTM that uses all three improvement techniques has achieved the best performance when compared with these state-of-the-art methods.

Table 2. Empirical accuracy of the MaxCosine-LSTM model compared with previous results

Model	k	Test accuracy
Separate-LSTM	100	77.8
Sequential-LSTM	100	80.9
Attention-LSTM	100	82.3
Word-by-Word Attention-LSTM (our implementation)	100	83.3
mLSTM with word embedding (our implementation)	300	84.2
MaxCosine-LSTM-biEmb-biWay-Ensemble	300	**85.0**

We also conduct experiments to study the effectiveness of the three techniques in Sect. 4. With the dropout ratio set as 0.3, 0.4, and 0.5, Table 3 shows the results of using different combinations of the three techniques. The accuracies of all the non-ensemble methods (MaxCosine-LSTM, MaxCosine-LSTM-biEmb, MaxCosine-LSTM-biWay, MaxCosine-LSTM-biEmb-biWay) are reported as the average over 5 runs.

It can be easily observed that all the three techniques have their own contributions to the final predictive ability. Discarding any technique would lead to some decrease in the prediction accuracy. It is also evident that the ensemble technique is the most important one, the biway integration goes next, and the bi-embedding integration is relatively less significant.

In addition, we can also observe the following two facts:

- The effect of ensemble technique on MaxCosine-LSTM-biWay is more significant than its effect on MaxCosine-LSTM or MaxCosine-LSTM-biEmb.
- The effect of ensemble is more significant with smaller dropout ratio.

These observations can be explained by the fact that the ensemble technique based on model averaging works better on the method of higher model complexity which usually has higher variance in bias-variance decomposition. Intuitively, the model complexity of MaxCosine-LSTM-biWay is much higher than that of MaxCosine-LSTM and that of MaxCosine-LSTM-biEmb. Another fact is that the dropout ratio can control the model complexity: lower dropout ratio means higher model complexity, because dropout is one kind of regularization.

Table 3. Effects of the three techniques for MaxCosine-LSTM model

Dropout ratio	Techiques used	Accuracy
0.3	MaxCosine-LSTM	82.45%
	MaxCosine-LSTM-biEmb	82.50%
	MaxCosine-LSTM-biWay	82.93%
	MaxCosine-LSTM-Ensemble	83.65%
	MaxCosine-LSTM-biEmb-biWay	83.26%
	MaxCosine-LSTM-biEmb-Ensemble	83.92%
	MaxCosine-LSTM-biWay-Ensemble	84.78%
	MaxCosine-LSTM-biEmb-biWay-Ensemble	84.98%
0.4	MaxCosine-LSTM	82.07%
	MaxCosine-LSTM-biEmb	82.56%
	MaxCosine-LSTM-biWay	83.05%
	MaxCosine-LSTM-Ensemble	83.09%
	MaxCosine-LSTM-biEmb-biWay	83.31%
	MaxCosine-LSTM-biEmb-Ensemble	83.57%
	MaxCosine-LSTM-biWay-Ensemble	84.23%
	MaxCosine-LSTM-biEmb-biWay-Ensemble	84.74%
0.5	MaxCosine-LSTM	81.68%
	MaxCosine-LSTM-biEmb	82.18%
	MaxCosine-LSTM-biWay	82.77%
	MaxCosine-LSTM-Ensemble	82.37%
	MaxCosine-LSTM-biEmb-biWay	83.44%
	MaxCosine-LSTM-biEmb-Ensemble	83.05%
	MaxCosine-LSTM-biWay-Ensemble	83.89%
	MaxCosine-LSTM-biEmb-biWay-Ensemble	84.43%

6   Conclusion and Outlook

We proposed a simple neural method to determine the relation between a hypothesis and a premise. It relies on word semantic matching from the hypothesis to the premise (or vice versa), then makes use of the LSTM to do the sequence modeling, and finally feed the final output from the LSTM into a softmax layer to make the classification decision. After equipped with three techniques to improve its performance, experimental results have shown that our method has achieved better accuracies than state-of-the-art systems. In addition, it is also shown that the three techniques all have their own contribution to the accuracy obtained.

Acknowledgments. This work is supported by National High-Tech R&D Program of China (863 Program) (No. 2015AA015404), and Science and Technology Commission of Shanghai Municipality (No. 14511106802). We are grateful to the anonymous reviewers for their valuable comments.

References

1. Aharon, R.B., Szpektor, I., Dagan, I.: Generating entailment rules from FrameNet. In: Proceedings of the 48th Annual Meeting of the Association for Computational Linguistics (Short Papers), pp. 241–246. Association for Computational Linguistics (2010)
2. Baker, C.F., Fillmore, C.J., Lowe, J.B.: The Berkeley FrameNet project. In: Proceedings of the 36th Annual Meeting of the Association for Computational Linguistics and 17th International Conference on Computational Linguistics, vol. 1, pp. 86–90. Association for Computational Linguistics (1998)
3. Bengio, Y., Simard, P., Frasconi, P.: Learning long-term dependencies with gradient descent is difficult. IEEE Trans. Neural Netw. **5**(2), 157–166 (1994)
4. Bowman, S.R., Angeli, G., Potts, C., Manning, C.D.: A large annotated corpus for learning natural language inference. In: Proceedings of the 2015 Conference on Empirical Methods in Natural Language Processing, pp. 632–642 (2015)
5. Bowman, S.R., Gauthier, J., Rastogi, A., Gupta, R., Manning, C.D., Potts, C.: A fast unified model for parsing and sentence understanding. In: Proceedings of the 54th Annual Meeting of the Association for Computational Linguistics, pp. 1466–1477 (2016)
6. Cheng, J., Dong, L., Lapata, M.: Long short-term memory-networks for machine reading. In: Proceedings of the 2016 Conference on Empirical Methods in Natural Language Processing, pp. 551–561 (2016)
7. Dagan, I., Dolan, B., Magnini, B., Roth, D.: Recognizing textual entailment: rational, evaluation and approaches. Nat. Lang. Eng. **15**(4), i–xvii (2009)
8. Dagan, I., Glickman, O.: Probabilistic textual entailment: generic applied modeling of language variability. In: Learning Methods for Text Understanding and Mining, pp. 26–29 (2004)
9. Fellbaum, C. (ed.): Wordnet: An Electronic Lexical Database. MIT Press, Cambridge (1998)
10. Giampiccolo, D., Magnini, B., Dagan, I., Dolan, B.: The third pascal recognizing textual entailment challenge. In: Proceedings of the ACL-PASCAL Workshop on Textual Entailment and Paraphrasing, pp. 1–9. Association for Computational Linguistics (2007)
11. Glickman, O., Dagan, I.: Acquiring lexical paraphrases from a single corpus. In: Recent Advances in Natural Language Processing III, pp. 81–90. John Benjamins Publishing, Amsterdam (2004)
12. Harris, Z.S.: Distributional structure. Word **10**(2–3), 146–162 (1954)
13. Harabagiu, S., Hickl, A.: Methods for using textual entailment in open-domain question answering. In: Proceedings of the 21st International Conference on Computational Linguistics and 44th Annual Meeting of the Association for Computational Linguistics, pp. 905–912 (2006)
14. Hochreiter, S.: Untersuchungen zu dynamischen neuronalen netzen. Technische Universität München, Diploma (1991)
15. Hochreiter, S., Schmidhuber, J.: Long short-term memory. Neural Comput. **9**(8), 1735–1780 (1997)
16. Kingma, D., Ba, J.: Adam: a method for stochastic optimization. arXiv preprint arXiv:1412.6980 (2014)
17. Landauer, T.K., Dumais, S.T.: A solution to Platos problem: the latent semantic analysis theory of acquisition, induction, and representation of knowledge. Psychol. Rev. **104**(2), 211–240 (1997)

18. Lin, D., Pantel, P.: DIRT - Discovery of inference rules from text. In: Proceedings of the Seventh ACM SIGKDD International Conference on Knowledge Discovery and Data Mining, pp. 323–328 (2001)
19. Lloret, E., Ferrández, O., Munoz, R., Palomar, M.: A text summarization approach under the influence of textual entailment. In: Proceedings of the 5th International Workshop on Natural Language Processing and Cognitive Science, pp. 22–31 (2008)
20. Mikolov, T., Sutskever, I., Chen, K., Corrado, G.S., Dean, J.: Distributed representations of words and phrases and their compositionality. In: Advances in Neural Information Processing Systems (NIPS), pp. 3111–3119 (2013)
21. Mou, L., Men, R., Li, G., Xu, Y., Zhang, L., Yan, R., Jin, Z.: Natural language inference by tree-based convolution and heuristic matching. In: Proceedings of the 54th Annual Meeting of the Association for Computational Linguistics (Short Papers), pp. 130–136 (2016)
22. Pekar, V.: Acquisition of verb entailment from text. In: Proceedings of the Main Conference on Human Language Technology Conference of the North American Chapter of the Association of Computational Linguistics, pp. 49–56. Association for Computational Linguistics (2006)
23. Pennington, J., Socher, R., Manning, C.D.: Glove: global vectors for word representation. In: Proceedings of the 2014 Conference on Empirical Methods on Natural Language Processing, pp. 1532–1543 (2014)
24. Rocktäschel, T., Grefenstette, E., Hermann, K.M., Kočiský, T., Blunsom, P.: Reasoning about entailment with neural attention. ArXiv Preprint arXiv:1509.06664 (2016)
25. Sekine, S.: Automatic paraphrase discovery based on context and keywords between NE pairs. In: Proceedings of International Workshop on Paraphrase (IWP 2005), pp. 4–6 (2005)
26. Srivastava, N.: Improving neural networks with dropout. Ph.D. Thesis, University of Toronto (2013)
27. Szpektor, I., Dagan, I.: Augmenting wordnet-based inference with argument mapping. In: Proceedings of the 2009 Workshop on Applied Textual Inference, pp. 27–35. Association for Computational Linguistics (2009)
28. Szpektor, I., Tanev, H., Dagan, I., Coppola, B., Kouylekov, M.: Unsupervised acquisition of entailment relations from the web. Nat. Lang. Eng. **21**(01), 3–47 (2015)
29. Wang, S., Jiang, J.: Learning natural language inference with LSTM. In: Proceedings of the 15th Annual Conference of the North American Chapter of the Association for Computational Linguistics, pp. 700–704 (2016)
30. Zaremba, W., Sutskever, I., Vinyals, O.: Recurrent neural network regularization. ArXiv Preprint arXiv:1409.2329 (2014)

An Intelligent Field-Aware Factorization Machine Model

Cairong Yan[✉], Qinglong Zhang, Xue Zhao, and Yongfeng Huang

School of Computer Science and Technology,
Donghua University, Shanghai 201620, China
cryan@dhu.edu.cn

Abstract. The widely-used field-aware factorization machines model (FFM) takes the interactions of all the text features into consideration which will lead to a large number of invalid calculations. An intelligent field-aware factorization machine model (iFFM) is proposed in this paper. In the model, the key attributes are promoted and the factor selection operations are embedded into the computation process intelligently by using the auto feature engineering technology. Meanwhile, Markov Chain Monte Carlo (MCMC) and stochastic gradient descent (SGD) methods are applied to optimize the loss function and improve the recommendation accuracy. In order to get better diversity, a new model iFFM-2 is put forward, which is the linear weighted combination of iFFM and a model built based on the heat-spreading algorithm. The experimental results show that iFFM can obtain higher accuracy and computation efficiency compared with FFMs, iFFM-2 inherits the accuracy of iFFM, and it can provide better diversity.

Keywords: FFMs · Recommender systems · MCMC · Heat-spreading algorithm

1 Introduction

As an effective information filtering method, recommendation system is one of the most effective methods to solve the problem of information overload and realize personalized information service. The current mainstream recommender system has been applied in many fields, including online e-commerce [1] (e.g., Netflix, Amazon, eBay, Alibaba, and Douban), Information retrieval [2] (e.g., iGoogle, MyYahoo, GroupLens, and Baidu), mobile applications [3] (e.g., Daily Learner and Appjoy) and life services [4] (e.g., Compass and M-CRS).

As a typical collaborative filtering algorithm, non-negative matrix factorization (NMF) [5] has been paid more attention due to its simple thought, easy to understand, high recommendation accuracy, and wide application range. But its extensibility is greatly limited to some extent because NMF is mainly confined to the user and item characteristic. For this reason, scholars have put forward a lot of solutions, e.g., SVD++ [6], pairwise interaction tensor factorization (PITF) [7] and factorization machines (FMs) [8, 9], etc. Among these ways, FMs is widely used since it synthesizes the advantages of the support vector machine (SVM) [10] and MF, and can effectively

© Springer International Publishing AG 2017
S. Candan et al. (Eds.): DASFAA 2017, Part I, LNCS 10177, pp. 309–323, 2017.
DOI: 10.1007/978-3-319-55753-3_20

solve the sparse problem of high dimensional feature combination with linear time complexity.

Learned from Michael Jahrer's paper [11], Yu-Chin Juan and his team put forward a new model, field-aware factorization machines model (FFMs) [12]. It uses the concept of "field" and attributes the same properties into the same field. Since its inception, FFMs has made significant headway in the recommender system and outperforms existing models in the worldwide click-through rate (CTR) prediction competitions held by Criteo[1] and Avazu[2].

Like FMs, FFMs takes the interactions of all the text features into consideration, by sharing a specific feature implicit vector to calculate parameters of factorization. However, in the real application scene, the text features are usually too many and not all interactions are valid. In order to improve the calculation efficiency, a feasible solution is to manually specify interactive features and the order of the features interact.

In this paper, an intelligent field-aware factorization machine model (iFFM) is proposed. In the model, the key attributes are promoted and the factor selection operations are embedded into the computation process intelligently by using the auto feature engineering technology. To achieve better diversity with short computation time, we introduce the heat-spreading algorithm [13], build a new model HeatS and combine it with iFFM to form a hybrid model iFFM-2 by using weighted linear fusion method.

The rest of the paper is organized as follows. Section 2 introduces preliminary of our models. We propose an intelligent field-aware factorization machine model and analyze its properties mathematically. We then use the heat-spreading algorithm to build a new model HeatS and combined with iFFM to gain better diversity in Sect. 4. The experimental results are presented in Sect. 5. And finally, we conclude this paper in Sect. 6.

2   Preliminary

In this section, we will first give a concise overview of FMs, which forms the foundation of our solution. We then present its recent extension FFMs. Some preprocessing methods that we used are also introduced.

2.1   FMs

FMs is a generic machine learning model that firstly proposed by Steffen Rendle in 2010 [8], aiming to solve the problem of feature combination under sparse data. The second-order FMs is defined as:

$$\hat{y}(x) = \omega_0 + \sum_{i=1}^{n} \omega_i x_i + \sum_{i=1}^{n} \sum_{j=1}^{n} \langle \mathbf{v}_i, \mathbf{v}_j \rangle x_i x_j, \tag{1}$$

---

[1] https://www.kaggle.com/c/criteo-display-ad-challenge .

[2] https://www.kaggle.com/c/avazu-ctr-prediction .

where n is the number of instances, k is the dimensionality of the factorization and $\Theta = \{\omega_0, \omega_i, \omega_{ij}\}$ are the model parameters.

This model is closely related to nested polynomial regression of order d with the very important difference that FM use factorized interactions whereas polynomial regression has independent parameters per interaction [14].

Let $\langle \mathbf{v}_i, \mathbf{v}_j \rangle = \sum_{f=1}^{k} v_{i,f} \cdot v_{j,f}$, then we can simplify Eq. (1) to

$$y(x) = \omega_0 + \sum_{i=1}^{n} \omega_i x_i + \frac{1}{2} \sum_{f=1}^{k} \left(\left(\sum_{i=1}^{n} v_{i,f} x_i \right)^2 - \sum_{i=1}^{n} v_{i,f}^2 x_i^2 \right). \qquad (2)$$

For deriving a point estimator of the model parameters, the task is to minimize an objective consisting of loss function l and regularization

$$\text{OPT}(S) = \operatorname*{argmin}_{\Theta} \left(\sum_{(c,x,y) \in S} cl(\hat{y}(x|\Theta), y) + \frac{1}{2} \sum_{\theta \in \Theta} \lambda_\theta \|\theta\|^2 \right), \qquad (3)$$

where the loss function l can be the least-squares for regression or logistic loss for binary classification.

Then we can use stochastic gradient descent (SGD) method to alternate least squares (ALS) or Markov Chain Monte Carlo (MCMC) method to minimize Eq. (3) [9].

FMs can be trained and predicted with a complexity O(kn), making it a quite efficient model.

2.2   FFMs

As introduced in Sect. 1, FFMs is based on FMs, it attributes the same properties into the same field [12]. In FFMs, each feature has several latent vectors. Depending on the field of other features, one of them is used to do the inner product. According to the field-sensitive properties, we can derive the FFMs mathematical equation

$$\hat{y}(x) = \omega_0 + \sum_{i=1}^{n} \omega_i x_i + \sum_{i=1}^{n} \sum_{j=i+1}^{n} \langle \mathbf{v}_{i,f_j}, \mathbf{v}_{j,f_i} \rangle x_i x_j, \qquad (4)$$

where f_i and f_j are respectively the fields of i and j. If f is the number of fields, then the number of variables of FFMs is nfk, and the complexity to compute Eq. (4) is O(kn^2). FFMs are available in the large-scale machine learning libraries GraphLab Create[3]. For more details about FMs, the reader is referred to reference [12].

2.3   One-Hot Encoding

One-hot encoding[4], also known as one valid code, is a method using N status register to encode N state. Each state has its own register, and at any time, only one is effective.

---

[3] https://github.com/turi-code/python-libffm .

[4] https://en.wikipedia.org/wiki/One-hot .

One-hot encoding solves the problem that the classifier cannot deal with the attribute data effectively, and to a certain extent, expanded the number of features. In this paper, the categorical feature is transformed into a numerical feature by using one-hot encoding to facilitate the prediction of the model.

In this paper, implicit feature matrix will be learned step-by-step. For each iteration, an optimal feature will be selected greedily to join and update the model. In the process of adding the factors, we will try depth-first and breadth-first principle respectively, calculate $T(t)$ in Eq. (17) and find the best way to add factors.

2.4   Sample Data Generation

Consider the widely used FFMs data format, we define the iFFM data structure as a triplet, (namely: label field: feature: value), where each category is considered as a field. To explain how the iFFM format works, Table 1 shows an example.

Table 1. An example of iFFM data format

Clicked	Country	Day	Ad_type
1	USA	16/11/15	Movie

We generate the following iFFM format:

1, Country: USA:1, Day:16/11/15:1, Ad_type: Movie:1.

Since zero-valued features do not contribute to the model at all, in order to improve the training and prediction speed, we only store the non-zero features.

For numerical features, simply setting the feature value as 1 is not reasonable. In this paper, we discretize each numerical feature to a categorical one, and in order to reduce the effect of noise, we transform the feature to int (feature). Then the value will be the actual value of the data. Table 2 is an example to predict what score a user will rate a movie.

We generate the data into three records as follows:

5.6, User: XSH:1, Movie: I Belonged to You:1, Genre: Love:1, Rating: 5:5.6
5.6, User: XSH:1, Movie: I Belonged to You:1, Genre: Art:1, Rating: 5:5.6
5.6, User: XSH:1, Movie: I Belonged to You:1, Genre: Comedy:1, Rating: 5:5.6

Table 2. An example of numerical features

User	Movie	Genre	Rating
XSH	I Belonged to You	Love|Art|Comedy	5.6

3   IFFM

3.1   Attribute Boosting

During the calculation, FFMs take the same measures for all the attributes. But in the real scene, users and items are targeted, i.e., the relevant attributes will show a certain preference (e.g., the male users born in the 1980s and early 1990s may pay more attention to game adapted movies such as World of Warcraft, Resident Evil, and Tomb Raider, etc.). At the same time, the younger girls born after 1995 generally are more interested in love topics. In addition, the importance of different attributes is also different. Normally, a user's occupation has a greater impact on interests than home addresses. So, we boost certain particularly important attributes to better exploit the potential features of attributes, thereby improving prediction accuracy. The specific process is as follows.

Firstly, we map the feature x_i into $x_i + B_i$. So, x_i can be dynamically changed to accommodate different combinations of users and items during the learning process. Assume that the latent interest of user u is p_u, the corresponding weight matrix is u_i, then we can build the user's attribute boosting function:

$$\hat{y}'_{user} = \sum_{i=1, i\notin t(u)}^{t_1} w_{f_i}\left(x_i + u_i^T p_u\right) = \hat{y}_{user}(x) + \sum_{i=1, i\notin t(u)}^{t_1} w_{f_i} x_i, \tag{5}$$

where $t(u)$ is the type of users.

Similarly, we can gain the item's attribute boosting function:

$$\hat{y}'_{item}(x) = \sum_{i=1, i\notin t(v)}^{t_2} w_{f_i}\left(x_i + v_i^T q_v\right) = \hat{y}_{item}(x) + \sum_{i=1, i\notin t(v)}^{t_2} w_{f_i} x_i, \tag{6}$$

where q_v indicates the degree to which an item is associated with an attribute, v_i is the corresponding weight matrix, and $t(v)$ is the type of items.

Then we can simplify Eq. (4) to

$$\hat{y}(x) = \omega_0 + \sum_{i=1}^{n} \omega_i x_i + \hat{y}_{user}(x) + \hat{y}_{item}(x) + \sum_{i=1}^{n} \sum_{j=i+1}^{n} \left\langle v_{i,f_j}, v_{j,f_i} \right\rangle x_i x_j. \tag{7}$$

3.2   Model Trainning

Normally, the number of p_u, q_v and the type of attributes are fixed, so we can use MCMC to generate the distribution of $\hat{y}_{user}$ and $\hat{y}_{item}$ by sampling.

Assuming $\hat{y}_{user}$ is a Gaussian distribution with mean $w_{f_i} \cdot u_i^T p_u$ and variance Λ_{uy}^{-1}, the likelihood of the parameters w_{f_i}, u_i and p_u can be written as:

$$P\left(\hat{y}_{user} | w, u, p_u, \Lambda_{uy}\right) = \prod_{i=1}^{t_1} N\left(\hat{y}_{user}^{(i)} | w_{f_i} \cdot u_i^T p_u, \Lambda_{uy}^{-1}\right). \tag{8}$$

Similarly, w_{f_i}, u_i, and p_u can be assumed as Gaussian distributions with mean $\{\mu_f, \mu_p, \mu_u\}$ and variance $\{\Lambda_f^{-1}, \Lambda_p^{-1}, \Lambda_u^{-1}\}$. Like Bayesian PMF [15], we set the prior distribution of the hyperparameters $\Theta_f = \{\mu_f, \Lambda_f^{-1}\}$, $\Theta_p = \{\mu_p, \Lambda_p^{-1}\}$, and $\Theta_u = \{\mu_u, \Lambda_u^{-1}\}$ as the Gaussian-Wishart distribution,

$$
\begin{aligned}
P(\Theta_f|\Theta_0) &= N\left(\mu_f \middle| \mu_0, (\beta_0 \Lambda_f)^{-1}\right) W(\Lambda_f | W_0, v_0), \\
P(\Theta_p|\Theta_0) &= N\left(\mu_p \middle| \mu_1, (\beta_1 \Lambda_p)^{-1}\right) W(\Lambda_p | W_1, v_1), \\
P(\Theta_u|\Theta_0) &= N\left(\mu_u \middle| \mu_1, (\beta_1 \Lambda_u)^{-1}\right) W(\Lambda_u | W_1, v_1),
\end{aligned}
\tag{9}
$$

where $\Theta_0 = \{\mu_0, \mu_1, \beta_0, \beta_1, W_0, W_1, v_0, v_1, \Lambda_{uy}\}$. MCMC allows us to integrate the hyperparameters into the inference process, i.e., values for optimal parameters are found automatically by sampling from their corresponding conditional posterior distributions.

Gibbs sampling is a one of the simplest MCMC algorithms, which cycles through the latent variables, sampling each one from its distribution conditional on the current values of all other variables. Gibbs sampling is typically used when the conditional distributions can be sampled from easily [15]. We extracted K sample points, then the sample $\{w_f, p_u, u\}$ can be considered as the sample from the joint probability $P(w_f, p_u, u | \hat{y}_{user}, \Theta_f, \Theta_p, \Theta_u)$. At last, we use Eq. (10) to predict the result.

$$
P\left(\hat{y}_{user}^{(i)} | \hat{y}_{user}, \Theta_0\right) = \frac{1}{K} \sum_{k=1}^{K} P\left(\hat{y}_{user}^{(i)} \middle| w_{f_i}^{(k)}, p_u^{(k)}, u_i^{(k)}\right).
\tag{10}
$$

The specific process is described as follows:
Firstly, we sample w_f as follows:

$$
\begin{aligned}
P(w_{f_i} | \hat{y}_{user}, p_u, u, \Theta_f) &\propto \prod_{i=1}^{t1} N\left(\hat{y}_{user} \middle| w_{f_i} \cdot u_i^{\mathrm{T}} p_u, \Lambda_{uy}^{-1}\right) \cdot P\left(w_{f_i} \middle| \mu_f, \Lambda_f^{-1}\right) \\
&\propto N\left(w_{f_i} \middle| \mu_{f_i}^*, \left[\Lambda_{f_i}^*\right]^{-1}\right),
\end{aligned}
\tag{11}
$$

where $\Lambda_{f_i}^*$ can be calculated by $\Lambda_{f_i}^* = \Lambda_{f_i} + \Lambda_{uy} \sum_{u=1}^{t1} \left(u_i^{\mathrm{T}} p_u\right)^2$, $\mu_{f_i}^*$ can be calculated by $\mu_{f_i}^* = \left[\Lambda_{f_i}^*\right]^{-1} \cdot \left(\Lambda_{uy} \cdot \sum_{u=1}^{t1} u_i^{\mathrm{T}} p_u \cdot \hat{y}_{user} + \mu_{f_i} \Lambda_{f_i}\right)$
The posterior probability of Θ_f is

$$
P(\Theta_f | w_f, \Theta_0) = N\left(\mu_f \middle| \frac{\beta_1 \mu_0 + t_1 \bar{w}_f}{\beta_1 + t_1}, \frac{\Lambda_f^{-1}}{\beta_1 + t_1}\right) \cdot W\left(\Lambda_f \middle| W_f^*, v_0 + t_1\right),
\tag{12}
$$

where $\left[W_f^*\right]^{-1}$ can be calculated by $\left[W_f^*\right]^{-1} = W_0^{-1} + \sum_{i=1, i \notin (u)}^{t_1} w_{f_i}^2 + \frac{\beta_1 t_1}{\beta_1 + t_1} \cdot (\mu_0 - \bar{w}_f)^2$.
Secondly, we sample p_u as follows:

$$P\left(p_{ui}\middle|\hat{y}_{user}, w_f, u, \Theta_f\right) \propto \prod_{i=1}^{t1} N\left(\hat{y}_{user}\middle|w_{f_i} \cdot u_i^{\mathrm{T}} p_u, \Lambda_{uy}^{-1}\right) \cdot P\left(p_{ui}\middle|\mu_p, \Lambda_p^{-1}\right)$$

$$\propto N\left(p_{ui}\middle|\mu_{pi}^*, \left[\Lambda_{pi}^*\right]^{-1}\right), \tag{13}$$

where Λ_{pi}^* can be calculated by $\Lambda_{pi}^* = \Lambda_{pi} + \Lambda_{uy}w_{f_i}^2 \sum_{i=1}^{t1} u_i^{\mathrm{T}} u_i, \mu_{f_i}^*$ can be calculated by $\mu_{f_i}^* = \left[\Lambda_{f_i}^*\right]^{-1} \cdot \left(\Lambda_{uy}w_{f_i}^2 \cdot \sum_{u=1}^{t1} \frac{u_i \cdot \hat{y}_{user}}{w_{f_i}} + \mu_{pi}\Lambda_{pi}\right)$.

According to the symmetry, the parameters u and p_u have similar properties. The hyperparameters Θ_p and Θ_u can be calculated in the way like Eq. (12).

The overall procedure is presented in Algorithm 1.

Algorithm 1 Gibbs sampling for $\hat{y}_{user}$

Input: train set tr, instance num n, field num m, Gibbs sampling num K
Output: K samples
1: Initialize model parameters $\left(w_f^1, p_u^1, u^1\right)$;

2: for $k =1, 2, ..., K$ do
3:    Sample the hyperparameters by (12) and (13);
4:    $\Theta_f^k \sim P\left(\Theta_f\middle|w_f^k, \Theta_0\right)$, $\Theta_p^k \sim P\left(\Theta_p\middle|p_u^k, \Theta_0\right)$, $\Theta_u^k \sim P\left(\Theta_u\middle|u^k, \Theta_0\right)$;
5:    for $i = 1, 2, ..., n$ do
6:        $w_{f_i}^{k+1} \sim P\left(w_{f_i}\middle|\hat{y}_{user}, p_u^k, u^k, \Theta_f^k\right)$, $p_{ui}^{k+1} \sim P\left(p_{ui}\middle|\hat{y}_{user}, w_f^k, u^k, \Theta_p^k\right)$, $u_i^{k+1} \sim P\left(u_i\middle|\hat{y}_{user}, p_u^k, w_f^k, \Theta_u^k\right)$;
7: return $\left(w_f^1, p_u^1, u^1\right), ..., \left(w_f^K, p_u^K, u^K\right)$.

The Gibbs sampling for $\hat{y}_{item}$ is similar to Algorithm 1.

3.3 Model Optimization

The optimization problem is the same as Eq. (3) and the loss function can be represented as $l = (y - \hat{y})^2$.

Firstly, we normalize the sample data and feature, the normalized coefficient of sample i is

$$R(i) = \frac{1}{\|X(i)\|}. \tag{14}$$

In this paper, iFFM is trained by SGD, in each step, the loss function of a single sample is used to calculate the sub gradient:

$$g_{i,f_j} = \nabla_{v_{i,f_j}} f(\theta) = \lambda_\theta \cdot v_{i,f_j} + \kappa \cdot v_{j,f_i}, \tag{15}$$

$$g_{j,f_i} = \nabla_{v_{j,f_i}} f(\theta) = \lambda_\theta \cdot v_{j,f_i} + \kappa \cdot v_{i,f_j}, \tag{16}$$

where $\kappa = \frac{\partial E}{\partial \hat{y}} = y - \hat{y}$.

Then the factors are added gradually according to auto feature engineering technology, we calculate

$$T(t) = \frac{g_{i,f_j}^{(t)} \cdot g_{j,f_{ij}}^{(t)}}{g_{i,f_j}^{(t)} + g_{j,f_{ij}}^{(t)}}, \tag{17}$$

and choose the factor which maxes the $T(t)$.

We use AdaDelta [16] to calculate gradient:

$$RMS(g_{i,f_j})_d = \sqrt{\rho E\left(g_{i,f_j}^2\right)_{d-1} + (1-\rho)\left(g_{i,f_j}^2\right)_d + \varepsilon,} \tag{18}$$

$$RMS(g_{j,f_i})_d = \sqrt{\rho E\left(g_{j,f_i}^2\right)_{d-1} + (1-\rho)\left(g_{j,f_i}^2\right)_d + \varepsilon,} \tag{19}$$

And lastly, we update v_{i,f_j}, v_{j,f_i} :

$$\left(v_{i,f_j}\right)_d = \left(v_{i,f_j}\right)_d - \frac{\eta}{RMS(g_{i,f_j})_d} \cdot \left(g_{i,f_j}\right)_d, \tag{20}$$

$$\left(v_{j,f_i}\right)_d = \left(v_{j,f_i}\right)_d - \frac{\eta}{RMS(g_{j,f_i})_d} \cdot \left(g_{j,f_i}\right)_d, \tag{21}$$

The overall procedure is presented in Algorithm 2.

Algorithm 2 Training iFFM using SGD and Gibbs sampling

Input: train set *tr*, test set *va*, instance num *n*, field num *m*, Gibbs sampling num *K*, train parameter *pa*
Output: average prediction error OPT
1: model = init (*tr. n, tr.m, pa*);
2: normalize tr and va by (14);
3: do Gibbs sampling for p_u and p_v ;
4: for $i \in$ non-zero items in $\{1, ..., tr. n\}$ do
5: for $j \in$ non-zero items in $\{i+1, ..., tr. n\}$ do
6:    calculate OPT by (3);
7:    calculate sub gradient by (15) and (16);
8:    add feature factor by (17);
9:    for $d \in \{1, ..., k\}$ do
10:    update gradient by (18) and (19);
11:    update model by (20) and (21);
12: for $i \in$ non-zero items in $\{1, ..., va. n\}$ do
13: for $j \in$ non-zero items in $\{i+1, ..., va. n\}$ do
14:    calculate OPT by (3).

The time complexity of Algorithm 2 is $O(kn^2)$.

4 iFFM-2

Similar to mainstream recommendation systems, iFFM pays more attention to the accuracy of recommended results. However, accuracy is not the only evaluation index of the recommendation system. Although the user's point of interest (POI) is consistent over a long-time span, the POI is often singular when the user visits the recommendation system. If the recommendation system can only cover a single POI of the user, and the POI is not the moment the user is interested, the recommendation result is difficult to make the user satisfied. Therefore, to improve the diversity, we introduce heat-spreading algorithm, build a new model HeatS, combined with iFFM in linear weighted. Reference [17] has shown that the fusion of multiple models can lead to better results than a single model.

4.1 Heat-Spreading Algorithm

It has been shown in reference [13] that the use of heat-spreading algorithm for resource allocation can lead to a better diversity of recommender systems. In this paper, the iFFM is coupled in a highly efficient hybrid with HeatS. The process is as follows:

Establish the graph of Heats: G = (V, E) where V contains all the nodes of users and items and E contains link between users and items.

Denote that f is the initial resource (where f_β is the resource possessed by item β), and then f is redistributed via the transformation $f = W^H f$,

$$W^H_{\alpha\beta} = \frac{1}{k_\alpha} \sum_{i=1}^n \frac{a_{\alpha i} a_{\beta i}}{k_i}, \tag{22}$$

where W^H is a row-normalized matrix representing a discrete analogy of a heat diffusion process, k_α and k_i represent respectively the number of users who have collected item α and the number of items collected by user i. We set $a_{\alpha i} = 1$ if item α is collected by user i, otherwise, $a_{\alpha i} = 0$.

A visual representation of the resource spreading processes of HeatS is given in Fig. 1.

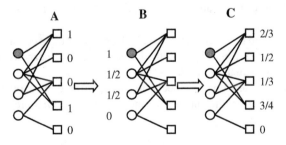

Fig. 1. The processing flow of HeatS

Resource in HeatS is redistributed via an averaging procedure, with users receiving a level of resource equal to the mean amount possessed by their neighboring items, and items then receiving back the mean of their neighboring users' resource levels.

For a given user i, the recommender score can be handled as follows.

A well-known and widely used technique is that the recommending items can be collected frequently by a given user's "taste mates" [13]. The taste overlap between users i and j is measured by the adjust cosine similarity [18]:

$$s_{ij} = \frac{\sum_{u \in U} (a_{\alpha i} - \overline{a_i}) \cdot (a_{\alpha j} - \overline{a_j})}{\sqrt{\sum_{u \in U} (a_{\alpha i} - \overline{a_i})^2 \sum_{u \in U} (a_{\alpha j} - \overline{a_j})^2}}, \tag{23}$$

and if user i has not yet collected item α, its recommendation score is given by:

$$v_{\alpha i} = \frac{\sum_{j=1}^{n} s_{ij} a_{\alpha j}}{\sum_{j=1}^{n} s_{ij}}. \tag{24}$$

We then sort $v_{\alpha i}$ in descending order (the higher the score the greater the degree of interest to the user), and finally exclude the user has selected items, get the final recommendation list.

4.2   Fusion of Two Models

In this paper, we use weighted linear aggregation to combine iFFM and HeatS. Assuming iFFM and Heats produces the result y_a and y_b respectively, then a hybrid result can be given by

$$z = (1 - \lambda) \frac{y_a}{|y_a|} + \lambda \frac{y_b}{|y_b|}. \tag{25}$$

In order to avoid the over-fitting caused by the fusion of two models, the dataset is treated as follows:

1. Divide the dataset into training set tr and test set va according to a certain proportion.
2. Divide tr into tr1 and tr2 in the same way, and the size of tr2 is equal to the size of va.
3. Train iFFM and HeatS on tr1 respectively, calculate the linear fusion coefficient λ on tr2 by the least square method.
4. Train iFFM and HeatS on tr respectively, make a prediction on va, process the prediction results by coefficient fusion, and then get the final prediction results.

5   Experiments

5.1   Experimental Setup

Datasets. We use R4-Yahoo! Movies User Ratings and Descriptive Content Information, v.1.0 (R4)[5], MovieLens 1 M (ml-1m)[6], and 2012 KDD Cup track2 (Track2)[7] to evaluate the performance of iFFM and iFFM-2.

R4 contains 221367 anonymous ratings of approximately 11915 movies made by 7642 Yahoo users. ml-1m contains 1000209 anonymous ratings of approximately 3900 movies made by 6,040 MovieLens users. Track 2 consisted of 149639105 records of user queries, a total of 10 Gb of data. Multiple queries with the same properties and their outputs were rolled up into one record in the training data.

Each record of datasets is defined as an instance, and each attribute is defined as a field. To make the attribute more reasonable, we modify some attributes (e.g., we modify the birth_year in R4 to age and divide the age into 6 values). Table 3 shows the detailed information about the modified three datasets.

Table 3. Information about the datasets

Dataset	# instances	# features	# fields
R4	221,367	575,584	38
ml-1 m	1,000,209	468,521	9
Track2	20,950,284	19,147,857	11

Platform. All experiments are conducted on a computer with 4 physical cores on Inter® Core™ i5-4460 CPU @ 3.20 GHz processors and 15.6 GB memory. The operating system is Linux.

Evaluation metrics. To evaluate the forecasting accuracy, we adopt two metrics, Root Mean Square Error (RMSE) and Area Under Roc Curve (AUC). They are defined as

$$\text{RMSE} = \frac{\sqrt{\sum_{i,j \in \text{T}} \left(r_{ij} - \hat{r}_{ij} \right)^2}}{|\text{T}|}, \tag{26}$$

$$\text{AUC} = \frac{\sum_{ans \in \text{pctr}} rank_{ans} - M \cdot (M+1)/2}{M \cdot N}, \tag{27}$$

---

[5] http://webscope.sandbox.yahoo.com/myrequests.php .

[6] http://grouplens.org/datasets/movielens/1m/ .

[7] http://www.kddcup2012.org/c/kddcup2012-track2/data .

where M is the number of clicks, N is the number of impressions and pctr is a list containing the predicted click-through rates. The greater the AUC, the better performance the model.

The metric of diversity is evaluated by the Hamming distance. Given two users i and j, the Hamming distance of recommendation lists can be defined as

$$h_{ij}(L) = 1 - \frac{q_{ij}(L)}{L}, \tag{28}$$

where $q_{ij}(L)$ is the number of common items in the top L places of both lists. Averaging $h_{ij}(L)$ over all pairs of users with at least one deleted link we obtain the mean distance $h(L)$, for which greater or less values mean, respectively, greater or less personalization of users' recommendation lists.

Implementation. We implement libFM, FFMs, iFFM, and iFFM-2 all in C++. For FFMs and iFFM, we use Streaming SIMD Extensions 3 (SSE3) instructions to boost the efficiency of inner products. The data preprocessing part we use python programming, the time is not included in the experimental results. For Track2 to make experiments faster, we randomly select 10% instances as the dataset.

5.2   Results

The impact of parameters. We conduct experiments to investigate the impact of k and η on iFFM. Table 4 shows the average running time per epoch and best AUC with different values of k. From it, we can see that the changes of k do not affect the AUC much. For the parameter η, Fig. 2 shows that bigger η can bring better performance, however, it may not that stable.

Table 4. The impact of k on iFFM

k	Time/s	AUC
1	27.236	0.7969
2	26.384	0.79696
4	27.875	0.79715
8	40.331	0.79773
16	70.164	0.79725

Performance comparison of libFM, FFMs, and iFFM. We conduct experiments to compare iFFM with libFM and FFMs on R4, ml-1m and Track2. The results can be found in Table 2. On all three datasets, the time required for the iFFM to achieve the highest accuracy and the time costed are better than that of libFM and FFMs. On Track2, the FFMs and iFFM share the same parameters to get the best AUC, but the iFFM takes 32% less time. It can be concluded that the auto feature engineering technology used by iFFM can reduce the model's running time (Table 5).

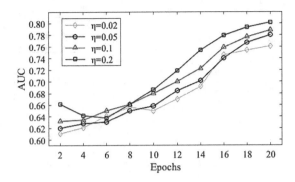

Fig. 2. The impact of η and k on iFFM

Diversity comparison of FFMs, iFFM, and iFFM-2. We conduct experiments to compare diversity of FFMs, iFFM and iFFM-2 on R4 and ml-1 m. The results can be found in Fig. 3. It can be seen from Fig. 3 that the Hamming distance of FFMs, iFFM and iFFM-2 decrease with the increase of the recommended list length L, and the diversity of iFFM-2 is much better than that of FFMs and iFFM. The diversity of iFFM on R4 is better than that of FFM. On ml-1 m, when the recommended list length is less than 30, the diversity of FFM is better than iFFM. Since both FFM and iFFM are focused on improving the accuracy of model predictions, it is difficult to determine the diversity between the two datasets.

Table 5. Performance comparison of libFM, FFMs, and iFFM

(a)Dataset R4			
Model	parameters	time/s	Best RMSE
libFM	$\lambda_0 = 12, k = 10$, Epochs = 8	142.24308	0.91986
FFMs	$\eta = 0.2$, k = 6, Epochs = 20	67.610396	0.87555
iFFM	$\eta = 0.2$, k = 2, Epochs = 10	20.178848	0.8541
(b)Dataset ml-1m			
Model	parameters	time/s	Best RMSE
libFM	$\lambda_0 = 12, k = 60$, Epochs = 80	1622.181313	0.753563
FFMs	$\eta = 0.2$, k = 50, Epochs = 40	463.589425	0.71362
iFFM	$\eta = 0.2$, k = 16, Epochs = 50	267.513094	0.70568
(c)Dataset Track2			
Model	parameters	time/s	Best RMSE
libFM	$\lambda_0 = 40, k = 100$, Epochs = 50	4765.43271	0.79011
FFMs	$\eta = 0.2$, k = 50, Epochs = 40	463.589425	0.71362
iFFM	$\eta = 0.2$, k = 16, Epochs = 50	267.513094	0.70568

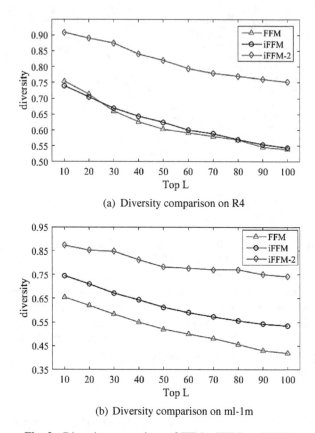

(a) Diversity comparison on R4

(b) Diversity comparison on ml-1m

Fig. 3. Diversity comparison of FFMs, iFFM, and iFFM

6   Conclusions and Future Works

In this paper, iFFM is proposed to solve the problem that FFMs cannot efficiently select valuable text features. We demonstrate that our model can achieve better performance with less time than other models. Meanwhile, to improve diversity, we combine the Heats with iFFM to get iFFM-2. The experimental results show that iFFM-2 inherits the accuracy of iFFM, and it can provide better diversity.

In the future, we will focus on how to deal with the memory leak problem for big scale datasets and parallel processing methods.

References

1. Xu, H., Zhang, R., Lin, C., Gan, W.: Construction of e-commerce recommendation system based on semantic annotation of ontology and user preference. Telkomnika Indonesian J. Electr. Eng. **12**(3), 2028–2035 (2014)

2. Gupta, Y., Saini, A., Saxena, A.K., Gupta, Y., Saini, A., Saxena, A.K.: A new fuzzy logic based ranking function for efficient information retrieval system. Expert Syst. Appl. **42**(3), 1223–1234 (2015)
3. Colombo-Mendoza, L.O., Valencia-García, R., Rodríguez-González, A., Alor-Hernández, G., Samper-Zapater, J.J.: RecomMetz: a context-aware knowledge-based mobile recommender system for movie showtimes. Expert Syst. Appl. **42**(3), 1202–1222 (2015)
4. Gavalas, D., Kenteris, M.: A web-based pervasive recommendation system for mobile tourist guides. Pers. Ubiquit. Comput. **15**(7), 759–770 (2011)
5. Lee, D.D., Seung, H.S.: Learning the parts of objects by non-negative matrix factorization. Nature **401**(6755), 788–791 (1999)
6. Koren, Y.: Factorization meets the neighborhood: a multifaceted collaborative filtering model. In: Proceedings of SIGKDD International Conference on Knowledge Discovery and Data Mining, pp. 426–434. ACM, New York (2008)
7. Rendle, S., Schmidt-Thieme, L.: Pairwise interaction tensor factorization for personalized tag recommendation. In: Proceedings of SIGKDD International Conference on Web Search and Web Data Mining, pp. 81–90. ACM, New York (2010)
8. Rendle, S.: Factorization machines. In: ICDM, pp. 995–1000 (2010)
9. Rendle, S.: Factorization machines with libFM. ACM Transactions on Intelligent Systems & Technology **3**(3), 219–224 (2012)
10. Platt, J.: A fast algorithm for training support vector machines. J. Inf. Technol. **2**(5), 1–28 (1998)
11. Jahrer, M., Toscher, A., Lee, J.Y.: Ensemble of collaborative filtering and feature engineered models for click through rate prediction. In: Proceedings of KDD 2012, vol. 1, pp. 1–23. ACM, New York (2008)
12. Juan, Y., Zhuang, Y., Chin, W.S., Lin, C.J.: Field-aware factorization machines for CTR prediction. In: Proceedings of the 10th ACM Conference on Recommender Systems, pp. 43–50. ACM, New York (2016)
13. Zhou, T., Kuscsik, Z., Liu, J.G., Medo, M., Wakeling, J.R., Zhang, Y.C.: Solving the apparent diversity-accuracy dilemma of recommender systems. Proc. Nat. Acad. Sci. **107**(10), 4511–4515 (2010)
14. Rendle, S.: Social network and click-through prediction with factorization machines. In: KDD Cup (2012)
15. Salakhutdinov, R., Mnih, A.: Bayesian probabilistic matrix factorization using Markov chain Monte Carlo. In: Proceedings of International Conference on Machine Learning, pp. 880–887. ACM, New York (2008)
16. Zeiler, M.D.: Adadelta: An adaptive learning rate method. arXiv preprint arXiv:1212.5701 (2012)
17. Jahrer, M., Scher, A., Legenstein, R.: Combining predictions for accurate recommender systems. In: Proceedings of SIGKDD International Conference on Knowledge Discovery and Data Mining, pp. 693–702. ACM, New York (2010)
18. Sarwar, B., Karypis, G., Konstan, J., Riedl, J.: Item-based collaborative filtering recommendation algorithms. In: Proceedings of the 10th International Conference on World Wide Web, vol. 4, pp. 285–295. ACM, New York (2001)

Query Processing and Optimization (I)

Beyond Skylines: Explicit Preferences

Markus Endres$^{(\boxtimes)}$ and Timotheus Preisinger

Department of Computer Science, University of Augsburg,
Universitätsstr. 6a, 86159 Augsburg, Germany
endres@informatik.uni-augsburg.de, t.preisinger@devnet.de
http://www.dbis.informatik.uni-augsburg.de

Abstract. Skyline queries are well-known in the database community and there are many algorithms for the computation of the Pareto frontier. But users do not only think of finding the Pareto optimal objects, they often want to find the best objects concerning an *explicit specified preference order*. While preferences themselves often are defined as *general strict partial orders*, almost all algorithms are designed to evaluate Pareto preferences combining weak orders, i.e., Skylines. In this paper, we consider general strict partial orders and we present a method to evaluate such explicit preferences by embedding any strict partial order into a complete lattice. This enables preference evaluation with specialized lattice based algorithms instead of algorithms relying on tuple-to-tuple comparisons and therefore speed-ups their computation as can be seen in our experiments.

Keywords: Skyline · Preference · Strict partial order · Lattice

1 Introduction

A Skyline query [1] selects those objects from a dataset that are not dominated by any others w.r.t. a given user preference. An example for a Skyline query is the search for a car that is *cheap* and has *high horsepower* (hp). Unfortunately, these two goals are complementary as cars with high power tend to be more expensive, cp. Table 1. The *Skyline* consists of all cars that are not worse than any other car in both dimensions. In our example these cars are id $\in \{3, 4, 7\}$.

In many approaches, preferences are modeled as *strict partial orders* (SPO), and therefore *transitivity* holds, cf. e.g., [2]. When evaluating a preference P on a dataset D, the tuples in D that are not dominated by any other tuple in D w.r.t. P are called the *maximal values* or *the Skyline* in the case of a Pareto preference query. The objective of a preference query is to find the tuple(s) in a dataset that are maximal with respect to a given set of preferences.

Figure 1 expresses a user preference on the domain of colors dom($color$) = {red, blue, green, yellow, purple, black, cyan} when searching for a car in Table 1. The colors *red, blue,* and *green* are preferred over *yellow* and *purple,* which are better than *black* and *cyan.* Thereby, all colors in the same set should be considered as equally good and as *substitutable* (we refer to this as *regular Substitutable Values (SV) semantics* later on in this paper [3]).

© Springer International Publishing AG 2017
S. Candan et al. (Eds.): DASFAA 2017, Part I, LNCS 10177, pp. 327–342, 2017.
DOI: 10.1007/978-3-319-55753-3_21

Table 1. Sample data.

car	id	Make	Color	Price	hp
	1	Ford	black	70 K	180
	2	Mercedes	purple	75 K	200
	3	BMW	red	50 K	230
	4	Audi	blue	45 K	170
	5	Mercedes	cyan	55 K	190
	6	GMC	yellow	70 K	150
	7	BMW	green	48 K	220

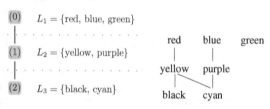

Fig. 1. Color preference.   **Fig. 2.** General SPO.

A *Better-Than Graph* (BTG) is a visual representation of the domination of domain elements for a preference as can be seen in Fig. 1. The *nodes* in this BTG represent *equivalence classes*. Each equivalence class contains objects which are mapped to the same level by a *utility function*. The *edges* in the BTG state dominance. The preference order in Fig. 1 forms a *weak order*, because all values in the same equivalence class are considered as substitutable and for each equivalence class we can specify a numerical value which represents a node in the BTG. However, in real life a user does not necessarily have such simple preferences, but specifies his wishes in a more explicit way.

A general strict partial order which does *not* form a weak order is given in Fig. 2. In fact, *red*, *blue*, and *green* are the best values, but they are not considered as *substitutable* (*trivial SV-semantics*). In addition, *green* is not better than *yellow* or *purple*, i.e., the 'best values' cannot lie in the same class, we cannot represent *red*, *blue*, and *green* with one single numerical value as in Fig. 1.

There are many algorithms for the computation of preference queries (see [4] for an overview). Many of them rely on a tuple-to-tuple comparison approach (e.g., BNL [1], SFS [5], SaLSa [6]). The major advantage of the nested-loop style algorithms (BNL, SFS, SaLSa) are their simplicity and suitability for computing the maxima of *general strict partial orders* due to its tuple-to-tuple comparison. However, most of these algorithms have a *quadratic* worst-case time complexity, $\mathcal{O}(n^2 z)$ – where n is the size of the z-dimensional input data, cp. [7].

On the other hand there are algorithms which exploit the *lattice structure* induced by a Pareto preference for efficient preference evaluation, e.g., [8,9]. Instead of direct comparisons of tuples, a lattice structure represents the better-than relationships. These algorithms offer excellent performance for domains with low cardinalities. They have a *linear runtime complexity*, $\mathcal{O}(zV+zn)$, where V is the product of the cardinalities of the z low-cardinality domains. Apart from the domain size restriction, lattice algorithms can only deal with *Pareto preferences consisting of preferences that form weak orders* on their domains [9,10]. But this is not always suitable as seen in Fig. 2.

In this paper, we present a method to embed any strict partial order into a lattice. That means, we overcome the weak order restriction and thus *make lattice-based algorithms capable of dealing with general strict partial orders*. In addition we show how to construct smaller lattices in the case of special base preferences. After such a lattice embedding we are able to apply well-known lattice algorithms like [8,9,11].

Referring to order structures, we have to mention the work of [12,13], where the authors consider the embedding of orders into small products of chains. The authors of [14] represent any order as union of interval orders, and [15] create a spanning tree on a direct acyclic graph where each node is associated with an interval. Non of these papers produce complete lattices, but 'arbitrary' graphs. Hence, to the best of our knowledge this is the first work which considers the embedding of general strict partial orders into complete lattices.

The rest of this paper is organized as follows: Sect. 2 contains the formal background. In Sect. 3 we discuss the embedding of general database preferences into lattices, and in Sect. 4 we present a method to construct smaller lattices for base preferences. Section 5 shows how to combine weak order preferences and general strict partial orders. In Sect. 6 we show our experimental results. Section 7 contains our concluding remarks.

2   Background

A database preference is defined as $P := (A, <_P)$, where A is a set of attributes and $<_P$ is a *strict partial order* (SPO, irreflexive and transitive) on $\text{dom}(A)$ [2]. The term $x <_P y$ for $x, y \in \text{dom}(A)$ is interpreted as *"I like y more than x"*. The *maximal values* of $P = (A, <_P)$ are defined as $M(P) := \{v \in \text{dom}(A) | \nexists w \in \text{dom}(A) : v <_P w\}$. The *indifference* relation is defined as $x \parallel_P y \Leftrightarrow \neg(x <_P y \vee y <_P x)$. In general, $\parallel_P$ is reflexive, symmetric, but *not transitive*. A weak order preference (WOP) is a SPO in which *indifference is transitive* [16]. For any WOP $P = (A, <_P)$ we can define a *utility function* $u_P : \text{dom}(A) \to \mathbb{R}_0^+$ mapping each domain value to a number to determine dominance between two values. The utility function depends on the type of preference as can be seen in Sect. 2.1. It holds that $x <_P y \Longleftrightarrow u_P(x) > u_P(y)$. For weak order preferences indifferent values belong to the same *equivalence class* identified by $u_P(x)$. Note that if P is a weak order then $\parallel_P$ is an equivalence relation.

Definition 1 (max(P)). $\max(P) \in \mathbb{R}_0^+$ *is the* maximum u_P *value for a weak order preference P:* $\max(P) := \max\{u_P(v) \mid \forall v \in \text{dom}(A)\}$

2.1   Preferences Constructors

Base preferences are defined on single attributes like *discrete* (*categorical*) or *continuous* (*numerical*) domains and are defined as WOPs having a utility function, such that $u_P(v) = f(v)$ if $d = 0$ and $u_P(v) = \lceil f(v)/d \rceil$ if $d > 0$, where $f : \text{dom}(A) \to \mathbb{R}_0^+$ is a score function and $d \in \mathbb{R}_0^+$ [2]. In the case of $d = 0$ the function $f(v)$ models the *distance* to the best value. A d-parameter $d > 0$ represents a discretization, which is used to group ranges of scores together. Choosing $d > 0$ has the effect that attribute values with identical $u_P(v)$ value become *indifferent* and stay in the same *equivalence class*.

The $\text{BETWEEN}_d(A, [low, up])$ preference for example expresses the wish for a value between a *lower* and an *upper* bound. If this is infeasible, values having the smallest distance to $[low, up]$ are preferred, where the distance is discretized by the parameter d. The scoring function is $f(v) = \max\{low - v, 0, v - up\}$. The $\text{AROUND}_d(A, v)$ is a special case of the former, where $low = up =: v$.

In a categorical domain $\text{LAYERED}_m(A, (L_1, \ldots, L_m))$ expresses that a user has a set of preferred values given by the disjoint sets L_i, which form a partition of $\text{dom}(A)$. Thereby the L_i form a sequence in order to know the preference order. The values in L_1 are the most preferred values. The scoring function equals $f(v) = i - 1 \Leftrightarrow v \in L_i$. Figure 1 shows an example for such a preference.

Complex preferences combine other preferences and determine the relative importance of these. But we need a notion of equality w.r.t. a complex preference. *Equivalence classes* can be applied here. For example, if for two values x, x', $u_P(x) = u_P(x')$ holds, then they belong to the same *equivalence class*. The *substitutable value* (SV) semantics has been introduced in [3] to have *indifference as a transitive relation* and every preference is associated with an SV-relation $\cong_P$ on $\text{dom}(A)$, where the equivalence classes contain "equally good" objects [16].

Definition 2 (Substitutable Values (SV)). *Let* $P = (A, <_P)$ *be a preference.* $\cong_P$ *is a substitutable values relation (SV) for* P *iff* $\forall x, y, z \in \text{dom}(A)$

(a) $x \cong_P y \Rightarrow x \parallel_P y$
(b) $(x <_P y \wedge y \cong_P z) \vee (x \cong_P y \wedge y <_P z) \Rightarrow x <_P z$
(c) $\cong_P$ *is an equivalence relation*

The identity relation on the attribute A *is called* trivial SV-relation $(=_P)$. *The regular SV-relation* $\sim_P$ *is the equivalence relation induced by the equivalence classes computed by* u_P.

The intuition behind SV-relations is that a tuple x can be substituted by x', if $x \cong_P x'$ holds. For *base preferences* regular SV-semantics $\sim_P$ does not affect $<_P$ itself, but expresses that it is admissible to substitute values for each other. The difference occurs when such preferences are combined to *complex preferences*, e.g., a Pareto preference, where $\sim_P$ does affect $<_P$. Then, the value of the utility function u_P alone is not sufficient to determine domination.

Definition 3 (Pareto Preference). *Given* z *preferences* $P_i = (A_i, <_{P_i})$ *and objects* $x = (x_1, \ldots, x_z)$, $y = (y_1, \ldots, y_z) \in \text{dom}(A_1 \times \cdots \times A_z)$. *A Pareto preference* $P := P_1 \otimes \ldots \otimes P_z$ *is defined as:*

$$x <_P y \iff \exists i : x_i <_{P_i} y_i \wedge \left(\forall j \in \{1, \ldots, z\}, j \neq i : (x_j <_{P_j} y_j \vee x_j \cong_{P_j} y_j)\right)$$
$$x \cong_P y \iff \forall i \in \{1, \ldots, z\} : x_i \cong_{P_i} y$$

Note that Pareto forms a SPO and not a WOP anymore. All input values for a Pareto preference P leading to the same utility value combination $(u_{P_1}(v), \ldots, u_{P_z}(v))$ for the WOPs P_i belong to the same class $(u_{P_1}(v), \ldots, u_{P_z}(v))$.

Example 1. The preference P_1 on colors (Fig. 1) should be equally important to $P_2 := \text{AROUND}_5(price, 50\,K)$ (Pareto). Using *trivial* SV-semantics the tuple $(red, 50\,K)$ is not better than $(blue, 45\,K)$, although a price of $50\,K$ $(u_{P_1}(50\,K) = 0)$ is better than $45\,K$ $(u_{P_1}(45\,K) = 1)$. Due to the trivial SV-semantics 'red' and 'blue' are *not* substitutable.

Having *regular* SV-semantics, 'red' and 'blue' become substitutable in the preference on the colors. Hence, $(red, 50\,K)$ is equally good as $(blue, 45\,K)$ concerning the color, but a price of $50\,K$ is better than $45\,K$ concerning AROUND$_5$. This means, $(red, 50\,K)$ domaintes $(blue, 45\,K)$.

2.2 Lattice Skyline

Lattice based algorithms like *LS-B* [9] and *Hexagon* [8] exploit the observation that the BTG for a Pareto preference constitutes a lattice [10] (e.g., Fig. 3e) to find the maximal values of a dataset. The elements of a dataset D that compose the Skyline set is build up by those nodes in the BTG that have no path leading to them from another non-empty node. For the implementation of such algorithms the lattice is usually represented by an *array*, where each position stands for one node in the lattice [9]. The array stores the *empty*, *non-empty*, and *dominated* state of a node. For each element $t \in D$ the algorithms compute the unique position in the array and mark this position as *non-empty*. Next, the nodes are visited in a breadth-first order (BFT). Non-empty nodes cause a depth-first traversal (DFT) where the dominance flags are set. Finally those nodes represent the maximal values which are both *non-empty and non-dominated*. Note that in general strict partial orders do not form lattices (e.g., Fig. 2) and therefore the above approach cannot be applied.

3 General Strict Partial Orders

In this section we will use lattices as an abstraction from the underlying preference to integrate *general strict partial orders*. This will enable us to handle arbitrary preferences in the same way as Pareto preferences. To define general strict partial orders the preference constructor EXPLICIT(A, E) was introduced in [2]. It constructs a preference from a given set of edges E. Unmentioned values are considered worse than any value in some element of E. The transitive hull of E is the Better-Than-Graph of the strict partial order expressed by EXPLICIT. In contrast to other base preferences it does *not* construct a weak order. A typical simple general strict partial order expressed as an EXPLICIT preference is known from the introduction, cp. Fig. 2.

Unfortunately there is no efficient algorithm to evaluate arbitrary database preferences but tuple-to-tuple comparison based algorithms like BNL, SFS, or SaLSa (worst case complexity $\mathcal{O}(zn^2)$). To be able to apply efficient algorithms like [8,9], we have to embed a strict partial order into a complete lattice.

For this embedding, we start with a BTG representing a general strict partial order. Since lattices need a least upper bound and a greatest lower bound (cp. [10]) we just add virtual nodes to the existing BTG. Then, we will assign a so-called *signature* to each node. This signature is a combination of integers that is *unique for each node* and hence can be used to identify it. For nodes in the same level (which are indifferent) we construct Pareto incomparable signatures. A node o which is directly dominated by a node $\hat{o}$ needs a "worse" signature than the dominator. For this we just increase one position in the signature of $\hat{o}$ and assign it to o. Which position we use is determined by a depth-first traversal. When the construction of the complete lattice structure is complete, the signature of a node is identical to the *integer combination* of the BTG node it is mapped to. Algorithm 1 describes our approach in detail. The proof that Algorithm 1 preserves the original strict partial order is given in our Technical Report [17]. The proof shows that the signatures can be used to determine a supremum and an infimum for each pair of nodes, which is the basic characteristic of a lattice. Apart from that, the relation between any two nodes of the SPO has to be preserved.

Algorithm 1 (Embedding SPOs into Lattices).[1]

1) *Identify non-dominated nodes and generate a virtual top node $\triangle$ for them. Add edges from $\triangle$ to the non-dominated nodes. Also add a virtual bottom node $\triangledown$ that is dominated by all the nodes not dominating other values in the graph. The top node will be labeled $(0, \ldots, 0)$ and the bottom node corresponds to the maximum values of the signatures.*

2) *Do a depth-first search beginning at the top node. The algorithm used for the depth-first search is irrelevant, but the following issues have to be kept in mind:*
   a) *Keep a counter. Each time the search finds a dead end, increase the counter by one.*
   b) *Annotate each edge during the search with the counter value.*
   c) *Do not follow annotated edges.*

3) *Do a breadth-first search on the graph, starting at the top node again. There are two possibilities in each node o:*
   a) *o is directly dominated by exactly one node and reached by an edge with an annotated value of v. The signature value of o is the signature value of its dominating node increased by one at position v.*
   b) *o is directly dominated by a number of nodes $\hat{o}_1, \hat{o}_2, \ldots, \hat{o}_x$. The signature of o at each position i is given by the maximum value of the $\hat{o}_i$ at the same position. If two nodes o and $\bar{o}$ (or more) are dominated directly by exactly the same set of nodes $\hat{o}_1, \ldots, \hat{o}_x$, this yields the same signature. In this case, increase the value of o (resp. $\bar{o}$) at the position of the edge on which it was reached first by the depth-first search.*

4) *Check the maximum values in use at each position of the node signature. Remove all those positions with a maximum value of zero. The remaining maximum signature values characterize the lattice.*

Note that Step 4 is unnecessary for the correctness of the algorithm. It is simply removing elements not containing any information for any node to reduce the signature length. For an EXPLICIT preference, all domain values not mentioned in its constructor are mapped to the virtual bottom node.

---

[1] Implementation available at https://github.com/endresma/ExplicitPreference.git.

We are now able to embed any strict partial order into a lattice. After that, we apply a lattice based algorithm like [8,9] (cp. Sect. 2.2), where the remaining nodes contain the maximal values. Our approach still has a linear runtime complexity. The construction of the lattice in Algorithm 1 only relies on DFT and BFT, both are linear in the number of nodes and edges. The exponential lattice size which may occur is not a major limitation, because in most cases preferences are specified only over a few items, all others can be considered as worse then the mentioned objects and hence can reside in the virtual bottom node of the lattice. That means, it is not necessary to construct the lattice on the complete domain, but only on the objects specified by the user. Therefore we create small lattices which can be handled efficiently.

Example 2. Consider the BTG presented in Fig. 3a (the same as Fig. 2), where we use the abbreviations red (r), blue (b), green (g), black (k), etc.

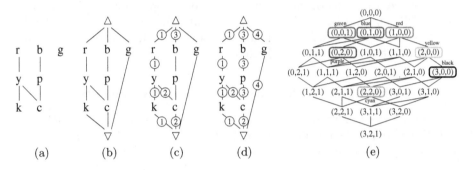

Fig. 3. Embedding a SPO into a lattice. (Color figure online)

Following Algorithm 1, we add virtual top and bottom nodes in Fig. 3b. Then, the depth-first marking of the edges begins. Figure 3c shows the moment when the counter value is "3" and the first edge has been marked with it. In Fig. 3d, all edges are marked. The depth-first orders leads to annotations on the edges which make each path from the top node to any other node in the BTG unique. Then, we determine the node signatures, starting at $\triangle$. As the highest number assigned to an edge is 4, the node signatures consists of four integer values. The top node has a signature of $\triangle \rightarrow (0,0,0,0)$. For r, we get the signature $r \rightarrow (1,0,0,0)$, as it is dominated by the top node by an edge marked with 1 and so the signature value of $\triangle$ is increased by one at position 1. Table 2 shows the signatures for all nodes after step 3 of the algorithm.

In this case the signature of the virtual bottom node is $\triangledown \rightarrow (3,0,2,1)$. We see that the maximum value at position 2 is 0. Hence it is removed. So we keep the maximum signature values 3, 2, and 1. The resulting lattice can be seen

Table 2. Signatures for all nodes after step 3.

node	r	y	k	b	p	c	g
signature	$(1,0,0,0)$	$(2,0,0,0)$	$(3,0,0,0)$	$(0,0,1,0)$	$(0,0,2,0)$	$(2,0,2,0)$	$(0,0,0,1)$

in Fig. 3e showing all signature values and the framed nodes connected to a value of the original order. Note that a lattice has $\Pi_{i=1}^{z}(\max(P_i)+1)$ nodes, i.e., $4 \cdot 3 \cdot 2 = 24$ nodes in this case.

4   WOPs with Trivial SV-Semantics

Our approach of embedding general SPOs does not necessarily construct minimal lattices. Therefore, in this section we show a method which in general constructs smaller lattices than Algorithm 1. However, this only applies for base preferences like LAYERED$_m$ and BETWEEN$_d$ and their sub-constructors with trivial SV-semantics $(=_P)$, where each object in a dataset forms its own (single valued) class. All proofs can be found in our Report [17].

4.1   Categorical Base Preferences with Trivial SV-Semantics

Modeling incomparability of values in categorical base preferences is straightforward. Using trivial SV-semantics in LAYERED$_m(A, (L_1, \ldots, L_m))$, all values in one of the L_i are incomparable.

Theorem 1. *Let $P = LAYERED_m(A, (L_1, \ldots, L_m))$ and P' derived from P by replacing regular SV-semantics with trivial SV-semantics. Let the elements of the L_i are labeled with indexes: $L_i := \{l_{i,1}, l_{i,2}, \ldots, l_{i,|L_i|}\}$.*
Then each value in $\mathrm{dom}(A)$ can be mapped to a pair of integer values as follows:

$$l_{i,j} \rightarrow (l_l, l_r), \quad where$$

$$l_l := \left| \bigcup_{x=1}^{i-1} L_x \right| - i + j$$
$$l_r := \left| \bigcup_{x=1}^{i} L_x \right| + 1 - (i+j) + |\{x \mid x \leq i \wedge |L_x| = 1 \wedge |L_{x-1}| = 1\}|$$

Example 3. Consider the color preference in Fig. 1. We derive a preference P' with the same sets but trivial instead of regular SV-semantics. This means for example that 'red', 'blue', and 'green' are *not* substitutable. Table 3 shows the integer combinations (l_l, l_r) for each color. For example, for 'red' we compute $l_l = 0 - 1 + 1 = 0$ and $l_r = 3 + 1 - (1+1) + 0 = 2$, i.e., 'red' $\rightarrow (0,2)$.

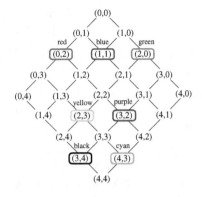

Table 3. Mappings for Example 3.

Color	L_i	u_P	$l_{i,j}$	(l_l, l_r)
red	L_1	0	$l_{1,1}$	$(0,2)$
blue	L_1	0	$l_{1,2}$	$(1,1)$
green	L_1	0	$l_{1,3}$	$(2,0)$
yellow	L_2	1	$l_{2,1}$	$(2,3)$
purple	L_2	1	$l_{2,2}$	$(3,2)$
black	L_3	2	$l_{3,1}$	$(3,4)$
cyan	L_3	2	$l_{3,2}$	$(4,3)$

Fig. 4. Lattice for LAYERED$_m$ with trivial SV. (Color figure online)

Figure 4 shows the lattice for P'. Nodes with invalid level combinations are white. All other nodes are labeled with the color and its assigned integer combination. Note that the number of occupied nodes is the size of the categorical domain, $|dom(A)| = 7$.

4.2 Numerical Base Preferences with Trivial SV-Semantics

In the case of numerical preferences it is sufficient to analyse the problem for BETWEEN$_d$, because all other numerical preferences are special cases. However, embedding BETWEEN$_d$ into a lattice is not a trivial task, because if two values exist in the same equivalence class, they are not substitutable when using trivial SV-semantics and must be modeled incomparable in the lattice structure.

Example 4. Let $P := $ AROUND$_5(price, 50\,K)$ be a preference with $\operatorname{dom}(price) = \{45\,K, 48\,K, 50\,K\}$. The value $50\,K$ is the *maximal value*, whereas $45\,K, 48\,K$ lie in a distance of $d = 5$ from the best value. Since $45\,K, 48\,K \in [45\,K, 50\,K]$, they share the same u_P value 1, but are not substitutable and must be modeled as incomparable in the lattice which they should be embedded in.

Since the domain of an attribute could be infinite, this makes the embedding of numerical base preferences with trivial SV-semantics difficult. However, in database relations we generally assume a *closed-world*, hence we have a finite number of objects in an equivalence class. In this case, BETWEEN$_d$ could be modeled in the same way as LAYERED$_m$.

Under some assumptions we can produce smaller lattices, e.g., when we have single occupied equivalence classes. In this case we avoid the problem of several indifferent objects in the same equivalence class; objects having the same u_P function value are either identical or lie "left and right" of the maximal values.

Theorem 2. *Given $P := BETWEEN_d(A, [low, up])$ and let P' be derived from P by replacing regular with trivial SV-semantics. Assume each equivalence class contains at most one element. We map $x \in \operatorname{dom}(A)$ to the integer combination (l_1, l_2) in P' as follows:*

$$x \to (l_1, l_2) = \begin{cases} (u_P(x) \quad, \ u_P(x) - 1) \ if \ x > up \\ (u_P(x) - 1 \ , \ u_P(x)) \quad if \ x < low \end{cases}$$

For $x \in [low, up]$ we set $x \to (0,0)$. Then, P' models the same order as P w.r.t. dom(A).

This can be interpreted as two WOPs being connected and used to model a strict partial order. A "virtual" Pareto preference is constructed by the numerical base preference.

Lemma 1 (#Nodes and #Used Nodes). *Consider a preference P' which is defined as a BETWEEN$_d$ preference P with trivial SV-semantics.*

The number of nodes in the lattice of P' is:

$$\#nodes_{P'} = (\max(P) + 1)^2 \tag{1}$$

The number of used nodes (i.e. the number of nodes that can be matched by values evaluated by P') in the lattice of P' is given by

$$\#used\_nodes_{P'} = 2 \cdot \max(P) + 1 \tag{2}$$

Note that Lemma 1 only holds when considering equivalence classes which contain at most one object.

Example 5. Let $P := \text{AROUND}_5(price, 50K)$ in the domain dom($price$) = $\{45\,\text{K}, 50\,\text{K}, 55\,\text{K}, 70\,\text{K}, 75\,\text{K}\}$. Then $\max(P) = 5$. No two domain values lie in the same equivalence class, we derive P' with trivial SV-semantics and create pair mappings: A perfect value of $50\,\text{K}$ is mapped to $(0,0)$, $45\,\text{K}$ and $55\,\text{K}$ (with $u_P(45\,K) = u_P(55\,K) = 1$) are mapped to incomparable value combinations $(0,1)$ and $(1,0)$, respectively. All combinations can be found in Table 4.

Table 4. Mappings for P'.

Price	u_P	(l_1, l_2)
45 K	1	(0,1)
50 K	0	(0,0)
55 K	1	(1,0)
70 K	4	(4,3)
75 K	5	(5,4)

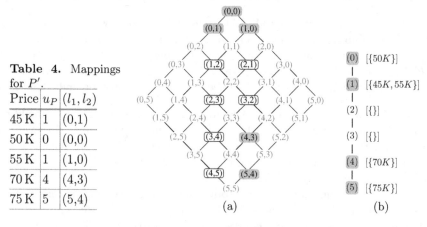

Fig. 5. Lattice for AROUND$_d$ with (a) trivial SV-semantics and (b) regular SV-semantics.

Figure 5a shows the lattice of P' with *trivial SV-semantics*. The filled gray nodes have tuples belonging to them, the framed white nodes represent valid integers for l_1 and l_2, while the gray scripted nodes are unused dummy nodes given by the graph structure. The number of nodes is $(5+1)^2 = 36$ from which $2 \cdot 5 + 1 = 11$ might be used. In Fig. 5b we present the BTG for P with *regular SV-semantics* for comparison only. Values with the same u_P value are substitutable and reside in the same equivalence class. The BTG forms a chain.

5   Combining SPOs and WOPs

The combination of a strict partial order S embedded into a lattice G_S and a Pareto preference P with its corresponding lattice G_P is just constructed by the combination of P and S. It is visualized by the product of G_S and G_P. Such a embedding can be very useful as it enables us to evaluate base preferences that are strict partial orders just like Pareto preferences and Pareto preferences consisting of strict partial orders just as if they only used standard WOPs as input preferences.

Example 6. Consider the SPO with its lattice embedding in Fig. 6a. Now we combine this order with a WOP with a maximum value $\max(P) = 3$, e.g., $AROUND_{10}(price, 50\,K)$ with regular SV-semantics as in Fig. 6b. The lattice for the combined (Pareto) order can be seen in Fig. 6c. As the resulting lattice is the product of the lattices of the two underlying preferences, each of the original nodes in Fig. 6a is multiplied. For example, the multiplication of $(0,1)$ by $\{0,1,2,3\}$ results in $(0,0,1)$, $(1,0,1)$, $(2,0,1)$, $(3,0,1)$. Nodes representing no reachable integer combination (due to the strict partial order) are printed in gray.

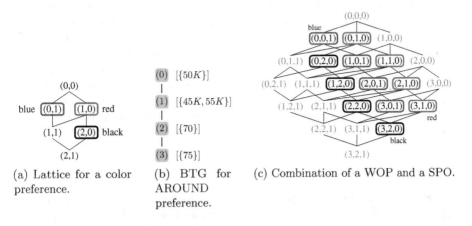

(a) Lattice for a color preference.

(b) BTG for AROUND preference.

(c) Combination of a WOP and a SPO.

Fig. 6. Combining SPOs and WOPs. (Color figure online)

6 Experiments

This section provides some experiments on synthetic and real world data to reveal the applicability of our approach.

6.1 Benchmark Framework

We implemented a Java 7.0 prototype, which is available as open source project on GitHub (See footnote 1). The experiments were performed on a single node running Debian 7.1 on a Intel Xeon 2.53 GHz processor. We used JGraphT[2], a free Java class library that provides graph-theory objects and algorithms, to model the EXPLICIT preferences. Since all lattice-based algorithms run in main memory, we used a buffer size large enough for SaLSa to hold all objects in memory, too.

For our synthetic datasets we used the data generator commonly used in preference research [1]. For the experiments on real-world data, we used the well-known Zillow dataset (www.zillow.com). It contains more than 2 M entries about real estate in the United States.

We compare our approach (named *Lattice Skyline Explicit*, **LSE**) to **SaLSa**, because SaLSa is the state-of-the-art tuple-comparison based algorithm which can evaluate arbitrary strict partial orders (worst case $\mathcal{O}(n^2)$, best case $\mathcal{O}(n \log n)$). Within SaLSa we use the *maximum coordinate sort* since this has been shown as best [6]. Note that we do not compare our algorithm to index-based approaches because we do not perform any pre-processing in LSE.

For our algorithm LSE we used different data structures to store the lattice: **A** (Array), **HM** (HashMap), and **SL** (SkipList) to show the different behavior concerning runtime and memory requirements. The advantage of using SkipLists in contrast to HashMaps lies in the reduced memory requirements, because we do not have to initialize the whole data structure in main memory. A node is initialized on-the-fly if it is marked as *non-empty* or *dominated*. Using a HashMap, we have $\mathcal{O}(1)$ for the look-up in the HashMap, since we can use a perfect hash function due the known width of the lattice, cf. [8]. In summary this leads to a runtime complexity of $\mathcal{O}(zV + zn)$, too (cp. Sect. 1). For the SkipList based implementation we have $\mathcal{O}(zV + zn \log w)$, since operations on SkipLists are $\mathcal{O}(\log w)$ [18], where w is the number of elements in the SkipList, i.e., the width of the lattice in the worst case. For more implementation details we refer to [17].

6.2 Experimental Results

Test Setting 1. Our first experiments follows Example 3 where we consider a LAYERED$_m$ preference on the domain of colors having trivial SV-semantics. As in the example, the domain size is $|\operatorname{dom}(color)| = 7$, which is realistic when considering colors of a product. We varied the number of input objects from 1 K to 5 M and measured the complete runtimes of the algorithms. The results can

---

[2] https://github.com/jgrapht/jgrapht.

be found in Fig. 7a. We print error bars based on 100 runs for each experiment on the same input, i.e., the mean and the best and worst runtime. For clarity we do this only for SaLSa and LSE-A. All LSE algorithms are nearly linear in the size of the lattice and the input size and outperform SaLSa clearly. Note that for the LSE approach the constructed lattice is always the same, the algorithms only differ in the number of input objects and the lattice representation. The time to create the lattice from the preference (Algorithm 1) is less than 0.03 s.

LSE-HM and LSE-SL use dynamic data structures and therefore are a little bit worse than LSE-A using an static array having constant access time. However, they differ in the amount of memory requirements: In the 1 M input test, LSE-HM uses 54 MB of memory, whereas LSE-SL can reduce the memory usage to 49 MB due to the fact of dynamic adding and removal of single nodes of the lattice. LSE-A needs about 53 MB of memory. Note that the memory usage includes all information on empty/non-empty/dominated nodes, references to the objects which belong to a node, additional memory requirements for the Java implementation, ... [3] The result size in this case is 424,428 objects (which are either 'red', 'blue', or 'green'). SaLSa needs 1,713,646 tuple comparisons for computing the result.

Test Setting 2. In this test setting we evaluate the SPO given in Fig. 3a. We compare all algorithms on a single attribute domain. We want to compare different domain sizes, therefore we map the strict partial order to integer numbers, i.e., $r \to 0$, $b \to 1$, ... We fixed the input size to 1 M. Remember that in EXPLICIT non-mentioned nodes are worse than all others. The results can be found in Fig. 7b. In the SPO there are only seven objects, i.e., all others are worse than the mentioned objects, hence SaLSa becomes worse with increasing domain size due to many tuple comparisons between all non-mentioned tuples. All LSE algorithms have similar runtime, whereas LSE-A is again a little bit better than LSE-HM and LSE-SL. The lattice construction from the SPO in Fig. 3a needs round about 0.05 s.

Test Setting 3. In our third experiment, we evaluate Example 6, which combines a general SPO (trivial SV-semantics) with a WOP (*regular SV-semantics*). We varied the number of input objects from 1 K to 5 M objects. Note that this strict partial order builds a Pareto preference since both preferences are equally important during the combination of them. However, the color preference is evaluated with trivial SV-semantics, which is not possible in common Pareto algorithms, cp. [4]. Figure 7c shows our results. The result size is 1420. SaLSa is nearly as good as LSE when considering small input sizes. However, from 50 K objects on all LSE algorithms outperform SaLSa clearly. Note, that in this experiment LSE-A (53 MB, 1 M input) is again better than LSE-SL (47 MB) and LSE-HM (54 MB), but needs more memory. The lattice construction time is less than 0.05 s.

---

[3] We used the Java `Runtime` object with the methods `totalMemory()` and `freeMemory()` to determine the total amount of used memory in the JVM.

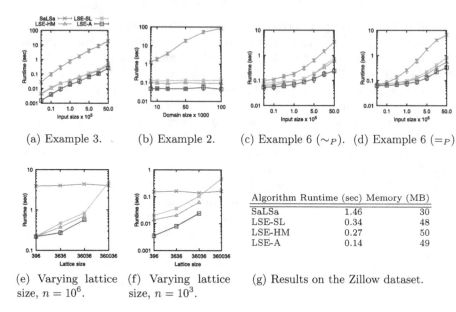

(a) Example 3.   (b) Example 2.   (c) Example 6 ($\sim_P$). (d) Example 6 ($=_P$)

(e) Varying lattice (f) Varying lattice   (g) Results on the Zillow dataset.
size, $n = 10^6$.   size, $n = 10^3$.

Algorithm	Runtime (sec)	Memory (MB)
SaLSa	1.46	30
LSE-SL	0.34	48
LSE-HM	0.27	50
LSE-A	0.14	49

Fig. 7. Experimental results (log scale).

Test Setting 4. In this experiment we use trivial instead of regular SV-semantics for the AROUND_d preference above, i.e., AROUND_d builds a lattice which is characterized by the values $(3, 3)$. The combination of both preferences (the preference on colors and AROUND_d) builds up the lattice structure $(2, 1, 3, 3)$ which has 96 nodes. Figure 7d presents our results, which are very similar to Fig. 7c. In the 1M case LSE-HM occupies 55 MB, LSE-SL 52 MB, and LSE-A uses 54 MB of memory to store all informations in memory. Algorithm 1 needs less than 0.06 s for the construction of the lattice. The result size is 568 objects, SaLSa needs 1,167,492 tuple comparisons.

Test Setting 5. This experiment considers the size of the lattice, because for larger preference sets the cardinality terms could dominate the runtime and hence LSE could be worse than SaLSa. A similar behavior was already shown in [8,9]. We use a numerical WOP (regular SV-semantics) which produces a chain with $\max(P) \in \{10, 100, 1000, 10000\}$ and combine it with the SPO given in Fig. 3a. This constructs a Pareto preference as described in Sect. 5. Note that this experiment is only to investigate the influence of the lattice size which is 396, 3636, 36036, and 360036 respectively. In practice, nobody would express a preference over 10000 objects, hence lattices normally are "small".

First, we fix the input size to 1M objects. In Fig. 7e we can see that LSE is faster than SaLSa, as long as the lattice size is "small". From 360.000 nodes on, we could not evaluate LSE-A and LSE-HM, because the number of nodes did not fit into main memory as necessary for these data structures. Here, LSE-SL shows its advantage as it initializes lattice nodes on-the-fly and not at once as the array-based or HashMap-based implementations. In Fig. 7f we use the same setting as before, but fix the input size to 1 K objects. This should favor the SaLSa algorithm, because of less tuple comparisons. However, LSE still outperforms SaLSa, but with increasing lattice size LSE takes too much time for the DFT and BFT search in the *dominance* and *removal* phase (cp. Sect. 2.2). Hence, it is outperformed by SaLSa for more than 360.000 nodes.

Test Setting 6. For the experiment on the Zillow dataset we use four preferences on the attributes *bedrooms, bathrooms, living area* (in m^2), and *age* of the building. The preference orders are shown in Fig. 8. The first two have trivial SV-semantics whereas the last two preferences have regular SV-semantics on the given intervals.

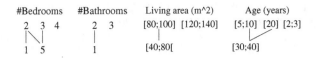

Fig. 8. EXPLICIT preferences for Zillow.

We combine these SPOs as described in Sect. 5 to build one complex preference. The complete lattice has 4608 nodes. The runtimes and memory requirements are depicted in Fig. 7g. Note that SaLSa only uses 30 MB of memory. This is because of the tuple-to-tuple comparison where objects can be 'removed' immediately after domination.

7   Conclusion and Future Work

The principal point of this paper is to show that it is possible to embed all kinds of strict partial orders into complete lattices. As a consequence, existing lattice based algorithms can be applied to general strict partial orders and prior restrictions on weak order preferences and their combinations in Pareto preferences are resolved. Our experiments confirm the applicability of lattice algorithms and the benefit in comparison to tuple-to-tuple comparison approaches like SaLSa.

As we have mentioned, the lattices we construct with our algorithm are not minimal. Nevertheless a reduction of the lattice size is wise as the size of a lattice for a Pareto preference grows exponentially with the lattice sizes of the underlying preferences. Hence our next step will be to address this and improve our algorithm so that it produces minimal lattices embedding strict partial orders. However, this could be a challenging task.

References

1. Börzsönyi, S., Kossmann, D., Stocker, K.: The skyline operator. In: Proceedings of ICDE 2001, pp. 421–430. IEEE, Washington, DC (2001)

2. Kießling, W.: Foundations of preferences in database systems. In: Proceedings of VLDB 2002, pp. 311–322. VLDB, Hong Kong (2002)
3. Kießling, W.: Preference queries with SV-semantics. In: Proceedings of COMAD 2005, pp. 15–26. Computer Society of India, Goa (2005)
4. Chomicki, J., Ciaccia, P., Meneghetti, N.: Skyline queries, front and back. SIGMOD **42**(3), 6–18 (2013)
5. Chomicki, J., Godfrey, P., Gryz, J., Liang, D.: Skyline with presorting. In: Proceedings of ICDE 2003, pp. 717–816 (2003)
6. Bartolini, I., Ciaccia, P., Patella, M.: SaLSa: computing the skyline without scanning the whole sky. In: Proceedings of CIKM 2006, pp. 405–414. ACM, New York (2006)
7. Godfrey, P., Shipley, R., Gryz, J.: Maximal vector computation in large data sets. In: VLDB 2005: Proceedings of the 31st International Conference on Very Large Data Bases, pp. 229–240. VLDB Endowment (2005)
8. Preisinger, T., Kießling, W.: The Hexagon algorithm for evaluating pareto preference queries. In: Proceedings of MPref 2007 (2007)
9. Morse, M., Patel, J.M., Jagadish, H.V.: Efficient skyline computation over low-cardinality domains. In: Proceedings of VLDB 2007, pp. 267–278 (2007)
10. Davey, B.A., Priestley, H.A.: Introduction to Lattices and Order, 2nd edn. Cambridge University Press, Cambridge (2002)
11. Lee, J., Hwang, S.-W.: BSkyTree: scalable skyline computation using a balanced pivot selection. In: Proceedings of EDBT 2010, pp. 195–206. ACM (2010)
12. Raynaud, O., Thierry, E.: The complexity of embedding orders into small products of chains. Order **27**(3), 365–381 (2009)
13. Zhang, S., Mamoulis, N., Cheung, D.W., Kao, B.: Efficient skyline evaluation over partially ordered domains. Proceedings of VLDB 2010 **3**(1–2), 1255–1266 (2010)
14. Capelle, C.: Representation of an order as union of interval orders. In: Bouchitté, V., Morvan, M. (eds.) ORDAL 1994. LNCS, vol. 831, pp. 143–161. Springer, Heidelberg (1994). doi:10.1007/BFb0019432
15. Fishburn, P.C.: Interval graphs and interval orders. Discrete Math. **55**(2), 135–149 (1985)
16. Fishburn, P.: Preference structures and their numerical representation. Th. Comp. Sci. **217**(2), 359–383 (1999)
17. Endres, M., Preisinger, T.: Preference structures and their lattice representation. Report–02, University of Augsburg (2016)
18. Pugh, W.: Skip lists: a probabilistic alternative to balanced trees. Commun. ACM **33**(6), 668–676 (1990)

Optimizing Window Aggregate Functions in Relational Database Systems

Guangxuan Song[1], Jiansong Ma[1], Xiaoling Wang[1(✉)], Cheqing Jin[1], and Yu Cao[2]

[1] Shanghai Key Laboratory of Trustworthy Computing,
MOE International Joint Lab of Trustworthy Software,
East China Normal University, Shanghai, China
guangxuan_song@163.com, ecnumjs@163.com, {xlwang,cqjin}@sei.ecnu.edu.cn
[2] EMC Labs, Beijing, China
yu.cao@emc.com

Abstract. The window function has become an important OLAP extension of SQL since SQL:2003, and is supported by major commercial RDBMSs (e.g. Oracle, DB2, SQL Server, Teradata and Pivotal Greenplum) and by emerging Big Data platforms (e.g. Google Tenzing, Apache Hive, Pivotal HAWQ and Cloudera Impala). Window functions are designed for advanced data analytics use cases, bringing significant functional and performance enhancements to OLAP and decision support applications. However, we identify that existing window function evaluation approaches are still with significant room for improvement. In this paper, we revisit the conventional two-phase evaluation framework for window functions in relational databases, and propose several novel optimization techniques which aim to minimize the redundant data accesses and computations during the function calls invoked over window frames. We have integrated the proposed techniques into PostgreSQL, and compared them with both PostgreSQL's and SQL Server's native window function implementation over the TPC benchmark. Our comprehensive experimental studies demonstrate significant speedup over existing approaches.

Keywords: Window function · Query optimization · Relational database

1 Introduction

Most of the queries or reports that arise in business intelligence (BI) and analytical applications can be expressed naturally in terms of window functions, which represent the state-of-the-art way of performing complex data analytics within a single SQL statement. Window functions perform common analyses such as ranking, percentiles, moving averages and cumulative sums in a flexible, intuitive and efficient manner, overcoming shortcomings of the traditional alternatives such as grouped queries, correlated subqueries and self-joins [1,2]. As one of the most

© Springer International Publishing AG 2017
S. Candan et al. (Eds.): DASFAA 2017, Part I, LNCS 10177, pp. 343–360, 2017.
DOI: 10.1007/978-3-319-55753-3_22

useful standardized extensions to SQL since the SQL:2003 standard, window functions have been widely implemented in most of the major commercial and open-source relational database systems. (e.g. Oracle, DB2, SQL Server, Teradata, Pivotal Greenplum and PostgreSQL), as well as in some emerging Big Data systems (e.g. Google Tenzing, SAP HANA, Amazon Redshift, Pivotal HAWQ and Cloudera Impala).

In this paper, we focus on an important class of window functions called *window aggregate functions*. The general form of a window aggregate function is as follows:

```
Agg_func(expression) OVER(
  [PARTITION BY expr_list]
  [ORDER BY order_list [frame_clause]])
```

Agg_func is a normal aggregate function (e.g. SUM, AVG and COUNT), except that it operates on a partition or "window" of a base or derived table, and returns a value for every row in that window. Unlike group functions that aggregate result rows, all rows in the table are retained. The values returned are calculated by invoking function calls over the sets of rows in that window. A window is defined using a window specification (the OVER clause), and is based on three main concepts:

- *Window partitioning*, which forms groups of rows (PARTITION BY clause);
- *Window ordering*, which defines an order or sequence of rows within each partition (ORDER BY clause);
- *Window frames*, which are defined relative to each row to further restrict the set of rows when using ORDER BY (frame_clause specification).

In current database systems [3,7,8], window function is evaluated over the windowed table in a two-phase manner. In the first phase, the windowed table is reordered into a set of physical window partitions, each of which has a distinct value of the PARTITION BY key and is sorted on the ORDER BY key. Then the generated window partitions are pipelined into the second phase, where the window function is sequentially invoked for each row over its window frame within each window partition.

In this paper, we focus on the second phase and propose our optimization techniques based on two important observations:

- According to the window function definition, different window frames may overlap, i.e. they may contain common rows;
- If a window aggregate function is exchangeable, which means that all operations involved in its transition function are commutative, then the aggregation sequence of the rows in a window frame does not affect the final function result. It turns out that most of the common window aggregate functions are exchangeable.

As a result, we propose three efficient algorithms. (1) FL(Fast Locating) is used to save the cost of locating the frame head pointer. (2) TF(Temporary Frame) can reduce the number of re-calculation by holding the previous transition value. (3) HTF(Hierarchical Temporary Frames) makes use of child TFs to calculate the value of the parent TFs in a recursive way. In comparison with native window function implementations of PostgreSQL and SQL Server, the experimental results demonstrate solid performance advantages of our proposed techniques.

The rest of this paper is organized as follows. In Sect. 2, we explain in details the motivation of our proposed optimization techniques and the related work. In Sect. 3, we describe our proposed approaches to optimize the evaluation of window aggregate functions. Section 4 empirically validates the effectiveness of our proposed techniques. Finally we draw conclusions in Sect. 5.

2   Motivation and Premilinary

2.1   Premilinary

Definition 1 (Frame f). *The frame is a window, containing a set of rows. The start position for the Frame is named as the frame head, and the frame tail is for the end position. The frame is defined as the triple (h, t, TV). h is the frame head and t is the frame tail. TV is the transition value for the frame. f_i is the i'th frame in a partition. $f_i \cap f_j$ is the largest common sub-frame for f_i and f_j.*

In Definition 1, transition value is the temporary result of rows in the frame, which is obtained by invoking correlated aggregation function, such as SUM, AVG, COUNT and so on. In this paper, we utilize **N** to denote the Table size and F_i to denote the Frame size of f_i (we usually regard all the frames' size as the same **F**).

A window frame is a set of consecutive rows within a window partition. For two consecutive rows in a partition, their corresponding window frames are also consecutive. Moreover, according to the window frame definition, it is obvious that, for a row r_1 (with frame head $f_1.h$ and frame tail $f_1.t$) and its following row r_2 (with frame head $f_2.h$ and frame tail $f_2.t$), $f_1.h \leq f_2.h$ and $f_1.t \leq f_2.t$. In case that $f_1.h = f_2.h$, it is feasible to directly reuse r_1's final aggregate transition value and incrementally process the rows between $f_1.t$ and $f_2.t$, which is called the *incremental strategy*.

However, if $f_1.h < f_2.h$, the aggregate transition values for r_2 need to be completely re-calculated, incurring the redundant data accesses and computations. Moreover, the calculation of aggregate transition values for r_2 starts with the row at h_2, which requires relocating the row reader position to $f_2.h$. However, since the row sizes may not be constant especially when attributes of non-fixed size exist, the reader needs to go from $f_1.t$ back to $f_2.h$ step by step, as illustrated in the example below.

Example 1. The following window query example calculates the average salary over a window frame covering the rows inclusively between the current row and the 10000th row preceding it.

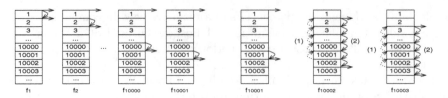

(a) In Example 1, the window function calcula- (b) In Example 1, the situation when the frame
tions of the first 10001 frames sharing the same head changes
frame head

Fig. 1. Illustrations of Example 1

```
SELECT
    empno, depno, salary,
    AVG(salary) OVER (PARTITION BY depno ORDER BY salary ROWS
        BETWEEN 10000 PRECEDING AND CURRENT ROW) as avg
FROM empsalary;
```

For the calculation of the first 10001 frames of Example 1, the frame head doesn't change. Figure 1a provides an illustration that depicts the applicable frame with respect to each row using two parallel arrows, the curved arrows show the detailed calculation process. The transition value of f_{i+1} can be obtained by an incremental aggregation from previous frame f_i.

Figure 1b illustrates the situation where frame head position changes. As such, each window function calculation consists of two subsequent steps: first, relocate the frame head position showed by dashed curved arrows; second, aggregate the final transition value showed by solid curved arrows.

From the above discussion and Example 1, we can figure out that when both the window frames and the window partitions are large enough, the overlaps among consecutive window frames are probably very large too. These overlaps result in redundant data accesses for relocating frame head positions and for fetching a row from either the buffer or disk and computations of aggregate transition values. On the other hand, as most of the common window aggregate function is exchangeable, the aggregation sequence of the rows in a window frame does not affect the final function result, it is possible for us to optimize the exchangeable window aggregation function calls by amortizing as much as possible the data access and computation costs of the overlaps among consecutive window frames.

2.2   Related Work

The sliding window technique in data stream, such as [10, 11], has few similarities to the window function in the traditional database SQL query language. In data stream applications, because the data comes as a sequence of tuples or items, it is difficult to store the entire data in memory. So the sliding window stores some approximate data, i.e., statistics, sketches or samples. [9] focuses on the top-k problem. However, different from the sliding window technique in data stream, the window query in SQL must provide exact query results, rather than approximate values. So the sliding window technique can't be used in the window query evaluation.

In recent years, more and more researches have focused on window function in database. Existing techniques [2–5] are available to optimize the table reordering operation in the first phase. Bellamkonda et al. [2] make use of the full sort technique and Cao et al. [5] promote it by the more competitive hash sort and segment sort. Other earlier work, such as [6,12–14], make many contributes from the point of the "ORDER BY" clause and "GROUP BY" clause and they propose the optimization framework based on either the function dependency or the reusing of the intermediate results. Cao et al. [6] proposes a novel method for the multi sorting operations in one query. They find that most of the time is spent in sorting and partitioning phase when the size of window is small. So it's a good idea to avoid repeated sorting. Cao et al. also prove it's an NP-hard problem to find the optimal sequence and provide a heuristic algorithm for this problem. Besides, there are several recent work about window operators and their functions. Leis et al. [15] propose a general execution framework which is more suitable for memory database. They propose an additional data structure named Segment Tree to store aggregates for sub ranges of the entire group. Therefore, it is possible to reduce redundant calculations by using Segment Tree. The work of Leis et al. is efficient enough, but it's only for distributive and algebraic aggregates. So Wesley et al. [16] propose an incremental method for three holistic windowed aggregates.

This paper improves the second phase of window function execution and proposes three novel optimization algorithms: FL, TF and HTF. Unlike [15] which needs construct a segment tree, not too much extra space are needed to store temporary values in our optimization methods. And the proposed algorithms are used to deal with distributive and algebraic aggregates, which are more general than [16].

3   Optimization

In this section, we consider the optimization techniques for window aggregation functions. The goal is to reduce the cost of locating the frame head C_{loc}, the cost of accessing data C_{da} and the cost of computing C_{cal} in the two-phase evaluation framework for window functions in relational database systems.

3.1   FL(Fast Locating) Approach

During the sequencing phase for evaluating aggregation function of a window query, if the frame head position changes, the reader position has to move back to the new frame head step by step, which is time-consuming.

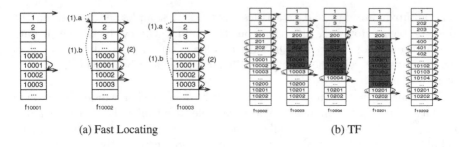

(a) Fast Locating   (b) TF

Fig. 2. Examples of fast locating and TF (Color figure online)

In order to reduce this cost of locating the frame head of a frame C_{loc}, we introduce a temporary reader pointer to keep track of memory or disk address for frame head position, and then the locating process can be divided into two steps, as showed in Fig. 2a. Firstly, the dashed curved arrow tagged with "(1).a" means jumping the temporary reader pointer forward until it reaches the frame head. Secondly the dashed curved arrow tagged with "(1).b" means copying the temporary reader pointer to the aggregate reader pointer directly. Since the temporary pointer always jumps forward, the cost of locating C_{loc} is at most $p \times N \times T_{DA}$ in our proposed FT approach, which is much smaller than the implementation in PostgreSQL, i.e., $p \times N \times (F - 1) \times T_{DA}$.

3.2   TF(Temporary Frame) Approach

Based on the analysis of the window query executing process, the bottleneck is the re-calculation of the overlap among the frames. In order to reduce the number of re-calculation, we introduce the TF-based optimization algorithm. First, we give the definitions of TF and Group.

Definition 2 (TF). *TF is a Temporary Frame in the process of sequencing and is defined as the triple* (s, e, TTV), *where s is the starting row position of the temporary frame, e is the ending row position, and TTV is the transition value for the temporary frame TF.*

For example, in Fig. 2b, when we calculate f_{10002}, we set a TF (The TF is shown in f_{10003} marked by the deep color). So, for this TF, its start position s is 201 and end position e is 10002.

Definition 3 (Group). *The frames,* $f_i, f_{i+1}, f_{i+2}, \dots, f_{i+k} (k > 0)$ *share the same TF, if the start position s of TF they associate with is the same. If the frames share the same TF, we call they are in a Group.*

Algorithm 1. TF(Temporary Frame)

```
1   for each partition in the TABLE do
2   │   init s;
3   │   for each frame fᵢ in the partition do
4   │   │   if fᵢ.h=fᵢ₋₁.h then
5   │   │   │   fᵢ.TV ← fᵢ₋₁.TV;
6   │   │   │   for each row r in (fᵢ₋₁.t, fᵢ.t] do
7   │   │   │   └   fᵢ.TV ← transfunc(fᵢ.TV, r);

8   │   │   else if s ≥ fᵢ.h then
9   │   │   │   fᵢ.TV ← TTV;
10  │   │   │   for each row r in [fᵢ.h, s) do
11  │   │   │   └   fᵢ.TV ← transfunc(fᵢ.TV, r);
12  │   │   │   skip interval[s, e];
13  │   │   │   for each row r in (e, fᵢ.t] do
14  │   │   │   │   fᵢ.TV ← transfunc(fᵢ.TV, r);
15  │   │   │   └   TTV ← transfunc(TTV, r);
16  │   │   └   e← fᵢ.t;

17  │   │   else
18  │   │   │   init fᵢ.TV, TTV and s= determinenext(case, F);
19  │   │   │   for each row r in [fᵢ.h, fᵢ.t] do
20  │   │   │   │   fᵢ.TV ← transfunc(fᵢ.TV, r);
21  │   │   │   │   if row r in [s, fᵢ.t] then
22  │   │   │   │   └   TTV ← transfunc(TTV, r);
23  │   │   └   e← fᵢ.t;
```

If we know the start position and the end position of TF and its transition value, we can calculate the aggregation value for each frame with the TF-based algorithm.

In the partition, if the frame head for the following frames doesn't change comparing with the last one, there is no need to traverse from the frame head again, and the previous frame's transition value could be reused. In other words, there is no need to calculate with TF. If we use the TF to calculate the TV instead of the previous frame's TV, some rows will be calculated twice or even more. So, the procedure for evaluating aggregates for the frames whose frame head does not change is as follows: the transition function is invoked for each row added to the frame, and the final function is invoked whenever we need the current aggregate value.

When the frame head has changed, the previous transition value is not applicable for the current frame. In order to reduce the rate of re-calculation, we design the TF-based approach.

TF algorithm is shown as Algorithm 1. First, we initialize the start position of TF (line 2). When the frame head $(f_i.h)$ of the current frame is the same as that of the last frame, we can use the incremental strategy (line 4–7). We first assign the transition value of last frame $(f_{i-1}.TV)$ to the transition value of current frame ($f_i.TV$) (line 5). Then we scan the new entering rows to update the value of $f_i.TV$ (line 6–7). If the frame head changes, there are two cases: (1) The frame head does not exceed the start position of TF (line 8–16). We first assign the transition value of temporary frame (TTV) to $f_i.TV$ (line 9), then scan the rows which are before s and belong to f_i to update the value of $f_i.TV$ (line 10–11). Next, we skip the interval that the TF contains (line 12). Then we scan the new entering rows and update the value of $f_i.TV$ and TTV (line 13–15). Finally, we update the end position of TF (line 16). (2) The frame head exceeds the start position of TF (line 17–23). Firstly, we initialize $f_i.TV$, TTV and determine s (line 18). Next we scan the rows in [$f_i.h$, $f_i.t$] which belong to the current frame f_i to update $f_i.TV$ and TTV with the rows which are in [s, $f_i.t$] (line 19–22). Finally, we set e (line 23).

Figure 2b illustrates the TF approach for Example 2. Before f_{10002}, the incremental strategy is used, and the process is the same as Fig. 1a. The black and purple curve arrows show the calculation process for the aggregation value of frame and temporary frame respectively. The black dashed arrow means the corresponding region is skipped. The purple shaded area is the rows which the temporary frame contains. It is updated to the frame tail after a frame is completed. When calculating f_{10002}, frame head changes compared with f_{10001}, so there is no appropriate TTV that can be used. Then the whole frame has to be aggregated. Meanwhile, redefine the start and end position of TF. TTV is calculated when calculating f_{10002}, which means each row in interval [201, 10002] is aggregated to both $f_{10002}.TV$ and TTV. For frame f_{10003}, TTV is assigned to $f_{10003}.TV$ first. Then each row in [3, 201) is aggregated to $f_{10003}.TV$. The overlap region is skipped. Finally, the rows that enter the frame is aggregated to both $f_{10003}.TV$ and TTV. Similarly, the four steps strategy (assign, aggregate, skip, aggregate) is also adopted by other following consecutive frames whose frame head don't exceed the start position of TF. f_{10201} is a special frame whose frame head is the same as TF. When calculating f_{10202}, $f_{10202}.h > s$, so the start position of TF is redefined and need to be recalculated over the interval [s, $f_{10202}.t$] into TTV. The process is similar to that of f_{10002}.

It is obvious that TF algorithm brings about performance improvement because of the TF. We will discuss how to determine the (s, e, TTV) in Appendix A. And Appendix A also shows the analysis of TF.

3.3   Optimizations for Mutilple TFs

By now, the overlap among frames has not disappeared, the reduction is a transformation of one problem into another problem, the overlap among the TFs still exists. So we propose another optimization algorithm with mutilple TFs. Since C_{loc} is reduced to minimum, the room for optimization focuses on C_{da} and C_{cal}.

In order to reduce the cost of both C_{cal} and C_{da}. An intuitive idea is to use HTF(Hierarchical Temporary Frames) approach. In this way, we can share the data access and calculation as much as possible.

If there are l levels of TF as shown in Fig. 3, every time only one path from the root to leaf node exists in the tree structure, and the calculation of TF_{i-1}

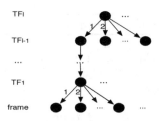

Fig. 3. Structure of l levels of TF-tree

depends on TF_i $(1 \leq i \leq l)$. At first, we calculate the aggregation value for the front frames with TF_1. When TF_1 is out of date, we first find the smallest level j $(1 < j \leq l)$ where the frame head of current frame doesn't exceed the start position of TF_j. Absolutely it is the TF_2 which is in the level 2. Then we calculate the aggregation value of current frame with TF_2. At the same time, we update the TF_1 (containing the initialization of the start and end position of TF_1 and the calculation of the TTV value of TF_1). Then we calculate the aggregation value of the following frames with the new TF_1. When the TF_1 and TF_2 are both out of date, we calculate the aggregation value of current frame with TF_3, and we update the TF_1 and TF_2 at the same time. Then we can calculate the aggregation value with the new TF_1 and TF_2. When TF_1 is out of date, first find the smallest level j $(i < j \leq l)$ where the frame head of current frame doesn't exceed the start position of TF_j. Then we calculate the aggregation value of current frame and reconstruct the path from TF_{j-1} to TF_1.

Algorithm 2 shows the evaluating of HTF. TF_j is the j-level TF. TF_{act} is the active TF and act is the active level of the TF-tree. For each partition, we initialize the active level first (line 2). When the frame head of the frame (f_i) is the same as that of the last frame (f_{i-1}), we can use the incremental strategy (line 4–7). When the frame head of the frame (f_i) is different from that of the last frame (f_{i-1}), we select the active and lowest level active TF (line 9). If we can find an active TF, we assign the value of $TF_{act}.TTV$ to $f_i.TV$ (line 11). If the value of act is not 1, then the TFs from level 1 to act-1 is not available. We need to reconstruct the path from TF_{act-1} to TF_1 based on TF_{act} (line 12–13). If act is -1, we initialize the $f_i.TV$ and the $TF - tree$ (line 15). Then, we scan the rows in the frame to update $f_i.TV$ and the $TTV - tree$ (line 16–20). If act is not -1, we scan the rows in the interval $[f_i.h, TF_{act}.s)$ to calculate $f_i.TV$ and update the non-available TFs at the same time (line 22–25). We then skip the interval which TF_{act} contains (line 26). Finally, we scan the rows which follow $TF_{act}.e$ and belong to f_i in order to calculate the $f_i.TV$ and update the TFs at the same time (line 27–31).

We first take two levels of TF into consideration, and then extend to arbitrary levels. We have showed each TF_1 in the first level is shared by $2\sqrt{F}$ frames. Here we assume each TF_2 in second level is shared by t TF_1 in first level. Since there are $N/2\sqrt{F}$ TF_1 in first level, and the start position of two adjacent

---

Algorithm 2. HTF(Hierarchical Temporary Frames)

---

1 **for** *each partition in the TABLE* **do**
2 $\quad$ $act \leftarrow -1$; (*act : the active level of the* $TTV - tree$)
3 $\quad$ **for** *each frame* f_i *in the partition* **do**
4 $\quad\quad$ **if** $f_i.h = f_{i-1}.h$ **then**
5 $\quad\quad\quad$ $f_i.TV \leftarrow f_{i-1}.TV$;
6 $\quad\quad\quad$ **for** *each row* r *in* $(f_{i-1}.t, f_i.t]$ **do**
7 $\quad\quad\quad\quad$ $f_i.TV \leftarrow$ transfunc$(f_i.TV, r)$;

8 $\quad\quad$ **else**
9 $\quad\quad\quad$ select the active level of the $TF - tree$;
10 $\quad\quad\quad$ **if** $act \geq 1$ **then**
11 $\quad\quad\quad\quad$ $f_i.TV \leftarrow TF_{act}.TTV$;
12 $\quad\quad\quad\quad$ **for** j *from 1 to act - 1* **do**
13 $\quad\quad\quad\quad\quad$ update the TF_j;

14 $\quad\quad\quad$ **if** $act = -1$ **then**
15 $\quad\quad\quad\quad$ initialize $f_i.TV$ and $TF - tree$;
16 $\quad\quad\quad\quad$ **for** *each row* r *in* $[f_i.h, f_i.t]$ **do**
17 $\quad\quad\quad\quad\quad$ $f_i.TV \leftarrow$ transfunc$(f_i.TV, r)$;
18 $\quad\quad\quad\quad\quad$ **for** r *in* $[TF_j.s, f_j.t]$ *of each* TF_j *which level* j *is in* $[1, l]$ **do**
19 $\quad\quad\quad\quad\quad\quad$ $TF_j.TTV \leftarrow$ transfunc$(TF_j.TTV, r)$;

20 $\quad\quad\quad\quad$ update the end position of each TF in the TF-tree;

21 $\quad\quad\quad$ **else**
22 $\quad\quad\quad\quad$ **for** *each row* r *in* $[f_i.h, TF_{act}.s)$ **do**
23 $\quad\quad\quad\quad\quad$ $f_i.TV \leftarrow$ transfunc$(f_i.TV, r)$;
24 $\quad\quad\quad\quad\quad$ **for** r *in* $[TF_j.s, TF_{act}.s)$ *of each* TF_j *which level* j *in* $[1, act-1]$ **do**
25 $\quad\quad\quad\quad\quad\quad$ $TF_j.TTV \leftarrow$ transfunc$(TF_j.TTV, r)$;

26 $\quad\quad\quad\quad$ ship interval $[TF_{act}.s, TF_{act}.e]$;
27 $\quad\quad\quad\quad$ **for** *each row* r *in* $[TF_{act}.e, f_i.t]$ **do**
28 $\quad\quad\quad\quad\quad$ $f_i.TV \leftarrow$ transfunc$(f_i.TV, r)$;
29 $\quad\quad\quad\quad\quad$ **for** r *in* $[TF_j.s, f_j.t]$ *of each* TF_j *which level* j *is in* $[1, l]$ **do**
30 $\quad\quad\quad\quad\quad\quad$ $TF_j.TTV \leftarrow$ transfunc$(TF_j.TTV, r)$;

31 $\quad\quad\quad\quad$ update the end position of each TF in the TF-tree;

TF_1s differs by $2\sqrt{F}$. Assuming that TF_1 contains m rows, then TF_2 contains $m - (t - 1) \times 2\sqrt{F}$ rows.

The cost of constructing TF_1 from TF_2 is :

$$cost_2 = \frac{N}{2\sqrt{F} \times t} \times (m - (t - 1) \times 2\sqrt{F}) \times T_{cpu}$$
$$+ \frac{N}{2\sqrt{F}} \times ((t - 1) \times 2\sqrt{F}) \times T_{CPU} \tag{1}$$

Equal to $\dfrac{N}{2\sqrt{F}} \times (\dfrac{m}{t} - \dfrac{(t-1)^2}{t} \times 2\sqrt{F}) \times T_{CPU}$.

The cost of constructing TF_1 without TF_2 is :

$$cost_1 = \frac{N}{2\sqrt{F}} \times m \times T_{CPU} \tag{2}$$

Then we can get $cost_1 > cost_2(t \geq 2)$. In a similar way, we can get that if the TF_3 can be shared by more than 2 TF_2s, the 3-level TF is superior to 2-level TF. By extension, the l-level TF is superior to $(l\text{-}1)$-level TF, if the TF_l can be shared by more than 2 TF_{l-1}s.

4   Experiments

We conduct a series of experiments. We have integrated the proposed techniques into PostgreSQL, and compared them with both PostgreSQL's and SQLServer's native window function implementation over both the TPC benchmark and generated data.

4.1   Experiment Environment

We experimented with the PostgreSQL 9.6.1. We changed the kernel of PostgreSQL by implementing the FL, TF and HTF. We ran the PostgreSQL system on a computer of Lenovo. The cpu is Intel(R) Core(TM) i5-4460M CPU@3.20 Hz. The memory is 24 GB 1600 MHz DDR3.

4.1.1   Data Sets

Two data sets are used to evaluate our methods. The first is TPC-H benchmark. We use the following TPC-H DBGEN instruction to generate 1.5 million rows of Table "order":

```
./dbgen -S 1 -T 0
```

The second data set is generated data. We created an "employee" table which contains three attributes, empno, depno and salary. Then we inserted 1.3 million rows into this table. The values of "salary" distributes between 0 and 10000.

```
CREATE TABLE empsalary(
  empno bigserial primary key,
  depno int,
  salary int
) ;

INSERT INTO empsalary ( depno, salary)
SELECT 1, random() * 10000 FROM generate_series (1, 1300000);
```

4.1.2   Queries

We use the following query template. Four aggregation functions, i.e., sum, avg, min and max, are tested.

```
SELECT *, aggfun(salary),
OVER (Rows BETWEEN frameoffset PRECEDING AND frameoffset
    FOLLOWING)
FROM table_name;
```

4.1.3   Comparison Methods

we compared our proposed methods with both PostgreSQL and SQLServer:

- **PG:** The implementation of the PostgreSQL itself.
- **SQLSERVER:** The implementation of the MS SQL Server 2012.
- **FL:** The fast locating method as shown in IV. (The following algorithms all include this method.)
- **TF:** We implemented the TTV algorithm as shown in IV.
- **HTF:** This is the HMTTV algorithm for the aggregation functions.

4.2   Experimental Results

In this subsection, we will show the experiment results with two data sets: empsalary and orders. The working memory can also influence the execution time. We set the working memory as 1 MB (data access is disk state) and 128 MB (data access is memory state).

4.2.1   Performance of Our Methods

For experimental convenience, we reduce the data size of table "orders" to 0.15 million and set the working memory as 1 MB. Figure 4a has shown the influence of FL. In Fig. 4a, as the frameoffset increases from 5000 to 10000, the frame size increases from 10000 to 20000. With PG, due to a large number of backward processing steps, the execution time of query grows from 141 s to 3733 s. It cost a lot of time to relocate the frame head. With FL, the execution time of query grows from 107 s to 285 s. It has a large degree of reduction of time for relocating the frame head with FL. And the effect of FL is more obvious as the window becomes larger.

We compared the efficiency between FL and TF by using the table "empsalary". Here, we set the working memory as 1 MB. Figure 4b shows the efficiency of the TF algorithm on the table "empsalary".

In Fig. 4b, when the FL grows from 460 s to 3769 s as the frame size increases from 5000 to 30000, the TF just grows from 36 s to 67 s. The execution time of TF is far less than FL. For example, when the frame size is 30000, speedup is over 50 times. And Fig. 4b shows that, as the frame size increases, the execution time of FL grows quickly comparing with TF algorithm. Because FL just helps to reduce the cost of locating the frame head and the TF algorithm reduces the times of recomputing as many as possible. As we discuss in Sect. 2, the bottleneck

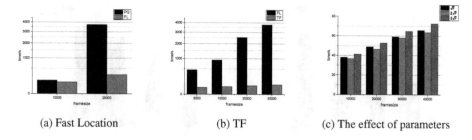

(a) Fast Location   (b) TF   (c) The effect of parameters

Fig. 4. Experimental resuls for singal-TF

is the re-calculation of the overlap among the frames. This means reducing the times of recomputing has a bigger impact on efficiency than locating the frame head.

We set the working memory as 1 MB and frame size from 10000 to 40000. The number of Group members has three values, $\sqrt{F}$, $2\sqrt{F}$ and $3\sqrt{F}$. Figure 4c shows the differences with different frame sizes. In all the cases, the value of $2\sqrt{F}$ performances best. It proves that, at the position of $2\sqrt{F}$, it has an minimal value for the execution time. The results are consistent with the analysis in Appendix A.

4.2.2 Comparison Among Different Optimizations

Figure 4b shows that the TF algorithm is more efficient than FL. But it is not the best algorithm for reducing the recomputing cost. Figures 5a, b, c and d shows the comparing result among TF and HTF.

Figure 5a is the execution result of the table "orders" with the 1 MB working memory and Fig. 5b is the result under the 128 MB working memory. Here, we use the MAX function. The frame size varies from 10000 to 70000. In all the four cases no matter what the working memory is, the HTF performances better. The overall speedup is over 2 times for all distributions.

Figure 5c is the execution result of the table "empsalary" with the 1 MB working memory and Fig. 5d is the result under the 128 MB working memory. We use the SUM function here. The frame size varies from 10000 to 70000, too. Figures 5c and d show the same performance of the three algorithms as the result of table "orders". Figure 5a, b, c and d all prove that the HTF algorithm is more efficient and more insensitive to the size of frame than TF.

4.2.3 Comparison with SQLServer

The SQLServer runs on Lenovo 90AU0010CP, with the operating system Windows 7. The cpu is Intel(R) Core(TM) i5-4460 M CPU@3.20 Hz. The memory is 24 GB 1600 MHz DDR3. In this subsection, we all use the table "orders". The SQLServer is SQLServer 2012 Express edition.

Figure 6a shows the results of function MAX under the 300 MB working memory (the minimal server memory of SQLServer). As the frame size increases from 1000 to 4000 (Here, we choose the smaller frame size, because it costs a lot of time to execute a query for larger frame size with SQLServer), the execution time

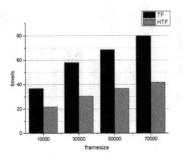

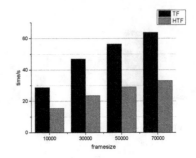

(a) The execution time on the Table "orders" with 1MB working memory

(b) The execution time on the Table "orders" with 128MB working memory

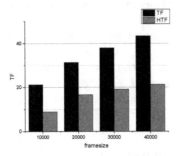

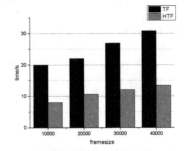

(c) The execution time on the Table "emp-salary" with 1MB working memory

(d) The execution time on the Table "emp-salary" with 128MB working memory

Fig. 5. Comparison among different optimizations

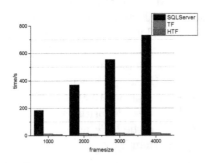

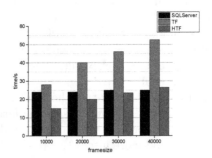

(a) The execution time of MAX aggregation function on the Table "orders" with 300MB working memory

(b) The execution time of SUM aggregation function on the Table "orders" with 300MB working memory

Fig. 6. Comparison with SQLServer

of SQLServer grows from 184 s to 782 s, much more than the two optimization algorithms.

Figure 6b shows the result of function SUM. As the frame size increases from 10000 to 40000, the execution time of SQLServer is all about 25 s, which is slightly longer than HTF algorithm but less than TF. So we can see that the average execution time of function SUM is much smaller than MAX with SQLServer from Figs. 6a and b. The possible reason is that SQL Server uses the native algorithm for MAX and an incremental method [16] for SUM. In other words, the optimization method for function SUM in SQLServer is much better than MAX.

Figure 6a and b both shows that HTF gets the best performance among the proposed algorithms. However, SUM function in SQL Server is not sensitive to the size of frame, and it may perform better than HTF as the size of frame increases.

5   Conclusion

This paper studies the problem of SQL query optimization for window function. Firstly, we introduce the two-phase window function evaluation framework systematically, and then built a cost model to analyse the sequencing phase. In order to minimize the redundant data accesses and computations during the function calls invoked over window frames, four optimization methods are proposed. We have prototyped our techniques in PostgreSQL. We conduct a series of experiments over TPC-H benchmark and generated data by comparing our methods with PostgreSQL and MS SQLServer. The experimental results show that our techniques bring substantial performance gains.

Acknowledgement. This work was supported by NSFC grants (No. 61472141, U1501252 and 61021004), Shanghai Knowledge Service Platform Project (No. ZF1213), Shanghai Leading Academic Discipline Project (Project NumberB412) and Shanghai Agriculture Science Program(2015) Number 3-2.

A   Analysis of TF

Lemma 1. *In ROWS model, the size of two adjacent frames in a partition differs at most one.*

Proof: Let f_i and f_{i+1} $(0 < i < N)$ be two adjacent frames in a partition. The size difference can be expressed as $sd_i = |(f_{i+1}.t - f_i.t) - (f_{i+1}.h - f_i.h)|$. Since frame head and frame tail moves at most one step ahead at one time, $0 \leqslant f_{i+1}.h - f_i.h \leqslant 1$ and $0 \leqslant f_{i+1}.t - f_i.t \leqslant 1$, then $0 \leqslant sd_i \leqslant 1$, Lemma 1 is proved.

- G_j: The j'th group in a partition.
- k_j: The number of frames which use the same TF in the group G_j.

The frames are divided into groups and frames in each group share the same TTV value, which is calculated when calculating the TV of the first frame in the group. Since two partitions share nothing, their cost could be analysed separately.

For purpose of simplicity to build the cost model, we set the whole table as a partition. Then the partition size is N. We first assume each frame has the same frame size F and frame head changes for each frame. Group G_j contains the $k_j + 1$ frames, f_i, f_{i+1}, f_{i+2}, ... , f_{i+k_j}. Let o denotes the overlap of the group of frames from f_i to f_{i+k_j} ($k_j > 0$), then $o = f_i \cap f_{i+1} \cap ... \cap f_{i+k_j}$, it's not hard to get that o starts from $f_{i+k_j}.h$ and ends at $f_i.t$. And apparently $f_i.t$ is bigger than $f_{i+k_j}.h$, because it is utterly useless if the overlap is empty. So $s = f_{i+k_j}.h$ and it will not be changed until it's out of the range of current frame. Besides, e is initialized to $f_{i_j}.t$ and assigned to the tail position of current frame in the following calculation. So in the end, e will be equal to $f_{i+k_j}.t$.

We can get that, the start position s of TF is associated with k_j. So, if we want to know s, we must confirm k_j firstly. Next, we will give the analysis of k_j.

For a group G_j which contains $k_j + 1$ frames, the calculation cost consists of two components, one is related to the calculation and updating of TTV, the other one is related to the calculation of each transition value (TV). Since the TF is finally identical with the last frame, the cost on TTV is $F \times T_{CPU}$. For the first frame, all rows need to be aggregated, for other frames, approximatively only rows before the overlap (the number of the rows for f_{i+1} is k-1, the number of the rows for f_{i+2} is k-2, ... , the number of the rows for f_{i+k} is 0. So, the sum is k(k-1)/2) and the new entering row (For every frame, there is just one row entering the frame) needs to be aggregated. So the calculation cost related to TV is $(F + k \times (k-1)/2 + k) \times T_{CPU}$. For each group G_j, the calculation cost of TF-based scheme is:

$$C_{cal} = (2F + k_j \times \frac{k_j + 1}{2}) \times T_{CPU} \qquad (3)$$

So, for the partition, the total calculation cost is:

$$C_{cal} = \sum_{j=1}^{t} [(2F + k_j \times \frac{k_j + 1}{2}) \times T_{CPU}] \qquad (4)$$

Since the number of frames in the partition is N, we get this:

$$\sum_{j=1}^{t} (k_j + 1) = N \qquad (5)$$

By the Formulas 4 and 5, we get this:

$$f = \frac{1}{T_{CPU}} \times C_{cal} = 2Ft + \frac{1}{2} \times \sum_{j=1}^{t} (k_j^2 + k_j) \qquad (6)$$

$$\varphi = t + \sum_{j=1}^{t} k_j - N = 0 \qquad (7)$$

f takes the minimum value when C_{cal} takes the minimum value. In order to minimize the C_{cal}. We use the *Lagrange Multiplier* with Formulas 6 and 7 (λ is the new variable):

$$L(k_1, k_2, ..., k_t) = f + \lambda\varphi \tag{8}$$

As well as:

$$L(k_1, k_2, ..., k_t) = 2Ft + \frac{1}{2} \times \sum_{j=1}^{t}(k_j{}^2 + k_j) + \lambda(t + \sum_{j=1}^{t} k_j - N) \tag{9}$$

Then we get the simultaneous formulae with a step partial derivative of Formula 9 by $k_1, k_2, ..., k_t$ and Formula 7:

$$\begin{cases} k_1 + \frac{1}{2} + \lambda = 0 \\ k_2 + \frac{1}{2} + \lambda = 0 \\ k_3 + \frac{1}{2} + \lambda = 0 \\ . \\ . \\ k_t + \frac{1}{2} + \lambda = 0 \\ t + \sum_{j=1}^{t} k_j - N = 0 \end{cases} \tag{10}$$

By the Formula 10, we get this:

$$k_1 = k_2 = ... = k_t = k = \frac{N}{t} - 1 \tag{11}$$

We can get that, if we give the partition and it is divided into t groups (N, F, T_{CPU} and t have been given), $k_1, k_2, ..., k_t$ is all the same when the cost C_{cal} is taking the minimum. Then we take the Formula 11 into Formula 6 and get this:

$$f = 2Ft + \frac{N^2}{2t} - \frac{N}{2} \tag{12}$$

By the Formula 12, we can get that the cost(C_{cal}) is associated with the number of groups in a partition. So, if we want to get a minimum cost, we should set a appropriate value of t. We get the first-order derivative of the Formula 12:

$$f' = 2F - \frac{N^2}{2} \times t^{-2} \tag{13}$$

Let $f' = 0$, we get that $t = \frac{N}{2\sqrt{F}}$. Then we take $t = \frac{N}{2\sqrt{F}}$ into Formula 11 and get that $k = 2\sqrt{F} - 1$.

So, if we have been given a partition and it's size is P, we set the number of groups in the partition as $\frac{P}{2\sqrt{F}}$. That means that each group contains $2\sqrt{F}$ frames. For each group, the start position of TF is $f_{i+\lfloor 2\sqrt{F}\rfloor-1}.h$, $s = f_i.h + \lfloor 2\sqrt{F}\rfloor - 1$.

In the experiments, we tried different frame sizes. From the experiment results, we can see that the TF algorithm have a good performance.

References

1. Zuzarte, C., Pirahesh, H., Ma, W., Cheng, Q., Liu, L., Wong, K.: Winmagic: subquery elimination using window aggregation. In: Proceedings of the ACM SIGMOD International Conference on Management of Data, SIGMOD 2003, pp. 652–656. ACM, New York (2003)
2. Bellamkonda, S., Ahmed, R., Witkowski, A., Amor, A., Zait, M., Lin, C.-C.: Enhanced subquery optimizations in oracle. Proc. VLDB Endow. **2**(2), 1366–1377 (2009)
3. Ben-Gan, I.: Microsoft SQL Server High-Performance T-SQL Using Window Functions. Microsoft Press (2012)
4. Cao, Y., Bramandia, R., Chan, C.Y., Tan, K.-L.: Optimized query evaluation using cooperative sorts. In: Proceedings of the 26th International Conference on Data Engineering, ICDE, Long Beach, California, USA, pp. 601–612. IEEE (2010)
5. Cao, Y., Chan, C.-Y., Li, J., Tan, K.-L.: Optimization of analytic window functions. Proc. VLDB Endow. **5**(11), 1244–1255 (2012)
6. Cao, Y., Bramandia, R., Chan, C.-Y., Tan, K.-L.: Sort-sharing-aware query processing. VLDB J. **21**(3), 411–436 (2012)
7. Window functions for postgresql desgin overview (2008). http://www.umitanuki.net/pgsql/wfv08/design.html
8. Bellamkonda, S., Bozkaya, T., Ghosh, B., Gupta, A., Haydu, J., Subramanian, S., Witkowski, A.: Analytic functions in oracle 8i. Technical report (2000)
9. Jin, C.Q., Yi, K., Chen, L., Yu, J.X., Lin, X.M.: Sliding-window top-k queries on uncertain streams. VLDB J. **19**(3), 411–435 (2010)
10. Li, J., Maier, D., Tufte, K., Papadimos, V., Tucker, P.A.: Semantics and evaluation techniques for window aggregates in data streams. In: Proceedings of the ACM SIGMOD International Conference on Management of Data, SIGMOD 2005, pp. 311–322. ACM, New York (2005)
11. Li, J., Maier, D., Tufte, K., Papadimos, V., Tucker, P.A.: No pain, no gain:efficient evaluation og sliding-window aggregates over data streams. SIGMOD Rec. **34**(1), 39–44 (2005)
12. Neumann, T., Moerkotte, G.: A combined framework for grouping, order optimization. In: Proceedings of the 30st International Conference on Very Large Data Bases, VLDB 2004, pp. 960–971. VLDB Endowment (2004)
13. Simmen, D., Shekita, E., Malkemus, T.: Fundamental techniques for order optimization. In: Proceedings of the 1996 ACM SIGMOD International Conference on Management of Data, SIGMOD 1996, pp. 57–67. ACM, NY (1996)
14. Wang, X.Y., Chernicak, M.: Avoiding sorting, grouping in processing queries. In: Proceedings of the 29th International Conference on Very Large Data Bases, VLDB 2003, pp. 826–837. VLDB Endowment (2003)
15. Leis, V., Kan, K., Kemper, A., et al.: Efficient processing of window functions in analytical SQL queries. Proc. VLDB Endow. **8**(10), 1058–1069 (2015)
16. Wesley, R., Xu, F.: Incremental computation of common windowed holistic aggregates. Proc. VLDB Endow. **9**(12), 1221–1232 (2016)

Query Optimization on Hybrid Storage

Anxuan Yu, Qingzhong Meng, Xuan Zhou$^{(\boxtimes)}$, Binyu Shen,
and Yansong Zhang

DEKE Lab, Renmin University of China, Beijing, China
zhou.xuan@outlook.com

Abstract. Thanks to the rapid growth of memory capacity, it is now
feasible to perform query processing completely in memory. Nevertheless,
as main memory is substantially more expensive than most secondary
storage equipments, including HDD and SSD, it is not suitable for storing
cold data. Therefore, a hybrid data storage composed of both memory
and secondary storage is expected to stay popular in the foreseeable
future. In this paper, we introduce a query optimization model for hybrid
data storage. Different from traditional query processors, which treat
either main memory as a cache or secondary storage as an anti-cache,
our model performs semantic data partitioning between memory and
secondary storage. Query optimization can thus take the partitioning
of data into account, to achieve enhanced performance. We conducted
experimental evaluation on a columnar query engine to demonstrate the
advantage of the proposed approach.

1 Introduction

As the size of RAM grows rapidly, it is already affordable to store an entire
database in memory. This can substantially speed up data processing, making
in-memory DBMS an popular tool for data intensive applications. Nevertheless,
a choice between an in-memory database and a disk resident one does not solely
depends on speed. Instead, it is determined by consideration of cost-effectiveness
most of the time. While memory appears more cost-effective than secondary
storage in speed and throughput, it is much less cost-effective in storage space.
Consequently, a practical choice is to store hot data in memory while keeping
cold data in more economical devices, such as HDD or SSD. The combination
of memory and secondary storage is expected to stay popular in future's data
processing infrastructures.

To live with a hybrid data storage, data processing technology is faced with
an essential question: how to partition data between memory and secondary
storage and how to conduct query processing over the two parts of data? In a
conventional DBMS, disk plays the leading role of storing and processing data.
Memory works as a cache, an auxiliary device regarded unimportant by the
query optimizer, which picks query plans by counting disk I/Os. However, when
memory is spacious, this approach becomes ineffective. If a significant proportion of frequently used data is cached in memory, the I/O cost estimated by a

© Springer International Publishing AG 2017
S. Candan et al. (Eds.): DASFAA 2017, Part I, LNCS 10177, pp. 361–375, 2017.
DOI: 10.1007/978-3-319-55753-3_23

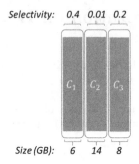

Selectivity: 0.4 0.01 0.2

Size (GB):   6   14   8

Fig. 1. Correlation between data placement and query optimization

traditional query optimizer will be inaccurate, which may result in suboptimal query plans. Modern in-memory DBMSs choose to treat memory as the main storage and disk as a backup or cold storage. Thus, they consider only memory access cost when performing query optimization. This approach works well when hot data can completely fit in memory. However, if there is a shortage of memory space, the resulting query plans will be problematic too.

The two opposite design choices made by today's database systems intend to conduct data placement and query optimization independently. While this separation of concerns can reduce the complexity of implementation, it may result in inefficiency of query processing, especially when memory and second storage are regarded equally important. In this paper, we consider the possibility of correlating data placement and query optimization. On the one hand, we enable a query optimizer to take the location of data into account, to improve the accuracy of query plan evaluation. On the other hand, we optimize the placement of data blocks by inferring their access frequencies from the common query plans generated by the query optimizer.

Figure 1 illustrates that correlating query optimization and data placement can result in enhanced performance. Consider a columnar database consisting of a single relational table. The table is composed of three columns (i.e., C_1, C_2 and C_3), whose sizes are 6 GB, 14 GB and 8 GB respectively. Suppose users issue only simple selection queries to the database. The database executes each query by performing column-wise scan – the three columns are scanned consecutively, to gradually filter out the tuples not satisfying the selection predicates. (For in-memory DBMSs, column-wise scan is a well used operator.) As illustrated in Fig. 2, during a column-wise scan, the first column needs to be fully scanned, while the subsequent columns can be partially accessed, since the tuples filtered out in the previous scans can be skipped. Therefore, the order of scan can have significant impact on the performance.

A query optimizer is responsible for picking the scanning order yielding the best performance. Suppose the expected selectivity of each query on C_1, C_2 and C_3 are 0.4, 0.01 and 0.2 respectively. If we do not consider the placement of data, the optimal order for scanning the three columns will be $C_1 \rightarrow C_3 \rightarrow C_2$, as this order can minimize the amount of data an average query should access.

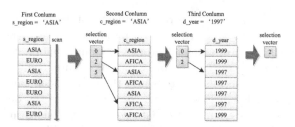

Fig. 2. Column-wise scan: the first column is fully scanned, and the subsequent columns are partially accessed.

Suppose the capacity of the main memory is 14 GB, which is only enough to accommodate either C_2 or the combination of C_1 and C_3. A traditional DBMS usually decides the placement of the columns based on their chance to be accessed. Since the optimal scanning order is $C_1 \rightarrow C_3 \rightarrow C_2$, C_1 and C_3 will be accessed more frequently than C_2. Thus, the DBMS will place C_1 and C_3 in memory. This is unfortunately not the optimal solution. As we know, the data access cost of disk is much higher than that of memory. For the scanning order $C_1 \rightarrow C_3 \rightarrow C_2$, about 8 percent (0.4×0.2) of C_2 will be accessed from the disk, resulting in an I/O cost of $0.08 \times 14 = 1.12$. If we place C_2 in memory and apply the scanning order $C_2 \rightarrow C_1 \rightarrow C_3$ instead, the I/O cost can be reduced to $0.01 \times 6 + 0.01 \times 0.4 \times 8 = 0.092$. This example shows that data placement and query plans should be considered simultaneously to achieve the optimal performance.

This paper addresses the problem of query optimization on a hybrid storage. Instead of conducting data placement and query optimization separately, we propose to integrate these two issues into one co-optimization problem. We formulate and solve this co-optimization problem in the context of A-Store [25], a columnar query processing engine specialized for star and snowflake schemas. We conducted extensive experiments on the queries of SSB and TPC-DS. The results show that our query optimizer can achieve significant performance improvement over traditional optimizers.

The rest of the paper is organized as follows. Section 2 reviews the related work. Section 3 provides an overview of A-Store. Section 4 introduces a data partitioning model that enables co-optimization of data placement and query plans. Section 5 presents our solution to the co-optimization problem. Section 6 evaluates our query optimizer using standard benchmarks. Section 7 concludes the paper and discusses future research opportunities.

2   Related Work

Research on query optimization dates back to several decades ago. A variety of query optimization models and algorithms have been proposed for different types of database systems, including traditional disk resident database [5], in-memory database [17,23], distributed and parallel databases [2,10], etc. To achieve effectiveness, the cost model of a query optimizer should reflect the dominant factor of performance. For instance, to a disk resident database, I/O cost should be the dominant factor. To a distributed or parallel database, the cost of data transmission should be a dominant factors. As main memory starts to work as data storage, memory access cost becomes increasing important to performance. This involves several additional factors into the cost model, such as cache efficiency [16,20] and code efficiency [3,18]. Nevertheless, as the query optimizers of in-memory databases assume that data completely resides in memory, they normally ignore the possibility of accessing cold data. This assumption is not necessarily true for a hybrid storage. In this paper, we assume that cold data is still required in query processing, so that a query optimizer should consider both memory and secondary storage.

In recent years, J. DeBrabant et al. have coined the concept of anti-cache [7], to deal with cold data management in an in-memory database. As it is uneconomical to store cold data in memory, cold data should be evicted to disks periodically. In such a case, disk plays an opposite role of cache, so it is called anti-cache. (Some in-memory systems [8] also call it cold storage directly.) There has been extensive discussion about how to ensure the transactional safety [7] and the efficiency [24] of anti-cache. However, little work is devoted to data access efficiency of anti-cache. We address this problem in this work.

In the case of hybrid storage, data partitioning is a sub-problem of query optimization. There has been a significant body of work on query optimization over partitioned data [13,19], especially in the context of parallel database [4,26]. It is commonly accepted that the knowledge of data partitioning can help to speed up query execution. However, existing approaches to partition-aware optimization all assume that data resides on symmetric storage devices. They cannot be directly applied to asymmetric storage composed of both memory and disk. In this paper, we consider asymmetric storage.

Semantic caching [6,21] is a technology that shares our goal of this paper. Different from the caches or buffers used in a conventional database systems, the contents in a semantic cache are describable in the database's query language. By keeping track of the contents in a semantic cache, a database can answer or partially answer a query directly using the cache, so as to relieve the burden of the backend storage. Some techniques on semantic cache consider data placement and query optimization simultaneously. For instance, the technique of cache investment [15] will intentionally choose suboptimal query plans to enhance the hit rate of a semantic cache. The same techniques can be applied to our case of hybrid storage, by treating the memory as a semantic cache. However, such an indirect solution does not seem efficient. First, the data granularity of semantic caching is at the level of queries and views, which could be too coarse to achieve good memory utilization. Second, as query answering based on views is a complex problem, it might impair the practicality of semantic caching. In contrast, our approach treats column chunks as the units of data placement. This makes our solution more straight forward and practical.

To enable co-optimization of data placement and query plans, our query optimizer needs to optimize multiple queries simultaneously. Multi-query

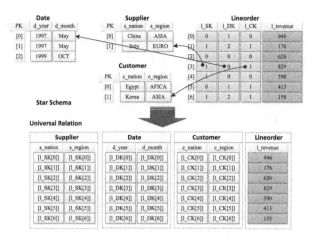

Fig. 3. Virtual denormalization of A-Store

optimization has been an intensively investigated area [9,22]. Usually, a multi-query optimizer [11] attempts to create a joint query plan for multiple queries, so that they can share intermediate results and save cost. Our work does not consider joint query plans, but focus on the efficiency of data placement and individual query plans. Joint query optimization can be scheduled in our future research.

3   Preliminary of A-Store

We build our query optimizer on A-Store [25], a columnar query processing engine specialized for star and snowflake schemas. A star or snowflake schema can be regarded as a normalized relational schema of multidimensional data [12]. While normalization can save storage space, it requires expensive joins to accomplish query processing. To achieve good performance, A-Store applies the strategy of virtual denormalization – it encodes database columns into arrays and treats the array indexes as the primary keys; then, join on foreign key can be accomplished by positional tuple access based on array indexes.

As illustrated in Fig. 3, A-Store stores a relational table in an array family, which is composed of a set of arrays of equal length, each representing a column of the table. The arrays in an array family are completely aligned with one another, such that all the ith elements in the arrays constitute the ith tuple of the table. As array indexes can be used to directly locate the tuples in a table, A-Store treats the array index as the primary key of a table. As a result, a foreign key can be used to directly address the tuples in the reference table.

With such as storage model, each star or snowflake schema is turned into a universal relation. As shown in the bottom of Fig. 3, we can obtain the universal relation by scanning the fact table and at the same time fetching the referred tuples in the dimension tables using the array indexes. Utilizing such a virtual universal relation, A-Store is able to process Selection-Projection-Join-Grouping-Aggregation (SPJGA) queries efficiently – it simply scans the universal relation, filters out the tuples that do not satisfy the selection criteria, and feeds the resulting records to grouping and aggregation operators. Performance study [25] showed that A-Store is competitive to the most prestigious in-memory DBMSs, including Vectorwise [27] and Hyper [14], in OLAP.

To achieve further performance improvement, a number of optimization techniques have been applied to A-Store. For instance, the dimension tables can be processed prior to the scan of the fact table, so as to reduce the cost of probing. Column-wise table scan can be performed to save memory bandwidth (see Fig. 2). Details of the optimization techniques can be found in [25]. As a result, each query execution plan of A-Store can always be divided into three stages – (1) evaluation of the dimension tables; (2) a column-wise scan of the fact table, to prune and group the tuples; (3) a scan of the measure columns to finish the aggregation. In general, the first and third stages are relatively cheap and usually unamendable for each particular query. Therefore, they are not subject to optimization. In contrast, the scanning order in the second stage can impact the overall performance significantly and must be optimized. Therefore, in this paper, we focus on the optimization of the scanning order over the fact table.

The query model of A-Store is limited to SPJGA queries. While SPJGA queries can cover the majority of the cases in MOLAP [12], they are only a subset of queries expressible in relational algebra. Therefore, we do not expect to apply our query optimizer to a generic relational database. To generalize our query optimizer, further extension is required.

4   The Data Placement Model

To enable dynamic placement of data, we need to break a relational table into pieces, which can be moved between memory and disk conveniently. In a traditional database system, a table can be partitioned horizontally or vertically, which would result in a row oriented or column oriented database. To enable effective query optimization, we choose to apply vertical and horizontal partitioning simultaneously. On the one hand, we expect that data partitions are semantically describable, such that the knowledge of data placement can be exploited by query optimization. Vertical data partitioning seems to meet this criterion, as columns are semantically describable. Moreover, as A-Store performs columnar query processing, it requires data to be vertically partitioned anyway. On the other hand, we expect that data partitioning can separate hot and cold data. In this sense, horizontal partitioning appears more effective than vertical partitioning, as the utility of data is often determined by its freshness or age, which varies from row to row. Horizontal partitioning also enables A-Store to parallelize query processing – we can assign each horizontal partition of the fact table to a worker thread, which works independently to generate a subset of query results.

Our data placement model first horizontally partitions a table into a number of *row groups*, and then vertically partition each row group into *column chunks*. This is illustrated in Fig. 4.

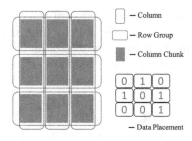

Fig. 4. Partitioning a relational table

Conventionally, a horizontal partitioning of a table can be performed based on the value ranges of a set of user defined attributes. As A-Store uses array indexes as the primary keys, a dimension table has to be partitioned based on its array index. Otherwise, insertion in a dimension table may result in rearrangement of the array index and cascading updates over the entire database [25]. In contrast, the attributes for partitioning the fact table can be chosen by the user. In most cases, we can partition a fact table using its timeline attributes (e.g., *date* or *time*). As fresh data is more likely to be used than dated data, it allows us to separate hot and cold data more clearly.

After the horizontal partitioning, each row group is further partitioned into a number of column chunks, one for each column. Each column chunk is stored as an array residing either in memory or on disk. As the data types of the columns differ from one another, the column chunks can vary in size. Column chunk is the basic unit of data placement. Our system can dynamically decide whether a column chunk should stay in memory or on disk. As shown in Fig. 4, a bitmap is used to keep track of the location of each column chunk. Using the bitmap, the system knows where to retrieve the required data during query processing.

5   Co-optimization of Data Placement and Query Plans

With the knowledge of data placement, a query optimizer can make informed decision about the optimal query plans. When frequent query plans are known, we can estimate the access frequency of each column chunk, which in turn help us determine the best data placement. The mutual reinforcement between data placement and query plans can continue indefinitely. Therefore, we expect that the query optimizer can co-optimize them simultaneously in a single round. As mentioned earlier, the query optimizer of A-Store is only concerned with the scanning order of the fact table. In this section, we focus on the co-optimization of the data placement and the scanning orders.

5.1   Data Placement Aware Query Optimization

When the knowledge of data placement is unavailable, we can estimate the cost of a query plan by the amount of the data it is going to process. This is how the state-of-the-art optimizers operate. In the case of A-Store, we can estimate the cost of a column-wise scan by the proportion of the data it is going to access. Given n columns $\{C_1, C_2, ..., C_n\}$, suppose the size of the column C_i is W_i. Suppose the scanning order of a query on the columns is $C_1 \rightarrow C_2 \rightarrow ... \rightarrow C_n$, and the selectivity of the query on C_i is S_i. Then, the cost of the column-wise scan, i.e., the total amount of data to be accessed during the scan, will be

$$Cost = W_1 + \sum_{k=2}^{n} \left(W_k \times \prod_{t=1}^{k-1} S_t\right). \tag{1}$$

A brute force method of query optimization is to calculate the cost of every possible scanning order using Formula (1) and select the order with the minimum cost. This can be costly when confronted with a large number of columns. In fact, there is a much simpler method. Consider two scanning orders, $... \rightarrow C_i \rightarrow ... \rightarrow C_j \rightarrow ...$ and $... \rightarrow C_j \rightarrow ... \rightarrow C_i \rightarrow ...$. Suppose the only difference between the two orders is that C_i and C_j are swapped. Based on Function (1), we can conclude that the first order is cheaper than the second order, if and only if

$$\frac{W_i}{1 - S_i} < \frac{W_j}{1 - S_j}.$$

In other words, if this inequation holds, C_i should always be scanned prior to C_j. Therefore, we can determine the optimal scanning order using the following formula alone.

$$\frac{W_i}{1 - S_i}. \tag{2}$$

The smaller the value of $\frac{W_i}{1-S_i}$, the earlier the column C_i should be scanned. As a result, the query optimizer only needs to sort the columns using Formula (2).

Given the example in Fig. 1, the $\frac{W_i}{1-S_i}$ values of C_1, C_2 and C_3 are 10, 14.14 and 10 respectively. Therefore, their optimal scanning order is $C_1 \rightarrow C_3 \rightarrow C_2$ or $C_3 \rightarrow C_1 \rightarrow C_2$.

The case will be different when the columns are distributed among the memory and the secondary storage. A secondary storage device, be it either HDD or SSD, is usually significantly slower than main memory. Thus, the objective of a query optimizer turns to minimize the data accesses on the secondary storage. Suppose the columns located in memory is $\{C_1, ..., C_i\}$ and those located in the secondary storage is $\{C_{i+1}, ..., C_n\}$. As data access in memory is no longer considered, a optimal query plan will certainly first scan the data in memory. To decide the optimal scanning order on the secondary storage, we can still use Formula (2). Suppose the scanning order on the secondary storage is $C_{i+1} \rightarrow ... \rightarrow C_n$. Then, the cost of the scan can be summarized as follows, which considers only the cost on the secondary storage:

$$Cost' = \prod_{t=1}^{i} S_t \times \sum_{k=i+1}^{n} \left(W_k \times \prod_{t=i}^{k-1} S_t\right). \tag{3}$$

For the case in Fig. 1, if C_2 is located in memory, the optimal scanning order will be $C_2 \rightarrow C_1 \rightarrow C_3$.

5.2   The Algorithm for Co-optimization

Co-optimization of data placement and query plans is a problem of much higher complexity. First, as data has been horizontally partitioned, the selectivity of a query may vary from one row group to another. In other words, instead of assigning a selectivity to an entire column, we have to assign a unique selectivity to each individual column chunk. Second, it is no longer sensible to optimize a single query. Instead, we need to consider all possible queries that could be issued to the database within a certain period of time.

To simplify the problem, we classify queries into a set of abstract *query templates*. Queries of the same template share the same query structure that aims to retrieve the same type of information. They differ only in selection predicates. For instance, one can use a query template to retrieve the orders of Product A. By changing the predicates, he can use the same template to retrieve the orders of Product B. Such an abstraction appears reasonable, as users usually access a database through a limited number of predefined programs or stored procedures, which can be regarded as templates. The goal of query optimization is to find a query plan for each template. To further simplify the problem, we assume that the expected selectivity of each query template on a column chunk is predetermined. While this assumption is not always true, it is necessary to make the computation feasible.

Finally, the problem of co-optimization can be defined as follows.

Problem Definition:

Given: (1) A relational table that has been partitioned into a set of column chunks. (2) A set of query templates, each imposing a predetermined selectivity on each column chunk. (If a query template does not need a column chunk, it sets the corresponding selectivity to 0.) (3) The weight (i.e., the frequency of invocation) of each query template. (4) The size of the main memory.

Goal: (1) A placement of the column chunks among the main memory and the secondary storage (the data in the memory should not exceed its capacity limit). (2) A query plan (i.e., a scanning order) for each query template on each row group. The combination of (1) and (2) should ensure that the expected data access cost on the secondary storage is minimized.

Optimization on a Single Row Group. To solve the problem, we first consider the optimization problem on a single row group. As discussed in Sect. 5.1, if the data placement is known, the optimal scanning order of a query template on a single row group can be easily determined. When the optimal scanning orders of all the templates are determined, we can calculate the *expected scanning cost* of a data placement, which is the weighted average of the scanning costs of all the templates based on Formula (3). If the data placement is undecided, we need to determine the optimal data placement that yields the minimum *expected scanning cost*. A brute force solution is to exhaustively enumerate all possible data placements, compute the optimal scanning orders for each data placement and select the data placement with the minimum expected scanning cost.

Data:
ml: memory capacity;
$rgCount$: number of row groups;
$msCount$: number of data placements for each row group;
$ms[i]$: memory usage of the ith data placement;
$cost[i][j]$: minimum cost of the ith row group if its memory size is $ms[j]$;
$minCost[i]$: minimum total cost when the total memory capacity is i;
$choice[i][j]$: optimal memory size for the jth row group if the total memory capacity is i;

Result: for each i ($1 \leq i \leq rgCount$), compute $choice[ml][i]$;

```
for i ← 1 to ml + l do
    minCost[i] ← MaxInteger;
    choice[i] ← an empty array;
    for j ← 1 to ccCount do
        | choice[i][j] ← −1;
    end
end
for i ← 1 to rgCount do
    for j ← ml + 1 downto 1 do
        for k ← 1 to msCount do
            if j ≥ ms[k] ∧ minCost[j − ms[k]] + cost[i][k] < minCost[j] then
                minCost[j] ← minCost[j − ms[k]] + cost[i][k];
                choice[j] ← choice[j − ms[k]];
                choice[j][i] ← ms[k];
            end
        end
    end
end
```

Algorithm 1: Algorithm for Optimizing Data Placement

Optimization on an Entire Database. So far, we can compute the optimal data placement for a single row group, as long as the size of the memory allocated to the row group is known. To optimize the data placement on multiple row groups, we only need to find the optimal memory allocation among different row groups. Intuitively, this can be reduced to the Multiple-Choice Knapsack Problem (MCKP) [1]. To achieve the reduction, each row group can represent a class of objects in MCKP – the size of an object corresponds to the size of memory allocated to the row group, and the weight of the object corresponds to the expected scanning cost of the optimal data placement for the memory size. With the reduction, the problem becomes to find an object from each class, such that their total weight is minimized and their aggregated size is within the limit of the memory capacity.

According to [1], there is a pseudo-polynomial algorithm for MCKP. We use the same algorithm (Algorithm 1) to find the optimal memory allocation. When the memory allocation is decided, the optimal data placement of each row group is automatically determined. So is the optimal query plan of each query template. The total complexity of Algorithm 1 is $O(N \times M \times D)$, where N represents the number of row groups, M represents the size of the memory, and D represents the number of possible data placements for a row group. The optimization is not invoked for each query. Instead, it is executed only once when there is dramatic change in the workload or the database structure. Therefore, such a complexity is acceptable in practice. A regular database usually contains a limited number of columns. The frequent query templates being executed on a database are usually limited in number too. Thus, it is unlikely that M and D will boost the

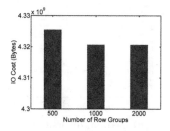

Fig. 5. How N influences the effectiveness of optimization

execution time of the algorithm. In contrast, N is the main influential factor to the performance. To ensure the feasibility of the algorithm, we can set a lower bound to the size of each row group to restrict N. According to our experimental study, when N is set to 1,000 to 100,000, we can achieve a good tradeoff between effectiveness and feasibility.

6   Evaluation

We implemented our optimizer in A-Store and conducted experiments to evaluate its performance. We selected a set of queries from the benchmarks of SSB and TPC-DS. By putting the queries into different combinations, we studied when and how the optimizer can achieve enhanced performance. The experiments were conducted on a HP Z820 workstation, equipped with two 2.60 GHz Intel Xeon processors E5-2670 and 256 GB DDR3 RAM. Its secondary storage is a PCIe SSD of 1TB. The operating system installed on the workstation was CentOS 7.1. Without otherwise mentioned, the scale factors of SSB and TPC-DS were both set to 100.

We compared our optimizer against a classical approach, which represents the "optimal" performance achievable by the traditional approaches that conduct data placement and query optimization independently. Basically, the classical approach first determines the query plan for each query template. Then, it deduces the access frequencies of the column chunks based on the query plans, and places the most frequently accessed chunks in memory.

6.1   Influence of N

Figure 5 shows how the number of row groups (i.e., N) influences the results of optimization on $Q3.4$ and $Q4.4$ of SSB. In principle, a fine-grained partitioning can achieve better optimization results than a coarse-grained partitioning. However, the marginal benefit of increased partitioning granularity keeps shrinking. As shown in Fig. 5, when $N = 1000$, the effectiveness of optimization has already approached the optimal. Therefore, in the rest of the experiments, we always set $N = 1000$.

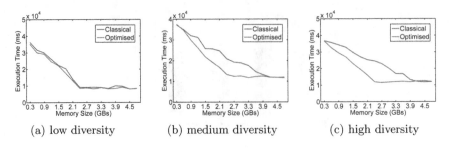

Fig. 6. Performance of uniform queries on SSB

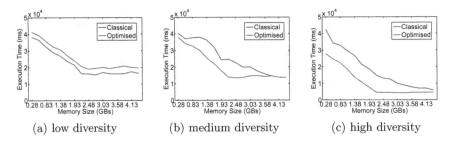

Fig. 7. Performance of uniform queries on TPC-DS

6.2 Experiments on Uniform Queries

In the first set of experiments, we considered the cases where all the query predicates follow an uniform distribution. In other words, we randomly set the parameters of a query template, such that the resulting queries have equal chance to access each row group. We used three combinations of the SSB queries and three combinations of the TPC-DS queries. For each of the queries, we turned it into a query template by parameterizing its selection predicates. Then, we used each template to generate a number of queries to execute. Figures 6 and 7 show the average response time of the queries.

In the experiments on SSB, we tested three combinations of query templates, which included $\{Q3.1, Q3.4\}$, $\{Q3.4, Q4.3\}$ and $\{Q2.2, Q2.3, Q3.1, Q3.4, Q4.3\}$. As $Q3.1$ and $Q3.4$ apply the same scanning order in query processing, they represent the case that lacks diversity. In contrast, $\{Q3.4, Q4.3\}$ represents the case of high diversity, as they tend to use opposite scanning orders. The combination of 5 queries represents the case of medium diversity. We compared our optimizer against the traditional approach, which determines query plans prior to data placement. We varied the size of the memory, and observed how the performance varies with the memory size. As Fig. 6 shows, the co-optimization of query plans and data placement can achieve significant performance enhancement, especially when there is a large diversity among the queries. This is expected, as diverse query plans imply low data locality, which hurts the utility of the memory. Through co-optimization, we increased the locality of data access, and thus enhancedthe performance.

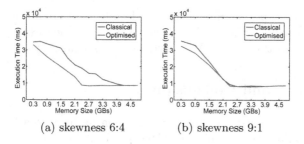

(a) skewness 6:4   (b) skewness 9:1

Fig. 8. Performance of skewed queries on SSB

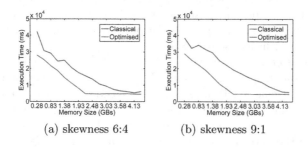

(a) skewness 6:4   (b) skewness 9:1

Fig. 9. Performance of skewed queries on TPC-DS

In the experiments on TPC-DS, we tested three combinations of query templates too. They are $\{Q7, Q43\}$, $\{Q10, Q42\}$ and $\{Q7, Q10, Q42, Q43\}$, representing the cases of low, high and medium diversity respectively. The same trend can be found in Fig. 7. The higher the diversity among queries, the more improvement can be made by the co-optimization.

6.3   Experiments on Skewed Queries

In real-world scenarios, the distribution of query predicates is usually skewed. In our second set of experiments, we sorted each table based on the timeline, and assumed that queries prefer new data to dated data. We selected the query combinations of high diversity (i.e., $\{Q3.4, Q4.3\}$ of SSB and $\{Q10, Q42\}$ of TPC-DS), and tested how they perform with skewed predicates. We used two types of skewness – 6 : 4 indicates that 60% of queries are directed to 40% of the row groups; 9 : 1 indicates that 90% of queries are directed to 10% of the row groups. The results are shown in Figs. 8 and 9.

As we can see, the skewness of predicates does have impact on the absolute performance of query processing. However, its influence on the effectiveness of optimization is unpredictable. In the experiments on SSB, it seems that increased skewness significantly reduced the performance enhancement made by the optimizer. In contrast, increased skewness visibly amplified the performance enhancement made by the optimizer in the case of TPC-DS. The co-optimization of query plans and data placement aims to improve the locality of data access. The skewness of selection predicates tends to increase data locality too.

However, the two types of enforcement on data locality are not necessarily correlated. They can be orthogonal to each other. A detailed experimental study seems necessary in this regard. We schedule it in our future work.

7 Discussion and Conclusion

In this paper, we studied the query optimization problem in a hybrid storage, composed of main memory and secondary storage. We showed that the query plans and data placement should be optimized together to achieve the best performance. We presented a query optimization model and a solution in the context of A-Store. Experiments showed that the co-optimization of query plans and data placement can achieve significant performance improvement, especially when confronted with diverse workload.

As mentioned earlier, A-Store is a specialized query processing engine and does not represent the generic cases in a DBMS. Hard work is required to apply this technique to a generic DBMS. In a generic DBMS, data placement does influence the performance of query processing. For instance, index-scan and index-join are common operations in query processing. The placement of the indexes is important to the performance of the operations. Thus, to decide which index to use or when to use an index, the query optimizer should take the location of the index into account. This issue is an interesting topic for future study.

Acknowledgement. This work is partially supported by Chinese National High-tech R&D Program (863 Program) (2015AA015307) and the NSFC Porject (No. 61272138).

References

1. Akbar, M.M., Rahman, M.S., Kaykobad, M., Manning, E.G., Shoja, G.C.: Solving the multidimensional multiple-choice knapsack problem by constructing convex hulls. Comput. Oper. Res. **33**(5), 1259–1273 (2006)
2. Bernstein, P.A., Goodman, N., Wong, E., Reeve, C.L., Rothnie Jr., J.B.: Query processing in a system for distributed databases (sdd-1). ACM TODS **6**(4), 602–625 (1981)
3. Boncz, P.A., Zukowski, M., Nes, N.: Monetdb, x100: hyper-pipelining query execution. In: CIDR, pp. 225–237 (2005)
4. Ceri, S., Gottlob, G.: Optimizing joins between two partitioned relations in distributed databases. J. Parallel Distrib. Comput. **3**(2), 183–205 (1986)
5. Chaudhuri, S.: An overview of query optimization in relational systems. In: Proceedings of the Seventeenth ACM SIGACT-SIGMOD-SIGART Symposium on Principles of Database Systems, pp. 34–43. ACM (1998)
6. Dar, S., Franklin, M.J., Jonsson, B.T., Srivastava, D., Tan, M., et al.: Semantic data caching and replacement. In: Proceedings of VLDB, vol. 96, pp. 330–341. Citeseer (1996)
7. DeBrabant, J., Pavlo, A., Tu, S., Stonebraker, M., Zdonik, S.: Anti-caching: a new approach to database management system architecture. Proc. VLDB Endow. **6**(14), 1942–1953 (2013)

8. Eldawy, A., Levandoski, J., Larson, P.-Å.: Trekking through siberia: managing cold data in a memory-optimized database. Proc. VLDB Endow. **7**(11), 931–942 (2014)
9. Finkelstein, S.: Common expression analysis in database applications. In: Proceedings of SIGMOD, pp. 235–245. ACM (1982)
10. Ganguly, S., Hasan, W., Krishnamurthy, R.: Query optimization for parallel execution. In: Proceedings of the SIGMOD, pp. 9–18 (1992)
11. Giannikis, G., Alonso, G., Kossmann, D.: Shareddb: killing one thousand queries with one stone. Proc. VLDB Endow. **5**(6), 526–537 (2012)
12. Gray, J., Chaudhuri, S., Bosworth, A., Layman, A., Reichart, D., Venkatrao, M., Pellow, F., Pirahesh, H.: Data cube: a relational aggregation operator generalizing group-by, cross-tab, and sub-totals. Data Mining Knowl. Discov. **1**(1), 29–53 (1997)
13. Herodotou, H., Borisov, N., Babu, S.: Query optimization techniques for partitioned tables. In: Proceedings of the SIGMOD, pp. 49–60. ACM (2011)
14. Kemper, A., Neumann, T.: Hyper: a hybrid OLTP & OLAP main memory database system based on virtual memory snapshots. In: Proceedings of ICDE, pp. 195–206. IEEE (2011)
15. Kossmann, D., Franklin, M.J., Drasch, G., Ag, W.: Cache investment: integrating query optimization and distributed data placement. ACM TODS **25**(4), 517–558 (2000)
16. Manegold, S., Boncz, P., Kersten, M.L.: Optimizing main-memory join on modern hardware. IEEE TKDE **14**(4), 709–730 (2002)
17. Manegold, S., Boncz, P., Kersten, M.L.: Generic database cost models for hierarchical memory systems. In Proceedings of VLDB, VLDB 2002, pp. 191–202. VLDB Endowment (2002)
18. Neumann, T.: Efficiently compiling efficient query plans for modern hardware. Proc. VLDB Endow. **4**(9), 539–550 (2011)
19. Polyzotis, N.: Selectivity-based partitioning: a divide-and-union paradigm for effective query optimization. In: Proceedings of CIKM, pp. 720–727. ACM (2005)
20. Rao, J., Ross, K.A.: Making b+-trees cache conscious in main memory. ACM SIGMOD Record **29**, 475–486 (2000)
21. Ren, Q., Dunham, M.H., Kumar, V.: Semantic caching and query processing. IEEE TKDE **15**(1), 192–210 (2003)
22. Sellis, T.K.: Multiple-query optimization. ACM TODS **13**(1), 23–52 (1988)
23. Zhang, H., Chen, G., Ooi, B.C., Tan, K.-L., Zhang, M.: In-memory big data management and processing: a survey. IEEE TKDE **27**(7), 1920–1948 (2015)
24. Zhang, H., Chen, G., Ooi, B.C., Wong, W.-F., Wu, S., Xia, Y.: Anti-caching-based elastic memory management for big data. In: Proceedings of ICDE, pp. 1268–1279. IEEE (2015)
25. Zhang, Y., Zhou, X., Zhang, Y., Zhang, Y., Su, M., Wang, S.: Virtual denormalization via array index reference for main memory OLAP. IEEE TKDE **28**(4), 1061–1074 (2016)
26. Zhou, J., Larson, P.-A., Chaiken, R.: Incorporating partitioning and parallel plans into the scope optimizer. In Proceedings of ICDE, pp. 1060–1071. IEEE (2010)
27. Zukowski, M., van de Wiel, M., Boncz, P.: Vectorwise: a vectorized analytical dbms. In: Proceedings of ICDE, pp. 1349–1350. IEEE (2012)

Efficient Batch Grouping in Relational Datasets

Jizhou Sun[✉], Jianzhong Li, and Hong Gao

School of Computer Science and Technology, Harbin Institute of Technology,
Harbin, China
{sjzh,lijzh,honggao}@hit.edu.cn

Abstract. Data Grouping is an expensive and frequently used operator in data processing, meanwhile data is often too big to fit in memory, where disk sorting based method is often employed. Disk sorting
reads and writes the entire dataset for many times, which is very time-
consuming, so reducing I/O costs is of great significants. In many applications, grouping a set of records multi-times on different keys is very
common. Grouping in batch manner and techniques of sharing intermediate results are studied in this paper for efficiency. In batch grouping
settings, different grouping orders may result in different I/O costs. To
minimize I/O costs, we formalize the group-order scheduling problem as
an optimization problem which can be proven in NP-Complete, and then
propose a heuristic algorithm. Experimental results on TPC-H as well as
synthetic datasets show the efficiency and robustness of our techniques.

Keywords: Batch grouping · I/O efficiency · Sharing · Scheduling

1 Introduction

In the database area, many applications need grouping the input data on a group-
key. In OLAP for example, the aggregation, join and set operations occur very
frequently, and they are often implemented by grouping. For another example,
when detecting data errors with functional dependencies, the dataset are grouped
by using the left-hand-side attributes of the dependency as the grouping key, and
checking whether all tuples in the same sub-group having the same right-hand-
side value [9].

There are mainly two ways to implement the grouping operator, hashing and
sorting. The dataset is always too big to fit in main memory, the two operators
are both completed on disk. Since hash based methods have an inherent defect
of imbalance, we focus on the sorting based ones. The most commonly used
technique for external sorting is the *external sort-merge algorithm*, where two or
even more times of I/O operations are required, which is known very expensive.

In many situations, grouping a set of records multiple times on different keys
respectively is common. For instance, the 22 queries in the TPC-H benchmark[1],

---

[1] The TPC-H specifications can be found in www.tpc.org/tpc_documents_current_
versions/current_specifications.asp.

© Springer International Publishing AG 2017
S. Candan et al. (Eds.): DASFAA 2017, Part I, LNCS 10177, pp. 376–390, 2017.
DOI: 10.1007/978-3-319-55753-3_24

especially the sub-query of Q9, often join a table with several other tables on different attributes. The join operations can be implemented by grouping the tuples multiple times on different join keys. For a more intuitive example, when detecting data errors according to several different functional dependencies, the dataset will be grouped multiple times. We consider batch grouping, where sharing intermediate results among grouping tasks is possible, even when where is no dependency relationship between the grouping keys.

When two grouping tasks a considered, pair-wise sharing is analysed mainly in three cases. If one grouping key k_2 is strongly decided by the other one k_1 as defined in [2], i.e., for any two tuples t_1 and t_2, if they equal on k_1, they must also equal on k_2, denoted as $k_1 \rightarrow k_2$. It can be shown that there exists such a grouping manner that when the tuples are grouped on k_1, they are also grouped on k_2 [1,12,15,17,19]. The grouping task on k_2 is totally reduced and we call this *total sharing*. In the second case, partial information of k_2 is decided by k_1, when tuples are grouped on k_1 in a certain manner, further sub-groupings on k_2 will create a k_2-grouped result. The I/O operations for k_2 is partially reduced and it is called *partial sharing*. In the third case, one of the two grouping keys is expected to have as few distinct values as possible. If k_1 is of fewer distinct values, when the tuples are grouped on k_1, and tuples in each sub-group are sorted on k_2, then merging all sub-groups together on k_2 can create a k_2-grouped result, which can save I/Os. Note that no dependency relationship between the two keys is required, and it is called *general sharing*. The numbers of distinct values on different keys are required, and the statistical information in hand or estimating techniques such as [6,10] can be helpful. To the best of our knowledge, no work has studied the latter two sharing techniques in batch grouping.

As for the case of more than two grouping tasks, there may be many different grouping orders, with different I/O costs. To this end, we formalize the group-order scheduling problem (GOS) as an optimization problem, and reduce the *set cover problem* to GOS to prove its NP-Completeness. Due to the intractableness, we propose a heuristic algorithm to schedule the tasks wisely.

We summarize our contributions as follows:

- Batch grouping is studied, where two types of sharing, partial sharing and general sharing, are proposed, which can be applied more broadly.
- The multi-task scheduling problem aiming at minimizing I/O costs is formalized and proven in NP-Complete.
- Due to the intractable of the scheduling problem, a heuristic algorithm is proposed and analysed.
- Various experiments on TPC-H and synthetic datasets are conducted to validate the efficiency and robustness of our methods.

In the rest of this paper, related works are reviewed in Sect. 2 and in Sect. 3 are some preliminaries. Then in Sect. 4 we discuss the sharing techniques. In Sect. 5, the GOS problem is studied. Experiments are conducted in Sect. 6 and in Sect. 7 is our conclusion.

2 Related Works

The most related works to ours are in [4], which studied three kinds of sort-sharing techniques similar to those in our study, but there are several important differences. Firstly, attributes in a grouping key are allowed to be of arbitrary orders. On the other hand, tuples are grouped other than strictly sorted in our settings, therefore the groups can be arranged in any order. For instance, values $1, 1, 2, 2, 6, 6, 6, 4, 4$ are grouped but not sorted. Furthermore, when functional dependencies are known satisfied, or one grouping key is a subset (not necessarily prefix) of the other one, the I/O costs can be totally saved in grouping tasks, while it is not the case in sorting tasks. Finally, the scheduling problems are different, and the NP-Completeness is proved in this paper, while in [4] it wasn't. All the first three differences make sharing techniques in grouping more applicable than in sorting.

Some other works about intermediate results sharing include [8,11,12,16]. [12] is mainly about sort sharing, and it has discussed the total sharing techniques in grouping. In [12,16], the sharing scheduling problem was defined on the query executing plans in DBMS. [8] is a review about sort-sharing.

In [1,15,17,19] the attributes in a key can be rearranged, just as it is in this paper, but limited to the total sharing technique. [1] focused on CUBE and optimization techniques in the storage level, techniques in [15,17] were designed for general SQL queries.

Recent years, grouping and sorting have still received a lot of attentions. In [3], sorting was compared with hashing in the parallel and in-memory environments experimentally. [18] proposed new sorting and join algorithms, to adapt to new occurred I/O asymmetric memory. [13] proposed new sorting algorithms to make full use the SIMD instructions and cache memory of modern CPUs. In [5], a long-forgotten technique, called *Patience Sort Scan*, was shown competitive with todays best comparison-based sorting techniques, by some key modifications. [20] studied sort-sharing in memory and improving the parallel degree. In [7], new sort algorithms were proposed on a hybrid storage system with both precise storage and approximate storage.

3 Preliminaries

A tuple t is a record with n attributes $\{A_1, A_2, ..., A_n\}$ and a table T is a finite set of tuples. A grouping key k is a set of attributes. A grouping operation (or task) of T on k is a partition of T, denoted as $\mathcal{G}(T, k) = \{G_1, G_2, ..., G_m\}$, satisfying 1) all tuples from the same subset equal on k, and 2) all tuples from different subsets differ on k.

Example 1. In Fig. 1(a), T contains 10 tuples on 4 attributes $\{A, B, C, D\}$. In Fig. 1(b), $k = \{A\}$, and the grouping operation $\mathcal{G}$ partitions T into four subsets, with tuples in each subset equal on A. In Fig. 1(c), $k = \{B, A\}$, which contains more than one attributes, and the grouping operation $\mathcal{G}$ partitions T into three subsets, with tuples in each subset equal on $\{B, A\}$.

tid	A B C D
t_0	3 1 2 1
t_1	4 2 1 1
t_2	2 2 2 2
t_3	3 1 2 2
t_4	4 2 2 1
t_5	2 2 1 1
t_6	3 1 2 2
t_7	3 1 2 2
t_8	1 1 2 1
t_9	1 1 1 1

(a) Table T

tid	A B C D
t_8	1 1 2 1
t_9	1 1 1 1
t_2	2 2 2 2
t_5	2 2 1 1
t_0	3 1 2 1
t_3	3 1 2 2
t_6	3 1 2 2
t_7	3 1 2 2
t_1	4 2 1 1
t_4	4 2 2 1

(b) $k = \{A\}$

tid	A B C D
t_9	1 1 1 1
t_8	1 1 2 1
t_0	3 1 2 1
t_3	3 1 2 2
t_6	3 1 2 2
t_7	3 1 2 2
t_5	2 2 1 1
t_2	2 2 2 2
t_1	4 2 1 1
t_4	4 2 2 1

(c) $k = \{B, A\}$

tid	A B C D
t_9	1 1 1 1
t_8	1 1 2 1
t_0	3 1 2 1
t_3	3 1 2 2
t_6	3 1 2 2
t_7	3 1 2 2
t_5	2 2 1 1
t_1	4 2 1 1
t_2	2 2 2 2
t_4	4 2 2 1

(d) $k = \{B, C\}$

Fig. 1. Relation T and grouped results on different keys

A sorting key is an ordered set of attributes $(A_1, A_2, ..., A_s)$, enclosed by parentheses instead of braces, to distinguish from grouping keys. Note that different from sorting, sub-groups in $\mathcal{G}(T, k)$ can be arranged in any order, not necessarily in ascending or descending. In this paper however, sorting is employed as an efficient method for grouping. The most common situation is that T is too big to fit in memory, and a popular method is called *external sort-merge algorithm*.

To evaluate the efficiency of the sorting method, several notations are necessary. The input data and work memory are measured in *blocks*, which often correspond to disk blocks or memory pages. The size of T is denoted as $B(T)$, and the work memory size can accommodate M blocks. The time cost for reading and writing a block of data from or to disk is denoted as $CostR$ and $CostW$ respectively. For the table T and the grouping key k, by $\delta(T, k)$ we denote the number of distinct values of T on k. Because the cpu operations are much faster compared with I/O operations, only the I/O costs are considered.

Now we use an example to demonstrate the sorting process.

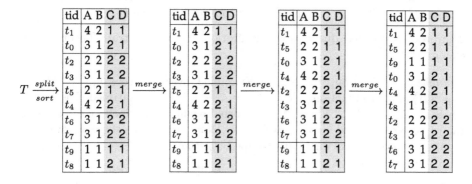

Fig. 2. The sorting process of T

Example 2. For table T in Fig. 1(a), if $M = 2$, $B(T) = 10$ and $k = \{C, D\}$. The sorting key can be (C, D) (note that (D, C) is also feasible). Assume that only 2 tuples can be loaded into memory at the same time.

In the first phase, the 10 tuples are split into 5 subsets, each of which contains 2 tuples and can be sorted in memory as shown in the leftmost table of Fig. 2. Then, each two (because only two tuples can be accommodated in memory) subsets are merged together and create three sorted subsets, then continue the merging process until a single sorted record set is obtained, as shown in the rest three tables in Fig. 2. By now, the sorted tuples are naturally grouped according to the key $\{C, D\}$.

The grouping task in Example 2 reads and writes T for four times respectively. Generally, the I/O cost can be calculated by

$$cost(\mathcal{G}(T, k)) = B(T) \times (\lceil \log_M \frac{B(T)}{M} \rceil + 1) \times (CostR + CostW) \qquad (1)$$

4 Group Sharing Techniques

In this section, we discus some sharing techniques between two grouping tasks. Depending on different relationships between the grouping keys, there are three different levels of sharing: total sharing, partial sharing and general sharing.

4.1 Total Sharing

Consider a table T and two grouping keys k_1 and k_2, if one key is decided by another, total sharing is applicable. Without loss of generality, we assume that k_2 is decided by k_1, denoted as $k_1 \rightarrow k_2$, which indicates that for any two tuples t_1 and t_2 in T, whenever $t_1[k_1] = t_2[k_1]$, we have $t_1[k_2] = t_2[k_2]$.

When both $\mathcal{G}(T, k_1)$ and $\mathcal{G}(T, k_2)$ are required, it can be shown that sorting T only once with key (k_2, k_1) is sufficient.

Theorem 1. *Given table T and two grouping keys k_1 and k_2. If $k_1 \rightarrow k_2$, sorting T by key (k_2, k_1) creates a table grouped both on k_1 and on k_2.*

The proof is omitted due to space limitation and we illustrate Theorem 1 by example.[2]

Example 3. If we group table T in Example 1 over two keys $k_1 = \{A\}$ and $k_2 = \{B\}$, and the functional dependency $A \rightarrow B$ is known satisfied. The sorting key is $(k_2, k_1) = (B, A)$, Fig. 1(c) is the sorted result, from which we can see that the tuples are grouped on k_1 as well as k_2. Note that for k_2 there are two sub-groups, all tuples equal on k_2 are grouped together.

---

[2] All proofs of theorems in this paper can be found in the technical report: www.researchgate.net/publication/312070129.

4.2   Partial Sharing

If none of the two keys depends on the other one, the sharing techniques may still be available to a certain degree. Given a table T and two grouping keys k_1 and k_2, this subsection consider the case that there is such a key k' that both $k_1 \to k'$ and $k_2 \to k'$ hold.

Suppose that T is required to be grouped with grouping keys k_1 and k_2 respectively. We firstly sort T with key (k', k_1). From Theorem 1 it is easy to assure that the sorted tuples are grouped on both k_1 and k'. Because the tuples in T are already grouped on k', and $k_2 \to k'$, tuples equal on k_2 must also equal on k' and in the same sub-group of $\mathcal{G}(T, k')$. To this end, by further grouping each subset $G_i \in \mathcal{G}(T, k')$ on k_2 we have all tuples in T grouped on k_2.

Example 4. Given table T in Example 1, two grouping keys $k_1 = \{A\}$ and $k_2 = \{B, C\}$, and the dependency $A \to B$. We have $k_1 \to B$, and $k_2 \to B$ (because $B \in k_2$), thus partial sharing is feasible and $k' = \{B\}$.

Table T is firstly sorted on $(k', k_1) = (B, A)$, as shown in Fig. 1(c), which is grouped on k_1 and k'. $\mathcal{G}(T, k') = \{\{t_9, t_8, t_0, t_3, t_6, t_7\}, \{t_5, t_2, t_1, t_4\}\}$. We further group each sub-group of $\mathcal{G}(T, k')$ on k_2: $\mathcal{G}(\{t_9, t_8, t_0, t_3, t_6, t_7\}, k_2) = \{\{t_9\}, \{t_8, t_0, t_3, t_6, t_7\}\}$, and $\mathcal{G}(\{t_5, t_2, t_1, t_4\}, k_2) = \{\{t_5, t_1\}, \{t_2, t_4\}\}$. By combining them, we get $\mathcal{G}(T, k_2) = \{\{t_9\}, \{t_8, t_0, t_3, t_6, t_7\}, \{t_5, t_1\}, \{t_2, t_4\}\}$, as shown in Fig. 1(d).

Note that the second grouping task is implemented on much smaller datasets, the number merge-phases may be fewer according to Eq. 1, and the overall I/O costs can be reduced.

4.3   General Sharing

In general cases, the two grouping keys may be totally irrelevant. Fortunately, sharing techniques are still possible, as long as the number of distinct values on one key is few enough.

For a table T, and two grouping keys k_1 and k_2, with $\delta(T, k_1)$ and $\delta(T, k_2)$ the corresponding numbers of distinct values respectively. Without loss of generality, we assume that $\delta(T, k_1) \leq \delta(T, k_2)$. Firstly, we sort the tuples in T on key $k = (k_1, k_2)$. Because k_1 is the first sorting key, the sorted tuples are grouped on k_1. However, the dependency $k_1 \to k_2$ no longer holds, and the sorted tuples are not grouped on k_2. Fortunately, tuples in each sub-group of $\mathcal{G}(T, k_1)$ are sorted, thus also grouped, on k_2 locally because k_2 is the second sorting key. So in the second step we merge the subsets in $\mathcal{G}(T, k_1)$ into fewer and bigger ones repeatedly on k_2, until a single set is obtained, just as the merging process shown in Fig. 2. By now, the merged tuples are grouped on k_2.

Example 5. Figure 3 illustrates an example of general sharing. In Fig. 3(a) is the original table T, with two grouping keys $k_1 = \{D\}$ and $k_2 = \{A\}$. No dependency relationship is available between the keys, but we know that $\delta(T, k_1) = 2$ and $\delta(T, k_2) = 4$. Obviously $\delta(T, k_1) \leq \delta(T, k_2)$, so the sorting key is (k_1, k_2), or

tid	A	B	C	D
t_0	3	1	2	1
t_1	4	2	1	1
t_2	2	2	2	2
t_3	3	1	2	2
t_4	4	2	2	1
t_5	2	2	1	1
t_6	3	1	2	2
t_7	3	1	2	2
t_8	1	1	2	1
t_9	1	1	1	1

(a) Table T

$\xrightarrow{\text{sort by } (D,A)}$

tid	A	B	C	D
t_8	1	1	2	1
t_9	1	1	1	1
t_5	2	2	1	1
t_0	3	1	2	1
t_1	4	2	1	1
t_4	4	2	2	1
t_2	2	2	2	2
t_3	3	1	2	2
t_6	3	1	2	2
t_7	3	1	2	2

(b) $\mathcal{G}(T, \{D\})$

$\xrightarrow{\text{merge by } A}$

tid	A	B	C	D
t_8	1	1	2	1
t_9	1	1	1	1
t_5	2	2	1	1
t_2	2	2	2	2
t_0	3	1	2	1
t_3	3	1	2	2
t_6	3	1	2	2
t_7	3	1	2	2
t_1	4	2	1	1
t_4	4	2	2	1

(c) $\mathcal{G}(T, \{A\})$

Fig. 3. A general sharing example

(D, A). In Fig. 3(b) are the sorted tuples and they are grouped on $k_1 = \{D\}$. More over, tuples in each sub-group, for example the first 6 tuples, are sorted on A, thus merging the groups together can generate a tuple-set grouped on A as shown in Fig. 3(c).

From Fig. 2 we know that a grouping task on T reads and writes the whole table for four times respectively. By general sharing technique however, grouping T on A on the basis of $\mathcal{G}(T, \{D\})$ reads and writes T only once.

It is not hard to see that, the fewer $\delta(T, k_1)$ is, the less merging phases it requires. That is why the key with fewer distinct values is chosen as the first sorting key. Generally, the I/O cost in the merging phases is calculated by

$$Merging\ cost = B(T) \times (\lceil \log_M \delta(T, k_1) \rceil) \times (CostR + CostW) \qquad (2)$$

5 Group-Order Scheduling

In the previous section, three different types of pairwise sharing techniques in grouping are discussed. In this section we focus on the batch grouping, where two or more grouping tasks are required. Firstly, each kind of sharing technique among multiple grouping tasks is discussed. And then, when all of the three techniques are considered, to minimize the I/O cost, a group-order scheduling problem is proposed proven in NP-Complete. Finally, a heuristic algorithm is designed.

5.1 Generalize Sharing Techniques to Multiple Tasks

We give three theorems without proof due to space limitation, to generalize the sharing techniques.

Theorem 2. *For a table T and n grouping keys $k_1, k_2, ..., k_n$, if $k_1 \rightarrow k_2 \rightarrow ... \rightarrow k_n$, sorting T by key $(k_n, k_{n-1}, ..., k_1)$ creates a table grouped on all of the n keys.*

Theorem 3. *For table T and n grouping keys $k_1, k_2, ..., k_n$, if there exists such a key k' that $k_1 \rightarrow k'$, $k_2 \rightarrow k'$, ... and $k_n \rightarrow k'$, sorting T by (k', k_1) creates a table grouped on both k' and k_1, and further grouping each subset of $\mathcal{G}(T, k')$ on k_i creates a table grouped on k_i for $i = 2, 3, ..., n$.*

Theorem 4. *For a table T and n grouping keys $k_1, k_2, ..., k_n$, sorting T by key $(k_1, k_2, ..., k_n)$ creates a table grouped on k_1, further merging sub-groups in $\mathcal{G}(T, k_1)$ on $(k_2, ..., k_n)$ creates a table grouped on k_2, ..., and further merging sub-groups in $\mathcal{G}(T, k_{n-1})$ on k_n creates a table grouped on k_n.*

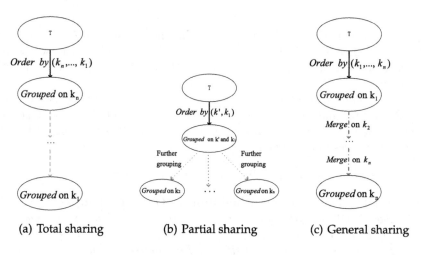

(a) Total sharing   (b) Partial sharing   (c) General sharing

Fig. 4. Sharing in multiple grouping tasks

For intuitiveness, we illustrate them graphically, as shown in Fig. 4. Solid edges indicate that the grouping tasks are accomplished by traditional sorting method. Dash edges indicate that the subsequent grouping tasks are directly finished by doing nothing, corresponding to Theorem 2. Dot edges indicate that the subsequent grouping tasks are finished by further grouping operations, corresponding to partial sharing in Theorem 3. Dash-dot edges indicate that the subsequent grouping tasks are finished with general sharing, corresponding to Theorem 4.

5.2 Group-Order Scheduling Problem

Actually, usually all of the 3 types of sharing are available meanwhile in batch grouping, and different orders of executing the tasks result in different I/O costs.

Example 6. In the previous examples, there are totally four grouping keys, i.e., $\{A\}, \{B\}, \{B, C\}$ and $\{D\}$, and an explicit dependency $A \rightarrow B$. The sharing

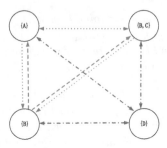

Fig. 5. Sharing in multiple grouping tasks

graph is shown in Fig. 5. The numbers of distinct values are $\delta(T, \{A\}) = 4$, $\delta(T, \{B\}) = 2$, $\delta(T, \{B, C\}) = 4$ and $\delta(T, \{D\}) = 2$. Consider two strategies, with grouping orders $(\{A\}, \{B\}, \{B, C\}, \{D\})$ and $(\{B\}, \{D\}, \{B, C\}, \{A\})$ respectively.

From the sharing graph it can be shown that in the first strategy, each of the three sharing techniques is employed once, and in the second strategy, general sharing is used twice and partial sharing is used once. It is very likely that the first strategy saves more I/O cost.

Now, the group-order scheduling problem can be defined.

Definition 1 (Group-Order Scheduling (GOS) Problem). *Given a relational table T with an attribute set Attrs, a set of functional dependencies $\mathcal{F}$ satisfied by T, a set of grouping keys $\mathcal{K}$, the number of distinct values $\delta(T, k)$ for each key k, and the work memory size M. Find out such a grouping order that all keys in $\mathcal{K}$ are covered and has the least I/O costs.*

Unfortunately, the GOS problem is intractable.

Theorem 5. *The GOS problem is in NP-Complete.*

To prove Theorem 5, the basic idea is reducing the *set cover problem*, which is known in NP-Complete, to the GOS problem. The detailed proof is omitted here due to space limitation.

5.3   A Heuristic Algorithm

Due to the intractability of the GOS problem, we consider a HEuristic Group-Order Scheduling (HEGOS) algorithm.

The main idea behind HEGOS is: because total sharing reduces the entire work load of the subsequent task, we make full use of total sharing as much as possible, and then the rest kinds of sharing are scheduled one by one in a greedy manner, as shown in Algorithm 1.

In step 1, E_{total} corresponds to total sharings, similarly for $E_{partial}$ and $E_{general}$.

Algorithm 1. HEGOS

Input: A relational table T with an attribute set $Attrs$, a set of functional dependencies $\mathcal{F}$ satisfied by T, a set of grouping keys $\mathcal{K}$, the number of distinct values $\delta(T, k)$ for each key k, and the work memory size M.

Output: Sets of sharing strategies which can reducing I/O costs in grouping tasks on $\mathcal{K}$.

1: $V := \emptyset$, $E := \emptyset$, $E_{total} := \emptyset$, $E_{partial} := \emptyset$, $E_{general} := \emptyset$.
2: For each $k \in \mathcal{K}$, add v_k into V.
3: Using $\mathcal{F}$, check all pairwise dependent relationships, if $k_j \rightarrow k_i$, $E := (v_{k_i}, v_{k_j})$.
4: In DAG (V, E), find out the maximum set of non-conflicting edges E_{total} using Hungarian algorithm. Each edge in E_{total} corresponds to a total sharing.
5: **repeat**
6:    Find out a pair of vertices v_{k_i} and v_{k_j}, such that v_{k_j} is not covered by any edge in $E_{total} \cup E_{partial} \cup E_{general}$, and perform the grouping task on k_j after k_i by partial sharing or general sharing saves the most I/O costs.
7:    **if** The costs saved in step 6 is by partial sharing **then**
8:       $E_{partial} := E_{partial} \cup \{(v_{k_i}, v_{k_j})\}$.
9:    **else**
10:       $E_{general} := E_{general} \cup \{(v_{k_i}, v_{k_j})\}$.
11:    **end if**
12: **until** There is no further changes or only one uncovered task remains.
13: **return** E_{total}, $E_{partial}$ and $E_{general}$;

In steps 2–4, all grouping tasks are connected into a directed acyclic graph (DAG) by total sharings. Finding a maximum set of non-conflicting edges is equivalent to the *maximum bipartite matching problem*, for which there is an algorithm called the *Hungarian Algorithm* with time complexity of $O(|V||E|)$, which can be found in [14], where $|V|$ is the number of vertices (corresponding to tasks) and $|E|$ is the number of edges (corresponding to sharings). By *conflicting* we mean that, total sharings can not be used simultaneously. For instance, in Fig. 5, the two green edges (total sharing) requires the table to be sorted on (B, A) and (B, C) respectively, which can not be satisfied at the same time.

In steps 5–12, HEGOS repeatedly select the task that is not covered and saves the most I/O costs, which is done in step 6. By covering, we mean that a vertex is not at the end of the edge. For instance, in Fig. 5, vertex B is covered by the blue edges, but not covered by the green ones. If the cost saving is contributed by partial sharing, then add it into $E_{partial}$, else $E_{general}$ (steps 7–11). Termination condition is no more changes or only one uncovered task rest, because an initial grouping is necessary.

In step 13, the algorithm returns three sets of edges, standing for the types of sharings respectively. The grouping tasks are then executed as scheduled.

The complexity of HEGOS is dominated by the Hungarian algorithm (with complexity $O(|V| \times |E|)$) and the repeat selecting process (with complexity $O(|V|^3)$), so HEGOS's complexity is $O(|V|^3)$.

It should be noted that in step 3, the dependent relationships can be determined by rules in *Armstrongs axioms* [2]. Moreover, in step 6, when calculating the I/O costs, some statistics such as number of distinct values are necessary, and related techniques can be found in [6,10] etc.

6   Experimental Evaluation

All experiments are implemented by JAVA, and performed on a Dell desktop with an Octa-Core Intel i7 3.60 GHz processor, 8 GB of memory, 1T SATA disk. The underlying OS is Windows 7.

Two baseline algorithms are implemented to study the efficiency of our methods, including *No Sharing* algorithm and *Total Sharing Only* algorithm. In *No Sharing* algorithm, all grouping tasks are performed independently, no sharing techniques are used. In *Total Sharing Only* algorithm, only the total sharing technique is employed, which has been studied in several previous works.

The experiments are conducted on the TPC-H benchmark generated dataset, as well as a synthetic dataset. The TPC-H dataset contains 8 tables, and the grouping tasks are conducted on the largest table *lineitem*, which occupies 70% of the total size and contains 16 attributes. The dataset size is $SF \times 1\,GB$, where SF is the specifiable *scale factor*. The synthetic dataset contains 20 fixed-length attributes, where the length of each attribute is randomly selected from $1, ..., 5$. All attribute values are strings of lower-case letters, and thus the number of distinct values differs. For both TPC-H and synthetic datasets, the number of attributes in each grouping key is randomly chosen from geometric distribution.

Table 1. Experiment parameter configuration ranges.

Parameter	Range
Data size in MB	$[100,...,\mathbf{500},...,1000]$
Scale factor (SF)	$[0.1,...,\mathbf{0.5},...,1]$
Work memory size in MB	$[\mathbf{1},...,10]$
Number of tasks	$[2,4,...,\mathbf{20}]$
Estimation Error (σ, in percent)	$[\mathbf{0},1,...,10]$

Table 1 shows some parameters considered in the experiments. When not explicitly stated, we use the default configuration value (highlighted in bold). The estimation error is to indicate inaccuracy of statistic information, see details in Sect. 6.2. The block size B is set 64 KB in all experiments.

6.1   Efficiency Results

We firstly ran 20 tasks on both the TPC-H dataset and the synthetic dataset, under different data sizes (or scale factors for TPC-H). Figure 6 illustrates the total running time to finish the 20 tasks. It can be shown that the *Total Sharing*

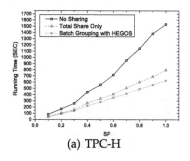
(a) TPC-H

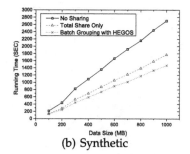
(b) Synthetic

Fig. 6. Performance comparison with changing data size

Only saved 35% to 50% running time compared with the naive *No Sharing* algorithm. And HEGOS is 10% to 20% faster than *Total Sharing Only* algorithm.

Considering that the results above depend greatly on relationships between tasks, we illustrate the average running time for per task of different sharing techniques in Fig. 7. In TPC-H dataset, the partial sharing technique saved about one-third time and the general sharing saved even more (about 60%). In the synthetic dataset, two of the sharing techniques consumed approximately the equal time, which make the grouping task much faster (by about 50%). Note that total sharing technique saved the whole grouping task and we omitted the discussion here.

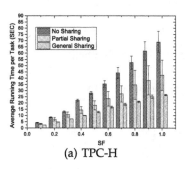
(a) TPC-H

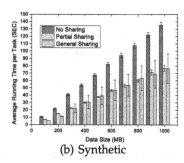
(b) Synthetic

Fig. 7. Performance comparison of different sharing techniques

Then, we study the efficiency under different sizes of working memory. In Fig. 8 it can be shown that the running time increase locally, with sudden decreases. That is because when the memory becomes big enough, fewer times of merging are needed, causing the running time's decreasing. While if the memory size grows slightly, the same number of merging process are required, but more blocks are loaded into memory, causing more random accesses of the underlying disk and the running time's increasing.

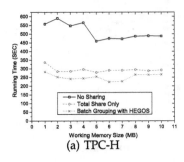
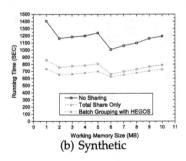

(a) TPC-H   (b) Synthetic

Fig. 8. Relationship between running time and working memory size

Different number of tasks also influent the effectiveness of sharing. In Fig. 9, the number of tasks changes from 2 to 20. The running time curve of *No Sharing* is nearly linear, while sharing techniques make it sub-linear, because more tasks brings more chance of sharing, and the average running time per task decreases.

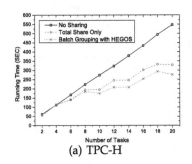
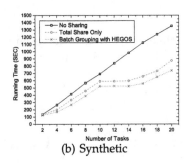

(a) TPC-H   (b) Synthetic

Fig. 9. Running time under different number of tasks

6.2   Influence of Estimating Inaccuracy

In partial and general sharing techniques, some statistics information should be estimated, such as number of distinct values on a grouping key. Techniques in [6,10], etc., can be helpful, but with a certain degree of inaccuracy. To study the influence of inaccuracy, we firstly scan the whole table to obtain the accurate statistics information, then multiply it by a Gaussian random variable, with mean $\mu = 1$ and standard deviation σ ranging from 1% to 10%. From Fig. 10 it can be shown that HEGOS is robust to the estimating inaccuracy, the time-curve is almost parallel with that of *Total Sharing Only*, which doesn't depend on the statistical information, and is immune to inaccuracy.

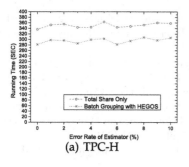

(a) TPC-H

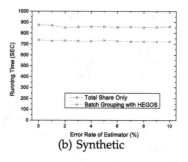

(b) Synthetic

Fig. 10. Influence of estimating inaccuracy

7   Conclusions and Future Works

In this paper, we have studied batch grouping, and proposed two new sharing techniques, i.e., partial sharing and general sharing techniques. In the case of multiple grouping tasks, the problem of scheduling with the sharing techniques in batch grouping, is formalized and proven in NP-Complete, for which a heuristic algorithm HEGOS is designed. We have implemented our techniques and shown the efficiency and robustness experimentally.

There are several further directions to discuss in future. Firstly, one grouping task may depend on another, and in parallel environments, the scheduling problem becomes even more complicated. Furthermore, we have only consider the sorting-based methods, and the probability of sharing in hash-based methods should be studied. Additionally, in SQL, the grouping tasks may be required with some order, which should be considered into our scheduling algorithm. Finally, when some sorting tasks are required besides grouping tasks, corresponding sharing techniques and scheduling algorithms should be reconsidered.

Acknowledgments. This work is supported in part by the Key Research and Development Plan of National Ministry of Science and Technology under grant No. 2016YFB1000703, and the Key Program of the National Natural Science Foundation of China under Grant No. 61190115, 61632010 and U1509216.

References

1. Agarwal, S., Agrawal, R., Deshpande, P. et al.: On the computation of multidimensional aggregates. In: Proceedings of 22th International Conference on Very Large Data Bases (1996)
2. Armstrong,W.W.: Dependency structures of data base relationships. In: IFIP Congress, pp. 580–583 (1974)
3. Balkesen, C., Alonso, G., Teubner, J., et al.: Multi-core, main-memory joins: sort vs. hash revisited. PVLDB **7**(1), 85–96 (2013)
4. Cao, Y., Bramandia, R., Chan, C., et al.: Sort-sharing-aware query processing. VLDB J. **21**(3), 411–436 (2012)

5. Chandramouli, B., Goldstein, J.: Patience is a virtue: revisiting merge and sort on modern processors. In: Proceedings of 33rd International Conference on Management of Data, Snowbird, USA, pp. 731–742 (2014)
6. Charikar, M., Chaudhuri, S., Motwani, R. et al.: Towards estimation error guarantees for distinct values. In: Proceedings of 19th ACM SIGMOD-SIGACT-SIGART Symposium on Principles of Database Systems, Dallas, USA, pp. 268–279 (2000)
7. Chen, S., Jiang, S., He, B. et al.: A study of sorting algorithms on approximate memory. In: Proceedings of 35th International Conference on Management of Data, SIGMOD 2016, San Francisco, USA, pp. 647–662. ACM (2016)
8. Estivill-Castro, V., Wood, D.: A survey of adaptive sorting algorithms. ACM Comput. Surv. **24**(4), 441–476 (1992)
9. Fan, W., Geerts, F., Jia, X., et al.: Conditional functional dependencies for capturing data inconsistencies. ACM Trans. Database Syst. **33**(2), 6 (2008)
10. Gibbons, P.B.: Distinct sampling for highly-accurate answers to distinct values queries and event reports. In: Proceedings of 27th International Conference on Very Large Data Bases, Roma, Italy, pp. 541–550 (2001)
11. Graefe, G.: Implementing sorting in database systems. ACM Comput. Surv. **38**(3), 10 (2006)
12. Guravannavar, R., Sudarshan, S.: Reducing order enforcement cost in complex query plans. In: Proceedings of 23rd International Conference on Data Engineering, Istanbul, Turkey, pp. 856–865 (2007)
13. Inoue, H., Taura, K.: SIMD- and cache-friendly algorithm for sorting an array of structures. PVLDB **8**(11), 1274–1285 (2015)
14. Jünger, M. (ed.): 50 Years of Integer Programming 1958–2008: From the Early Years to the State-of-the-Art. Springer, Heidelberg (2010)
15. Neumann, T., Moerkotte, G.: A combined framework for grouping and order optimization. In: Proceedings of 30th International Conference on Very Large Data Bases, Toronto, Canada, pp. 960–971 (2004)
16. Neumann, T., Moerkotte, G.: An efficient framework for order optimization. In: Proceedings of 20th International Conference on Data Engineering, Boston, USA, pp. 461–472 (2004)
17. Simmen, D.E., Shekita, E.J., Malkemus, T.: Fundamental techniques for order optimization. In: Proceedings of 15th International Conference on Management of Data, Montreal, Canada, pp. 57–67 (1996)
18. Viglas, S.: Write-limited sorts and joins for persistent memory. PVLDB **7**(5), 413–424 (2014)
19. Wang, X., Cherniack, M.: Avoiding sorting and grouping in processing queries. In: Proceedings of 29th International Conference on Very Large Data Bases, VLDB 2003, Berlin, Germany, pp. 826–837. VLDB Endowment (2003)
20. Xu, W., Feng, Z., Lo, E.: Fast multi-column sorting in main-memory column-stores. In: Proceedings of 35th International Conference on Management of Data, SIGMOD 2016, San Francisco, USA, pp. 1263–1278. ACM (2016)

Text Mining

Memory-Enhanced Latent Semantic Model: Short Text Understanding for Sentiment Analysis

Fei Hu[1,2], Xiaofei Xu[1], Jingyuan Wang[1], Zhanbo Yang[1], and Li Li[1(✉)]

[1] School of Computer and Information Science, Southwest University,
Chongqing, China
lily@swu.edu.cn
[2] Network Centre, Chongqing University of Education, Chongqing, China

Abstract. Short texts, such as tweets and reviews, are not easy to be processed using conventional methods because of the short length, the irregular syntax and the lack of statistical signals. Term dependencies can be used to relax the problem, and to mine latent semantics hidden in short texts. And Long Short-Term Memory networks (LSTMs) can capture and remember term dependencies in a long distance. LSTMs have been widely used to mine semantics of short texts. At the same time, by analyzing the text, we find that a number of key words contribute greatly to the semantics of the text. In this paper, we propose a LSTM based model (MLSM) to enhance the memory of the key words in the short text. The proposed model is evaluated with two datasets: IMDB and SemEval2016, respectively. Experimental results demonstrate that the proposed method is effective with significant performance enhancement over the baseline LSTM and several other latent semantic models.

Keywords: Long Short-Term Memory · Deep learning · Sentiment analysis · Neural network · Semantic understanding

1 Introduction

Short texts are very prevalent on today's web sites, such as forums, tweets, microblogs, internet ads, commodity reviews and cell phone messages. They are playing an increasingly important role in our daily lives. Compared with long texts, dealing with short texts is a big challenge according to the following characteristics.

1. Short texts do not always observe the syntax. Methods that work on long and well-organized texts fail on short texts, such as part-of-speech (POS) tagging and dependency parsing methods [16,34,39];
2. Short texts are less than 140 words, having limited, even ambiguous context. Thus, short texts usually do not contain sufficient statistical signals. For example, Bag of Words (BOW) based probabilistic latent semantic models do not per-form better in tasks with short texts than with long texts [5,28].

© Springer International Publishing AG 2017
S. Candan et al. (Eds.): DASFAA 2017, Part I, LNCS 10177, pp. 393–407, 2017.
DOI: 10.1007/978-3-319-55753-3_25

These characteristics bring a significant amount of ambiguity for understanding the semantics of short texts. Many approaches were introduced to relax them [7,16,20,31,33,36]. In [16], models were improved from three aspects: text segmentation, POS tagging and concept labeling. In the aspect of text segmentation, short text was divided into a sequence of meaningful components, and this method was improved from the Longest-Cover method [19,32]; in the aspect of POS tagging, not only lexical features but also semantics were considered as tag POS; in the aspect of concept labeling, type detection was incorporated into the framework of short text understanding, and instance disambiguation was conducted on the basis of all types of contextual information [19,32]. In [20], an n-gram model was suggested. In [20,31,33], topic models were used to analyze the latent topics of short texts that helped to extract potential semantic structures. In [7,36], features were extended and carefully selected.

All these improved methods mentioned above are still BOW based methods. They did not consider the term dependencies, or just considered adjacent term dependencies like n-gram models [17,30]. Term dependencies contribute significantly to the semantics of short texts. For example in Table 1, *text 1* and *text 2* have the same vocabularies, yet the wording of the sentences is shuffled. As a result, the meanings of them are totally different. On the other hand, *text 3* and *text 4* have no single word occurrence in two texts, but they imply the same thing. Term dependencies hide in sequences, but BOW based models fail to capture the hidden dynamics in sequences. On the contrary, the LSTM network, an extension of Recurrent Neural Networks (RNNs), which was proposed by Hochreiter et al. [14], is promising. It is explicitly designed to capture the dependencies between words, even in a long distance. Unlike n-gram models that suffer from long term memory property, LSTM remembers information for long periods of time, which is consistent with long term memory requirement. The LSTM model takes every single word as the input for processing of temporal patterns rather than BOW tokens or sentences. Imitating human memory mechanism, it memorizes words one by one.

By analyzing texts, we found that a number of key words contribute greatly to the semantics of them. It is similar to the process of human memorization. When reading a text, people are always impressed by a few words or phrases. For example, from the sentence *"It is a wonderful day. I am very happy."*, we can see that the words *"wonderful"* and *"happy"* contribute more to understanding the semantics than the other words.

Table 1. Sample texts, texts are lower cased.

text	
text 1	tom help a dog win the game
text 2	the game help tom win a dog
text 3	this novel is very interesting
text 4	i was moved by the book

In this paper, we developed a new latent semantic model based on LSTM, called memory-enhanced latent semantic model (MLSM), as shown in Fig. 2. The proposed model pays more attention on the key words, and leaves more room in memory block for them. As a result, semantics are biased to those words, and short texts can be better understood.

2   Related Work

2.1   Latent Semantic Models

Latent semantic models (LSMs) can measure similarity between documents, no matter they have the same words or not. They address the language discrepancy between documents by grouping different terms which frequently occur in similar contexts into the same semantic cluster [17]. For example, latent semantic analysis (LSA) projects high-dimensional vector space representation of a document into a lower dimensional vector space [15]. The low-dimensional representation is full of semantics and can be understood by the system. The singular value decomposition (SVD) is introduced to realize this process by retaining a certain number of the largest characteristic values and eliminating the noises of the middle matrix. The refactored middle matrix more exactly describes the semantics [35]. Extending from LSA, probabilistic topic models such as probabilistic LSA (PLSA) [18], Latent Dirichlet Allocation (LDA) [1] and Bi-Lingual Topic Model (BLTM) [8] have been proposed and successfully been used to mine semantics of documents. In recent years, neural networks, especially deep learning networks [12], have been introduced to promote the development of LSMs [3]. Salakhutdinov et al. demonstrated that hierarchical semantic representations of a document can be extracted via a deep learning structure [29]. RNN based models are a kind of deep learning structure. They utilized the contextual information of a document to help understand the document. Our works are based on an extension of RNN, known as LSTM.

2.2   Long-Term Dependencies

RNN has a hidden layer which is capable of memorizing recent words. The hidden layer connects previous information to the present task that is like a memory block flowing through time. The memory mechanism helps capture term dependencies, and further to understand a document. It has been successfully applied in many natural language processing (NLP) tasks, such as predicting next word given contextual information [23]. Compared with feed-forward neural networks, RNNs are characterized by the ability of encoding past information, and hence are suitable for sequential models. The backpropagation through time algorithm (BPTT) and the Real-Time Recurrent Learning algorithm (RTRL) [37] extend the ordinary BP algorithm to suit the recurrent neural architecture; however, RNNs have the problem of memorizing past information in a long distance. With BPTT or RTRL, error signals tend to vanish after steps of flowing, incapable of changing weights. As a result, long-term dependencies are hard to learn [11]. See detailed analysis in following equations [13]. The error signal for the j-th unit at time step t-1 can be formulated as follows:

$$\theta_j(t-1) = f_j'(net_j(t-1)) \sum_i w_{ij}\theta_i(t) \tag{1}$$

where θ is the error signal, $f'()$ is the partial derivative of the activation function, $net()$ is the input, w_{ij} is a weight parameter for two neurons from the $t-1$ and t time step. The target error has been removed from the equation, because it is zero when the unit belongs to the non-output layer.

Derived from the above equation, we get the error signal for the j-th neuron at the time step of $t - q(q >= 1)$. The error is scaled by:

$$\frac{\partial \theta_j(t-q)}{\partial \theta_i(t)} = \begin{cases} f_j'(net_j(t-1))w_{ij} & q = 1 \\ f_j'(net_j(t-q)) \sum_{l=1}^n \frac{\partial \theta_l(t-q+l)}{\partial \theta_i(t)} w_{lj} & q > 1 \end{cases} \tag{2}$$

The above equation is expanded as follows:

$$\frac{\partial \theta_j(t-q)}{\partial \theta_i(t)} = \sum_{l_1} \cdots \sum_{l_{q-1}} \prod_{m=1}^q f_{l_m}'(net_{l_m}(t-m))w_{l_m l_{m-1}}$$

$$T = \prod_{m=1}^q f_{l_m}'(net_{l_m}(t-m))w_{l_m l_{m-1}} \tag{3}$$

If $|T| > 1$, the error will increase exponentially with the increase of q; if $|T| < 1$, the error will vanish with the increase of q. In practice, we will always meet the error vanishing problem.

LSTM has fixed this problem by introducing a memory cell, named as constant error carrousel (CEC). It is a self-connected device that keeps constant error signals, even from a far distance.

2.3   Long Short-Term Memory Networks

The LSTM network [14], as an extension of RNNs, plays an important role in text understanding. It is born with the ability to retain information over a much longer period of time than 10 to 12 time steps which is the limit of RTRL or BPTT models. Figure 1 depicts the LSTM structure [27], and the symbols are illustrated in Table 2, where subscripts t and $t - 1$ are the current time step and the previous time step, respectively. C is like a conveyor belt. Along the belt, information flows. Through f and i, old information is attenuated and new information is inserted. Through o, the model produces the output for current time step. C is the key of the LSTM structure. It keeps the information for long periods of time. Thus this structure supports long-term dependencies. The following equations mathematically abstract this process [27].

Table 2. Symbols for LSTM.

Symbol	Description
W_f, W_i and W_o	Weight matrices for the forget, input and output gates
W_C	The weight matrix for the input
b_f, b_i, b_o and b_C	Offset values
h	The output
σ	The sigmoid function
tanh	The hyperbolic tangent function
x	The input
f, i and o	The forget, input and output gates
C	The memory cell (CEC)
C'	The new candidate value

$$f_t = \delta(W_f \cdot [h_{t-1}, x_t] + b_f) \tag{4}$$

$$i_t = (W_i \cdot [h_{t-1}, x_t] + b_i) \tag{5}$$

$$o_t = (W_o \cdot [h_{t-1}, x_t] + b_o) \tag{6}$$

$$C'_t = \tanh(W_C \cdot [h_{t-1}, x_t] + b_C) \tag{7}$$

$$C_t = f_t * C_{t-1} + i_t * C'_t \tag{8}$$

$$h_t = o_t * \tanh(C_t) \tag{9}$$

LSTM is born to capture long term dependencies, and is able to memorize previous word in a long distance [14]. The long-distance memory is helpful to mining the semantics of the short text which might has related words in a long distance. However, it cannot distinguish memory of different words effectively, and the disginguishing memory is especially important in sentiment analysis tasks. For example, the sentence "*It is a wonderful day. I am very happy.*" has two key words "*wonderful*", "*happy*" which contribute more to the sentiment expressed by the user. In this paper, we developed a novel variant of LSTM to enhance key information with long-term dependencies.

3   Our Model

To capture semantic information from short texts, we proposed a latent semantic model. This model enhances LSTM by revising the input and forget gates, the former controls the flows of newly-incoming information and the latter retains old information. This improvement emphasizes the contribution from key words. Section 3.1 illustrates the proposed model, and Sect. 3.2 discusses the selection of the key words.

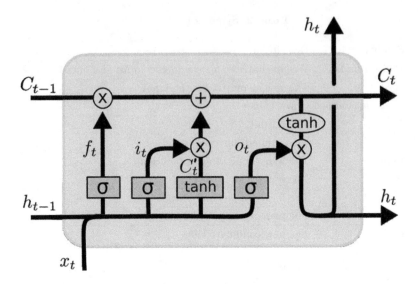

Fig. 1. LSTM structure.

3.1 Enhancing Memory of Key Words

Figure 2 depicts our model whose pseudo code will be presented below. Key words are introduced to affect the forget and input gates. The red line illustrates this affection. Mat is matching function. E is key-memory-enhancing function. E' is context-weakening function. They are formulated as follows:

$$Mat(x_t, K) = \begin{cases} 0 & x_t \notin K \\ k_j^v \mid x_t = k_j & k_j \in K \end{cases} \tag{10}$$

where K is a dictionary whose keys make a list of key words and whose values are levels of importance. $K = \{k_1 : k_1^v, k_2 : k_2^v, \cdots, k_j : k_j^v, \cdots\}$. In our task of sentiment analysis, a key word is a sentiment word, and the value represents the corresponding word's sentiment polarity. k_j is the j-th key word, k_j^v is the value of sentiment polarity of k_j. k_j^v is on the interval between 0.0 to 1.0. The larger the value, the more the word contributes to the semantics. If current word does not belong to K, Eq. 10 returns zero. The construction of the key word dictionary will be discussed in the next subsection.

$$E(i, Mat) = (1. + Mat) \cdot i$$
$$E'(i, Mat) = (1. - \frac{Mat}{2}) \cdot i \tag{11}$$

where i is the input gate of baseline LSTM, see Eq. 5 for more detail.

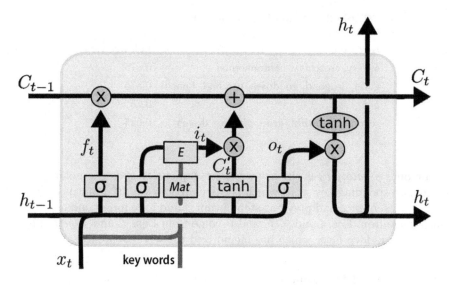

Fig. 2. MLSM structure based on LSTM. (Color figure online)

Then we get a revised input and forget gates:

$$i_t = (1. + Mat) \cdot (W_i \cdot [h_{(t-1)}, x_t] + b_i) \tag{12}$$

$$f_t = (1. - \frac{Mat}{2}) \cdot (W_f \cdot [h_{(t-1)}, x_t] + b_f) \tag{13}$$

This revised input gate carefully controls the flows of newly-incoming information, the more important the newly-incoming word, the more information getting through. The revised forget gate weakens the reserved old information, the more important the newly-incoming word, the less old information retained, i.e., the more memory blocks emptied for newly-incoming information. Thus, key word information flows are enhanced and normal word information flows are weakened. This model enhances the memory of key words.

The following pseudo code is presented to describe the LSTM-based MLSM which is depicted in Fig. 2.

```
Initialize inputsentence, inputtopics
state = self.initial_state
for time_step in range(num_steps):
    if inputsentence[time_step] in inputtopics:
        state = state * f' + input * i'
    else:
        state = state * f + input * i
return state
```

Here, *num_steps* is the loop count of the recurrent neural network, *inputsentence* is a sentence of the text whose words will be fed into the model

Table 3. Key word dictionary for sentiment analysis.

Key words	Polarity value
ornery, crotchety, cantankerous	0.875
good, do_good, goodness, not_bad, benefit	0.875
hot	0.75
sorry, sad, pitiful, grim, gloomy, dreary	0.625

one by one, *inputtopics* is the key-word set, *topicpolarity* is the polarity value of the corresponding key word which is calculated by Eq. 10, input is generated using *inputsentence* and previous-step output, f and i are forget and input gates which come from Eqs. 4 and 5, f' and i' come from Eqs. 12 and 13, state is the state of the hidden layer which has memorized the whole sentence.

3.2 Key Word Dictionary

Key words make the most useful part for understanding a document. Especially with a short text which does not contain sufficient statistical signals, key words emphasize the intent of a user who has presented the text. So key words can be utilized to better interpret what the user wants to express. The sentiment analysis task is asked to pick key words that would be good indicators of sentiment polarity. Many pairs of word and polarity make a sentiment dictionary, which is also called key word dictionary. Table 3 lists several sentiment words and the corresponding polarity values, where polarity values have been normalized on the interval between 0.0 to 1.0. This table only considers the polar intensity, ignoring positive and negative factors.

In this paper, we establish a key-word dictionary from SentiWordNet 3.0 [2], and use the dictonary to help enhancing memory of the essential words in the text. SentiWordNet is a lexical resource for opinion mining. It assigns to each synset of WordNet [24] three sentiment scores: positivity, negativity and objectivity, where the object score is derived from the other two scores by the equation "ObjScore = 1 − (PosScore + NegScore)". We use the larger one of positive and negative scores as the value of sentiment polarity of certain word. Table 3 lists several words with their polar values.

4   Experiments and Results

4.1 Experimental Settings

Two datasets were introduced to evaluate our model which were shown as follows:

1. One is a large movie review dataset (IMDB) [22]. The dataset has a set of 25000 highly polar movie reviews for training and 25000 for testing, totally 50000 reviews, and the polarity labels are positive and negative. All reviews

were grouped into four subsets according to the length of text: short text with the length of less than (including) 140; medium length text between 141 and 200; ordinary length text between 201 and 400; long text greater than 400; in addition, a group of total reviews is considered. Of each subset, 80% is used for training and the rest 20% for testing.

2. The other one is from SemEval-2016 Task 4 [25]. The dataset consists of 6000 tweets, and three polarity labels are considered: positive, negative and neutral. All tweets are short because of the length of less than 140 words. 80% of the dataset is used for training and the rest 20% for testing.

In order to evaluate our model with a sentiment analysis task, we introduced our model into a sentiment analysis structure, as shown in Fig. 3. The input sequence $\{x(0), x(1), \cdots, x(n)\}$ represents a review. And each $x(i)$, $i = 0, 1, \cdots, n$, is a word encoded using one-hot representation (One-hot representation is a word-encoding method that encodes every word into a long sparse vector. Each bit of this vector represents a unique word, and one bit will be set to 1 if a certain word is encoded but the rest bits will be set to 0. The length of the vector is the size of the vocabulary which usually includes all unique words in a dataset.). The embedding layer projects x(i) from a high dimensional vector space into a lower dimensional space which has 128 dimensions in our study. The CECs in the MLSM layer produce a sequence $\{h(0), h(1), \cdots, h(n)\}$. Then h is obtained by averaging this sequence over all time steps. At last, h is fed to a logistic regression layer whose output is a binary classification label which represents positivity with 1 or negativity with 0.

Comparative study was performed with the baseline of LSTM and other non-LSTM LSMs. Each LSTM based model has 32, 128 or 1024 neurons in the hidden layer and uses the optimization algorithm Adadelta [38]. IMDB and SemEval-2016 datasets are separately used with these models. The comparison of their accuracies is shown in Table 4. Table 5 lists the comparison of the six LSTM based models (they are MLSM@32, MLSM@128, MLSM@1024, LSTM@32, LSTM@128 and LSTM@1024, respectively, where MLSM is the sentiment analysis structure into which our proposed model is incorporated, basic LSTM is incorporated into LSTM, and @k is the number of neurons in the hidden layer) in Mean Squared Error (MSE) metric which is the lost function in the experiments. The MSE metric reflects the error between the real output and the expected output. The less the error, the more robust the model.

Furthermore, several optimization algorithms were used respectively in the MLSM@128 and the LSTM@128. And for simplicity, only datasets with short texts are considered, i.e., IMDB 140 and SemEval-2016. Comparison results are shown in Table 6 (for accuracy metric) and Table 7 (for MSE metric), respectively.

4.2   Experimental Results

In the accuracy study of LSMs, we evaluated the accuracy metric within ten models which are the six LSTM based models, three LDA based models [26]

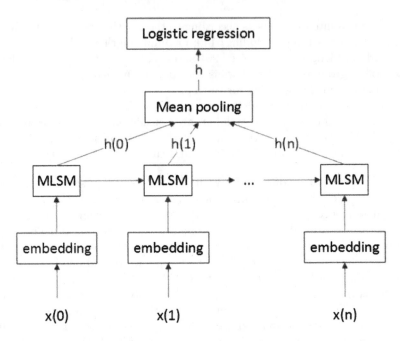

Fig. 3. Illustration of MLSM used in the sentiment analysis task.

and one word2vec based model [4]. Three LDA based models are different in the vector representation of topic words: Words are projected into a 30-dimensional vector space in lda1, 100-dimensional vector space in lda2 and 200-dimensional vector space in lda3. Word2vec uses a 100-dimensional vector space to represent words. Then the LDA and word2vec based models utilize support vector machine (SVM) to classify documents. As we can see in Table 4: The LSTM based models have higher accuracy, i.e. they work better than other non-LSTM LSMs, especially with short texts; and our improvements enhance the effectiveness of baseline LSTM model.

In the loss study of the LSTM based models, we evaluated the MSE metric within the six LSTM based models. Table 5 shows the results: The MLSM based models have less MSE that means they work better than the baseline models.

Furthermore, we evaluated the MLSM@128 and the LSTM@128 with several optimization algorithms to demonstrate the effectiveness of our improvements. Five optimization algorithms were used, they are: Adagrad [6], Adadelta [38], GradientDesent, Momentum and Adam [21]. For simplicity, only datasets with short texts are considered, i.e., IMDB 140 and SemEval2016. Tables 6 and 7 show the comparative results in terms of accuracy and MSE, respectively: models with Adadelta and Adam perform better; with any optimization algorithm, the MLSM based model has higher accuracy than the LSTM based model, which shows the generalizability of our method.

Table 4. Accuracy study of LSMs with different length of documents.

Models	Datasets					
	IMDB					SemEval2016
	<=140	<=200	<=400	>400	All	
lda1-svm	0.7857	0.8049	0.8169	0.8168	0.8008	-
Lda2-svm	0.7783	0.8098	0.8252	0.8191	0.8088	-
Lda3-svm	0.7739	0.8074	0.8161	0.8084	0.8089	-
word2vec-svm	0.802	0.826	0.847	0.822	0.828	-
LSTM@32	0.8299	0.8293	0.8288	0.8294	0.8288	-
LSTM@128	0.8651	0.8644	0.8648	0.8646	0.8655	0.7076
LSTM@1024	0.8645	0.8638	0.8651	0.8648	0.8652	0.7079
MLSM@32	0.8415	0.8422	0.8415	0.8418	0.8420	-
MLSM@128	0.8787	**0.8788**	**0.8788**	0.8783	0.8790	0.7243
MLSM@1024	**0.8798**	0.8787	0.8785	**0.8796**	**0.8793**	**0.7256**

Table 5. MSE of LSTM-based models with different length of documents.

Models	Datasets					
	IMDB					SemEval2016
	<=140	<=200	<=400	>400	All	
LSTM@32	0.1242	0.1244	0.1238	0.1242	0.1241	-
LSTM@128	0.0998	0.0992	0.1001	0.0996	0.0998	0.1384
LSTM@1024	0.1001	0.0996	0.0995	0.0994	0.0995	0.1386
MLSM@32	0.1110	0.1111	0.1106	0.1113	0.1110	-
MLSM@128	**0.0955**	**0.0948**	0.0956	**0.0955**	0.0958	0.1351
MLSM@1024	0.0959	0.0955	**0.0953**	0.0955	**0.0954**	**0.1345**

Table 6. Accuracy study of LSTM-based models with different optimization algorithms.

Datasets	Models	Optimization algorithms				
		Adagrad	Adadelta	GD	Momentum	Adam
IMDB 140	LSTM@128	0.8476	**0.8651**	0.8585	0.8623	0.8648
	MLSM@128	0.8572	**0.8787**	0.8658	0.8768	0.8780
SemEval2016	LSTM@128	0.7018	0.7076	0.7000	0.7073	**0.7080**
	MLSM@128	0.7181	**0.7243**	0.7189	0.7238	0.7238

Table 7. MSE of LSTM based models with different optimization algorithms.

Datasets	Models	Optimization algorithms				
		Adagrad	Adadelta	GD	Momentum	Adam
IMDB 140	LSTM@128	0.1072	**0.0998**	0.1042	0.1011	0.1001
	MLSM@128	0.1038	**0.0955**	0.0996	0.0964	0.0958
SemEval2016	LSTM@128	0.1398	**0.1384**	0.1401	0.1387	0.1388
	MLSM@128	0.1373	**0.1351**	0.1373	0.1360	0.1357

Results above show that the proposed model is effective and robust, especially in dealing with short texts. The other non-LSTM LSMs have declined in effectiveness with short texts, but the proposed model has worked stably and outperformed the baseline LSTM.

5   Discussions

Compared with traditional LSMs, LSTM based models take into account the dependencies between words. The dependencies are conducive to understanding texts, and they are significantly effective in dealing with short texts. As we can see in Table 4, the LSTM models have higher accuracy than the other four non-LSTM LSMs. Especially with short texts, the accuracy of the non-LSTM LSMs declines sharply, and the effectiveness of the LSTM based models remains stable. Term dependencies could be utilized through memory of words. The farther and the more words memorized, the more contextual information took into account. Therefore, previous researches have focused on how to extend the length of memory [9,10].

Actually, memory length is not the sole factor to be considered in understanding the texts. That is, every single word in a text does not contribute the same weight to the semantics. People's focus is mostly on scanning those high-weighted words, and the other words are skimmed through or just skipped through. Through long-term training, people can enhance the semantic understanding by showing different memory for different words in the text. For example, when reading the sentence mentioned in the section of Introduction: "*It is a wonderful day. I am very happy.*", we are impressed by the two words "*wonderful*" and "*happy*", and relatively weaken the memory of the other words. The mechanism of distinguishing memory is helpful for people to grasp the key points, thus enhancing the understanding of the text.

As a kind of memory machine, LSTM can memorize contextual information in a long distance, and it is supposed to imitate human distinguishing memory. We dream that the machine would automatically figure out which parts of the text are important and read them more carefully. To simplify the problem and facilitate the application, we have collected the key words, and the automation will be studied in the future.

6   Conclusions

LSTM based models are sequential models, capable of memorizing sequential data such as texts. In this paper, we have shown the importance of key words in the text. They are beneficial for understanding the semantics. The proposed model (MLSM) is further used to perform sentiment analysis. The MLSM outperforms the baseline LSTM by 1–2% in terms of accuracy. Especially in dealing with short texts, the performance upgrade still exists. In addition, we obtained quite similar gains with different optimization algorithms. It shows the generalizability of our method. In other words, our method is general enough to be applied to different LSTM based models. Our work is promising in dealing with NLP related tasks, such as query suggestion, web search recommendation, and movie recommendation.

Acknowledgements. This work was supported by Scientific and Technological Research Program of Chongqing Municipal Education Commission (No. KJ1501405, No. KJ1501409); Scientific and Technological Research Program of Chongqing University of Education (No. KY201522B); Fundamental Research Funds for the Central Universities (No. XDJK2016E068) and NSFC (No. 61170192).

References

1. Anoop, V.S., Prem, S.C.: Generating and visualizing topic hierarchies from microblogs: an iterative latent dirichlet allocation approach. Proc. IEEE (2015)
2. Baccianella, S., Esuli, A., Sebastiani, F.: Sentiwordnet 3.0: an enhanced lexical resource for sentiment analysis and opinion mining. In: International Conference on Language Resources and Evaluation, LREC 2010, Valletta, Malta, 17–23 May 2010, pp. 83–90 (2010)
3. Bengio, Y., Schwenk, H., Sencal, J., Morin, F., Gauvain, J.L.: Neural probabilistic language models. J. Mach. Learn. Res. **3**(6), 1137–1155 (2003)
4. Chen, P., Fu, X., Teng, S., Lin, S., Lu, J.: Research on micro-blog sentiment polarity classification based on SVM. In: Zu, Q., Hu, B., Gu, N., Seng, S. (eds.) HCC 2014. LNCS, vol. 8944, pp. 392–404. Springer, Cham (2015). doi:10.1007/978-3-319-15554-8_32
5. Cheng, X., Yan, X., Lan, Y., Guo, J.: BTM: topic modeling over short texts. IEEE Trans. Knowl. Data Eng. **26**(12), 2928–2941 (2014)
6. Duchi, J., Hazan, E., Singer, Y.: Adaptive subgradient methods for online learning and stochastic optimization. J. Mach. Learn. Res. **12**(7), 257–269 (2010)
7. Fan, X., Hu, H.: Construction of high-quality feature extension mode library for chinese short-text classification. In: Wase International Conference on Information Engineering, pp. 87–90 (2010)
8. Gao, J., Toutanova, K., Yih, W.T.: Clickthrough-based latent semantic models for web search. In: Proceeding of the International ACM SIGIR Conference on Research and Development in Information Retrieval, SIGIR 2011, Beijing, China, pp. 675–684, July 2011
9. Graves, A., Wayne, G., Danihelka, I.: Neural turing machines. arXiv preprint arXiv:1410.5401 (2014)

10. Greff, K., Srivastava, R.K., Koutnik, J., Steunebrink, B.R., Schmidhuber, J.: LSTM: A search space odyssey. IEEE Trans. Neural Netw. Learn. Syst. (2015)

11. Gustavsson, A., Magnuson, A., Blomberg, B., Andersson, M., Halfvarson, J., Tysk, C.: On the difficulty of training recurrent neural networks. Comput. Sci. **52**(3), 337–345 (2012)

12. Hinton, G.E., Salakhutdinov, R.R.: Reducing the dimensionality of data with neural networks. Science **313**(5786), 504–507 (2006)

13. Hochreiter, S.: The vanishing gradient problem during learning recurrent neural nets and problem solutions. Int. J. Uncertainty Fuzziness Knowl.-Based Syst. **6**(2), 107–116 (1998)

14. Hochreiter, S., Schmidhuber, J.: Long short-term memory. Neural Comput. **9**(8), 1735–1780 (1997)

15. Hofmann, T.: Unsupervised learning by probabilistic latent semantic analysis. Mach. Learn. **42**(1), 177–196 (2001)

16. Hua, W., Wang, Z., Wang, H., Zheng, K., Zhou, X.: Short text understanding through lexical-semantic analysis. In: 2015 IEEE 31st International Conference on Data Engineering, pp. 495–506. IEEE (2015)

17. Huang, P.S., He, X., Gao, J., Deng, L., Acero, A., Heck, L.: Learning deep structured semantic models for web search using click through data. In: ACM International Conference on Conference on Information & Knowledge Management, pp. 2333–2338 (2013)

18. Ke, X., Luo, H.: Using LSA and PLSA for text quality analysis. In: International Conference on Electronic Science and Automation Control (2015)

19. Kim, D., Wang, H., Oh, A.: Context-dependent conceptualization. In: International Joint Conference on Artificial Intelligence, pp. 2654–2661 (2013)

20. Kim, K., Chung, B.S., Choi, Y., Lee, S., Jung, J.Y., Park, J.: Language independent semantic kernels for short-text classification. Expert Syst. Appl. **41**(2), 735–743 (2014)

21. Kingma, D., Ba, J.: Adam: a method for stochastic optimization. Comput. Sci. (2015)

22. Maas, A.L., Daly, R.E., Pham, P.T., Huang, D., Ng, A.Y., Potts, C.: Learning word vectors for sentiment analysis. In: Meeting of the Association for Computational Linguistics: Human Language Technologies, pp. 142–150 (2011)

23. Mikolov, T.: Statistical language models based on neural networks. Presentation at Google, Mountain View, 2 April 2012

24. Miller, G.A.: Wordnet: a lexical database for english. Commun. ACM **38**(11), 39–41 (1995)

25. Nakov, P., Ritter, A., Rosenthal, S., Sebastiani, F., Stoyanov, V.: Semeval-2016 task 4: sentiment analysis in twitter. In: International Workshop on Semantic Evaluation (2016)

26. Nalini, K., Sheela, L.J.: Classification of tweets using text classifier to detect cyber bullying. In: Satapathy S., Govardhan A., Raju K., Mandal J. (eds) Emerging ICT for Bridging the Future - Volume 2. AISC, vol. 338, pp. 637–645. Springer, Cham (2015)

27. Olah, C.: Understanding lstm networks. http://colah.github.io/posts/2015-08-Understanding-LSTMs/. Accessed 8 Jan 2016

28. Rao, Y., Lei, J., Wenyin, L., Li, Q., Chen, M.: Building emotional dictionary for sentiment analysis of online news. World Wide Web **17**(4), 723–742 (2014)

29. Salakhutdinov, R., Hinton, G.: Semantic hashing. Int. J. Approximate Reasoning **50**(7), 969–978 (2009)

30. Shen, Y., He, X., Gao, J., Deng, L., Mesnil, G.: Goire: a latent semantic model with convolutional-pooling structure for information retrieval. In: ACM International Conference on Conference on Information & Knowledge Management, pp. 101–110 (2014)

31. Song, G., Ye, Y., Du, X., Huang, X., Bie, S.: Short text classification: a survey. J. Multimedia **9**(5), 635–643 (2014)

32. Song, Y., Wang, H., Wang, Z., Li, H., Chen, W.: Short text conceptualization using a probabilistic knowledgebase. In: Proceedings of the International Joint Conference on Artificial Intelligence, IJCAI 2011, Barcelona, Catalonia, Spain, pp. 2330–2336, July 2011

33. Wang, B.K., Huang, Y.F., Yang, W.X., Li, X.: Short text classification based on strong feature thesaurus. J. Zhejiang Univ. Sci. C **13**(9), 649–659 (2012)

34. Wang, G., Zhang, Z., Sun, J., Yang, S., Larson, C.A.: Pos-rs: a random subspace method for sentiment classification based on part-of-speech analysis. Inf. Process. Manag. **51**(4), 458–479 (2015)

35. Wang, J., Peng, J., Liu, O.: A classification approach for less popular webpages based on latent semantic analysis and rough set model. Expert Syst. Appl. **42**(1), 642–648 (2015)

36. Wang, M., Lin, L., Wang, F.: Improving short text classification through better feature space selection. In: 2013 9th International Conference on Computational Intelligence and Security (CIS), pp. 120–124. IEEE (2013)

37. Williams, R.J., Zipser, D.: Gradient-Based Learning Algorithms for Recurrent Networks and Their Computational Complexity. L. Erlbaum Associates Inc., Hillsdale (1998)

38. Zeiler, M.D.: Adadelta: an adaptive learning rate method. Comput. Sci. (2012)

39. Zou, H., Tang, X., Xie, B., Liu, B.: Sentiment classification using machine learning techniques with syntax features. In: 2015 International Conference on Computational Science and Computational Intelligence (CSCI), pp. 175–179. IEEE (2015)

Supervised Intensive Topic Models for Emotion Detection over Short Text

Yanghui Rao[1]([⊠]), Jianhui Pang[1], Haoran Xie[2], An Liu[3], Tak-Lam Wong[2], Qing Li[4], and Fu Lee Wang[5]

[1] School of Data and Computer Science, Sun Yat-sen University, Guangzhou, China
raoyangh@mail.sysu.edu.cn
[2] Department of Mathematics and Information Technology,
The Education University of Hong Kong, Tai Po, Hong Kong
[3] School of Computer Science and Technology, SooChow University, Soochow, China
[4] Department of Computer Science, City University of Hong Kong,
Kowloon Tong, Hong Kong
[5] Caritas Institute of Higher Education, Tseung Kwan O, Hong Kong

Abstract. With the emergence of social media services, documents that only include a few words are becoming increasingly prevalent. More and more users post short messages to express their feelings and emotions through Twitter, Flickr, YouTube and other apps. However, the sparsity of word co-occurrence patterns in short text brings new challenges to emotion detection tasks. In this paper, we propose two supervised intensive topic models to associate latent topics with emotional labels. The first model constrains topics to relevant emotions, and then generates document-topic probability distributions. The second model establishes association among biterms and emotions by topics, and then estimates word-emotion probabilities. Experiments on short text emotion detection validate the effectiveness of the proposed models.

Keywords: Topic model · Emotion detection · Short text analysis

1 Introduction

With the broad availability of portable devices such as tablets, online users can now conveniently express their opinions through various channels, which generates large-scale short messages and accelerates the emotion detection research. Many studies on emotion detection have focused on classifying emotions via latent topics [3,17], because a topic represents the real-world event that indicates the subject or context of the emotion [13]. However, short messages, as indicated by the name, typically only include a few words and result in the sparsity of word co-occurrence patterns. Thus, traditional topic models such as the Latent Dirichlet Allocation (LDA) [4], would suffer from this severe data sparsity problem when inferring latent topics. Based on the idea that two frequently co-occurred words are more likely to belong to a same topic, Cheng et al. [6] proposed the biterm topic model to extract biterms for each document and alleviate

© Springer International Publishing AG 2017
S. Candan et al. (Eds.): DASFAA 2017, Part I, LNCS 10177, pp. 408–422, 2017.
DOI: 10.1007/978-3-319-55753-3_26

the problem of data sparsity. However, it was designed for unsupervised learning rather than labeled documents. Another stream of work focused on developing supervised topic models by using labels to constraint topics' probability distributions. For instance, Ramage et al. [15] proposed the labeled latent Dirichlet allocation that constrains LDA by defining a one-to-one mapping between topics and labels, which probably ignores latent topic features. Bao et al. [3] proposed the emotion topic model that introduces an intermediate layer into LDA to associate words and emotional labels with topics. However, these supervised topic models do not consider the sparsity of short messages.

In light of these considerations, we propose two supervised intensive topic models named the Weighted Labeled Topic Model (WLTM) and the Intensive Emotion Topic Model (IETM) for emotion detection over short messages. Both WLTM and IETM use intensive features by modeling multiple emotion labels, valence scored by numerous users, and the word pair, aka "biterm" [6] for short text emotion detection jointly. To tackle the issue of sparse words, the WLTM firstly extracts biterms for each short message. Then, we define a one-to-many mapping between emotion labels and latent topics. Specifically, we use emotion distributions of labeled documents to constraint documents' topic probabilities during training. After generating topic probabilities, testing documents' emotion distributions can be estimated by supervised learning algorithms. For IETM, we associate short documents' biterms and their emotion labels with topics. The IETM also alleviates the feature sparsity issue of short messages, in which the supervision of labeled documents' emotion distributions further improves the performance of short text emotion detection.

2   Related Work

To detect user emotions over documents, many studies increasingly consider the latent topics of documents rather than individual words [11,25]. For instance, the emotion topic model was proposed to integrate an emotion layer into LDA [2], which employed the emotional distribution to guide the topic generation. The sentiment latent topic model [18] was proposed to associate each topic with words and emotions jointly and to infer the emotion distribution of documents. Recently, the contextual sentiment topic model [16] was proposed to explicitly distinguish context-independent topics from nondiscriminative information such as some very common words, and a contextual theme which characterizes context-dependent information across different collections. The above methods are supervised topic models that consider the prior knowledge of labeled samples to detect emotions of unlabeled documents. However, these methods target on normal documents primarily rather than short messages.

The feature sparsity of short text brings new challenges to emotion detection and other related tasks. This is because short messages may not share any common words, but they could relate to each other semantically. To overcome this limitation, external documents from the web were first exploited to enrich the short content [20]. The idea of enriching short messages is valuable, but the enriched documents may not always be consistent to the original messages. Furthermore, the existing knowledge bases such as WordNet or Wikipedia were applied to identify the semantic association among words [1]. Although several topic models were developed to enrich the representation of short messages by exploiting the external knowledge [9,14], finding such auxiliary data could be expensive or time-consuming. To overcome these shortages, the biterm topic model was proposed based on the idea that if two words co-occurred more frequently, they are more likely to belong to a same topic [6]. For every short text, the model can extract a set of unordered word pair named "biterm" within the document's context. For example, a text with three distinct words will generate three biterms: $(\omega_1, \omega_2, \omega_3) \Rightarrow \{(\omega_1, \omega_2), (\omega_1, \omega_3), (\omega_2, \omega_3)\}$, where $(.,.)$ is unordered. After the pre-processing, each short text is represented as a biterm set, and every two words in a biterm are drawn independently from a topic. The limitation, however, of applying such unsupervised topic models to emotion detection is that the generated topics are not aligned to emotional labels.

3   Supervised Intensive Topic Models

In this section, we propose two supervised intensive topic models—WLTM and IETM, to address the feature sparsity and topic-emotion alignment issues of short text emotion detection. Intensive features, including multiple emotion labels, valence scored by numerous users, and biterms, are modeled jointly.

3.1   Problem Definition

For convenience of describing the proposed models, we here define terms and notations. Table 1 summarizes the notation of frequently used variables.

Assume that a short text collection consists of D documents $\{d_1, d_2, ..., d_D\}$ with words and emotions. In particular, each document d contains a sequence of N_d word tokens denoted by $\omega_d = \{\omega_1, \omega_2, \omega_3, ..., \omega_{N_d}\}$, and ratings by labelers over multiple emotion labels represented by $E_d = \{E_{d,1}, E_{d,2}, ..., E_{d,|E|}\}$. We then assume that the collection contains N_B biterms $\mathbf{B} = \{b_i\}_{i=1}^{N_B}$, which are generated from each document. The amount and categories of emotion labels are determined by the corpus. We denote the number of topics as K, the topic distribution of the collection as θ, and the word distribution of topic k as ϕ_k.

For the proposed WLTM (Sect. 3.2), θ is a K-dimensional vector that indicates the topic probability distribution over the whole collection. To ensure that all topic assignments are limited to the document's labels, we define a list of binary topic presence/absence indicators $\mathbf{\Lambda}_{b_i} = (\iota_1, \iota_2, ..., \iota_K)$ where each $\iota_k \in \{0, 1\}$ is determined by topic prior $\mathbf{\Psi}$. Since the number of topics may not equal to the number of emotions, we define a parameter τ to indicate the multiplier between topic and emotion numbers (ref. Fig. 1). Given two sentences— "The water of the polluted river is disgusting" and "The expired food is disgusting", both of them trigger the emotion of "disgusting". However, their embedded topics are totally different. The topic of the first sentence is about the water of

Table 1. Notations

Symbol	Descriptions
K	Number of topics
D	Number of documents
$|E|$	Number of emotion labels
z_i	The topic of the i-th biterm
ε_i	The emotion label of the i-th biterm
Ψ	The label prior for topics
Λ_{b_i}	The binary topic indicator of biterm b_i
λ_{b_i}	The vector of topics relative to biterm b_i
θ	Multinomial distributions of topics
ϕ	Multinomial distributions of words to topics
δ	Multinomial distributions of emotions to topics
τ	Multiplier between topic and emotion numbers
$\varepsilon_{d,m}$	The m-th emotion label of document d
$z_{d,m}$	The topic assigned to emotion label $\varepsilon_{d,m}$
γ	Prior emotion frequencies in the corpus
α	Dirichlet prior of θ and δ
β	Dirichlet prior of ϕ

the river, while the topic of the second sentence is about the expired food. Thus, a same emotion is probably related to several or more topics. We use constant mapping of each emotion to topics via multiplier τ, by following the method of mapping each document to a constant value of topics in LDA [4]. Although a hierarchical Dirichlet process can be used as the weighting over a topic probability vector of infinite length for each emotion or document, it requires time-consuming inference about high-dimensional word-topic memberships [24].

For the proposed IETM (Sect. 3.3), δ is a $|E| \times K$ matrix that indicates the emotion probability distribution conditioned to topics. The IETM associates biterms and emotion labels with topics, in which $\varepsilon_{d,m}$ is the m-th emotion in document d and $z_{d,m}$ is the topic assigned to emotion label $\varepsilon_{d,m}$.

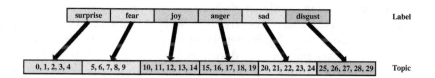

Fig. 1. Label-topic projection with $\tau = 5$

3.2 Weighted Labeled Topic Model

We here present the proposed WLTM's generative processes of topic probability distributions and detail its inference. Figure 2 presents the graphical representation of WLTM, where shaded nodes are observed data, blank ones are latent (not observed), and arrows indicate dependence. The WLTM extracts biterms and models the generation of biterms rather than words [15], and constraints topics that correspond to a document's label set only. As a complete generative model, the WLTM allows us to associate each topic with word tokens directly through weighting the topic probability for each word during training, in addition to infer/predict probabilities of emotions conditioned to unlabeled documents that only contain word tokens.

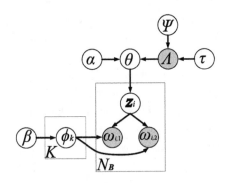

Fig. 2. Graphical representation of WLTM

As mentioned earlier, Λ_{b_i} is a list of binary topic indicators (presence/absence) related to labels of biterm b_i. Thus, Λ_{b_i} can be considered as topic indicators of each biterm in the collection. We then develop the label-topic projection (ref. Fig. 1) for each biterm. According to Table 1, we use an integer τ to indicate the multiplier between topic and emotion numbers. For example, given a dataset with two kinds of emotion labels ("joy" and "anger", $|E| = 2$) and $\tau = 2$, then the number of topics $K = |E| \times \tau = 4$. The first 2 topics are related to the first label "joy", and the last 2 topics are related to the second label "anger". In this way, the WLTM generates topic probability distribution from different numbers of topics, and enables us to tune topic number K by setting different values of multiplier τ via label-topic projection.

Figure 3 presents the generative process of WLTM that restricts multinomial distributions of topics θ to be corresponded to labels Λ_{b_i}. As shown in step 7, the biterm-topic assignment z_i is drawn from the above distribution so that this restriction limits all topic assignments to the biterm's labels. To this end, we firstly employ a Bernoulli coin toss for each topic k with a labeling prior probability to generate Λ_{b_i} (step 5). Then, we define the vector of biterms' topics as $\lambda_{b_i} = \{k|\Lambda_{b_i}^k = 1\}$. For the aforementioned example (i.e., $\tau = 2$, $|E| = 2$, and $K = 4$), if there is a labeled document that contains biterm b_i with emotion

1. Draw $\theta \sim$ Dirichlet (α)
2. For each topic $k \in [1, K]$, draw $\phi_k \sim$ Dirichlet (β)
3. For each biterm $b_i \in \mathbf{B}$:
4.   For each topic $k \in [1, K]$:
5.   Generate $\mathbf{\Lambda}_{b_i}^k \in \{0, 1\} \sim$ Bernoulli$(\cdot|\mathbf{\Psi}_k)$
6.   Generate $\lambda_{b_i} = \{k|\mathbf{\Lambda}_{b_i}^k = 1\}$
7.   Generate $z_i \in \lambda_{b_i} \sim$ Multinomial$(\cdot|\theta, \alpha)$
8.   Generate $\omega_{i,1}, \omega_{i,2} \in b_i \sim$ Multinomial(ϕ_{z_i})

Fig. 3. Generative process of WLTM

ratings $\{2, 0\}$, we can generate $\mathbf{\Lambda}_{b_i} = \{1, 1, 0, 0\}$ and $\lambda_{b_i} = \{1, 2\}$. According to our label-topic projection, the 2 user ratings over the first kind of emotion are mapped to the first and the second topic for biterm b_i. Thus, the vector of each biterm's topics λ_{b_i} can be generated by step 6.

The above variables are used to build our intensive topic model for supervised learning. As shown in Fig. 2, the only additional dependency is the dependency of θ on both α and $\mathbf{\Lambda}$, which is represented by directed edges from α and $\mathbf{\Lambda}$ to θ. To estimate parameters of WLTM, we use an approximate inference method based on Gibbs sampling which is also employed to estimate parameters of other topic models [2,8,12]. Given parameters θ and ϕ, the conditional probability of biterm b_i is estimated as follows:

$$P(b_i|\theta, \phi) = \sum_{k=1}^{K} P(\omega_{i,1}, \omega_{i,2}, z_i = k|\theta, \phi) = \sum_{k=1}^{K} \theta_k \phi_{k,\omega_{i,1}} \phi_{k,\omega_{i,2}}. \tag{1}$$

For all biterms, the proposed WLTM maximizes the likelihood function over the whole corpus, as follows:

$$P(\mathbf{B}|\theta, \phi) = \prod_{i=1}^{N_B} \sum_{k=1}^{K} \theta_k \phi_{k,\omega_{i,1}} \phi_{k,\omega_{i,2}}. \tag{2}$$

Given the observed document labels $\mathbf{\Lambda}_{b_i}$, the topic probability distribution of b_i is restricted to the set of labeled topics λ_{b_i}. Thus, the WLTM samples topic z_i according to the following conditional distribution for each biterm b_i:

$$P(z_i = k, k \in \lambda_{b_i}|z_{-i}, \mathbf{B}) \propto (n_{-i,k} + \alpha)$$
$$\times \frac{(n_{-i,\omega_{i,1}|k} + \beta)(n_{-i,\omega_{i,2}|k} + \beta)}{(n_{-i,\cdot|k} + W\beta + 1)(n_{i,\cdot|k} + W\beta)} \times \frac{\gamma_{d_i,|\frac{k}{\tau}|}}{\sum_{k'} \gamma_{d_i,|\frac{k'}{\tau}|}}, \tag{3}$$

where z_{-i} is topic assignments of all biterms, $n_{-i,k}$ is the number of biterms assigned to topic k, $n_{-i,\omega|k}$ is the number of times word ω assigned to topic k, and $n_{-i,\cdot|k}$ is the number of times that all words assigned to topic k. The subscript $-i$ indicates that the number count does not include the current assignment of biterm b_i, d_i denotes the document from which the current biterm b_i is sampled, and $|\frac{k}{\tau}|$

is the absolute value of k divides by τ, which maps the related emotion votes with topic k. An important feature of the above equation is that the target topic k is restricted to the corresponding labels of the document that contains b_i. Thus, topic k is assigned to the set of b_i's relative topics according to its emotion labels (i.e., λ_{b_i}). For the rest of topics that are not assigned to λ_{b_i}, the probability distribution of b_i equal to 0.

After a sufficient number of iterations, the WLTM counts the the number of biterms assigned to each topic n_k, and the number of times word ω assigned to each topic $n_{\omega|k}$ in the training set. Then, the topic-word probability distribution ϕ and the topic probability distribution of the whole corpus θ are estimated by

$$\phi_{k,\omega} = \frac{n_{\omega|k} + \beta}{n_{\cdot|k} + W\beta}, \text{ and } \theta_k = \frac{n_k + \alpha}{N_B + K\alpha}. \tag{4}$$

Based on the topic assigned to each biterm, the proposed WLTM can extract the topic proportion by estimating the posterior topic probability for each document $P(z|d)$. Specifically, we derive the topic proportion of a document by topics of biterms since the WLTM does not model words in the generative process.

Given a document d with N_d biterms, we assume that the topic of each biterm $b_i^{(d)} = (\omega_{i,1}^{(d)}, \omega_{i,2}^{(d)})$ is conditionally independent of document d. After generating biterms for each document, we have $P(z|d) = \sum_i P(z|b_i^{(d)})P(b_i^{(d)}|d)$, where $P(z|b_i^{(d)})$ can be estimated by Bayes' rule based on parameters learned in the WLTM, as follows:

$$P(z = k|b_i^{(d)}) = \frac{\theta_k \phi_{k,\omega_{i,1}^{(d)}} \phi_{k,\omega_{i,2}^{(d)}}}{\sum_{k'} \theta_{k'} \phi_{k',\omega_{i,1}^{(d)}} \phi_{k',\omega_{i,2}^{(d)}}}, \text{ and } P(b_i^{(d)}|d) = \frac{n(b_i^{(d)})}{\sum_{i=1}^{N_d} n(b_i^{(d)})}, \tag{5}$$

where $n(b_i^{(d)})$ represents the frequency of biterm b_i in document d.

According to Eq. 3, the Gibbs sampling algorithm based on biterms is employed for the proposed WLTM. After convergence, we get the topic probability distribution of each document $P(z|d)$. Finally, the Support Vector Regression (SVR) [5] is employed to predict testing documents' emotion distributions based on the above document-topic probability.

3.3   Intensive Emotion Topic Model

In this part, we present our IETM for short text emotion detection. To alleviate the problem of feature sparsity in short messages, the proposed IETM firstly generates biterms for each document, and then associates biterms and emotion labels with topics. The graphical model of IETM is shown in Fig. 4.

In the graphical representation of IETM, words $\omega_{i,1}$ and $\omega_{i,2}$ belong to a same biterm. The IETM is proposed to generate emotion probabilities conditioned to words. To achieve this, we firstly draw the emotion-topic probability because there are too few words in each short document. Extracting biterms for short messages can enrich features of each document, which enables us to alleviate the problem of feature sparsity.

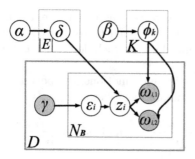

Fig. 4. Graphical representation of IETM

1. For emotion $e \in [1, |E|]$, draw $\delta_e \sim$ Dirichlet (α)
2. For each topic $k \in [1, K]$, draw $\phi_k \sim$ Dirichlet (β)
3. For each biterm $b_i \in \mathbf{B}$:
4.   Generate $\varepsilon_i \sim$ Multinomial(γ)
5.   Generate $z_i \sim$ Multinomial(δ_{ε_i})
6.   Generate $\omega_{i,1}, \omega_{i,2} \in b_i \sim$ Multinomial(ϕ_{z_i})

Fig. 5. Generative process of IETM

Figure 5 presents the generative process of IETM. The variables $\varepsilon_i \in E$ and $z_i \in Z$ are corresponding emotion and topic assignments for biterm b_i. In particular, ε is sampled from a multinomial distribution with prior observed emotion labels parameterized by γ, which is normalized and summed to 1. According to this generative process, the joint probability of all random variables for a document collection is given below:

$$P(\gamma, \varepsilon, \mathbf{z}, \mathbf{B}, \delta, \phi; \alpha, \beta) = P(\delta; \alpha) P(\phi; \beta) P(\gamma) \times P(\varepsilon|\gamma) P(\mathbf{z}|\varepsilon, \delta) P(\mathbf{B}|\mathbf{z}, \phi). \quad (6)$$

Specifically, we estimate the posterior probabilistic distribution of emotion ε and topic z for each biterm based on the following conditional probability, which is derived from Eq. 6:

$$P(\varepsilon_i = e|\gamma, \varepsilon_{-i}, \mathbf{z}, \mathbf{B}; \alpha, \beta) \propto \frac{\alpha + nz_{-i}^{e,z_i}}{K\alpha + \sum_z nz_{-i}^{e,z}} \times \frac{\gamma_{d_i,e}}{\sum_{e'} \gamma_{d_i,e'}}. \quad (7)$$

After generating emotion ε for biterms $\mathbf{B}$ conditioned to topic z, we generate the topic conditioned to biterms $\mathbf{B}$ as follows:

$$P(z_i = z|\mathbf{z}_{-i}, \gamma, \varepsilon, \mathbf{B}; \alpha, \beta) \propto \frac{\alpha + nz_{-i}^{\varepsilon_i,z}}{K\alpha + \sum_{z'} nz_{-i}^{\varepsilon_i,z'}}$$
$$\times \frac{\beta + n\omega_{-i}^{z,\omega_{i,1}}}{W\beta + \sum_\omega n\omega_{-i}^{z,\omega}} \times \frac{\beta + n\omega_{-i}^{z,\omega_{i,2}}}{W\beta + \sum_\omega n\omega_{-i}^{z,\omega}}, \quad (8)$$

where $\mathbf{B}$ denotes the current sampled biterms that each biterm b_i contains $\omega_{i,1}$ and $\omega_{i,2}$, ε and z are respectively the candidate emotion and topic for sampling,

$nz^{\varepsilon,z}$ is the number of times topic z assigned to emotion ε, and $nw^{z,\omega}$ denotes the number of times word ω assigned to topic z. The subscript $-i$ of nz and nw indicates that the count does not include the current assignment of emotions and topics for words.

Given the sampled topics and emotions, the posterior probabilistic distribution of δ and ϕ can be estimated as follows:

$$\delta_{e,z} = \frac{\alpha + nz^{e,z}}{K\alpha + \sum_{z'} nz^{e,z'}}, \text{ and } \phi_{z,w} = \frac{\beta + nw^{z,w}}{W\beta + \sum_{w'} nw^{z,w'}}. \tag{9}$$

With emotion-topic probability distribution δ and topic-word probability distribution ϕ derived above available, the emotion probability distribution for each testing document is estimated as follows:

$$P(e|d) = \frac{P(e) \prod_{w,w \in d} P(w|e)}{\sum_e P(e) \prod_{w,w \in d} P(w|e)}, \tag{10}$$

where $P(e)$ is the emotion probability distribution for the whole training set, and the probability of word ω conditioned to emotion e can be estimated by integrating the latent topic z: $P(w|e) = \sum_z \delta_{e,z} \phi_{z,w}$. Based on the values of $P(e|d)$ for the testing set, we can evaluate the IETM's emotion detection performance.

4   Experiments

In this section, we conduct experiments to achieve the following two goals: (i) to observe the influence of multiplier values on the task of short text emotion detection, and (ii) to compare the performance of our models with baselines.

4.1   Experimental Design

To verify the effectiveness of our models, we employ the following two datasets.

SemEval: This is a dataset used in the 14th task of the 4th International Workshop on Semantic Evaluations (SemEval-2007) [10], which contains 1,246 valid news headlines with the total score of the 6 emotions (anger, disgust, fear, joy, sad and surprise) larger than 0. These emotions are posited to be basic [7]. The dataset is officially divided into a trail-set with 246 documents and a test-set with 1,000 documents. We use the trail-set for training and the rest for testing.

ISEAR: This is a collection of 7,666 sentences annotated by 1,096 participants [21]. The emotional labels of this dataset are anger, disgust, fear, joy, sadness, shame, and guilt. We randomly select 60% of sentences as the training set, 20% as the validation set, and the rest as the testing set.

We implement the Labeled Latent Dirichlet Allocation (LLDA) [15], the Biterm Topic Model (BTM) [6], the Emotion Topic Model (ETM) [3], the Contextual Sentiment Topic Model (CSTM) [16], the Sentiment Latent Topic Model (SLTM) [18], and some classical approaches that do not exploit topics [3, 5, 10] as baselines. For BTM and our methods, all biterms are generated by taking each short text as an individual context unit. To take emotional distributions into account [10,16], we use two fine-grained evaluation metrics as indicators of performance: $AP_{document}$ and $AP_{emotion}$, where AP denotes the averaged Pearson's correlation coefficient. For each document, we firstly measure the correlation coefficient between predicted probabilities and actual votes over all emotion labels [16]. Then, $AP_{document}$ is calculated by averaging the above values over all documents. For each emotion label, we firstly measure the correlation coefficient between predicted probabilities and actual votes over all documents [10]. Then, $AP_{emotion}$ is calculated by averaging the above values over all emotions.

Since WLTM, LLDA, and BTM do not directly generate documents' emotion distributions, we employ SVR [5] using Radial Basis Function (RBF) as the kernel function to predict emotion distributions of testing documents based on the document-topic probability. Specifically, we perform 5-fold cross-validation on the training or validation data, and select parameters that achieve the lowest mean squared error during cross-validation for SVR. All cross-validations are performed on the training set or validation set only. To evaluate the IETM, we generate the emotion probability distribution $P(e|d)$ according to Eq. 10 directly. Similar to the previous studies [2,3,15], the hyperparameters α and β are respectively set to symmetric Dirichlet priors with values of $50/K$ and 0.1, and the number of Gibbs sampling iteration is 500 in the following experiments.

4.2   Influence of the Multiplier

The multiplier is used to determine the number of topics K that indicates how many latent aspects of documents can be derived, which may influence the performance of topic models. We therefore evaluate the performance of WLTM and IETM under different latent aspects.

As mentioned earlier, the topic number of WLTM depends on the multiplier (i.e., τ) between topic and emotion numbers. Thus, we vary τ from 1 to 15 for both datasets, and topic numbers of $SemEval$ and $ISEAR$ are set to be from $|E_{SemEval} * 1| = 6$ to $|E_{SemEval} * 15| = 90$ and from $|E_{ISEAR} * 1| = 7$ to $|E_{ISEAR} * 15| = 105$, respectively. These values are used to evaluate the influence

Table 2. Performance statistics of different models

(a) *SemEval*

Models	$AP_{document}$		$AP_{emotion}$	
	Mean	Variance	Mean	Variance
WLTM	0.20	**0.0006**	**0.24**	**0.0002**
IETM	**0.31**	0.0007	0.20	0.0004
LLDA [15]	0.00	0.0032	0.01	**6.77E-05**
BTM [6]	0.19	0.0011	**0.23**	0.0008
ETM [3]	**0.23**	0.0009	0.07	**0.0001**
CSTM [16]	**0.30**	**0.0001**	0.11	0.0009
SLTM [18]	0.17	0.0044	0.02	0.0005

(b) *ISEAR*

Models	$AP_{document}$		$AP_{emotion}$	
	Mean	Variance	Mean	Variance
WLTM	**0.43**	**4.12E-05**	**0.45**	**9.31E-05**
IETM	0.30	**1.93E-05**	0.34	**0.0001**
LLDA [15]	0.01	**3.41E-05**	0.02	**7.90E-05**
BTM [6]	**0.33**	0.0014	**0.36**	0.0015
ETM [3]	**0.35**	6.87E-05	**0.41**	0.0002
CSTM [16]	0.21	0.0004	0.23	0.0007
SLTM [18]	0.10	0.0012	0.09	0.0010

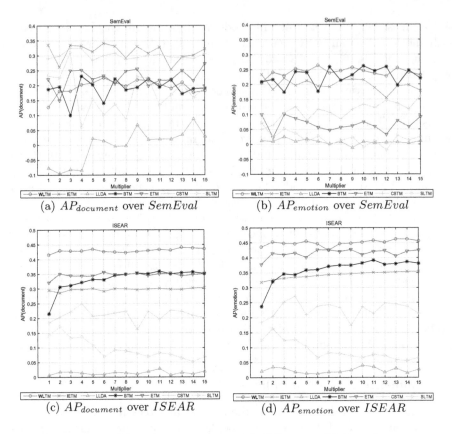

(a) $AP_{document}$ over $SemEval$

(b) $AP_{emotion}$ over $SemEval$

(c) $AP_{document}$ over $ISEAR$

(d) $AP_{emotion}$ over $ISEAR$

Fig. 6. Model performance curve with different multiplier values

of multiplier or topic numbers on WLTM, IETM, LLDA, BTM, ETM, CSTM and SLTM. The mean and variance of $AP_{document}$ and $AP_{emotion}$ under those multiplier values for different models are shown in Table 2, where top 3 values of each metric are highlighted in boldface. We also present the performance curve of different models in Fig. 6.

Experimental results over $SemEval$ show that the IETM yields competitive performance on $AP_{document}$ and the WLTM achieves the best $AP_{emotion}$ value. The performance of WLTM in terms of $AP_{document}$ is sightly worse than some other models, with a value of 0.20. A possible reason is that for the 1,000 testing documents, there are 28 documents' words not appeared in the training set. Due to the lack of samples in tuning parameters, WLTM, LLDA, and BTM that use SVR for prediction may underfit emotional distributions at the document level. Compared to LLDA, BTM, ETM, CSTM, and SLTM, the mean $AP_{document}$ of IETM improves 0.31, 0.12, 0.08, 0.01, and 0.14, and the mean $AP_{emotion}$ of WLTM improves 0.23, 0.01, 0.17, 0.13, and 0.22, respectively (ref. Table 2). Particularly, the variances of WLTM always rank top 3, which indicates that the WLTM performs stable over $SemEval$. On the other hand, the IETM's variances rank top 3 on $AP_{document}$ and rank top 4 on $AP_{emotion}$.

Experimental results over *ISEAR* indicate that the WLTM yields competitive performance on both $AP_{document}$ and $AP_{emotion}$ (ref. Fig. 6). Compared to LLDA, BTM, ETM, CSTM and SLTM, the proposed WLTM respectively improves 0.42, 0.10, 0.08, 0.22, and 0.33 on $AP_{document}$, and improves 0.43, 0.09, 0.04, 0.22, and 0.36 on $AP_{emotion}$, respectively (ref. Table 2). In particular, the variances of WLTM's performance rank top 3, which shows that the proposed WLTM performed stable. Although the proposed IETM can not achieve the best results in Pearson's correlation coefficient, its variances with different multiplier values also rank top 3.

Note that performance variances of WLTM and IETM over both *SemEval* and *ISEAR* almost rank top 3, which shows that these two models perform more stable than other baseline models. Particularly, the Pearson's correlation coefficients of WLTM are the best over *ISEAR*, and the WLTM's performance is the best over *SemEval* in terms of $AP_{emotion}$. Therefore, the WLTM is an accurate and stable supervised topic model for short text emotion detection. The IETM achieves the best $AP_{document}$ over *SemEval*. Although other Pearson's correlation coefficients of the IETM are not the best, they still indicate acceptable stable results. Experimental results over *ISEAR* show that the WLTM achieves best results on both metrics, which indicates that the WLTM is more efficient than IETM and baselines. Based on the generative process of the IETM, it samples a most-likely relative emotion ε for a biterm b_i, and then samples a most-likely relative topic z. This process ignores the other relative emotion of each biterm. On the other hand, the WLTM samples all relative topics in the generative process and reserves discriminative features. Statistical analysis on variances and mean values are conducted by the following F-test and t-test.

4.3 Comparison with Baselines

To compare the performance of our supervised intensive topic models on short text emotion detection statistically, we employ the analysis of variance F-test to test the assumption of homoscedasticity (the homogeneity of variance). The F-test is conducted on WLTM, IETM and baseline models of LLDA, BTM, ETM, CSTM and SLTM. Similarly, we conducted t-tests to test the assumption that the difference in performance between paired models has a mean value of zero.

Table 3. The p values of statistical tests on the WLTM and baselines

(a) *SemEval*

Models	$AP_{document}$		$AP_{emotion}$	
	$F-test$	$t-test$	$F-test$	$t-test$
LLDA [15]	0.0015	7.79E-11	0.0216	3.35E-30
BTM [6]	0.1184	0.2984	0.0078	0.0429
ETM [3]	0.2210	0.0018	0.0457	6.31E-19
CSTM [16]	0.0070	5.37E-13	0.0044	1.31E-12
SLTM [18]	0.0003	0.1358	0.0484	2.34E-21

(b) *ISEAR*

Models	$AP_{document}$		$AP_{emotion}$	
	$F-test$	$t-test$	$F-test$	$t-test$
LLDA [15]	0.3646	4.15E-45	0.3818	2.22E-40
BTM [6]	2.78E-08	2.32E-08	2.87E-06	8.14E-08
ETM [3]	0.1748	2.23E-23	0.1341	2.73E-09
CSTM [16]	4.35E-05	2.09E-18	0.0003	2.71E-17
SLTM [18]	5.40E-08	2.49E-16	3.29E-05	6.54E-19

As shown in Table 3(a), the F-test results on $AP_{document}$ over *SemEval* between the WLTM and baselines of LLDA, CSTM and SLTM are less than 0.05, yet results of that with BTM and ETM are larger than 0.05. The F-test results on $AP_{emotion}$ of *SemEval* between the WLTM and all baselines are less than 0.05. These results indicate that the WLTM is significantly more stable than baselines statistically. The p-values of t-test are almost less than 0.05, which indicates that the proposed WLTM outperforms baselines. As shown in Table 3(b), the F-test results between the WLTM and baselines over *ISEAR* are less than 0.05 except for LLDA and ETM. However, the t-test results are all less than 0.05, which indicates that the proposed WLTM is more effective than baseline models.

Table 4. The p values of statistical tests on the IETM and baselines

(a) *SemEval*

Models	$AP_{document}$		$AP_{emotion}$	
	$F-test$	$t-test$	$F-test$	$t-test$
LLDA [15]	0.0043	1.01E-14	0.0011	6.70E-19
BTM [6]	0.2121	5.21E-12	0.0922	0.0030
ETM [3]	0.3518	3.16E-09	0.2898	1.55E-16
CSTM [16]	0.0025	0.0674	0.0611	1.83E-10
SLTM [18]	0.0009	2.28E-07	0.2997	5.71E-20

(b) *ISEAR*

Models	$AP_{document}$		$AP_{emotion}$	
	$F-test$	$t-test$	$F-test$	$t-test$
LLDA [15]	0.1481	1.50E-42	0.2608	3.18E-36
BTM [6]	1.64E-10	0.0013	9.13E-06	0.0640
ETM [3]	0.0117	1.34E-15	0.2203	2.12E-16
CSTM [16]	0.0451	4.33E-11	0.0008	2.95E-12
SLTM [18]	3.25E-10	1.37E-12	9.83E-05	2.85E-16

Statistical tests over *SemEval* and *SemEval* between the proposed IETM and baselines are shown in Tables 4(a) and (b), respectively. The results indicate that the IETM is a stable and competitive model for short text emotion detection. Above all, the IETM outperforms others over *SemEval* on $AP_{document}$, and achieves stable performance over both datasets.

Furthermore, we compare our proposed models with some other classical models that do not exploit latent topics, i.e., SWAT [10,23], ET [3] and SVR [5]. The SWAT was one of the top-performing systems on the "affective text" task in SemEval-2007 [22], which made use of emotional ratings of news headlines and scored emotions of each word with a unigram model. The ET used the naïve Bayes method to model word-emotion associations, and considered emotion ratings when estimating parameters. The SVR was a baseline method on emotion detection, where the kernel of RBF was adopted [3]. We used the word frequency to construct the feature vector, and employed 5-fold cross-validation to tune parameters *gamma* and *cost*.

With respect to $AP_{document}$ and $AP_{emotion}$ values over *SemEval*, the SWAT achieves 0.26 and 0.22; the ET achieves 0.24 and 0.23; and the SVR achieves 0.01 and 0. Compared to the proposed WLTM, SWAT and ET achieve better or similar performance on $AP_{document}$, and the WLTM has the best performance on $AP_{emotion}$. For the IETM, it has the best performance on $AP_{document}$ and the third performance on $AP_{emotion}$. For $AP_{documents}$ and $AP_{emotion}$ over *ISEAR*, the SWAT achieves 0.21 and 0.21; the ET achieves 0.38 and 0.43; and the SVR achieves 0.05 and 0.07. The WLTM performs better than baselines of SWAT, ET and SVR on $AP_{document}$ and $AP_{emotion}$. The IETM obtains better performance than SWAT and SVR on both metrics except for ET. Due to space limit, we leave the comparison of different topic models' convergence speed and in-depth qualitative analysis on the generated topics to further research.

5   Conclusion

This paper proposed two supervised intensive topic models named WLTM and LETM, which aimed at addressing the challenging issues of feature sparsity in short messages and topic-emotion alignment. Experimental results indicated that the WLTM performed best over $ISEAR$ (the mean number of words is 22.0 for each sentence) in terms of both metrics, and the LETM achieved competitive performance over $SemEval$ (the mean number of words is only 6.6 for each instance), especially on $AP_{document}$. To conclude, the WLTM performs well and robustness on common short corpora, and the LETM is suitable for modeling extremely sparse features.

Although it is straightforward to take each short text as an individual context unit [6], extracting biterms using the length (context) of every document may be unreasonable for unbalanced datasets like $ISEAR$ (the least and the most numbers of words are one and 179 for each sentence). For example, assume that there are two documents d_1 and d_2, where d_1 contains 3 words and d_2 contains 10 words. After extracting biterms, d_1 has 3 biterms while d_2 has 45 biterms. In this case, document d_2 will dominate the training data and negatively influence the performance. The method of using a small, fixed-size window over a word sequence within a document to generate biterms [6] has the same shortage. To alleviate this problem, we plan to develop a dynamic approach for biterm extraction. Besides, all biterms are regarded as independent in the proposed methods, which can be extended to incorporate the correlation information via co-occurred words. We also plan to validate our models over other scenarios, such as a collection of tweets [19].

Acknowledgements. We are grateful to the anonymous reviewers for their valuable comments on this manuscript. The research has been supported by the National Natural Science Foundation of China (61502545, 61572336), two grants from the Research Grants Council of the Hong Kong Special Administrative Region, China (UGC/FDS11/E03/16 and UGC/FDS11/E06/14), the Start-Up Research Grant (RG 37/2016-2017R), and the Internal Research Grant (RG 66/2016-2017) of The Education University of Hong Kong.

References

1. Banerjee, S., Ramanathan, K., Gupta, A.: Clustering short texts using wikipedia. In: SIGIR, pp. 787–788 (2007)
2. Bao, S., Xu, S., Zhang, L., Yan, R., Su, Z., Han, D., Yu, Y.: Joint emotion-topic modeling for social affective text mining. In: ICDM, pp. 699–704 (2009)

3. Bao, S., Xu, S., Zhang, L., Yan, R., Su, Z., Han, D., Yu, Y.: Mining social emotions from affective text. IEEE Trans. Knowl. Data Eng. **24**(9), 1658–1670 (2012)
4. Blei, D.M., Ng, A.Y., Jordan, M.I.: Latent Dirichlet allocation. J. Mach. Learn. Res. **3**, 993–1022 (2003)
5. Chang, C., Lin, C.: LIBSVM: a library for support vector machines. ACM Trans. Intell. Syst. Technol. **2**(3), 389–396 (2011)
6. Cheng, X., Lan, Y., Guo, J., Yan, X.: BTM: topic modeling over short texts. IEEE Trans. Knowl. Data Eng. **26**(12), 2298–2941 (2014)
7. Ekman, P.: Facial expression and emotion. Am. Psychol. **48**(4), 384–392 (1993)
8. Griffiths, T.L., Steyvers, M.: Finding scientific topics. Proc. Nat. Acad. Sci. U.S.A **101**(suppl. 1), 5228–5235 (2004)
9. Jin, O., Liu, N.N., Zhao, K., Yu, Y., Yang, Q.: Transferring topical knowledge from auxiliary long texts for short text clustering. In: CIKM, pp. 775–784 (2011)
10. Katz, P., Singleton, M., Wicentowski, R.: SWAT-MP: the semeval-2007 systems for task 5 and task 14. In: SemEval, pp. 308–313 (2007)
11. Kim, S.B., Han, K.S., Rim, H.C., Myaeng, S.H.: Some effective techniques for naive bayes text classification. IEEE Trans. Knowl. Data Eng. **18**(11), 1457–1466 (2006)
12. Lau, R.Y.K., Xia, Y., Ye, Y.: A probabilistic generative model for mining cyber-criminal networks from online social media. IEEE Comput. Intell. Mag. **9**(1), 31–43 (2014)
13. Pang, B., Lee, L., Vaithyanathan, S.: Thumbs up? Sentiment classification using machine learning techniques. In: EMNLP, pp. 79–86 (2002)
14. Phan, X.H., Nguyen, L.M., Horiguchi, S.: Learning to classify short and sparse text & web with hidden topics from large-scale data collections. In: WWW, pp. 91–100 (2008)
15. Ramage, D., Hall, D., Nallapati, R., Manning, C.D.: Labeled LDA: a supervised topic model for credit attribution in multi-labeled corpora. In: EMNLP, pp. 248–256 (2009)
16. Rao, Y.: Contextual sentiment topic model for adaptive social emotion classification. IEEE Intell. Syst. **31**(1), 41–47 (2016)
17. Rao, Y., Lei, J., Liu, W., Li, Q., Chen, M.: Building emotional dictionary for sentiment analysis of online news. World Wide Web **17**, 723–742 (2014)
18. Rao, Y., Li, Q., Mao, X., Liu, W.: Sentiment topic models for social emotion mining. Inf. Sci. **266**, 90–100 (2014)
19. Roberts, K., Roach, M.A., Johnson, J., Guthrie, J., Harabagiu, S.M.: EmpaTweet: annotating and detecting emotions on twitter. In: LREC, pp. 3806–3813 (2012)
20. Sahami, M., Heilman, T.D.: A web-based kernel function for measuring the similarity of short text snippets. In: WWW, pp. 377–386 (2006)
21. Scherer, K.R., Wallbott, H.G.: Evidence for universality and cultural variation of differential emotion response patterning. J. Pers. Soc. Psychol. **66**(2), 310–328 (1994)
22. Snow, R., O'Connor, B., Jurafsky, D., Ng, A.Y.: Cheap and fast–but is it good? Evaluating non-expert annotations for natural language tasks. In: EMNLP, pp. 254–263 (2008)
23. Strapparava, C., Mihalcea, R.: Semeval-2007 task 14: affective text. In: SemEval, pp. 70–74 (2007)
24. Taddy, M.A.: On estimation and selection for topic models. In: AISTATS, pp. 1184–1193 (2012)
25. Wang, J., Yao, Y., Liu, Z.: A new text classification method based on HMM-SVM. In: ISCIT, pp. 1516–1519 (2007)

Leveraging Pattern Associations for Word Embedding Models

Qian Liu[1,2], Heyan Huang[1], Yang Gao[1(✉)], Xiaochi Wei[1], and Ruiying Geng[1]

[1] School of Computer Science and Technology,
Beijing Institute of Technology, Beijing, China
gyang@bit.edu.cn
[2] Faculty of Engineering and Information Technology,
University of Technology Sydney, Ultimo, Australia

Abstract. Word embedding method has been shown powerful to capture words association, and facilitated numerous applications by effectively bridging lexical gaps. Word semantic is encoded with vectors and modeled based on n-gram language models, as a result it only takes into consideration of words co-occurrences in a shallow slide windows. However, the assumption of the language modelling ignores valuable associations between words in a long distance beyond n-gram coverage. In this paper, we argue that it is beneficial to jointly modeling both surrounding context and flexible associative patterns so that the model can cover long distance and intensive association. We propose a novel approach to combine associated patterns for word embedding method via joint training objection. We apply our model for query expansion in document retrieval task. Experimental results show that the proposed method can perform significantly better than the state-of-the-arts baseline models.

1 Introduction

As an improvement of traditional one-hot word representation method, word embedding overcomes the data sparsity, high dimensional, and lexical gap problems by capturing both word semantics and syntactics with dense vectors. A great efforts have been conducted to construct prominent word embedding [4,10,16,20,21,23,25], and they are widely used in natural language processing tasks to compute word semantical similarity and regularity.

Most existing word embedding methods generate word vectors based on its surrounding context, encoding only shallow adjacency information. However, in real world, many valuable associative relationships between words actually exist in a longer linguistic distance instead of adjacent rules. A few examples are shown in Fig. 1(a). As can be seen from these sentences, *"programming languages"* and *"algorithm"* hold intensive relations, but they are distributed apart in long distance. And it's hard for a local slide window (i.e. the length of slide window defines as n = 5) to cover with.

It is worth mentioning that association rule mining [1,27], as a data analysis technique, is generally used for discovering frequently co-occurring data items

© Springer International Publishing AG 2017
S. Candan et al. (Eds.): DASFAA 2017, Part I, LNCS 10177, pp. 423–438, 2017.
DOI: 10.1007/978-3-319-55753-3_27

without the limitation of n-gram coverage. We employ association rule mining in text data processing to mine frequent patterns, and utilise the discovered association rules among them as trustful relationships between individual words. Figure 1(b) present several associated items mined from sentences using 5-gram and associated patterns respectively. As we can see, several intensive relations (i.e. *"programming languages"* with *"algorithm"*, *"language"* with *"express algorithm"*) cannot be discovered in 5-gram, while can be easily captured by pattern mining methods.

Associated patterns can discover rich words associations [3,7,9] but not yet well-utilised in generating word embedding. To illustrate this fact, we visually show the relationship between the number of word co-occurrences in n-gram windows and the number of same related words in associated patterns. The result is shown in Fig. 1(b). In the figure, the X-axis denotes the number of two words co-occurrence in n-gram model window (n = 5) and the Y-axis denotes the number of two words co-occurrence in association patterns. We find that the statistic follows a power-law distribution, which means that most frequently appeared patterns can hardly be fetched by a window with a fixed-length. Specially, great number of related words in associated patterns (marked as area1) cannot be found in local context word co-occurrence. This indicates that rich word associations in the corpus may be ignored if the word embedding methods only consider local contextual information, although area2 and area3 demonstrate that n-gram models can also cover reasonable number of related words.

In order to solve the urging problem of losing word associative relations of language models, in this paper, we propose an Associated Patterns enhanced Word Embedding model, APWE for short. In this model, associated patterns

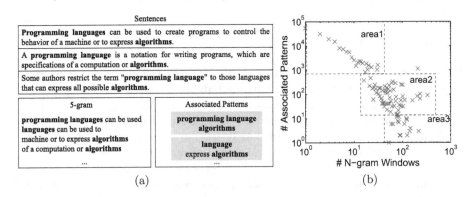

(a)   (b)

Fig. 1. Examples of associated words in long sentences. (a) shows several sentences, and associated words in the sentences are in bold. Related words are shown using n-gram (n = 5) and associated patterns respectively. (b) compares the ability of mining related words in a 5-gram context window and associated patterns. Relations in area1 can be discovered by associated patterns while hard by context. N-gram method is good at capturing relations in area3. Both methods work well in discover relations in area2.

are introduced and jointly used with context information to predict the target word, so that both local co-occurrence and long distance associated relations are considered in the generated word embedding. Further more, we apply APWE method for query expansion in document retrieval task, and the results demonstrate that our methods outperform state-of-the-arts methods.

The main contributions of our work are summarized as follows:

- Association patterns are wisely integrated in the process of generating semantic word embeddings. The new word embedding model can capture intensive word associations beyond n-gram limitation which is covered by a sentence level.
- To enable word relations discovery more sensitive for word embedding in a flexible and fast way, especially under circumstances of insufficient datasets.
- We conduct experiments to demonstrate the effectiveness of our method for document modelling in real application.

The rest of this paper is organized as follows. Section 2 summarizes the background of our methods. We then proposed associated patterns enhanced word embedding model in Sect. 3. Section 4 reports the experimental results. Section 5 surveys the related work. Finally, we conclude the paper in Sect. 6.

2 Background

2.1 Word2Vec Method

Recently, neural networks relevant methods have been introduced to model languages with promising results. Especially, Mikolov et al. [15,16] proposed Word2Vec method is an efficient method for learning high quality word embedding from large-scale unstructured text data.

The basic assumption behind Word2Vec is that the representation of co-occurrenced words have the similar representation in the semantic space. To this target, a sliding window is employed on the input text stream, where the central word is the target word and others are contexts. Word2Vec method contains two models: continuous bag-of-word model (CBOW) and Skip-gram model.

CBOW aims at predicting the target word using the context words in the sliding window. Formally, given a word sequence $\mathcal{D} = \{w_{i-k}, ..., w_{i-1}, w_i, w_{i+1}, ..., w_{i+k}\}$, the objective of CBOW is to maximize the average log probability

$$L(\mathcal{D}) = \frac{1}{T} \sum_{i=1}^{T} \log Pr(w_i \mid w_{i-k}, ..., w_{i-1}, w_{i+1}, ..., w_{i+k}). \tag{1}$$

where, w_i is the target word, T is the corpus size, and k is the context size of the target word, which indicates that the window size is $2k + 1$. CBOW formulates the probability $Pr(w_i \mid w_{i-k}, ..., w_{i-1}, w_{i+1}, ..., w_{i+k})$ with a softmax function as

$$Pr(w_i \mid w_{i-k}, ..., w_{i-1}, w_{i+1}, ..., w_{i+k}) = \frac{\exp(\mathbf{x}_i \cdot \mathbf{x}_c)}{\sum_{w \in \mathcal{W}} \exp(\mathbf{x} \cdot \mathbf{x}_c)}, \tag{2}$$

where $\mathcal{W}$ represents the vocabulary, $\mathbf{x}_i$ is the vector representation of the target word w_i, and $\mathbf{x}_c$ is the average of all context word vectors.

Different from CBOW, Skip-gram aims to predict context words given the target word. Therefore, the objective of Skip-gram is to maximize the average log probability

$$L(\mathcal{D}) = \frac{1}{T} \sum_{i=1}^{T} \sum_{-k \leqslant c \leqslant k, c \neq 0} \log Pr(w_{i+c} \mid w_i), \tag{3}$$

where, k is the context size of the target word, and the probability $Pr(w_{i+c} \mid w_i)$ is formulated with softmax function, which is denoted as

$$Pr(w_{i+c} \mid w_i) = \frac{\exp(\mathbf{x}_{i+c} \cdot \mathbf{x}_i)}{\sum_{w \in \mathcal{W}} \exp(\mathbf{x} \cdot \mathbf{x}_i)}, \tag{4}$$

where $\mathcal{W}$ represents the vocabulary, $\mathbf{x}_i$ is the vector representation of the target word w_i, and $\mathbf{x}_{i+c}$ is the vector of context word.

Word2Vec has been shown useful in many applications. Nevertheless, most existing works learn word representations mainly based on the word co-occurrences, therefore the obtained word embedding cannot capture associated words if either of them yield very little context information. On the other hand, in small and insufficient datasets, word co-occurrences may be sparse and unreliable, which may mislead the training process. In this paper, we extend Word2Vec method leveraging associated patterns to further improve the word representation.

2.2 Associated Patterns

In this paper, we employ association rule mining in text data processing to discover associated patterns. Association rule mining, as a data analysis technique, is generally used for discovering frequently co-occurring data items, and aims to discover hidden rules among enormous pattern combinations based on their individual and conditional frequencies.

An association rule contains two patterns, i.e. an antecedent pattern X and a consequent pattern Y. These two patterns are considered to be a rule if its frequency satisfies a minimum support threshold (T_s) and the conditional probability satisfies a minimum confidence threshold (T_c). Formally, this can be expressed as

$$\begin{cases} Supp(X \cup Y) \geq T_s, \\ Conf(X \rightarrow Y) = \dfrac{(Supp(X \cup Y))}{Supp(X)} \geq T_c, \end{cases} \tag{5}$$

where $Supp(\cdot)$ is the frequency support, and $Conf(X \rightarrow Y)$ indicates the conditional probability of X's occurrence implies Y's occurrence. The support can be considered as a global measure of being interesting, and the confidence is used as a localization measure.

Based on the association rule, in this paper we regard X and Y as associated patterns.

Definition (Associated Patterns). Let $\mathcal{I} = \{w_1, w_2, ..., w_n\}$ be a set of items, pattern X and Y are associated patterns, if: (1) $X \subseteq \mathcal{I}$, $Y \subseteq \mathcal{I}$, $X \bigcap Y = \emptyset$; (2) $Supp(X) \geq T_s$, $Supp(Y) \geq T_s$; (3) $Conf(X \rightarrow Y) \geq T_c$.

In an associated patterns pair, we denote the antecedent pattern as pat_A and the consequent pattern as pat_C. For example, in Fig. 1, *"program language"* and *"algorithm"* are regard as associated patterns, where $pat_A = $ *"program language"* and $pat_C = $ *"algorithm"*.

After generating associated patterns, we need to align target word with its associated words for training the model. We denote $\mathcal{PAT}(w_i)$ as the pattern set associated with word w_i. According to associated patterns' definition, there are two roles of pattern, antecedent and consequent. Therefore, we also define $\mathcal{PAT}(w_i)$ contains two subsets, word w_i's antecedent set $\mathcal{PAT}^A(w_i)$ and w_i's consequence set $\mathcal{PAT}^C(w_i)$. For each associated patterns, if pat_C contains w_i, we select pat_A as one of w_i associated patterns in $\mathcal{PAT}^A(w_i)$. While if pat_A contains w_i, we select pat_C as one of w_i associated patterns in $\mathcal{PAT}^C(w_i)$. Hence, $\mathcal{PAT}(w_i)$ contains two subsets, $\mathcal{PAT}^A(w_i)$ is the pattern collection to predict word w_i; $\mathcal{PAT}^C(w_i)$ is the pattern collection which can be predicted by w_i. In our method, different subsets are used for training different models. The details are described as following section.

3   Model

In this section, we describe the details of the proposed associated patterns enhanced word embedding (APWE) method. Follow basic structures of generating word embedding of the Word2Vec model, we propose our novel APWE model in terms of CBOW model and Skip-gram model, and theoretically demonstrate the flexibility and applicability of the integrating approach. At last, we describe how to apply the proposed embedding method for query expansion.

3.1   Pattern Enhanced Word Embedding Model

In this subsection, we detail the approach of generating newly word embedding with associated patterns. The basic idea is that associated patterns encode the words co-occurrence information in sentence level, from which we can extract hidden relations in long distance. We refine the word embedding model according to both local context and additionally long distance patterns, in order to consider word similarity and relatedness at the same time.

Model 1 (Associated Patterns Enhanced CBOW Model). In this model, we use both context and associated patterns to predict the target word. The associated patterns set is generated with the aforementioned rules, and we choose $\mathcal{PAT}^A(w_i)$ subset (we defined it in Sect. 2.2) to predict target word w_i.

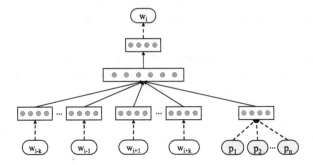

Fig. 2. Associated patterns enhanced CBOW model, we denote as A-APWE.

We named this mode as antecedent associated pattern enhanced word embedding models, and we use A-APWE model for easy of reference.

The model architecture is shown in Fig. 2, it contains two prediction tasks. The former prediction task captures word relations in local level, since words with similar context tend to have similar representation. The latter prediction task capture words association in long distance, since words frequently co-occure in the sentence tend to have similar representation. To joint context and associated patterns in the model, we encode the associated patterns using the following objection function

$$L(\mathcal{PAT}^A(w_i)) = \sum_{pat_j \in \mathcal{PAT}^A(w_i)} \log Pr(w_i \mid pat_j). \tag{6}$$

In this function, we present a pattern using its terms average vector, and $Pr(w_i \mid pat_j)$ is also a softmax funciton

$$Pr(w_i \mid pat_j) = \frac{\exp(\mathbf{x}_i \cdot \mathbf{x}_{pat_j})}{\sum_{w \in \mathcal{W}} \exp(\mathbf{x} \cdot \mathbf{x}_{pat_j})}. \tag{7}$$

Combining Eq. 6 with existing CBOW model (Eq. 1), we obtained the following objective function,

$$L = \frac{1}{T} \sum_{i=1}^{T} (\log Pr(w_i \mid w_{i-k}, ..., w_{i+k}) + \alpha \sum_{n=1}^{N} \log Pr(w_i \mid pat_n^{w_i})), \tag{8}$$

where α is the combination coefficient. We follow the similar optimization scheme as CBOW model and adopt the negative sampling technique for learning A-APWE model. The training objective function is defined as

$$\ell = \sum_{i=1}^{T} (log\sigma(\mathbf{w_i} \cdot \mathbf{w_c}) + k \cdot \mathcal{N}(w' \sim w_i) \cdot log\sigma(\mathbf{w'} \cdot \mathbf{w_c})$$

$$+ \sum_{n=1}^{N} (log\sigma(\mathbf{w_i} \cdot \mathbf{pat_n}) + k \cdot \mathcal{N}(w' \sim w_i) \cdot log\sigma(\mathbf{w'} \cdot \mathbf{pat_n}))), \tag{9}$$

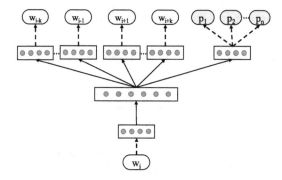

Fig. 3. Associated patterns enhanced Skip-gram model, we denote it as C-APWE

where $\sigma(x) = 1/(1 + exp(-x))$ and k is the number of negative samples. $\mathcal{N}(w' \sim w_i)$ denotes the sampled word collection of word w_i, and w' represents one sampled word. We use stochastic gradient descent (SGD) for optimization, and gradients are calculated using the back propagation neural networks.

Model 2 (Associated Patterns Enhanced Skip-gram Model). We use the target word to predict context and its associated patterns in this model. Similar with model 1, we need to mine the pattern collection with the aforementioned rules firstly. However, in this model, the basic idea is using target word to predict patternss. Therefore, for target word w_i, we select $\mathcal{PAT}^C(w_i)$ (we defined it in Sect. 2.2) as its associated patterns set. We denote this model as consequent associated patterns enhanced word embedding model, C-APWE model for short. Figure 3 shows the model architecture. The target word is used to predict its context, as well as its associated patterns. Formally, we encode the associated patterns using the following objection function

$$L(\mathcal{PAT}^C(w_i)) = \sum_{pat_j \in \mathcal{PAT}^C(w_i)} \sum_{w_j \in pat_j} \log Pr(w_j \mid w_i), \qquad (10)$$

where $Pr(pat_j \mid w_i)$ is also a softmax function

$$Pr(w_j \mid w_i) = \frac{\exp(\mathbf{x}_j \cdot \mathbf{x}_{w_i})}{\sum_{w \in \mathcal{W}} \exp(\mathbf{x} \cdot \mathbf{x}_{w_i})}. \qquad (11)$$

Combining Eq. 10 with existing Skip-gram model (Eq. 3), we obtained the following objective function

$$L = \frac{1}{T} \sum_{i=1}^{T} (\sum_{-k \leq c \leq k, c \neq 0} \log Pr(w_{i+c} \mid w_i) + \beta \sum_{n=1}^{N} \log Pr(pat_n^{w_i} \mid w_i)), \qquad (12)$$

where β is the combination coefficient. When it comes to optimization, C-APWE model adopts negative sampling method and follows the similar optimization scheme as Skip-gram model. The target objection can be represented as

$$\ell = \sum_{i=1}^{T} (\sum_{-k \leq c \leq k, c \neq 0} (log\sigma(\mathbf{w_{i+c}} \cdot \mathbf{w_i}) + k \cdot \mathcal{N}(w' \sim w_{i+c}) \cdot log\sigma(\mathbf{w'} \cdot \mathbf{w_i})$$

$$+ \sum_{n=1}^{N} \sum_{w_p \in pat_n} (log\sigma(\mathbf{w_p} \cdot \mathbf{w_i}) + k \cdot \mathcal{N}(w' \sim w_p) \cdot log\sigma(\mathbf{w'} \cdot \mathbf{w_i}))), \quad (13)$$

where k is the number of negative samples, $\mathcal{N}(w' \sim w_{i+c})$ denotes the negative samples collection of context word w_{i+c} and $\mathcal{N}(w' \sim w_p)$ denotes the negative samples collection of word w_p in pattern pat_n. We use SGD for optimization, and gradients are calculated using the back propagation neural networks.

3.2   Query Expansion for Document Retrieval

In information filtering tasks, document queries are formulated using several terms. Term-matching retrieval functions could fail at retrieving relevant documents if they cannot judge word semantic similarity, which also known as lexicon gap problem. Expanding queries based on words semantic meaning could enhance the likelihood of retrieving relevant documents. In this subsection, we detail how our proposed methods are used in query expansion.

Specifically, we obtained the word embedding with the proposed APWE models. Then, given a query q, we construct expansion set $Q^+ = \{q_1^+, ..., q_n^+\}$ by selecting top n most similar words with the cosine similarity. Each term q^+ is associated with a term weight according its cosine distance to query q. Then the final expansion of query q is represented as

$$Q = q \bigcup Q^+. \quad (14)$$

Formally, in the information ranking task, we define the computation method on Q as

$$f(Q) = f(q) \cdot (1 + \gamma \sum_{q_i^+ \in Q^+} \cos(v_{q_i^+}, v_q)), \quad (15)$$

where γ is the combination coefficient, and $f(q)$ is the origin function on query term in the retrieval model (i.e. word frequency, vector representation).

4   Experiments

In this section, we present experiments to evaluate the performance of our method in document filtering task. We discuss the experiments and evaluation in terms of dataset, baseline models and setting, measures and results. The results show that query expansion based on our APWE method significantly outperforms the-state-of-the-arts models in terms of effectiveness.

4.1 Dataset

To evaluate the performance of the proposed method with existing different baseline approaches, we conducted our experiments using the Reuters Corpus Volume 1 (RCV1) dataset, which is widely used in document ranking task.

Table 1. Statistic of RCV1 Dataset

# Documents	Corpus	Vocabulary size	# Sentences	# Associated patterns
806,791	70.1 M	111,257	20,300	88,564

In RCV1 dataset, there are a total of 806,791 documents that cover a variety of topics and a large amount of information. These documents are divided into 100 collections in total, and each collection is divided into a testing set and a training set. The first 50 collections were composed by human assessors and another 50 collections were constructed artificially from intersections collections. In this paper, only the first 50 collections used for experiments. Each document contains 'title' and 'text', and these parts were used by all the models in the experiments. We then tokenized all text in the dataset with the help of Stanford tokenizer tool and we at last converted every word into lower case.

To train word embedding, we combined all documents in RCV1 dataset as the training corpus, which includes 16 million words. In this paper, we mined the association patterns in the sentence level. We hence segmented 'text' of positive labeled documents into sentences and combined all sentences together for association rules mining. In total, we generated 88,564 associated patterns. More statistic of the dataset is given in Table 1.

4.2 Measures

In order to evaluate the performance, we apply four standard evaluation metrics: average precision of the top 10 documents ($P@10$), $F1$ measure, Mean Average Precision (MAP), break-even point (b/p). The precision is the proportion of labeled documents identified by the model which are correct. The recall is the proportion of labeled documents in the result which are correctly identified by the model. $F1$ measure is a criterion that assesses the effect involving both precision (p) and recall (r), which is defined as $F1 = \frac{2pr}{p+r}$. The larger the $Percision@10$, MAP, b/p, $F1$ score, the better the system performs.

4.3 Baseline Models and Settings

We choose BM25 [22], Word2Vec, Topical Word Embedding [14] (TWE), and GloVe [21] as baseline methods. The underlying idea of using these methods are highlighted below.

BM25 is one of the state-of-the-arts term-based document ranking approach. The term weights are estimated using the following equation:

$$W(t) = \frac{tf \cdot (k+1)}{k \cdot ((1-b) + b \cdot \frac{dl}{avgdl})} \cdot \log(\frac{N - n + 0.5}{n + 0.5}),\tag{16}$$

where N is the number of documents in the collection; n is the number of documents which contain term t; tf represents term frequency; dl is the document length and $avdl$ is the average document length.

Word2Vec, as mention above, is the-state-of-the-arts model which captures word relations based on local word co-occurrence information.

TWE method employs the widely used latent dirichlet allocation (LDA) [5] to refine Skip-gram model. TWE model can employ topic models to take advantages of all words as well as their context together to learn topical word embeddings.

GloVe method is an weighted least squares regression model that performs global matrix factorization with a local context window models. GloVe model leverages statistical information by training only on the nonzero elements in a word to word co-occurrence matrix, rather than on the entire sparse matrix or on individual context windows in a large corpus.

As to the document filtering task, for each collection, we generated document queries according to term's BM25 weight. We selected top 10 terms as collection's queries and expanded them using different word embedding methods. There are

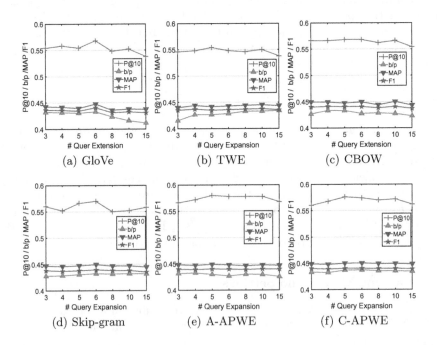

Fig. 4. The size of query expansion.

two parameters in BM25 method, b and k. Following the experimental setting in [22], k is set to 1.2 and b is set to 0.75. Another essential parameter is the size of Q^+ in Eq. 14, which indicates how many associated terms were used to expand the query. We carefully tuned this parameter for different word embedding methods in terms of $P@10$, b/p, MAP and $F1$. As Fig. 4 shows, the results with 5 expanding range in our method achieves the best performance for this particular dataset. Therefore, we set number of expanding terms as 5 in APWE method. We also carefully tuned the size of Q^+ in other baseline methods, and seleted 6,5,5 and 6 for GloVe, TWE, CBOW and Skip-gram mehtod respectively. Word embedding size is 300 in all these methods.

Besides, in our APWE models, there are four parameters, T_s, T_c, α and β. The first two are the threshold of *support* and *confidence* respectively, used in associated patterns mining. The last two are used for combination adjustment in different APWE models. We experiential set the T_s to 5 and T_c to 0.7 to take the best balance between accuracy and comprehension for extracting word associations. In order to guarantee the quality of word's associated pattern set, we select no more than ten words regarding to pattern's support for $\mathcal{PAT}^A$ and $\mathcal{PAT}^C$. The parameter α used in A-APWE model and β used in C-APWE model control the contribution of pattern information in the training process. Therefore, we do an experiment on the validation dataset to determine the best α and β in terms of $P@10$, b/p, MAP and $F1$. We carefully tuned α and β from 0.1 to 1.0, with the step size of 0.1, the parameters corresponding to the best $P@10$ value are used to report the final result. As the result shown in Fig. 5, we observed that both of our models reached the optimal performance when combination weight is 0.6. We find that the performance is relatively stable around the optimal parameter.

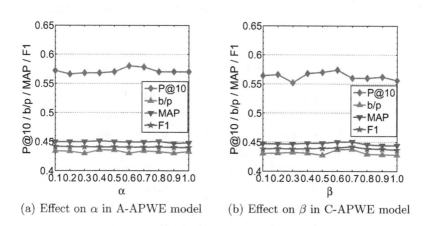

(a) Effect on α in A-APWE model   (b) Effect on β in C-APWE model

Fig. 5. Combination parameters.

4.4   Overall Performances

Tabel 2 shows the document ranking performances comparison between our model and baselines. From the table, we can observe that: (1) Using continuous word embedding as a foundation of expanding queries is an effective way for retrieving documents. (2) The proposed A-APWE model, which is updated according to CBOW model, is outperforming CBOW based Word2Vec model, and the proposed C-APWE model, which is updated according to Skip-gram model, is better than Skip-gram based Word2Vec model, respectively. From the both results of APWE models, we can imply that combing patterns is a flexible and applicable way to capture long distance word relationships rather than n-gram assumption. It also demonstrates that the association patterns are effectively boosting to capture semantical similarities, especially for document retrieval modelling. (3) The APWE models are consistantly superior to the GloVe and the TWE model. As aforementioned, the TWE model improved Word2Vec model by integrating topic assignment for generating word vectors, while the GloVe model considered global information for local calculations. However, compared with our proposed model, none of them take into consideration of those intensive and trustful words associations that are often distributed in a long range but actually semantically related. That is why our models are outperforming those state-of-the-arts high-quality word embeddings for document modelling.

We also find that the percentage of the A-APWE model outperforming CBOW model is higher than the counterpart percentage of C-APWE to Skip-gram model. Word prediction upon patterns and surrounding context is under more trustful and sufficient condition. However, conversely utilising single word is not that convincing to predict accurate patterns. Hence, A-APWE achieved the best performance. And associated patterns' quality promotion can further enhance models' performance.

Table 2. Overall performances

Methods	P@10	b/p	MAP	F1
BM25	0.446	0.406	0.408	0.415
GloVe+QE	0.562	0.434	0.449	0.440
TWE+QE	0.554	0.427	0.442	0.435
CBOW+QE	0.568	0.433	0.448	0.440
Skip-gram+QE	0.570	0.432	0.449	0.440
A-APWE+QE	**0.580**	0.430	0.449	**0.440**
C-APWE+QE	0.576	**0.437**	**0.450**	**0.440**

4.5   Case Study

A case study study was conducted to analyze in-depth reason why the APWE models surpass conventional candidate selection methods. Several examples are listed in Table 3. We present the most similar 5 words with cosine distance to a given target word under different word embedding approaches.

Two facts mainly accounted for the failure of traditional word embedding methods, according to the observation in our datasets. The first is that, in view of word similarity, unrelated words are selected in baseline methods, i.e. in GloVe method, *horizons* was regarded as a similar word to *computer*; in Skip-gram method, *credibity* was regarded as a similar word to *arms*, which is either unreasonable. Another problem is that, language semantic is comlicated which is formed by different granularities when expressing a general or specific level of semantic meaning. But the baseline models can hardly capture the structural levels simply by similarities. For example, *javastation* is more specific than *computer* at abstraction level, yet they are in a same list of similar words in the Skip-gram model, and same examples happen in the CBOW and the GloVe and so forth. From the results in Table 3, we can find that similar words discovered by our proposed models are mostly distributed at a same abstraction level, which can keep coherent semantics.

As can be seen, the most similar words of APWE methods and baseline models are in common, i.e. the most similar word for target word *arms* is *weapons* in all methods. This demonstrates that our models also take the important context information into consideration, and the wisely joint surrounding context and associative patterns while learning word embeddings.

Table 3. Target words and their 5 most similar words under different models.

Computer					
GloVe	assoc	navio	dell	laptop	horizons
TWE	software	computers	networking	macintosh	handheld
CBOW	motherson	audio	mediwar	disk	comint
Skip-gram	software	computers	legent	playback	javastation
A-APWE	computers	software	computing	pc	workstations
C-APWE	software	computers	internet	embed	hpcs
Arms					
GloVe	weapons	cache	decommissioning	morgane	proliferation
TWE	weapons	baghdads	arsenal	cooperates	nikita
CBOW	weapons	unsafeguarded	missile	weaponry	airacraft
Skip-gram	weapons	credibity	ballastic	churns	dustruction
A-APWE	weapons	missile	iraq	weaponry	haemorrhaging
C-APWE	weapons	ballastic	bilogical	churns	iraq

5   Related Work

Representing words using fixed-length vectors is an essential step in text processing tasks. In the early stage, one-hot representations have been widely used for its simplicity and efficiency. However, this traditional representation method suffers from data sparsity, the curse of dimensions and lexical gap, which make NLP and IR tasks difficult to use. Distributed word representation, also known as word embedding, is then introduced to solve these problems. In this method, words are represented as dense, low-dimensional, real-valued vectors, and each dimension represents latent semantic and syntactic features of words.

Recently, there is a surge of works focusing on Neural Network (NN) algorithms for word representations learning (Bengio et al. [4]; Mnih and Hinton [17]; Collobert et al. [8]; Mikolov et al. [15,16]; Mnih and Kavukcuoglu [18]; Lebret and Collobert [11]; Pennington et al. [21]). Most of these methods hold the assumption that words with similar context tend to have similar meanings.

Several researches focus on generating word representation beyond the context-based assumption, considering deeper relationships between linguistic items [13]. The Strudel system [2] represented a word using the clusters of lexical-syntactic patterns in which it occurs. Murphy et al. [19] represented words through their co-occurrence with other words in syntactic dependency relations, and then used the Non-Negative Sparse Embedding (NNSE) method to reduce the dimension of the resulted representation. Levy and Goldberg [12] extended the Skip-gram Word2Vec model with negative sampling [?] by basing the word co-occurrence window on the dependency parse tree of the sentence.

There are also several works have been done to use patterns enhanced performance of word embedding. Pattern has been suggested to be useful to capture word co-occurred relations in word representation [26]. Bollegala et al. [6] replaced bag-of-words contexts with various patterns (lexical, POS and dependency). Roy Schwartz et al. [24] proposed a symmetric pattern based approach to word representation which is particularly suitable for capturing word similarity. These works show the effectiveness of patterns in word representations. As mentioned above, local context also has shown its effective in capture word relatedness. To improve word representation, we consider both context and associated pattern, and leverage pattern associations for word embedding model.

6   Conclusion

This paper presents an innovative associated patterns enhanced word embedding method for covering word associations both in local and long distance. The proposed APWE method employs associated patterns to model associated words which occur in complex sentences without adjacency rules. APWE models extend the Word2Vec method by both using context information and associated patterns to capture word relatedness and semantics. Our method has been evaluated by using RCV1 for the task of document filtering. In comparison with the-state-of-the-arts models, the proposed models demonstrate excellent strength on query expansion and document ranking. As part of our future work, we will focus our efforts on dealing with those unseen words in large amount of data, since the APWE models in this stage can not well deal with the problem of the data size arises. We plan to collaborate conceptual assistance with the word associations for modelling unseen words and try to improve the robustness of the APWE models in the future.

Acknowledgments. The work was supported by National Nature Science Foundation of China (Grant No. 61132009), National Basic Research Program of China (973 Program, Grant No. 2013CB329303).

References

1. Agrawal, R., Srikant, R.: Fast algorithms for mining association rules in large databases. In: VLDB 1994, pp. 487–499 (1994)
2. Baroni, M., Murphy, B., Barbu, E., Poesio, M.: Strudel: a corpus-based semantic model based on properties and types. Cogn. Sci. **34**(2), 222–254 (2010)
3. Bastide, Y., Taouil, R., Pasquier, N., Stumme, G., Lakhal, L.: Mining frequent patterns with counting inference. SIGKDD **2**(2), 66–75 (2000)
4. Bengio, Y., Ducharme, R., Vincent, P., Janvin, C.: A neural probabilistic language model. JMLR **3**, 1137–1155 (2003)
5. Blei, D.M., Ng, A.Y., Jordan, M.I.: Latent dirichlet allocation. JMLR **3**, 993–1022 (2003)
6. Bollegala, D., Maehara, T., Yoshida, Y., Kawarabayashi, K.: Learning word representations from relational graphs. In: AAAI 2015, pp. 2146–2152 (2015)
7. Cheng, H., Yan, X., Han, J., Hsu, C.: Discriminative frequent pattern analysis for effective classification. In: ICDE 2007, pp. 716–725 (2007)
8. Collobert, R., Weston, J., Bottou, L., Karlen, M., Kavukcuoglu, K., Kuksa, P.P.: Natural language processing (almost) from scratch. JMLR **12**, 2493–2537 (2011)
9. Gao, Y., Xu, Y., Li, Y.: Pattern-based topics for document modelling in information filtering. TKDE **27**(6), 1629–1642 (2015)
10. Iacobacci, I., Pilehvar, M.T., Navigli, R.: Sensembed: learning sense embeddings for word and relational similarity. In: ACL 2015, pp. 95–105 (2015)
11. Lebret, R., Collobert, R.: Word embeddings through hellinger PCA. In: EACL 2014, pp. 482–490 (2014)
12. Levy, O., Goldberg, Y.: Dependency-based word embeddings. In: ACL 2014, pp. 302–308 (2014)
13. Li, J., Li, J., Fu, X., Masud, M.A., Huang, J.Z.: Learning distributed word representation with multi-contextual mixed embedding. KBS **106**, 220–230 (2016)
14. Liu, Y., Liu, Z., Chua, T., Sun, M.: Topical word embeddings. In: AAAI 2015, pp. 2418–2424 (2015)
15. Mikolov, T., Chen, K., Corrado, G., Dean, J.: Efficient estimation of word representations in vector space. CoRR abs/1301.3781 (2013)
16. Mikolov, T., Sutskever, I., Chen, K., Corrado, G.S., Dean, J.: Distributed representations of words and phrases and their compositionality. In: NIPS 2013, pp. 3111–3119 (2013)
17. Mnih, A., Hinton, G.E.: A scalable hierarchical distributed language model. In: NIPS 2008, pp. 1081–1088 (2008)

18. Mnih, A., Kavukcuoglu, K.: Learning word embeddings efficiently with noise-contrastive estimation. In: NIPS 2013, pp. 2265–2273 (2013)
19. Murphy, B., Talukdar, P.P., Mitchell, T.M.: Learning effective and interpretable semantic models using non-negative sparse embedding. In: COLING 2012, pp. 1933–1950 (2012)
20. Nam, J., Loza Mencía, E., Fürnkranz, J.: All-in text: learning document, label, and word representations jointly. In: AAAI 2016, pp. 1948–1954 (2016)
21. Pennington, J., Socher, R., Manning, C.D.: Glove: global vectors for word representation. In: EMNLP 2014, pp. 1532–1543 (2014)
22. Robertson, S.E., Zaragoza, H., Taylor, M.J.: Simple BM25 extension to multiple weighted fields. In: CIKM 2004, pp. 42–49 (2004)
23. Rothe, S., Schütze, H.: Autoextend: extending word embeddings to embeddings for synsets and lexemes. In: ACL 2015, pp. 1793–1803 (2015)
24. Schwartz, R., Reichart, R., Rappoport, A.: Symmetric pattern based word embeddings for improved word similarity prediction. In: CoNLL 2015, pp. 258–267 (2015)
25. Sun, F., Guo, J., Lan, Y., Xu, J., Cheng, X.: Learning word representations by jointly modeling syntagmatic and paradigmatic relations. In: ACL 2015, pp. 136–145 (2015)
26. Turney, P.D., Pantel, P.: From frequency to meaning: vector space models of semantics. JAIR 37, 141–188 (2010)
27. Vaidya, J., Clifton, C.: Privacy preserving association rule mining in vertically partitioned data. In: KDD 2002, pp. 639–644 (2002)

Multi-Granularity Neural Sentence Model for Measuring Short Text Similarity

Jiangping Huang[1], Shuxin Yao[2], Chen Lyu[1], and Donghong Ji[1(✉)]

[1] Computer School, Wuhan University, Wuhan 430072, China
{hjp,lvchen1989,dhji}@whu.edu.cn
[2] Language Technologies Institute, Carnegie Mellon University,
Pittsburgh, PA 15213, USA
shuxiny@cs.cmu.edu

Abstract. Measuring the semantic similarities between short texts is a critical and fundamental task because it is the basis for many applications. Although existing methods have explored this problem through enriching the short text representations based on the pre-trained word embeddings, the performance is still far from satisfaction because of the limited feature information. In this paper, we present an effective approach that combines convolutional neural network and long short-term memory to exploit from character-level to sentence-level features for performing the semantic matching of short texts. The proposed approach nicely models the feature information of sentences with the multiple representations and captures the rich matching patterns at different levels. Our model is rather generic and can hence be applied to matching tasks in different language. We use both paraphrase identification and semantic similarity tasks for evaluating our approach. The experimental results demonstrate that the proposed multiple-granularity neural sentence model obtains a significant improvement on measuring short texts similarity compared with the existing benchmark approaches.

Keywords: Neural sentence model · Semantic similarity · Paraphrase identification · Convolutional neural network · Long short-term memory · Embedding

1 Introduction

Modeling short texts for measuring the semantic relatedness is a fundamental problem in natual language processing applications like paraphrase identification [12,30], sentence completion [23], automatic summarization [1] and question answering [22]. Taking paraphrase recognition as an example, given a pair of sentences, a measuring function is required to determine the matching degree between these two sentences. Figure 1 shows the schematic illustration of paraphrase identification on two short texts. Existing approaches have been proposed for semantic similarity that use lexical matching and linguistic analysis, next to semantic features. Methods for lexical matching aim to determine whether the

© Springer International Publishing AG 2017
S. Candan et al. (Eds.): DASFAA 2017, Part I, LNCS 10177, pp. 439–455, 2017.
DOI: 10.1007/978-3-319-55753-3_28

words or string in two short text look alike such as in terms of edit distance. The main problem when using edit distance is the lack of robust in morphologically rich languages or domains with dynamic vocabularies (e.g. social media). For example, *fruit* would be closer to *bruit* in this way, than it would be to *fruitful*. To better represent morphological variants and rare words, some work proposed to use character-based compositional models for encoding arbitrary character sequences into vectors [13,31], which has potential benefit for lexical matching. Features based on linguistic analysis, like dependency parses or syntactic trees, are often used for short text similarity [26]. Linguistic tools such as parsers are commonly available these days for many languages, though the quality might vary between languages. However, high-quality parses might be expensive to compute at run time.

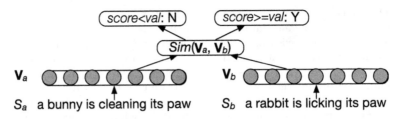

Fig. 1. The illustration of paraphrase identification. The short texts S_a and S_b are represented with vectors $\mathbf{V}_a$ and $\mathbf{V}_b$ respectively, then similarity layer $Sim(\mathbf{V}_a, \mathbf{V}_b)$ calculates the matching value *score*, when *score* is less than the threshold *val*, it should be non-paraphrase **N**, vice versa. Notice that in this work, we focus on sentence-level short texts, so we do not distinguish the difference of short texts and sentences.

Many approaches use external sources of structured semantic knowledge such as Wikipedia or WordNet for semantic features [18,25]. Wikipedia is structured around entities and as such is primarily of avail in settings where a focus on rather well-known persons and organizations can be assumed, such as news articles. However, such an assumption cannot always hold, and a drawback of using dictionaries or WordNet is that high-quality resources like these are not available for all languages, and proper names, domain-specific technical terms and slang tend to be underrepresented. In this work, we aim to make as few assumptions as possible and to propose a generic model, which requires no prior knowledge of natural language and no external resources of structured semantic information. Recent developments in distributional semantics, in particular neural network-based approaches only require a large amount of unlabelled text data, and this data is used to create a semantic space. Words are represented in the space as vectors that are called word embeddings. The geometric properties of this space prove to be semantically and syntactically meaningful, that is, words that are semantically or syntactically similar tend to be close in the semantic space.

There are two challenges for the semantic representation of short text. On the one hand, although word embedding can capture contextual semantic information and syntactic structure features for short text representation, there is an undoubted fact that numerous short texts contain out-of-vocabulary and non-standard words. Due to the length limited of short texts, when several these words exist in short text, the semantic representation for these texts is distorted and inaccurate. On the other hand, there is also a challenge for applying word embeddings to the task of determining semantic similarity of short texts, which is going from word level semantics to short text level semantics. Intuitively, it will be very helpful to further exploit the fine-grained features by utilizing subword information to solve the first challenge [31]. Meanwhile, because of the superior ability to preserve sequence information over time, long short-term memory (LSTM) with a more complex computational unit, has obtained strong results on a variety of sequence modeling tasks [16], so it is a natural choice to apply the LSTM model for representing short text level semantics based on word embeddings.

In this paper, we propose a novel neural network structure that combines convolutional neural network (CNN) and LSTM for short text semantic representation. Our model applies three neural network models for modeling different level features: the subword information for fine-grained features is captured with character-level CNN, the n-gram features are obtained via word-level CNN and the text-level semantic representation is achieved with LSTM, which is the variant of recurrent neural network [9] and can addresses the problem of exploding or vanishing gradients of learning long-term dependencies by introduction a memory cell that is able to perserve state over long periods of time. A novel feature of our approach is that an arbitary length short text can be represented with a fixed-length vector which is very important for understanding of short texts. The similarity layer calculates a score between the short text pairs which is used to judge whether two texts keep paraphrase relation or similarity. The proposed method is different from existing models [10,17,19] in that our approach can capture multiple-level feature information for short text representation.

The remainder of this paper is organized as follows. Section 2 presents an overview of the related work. The problem settings and our model is described in Sect. 3 and Sect. 4 respectively, Sect. 5 details our experimental setup and results. Finally, in Sect. 6 we conclude our final remarks.

2   Related Work

Existing methods have investigated the ways of automatically detecting semantic similarity on short text pairs, like newswire texts. For the Microsoft Research Paraphrase Corpus (MRPC) task, a complementary logistic regression model was proposed based on lexical overlap features [6], which chose a threshold for binary classification. Ji and Eisenstein [11] used matrix factorization techniques to obtain sentence representations, and combined them with fine-tuned sparse features using a SVM classifier for similarity prediction. Hu et al. [10] used convolutional neural network that combines hierarchical sentence modeling with layer-by-layer composition and pooling. Socher et al. [24] used a recursive neural network to model each sentence, which can recursively compute the representation for the sentence from the representations of its constituents in a binarized constituent parsing. Kim [15] proposed several modifications to the convolutional neural network architecture of Collobert and Weston [21], including the use of both fixed and learned word vectors and varying window sizes of the convolution filters. Yin and Schütze [30] also developed a convolutional neural network architecture for paraphrase identification. Tai et al. [26] and Zhu et al. [28] concurrently proposed a tree-based LSTM neural network architecture for sentence modeling. Kalchbrenner et al. [14] introduced a convolutional neural network for sentence modeling that uses dynamic k-max pooling to better model inputs of varying sizes.

Xu et al. [27] proposed a multi-instance learning paraphrase model for detecting paraphrase within the short messages in Twitter. In addition, Xu et al. [29] also proposed an evaluation on two related tasks, paraphrase identification and semantic textual similarity, for the Twitter data. They indicated that the very informal language, especially the high degree of lexical variation, used in social media has posed serious challenges to both tasks. The task is distinct from previous paraphrase identification dataset in three aspects. First of all, it contains sentence that are opinionated and colloquial, representing realistic informal language usage. Secondly, it contains paraphrases that are lexically diverse. And thirdly, it contains sentences that are lexically similar but semantically dissimilar. It raises many interesting research questions and could lead to a better understanding of our daily used language and how semantics can be captured in such language. Zarrella et al. [32] proposed an approach that explored various features including mixtures of string matching metrics, alignments using tweet-specific distributed word representations, and recurrent neural network is used for modeling similarity with those alignments and distance measurement on pooled latent semantic features, the effect of proposed method placed first in semantic similarity and second in paraphrase identification. Eyecioglu and Keller [7] proposed an approach that utilize features based on an overlap of word and character n-gram and train support vector machine to identify Twitter paraphrases, and the proposed method achieve the highest F-score. However, these methods depend on hand-craft feature engineering and the simple overlap of characters and words features.

In this work, we propose an approach for paraphrase identification and semantic similarity in short texts, such as tweets. The proposed method combined convolutional neural networks and long short-term memory for modeling multiple-granularity features. We use CNN to extract character-level and word-level features for obtaining the fine-grained features and semantic representation, which are proved to be very effective in existing methods [7,11]. In addition, we use LSTM for modeling sentence-level semantic representation, which keeping important contextual and syntactic features. Our method overcomes the shortcoming of single neural network model and adopts three kinds of models for capturing multiple-level features to represent the semantic vectors of short text pairs, which fine-grained feature information is very helpful for the semantic representation of short texts.

3   Problem Settings

In this section, we begin to describe the notation used in this work. Take the sentences S_a and S_b in Fig. 1. as a running example, so the input to the proposed model is a pair of texts, which denoted by S_a and S_b. Because the short texts are composed with words, thus the input S_a and S_b are two sequences of words $S_a = w_1, \cdots, w_m$ and $S_b = w_1, \cdots, w_n$ respectively. To transform words into real-valued feature vector (embeddings) that capture morphological, syntactic and semantic information about the words, we use a fixed-sized word vocabulary V^{word}, and we consider that words are composed of characters from a fixed-sized character vocabulary V^{char}. So for every word in S_a and S_b is converted into a vector $\mathbf{v} = [\mathbf{v}^{word}; \mathbf{v}^{char}]$, which is composed of two sub-vectors: the word-level embedding $\mathbf{v}^{word} \in \mathbb{R}^{d^{word}}$ and the character-level embedding $\mathbf{v}^{char} \in \mathbb{R}^{d^{char}}$. The word-level embeddings are applied to capture syntactic and semantic information, and character-level embeddings are used to capture morphological and shape information. When the short texts are represented, the proposed approach will further extract multiple-level feature information for short text representation based on word-level and character-level embeddings. In addition, the output of our model should be the score of semantic textual similarity, which can be used to detect the binary paraphrase relation between short text pairs.

Short Text Matrix. To capture and compose the features of individual word in a given text from low level word embeddings into higher level semantic concepts, the neural network should apply a series of transformations to the input short text matrix using convolution, non-linearity and pooling operations. To represent short text well, we adopt short text matrix for describing each input sentence S, which treated as a sequence of words: $[w_1, \cdots, w_{|S_w|}]$, where each word is drawn from a vocabulary V^{word}. Words are represented by distributional vectors $\mathbf{w} \in \mathbb{R}^{d^{word}}$ looked up in a word embeddings matrix $\mathbf{W} \in \mathbb{R}^{d^{word} \times |V^{word}|}$ which is formed by concatenating embeddings of all words in V^{word}. For convenience and ease of lookup operations in $\mathbf{W}$, words are mapped to integer indices $1, \cdots, |V^{word}|$. For each input short text S, we build a short text matrix $\mathbf{S}_w \in \mathbb{R}^{d^{word} \times |S_w|}$, where each column i represents a word embedding $\mathbf{w}_i$ at the corresponding position i in the short text S:

$$\mathbf{S}_w = \begin{bmatrix} | & | & \cdots & | & | \\ \mathbf{w}_1 & \mathbf{w}_2 & \cdots & \mathbf{w}_{|S_w|-1} & \mathbf{w}_{|S_w|} \\ | & | & \cdots & | & | \end{bmatrix}$$

As the same as word-based short text representation, we adopt character-level short text representation to obtain fine-grained feature information. For each word in input text S, which should be treated as a sequence of characters: $[c_1^{w_1}, c_2^{w_1}, \cdots, c_l^{w_k}, \cdots, c_{|S_c|-1}^{w_{|S_w|}}, c_{|S_c|}^{w_{|S_w|}}]$, where $c_l^{w_k}$ means the l-th character within word w_k in S and each character is drawn from the character vocabulary V^{char}. Characters are represented by distributional vectors $\mathbf{c} \in \mathbb{R}^{d^{char}}$ looked up in a character embeddings matrix $\mathbf{C} \in \mathbb{R}^{d^{char} \times |V^{char}|}$ which is formed by concatenating embeddings of all characters in V^{char}. As the same as words embeddings, for convenience and ease of lookup operations in $\mathbf{C}$, characters are mapped to integer indices $1, \cdots, |V^{char}|$. For each input text S we build a sentence matrix $\mathbf{S}_c \in \mathbb{R}^{d^{char} \times |S_c|}$, where each column j represents a character embedding $\mathbf{c}_j$ at the corresponding position j in the short text S:

$$\mathbf{S}_c = \begin{bmatrix} | & | & \cdots & | & | \\ \mathbf{c}_1^{w_1} & \mathbf{c}_2^{w_1} & \cdots & \mathbf{c}_{|S_c|-1}^{w_{|S_w|}} & \mathbf{c}_{|S_c|}^{w_{|S_w|}} \\ | & | & \cdots & | & | \end{bmatrix}$$

The character-level short text representation can capture the subword information of individual word in a given sentence from low level character embeddings into higher level morphological representation. It is noted that we ignored whitespace in $\mathbf{S}_c$.

Short Text Similarity. To calculate semantic similarity between two short texts, we used character-level CNN (charCNN), word-level CNN (wordCNN) and sentence-level LSTM (sentLSTM) models for modelling different-level short text representations, the output of these models are higher-level vector representations, which can be used to compute the semantic similarity score of short texts. The pair of short texts is associated with a scalar score from zero to one, then the pearson correlation coefficient should further denote the short texts similarity for the input short text pairs.

4 Multi-Granularity Neural Sentence Model

This section presents the architecture of multiple-granularity neural sentence model for short texts representation and similarity measurement based on the character-level and word-level sentence matrices. Our framework includes four parts: charCNN, wordCNN, sentLSTM and semantic similarity interaction tensor layer, as shown in Fig. 2. In our model, both character embeddings matrix and word embeddings matrix of sentences are the input of model, and a semantic similarity score is the output of the proposed model. We firstly depict the initial representation, which including word-level embeddings and character-level embeddings. Then we detail the neural sentence model from charCNN, word-CNN to sentLSTM. Finally, we describe the semantic similarity measurement layer, which measures the matching degree between two sentence representations. The network takes the sequence of words and the sequence of characters of the short text pairs as input, and passes them through a sequence of layers where features with increasing levels of complexity are extracted. The network extracts features from the character-level up to the sentence-level. The main novelty in our network architecture is the inclusion of two convolutional models and a long short-term memory model, which allows our model to copy with characters, words and sentences of any size for capturing multiple-level features information for better short text representation.

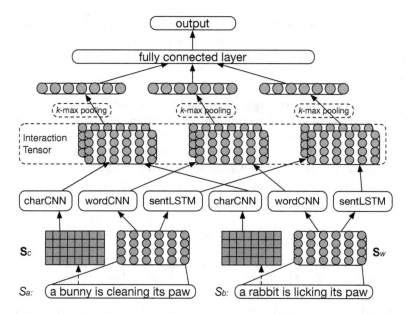

Fig. 2. The architecture of our model applied to short text pairs. Here the input layer of the network transforms words and characters in sentences S_a and S_b into character-level embeddings $\mathbf{S}_c$ (blue) and word-level embeddings $\mathbf{S}_w$ (green) respectively. Then the charCNN applies *wide* convolution operation on character-embedding matrix and wordCNN applies *narrow* convolution operation on word-embedding matrix, the sentLSTM also applies LSTM operation to capture semantic and syntactic features representation on word-embedding matrix. There are three layers convolution operation and k-max pooling inside charCNN and wordCNN models. A interaction tensor is applied for same models for obtaining similarity information. The k-max pooling extracts the k most activated features, and a fully connected layer is used to join the fixed-length vector representation. The matching score is finally outputted through a multi-layer perceptron. (Color figure online)

4.1   Initial Representation

The input layer of the network transforms words and characters into real-valued feature vectors (embeddings) that capture morphological, syntactic and semantic information about the words and fine-grained feature information about the characters respectively. The word-level embeddings and character-level embeddings for the input of network are described as follows.

Word-Level Embeddings. Word-level embeddings are encoded by column vectors in an embedding matrix $\mathbf{W}^{word} \in \mathbb{R}^{d^{word} \times |V^{word}|}$. Each column $\mathbf{W}_i^{word} \in \mathbb{R}^{d^{word}}$ corresponds to the word-level embedding of the i-th word in the vocabulary. We transform a word w_i into its word-level embedding $\mathbf{w}_i$ by using the matrix-vector product:

$$\mathbf{w}_i = \mathbf{W}^{word} u^i \tag{1}$$

where u^i is a vector of size $|V^{word}|$ which has value 1 at index i and zero in all other positions. The matrix $\mathbf{W}^{word}$ can be as parameters to be learnt, and the size of the word-level embedding d^{word} is a hyper-parameter to be chosen by the user.

Character-Level Embeddings. Due the limit of contextual length in short texts, it is essential to extract morphological and shape information from words robust and take into consideration all characters of the word. Given a word w composed of m characters $\{c_1, c_2, \cdots, c_m\}$, we firstly transform each character c_j into a character embedding $\mathbf{c}_j^{char}$. Character embeddings are encoded by column vectors in the embedding matrix $\mathbf{W}^{char} \in \mathbb{R}^{d^{char} \times |V^{char}|}$. Given a character c_j, its embedding $\mathbf{c}_j^{char}$ is obtained by the matrix-vector product:

$$\mathbf{c}_j^{char} = \mathbf{W}^{char} v^j \tag{2}$$

where v^j is a vector of size $|V^{char}|$ which has value 1 at index j and zero in all other positions. The input for the covolution layer is the sequence of character embeddings $\{\mathbf{c}_1^{char}, \mathbf{c}_2^{char}, \cdots, \mathbf{c}_m^{char}\}$. The convolution layer applies a matrix-vector operation to each window of size l^{char} of successive windows in $\{\mathbf{c}_1^{char}, \mathbf{c}_2^{char}, \cdots, \mathbf{c}_m^{char}\}$.

4.2   Neural Sentence Model

Various neural sentence models have been described. A general class of basic sentence models is that of neural bag-of-words models. These generally consist of a projection layer that maps words, sub-word unit or n-grams to high dimensional embeddings. The latter are then combined component-wise with an operation such as summation. The resulting combined vector is classified through one or more fully connected layers. In our work, we apply convolutional neural network and long short-term memory for modelling sentence, which can capture from character-level, word-level and sentence-level feature information.

Convolutional Neural Network. The aim of the convolutional layer is to extract patterns. In our model, we apply *wide* and *narrow* convolution for character and word embeddings respectively. More formally, the convolution operation $*$ between two vectors $\mathbf{m} \in \mathbb{R}^m$ and a vector of inputs viewed as a sequence $\mathbf{s} \in \mathbb{R}^s$. The vector $\mathbf{m}$ is the filter of the convolution. Concretely, we think of $\mathbf{s}$ as the input sentence and $\mathbf{s}_i \in \mathbb{R}$ is a single feature value associated with the i-th word in the sentence. The idea behind the one-dimensional convolution is to take the dot product of the vector $\mathbf{m}$ with each m-gram in the sentence $\mathbf{s}$ to obtain another sequence:

$$\mathbf{c}_j = \mathbf{m} * \mathbf{s}_{j-m+1:j} \tag{3}$$

The equation gives rise to two types of convolution depending on the range of the index j. The *narrow* type of convolution requires that $s \geq m$ and yields a sequence $\mathbf{c} \in \mathbb{R}^{s-m+1}$ with j ranging from m to s. The *wide* type of convolution does not have requirements on s or m and yields a sequence $\mathbf{c} \in \mathbb{R}^{s+m-1}$ where the index j ranges from 1 to $s + m$ - 1. Out-of-range input values $\mathbf{s}_i$ where i < 1 or $i > s$ are taken to be zero. The result of the narrow convolution is a subsequence of the result of the wide convolution. The trained weights in the filter $\mathbf{m}$ correspond to a linguistic feature detector that learns to recognize a specific class of n-grams. These n-grams have size $n \leq m$, where m is the width of the filter. Applying the weights $\mathbf{m}$ in a wide convolution has some advantages over applying them in a narrow one. A wide convolution ensures that all weights in the filter reach the entire sequence, including the words or characters at the margins. This is particularly significant when m is set to a relatively large value. In addition, a wide convolution guarantees that the application of the filter $\mathbf{m}$ to the input sentence $\mathbf{s}$ always produces a valid non-empty result $\mathbf{c}$, independently of the width m and the sentence length s. In our model, we apply *wide* convolution on character-level embeddings and *narrow* convolution on word-level embeddings.

The output from the convolutional layer are then passed to the pooling layer, whose goal is to aggregate the information and reduce the representation. There are three convolutional choices for the pool($\cdot$) operation: *average, max* and *min*. Each operation applies to columns of the feature map matrix, by mapping them to a single value. These pooling methods exhibit certain disadvantages: in *average* pooling operation, all elements of the input are considered, which may weaken strong activation values. This is especially critical with tanh non-linearity function, where strong positive and negative activations can cancel each other out. The *min* pooling chooses minimum in feature map matrix, which ignores the strong activation values. The *max* pooling is used more widely and dose not suffer from the drawbacks of *average* and *min* pooling. However, it can lead to strong overfitting on the training set and, hence poor generalization on the test data.

The *max* pooling has been generalized to k-max pooling, where instead of a single max value, k values are extracted in their original order. Given a value k and a sequence $\mathbf{p} \in \mathbb{R}^p$ of length $k \leq q$, k-max pooling selects the subsequence $\mathbf{p}_{max}^k$ of the k highest values of $\mathbf{p}$. The order of the values in $\mathbf{p}_{max}^k$ corresponds to their original order in $\mathbf{p}$. The k-max pooling operation makes it possible to pool the k most active features in $\mathbf{p}$ that may be a number of positions apart. It preserves the order of the features, but is insensitive to their specific positions. It can also discern more finely the number of times the feature is highly activated in $\mathbf{p}$ and the progression by which the high activations of the feature change across $\mathbf{p}$. The k-max pooling operator is applied in the convolutional network after the topmost convolutional layer. This guarantees that the input to the fully connected layers is independent of the length of the input sentence.

Long Short-Term Memory. A recurrent neural network is a type of neural network architecture particularly suited for modeling sequence phenomena. Given an input sequence $\mathbf{x} = (x_1, \cdots, x_T)$. At each time step t, the network takes the input vector $\mathbf{x}_t \in \mathbb{R}^n$ and the hidden state vector $\mathbf{h}_{t-1} \in \mathbb{R}^m$ and produces the next hidden state $\mathbf{h}_t$ by applying the following recursive operation:

$$\mathbf{h}_t = f(\mathbf{W}\mathbf{x}_t + \mathbf{U}\mathbf{h}_{t-1} + \mathbf{b}) \tag{4}$$

Here $\mathbf{W} \in \mathbb{R}^{m \times n}$, $\mathbf{U} \in \mathbb{R}^{m \times m}$, $\mathbf{b} \in \mathbb{R}^m$ are the parameters of an affine transformation and f is an element-wise nonlinearity function. Though, in theory, recurrent neural networks are capable to capturing long-distance dependencies, in practice, they fail due to the gradient vanishing or exploding problems.

Long short-term memory addresses the problem of learning long range dependencies by augmenting the recurrent neural network with a memory cell vector $\mathbf{c}_t \in \mathbb{R}^n$ at each time step. Basically, a LSTM unit is composed of three multiplicative gates which control the proportions of information to forget and to pass on the next time step t. Concretely, one step of LSTM takes an input $\mathbf{x}_t$, $\mathbf{h}_{t-1}$, $\mathbf{c}_{t-1}$ and produces $\mathbf{h}_t$, $\mathbf{c}_t$ via the following intermediate calculations:

$$\mathbf{i}_t = \sigma(\mathbf{W}_i \mathbf{x}_t + \mathbf{U}_i \mathbf{h}_{t-1} + \mathbf{b}_i) \tag{5}$$

$$\mathbf{f}_t = \sigma(\mathbf{W}_f \mathbf{h}_{t-1} + \mathbf{U}_f \mathbf{x}_t + \mathbf{b}_f) \tag{6}$$

$$\mathbf{o}_t = \sigma(\mathbf{W}_o \mathbf{h}_{t-1} + \mathbf{U}_o \mathbf{x}_t + \mathbf{b}_o) \tag{7}$$

$$\mathbf{g}_t = \tanh(\mathbf{W}_g \mathbf{h}_{t-1} + \mathbf{U}_g \mathbf{x}_t + \mathbf{b}_g) \tag{8}$$

$$\mathbf{c}_t = \mathbf{f}_t \odot \mathbf{c}_{t-1} + \mathbf{i}_t \odot \mathbf{g}_t \tag{9}$$

$$\mathbf{h}_t = \mathbf{o}_t \odot \tanh(\mathbf{c}_t) \tag{10}$$

Here $\sigma(\cdot)$ and $\tanh(\cdot)$ are the element-wise sigmoid and hyperbolic tangent functions, $\odot$ is the element-wise multiplication operator, and $\mathbf{i}_t$, $\mathbf{f}_t$, $\mathbf{o}_t$ are referred to as *input*, *forget*, and *output* gates. At $t = 1$, $\mathbf{h}_0$ and $\mathbf{c}_0$ are initialized to zero vectors. Parameters of the LSTM are $\mathbf{W}_j$, $\mathbf{U}_j$, $\mathbf{b}_j$ for $j \in \{i, f, o, g\}$. Memory cells in the LSTM are additive with respect to time, alleviating the gradient vanishing problem. Gradient exploding is still an issue, though in practice simple optimization strategies work well. LSTM has been shown to outperform vanilla recurrent neural network on many tasks, including on sentence representation [16] and semantic similarity [26]. It is easy to extend LSTM to more layers by having another network whose input at t is $\mathbf{h}_t$. Indeed, having multiple layers is often crucial for obtaining competitive performance on various tasks [5,28].

For the proposed sentLSTM model, let $\mathbf{S}_w$ be the words embeddings matrix for the sequence $w_i^1, \cdots, w_i^N$ in short text s_i, where N is the number of words in the short text. At each time step t, the sentLSTM produces a hidden state $\mathbf{h}_i^t$ which can be interpreted as the representation of the word sequence $w_i^1, \cdots, w_i^t$. The hidden state $\mathbf{h}_i^N$ thus represents the full sentence, which is a fixed-length vector.

4.3 Similarity

Many kinds of similarity functions can be used for modeling the interactions between p_{a_i} and p_{b_j}, where p_{a_i} and p_{b_j} stand for the i and j-th positional sentence representations for two sentences S_a and S_b, respectively. We use three similarity functions, including cosine, bilinear and tensor layer, for similarity measurement in the composition of the proposed three models, and single model and the combination of two models only use the interaction tensor measurement. Given two vectors u and v, the three functions will output the similarity score $sim(u, v)$. Cosine is a common function to model interactions, the similarity score is viewed as the angle of two vectors. The bilinear further considers the interactions between different dimensions, and thus can capture more complicated interactions compared with cosine, and bilinear can capture more meaningful interactions between two positional sentence representations compared with cosine as well. We describe the detail of interaction tensor as following.

Tensor interaction is more powerful than the above two functions, which can roll back to other similarity metrics such as bilinear and dot product. It has also shown great superiority in modeling interactions between two vectors [24]. That is why we choose it as an interaction function in this paper. Other than outputing a scalar value as bilinear and cosine do, tensor layer outputs a vector, as described as follows.

$$sim(u, v) = f(u^T M^{[1:c]} v + W_{uv}[uv]^T + b) \tag{11}$$

where M^i, $i \in [1,...,c]$ is one slice of the tensor parameters, W_{uv} and b are parameters of the linear part. f is a non-linear function, and we use rectifier $f(z) = \max(0,z)$ in this work, since it always outputs a positive value which is compatible as a similarity.

5   Experiments

In this section, we demonstrate our experiments on two semantic matching tasks, paraphrase identification and semantic textural similarity for the Twitter data (PIT) [29], which is used for evaluating tweets semantic similarity.

5.1   Setting

Dataset. The PIT corpus uses a training and development set of 17,790 sentence pairs and test set of 972 sentence pairs with paraphrase annotations. Table 1 shows the basic statistics of the corpus. The dataset is more realistic and balanced, containing about 70% non-paraphrase vs. the 34% non-paraphrases in the benchmark MRPC.

Baselines. We adopted seven benchmark methods compared with the proposed model, including recursive auto-encoders [24] and a referential machine translation method [4] for paraphrase identification, a standard two-layer neural networks model [3] for semantic similarity, and a supervised logistic regression [6], a weighted matrix factorization [8], MITRE System [32] and a support vector machine classifier [2] for both paraphrase identification and semantic similarity tasks.

Table 1. The statistics of PIT-2015 Twitter Paraphrase Corpus. Debatable cases are those received a medium-score from annotators, which are excluded in our paraphrase identification.

Dataset	Unique sent	Sent pair	Paraphrase	Non-Paraphrase	Debatable
Train	13231	13063	3996 (30.6%)	7534 (57.7%)	1533 (11.7%)
Dev	4772	4727	1470 (31.1%)	2672 (56.5%)	585 (12.4%)
Test	1295	972	175 (18.0%)	663 (68.2%)	134 (13.8%)

Parameter Settings. We conduct two related experiments with same parameters. In convolutional networks, the slide window m is chosen in the range [1, 5] and [1, 3] for *wide* and *narrow* convolution operation respectively. We use the $dim = 300$-dimensional GloVe word embeddings trained on 840 billion tokens [20]. Embeddings for words not presented in the GloVe model are randomly initialized with the uniform distribution $U[-0.1, 0.1]$. We also use $dim = 100$-dimensional random initial vector for character embeddings, which each component sampled from the uniform distribution $U[-0.25, 0.25]$. We perform optimization using stochastic gradient descent. The back propagation algorithm is used to compute gradients for all parameters during training. We fixed the learning rate to 0.01 and regularization parameter $\lambda = 0.0001$.

Evaluation Metric. For paraphrase identification, *Precision*, *Recall* and F_1 are used for evaluating the performance of proposed model, the F-measure $F_1 = 2 * Precision * Recall/(Precision+Recall)$. And *Pearson* correlation coefficient, *maxF1*, *mPrec* and *mRecall* are used for semantic textual similarity measurement.

5.2   Results

We report the experimental results of paraphrase identification and semantic similarity measurement in Twitter respectively. There are six methods as baselines in the paraphrase identification and five methods as baselines in semantic similarity task.

Paraphrase Identification. In the first experiment, we validate the effective of the proposed multi-granularity sentence model on paraphrase identification, which the model combined charCNN, wordCNN and sentLSTM. We used cosine, bilinear and interaction tensor similarity measurement methods in the proposed model. The *Precision*, *Recall* and F_1 of the baselines from prior work and our experimental results are reported in Table 2. As can be seen from Table 2, our model with interaction tensor method obtained the best performance than both cosine and bilinear. That's why we choose it as an interaction function in this paper. Other than outputing a scalar value as bilinear and cosine do, tensor interaction outputs a vector, as described in this work.

Table 2. Test set results on PIT for paraphrase identification.

Model	Precision(%)	Recall(%)	F_1(%)
Logistic Regeression [6]	67.9	52.0	58.9
Recursive Auto-Encoder [24]	54.3	39.4	45.7
Weighted Matrix Factorization [8]	45.0	66.3	53.6
Referential Machine Translation [4]	**85.9**	41.7	56.2
MITRE System [32]	56.9	**80.6**	66.7
Support Vector Machine [2]	68.0	66.9	**67.4**
Multi-Granularity Sentence Model (Cosine)	60.1	55.4	57.7
Multi-Granularity Sentence Model (Bilinear)	59.2	63.9	61.5
Multi-Granularity Sentence Model (Tensor)	64.3	65.7	65.0

We can observe that the performance of our model is lower than both support vector machines and MITRE system as shown in Table 2. Although support vector machine obtained the highest F_1-score, they used hand-craft feature engineering for paraphrase identification. Moreover, MITRE system also combined seven systems, including normalization and alignment, for the paraphrase detection task. Our multi-granularity sentence model, which only used pre-trained word embeddings, still obtained 65.0% F_1-score and surpassed other four benchmark methods as well. The experimental results indicate that the proposed multiple-level features extraction method is very helpful for the paraphrase identification in Twitter.

Semantic Similarity. In the second experiment, the performance of the proposed multi-granularity sentence model with three semantic similarity methods and five baselines are shown in Table 3. We can see that for semantic similarity task, the proposed model with tensor interaction still performed better than the cosine and bilinear similarity measurement methods. In addition, we observe that the proposed model outperforms other benchmark methods in *Pearson, maxF1* and *mRecall*.

The support vector machine obtained the best performance on paraphrase identification in Table 2, however, the pearson correlation coefficient of the approach is 0.475, which is much less than the performance of our model on pearson. Meanwhile, our method with tensor interaction also outperforms the MITRE system in *Pearson, maxF1* and *mRecall*, the system contains seven components for semantic similarity and our model only uses pre-trained word embeddings. It indicates that the proposed neural sentence model can capture multiple-granularity semantic information for short texts.

Table 3. Test set results on PIT for semantic similarity.

Model	Pearson	maxF1	mPrec	mRecall
Logistic Regeression [6]	0.511	0.601	0.674	0.543
Two-layer Neural Network [3]	0.545	0.669	0.738	0.661
Weighted Matrix Factorization [8]	0.350	0.587	0.570	0.606
MITRE System [32]	0.619	0.716	**0.750**	0.686
Support Vector Machine [2]	0.475	0.616	0.732	0.531
Multi-Granularity Sentence Model (Cosine)	0.583	0.687	0.701	0.673
Multi-Granularity Sentence Model (Bilinear)	0.611	0.702	0.727	0.684
Multi-Granularity Sentence Model (Tensor)	**0.626**	**0.720**	0.748	**0.694**

5.3   Analysis

We analyse the performance of different neural network methods in the proposed multi-granularity neural sentence model for paraphrase identification and semantic similarity as shown in Fig. 3. We observe that charCNN is more superior than wordCNN and sentLSTM on both paraphrase identification and semantic similarity. Although the effectiveness of the wordCNN does not seem significant compared with sentLSTM, the combination of charCNN and wordCNN obtained better performance than the other combinations with only two methods on paraphrase identification as shown in Fig. 3(a). For semantic similarity, the sentLSTM has a higher pearson correlation coefficient than wordCNN model as shown in Fig. 3(b). However, the combination of charCNN and sentLSTM outperformed the compositions of two models. The model with charCNN, wordCNN and sentLSTM obtained the best performance on two tasks. However, the effectiveness of charCNN is more obvious than wordCNN and sentLSTM. The experiments indicate that character-level feature information is essential for modeling tweets

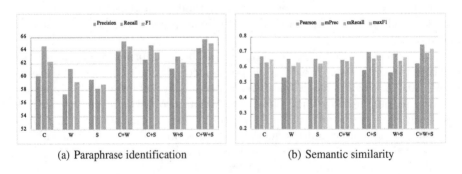

(a) Paraphrase identification   (b) Semantic similarity

Fig. 3. The performance of different models in the proposed approach for paraphrase identification and semantic similarity. C ← charCNN, W ← wordCNN, S ← sentLSTM. C+W ← the proposed multi-granularity sentence model with charCNN and wordCNN, and so on.

and the combination of both character-level and word-level features will obtain preferable performance on semantic similarity tasks. The experiments also indicate that our approach has more advantages when considering both tasks at the same time.

6 Conclusion

In this paper, we presented a novel approach for paraphrase identification and semantic similarity in short texts, such as tweets. Further, we showed how character-level, word-level and sentence-level features could be extracted for modeling sentence semantic representations. The character-level convolutional neural network has been used for extracting character n-gram features, and word-level CNN for word n-gram features. We also used long short-term memory for extracting sentence-level feature information, which contains more contextual knowledge. We used tweets for evaluating the proposed model on two semantic similarity tasks. The experimental results indicate that character-level features are effective in paraphrase detection and semantic similarity for tweets. We also can conclude that the composition of different feature extraction model can obtain better performance than single model, so the proposed multi-granularity neural sentence modeling is an efficient method for representing short texts. In future work, we will investigate some methods of the normalization of short texts in social media and amplify the applicability of our approach for other text representation tasks.

Acknowledgments. This work was supported by the National Natural Science Foundation of China under grant 61133012 and 61373108, and supported by Humanities and Social Science Foundation of Ministry of Education of China (16YJCZH004).

References

1. Aliguliyev, R.M.: A new sentence similarity measure and sentence based extractive technique for automatic text summarization. Expert Syst. Appl. **36**(4), 7764–7772 (2009)
2. Asli, E., Keller, B.: ASOBEK: twitter paraphrase identification with simple overlap features and SVMs. In: Proceedings of SemEval, pp. 64–69. ACL, Denver (2015)
3. Bertero, D., Fung, P.: HLTC-HKUST: a neural network paraphrase classifier using translation metrics, semantic roles and lexical similarity features. In: Proceedings of SemEval, pp. 23–28. ACL (2015)
4. Bicici, E.: RTM-DCU: predicting semantic similarity with referential translation machines. In: Proceedings of SemEval, pp. 56–63. ACL (2015)
5. Byeon, W., Breuel, T.M., Raue, F., Liwicki, M.: Scene labeling with LSTM recurrent neural networks. In: 2015 IEEE Conference on CVPR, pp. 3547–3555 (2015)
6. Das, D., Smith, N.A.: Paraphrase identification as probabilistic quasi-synchronous recognition. In: Proceedings of ACL and IJCNLP, pp. 468–476. ACL, Suntec, Singapore (2009)
7. Eyecioglu, A., Keller, B.: Twitter paraphrase identification with simple overlap features and SVMs. In: Proceedings of SemEval, pp. 64–69. ACL (2015)

8. Guo, W., Diab, M.: Modeling sentences in the latent space. In: Proceedings of ACL, pp. 864–872. ACL, Jeju Island (2012)
9. Hochreiter, S., Schmidhuber, J.: Long short-term memory. Neural Comput. **9**(8), 1735–1780 (1997)
10. Hu, B., Lu, Z., Li, H., Chen, Q.: Convolutional neural network architectures for matching natural language sentences. In: Proceedings of NIPS, pp. 2042–2050. Curran Associates Inc. (2014)
11. Ji, Y., Eisenstein, J.: Discriminative improvements to distributional sentence similarity. In: Proceedings of EMNLP, pp. 891–896. ACL, Seattle (2013)
12. Huang, J., Ji, D., Yao, S., Huang, W.: Character-aware convolutional neural networks for paraphrase identification. In: Hirose, A., Ozawa, S., Doya, K., Ikeda, K., Lee, M., Liu, D. (eds.) ICONIP 2016. LNCS, vol. 9948, pp. 177–184. Springer, Cham (2016). doi:10.1007/978-3-319-46672-9_21
13. Huang, J., Ji, D., Yao, S., Huang, W., Chen, B.: Learning phrase representations based on word and character embeddings. In: Hirose, A., Ozawa, S., Doya, K., Ikeda, K., Lee, M., Liu, D. (eds.) ICONIP 2016. LNCS, vol. 9950, pp. 547–554. Springer, Cham (2016). doi:10.1007/978-3-319-46681-1_65
14. Kalchbrenner, N., Grefenstette, E., Blunsom, P.: A convolutional neural network for modelling sentences. In: Proceedings of ACL, pp. 655–665. ACL, Baltimore (2014)
15. Kim, Y.: Convolutional neural networks for sentence classification. In: Proceedings of EMNLP, pp. 1746–1751. ACL, Doha (2014)
16. Li, J., Luong, T., Jurafsky, D.: A hierarchical neural autoencoder for paragraphs and documents. In: Proceedings of ACL, pp. 1106–1115. ACL, Beijing (2015)
17. Lyu, C., Lu, Y., Ji, D., Chen, B.: Deep learning for textual entailment recognition. In: Tools with Artificial Intelligence (ICTAI), pp. 154–161. IEEE (2015)
18. Mihalcea, R., Corley, C.D., Strapparava, C.: Corpus-based and knowledge-based measures of text semantic similarity. In: Proceedings of AAAI, pp. 775–780 (2006)
19. Milajevs, D., Kartsaklis, D., Sadrzadeh, M., Purver, M.: Evaluating neural word representations in tensor-based compositional settings. In: Proceedings of EMNLP, pp. 708–719. ACL, Doha (2014)
20. Pennington, J., Socher, R., Manning, C.: GloVe: global vectors for word representation. In: Proceedings of EMNLP, pp. 1532–1543. ACL, Doha (2014)
21. Ronan, C., Jason, W.: A unified architecture for natural language processing: deep neural networks with multitask learning. In: Proceedings of ICML, pp. 160–167 (2008)
22. Severyn, A., Moschitti, A.: Learning to rank short text pairs with convolutional deep neural networks. In: Proceedings of SIGIR, pp. 373–382 (2015)
23. Shengxian, W., Yanyan, L., Jiafeng, G., Jun, X., Liang, P., Xueqi, C.: A deep architecture for semantic matching with multiple positional sentence representations. In: Proceedings of AAAI, Phoenix, Arizona, pp. 2835–2841 (2016)
24. Socher, R., Huang, E.H., Pennin, J., Manning, C.D., Ng, A.Y.: Dynamic pooling and unfolding recursive autoencoders for paraphrase detection. In: Proceedings of NIPS, pp. 801–809. Curran Associates Inc. (2011)
25. Stefanescu, D., Rus, V., Niraula, N., Banjade, R.: Combining knowledge and corpus-based measures for word-to-word similarity. In: Proceedings of the Twenty-Seventh International Florida Artificial Intelligence Research Society Conference, pp. 87–90 (2014)
26. Tai, K.S., Socher, R., Manning, C.D.: Improved semantic representations from tree-structured long short-term memory networks. In: Proceedings of ACL. pp. 1556–1566. ACL, Beijing (2015)

27. Wei, X., Alan, R., Chris, C.B., William, B., Yangfeng, J.: Extracting lexically divergent paraphrases from twitter. TACL **2**(1), 435–448 (2014)
28. Xiaodan, Z., Parinaz, S., Honhyu, G.: Long short-term memory over recursive structures. In: Proceedings of ICML, Beijing, China, pp. 1604–1612 (2015)
29. Xu, W., Callison-Burch, C., Dolan, B.: SemEval-2015 task 1: paraphrase and semantic similarity in twitter (PIT). In: Proceedings of SemEval, pp. 1–11. ACL (2015)
30. Yin, W., Schütze, H.: Convolutional neural network for paraphrase identification. In: Proceedings of NAACL, pp. 901–911. ACL (2015)
31. Yoon, K., Yacine, J., David, S., Alexander, M.: Character-aware neural language models. In: Proceedings of AAAI, pp. 2741–2749. AAAI (2016)
32. Zarrella, G., Henderson, J., Merkhofer, E.M., Strickhart, L.: MITRE: seven systems for semantic similarity in tweets. In: Proceedings of SemEval, pp. 12–17. ACL (2015)

Recommendation

Leveraging Kernel Incorporated Matrix Factorization for Smartphone Application Recommendation

Chenyang Liu[1], Jian Cao[1(✉)], and Jing He[2]

[1] CIT Lab, Department of Computer Science and Engineering,
Shanghai JiaoTong University, No. 800, Dongchuan Road, Shanghai, China
Schumeichel_2003@163.com, cao-jian@sjtu.edu.cn
[2] College of Engineering and Science, Victoria University,
International PO Box 14428, Melbourne, VIC 8001, Australia
jing.he@vu.edu.au

Abstract. The explosive growth in the number of smartphone applications (apps) available on the market poses a significant challenge to making personalized recommendations based on user preferences. The training data usually consists of sparse binary implicit feedback (i.e. user-app installation pairs), which results in ambiguities in representing the users interests due to a lack of negative examples. In this paper, we propose two *kernel incorporated matrix factorization* models to predict user preferences for apps by introducing the categorical information of the apps. The two models extends Probabilistic Matrix Factorization (PMF) by constraining the user and app latent features to be similar to their neighbors in the app-categorical space, and adopts Stochastic Gradient Decent (SGD)-based methods to learn the models. The experimental results show that our model outperforms the baselines, in terms of two ranking-oriented evaluation metrics.

1 Introduction

With the rapid growth in the number of smartphone application available on the app markets (e.g. more than 1 million on Google Play Store and Apple App Store respectively), it is a significant challenge for users to find apps that they prefer. Most app markets provide search engines to match the keywords in user queries and app features, and simple recommendations are made based on the apps popularity. These methods suffer from low recommendation accuracy due to the ambiguous descriptions in the user queries and the lack of personalization.

Collaborative Filtering (CF) [1], which predicts user preferences for items by leveraging the information of similar users or items, has been proved to be effective in rating-oriented recommendation systems. CF-based methods are categorized as either memory-based or model-based. Memory-based methods (e.g. user-oriented [2] and item-oriented [3]) predict the ratings of a user-item matrix based on the similarity-weighted average of the ratings from k-nearest users or

© Springer International Publishing AG 2017
S. Candan et al. (Eds.): DASFAA 2017, Part I, LNCS 10177, pp. 459–474, 2017.
DOI: 10.1007/978-3-319-55753-3_29

items. The similarities between the users or items are computed by Pearson Correlation, Spearman Rank Correlation, Vector Cosine Similarity, etc. However, memory-based methods suffer from data sparsity and the scalability problem. More importantly, for app recommendations, conventional similarity computation results in low similarity values due to the sparsity of the training data and a lack of awareness of the inherent similarity between different apps. For example, two users who install Twitter and Facebook respectively on their smartphones are regarded to be similar to some extent, because they both prefer social apps.

Model-based methods (e.g. cluster-based CF [4] and latent semantic models [5]) predict missing ratings by modeling observed ratings, which captures latent relationships between users and items. *Matrix Factorization*-based models [6] are the most popular methods in this category and have been widely studied recently. These models approximate the original matrix through the products of multiple low-rank matrices, which are scalable to large data sets and best describe user and item profiles. However, the training observations in app recommendations are usually implicit feedback from users, e.g. users historical data of downloading and installing apps. This is typical *One-Class Collaborative Filtering* (OCCF) [7], where only sparse positive observations exist. Conventional strategies, e.g. all missing as negative (AMAN) and all missing as unknown (AMAU), suffer from low prediction accuracy due to the lack of negative examples. Several studies [7,8] propose a weighted low-rank matrix factorization, which measures the credibility of the loss functions by predetermined weights. The predetermined credibility is based on the assumption that unobserved user-item pairs are regarded as negative examples with high probability proportional to the number of observed pairs for each user or item. This assumption ignores the relationship between the content information of positive and 'negative' examples.

In this paper, we propose two kernel incorporated PMF models to enhance the prediction accuracy of user preferences for apps. The first model extends the conventional PMF by decomposing the user and app latent feature vectors as the summation of personalized offsets and the weighted average effects of neighbors. The second model extends PMF by regularizing the user and app latent factors with the weighted linear combination of their neighbors. The weights in the two models are computed by two kernel functions for users and apps over categorical labels, with a couple of predefined categorical weight parameters. To ensure there is a trade-off between learning efficiency and prediction accuracy, we constrain the neighborhood sizes of latent user and app features by two parameters. The experimental results on a real-world user-app dataset prove that our models have advantages over the baselines, in terms of *Mean Average Precision* (MAP) and *Normalized Discounted Coverage Gain* (NDCG). In summary, our work makes the following contributions:

- We model the hierarchical similarities among the users and the apps through a kernel function, i.e. a weighted cosine similarity based on the app-categorical labels.
- We invent two extended PMF models by constraining the latent user and app factors to be similar to their neighbors in the kernel-based neighborhoods.

– We introduce SGD-based algorithms to learn the factors, and invent a user-oriented sampling scheme to generate the training examples.

The remainder of this paper is organized as follows. Section 2 introduces the work related to matrix factorization and the application of machine learning methods for app recommendation. Section 3 presents the definition of the kernel function and the two extended PMF models, and the user-oriented sampling scheme. The recommendation effectiveness of our models compared with the baselines is presented in Sect. 5. Finally, we conclude the paper with a discussion on future work in Sect. 6.

2   Related Work

In this section, we first present two main types of PMF-based models, distinguished by model construction and learning methods. Then we introduce the work of OCCF for datasets with implicit feedback, followed by an overview of recent research on app recommendation.

2.1   Probabilistic Matrix Factorization (PMF)

The PMF model, arguably first proposed in [9], models the observations in a user-item rating matrix as the inner products of two D-dimensional user and item latent factor matrices. Through maximizing the posterior over the ratings with Gaussian priors over the latent matrices, the problem is converted to minimizing the squared errors with fixed penalty parameters, and is solved by *Gradient Descent* (GD)-based methods. Based on this model, [10] extends the PMF by incorporating the trust information between the users, which is constructed by a *Radius Basis Function* (RBF) kernel based on *Benevolence, Integrity, Competence,* and *Predictability*. [11] substitutes the multiplicative rules for the additive updates in PMF by introducing non-negative constraints, which accelerates the convergence of the GD-based optimization. The work in [12] proposes a Kernelized PMF, which explores the underlying covariance structures in the latent factor matrices by introducing zero-mean *Gaussian Process* based on side information (e.g. user social relationship and item attributes).

Different from the aforementioned pointwise MAP estimate of parameters, [13] proposes a *Bayesian* treatment for PMF (BPMF), which generalizes PMF to handle non-zero mean and non-spherical Gaussian priors. This method places *Gaussian-Wishart* priors on the mean and precision parameters of the latent factor matrices, and approximates the integration over the model parameters and the hyperparameters through a *Markov Chain Monte Carlo*-based method (e.g. Gibbs Sampling) to predict user-item ratings. The work in [14] extends BPMF by incorporating the linear regression of the users and items against the features of the other or the rating-specific information. A Dirichlet process mixture prior is placed on the user and item latent features, and a Gibbs sampler marginalizes over the parameters by a cyclic conditional sampling to find the posterior of the rating predictions. The work in [15] further proposes a hierarchical BPMF with side information, which places individual mean and precision parameters for the Gaussian-Wishart priors over the latent features. The hyperparameters are updated through *Bayesian Cramer-Rao Bound* to perform holdout prediction for missing entries and fold-in prediction for new users or items. The work in [16] develops a variational matrix co-factorization for the hierarchical Bayesian model, which computes the variational posterior distributions over the latent factor matrices and the hyperparameters by maximizing a variational low bound. In summary, the advantage of BPMF-based methods is their less proneness to overfitting due to its approximation to the true posterior distribution, however these models suffer from high computing complexity for parameter estimation.

2.2   One-Class Collaborative Filtering (OCCF)

Most web information systems only provide implicit feedback consisting of binary data, e.g. users' action or inaction in relation to items. This *one-class* setting results in difficulties in distinguishing the positive and negative examples from large amounts of unobserved user-item pairs.

To leverage the abundant implicit feedback, [7] proposes a weighted low rank approximation, which places different credibility weights on the Frobenius loss-based objective of PMF, according to a uniform, user-oriented or item-oriented weighting scheme. A sampling-based weighted ALS is proposed to train the model. The sampling strategy retrieves some missing values as negative examples from the original data, which are assumed to be negative with high probability. The work in [8] formulates a weighted NMF, which divides the loss function into two parts corresponding to the positive and non-positive examples, respectively. The non-positive part is further divided into two components, which denotes the probably positive and negative classes respectively. A NMF-based alternative optimization is adopted to solve the latent factor matrices and a discrete optimization method combining the Newton-Raphson iterations and the bisection method computes the positivity probability for non-positive examples. The shortage of both models is that the rough assignment of the credibility weights reduces the prediction accuracy.

2.3   Smartphone Application Recommendation

Many researchers have focused on smartphone application recommendation over the last decade. For example, the *GetJar* project in [17] proposes a PCA-based model, *EigenApp*, which performs *Eigen Decomposition* on the normalized matrix of the users' daily usage frequency on the apps. The k eigenvectors corresponding to the k highest eigenvalues are extracted to compute the app-app similarities, which is adopted in a top-N recommendation for each user. To relieve the cold-start problem in app recommendation, the work in [18] leverages the information on app followers available on *Twitter* and introduces *Latent Dirichlet Allocation* (LDA) to generate the latent user groups which share similar Twitter-ID follownership information, and the probability that represents a user's interest in an app is computed through the marginalization over these groups. The semi-supervised model in [19] extends LDA to generate the latent topics over app version metadata and textual description, and defines the weight of a genre-topic pair to identify important topics in the genres. The recommendation is based on the version-snippet score, i.e., a weighted marginalization over the latent topics.

However, these approaches require users' behavior on apps (e.g. daily app usage and app downloading sequence) or users' social information to be collected, which is privacy-sensitive information. By contrast, our models depend only on the information of users' app installation, therefore it is more practical for app recommendation than these approaches.

3   Kernel Incorporated Matrix Factorization

In our scenario, we have an $N \times M$ relationship matrix R on N users and M apps, where $R_{ij} = 1$ denotes user i installing app j on his smartphone; otherwise $R_{ij} = 0$. Conventional PMF factorizes this matrix as an inner product of two D-dimensional matrices, $U \in R^{N \times D}$ and $V \in R^{M \times D}$, and models the conditional distribution over the observed ratings as in Eq. (1),

$$\prod_{i=1}^{N} \prod_{j=1}^{M} \left(\mathcal{N} \left(R_{ij} | g \left(U_i^T V_j \right), \sigma_R^2 \right) \right)^{I_{ij}} \mathcal{N}(U_i | 0, \sigma_U^2 I) \mathcal{N}(V_j | 0, \sigma_V^2 I) \qquad (1)$$

where $\mathcal{N}(x | \mu, \sigma^2)$ refers to a Gaussian distribution with mean μ and variance σ^2, and I_{ij} is a binary indicator which represents the existence of the installation relationship between user i and app j. Through a Bayesian inference, the objective is converted to minimizing the square loss with quadratic regularization terms, as in Eq. (2),

$$\mathcal{L} = \frac{1}{2} \sum_{i=1}^{N} \sum_{j=1}^{M} I_{ij} \left(R_{ij} - g \left(U_i^T V_j \right) \right)^2 + \frac{\lambda_U}{2} \sum_{i=1}^{N} \|U_i\|_F^2 + \frac{\lambda_V}{2} \sum_{j=1}^{M} \|V_j\|_F^2 \qquad (2)$$

where $\|\cdot\|_F$ is the Frobenius norm, λ_U and λ_V are regularization parameters to prevent the model from overfitting.

To overcome the less effectiveness of PMF on user-app preference prediction due to data sparsity and one-class setting, we extend PMF by fusing latent neighboring information between users and apps, which is obtained through introducing a kernel function over additional information. We propose two models, i.e. *Kernel Constrained PMF* (KCPMF) and *Kernel Regularized PMF* (KRPMF), to preserve the neighboring information in the user-specific and app-specific latent feature representation.

3.1   Kernel Function Definition

An intuitive inspiration is to leverage the hierarchical app-category information available on the Web to measure the similarities between users and apps.

For example, the apps on www.wandoujia.com are categorized into *Travelling, News Reading, Social*, etc. Therefore, we assume a $M \times L$ matrix Q on M apps and L categorical labels, where $Q_{ij} = 1$ denotes app i is categorized by label j; otherwise $Q_{ij} = 0$. We construct another $N \times L$ matrix P on N users and L labels by multiplying the user-app matrix R by Q. The entries of P could be interpreted as the user preference degrees for the categorical labels.

With matrices P and Q, we introduce a kernel function to measure the similarities between the users and the apps in the categorical label space. Commonly used kernel functions include RBF kernels and cosine similarity. For simplicity, we define a weighted cosine similarity as in Eqs. (3) and (4),

$$\mathcal{K}(P_i, P_j) = \frac{(W_L \odot P_i)^T (W_L \odot P_j)}{\parallel W_L \odot P_i \parallel_2 \cdot \parallel W_L \odot P_j \parallel_2} \tag{3}$$

$$\mathcal{K}(Q_i, Q_j) = \frac{(W_L \odot Q_i)^T (W_L \odot Q_j)}{\parallel W_L \odot Q_i \parallel_2 \cdot \parallel W_L \odot Q_j \parallel_2} \tag{4}$$

where P_i, P_j, Q_i, Q_j are the row vectors in P and Q, $\odot$ is the *Hadamard* production of two vectors (e.g. $x \odot y = [x_1 y_1, x_2 y_2, \cdots, x_i y_i]$), and W_L is a predetermined weight vector for the labels of different hierarchies. Since a label of higher hierarchy means that more apps belong to the corresponding category, and thus a low similarity between users and apps, the weights could be empirically set to be inversely proportional to the numbers of the apps owning the categorical labels.

3.2   Kernel Constrained PMF (KCPMF)

In this model, we assume that every user-specific or item-specific latent features should be the combination of personal features and effects from similar users and apps. Therefore, we decompose the user and app latent factors in conventional PMF into two components, as in Eqs. (5) and (6),

$$U_i = Y_i^u + \sum_{k=1}^{N_u} W_{ik}^u E_k^u, V_j = Y_j^v + \sum_{l=1}^{N_v} W_{jl}^v E_l^v \tag{5}$$

$$W_{ik}^u = \frac{\mathcal{K}(P_i, P_k)}{\sum_{k'=1}^{N_u} \mathcal{K}(P_i, P_{k'})}, W_{jl}^v = \frac{\mathcal{K}(Q_j, Q_l)}{\sum_{l'=1}^{N_v} \mathcal{K}(Q_j, Q_{l'})} \tag{6}$$

where Y_i^u and Y_j^v are the D-dimensional *personalization offsets* of user i and app j, and E_k^u and E_l^v are the D-dimensional *neighboring effects* of similar user k and app l on current latent factors, which constrain the factors to be close to the weighted average of the effects. The parameters N_u and N_v control the neighborhood size of the users and the apps in the categorical label spaces, which results in a trade-off between computing complexity and accuracy.

Substituting the definition of user and app latent factor in Eq. (5) into the conventional PMF objective, i.e. Eq. (1), and placing zero-mean spherical Gaussian distributions on Y_i^u, Y_j^v, E_k^u and E_l^v, we get the following posterior distribution over the observed user-app installation relationship through *Bayes Inference*,

$$
\prod_{i=1}^{N}\prod_{j=1}^{M}\mathcal{N}\left(R_{ij}\Big|g\left(\left(Y_i^u+\sum_{k=1}^{Nu}W_{ik}^uE_k^u\right)^T\left(Y_j^v+\sum_{l=1}^{Nv}W_{jl}^vE_l^v\right)\right),\sigma_R^2\right)
$$

$$
\prod_{i=1}^{N}\mathcal{N}(Y_i^u|0,\sigma_{Yu}^2I)\prod_{j=1}^{M}\mathcal{N}(Y_j^v|0,\sigma_{Yv}^2I)\prod_{k=1}^{N}\mathcal{N}(E_k^u|0,\sigma_{Eu}^2I)\prod_{l=1}^{M}\mathcal{N}(E_l^v|0,\sigma_{Ev}^2I) \quad (7)
$$

where $g(x) = 1/(1+\exp(-x))$ is the logistic function that bounds the prediction within the interval [0,1]. The graphical model of KCPMF is shown in Fig. 1(a). Maximizing the log-posterior in Eq. (7) with hyperparameters is equivalent to minimizing the regularized squared error loss function,

$$
\mathcal{L}_1 = \frac{1}{2}\sum_{i=1}^{N}\sum_{j=1}^{M}I_{ij}\left\{R_{ij}-g\left(\left(Y_i^u+\sum_{k=1}^{Nu}W_{ik}^uE_k^u\right)^T\left(Y_j^v+\sum_{l=1}^{Nv}W_{jl}^vE_l^v\right)\right)\right\}^2
$$

$$
+\frac{\lambda_{Yu}}{2}\sum_{i=1}^{N}\|Y_i^u\|_F^2+\frac{\lambda_{Yv}}{2}\sum_{j=1}^{M}\|Y_j^v\|_F^2+\frac{\lambda_{Eu}}{2}\sum_{k=1}^{N}\|E_k^u\|_F^2+\frac{\lambda_{Ev}}{2}\sum_{l=1}^{M}\|E_l^v\|_F^2 \quad (8)
$$

where $\lambda_{Yu}=\sigma_R^2/\sigma_{Yu}^2$, $\lambda_{Yv}=\sigma_R^2/\sigma_{Yv}^2$, $\lambda_{Eu}=\sigma_R^2/\sigma_{Eu}^2$, $\lambda_{Ev}=\sigma_R^2/\sigma_{Ev}^2$ are the regularization parameters.

Taking the partial derivatives of the objective function in Eq. (8) with respect to Y_i^u, Y_j^v, E_k^u and E_l^v, we get the gradients of the latent factors as in Eqs. (9)–(12),

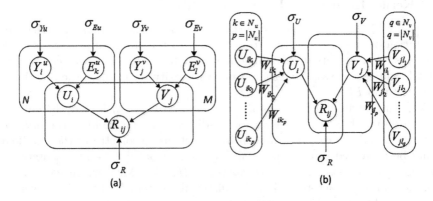

Fig. 1. The Graphical Model of Kernel Integrated MF. (a) KCPMF. (b)KRPMF.

$$\frac{\partial \mathcal{L}_1}{\partial Y_i^u} = - \sum_{j=1}^{M} ERR_{ij} g'_{ij} \left(Y_j^v + \sum_{l=1}^{Nv} W_{jl}^v E_l^v \right) + \lambda_{Y_u} Y_i^u \tag{9}$$

$$\frac{\partial \mathcal{L}_1}{\partial Y_j^v} = - \sum_{i=1}^{N} ERR_{ij} g'_{ij} \left(Y_i^u + \sum_{k=1}^{Nu} W_{ik}^u E_k^u \right) + \lambda_{Y_v} Y_j^v \tag{10}$$

$$\frac{\partial \mathcal{L}_1}{\partial E_k^u} = - \sum_{i \in \mathcal{E}u(k)} \sum_{j=1}^{M} ERR_{ij} g'_{ij} \left(Y_j^v + \sum_{l=1}^{Nv} W_{jl}^v E_l^v \right) W_{ik}^u + \lambda_{Eu} E_k^u \tag{11}$$

$$\frac{\partial \mathcal{L}_1}{\partial E_l^v} = - \sum_{j \in \mathcal{E}v(l)} \sum_{i=1}^{N} ERR_{ij} g'_{ij} \left(Y_i^u + \sum_{k=1}^{Nu} W_{ik}^u E_k^u \right) W_{jl}^v + \lambda_{Ev} E_l^v \tag{12}$$

where

$$ERR_{ij} = I_{ij} \left\{ R_{ij} - g \left(\left(Y_i^u + \sum_{k=1}^{Nu} W_{ik}^u E_k^u \right)^T \left(Y_j^v + \sum_{l=1}^{Nv} W_{jl}^v E_l^v \right) \right) \right\},$$

$$g'_{ij} = g' \left(\left(Y_i^u + \sum_{k=1}^{Nu} W_{ik}^u E_k^u \right)^T \left(Y_j^v + \sum_{l=1}^{Nv} W_{jl}^v E_l^v \right) \right),$$

$\mathcal{E}u(k)$ and $\mathcal{E}u(l)$ are the sets of users and apps that include user k and item l in their N_u and N_v nearest neighbor sets, respectively. Since the gradient-based updates of the factors for the minimization of the objective is analytically intractable, we perform an SGD-based alternative optimization in the factors to obtain a local minima.

3.3   Kernel Regularized PMF (KRPMF)

The KCPMF captures the personalization factors while constraining the representation of the latent factors to be similar to the weighted average of their neighbors. However, its disadvantage is that it has twice the space complexity of PMF and the linearly increasing time complexity for training the additional effects E_k^u and E_l^v.

A natural assumption (i.e. *local invariance assumption*) is that the data in the user-app matrix is regarded as a low-dimensional manifold in the original data space, and the closeness hidden in the intrinsic geometry of the manifold should be preserved in corresponding low-dimensional feature representations. Therefore, we propose *Kernel Regularized PMF* (KRPMF), which models the latent user and app factors as Gaussian distribution around the weighted average of their neighbors in the app-categorical space and incorporate them into the conventional PMF, as in Eq. (13).

$$\prod_{i=1}^{N} \prod_{j=1}^{M} \left(\mathcal{N} \left(R_{ij} | g \left(U_i^T V_j \right), \sigma_R^2 \right) \right)^{I_{ij}} \mathcal{N} \left(U_i | \sum_{k=1}^{Nu} W_{ik} U_k, \sigma_{Nu}^2 \right)$$
$$\mathcal{N} \left(V_j | \sum_{l=1}^{Nv} W_{jl} V_j, \sigma_{Nv}^2 \right) \mathcal{N}(U_i | 0, \sigma_U^2 I) \mathcal{N}(V_j | 0, \sigma_V^2 I) \tag{13}$$

where W_{ik}, W_{jl} refer to the definition in Eqs. (6), and N_u, N_v are neighborhood-size parameters. Similarly, maximizing the posterior in Eq. (13) is converted to minimizing the squared loss function with regularization terms in Eq. (14),

$$
\mathcal{L}_2 = \frac{1}{2} \sum_{i=1}^{N} \sum_{j=1}^{M} I_{ij} \left(R_{ij} - g \left(U_i^T V_j \right) \right)^2 + \frac{\lambda_{Nu}}{2} \sum_{i=1}^{N} \left(U_i - \sum_{k=1}^{Nu} W_{ik} U_k \right)^2
$$

$$
+ \frac{\lambda_{Nv}}{2} \sum_{j=1}^{M} \left(V_j - \sum_{l=1}^{Nv} W_{jl} V_j \right)^2 + \frac{\lambda_U}{2} \sum_{i=1}^{N} \|U_i\|_F^2 + \frac{\lambda_U}{2} \sum_{i=1}^{N} \|U_i\|_F^2 \quad (14)
$$

where $\lambda_{Nu} = \sigma_R^2/\sigma_{Nu}^2$ and $\lambda_{Nv} = \sigma_R^2/\sigma_{Nv}^2$ are the trade-off between the squared loss term and the prior of neighboring information in local invariance assumption, $\lambda_U = \sigma_R^2/\sigma_U^2$ and $\lambda_V = \sigma_R^2/\sigma_V^2$ are the regularization parameters to penalize the model complexity.

Similarly, a local minima of the objective in Eq. (14) can be obtained by performing a gradient descent-based iterative optimization in U_i and V_j,

$$
\frac{\partial \mathcal{L}_2}{\partial U_i} = - \sum_{j=1}^{M} I_{ij} \left(R_{ij} - g \left(U_i^T V_j \right) \right) g' \left(U_i^T V_j \right) + \lambda_{Nu} \left(U_i - \sum_{k=1}^{Nu} W_{ik} U_k \right)
$$

$$
+ \sum_{p \in \mathcal{E}u(i)} \left(U_p - \sum_{m=1}^{Nu} W_{pm} U_m \right) (-W_{pi}) + \lambda_U U_i \quad (15)
$$

$$
\frac{\partial \mathcal{L}_2}{\partial V_j} = - \sum_{i=1}^{N} I_{ij} \left(R_{ij} - g \left(U_i^T V_j \right) \right) g' \left(U_i^T V_j \right) + \lambda_{Nv} \left(V_j - \sum_{l=1}^{Nv} W_{jl} V_l \right)
$$

$$
+ \sum_{q \in \mathcal{E}v(j)} \left(V_q - \sum_{n=1}^{Nv} W_{qn} U_n \right) (-W_{qj}) + \lambda_V V_j \quad (16)
$$

where $\mathcal{E}u(i)$ and $\mathcal{E}v(j)$ refer to the definition in Eqs. (11) and (12), respectively.

3.4   Time Complexity Analysis

As presented in Sects. 3.2 and 3.3, the model learning of KCPMF and KRPMF comprise two stages, with a shared previous stage that computes the user and app kernel matrices over the categorical labels and takes the time complexity of $O\left(N^2 L\right)$ and $O\left(M^2 L\right)$ respectively, where L are the numbers of labels. Notice that $L \ll N, L \ll M$, therefore the time complexities are approximately linear to the square of N and M.

The second stage of KCPMF and KRPMF executes an iterative optimization of evaluating the objectives in Eqs. (8) and (14), and the gradients in Eqs. (9)–(12) and Eqs. (15)–(16) against the latent factors. For KCPMF, the complexity of evaluating the objective is $O\left(\rho_r \left(N_u + N_v\right) d\right)$, where ρ_r is the total number of user-app installation records, and d is the latent dimension. The complexity of computing the gradients is $O\left(\left(\overline{\rho_k \rho_u} N_v + \overline{\rho_l \rho_v} N_u\right) \left(N_u + N_v\right) d\right)$, where $\overline{\rho_k}$ and $\overline{\rho_l}$

are the average numbers of users and apps that user k and app l have effects on respectively, $\overline{\rho_u}$ and $\overline{\rho_v}$ are the average number of apps per user and the average number of users per app respectively. For KRPMF, the complexity of evaluating the objective is $O\left((\rho_r + N_u N + N_v M)d\right)$, while the complexity of computing the gradients is $O\left((\rho_u + \rho_v + \overline{\rho_p}N_u + \overline{\rho_q}N_v)\,d\right)$, where $\overline{\rho_p}$ and $\overline{\rho_q}$ amount to $\overline{\rho_k}$ and $\overline{\rho_l}$ in the complexity of KCPMF, respectively.

3.5   Negative Sampling Scheme

To relieve the problem of data sparsity and the lack of negative examples in one-class setting, [7] proposes a negative sampling scheme based on a sampling probability matrix S and a negative sample size q to model the uncertainty on missing data, where the entries in S are generated by a *user-oriented* or *item-oriented* scheme. The two schemes assume that the probability of being a negative example for a missing user-item pair is correlated to the corresponding row count or column count in the training matrix R. However, the constraint on q cannot guarantee that enough and accurate negative examples for the active users (i.e. the users installing more apps) are sampled.

We include all the positive examples in the training set. For the missing user-item pairs, we invent a user-oriented sampling scheme as follows. We introduce a negative sampling size q_i for each user i that is proportional to the count of the apps installed by this user, i.e. $q_i \propto \sum_j I\left(R_{ij} = 1\right)$. This sampling setting is based on the assumption that the missing pairs of the active users are more valuable for the model training than those of other users. The independent q_i for each user guarantees that more probably negative examples are sampled, and models our uncertainty on the missing user-app installation data based on the implicit feedback.

4   Performance Evaluation

We conduct the experiments on a real-world dataset consisting of user-app installation records, to verify the effectiveness of KCPMF and KRPMF compared with conventional methods. The comparison is based on two common evaluation metrics, MAP and NDCG. We also study the parameter sensitivity of KCPMF and KRPMF under different settings.

4.1   Experiment Dataset

Our original dataset contains more than 56000 users in a large city and more than 7700 apps installed on their smartphones. We filter the dataset by selecting the users who have installed at least 10 apps. The experimental dataset contains the installation relationship between 11166 users and 6257 apps. We extract 211 hierarchical categorical labels of the apps from an online app store, www.wandoujia.com, which divide the apps into categories by three hierarchies. Each app is labeled by at least one label per level. The final experimental dataset is listed in Table 1.

Table 1. Description of the experimental dataset

# of users	11166	# of level 1 labels	2
# of apps	6257	# of level 2 labels	29
# of user-app installations	157782	# of level 3 labels	180
user-app data sparsity	2.25%	# of app-label relations	18654
avg # of apps by user	14.1	App-label data sparsity	1.41%
avg # of users by app	25.2		

4.2 Experiment Setup

We randomly choose 80% user-app installation data for each user as the training set, and the remaining 20% of the data as the testing set. In our KCPMF and KRPMF model, we empirically set the weights of the categorical labels to be inversely proportional to the proportions of the apps that have the labels. We adopt cross-validation to determine latent dimension $d \in \{20, 40, 60, 80\}$, and regularization parameters $\lambda_{Yu}, \lambda_{Yv}, \lambda_{Eu}, \lambda_{Ev}$ in KCPMF and $\lambda_{Nu}, \lambda_{Nv}, \lambda_{U}, \lambda_{V}$ in KRPMF $\in \{0.001, 0.01, 0.1, 1, 10\}$. Learning rate η is set to 0.05, maximum number of epochs is set to 1000, and the threshold value for convergence is set to 0.00001.

We compare our models against the following baselines in the experiments: *User-based CF* (UCF) [2], *Item-based CF* (ICF) [3], *Probabilistic Matrix Factorization* (PMF) [9], *constrained PMF* (cPMF) [9], *weighted PMF* (wPMF) [7]. In the experiments, we set $k = 10$ nearest neighbors for UCF and ICF. The setting of latent dimension, regularization parameters and learning rate in PMF, cPMF, wPMF are the same as KCPMF and KRPMF. PMF and cPMF adopt a uniform sampling for the generation of negative examples, while wPMF adopts a user-oriented negative sampling and weighting scheme. SGD-based learning is adopted for learning the model parameters. After these models compute the pointwise user preferences for the apps, a top-N ranking-based recommendation is executed to present a personalized app list for each user, which is evaluated by the following two metrics.

We use two ranking-oriented evaluation metrics, *Mean Average Precision* (MAP) and *Normalized Discounted Cumulative Gain* (NDCG), to measure the performance of our models and the other methods. MAP evaluates the mean of the average precision (AP) over the users. AP for user u is computed at the point of each installed app at the ranked list,

$$AP_u = \frac{\sum_{i=1}^{N} prec(i) \times pref(i)}{\# \text{ of installed apps}} \quad (17)$$

where $prec(i)$ is the precision at ranked position i, and $pref(i)$ is a binary relevance indicator at position i. NDCG evaluates the discounted cumulative gain of a ranking list for user u,

$$NDCG_u@N = Z_u \sum_{k=1}^{N} \frac{2^{Y_u(k)} - 1}{\log_2(1 + k)} \qquad (18)$$

where Z_u is the normalization factor, and $Y_u(k)$ denotes the binary relevance of the app ranked in position k.

4.3　Experiment Results

We conduct the experiments on the top-N recommendations with N set to {5, 10, 15, 20}. The performance comparison between our models and the other baselines is shown in Table 2, where d is set to 80; $\lambda_U = \lambda_V = 0.001$ for PMF and wPMF; $\lambda_Y = \lambda_V = 0.001$, $\lambda_W = 0.01$ for cPMF. $\lambda_{Yu} = \lambda_{Yv} = 0.001$, $\lambda_{Eu} = \lambda_{Ev} = 0.01$ for KCPMF, $\lambda_{Nu} = \lambda_{Nv} = 0.01$, $\lambda_U = \lambda_V = 0.001$ for KRPMF, and $N_u = N_v = 10$ for both models. The results show that both KCPMF and KRPMF outperform the other baselines, with an appropriate 40~50% gain in recommendation accuracy over UCF and ICF, and 15~20% over PMF, cPMF and wPMF.

Table 2. The comparison of MAP & NDCG among the baselines

Metrics	MAP@N				NDCG@N			
Model	N = 5	N = 10	N = 15	N = 20	N = 5	N = 10	N = 15	N = 20
UCF	0.286	0.314	0.323	0.327	0.372	0.421	0.442	0.455
ICF	0.172	0.197	0.206	0.209	0.245	0.295	0.317	0.329
PMF	0.489	0.497	0.503	0.507	0.473	0.508	0.518	0.523
cPMF	0.493	0.523	0.525	0.531	0.528	0.543	0.552	0.565
wPMF	0.554	0.563	0.571	0.576	0.579	0.595	0.604	0.613
KCPMF	**0.663**	**0.684**	**0.695**	**0.701**	**0.642**	**0.663**	**0.675**	**0.686**
KRPMF	**0.672**	**0.681**	**0.689**	**0.705**	**0.648**	**0.659**	**0.682**	**0.691**

Compared with UCF and ICF, PMF, cPMF, and wPMF improve the performance by modeling the latent features of users and items. However, they are inferior to KCPMF and KRPMF due to the binary training examples and the mechanisms of handling the missing observations in the user-app matrix. PMF and cPMF train the model over the positive examples and a group of uniformly generated negative examples with a fixed sampling size. The binary examples are ineffective for revealing the graded user preferences for the apps. The training process may mistake the latent positive examples for the negative ones due to the uniform negative sampling. Although cPMF is superior to PMF due to the constraints of the similar selections on the apps, the experimental results show that the outperformance is limited. The outperformance of wPMF over PMF and cPMF is attributed to the incorporation of the credibility weights into the objective and the introduction of a negative sampling probability matrix. However, the user-oriented weighting and sampling schemes may distinguish the latent positive examples from the negative ones in the missing user-app observations incorrectly, which results in its inferiority to our models.

Another observation worth noting is that extending the recommendation length for all the baselines brings limited gain in MAP and NDCG metrics. As presented in Table 2, the improvement gained by changing the N in the top-N recommendation from 5 to 10 is more evident than that by changing the N from 15 to 20. Since 80% of users install 10~20 apps in our dataset, the train-test split scheme determines that the number of apps for most users in the testing set are no more than 5. Therefore, the recommendation performance of the baselines benefits few from recommending more apps for users. This conforms to the situation in the real world that most users only install very few apps with respect to the existing huge amounts of apps, and they pay less attention to the apps ranked in higher positions in recommendation lists.

We also conduct experiments to study the effects of the kernel-based neighborhood size parameters N_u and N_v in our models on computing complexity and prediction accuracy. We make trials of different N_u and N_v in {5, 10, 15, 20} for KCPMF and KRPMF. The evaluation metrics are the time consumed in each epoch, and *Root Mean Square Error* (RMSE),

$$RMSE = \sqrt{\left(\sum_{(u,i)\in R} (r_{u,i} - \widehat{r}_{u,i})^2 \right) \Big/ |R|} \qquad (19)$$

where R denotes the training set. We adopt this metric to evaluate the rate of the convergence to the known observations, i.e. the approximation to the binary values of the training examples.

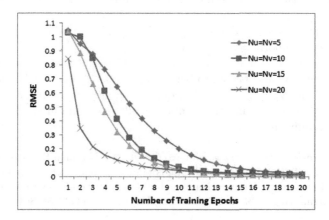

Fig. 2. The training process of KCPMF with different settings of N_u and N_v

Figure 2 shows the results of RMSE in the first 20 epochs of KCPMF on an experimental machine with a 3.20 GHz qual core CPU and 16 GB memory, while we omit the results of KRPMF since they present similar observations. With N_u and N_v set to higher values (e.g. 15 and 20), KCPMF obtains higher convergence rates, with the improvement on RMSE more evident. However, higher values for N_u and N_v result in higher computing complexity for each training epoch, i.e. the increasing iteration time for updating the latent factors and evaluating the objective functions. In our experiments, the average training time per epoch rises from 18.774 s to 65.359 s for KCPMF and 19.658 to 67.485 s for KRPMF when we raise N_u and N_v from 5 to 20. This experimental observation may lead to the inspiration of a variable parameter setting scheme for N_u and N_v based on the enhancement of the convergence rate and the computing complexity. We also notice that the overall training epoch time of KRPMF is higher than that of KCPMF. This observation is mainly attributed to the linear time complexity of KRPMF w.r.t the number of users and items (i.e. N and M) for evaluating the objective function, while the time complexity of KCPMF is quadratic to N_u and N_v for computing the gradient of the latent factors. $N_u^2 \ll N$ and $N_v^2 \ll M$ result in the advantage of KCPMF over KRPMF in time complexity.

5   Conclusion

In this paper, we propose two extended PMF models (i.e. KCPMF and KRPMF) for smartphone application recommendation, which extends the conventional PMF by incorporating the kernel-based user and app similarities over the app-categorical label space. KCPMF decomposes the latent factors into the personalized offsets and the linear combination of the effects of neighbors, while KRPMF regularizes the latent factors with their closeness to the linear combination of their neighbors in the latent space. Both models are optimized through SGD-based learning with a revised user-oriented sampling scheme for the generation of negative examples. The experiments verify the effectiveness and outperformance of our models compared to the baselines, and study the sensitivity of the models on the neighborhood-size parameters.

In the future, we will focus on exploiting the relationship between users, apps and app-genre information, to better capture user latent preferences. More techniques for analyzing the implicit feedback, e.g. users' daily behavior on apps, should be adopted to enhance the computing efficiency and prediction accuracy in user-app recommendation.

Acknowledgement. This work is supported by China National Science Foundation (Granted Number 61472253), Research Funds of Science and Technology Commission of Shanghai Municipality (Granted Number 15411952502) and Cross Research Fund of Biomedical Engineering of Shanghai JiaoTong University (YG2015MS61).

References

1. Shi, Y., Larson, M., Hanjalic, A.: Collaborative filtering beyond the user-item matrix: a survey of the state of the art and future challenges. ACM Comput. Surv. (CSUR) **47**(1), 1–45 (2014)
2. Cai, Y., Leung, H., Li, Q., Min, H., Tang, J., Li, J.: Typicality-based collaborative filtering recommendation. IEEE Trans. Knowl. Data Eng. **26**(3), 766–779 (2014)
3. Wang, J., de Vries, A.P., Reinders, M.J.T.: Unifying user-based and item-based collaborative filtering approaches by similarity fusion. In: 29th Annual International ACM SIGIR Conference on Research and Development in Information Retrieval, pp. 501–508. ACM (2006)
4. Wu, J., Chen, L., Feng, Y., Zheng, Z., Zhou, M., Wu, Z.: Predicting quality of service for selection by neighborhood-based collaborative filtering. IEEE Trans. Syst. Man Cybern. Syst. **43**(2), 428–439 (2013)
5. Hofmann, T.: Latent semantic models for collaborative filtering. ACM Trans. Inf. Syst. (TOIS) **22**(1), 89–115 (2004)
6. Koren, Y., Bell, R., Volinsky, C.: Matrix factorization techniques for recommender systems. Computer **8**, 30–37 (2009)
7. Pan, R., Zhou, Y., Cao, B., Liu, N.N., Lukose, R., Scholz, M., Yang, Q.: One-class collaborative filtering. In: Proceedings of the 8th IEEE International Conference on Data Mining, pp. 502–511 (2008)
8. Sindhwani, V., Bucak, S.S., Hu, J., Mojsilovic, A.: One-class matrix completion with lowdensity factorizations. In: Proceedings of the 10th IEEE International Conference on Data Mining, pp. 1055–1060 (2010)
9. Salakhutdinov, R., Mnih, A.: Probabilistic matrix factorization. Citeseer (2011)
10. Bao, Y., Fang, H., Zhang, J.: Leveraging decomposed trust in probabilistic matrix factorization for effective recommendation. In: Proceedings of the 28th AAAI Conference on Artificial Intelligence, pp. 30–36 (2014)
11. Luo, X., Zhou, M., Xia, Y., Zhu, Q.: An efficient non-negative matrix-factorization-based approach to collaborative filtering for recommender systems. IEEE Trans. Indus. Inf. **10**(2), 1273–1284 (2014)
12. Zhou, T., Shan, H., Banerjee, A., Sapiro, G.: Kernelized probabilistic matrix factorization: exploiting graphs and side information. Proc. SIAM Int. Conf. Data Mining **12**, 403–414 (2012)
13. Salakhutdinov, R., Mnih, A.: Bayesian probabilistic matrix factorization using Markov chain Monte Carlo. In: Proceedings of the 25th International Conference on Machine Learning, pp. 880–887. ACM (2008)
14. Porteous, I., Asuncion, A.U., Welling, M.: Bayesian matrix factorization with side information and Dirichlet process mixtures. In: Proceedings of 24th AAAI Conference on Artificial Intelligence (2010)
15. Park, S., Kim, Y., Choi, S.: Hierarchical Bayesian matrix factorization with side information. In: Proceedings of International Joint Conference on Artificial Intelligence, AAAI Press, pp. 1593–1599 (2013)
16. Yoo, J., Choi, S.: Hierarchical variational Bayesian matrix co-factorization. In: IEEE International Conference on Proceedings of Acoustics, Speech and Signal Processing, pp. 1901–1904 (2012)
17. Shi, K., Ali, K.: GetJar mobile application recommendations with very sparse datasets. In: Proceedings of the 18th ACM SIGKDD International Conference on Knowledge Discovery and Data Mining, pp. 204–212 (2012)

18. Lin, J., Sugiyama, K., Kan, M., Chua, T.: Addressing cold-start in app recommendation: latent user models constructed from twitter followers. In: Proceedings of the 36th International ACM SIGIR Conference on Research and Development in Information Retrieval, pp. 283–292 (2013)
19. Lin, J., Sugiyama, K., Kan, M., Chua, T.: New and improved: modeling versions to improve app. recommendation. In: Proceedings of the 37th International ACM SIGIR Conference on Research and Development in Information Retrieval, pp. 647–656 (2014)

Preference Integration in Context-Aware Recommendation

Lin Zheng[1] and Fuxi Zhu[1,2]([envelope])

[1] Computer School, Wuhan University, Wuhan 430072, China
{linzheng,fxzhu}@whu.edu.cn
[2] School of Computer Science and Technology,
Hankou University, Wuhan 430212, China

Abstract. In Recommender Systems, recommendation tasks are usually implemented by preference mining. Most existing methods in context-aware recommendation focus on learning user interests from all the records to implement preference mining. However, some records contribute to recommendations, whereas some others may reduce the performances. To address these limitations, we propose a division learning strategy to divide the original records into several groups based on regression tree techniques. Then, a two-layer preference mining process is carried out to produce group and local preferences. Finally, the two preferences are integrated by the preference integration (PRIN) approach to give recommendations. The experimental results demonstrated that our model outperformed other state-of-the-art methods, which illustrated the importance of targeted modeling.

Keywords: Recommender system · Context-aware recommendation · Division learning · Preference integration · Gradient boosted regression tree

1 Introduction

Background. Recommender systems are designed to filter information and mine preferences to benefit web users. Traditional recommendation algorithms such as Collaborative Filtering (CF) [15] focus on mining user preferences from user actions that are often termed feedback [9,14], e.g., clicking, purchasing, or rating. Hence, preference mining in recommender systems is usually carried out by feedback modeling. For example, Matrix Factorization (MF) [10,16] techniques treat feedback as training targets to model user-item interactions by low-rank factors. In addition to user and item aspects, contexts are introduced into MF methods that make recommendation algorithms be of practical significance [1,17]. Contexts can be user profiles, item attributes [18], or other ancillary information such as tags [17], locations [8], and social networks [12]. Thus, the research of context-aware recommendation is growing increasingly popular [1]. General factorization models [13,18] tend to introduce a large number of contextual factors

© Springer International Publishing AG 2017
S. Candan et al. (Eds.): DASFAA 2017, Part I, LNCS 10177, pp. 475–489, 2017.
DOI: 10.1007/978-3-319-55753-3_30

to simulate various contexts. Then, the recommendation results can be obtained from the combinations of such contextual parameters. For instance, the Factorization Machine (FM) [13] can capture single and pairwise feature interactions to mimic the original contexts. When there are huge numbers of contextual factors involved in the training, single factorization models may not be able to handle all of them appropriately. To compensate for this shortcoming, ensemble learning [19] is introduced into factorization methods for context-aware recommendation. As a successful ensemble learning tool, the Gradient Boosted Regression Tree (GBRT) [6,7,11] algorithm can be seamlessly integrated into many other algorithms because of the adaptability of boosting techniques [2,4,18].

Limitations and Motivations. The goal of both single and ensemble methods is to generate an item list that users are likely to prefer. However, most context-aware approaches [1,13,18] concentrate on mining user preferences from all the records without considering the benefits from specific records to some users. In other words, some records contribute to recommendations, whereas some others may reduce the performances. Specifically, there exist some records that have similar properties or categories, if such records can be gathered into one group according to some criterion, then learning respectively from different groups would be more targeted to further benefit the overall performances. This motivates us to divide original records into several groups to be learned separately, which we term *division learning*. Nevertheless, the division learning idea produces two sub questions: (1) How to divide records? (2) Do the preferences learned from groups really reflect the user interests? To solve the first problem, we can assume that records involve similar contexts and targets are reasonable to be placed in the same group. Here, target denotes user feedback, such as purchasing or rating, which is a strong indicator of user preference. Based on this assumption, the GBRT [6,7,11] algorithm is an appropriate choice to be improved for group division. Not only because regression trees consider both contexts and targets when dividing records, but also because the predicted values from tree leaves can represent the user interests in a specific group to some extent. Then, the key to solve the second problem is to compensate for the rough user tastes by mining personalized user preferences in each group precisely. The preference mining strategy should be efficient, because there are several groups and each group contains only a limited number of records to be learned. A small number of records means that the number of features is also small. This motivates us to utilize the entire small feature matrix rather than considering its factor dimension as existing complicated methods [13,18] do. Based on the small matrix, a lightweight algorithm can be created to capture personalized user interests termed *local preferences* that plays the role of compensation for *group preferences*. Finally, the division learning results (the two preferences) must be integrated to form a complete $dividing \Rightarrow mining \Rightarrow integrating$ learning process, which is described in detail as follows.

Our approach. In this paper, we propose a PReference INtegration (PRIN) approach to implement division learning and preference mining for context-aware recommendation. To summarize, the contribution of our work is fourfold:

1. To implement precise preference mining, we classified *preferences* into two categories: *group preferences* and *local preferences*. Group preferences indicate the overall user interests in a specific group, whereas local preferences reflect the personalized tastes of specific users.
2. For division learning, we improve the Gradient Boosted Regression Tree (GBRT) algorithm, transforming it into the Group Division Tree (GDT) algorithm. GDT leverages all the contexts and targets to divide the original records into several groups and generate group preferences.
3. We develop a lightweight regression method called the Local preference model (LPREF) to mine local preferences for records in each group. The role of local preferences is to compensate for the broad group preferences by capturing more individualized user tastes.
4. To fully reflect user interests, we integrate the two-layer preferences when providing a recommendation. The reasonableness of preference integration is discussed, moreover, we demonstrate that the combined results of LPREF converge to make preference integration reliable.

The remainder of the paper is organized as follows. In Sect. 2, we introduce the construction details of the preference integration approach, while Sect. 3 demonstrates the advantages of PRIN by analyzing the results of experiments. Section 4 lists related works to highlight similar studies. Finally, we give conclusions in Sect. 5 and provide some perspectives on further research.

2   Preference Mining and Integration

This session first introduces the framework of our approach by an example in Sect. 2.1. Then, the details of the two-layer preference mining follow in Sect. 2.2. Finally, in Sect. 2.3, we discuss reasonableness of the preference integration for recommendations.

2.1   Model Framework

Based on the idea of division learning, the first part of the framework is to implement group division by dividing original records into several groups, each of which contains similar records. An intuitive idea is to use similarity or distance metrics as the primary criteria for record selection. However, similar behaviors may lead to entirely different feedback, for example, the same clicking action on the same item may result in a purchase or only in browsing. The shortcoming of existing similarity criteria is that the similarity calculations do not include the targets (or called user feedback). In fact, it is the targets that play the strong indicator roles in presenting user preferences. A target can be a numerical value that denotes rating or a boolean value that indicates purchasing. In contrast to similarity measures, the Gradient Boosted Regression Tree (GBRT) [6,7,11] can divide the records according to both targets and contexts during the learning process. Most of the records in each leaf node have the same target after the tree

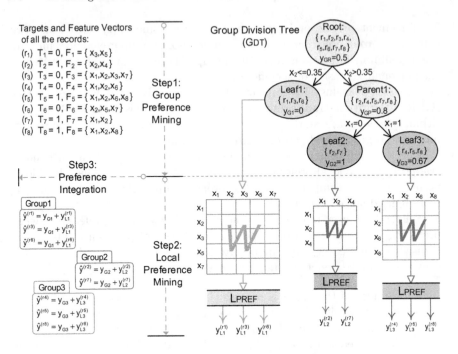

Fig. 1. Framework of the PRIN approach

was built. Hence, it is natural to adopt and improve GBRT for group division, which is implemented by a Group Division Tree (GDT) as shown in Fig. 1.

In step one, GDT has a simple and efficient structure by employing a binary regression tree for group division. Each record contains a target and a feature vector representing the relevant contexts, e.g., the target of record r_2 is $T_2 = 1$ and the feature vector of r_2 is $F_2 = \{x_2, x_4\}$. To simply explain the process, we use boolean values (0 or 1) to present targets, whereas feature values can be either numerical or boolean. First, all the records $\{r_1, r_2, \cdots, r_8\}$ in the *root* node are separated into two child nodes according to the best splitting feature x_2, which has a value of 0.35. The criterion for finding the best splitting feature and its value will be introduced in Sect. 2.2. Hence, the records where $x_2 \leqslant 0.35$ are assigned to the left child node and the records where $x_2 > 0.35$ are assigned to the right child node. Although record r_1 does not have feature x_2, it is assigned in the left child node because the default values of missing features are regarded as 0. Based on this setting, a binary tree does not require a third child to store records with missing feature values. The predicted value of each node is set to be the mean of all the record targets in that node, thus, we term the node predicted value *group preference* because this value indicates the overall user interest in a specific group (tree leaf). For instance, the group preference of the *root* node is $y_{GR} = 0.5$ and its left child has a pure group preference $y_{G1} = 0$. Because the left child contains all the records with 0 as the same target value, it is set to be the *leaf* node at which splitting stops. Then, the records in the right child continued to be split according to feature x_1 until the nodes in the third layer are changed into leaves, for the reason that the max tree depth is assumed to be 3. At this point, GDT growth is finished and the records have been divided into several groups.

In the second step, for each group (leaf node), the union set of features from all the records in that leaf are utilized to form a feature matrix W. For example, records $\{r_1, r_3, r_6\}$ in *Leaf1* have the union feature set $\{x_1, x_2, x_3, x_5, x_7\}$; thus, a 5×5 feature matrix is built. Based on these small feature matrices, the LPREF method described in Sect. 2.2 is called to implement local preference mining for each record in the leaves. Three predicted values of local preferences $y_{L1}^{(r_1)}$, $y_{L1}^{(r_3)}$, and $y_{L1}^{(r_6)}$ in *Leaf1* are learned for records r_1, r_3, and r_6, respectively. From this example, it can be observed that the advantage of local preferences is to refine the group preferences across a more precise range of records so that the tiny variations of user tastes can be captured.

Finally, the local preferences produced by LPREF are combined with the relevant group preferences to give recommendation. For instance, local preferences $\{y_{L2}^{(r_2)}, y_{L2}^{(r_7)}\}$ are relevant to y_{G2} because they are in the second group (*Leaf2*). Hence, the final predicted value of record r_7 is $\hat{y}^{(r_7)} = y_{G2} + y_{L2}^{(r_7)}$. This example simply illustrates one pair of group and local preference combination, a more general case of *Preference Integration* will be discussed in Sect. 2.3.

2.2   Two-Layer Preference Mining

The first layer. As discussed in Sect. 2.1, the first step (or layer) of preference mining employs a GDT to generate group preferences. GDT is a recursive partitioning method as shown in the first column of Table 1. The tree begins with a root node and grows by splitting the current node until stopping criterion met. We employ *node impurity* to be the splitting metric as GBRT does. More specifically, by traversing all the possible splitting features in the current node, the impurities of candidate child nodes are calculated. Then, the best split feature and its relevant value are found by minimizing the summation of child node impurities. There are a variety of criteria to estimate the node impurity of the current splitting. It was demonstrated that the Least-Squared Deviation (LSD) is an appropriate estimation of node impurity when both continuous and discrete features exist in the data set [5]. Hence, we adopt LSD as shown in Eq. 1 to compute the node impurity.

$$\mathcal{I_M} = \frac{1}{N} \sum_{r=1}^{N} (t_r - \bar{t}_r)^2 \tag{1}$$

where N is the record number in the current node. t_r indicates the distance between the predicted value $\hat{y}^{(r)}$ and the target $y^{(r)}$, which is usually set to be $\hat{y}^{(r)} - y^{(r)}$. Whereas $\bar{t}_r$ denotes the average value of t_r, meaning that $\bar{t}_r = \frac{1}{N} \sum_{r=1}^{N} t_r$. More children need more time to compute the impurity summation, thus we employ a binary tree that has only two children as the basic tree structure to reduce the calculation complexity.

Table 1. Two-layer preference mining algorithms

Group Division Tree (GDT)	Local Preference Model (LPREF)
1: **Input:** T_S - training set, d_t - depth of trees	1: **Input:** T_G - sub training set in a specific group
2: **procedure** GDT(T_S, d_t)	2: λ_w - regularization factor
3: $\hat{y}_G^{(r)} = \frac{1}{N} \sum_{r=1}^{N} y^{(r)}$	3: α - learning rate
4: **if** $d_t = max\_depth$ **then**	4: **function** LPREF(T_G, λ_w, α)
5: Set current node to be a leaf node	5: Generate $n \times n$ matrix W from T_G
6: Break	6: Initialize W with $\mathcal{N}(0,1)$ Gaussian
7: **end if**	7: **repeat**
8: Extract feature set F_S from T_S	8: Randomly choose row r in T_G with
9: **for each** f in F_S **do**	9: $target$ $y^{(r)}$ and feature vector $\mathbf{x}$
10: Sort T_S according to f value	10: $\hat{y}_L^{(r)} = 0$
11: $\mathcal{I}_\mathcal{M} = +\infty$	11: **for** $i = 1$ to n **do**
12: **for** $s = N$ **downto** 0 **do**	12: **for** $j = 1$ to n **do**
13: **if** $s > 0$ **then**	13: $\hat{y}_L^{(r)} \mathrel{+}= x_i w_{ji}$
14: $\mathcal{I}_\mathcal{L} = \frac{1}{s} \sum_{l=1}^{s} (t_l - \bar{t}_l)^2$	14: **end for**
15: **end if**	15: **end for**
16: **if** $s < N$ **then**	16: $\gamma = \hat{y}_L^{(r)} - y^{(r)}$
17: $\mathcal{I}_\mathcal{R} = \frac{1}{N-s} \sum_{r=s+1}^{N} (t_r - \bar{t}_r)^2$	17: **for** $i = 1$ to n **do**
18: **end if**	18: **for** $j = 1$ to n **do**
19: **if** $\mathcal{I}_\mathcal{L} + \mathcal{I}_\mathcal{R} < \mathcal{I}_\mathcal{M}$ **then**	19: $w_{ji} \mathrel{-}= \alpha \cdot (\gamma x_i + \lambda_w w_{ji})$
20: Get split value x_f by s	20: **end for**
21: $\mathcal{I}_\mathcal{M} = \mathcal{I}_\mathcal{L} + \mathcal{I}_\mathcal{R}$	21: **end for**
22: $best\_f = f$	22: **if** $Convergence$ **then**
23: $best\_x_f = x_f$	23: Break
24: **end if**	24: **end if**
25: **end for**	25: **until** stopping criterion met
26: **end for**	26: **return** W
27: Divide T_S into T_{SL}, T_{SR}	27: **end function**
28: according to $best\_f, best\_x_f$	
29: GDT($T_{SL}, d_t + 1$)	
30: GDT($T_{SR}, d_t + 1$)	
31: **end procedure**	

The main computation cost of GDT is searching the best splitting position that needs impurity calculation, which takes $\mathcal{O}(N)$ by considering both left child ($\mathcal{I}_\mathcal{L}$) and right child ($\mathcal{I}_\mathcal{R}$). Assume that there are N_f features, the total time for the selection of splitting position is in the order of $\mathcal{O}(N_f N^2)$. Because the sorting of records can be pre-computed and there are $N + 1$ candidate splitting position for each feature. For a small value of max depth ($d_t \leqslant 10$), the node number of a tree is small that can be regarded as a constant. Moreover, the record number in each node will be rapidly reduced when a tree is growing. Hence, the complexity of creating a regression tree is in the order of $\mathcal{O}(N_f N^2)$ if $d_t \leqslant 10$ is satisfied.

The second layer. The predicted group preference $\hat{y}_G^{(r)}$ of record r is automatically generated after GDT was built. The original training set T_S is divided into several sub sets as the inputs of the LPREF approach, which means that the number of local preference models equals to the number of groups (or tree leaves). Each LPREF method contains different training data T_G, whose union set of features gives birth to different feature matrix W. The goal of local preference mining is to fully utilize W for personalized taste prediction. W is much smaller than the original feature matrix that involves all the features, meanwhile, the record number of T_G is also small. Hence, traditional pairwise interactions

[13,18] of existing methods are complicated and not applicable in this scenario. In contrast, pure linear regression models are too simple to represent local interests. To address such limitations, we propose an extended regression method that is more lightweight than pairwise algorithms and more expressive than linear regression models as shown in Eq. 2.

$$\hat{y}_L(\mathbf{x}) = \sum_{i=1}^{n} \left(x_i \sum_{j=1}^{n} w_{ji} \right) \tag{2}$$

where the local preference $\hat{y}_L$ is derived from weighting combination of n dimensional feature vector $\mathbf{x} = \{x_1, x_2, \cdots, x_n\}$. Here, w_{ji} denotes the element in row j and column i of feature matrix W; the summation of w_{ji} is the extended regression coefficient of feature x_i. In this way, the feature interactions are implicitly expressed by their coefficients and are more robust than direct feature interactions, because the coefficients in W can still have an effect even when the values of two interactive features are missing. Based on an L-2 regression loss function $(\hat{y}_L^{(r)} - y^{(r)})^2 + \lambda_w w_{ji}^2$, the training of our local models can be implemented by Stochastic Gradient Descent (SGD) as illustrated in the second column of Table 1.

LPREF is highly efficient because its complexity is only $\mathcal{O}(n_a^2)$ for each iteration, where n_a is the average feature length of the records in T_G. Moreover, T_G is only a sub set of N training records. Hence, the running time to train the group records has an upper bound of $\mathcal{O}(N n_a^2)$, where N is the record number of T_S. We prefer such linear combination technique not only because of its high efficiency in mining partial records, but also because of its advantage of additive convergence that guarantees the entire convergence of the PRIN approach. The integrating of the two-layer preferences to complete our entire algorithm will be discussed in the next section.

2.3 Preference Integration

The purpose of preference integration is to combine each pair of group and local preference into one final score in an appropriate way. The group preferences and the local preferences reflect two different levels of user interests, which can compensate for each others to form a strong indicator of user tastes. Hence, the integration of them is reasonable. However, we have to demonstrate the convergence of the integrating result to make the recommendation reliable. Without loss of generality, consider M pairs of group and local preferences that can be integrated as follows.

$$\hat{y} = \sum_{p=1}^{M} \eta_p (\hat{y}_{G_p} + \hat{y}_{L_p}) \tag{3}$$

where $\hat{y}_{G_p}$ denotes the group preference and $\hat{y}_{L_p}$ indicates the local preference of the p-th pair, η_p is the shrinking factor of the p-th preference pair that satisfies $\sum_{p=1}^{M} \eta_p = 1$ and $\eta_p > 0$. The group preference part $\sum_{p=1}^{M} \eta_p \hat{y}_{G_p}$ is actually the additive leaf values of regression trees, the convergence of which had been discussed in the existing literature [7]. What we emphasize is the convergence of the additive local preference part $\sum_{p=1}^{M} \eta_p \hat{y}_{L_p}$. Consider the target value y and the p-th local predicted preference value $\hat{y}_{L_p}$ that can be trained by LPREF to satisfy the following inequality.

$$y - \epsilon \leqslant \hat{y}_{L_p} \leqslant y + \epsilon \tag{4}$$

where ϵ indicates a positive infinitesimal. Inequality 4 means that $\hat{y}_{L_p}$ can converge to y through local preference learning. In other words, the SGD training process of LPREF can theoretically guarantee that $\hat{y}_{L_p}$ converges to y. Then, we expand $\hat{y}_{L_p}$ according to Eq. 2 and convert it into the following format to benefit demonstration.

$$
\begin{aligned}
\hat{y}_{L_p}(\mathbf{x}_p) &= \sum_{i=1}^{n} \left(x_i^{(p)} \sum_{j=1}^{n} w_{ji}^{(p)} \right) = \left(\sum_{j=1}^{n} w_{j1}^{(p)} \right) x_1^{(p)} + \left(\sum_{j=1}^{n} w_{j2}^{(p)} \right) x_2^{(p)} + \cdots \\
&+ \left(\sum_{j=1}^{n} w_{jn}^{(p)} \right) x_n^{(p)} = \widetilde{w}_{.1}^{(p)} x_1^{(p)} + \widetilde{w}_{.2}^{(p)} x_2^{(p)} + \cdots + \widetilde{w}_{.n}^{(p)} x_n^{(p)} = \widetilde{\mathbf{w}}_p^T \mathbf{x}_p
\end{aligned}
\tag{5}
$$

here $\widetilde{w}_{.n}^{(p)}$ is the ensemble linear regression coefficient of $x_n^{(p)}$, which is the element summation in the n-th column of feature matrix W. The coefficients $\{\widetilde{w}_{.1}^{(p)}, \widetilde{w}_{.2}^{(p)}, \cdots, \widetilde{w}_{.n}^{(p)}\}$ are the elements of the row vector $\widetilde{\mathbf{w}}_p^T$. By substituting the result of Derivation 5 into Inequality 4 and multiplying both sides of the inequality with a shrinking factor η_p in Eq. 3, we obtain a new inequality as follows.

$$\eta_p(y - \epsilon) \leqslant \eta_p(\widetilde{\mathbf{w}}_p^T \mathbf{x}_p) \leqslant \eta_p(y + \epsilon) \tag{6}$$

Note that the shrinking factor $\eta_p > 0$ and $\sum_{p=1}^{M} \eta_p = 1$. Thus, the integration of all the local preferences $\hat{y}_{L_p}$ results in convergence as shown by the following inequality.

$$y - \epsilon \leqslant \sum_{p=1}^{M} \eta_p(\widetilde{\mathbf{w}}_p^T \mathbf{x}_p) \leqslant y + \epsilon \tag{7}$$

This is because both $y - \epsilon$ and $y + \epsilon$ are constants that stay unchanged after being multiplied by 1, for example, $\sum_{p=1}^{M} \eta_p(y - \epsilon) = (y - \epsilon) \sum_{p=1}^{M} \eta_p = y - \epsilon$. This linear system has the advantages of lightweight and additive convergence, which are the reasons why we utilize Eq. 2 to model local preferences. Consequently, the preference integration result is converged because the summation of finite pairs of group preferences and local preferences are also converged. In fact, the experiments will prove that one pair of preference integration can outperform other advanced methods as shown in Sect. 3.3. Algorithm 1 illustrates the entire learning process for one pair preference integration.

The creation of GDT needs $\mathcal{O}(N_f N^2)$, where N_f denotes feature number and N is the record number. The K group LPREF algorithms can be carried out in parallel, because each LPREF model has its own feature matrix that will not

Algorithm 1. Preference Integration (PRIN)

1: **Input:** T_S - training set, d_t - depth of trees, α - learning rate
2:           λ_w - regularization factor, η - shrinking factor of preference integration
3: **for** each record r in T_S **do**
4:     $\hat{y}^{(r)} = 0$ ▷ initialize predicted values
5: **end for**
6: TREE = GDT(T_S, d_t) ▷ learn group preferences
7: Divide T_S into K groups according to TREE leaves
8: **for all** $k \in \{1, \cdots, K\}$ **in parallel do**
9:     **for** each record r in $T_S[k]$ **do**
10:        $\hat{y}_G^{(r)}$ = TREE.LEAF$[k].value$
11:        $\hat{y}^{(r)} += \eta \cdot \hat{y}_G^{(r)}$ ▷ group preference integration
12:    **end for**
13:    W = LPREF$(T_S[k], \lambda_w, \alpha)$ ▷ learn local preferences
14:    **for** each record r in $T_S[k]$ **do**
15:        $\hat{y}_L^{(r)} = \sum_{i=1}^{n} \left(x_i \sum_{j=1}^{n} w_{ji} \right)$
16:        $\hat{y}^{(r)} += \eta \cdot \hat{y}_L^{(r)}$ ▷ local preference integration
17:    **end for**
18: **end for**

affect those of the others during the training process. Thus, the total complexity of local preference mining is equivalent to the consumption of only one LPREF method, which is $\mathcal{O}(Nn_a^2)$ and n_a is the average length of feature vectors in T_S. Based on the above discussion, the running time of PRIN is in the order of $\mathcal{O}(N(N_f N + n_a^2))$. Actually, recommendation datasets are often sparse, the feature number (length of feature vector) of each record is mush smaller than total feature number ($n_a \ll N_f$). Meanwhile, the feature length n_a is also smaller than the record number N ($n_a < N$) when N is large enough. Hence, in most cases, the condition $n_a^2 < N_f N$ holds; therefore the total complexity of PRIN is in the order of $\mathcal{O}(N_f N^2)$.

3   Experiments and Analysis

3.1   Datasets and Evaluation Metrics

We conducted experiments on two real world datasets derived from the Movie-Lens (ML) web-site[1] and the Yahoo! Webscope program[2]. The first dataset is the *ML-Latest* dataset that describes rating and tagging activities from Movie-Lens. *ML-Latest* contains $100,023$ records with 706 users and $8,552$ items. The movie contexts include *genre* and *tag*, whereas the user contexts involve *rating* and *rating timestamps* that can be extended to four sub contexts: *year*, *month*, *hour*, and *day*. In contrast to *ML-Latest*, the *Yahoo!-Movie* (R4) dataset contains $221,367$ records with $7,642$ users and $11,915$ items. The user profiles in *Yahoo!-Movie* involve *user id*, *birthyear*, and *gender*, whereas the contextual information of items is far richer than that of *ML-Latest*. Therefore, we chose all the user contexts and some representative item contexts, such as *actors*, *genres*,

---

[1] https://movielens.org/.
[2] http://webscope.sandbox.yahoo.com.

and *directors*, to carry out experiments. In each dataset, 80% observations were randomly chosen for model training, whereas the remaining 20% records were left for testing. To make the experimental results reliable, the ten-fold cross-validation strategy was employed throughout all the experiments.

To evaluate the experimental results, common metrics such as *Precision*, *Recall*, *F1-Measure*, and *AUC* (Area Under the ROC Curve) were adopted for evaluations. For each user, an item list that consisted of *top*-10 highest scored items was generated for recommendation. In the following experiments the influences of the model granularity were first investigated in Sect. 3.2. Then, we focused on method comparisons in Sect. 3.3 to finally analyze and demonstrate the advantage of the PRIN approach.

3.2   Influences of the Model Granularity

The model granularity of our algorithm is determined by the number of groups, that is, the number of local models. The settings of granularity directly affect the model performances. A small number of groups does not make a careful division of original records, thus, LPREF may not play its full role in local preference mining for specific users. In contrast, a large number of groups would provide insufficient records for each LPREF model to mine user interests. Hence, in each dataset, there exists an optimal group number for the PRIN method to achieve its best performances on most metrics. To investigate the influences of the model granularity, we changed the depth of GDT to adjust the number of groups. In fact, the GDT depth (denoted by d_t), is the only hyper-parameter that directly determines the group number. For example, in the *Yahoo!-Movie* dataset, a GDT with $d_t = 5$ gave birth to 16 groups, which means that 16 LPREF models were created. In comparison, a GDT with $d_t = 10$ generated 138 groups for LPREF to learn. This indicated that the number of local models grew rapidly when the tree depth was increasing, moreover, the minimum number of d_t was 2 because at least 2 groups were required for preference integration. Based on the above discussion, we gradually increased d_t from 2 to 10 in two datasets to observe the influences of model granularity as shown in Fig. 2 and Table 2.

Table 2. Granularity Influences in *Top*-10 Recommendations on *AUC*

Datasets	$d_t = 2$	$d_t = 3$	$d_t = 4$	$d_t = 5$	$d_t = 6$	$d_t = 7$	$d_t = 8$	$d_t = 9$	$d_t = 10$
ML-Latest	0.6931	0.7269	0.7655	0.8097	0.8331	**0.8736**	0.8335	0.7633	0.7426
Yahoo!-Movie	0.6561	0.6870	0.6941	0.7182	0.7393	0.7694	0.8163	**0.8246**	0.7618

The LPREF learning rates for *ML-Latest* and *Yahoo!-Movie* were set to be $1.6e^{-3}$ and $2.4e^{-3}$ respectively, whereas the regularizing factors of the two datasets were unified as $1.2e^{-3}$. From *top*-1 to *top*-10 recommendations, the *Recall* performances gradually increased whereas the *Precision* performances gradually reduced in the two datasets. This resulted in an optimal range on *F1-Measure*. For instance, the PRIN algorithm performed well from *top*-5 to

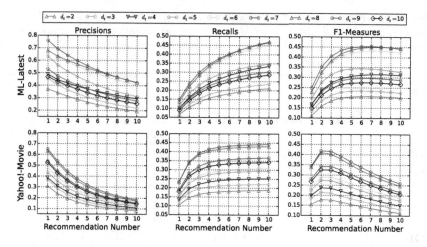

Fig. 2. Influences of the Model Granularity on *Precision*, *Recall*, and *F1-Measure*

top-8 recommendations in *ML-Latest*, where the best range of performances in *Yahoo!-Movie* was between *top*-2 and *top*-4 recommendations. The reason was that the two datasets had different characteristics such as record number, contexts, and average rating number, e.g., the average rating number per user in *ML-Latest* was much larger than that of *Yahoo!-Movie*. Consequently, the optimal granularities of the two datasets were also different. The PRIN model in *ML-Latest* required a GDT with $d_t = 7$ for its best granularity, while in *Yahoo!-Movie* the optimal model granularity was smaller ($d_t = 9$). Because bigger tree depth indicated more groups that represented smaller granularity. Specifically, the record and context numbers of *Yahoo!-Movie* were larger than those of *ML-Latest*, therefore *Yahoo!-Movie* needed more groups for preference learning. The *AUC* performances were consistent with those on the above three metrics, thus, we adopted the optimal model granularities in this experiments to compare our method with other advanced models in the next section.

3.3 Method Comparisons

The compared models were technically related to our approach. First, an improved GBRT method, the core algorithm belonging to RT-Rank [11], was employed for comparisons, because we borrowed the similar GBRT techniques to propose the GDT approach. The second and third methods were the Factorization Machine (FM) [13] and the Attribute Boosting (AB) [18] recommenders, both of them were state-of-the-art factorization models that were worth comparing with the LPREF part. The improved version of AB was the Attribute and Global Boosting (AGB) method, which leveraged GBRT as the global optimization part of AB. Consequently, AGB was also comparable because it combined factorization model with GBRT as the PRIN approach did. We conducted method comparisons on these algorithms and appropriately adjusted their hyperparameters according to two datasets as listed in Table 3.

Table 3. The hyper-parameter settings of compared methods. The notation M denotes *ML-Latest* and Y indicates *Yahoo!-Movie*. If the two notations are not specified, it means that the value is adopted in both datasets. *Dimension* is the abbreviation of *factor (or feature) dimension* indicating the dimension of low-rank factors in feature matrices, moreover, *Regularizing Factor* is denoted by λ whereas *Shrinking Factor* is represented by η.

Methods	Tree depth	Tree number	Dimension	Learning rate	Regularizing/ Shrinking factor
GBRT	$16(M), 25(Y)$	$6(M), 9(Y)$	-	-	$\eta = 1.0e^{-1}$
FM	-	-	64	$1.2e^{-4}(M),$ $0.9e^{-4}(Y)$	$\lambda = 0.8e^{-6}$
AB	-	-	64	$1.8e^{-3}(M),$ $2.8e^{-3}(Y)$	$\lambda = 1.2e^{-5}$
AGB	$4(M), 6(Y)$	$15(M), 25(Y)$	64	$1.8e^{-3}(M),$ $2.8e^{-3}(Y)$	$\lambda = 1.2e^{-5},$ $\eta = 3.0e^{-2}$
PRIN	$7(M), 9(Y)$	1	-	$1.6e^{-3}(M),$ $2.4e^{-3}(Y)$	$\lambda = 1.2e^{-3},$ $\eta = 1.0$

The *shrinking factor* η was inversely proportional to the tree number, e.g., GBRT required less than 10 tall trees, thus, we set $\eta = 1.0e^{-1}$. Whereas AGB needed more short trees so that η had a small value of $3.0e^{-2}$. In contrast to them, PRIN required only one tree to generate one pair of group and local preferences, hence, η was set to be 1.0. All the factorization models had the same *factor dimension*, however, it was not necessary to provide this hyper-parameter for LPREF because it utilized the entire feature matrix. We adjusted both *learning rate* and *regularizing factor* for each model and adopted SGD in training to prevent over-fitting. Based on these settings, the performances of method comparisons were shown in Fig. 3.

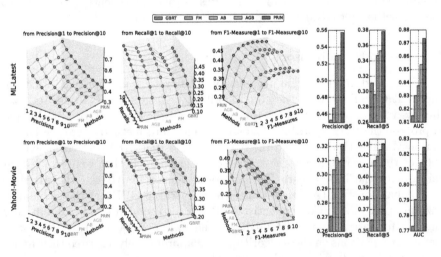

Fig. 3. Performances of method comparisons

In the *ML-Latest* dataset, FM had high *Precisions* and low *Recalls* in contrast to GBRT, while two attribute boosting approaches performed better than both FM and GBRT. In particular, AB held an advantage when the number of recommended items was relatively large, e.g., from *top*-5 to *top*-10 recommendations. In the *Yahoo!-Movie* dataset, four factorization models outperformed GBRT on all the measures. AB showed its advantages on *Precisions* and AGB had good performances on other metrics. It can be observed in two datasets that employing regression trees can bring improvements on the *Recall* measure. For example, GBRT outperformed FM on *Recalls* in *ML-Latest* and AGB enhanced the *Recall* performances based on AB in *Yahoo!-Movie*. This phenomenon existed in tree-based models except for the PRIN approach. Because our method not only leveraged trees for preference representations, more importantly, the GDT algorithm implemented division learning to accurately control the learning range, in which LPREF can carry out more targeted and personalized preference mining. Consequently, this new learning strategy made PRIN outperformed other methods on all the metrics, although the complexity of LPREF was much lower than those of FM and the AB family. As a matter of fact, division learning was a way of targeted modeling. This experiment illustrated that an appropriate targeted modeling was more important than simply increasing the model complexity, which was the main idea and highlight of our approach.

4   Related Works

Collaborative Filtering (CF) algorithms [15] usually capture user-item relations through Matrix Factorization (MF) techniques [10,16]. The user-item relations are commonly represented as user actions on items that are termed feedback. Therefore, feedback modeling [9,10,14] is quite popular in MF methods. Representative methods involve SVD++ [10], WRMF [9], and BPR-MF [14]. These matrix-based approaches lay a solid foundation for factorization models, including context-aware approaches.

In context-aware recommendation [1], one group of researchers integrated contextual factors into existing MF models to propose Tensor Factorization [17]. Another successful common methods, such as the Factorization Machine (FM) [13], treated all features equally to capture the single and pairwise interactions between them. This category of techniques that were concerned with modeling feature interactions [13,18], was particularly useful in preference mining. However, the efficiency of context-aware factorization approaches degrades when there are large numbers of features on which interactions are carried out. In these cases, GBRT [6,7,11] can benefit the factorization models. Because GBRT has the ability to handle thousands of features by distinguishing different targets efficiently, moreover, it can be applied in regression, classification, and ranking problems based on the flexible boosting techniques [2,4]. For instance, researchers in [2] showed that GBRT can be utilized as a general framework for functional matrix factorization. In other words, this means that the integration of regression trees and MF is seamless. For context-aware methods, some researchers have employed regression trees with existing methods [3,11,18]. For example, [3] improved FM by developing a tree-based greedy solution to identify useful feature pairs as its inputs. Another approach is to leverage GBRT as a part of existing models. The authors of [18] proposed an attribute boosting algorithm to implement preference mining and used regression trees as the global portion of their methods. In this work, we refine the regression tree approach not only using them for preference scoring, but also for division learning to implement targeted modeling.

5   Conclusions

In this work, we develop the PRIN approach to carry out preference mining for context-aware recommendation. The highlight of our model is combining tree-based methods and factorization algorithms together to implement a *dividing, mining, and integrating* learning process, in which the user tastes were precisely learned by a two-layer preference mining approach. As future work, we mainly focus on extending the PRIN approach into a multi-layer recommender by another tools such as deep learning. It is a challenge because the division learning strategy should be refined to apply a multi-layer learning process, which is interesting for us to investigate in further research.

Acknowledgments. This research is supported by the National Natural Science Foundation of China with Grant No. 61272277. We acknowledge the editors and other anonymous reviewers for insightful suggestions on this work.

References

1. Baltrunas, L., Ludwig, B., Ricci, F.: Matrix factorization techniques for context aware recommendation. In: Proceedings of the Fifth ACM Conference on Recommender Systems, RecSys 2011, pp. 301–304. ACM, Chicago, 23–27 October 2011. http://doi.acm.org/10.1145/2043932.2043988
2. Chen, T., Li, H., Yang, Q., Yu, Y.: General functional matrix factorization using gradient boosting. In: Proceedings of the 30th International Conference on Machine Learning (ICML 2013). JMLR Workshop and Conference Proceedings, vol. 28, pp. 436–444. Microtome Publishing, USA, 16–21 June 2013. http://jmlr.csail.mit.edu/proceedings/papers/v28/chen13e.pdf
3. Cheng, C., Xia, F., Zhang, T., King, I., Lyu, M.R.: Gradient boosting factorization machines. In: Proceedings of the 8th ACM Conference on Recommender Systems, RecSys 2014, pp. 265–272. ACM, NY, 6–10 October 2014. http://doi.acm.org/10.1145/2645710.2645730
4. Chowdhury, N., Cai, X., Luo, C.: BoostMF: boosted matrix factorisation for collaborative ranking. In: Appice, A., Rodrigues, P.P., Santos Costa, V., Gama, J., Jorge, A., Soares, C. (eds.) ECML PKDD 2015. LNCS (LNAI), vol. 9285, pp. 3–18. Springer, Cham (2015). doi:10.1007/978-3-319-23525-7_1
5. Larsen, D.R., Speckman, P.L.: Multivariate regression trees for analysis of abundance data. Biometrics **60**(2), 543–549 (2004). http://iras.lib.whu.edu.cn:8080/rwt/JSTOR/http/P75YPLUKPN4G86SPN73GH/stable/3695785

6. Elith, J., Leathwick, J.R., Hastie, T.: A working guide to boosted regression trees. J. Animal Ecol. **77**(4), 802–813 (2008). http://dx.doi.org/10.1111/j.1365-2656.2008.01390.x

7. Friedman, J.H.: Greedy function approximation: a gradient boosting machine. Ann. Stat. **29**(5), 1189–1232 (2001). http://www.jstor.org/stable/2699986

8. Hosseini, S., Li, L.T.: Point-of-interest recommendation using temporal orientations of users and locations. In: Navathe, S.B., Wu, W., Shekhar, S., Du, X., Wang, X.S., Xiong, H. (eds.) DASFAA 2016. LNCS, vol. 9642, pp. 330–347. Springer, Cham (2016). doi:10.1007/978-3-319-32025-0_21

9. Hu, Y., Koren, Y., Volinsky, C.: Collaborative filtering for implicit feedback datasets. In: Proceedings of the 8th IEEE International Conference on Data Mining, ICDM 2008, pp. 263–272. IEEE, Pisa, 15–19 December 2008

10. Koren, Y.: Factorization meets the neighborhood: a multifaceted collaborative filtering model. In: Proceedings of the 14th ACM SIGKDD International Conference on Knowledge Discovery and Data Mining, KDD 2008, pp. 426–434. ACM, NY (2008). http://doi.acm.org/10.1145/1401890.1401944

11. Mohan, A., Chen, Z., Weinberger, K., Chapelle, O., Chang, Y., Liu, T.Y.: Web-search ranking with initialized gradient boosted regression trees. In: Proceedings of the Learning to Rank Challenge. JMLR Workshop and Conference Proceedings, vol. 14, pp. 77–89. Microtome Publishing, Brookline, 25 June 2011. http://www.jmlr.org/proceedings/papers/v14/mohan11a.html

12. Pham, T.A.N., Li, X., Cong, G., Zhang, Z.: A general recommendation model for heterogeneous networks. IEEE Trans. Knowl. Data Eng. **28**(12), 3140–3153 (2016)

13. Rendle, S.: Factorization machines with libfm. ACM Trans. Intell. Syst. Technol. **3**(3), 57:1–57:22 (2012). http://doi.acm.org/10.1145/2168752.2168771

14. Rendle, S., Freudenthaler, C., Gantner, Z., Schmidt-Thieme, L.: Bpr: Bayesian personalized ranking from implicit feedback. In: Proceedings of the Twenty-Fifth Conference on Uncertainty in Artificial Intelligence, UAI 2009, pp. 452–461. AUAI Press, Arlington, 18–21 June 2009. http://dl.acm.org/citation.cfm?id=1795114.1795167

15. Ricci, F., Rokach, L., Shapira, B., Kantor, P.B.: Recommender Systems Handbook, vol. 1. Springer, US (2011). http://dx.doi.org/10.1007/978-0-387-85820-3_1

16. Salakhutdinov, R., Mnih, A.: Bayesian probabilistic matrix factorization using markov chain monte carlo. In: Proceedings of the 25th International Conference on Machine Learning, ICML 2008, pp. 880–887. ACM, NY, 5–9 July 2008. http://doi.acm.org/10.1145/1390156.1390267

17. Symeonidis, P., Nanopoulos, A., Manolopoulos, Y.: A unified framework for providing recommendations in social tagging systems based on ternary semantic analysis. IEEE Trans. Knowl. Data Eng. **22**(2), 179–192 (2010)

18. Zheng, L., Zhu, F., Mohammed, A.: Attribute and global boosting: a rating prediction method in context-aware recommendation. Comput. J. (2016). http://comjnl.oxfordjournals.org/content/early/2016/03/25/comjnl.bxw016.abstract

19. Zhou, Z.H.: Ensemble Methods: Foundations and Algorithms, 1st edn. Chapman & Hall/CRC, London (2012)

Jointly Modeling Heterogeneous Temporal Properties in Location Recommendation

Saeid Hosseini[1(✉)], Hongzhi Yin[1], Meihui Zhang[2], Xiaofang Zhou[1],
and Shazia Sadiq[1]

[1] The University of Queensland, Brisbane, Australia
saeid.hosseini@uq.net.au, h.yin1@uq.edu.au, {zxf,shazia}@itee.uq.edu.au
[2] Singapore University of Technology and Design, Singapore, Singapore
meihui_zhang@sutd.edu.sg

Abstract. Point-Of-Interest (POI) recommendation systems suggest interesting locations to users based on their previous check-ins via location-based social networks (LBSNs). Individuals visiting a location are partially affected by many factors including social links, travel distance and the time. A growing line of research has been devoted to taking advantage of various effects to improve existing location recommendation methods. However, the temporal influence owns numerous dimensions which deserve to be explored more in depth. The subset property comprises a set of homogeneous slots such as an hour of the day, the day of the week, week of the month, month of the year, and so on. In addition, time has other attributes such as the recency which signifies the newly visited locations versus others. In this paper, we further study the role of time factor in recommendation models. Accordingly, we define a new problem to jointly model a pair of heterogeneous time-related effects (recency and the subset feature) in location recommendation.

To address the challenges, we propose a generative model which computes the probability for the query user to visit a proposing location based on various homogeneous subset attributes. At the same time, the model calculates how likely the newly visited venues obtain a higher rank compared to others. The model finally performs POI recommendation through combining the effects learned from both homogeneous and heterogeneous temporal influences. Extensive experiments are conducted on two real-life datasets. The results show that our system gains a better effectiveness compared to other competitors in location recommendation.

Keywords: Heterogeneous time-related influence · Hybrid temporal location recommendation · Location-based service

1 Introduction

Recently, more and more people use GPS-enabled smart devices to easily socialize and share their visiting history through Location-based Social Networks (LBSNs). The LBSN geo-social platform involves the users, locations, and

© Springer International Publishing AG 2017
S. Candan et al. (Eds.): DASFAA 2017, Part I, LNCS 10177, pp. 490–506, 2017.
DOI: 10.1007/978-3-319-55753-3_31

contents[1]. The user performs the check-in at a particular venue and reports her current location at the time, enclosed with further contents (text, photo, video and etc.). In return, the check-in data is used by Point-Of-Interest (POI) recommendation systems to propose new appealing places to the users. A user-location matrix can be generated from a check-in dataset. Each entry represents the frequency of a user's visit at a location [1] or a binary value (e.g. [16,22]) indicating whether she has visited the location or not. Such a matrix is sparse because people mostly perform a limited number of check-ins at various locations. The main challenge of location recommendation in LBSN ecosystem is to mitigate the sparsity in user-location matrix. The prevalent *Collaborative Filtering (CF)* [2,9,16,21,22] classified into memory-based and model-based approaches have already been pervasively used to tackle the location recommendation problem. CF methods recommend locations visited by those who gain higher check-in similarity to the query user. In addition, alongside CF methods, influencing parameters such as *social links* [1,26], *spatial proximity* [12,22], and *content similarity* have been utilized to promote the effectiveness of POI recommender systems. Numerous works have included the *temporal influence* in location recommendation. Nevertheless, in this paper we aim to explore it more in depth.

Intuitively, the time dimension comprises multiple intervals with subset relation in between. Hence, we take this feature into consideration to model a set of homogeneous time-related attributes. We select four scales (i.e. $hour \subset day \subset week \subset month$) based on the density and the duration of our datasets (Sect. 3.3). However, our proposed model can include more intervals. The previous works solely consider one or two temporal scales such as hour [5,17,22,26] or day, and weekday/weekend periods [5,28] to circumvent overfitting concerns [26]. In addition, we model the recency attribute that is two-fold: Firstly, users with recent check-in activity must obtain higher weights in CF method's similarity metrics. Secondly, the locations which have recently absorbed more check-ins should be ranked higher in the recommendation list. To summarize, our model jointly incorporates two heterogeneous temporal attributes of the subset and recency in location recommendation. To this end, we employ the recency attribute in CF module of the framework. Moreover, we propose a latent generative module (Fig. 1) which fuses multiple homogeneous temporal scales of the subset property in location recommendation.

To start with, we associate a three dimension UTP (User-Time-POI) matrix with each temporal aspect to observe whether every user has visited a location (0 or 1) at a particular time (e.g. hour of the day) or not. UTP cubes are more sparse than user-location (or user-POI) matrices that merely report whether the query user has visited a location or not (disregarding the time factor). Accordingly, we devise a clustering method to decrease the sparsity in UTP cubes through merging similar slots (hours, days, weeks and etc.) and construct the temporal slabs. For instance, having a set of $\{11, 12\}$ as a temporal block constructed for the hour latent factor causes the check-ins between 11 am and 1 pm to be treated the same collectively. The Bayesian generative model will then intake the

---

[1] http://en.wikipedia.org/wiki/Geosocial_networking.

temporal slabs to recommend top K locations which are temporally correlated to the query user's existing check-in log. Moreover, we further utilize an *Expectation Maximization* approach to infer the latent parameters to compute the final time-related similarity metric between the query user and the proposed locations.

In short, this work concentrates on the problem of enhancing the location recommendation in LBSN environment. To the best of our knowledge, no prior work jointly considers the set of homogeneous and heterogeneous time-related features in LBSN based location recommendation systems. The proposed model in this paper specifically presents the following new contributions:

- We propose an approach which obtains a set of full similarity maps, each dedicated to a particular scale in temporal subset attribute.
- Rather than considering the effects of a sample temporal dimension in location recommendation [7], our proposed solution considers a set of concurrent homogeneous and heterogeneous temporal features in location recommendation.
- From one perspective the latent generative module predicts the query user's mobility patterns involving multiple homogeneous temporal effects denoted by the subset feature. From another perspective, the time decay effect integrates the recency influence in collaborative filtering module.

The rest of this paper is structured as follows. In Sect. 2, we provide the essential information about our framework which incorporates heterogeneous temporal effects (Recency + subset) in location recommendation. We explain how to employ the time-decay function in collaborative filtering module in Sect. 2.1. Moreover, in order to decrease the sparsity of the UTP cubes for all intervals, in Sect. 2.2 we present the method which exploits the temporal blocks out of similar slots. The parameter inference algorithm of our solution is elucidated in Sect. 2.3. The evaluation metrics are also explained in Sect. 3. Related research work is surveyed in Sect. 4. Finally, we conclude this paper in Sect. 5 to include promising future directions.

2   Modeling the Heterogeneous Temporal Effects

POI recommendation problem can be considered as a conditional visiting probability of the location l_j given the query user u_i. All users are treated the same, so the probability of location l_j to be visited by user u_i is proportionate to their joint probability (Eq. 1).

$$Pr(l_j|u_i) \propto Pr(u_i, l_j) \tag{1}$$

Talking about the time-related influences, a growing line of research [5,17, 22,26,28] has also utilized the time entities to foster the effectiveness of location recommendation systems. Nevertheless, most of them merely consider one or two temporal attributes of hour, day, or weekly periods. We include a mixture of heterogeneous temporal effects designated in a unified framework. From one perspective, we define four homogeneous temporal granularities. Relevant latent

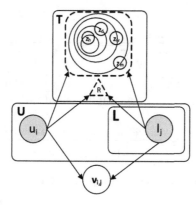

Fig. 1. The graphical representation of the model in location recommendation.

factors denoted as z_h, z_d, z_w, and z_m are respectively associated with hour, day, week and month of the year. In practice, such homogeneous temporal scales embody a time-related subdivision feature ($minute \subset hour \subset day \subset week$). In denser datasets, our solution can employ more latent factors including smaller scales (e.g. minutes). Adversely, if the dataset covers several years of check-in history, we can append the year latent factor (e.g. z_y). From another perspective, we also fuse the recency effect (indicated by R) in location recommendation.

As illustrated in Fig. 1, the query user $u_i \in U$ can visit any location $l_j \in L$ influenced by certain parameters. The parameter $\nu_{i,j}$ delegates primary impacts (Sect. 2.1). Accordingly, $Pr_\nu(l_j|u_i)$ is the probability for user u_i to visit location l_j. This can be computed through a mixture model [16] consisting of social and geographical effects which are built upon the CF module (Sect. 2.1). We continue with Sects. 2.1 and 2.2 which explain recency based CF module and the method in leveraging of the temporal slabs respectively. Both of these modules are consumed with our final location recommendation framework which is proposed in Sect. 2.3.

2.1   Integrating Recency in Collaborative Filtering

We integrate the recency effect in collaborative filtering module of our location recommendation framework. We take two directives into consideration: (1) The user who has performed recent check-ins must obtain higher similarity weights versus others. (2) The location which has been visited recently should gain higher rates in the recommendation. Subsequently, we define the time function $f(t_{k,j}) \in [0,1]$ in order to determine the recency significance of l_j visited by u_k at the time t. The value deduces by time monotonically. Each of the two rules adjusts CF module respectively. We obtain the first rule by multiplying the average temporal value of $f(t_{k,j})$ to the baseline similarity metric as formulated in Eq. 2.

$$w^*_{i,k,t} = \frac{\sum_{l_j \in L_k} f(t_{k,j}) \sum_{l_j \in L_i \cap L_k} c_{i,j} c_{k,j}}{|L_k| \sqrt{\sum_{l_j \in L_i} c^2_{i,j}} \sqrt{\sum_{l_j \in L_k} c^2_{k,j}}} \tag{2}$$

Likewise, in order to consider the second rule, we update the value of $c_{k,j}$ by including the time factor $(f(t_{k,j}))$ as implemented in Eq. 3.

$$c^*_{i,j,t} = \frac{\sum_{\{\forall u_k | w_{i,k} > 0\}} w_{i,k} c_{k,j} f(t_{k,j})}{\sum_{\{\forall u_k | w_{i,k} > 0\}} w_{i,k}} \tag{3}$$

We employ the time-decay model to compute the recency metric [4].

$$f(t) \in [0,1], \quad f(t_{1/2}) = 0.5, \quad t_{1/2} = \frac{1}{\Phi} \tag{4}$$

The value of the time function for u_k regarding the check-in at l_j is between 0 and 1 (Eq. 4). $t_{1/2}$ is the half-time parameter. Hence the highest value of freshness for u_k on the check-in date will be 1 $(f(0))$ and it will deduce to 0.5 on $t_{1/2}$ days. $t_{1/2}$ and the deduction rate (Φ) have reverse relation in time function (Eq. 5):

$$f(t) = e^{-\Phi.t} \tag{5}$$

The higher value of Φ grants lower significance to the old check-ins. Hence, in Sect. 3.4 we find an optimal value for $t_{1/2}$ to secure the best performance.

2.2   Leveraging Homogeneous Temporal Slabs

UTP cubes are more dispersed than User-location matrices. The smaller the scale, the sparsity extends higher. For example, the UTP cube associated with the minute scale is far scattered compared to the hours. We witness in our datasets (Sect. 3.3) that the minute's UTP cube is extremely scattered and the duration of the whole check-in histories is less than 5 years. Hence we select a set of four latent parameters between the minute and the year (i.e. hour:z_h, day:z_d, week:z_w, and month:z_m). The latent factors reflect the homogeneous scales of the temporal subdivision attribute (e.g. $z_h \subset z_d$). In other words, each user u_i visits a location l_j defined by four aspects as denoted by $\mathbb{T} = \{z_h, z_d, z_w, z_m\}$.

We can apply Cosine or Pearson similarity metrics to find similar hours, days, weeks and months [22]. The final value will be the average similarity metric for those users who have the check-ins in both of slots. Eventually, we can merge similar slots in each scale (e.g. similar hours). Table 1 reports exploited similar temporal slabs. For example, the set of {21,22,23} in Foursquare dataset is an hourly slab made up of 9, 10, and 11pm. Hence, as the check-ins during 21–24 will be counted together, the sparsity will decrease in the hour UTP cube. We can formulate the problem of leveraging the homogeneous temporal slabs as follows:

Table 1. Similar time-related slabs

	Foursquare	Brightkite
Hour slabs	{0,1}{2,3}{4,5}{6}{7}{8,9}, {10,11,12,13,14,15}{16,17,18,19,20}, {21,22,23}	{0,1,2}{3,4}{5}{6}{7}{8}{9,10} {11,12,13},{14,15}{16,17,18,19} {20,21,22,23}
Day slabs	{Mon,Wed,Fri,Sat}, {Tue,Thu},{Sun}	{Mon},{Sat,Sun}, {Tue,Wed}, {Thu,Fri}
Week slabs	{1,4,5}{2,3}	{1}{2,3,4}{5}
Month slabs	{3,8}{4,5,6,7}{1,9,10}{2,11,12}	{1,2,11,12}{3,4,5,6}{7,8,9,10}

Problem 1 (**Leveraging Similar Temporal slabs**). *Given the set of users* ($\mathbb{U}$), *their check-in logs* $\mathbb{L} = \{L_1, L_2, \ldots, L_n\}$ *(i.e.,* L_1 *is* u_1*'s check-in log) and predefined four-fold latent parameters (*$\mathbb{T} = \{z_h, z_d, z_w, z_m\}$*) representing hour, day, week and the month, our goal is to leverage all possible temporal slabs of similar times that are associated with each of the latent factors.*

Accordingly, we choose stratified sampling method which divides the dataset into certain portions including active and cold start users. We process $m\%$ of the users in every iteration and compute similarity values until we achieve an ideal n similarity samples for each pair in every latent factor. We then obtain the matrices of $s * s$ implemented by the LINQ queries. The s denotes the number of slots in the scale (e.g. 24 for z_h). After completing the whole process, we use matrix factorization [6] to find missing values in the similarity matrices. Figure 3 illustrate the similarity maps in both datasets.

We observe that: (1) The maps regarding z_h factor in both datasets are quite similar. But the temporal mobility pattern regarding other latent factors are different. (2) Proximate temporal slots are more similar as reported earlier by [22]. (3) Considering the check-in density, the z_h similarity map in Brightkite is more smooth compared to its Foursquare counterpart. We used Hierarchical Agglomerative Clustering (*HAC* [3]) to merge similar slots of the maps. We applied *complete linkage* function alongside the distance threshold to ensure sufficient similarity among all temporal slots in each slab. As Table 1 shows we attained a set of quadrilateral temporal slabs which comprise merged slots from all four temporal aspects (Fig. 2).

A sample set of Λ_i^s in Foursquare dataset possesses four vector features which represent all time-related blocks of $z_m^r = \{April, May, June, July\}$, $z_w^v = \{Week2, Week3\}$, $z_w^v = \{Tue, Thu\}$, and $\{21, 22, 23\}$. We can now verbalize the problem regarding the recommendation using homogeneous four-folded temporal slabs:

Problem 2 (**Recommendation Via Homogeneous Temporal Factors**). *Given the dataset* $\mathbb{D}$*, a set of four latent temporal factors* $\mathbb{T}$ *representing the time-related subset feature (hour* $\subset$ *day* $\subset$ *week* $\subset$ *month), set of leveraged*

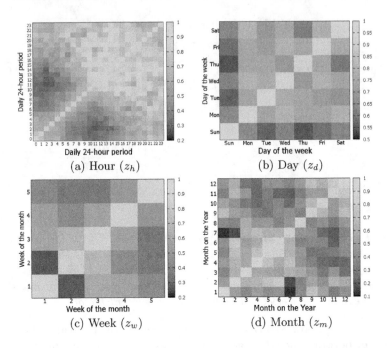

Fig. 2. Similarity between temporal slots in Foursquare dataset

quadrilateral temporal similarity slabs Λ^s based on $\mathbb{T}$, and the query user u_i, our goal is to suggest a list of new interesting locations where they are temporally correlated with u_i in four temporal aspects.

2.3   Recommendation via Recency and Homogeneous Scales

So far we have explained the method to integrate the recency impact in CF module as described in Sect. 2.1. Moreover, the value of $Pr_\nu(l_j|u_i)$ is jointly computed using CF module alongside the social and geographical influences. As reported in Sect. 2.2, we have also exploited temporal slabs through merging similar slots (e.g. hours, days, weeks and etc.). Finally, both $Pr_\nu(l_j|u_i)$ prior and temporal slabs are consumed by the final model that can further comprise quadrilateral homogeneous temporal factors.

Our proposed location recommendation model includes multiple homogeneous scales derived from time-related subset feature. The joint probability of u_i to visit every l_j (Eq. 6) is proportionate to the average value (denoted by $\sum$) of the joint probabilities of the user-location pair and four-fold factors ($Z = \{z_h, z_d, z_w, z_m\}$).

$$Pr(u_i, l_j) \propto \sum_{z_m} \sum_{z_w} \sum_{z_d} \sum_{z_h} Pr(u_i, l_j, z_h, z_d, z_w, z_m) \qquad (6)$$

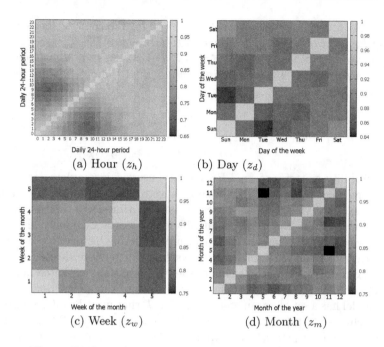

Fig. 3. Similarity between temporal slots in Brightkite dataset

Moreover, we don't take the query user and locations independently conditioned on time-related parameters. Hence, Eq. 7 represents the joint probability of the query user u_i to visit l_j influenced by all four time-related latent parameters:

$$Pr(u_i, l_j, z_h, z_d, z_w, z_m) \propto Pr(u_i)Pr_\nu(l_j|u_i)Pr(z_h|z_d, z_w, z_m, u_i, l_j)$$
$$Pr(z_d|z_w, z_m, u_i, l_j)Pr(z_w|z_m, u_i, l_j)Pr(z_m|u_i, l_j) \qquad (7)$$

We determine the log likelihood from both side of the Eq. 7 as multiplication of probabilities may reach less than the decimal minimum and get treated as zero. We continue with the proof of Eq. 7.

Proof. We model subset homogeneous temporal factors ($z_h \subset z_d \subset z_w \subset z_m$) in a generative process. The visiting probability of the query user at a proposing location is affected by some $\nu_{i,j}$ and homogeneous temporal preferences. Through applying the Bayesian theorem on $Pr(z_h, z_d, z_w, z_m|u_i, l_j)$ we can reach Eq. 7. We can also rearrange $Pr(u_i, l_j, z_h, z_d, z_w, z_m)$ as denoted in Eq. 8:

$$Pr(u_i, l_j, z_h, z_d, z_w, z_m) \propto Pr(u_i, l_j)Pr(z_h, z_d, z_w, z_m|u_i, l_j) \qquad (8)$$

$$Pr(u_i, l_j) \propto Pr(u_i)Pr_\nu(l_j|u_i) \qquad (9)$$

As Eq. 9 denotes, the general visiting probability (CF module + Recency + Non-temporal impacts) for location l_j to be visited by the query user u_i (i.e. $Pr_\nu(l_j|u_i)$)

is proportionate to the joint probability of u_i and l_j. We can equally treat all the users, so $Pr(u_i)$ can be set to 1. Equation 10 formulates $Pr(z_h, z_d, z_w, z_m | u_i, l_j)$ through the joint probability of z_h, z_d, z_w, z_m and assumes the user-location pair as a unique parameter. By substituting Eqs. 9 and 10 into 8, we can prove the correctness of Eq. 7.

$$Pr(z_h, z_d, z_w, z_m | u_i, l_j) = Pr(z_h | z_d, z_w, z_m, u_i, l_j) Pr(z_d | z_w, z_m, u_i, l_j)$$
$$Pr(z_w | z_m, u_i, l_j) Pr(z_m | u_i, l_j) \qquad (10)$$

The visibility pattern of an LBSN query user is incomplete for two reasons. Firstly, the majority of LBSN users have a limited number of check-ins. Moreover, as they visit various locations for few times, the visiting information regarding each of the locations at different times is extremely insufficient. Secondly, we eliminate a part of the query user's visiting history to evaluate how effectively each of the recommendation systems can retrieve deleted locations. Therefore, our novel approach consumes the temporal slabs exploited in Sect. 2.2 aiming to compensate incomplete mobility pattern of the query user.

The model has a set of parameters defined as Γ that must be inferred from the data. Γ includes $Pr(z_h | z_d, z_w, z_m, u_i, l_j)$, $Pr(z_d | z_w, z_m, u_i, l_j)$, $Pr(z_w | z_m, u_i, l_j)$, $Pr(z_m | u_i, l_j)$, and $Pr_\nu(l_j | u_i)$. The value of $Pr_\nu(l_j | u_i)$ can be calculated using social and geographical effects built upon the collaborative filtering module (Sect. 2.1). Intuitively, we pursue to maximize the log-likelihood of $\mathcal{F}(\Gamma)$.

$$\mathcal{F}(\Gamma) = \sum_{<u_i, l_j> \in <U, L>} log(Pr(u_i, l_j; \Gamma)) \qquad (11)$$

We employ Expectation-Maximization (EM) method to compute the best values for the parameters Γ that can maximize the log-likelihood of the historical data. E and M steps are explained below:

– **E-step:** Following the Bayesian theorem, in E-step we update the joint expectation of the latent variables for the given user-location pair (Eq. 12).

$$Pr(z_h, z_d, z_w, z_m | u_i, l_j) = \frac{Pr(u_i, l_j, z_h, z_d, z_w, z_m)}{\sum_{z_m} \sum_{z_w} \sum_{z_d} \sum_{z_h} Pr(u_i, l_j, z_h, z_d, z_w, z_m)} \qquad (12)$$

– **M-step:** We find the new Γ that can maximize the log-likelihood denoted in Eq. 13:

$$Pr(z_h | z_d, z_w, z_m, u_i, l_j) = \frac{Pr(z_h, z_d, z_w, z_m | u_i, l_j)}{\sum_{z'_h} Pr(z'_h, z_d, z_w, z_m | u_i, l_j)}$$

$$Pr(z_d | z_w, z_m, u_i, l_j) = \frac{\sum_{z_h} Pr(z_h, z_d, z_w, z_m | u_i, l_j)}{\sum_{z'_d} \sum_{z_h} Pr(z_h, z'_d, z_w, z_m | u_i, l_j)}$$

$$Pr(z_w | z_m, u_i, l_j) = \frac{\sum_{z_d} \sum_{z_h} Pr(z_h, z_d, z_w, z_m | u_i, l_j)}{\sum_{z'_w} \sum_{z_d} \sum_{z_h} Pr(z_h, z_d, z'_w, z_m | u_i, l_j)} \qquad (13)$$

$$Pr(z_m | u_i, l_j) = \frac{\sum_{z_w} \sum_{z_d} \sum_{z_h} Pr(z_h, z_d, z_w, z_m | u_i, l_j)}{\sum_{z'_m} \sum_{z_w} \sum_{z_d} \sum_{z_h} Pr(z_h, z_d, z_w, z'_m | u_i, l_j)}$$

The value for $\sum_{z'_m} \sum_{z_w} \sum_{z_d} \sum_{z_h} Pr(z_h, z_d, z_w, z'_m | u_i, l_j)$ is 1. Therefore, in implementation, we have $Pr(z_m | u_i, l_j) \propto \sum_{z_d} \sum_{z_h} \sum_{z_w} Pr(z_h, z_d, z_w, z_m | u_i, l_j)$.

3 Experiments

In this section, we release a comprehensive set of experiments to compare our proposed solution (described in Sect. 2.3) with other rival baselines. Firstly, we compute tuning metrics owned by all competitors to ensure they will achieve the best performance. We also find the best half-time value regarding the recency module of our model. Secondly, we study how the proposed model can promote the effectiveness of the state-of-the-art location recommendation methods.

3.1 Evaluation Metric

We perform the evaluation using 20% of users in each dataset. Top N highly ranked locations are initially suggested by each of the location recommendation methods. Rather than utilizing the survey-based evaluation approach (nDCG), we opt for the F1-score ratios. We firstly eliminate a portion (30%) of visited POIs by each test user. Accordingly, we compare the baselines to measure how successfully they can recover excluded locations. As Eq. 14 shows, R_p and E_p are the respective total Number of recovered POIs and the number of initially excluded POIs. Considering the Recommendation@N, Precision, Recall, and F1-score values will be computed for every test user. We then use the best F1-score average of all test users to determine the best recommendation approach.

$$Precision@N = \frac{R_p}{N}, Recall@N = \frac{R_p}{E_p}, F1 - score@N = \frac{2 \times Precision@N \times Recall@N}{Precision@N + Recall@N}$$
$$(14)$$

3.2 Recommendation Methods

The POI recommendation models compared in the experiments are listed below:

- **CF:** The collaborative filtering method excluding any temporal/non-temporal enhancing parameters.
- **CFT:** The temporal CF model proposed in [7].
- **USG:** The CF method promoted by the geo-social [16] effects.
- **USGT:** The model proposed in [7] which merely takes a single temporal aspect into consideration.
- **NH-JTI:** *Non Homogeneous-Joint Temporal Influence* is the proposed method in this paper. It is capable of combining multiple homogeneous and heterogeneous temporal effects in location recommendation.

USG's parameters such as α and β have been tuned to gain the best performance. UTP-based model [22] is not included in the evaluation process as it addresses a different problem of time-aware POI recommendation. We aim to prove the superiority of NH-JTI over other rivals.

3.3 Dataset

We perform the experiments on two real-life LBSN datasets as used in [7]. The statistics are reported in Table 2. Confirmed by low densities, both datasets are scattered. While more than half of the users in Brightkite dataset have less than 15 check-ins (Cold start scenario), merely less than 10% of foursquare user pairs share a minimum of six locations.

Table 2. Statistics of the datasets

	Brightkite	Foursquare
#users	58,228	4,163
#POIs	772,967	121,142
#check-ins	4,491,143	483,813
#Network links	214,078	32,512
Time span	Mar 2008 to Oct 2010	Dec 2009 to Jul 2013
User-POI density	2.7×10^{-5}	5.33×10^{-4}

3.4 Impact of Parameters

Each recommendation model explained in Sect. 3.2 owns its specific parameters. Similarly, our method (NH-JTI) takes various time-related properties into consideration which carry both benefits and obstructions. Hence, parameter tuning aims to find the best values which can maximize the effectiveness of competitors in performance analysis. Notable parameters of NH-JTI are two-fold. (i) **Subdivision**: The generative latent module in our system integrates various homogeneous temporal subset attributes and includes a set of parameters called as Γ. (ii) **Recency**: In order to analyze the significance of the recent check-ins versus others, we set up an experiment to obtain the best half-time value which can guarantee a better effectiveness for NH-JTI.

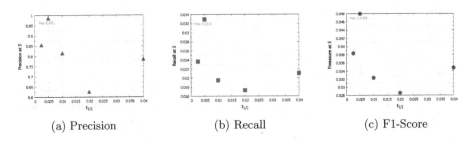

| (a) Precision | (b) Recall | (c) F1-Score |

Fig. 4. Studying the recency effect in Foursquare dataset

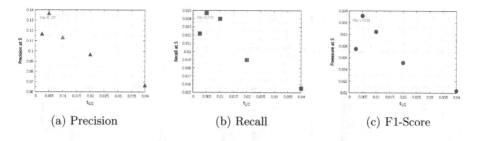

| (a) Precision | (b) Recall | (c) F1-Score |

Fig. 5. Studying the recency effect in Brightkite dataset

We use Expectation-Maximization method to find the optimum values of Γ for subdivision property. The inference algorithm is explained in Sect. 2.3. Moreover, regarding the recency effect, we choose a set of test users (20% of all users) and compute the CF probability (integrated with recency) for each query user u_i to visit every location l_j based on various half-time values (0.04, 0.02, 0.01, 0.005, 0.0025). Accordingly, we evaluate the CF method using F1-score (Sect. 3.1) for location Recommendation@5. The final metric is the average of all test users' F1-scores which can signify the best $t_{1/2}$ value (0.005 as illustrated in Figs. 4 and 5). As the best half-time value in both datasets is a small number, we can conclude that the recency property doesn't play a key role in mobility pattern of the LBSN users. Similarly, in order to exploit the best values for USG parameters [16], we tuned α and β between 0 and 1 to achieve the best F1-score@5 (Table 3). The final optimized values were chosen based on the highest F1-score@5 average.

Table 3. USG Optimized values

	F1-Score @5	
	α	β
Foursquare	0.2	0.6
Brightkite	0.3	0.4

3.5 Performance Comparison

We now elucidate the comparison results for the location recommendation models listed in Sect. 3.2. While, it is demanding to devise an approach to incorporate a mixture of homogeneous and heterogeneous time-related features in recommendation, we empirically prove that our model can surpass the state-of-the-art counterparts. Figures 6 and 7 illustrate that NH-JTI out-performs other models in the recommendation at 5, 10, and 20. Both user-location and UTP cubes have low densities in LBSN datasets which directly causes a reduction in the effectiveness of location recommendation systems. Giving an example, both precision and recall in [5,11,22] are less than 5%. Therefore, the relative improvement ratios are taken into account.

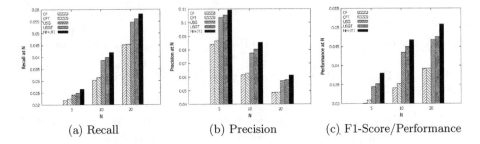

(a) Recall   (b) Precision   (c) F1-Score/Performance

Fig. 6. Comparing the methods - Foursquare dataset

The subset property is mostly useful for those who own richer check-in logs. In that case, NH-JTI can reclaim a more comprehensive mobility pattern and consequently, it can temporally correlate the query user with proposed locations more accurately. From another perspective, our model succeeds where other state-of-the-art models fail to offer interesting locations to the query user. Proposing a true suggestion can increase precision@5 by 20%. However, as most of the succeeded users possess a rich visiting history, relevant Recall@5 doesn't augment for them considerably (Sect. 3.1). Consequently, the overall F1-score will not increase substantially.

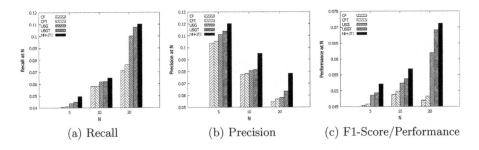

(a) Recall   (b) Precision   (c) F1-Score/Performance

Fig. 7. Comparing the methods - Brightkite dataset

4   Related Work

Nowadays, people use Location-based Social Networks on a daily basis to socialize and report their location enclosed with further artifacts[2] at the visiting time.

---

[2] Photos, Videos, and etc.

Accordingly, the pervasive POI recommendation systems utilize the data generated by such mediums (e.g. Foursquare, Yelp, and etc.) to suggest new appealing locations to the subscribers. Despite some traditional approaches [14], the prominent *Collaborative Filtering (CF)* [9,16,21,22] methods are used more pervasively. The CF approach relies on the user's check-in log to compute her interest regarding each of unvisited locations. In short, prior techniques employed for location recommendation include Collaborative Filtering [22], Matrix Factorization [5], Density estimation [24] and the Graph theory [23]. Moreover, a variety of features such as *geographical, social, context-oriented* (e.g. text contents and word-of-mouth) and *temporal* influences are combined with the CF algorithm to promote the performance of location recommendation task. In this paper, we aim to jointly include a set of temporal attributes to further enhance LBSN based location recommenders.

Collaborative Filtering (CF). CF-based methods [9,16,21,22] are commonly used in location recommendation and have two categories of model-based and memory-based [2]. Memory-based model has two sub categories of user-based [7,16] and item-based [4] which recommend new locations to a query user based on the similarity values(e.g. *Cosine* or *Pearson*) computed among users and items respectively. CF models have been used on diverse datasets such as Trajectory, LBSN, and semantics. Beside the CF core, social, temporal, and geographical influences are jointly utilized to enhance the effectiveness of location recommendation task.

Geographical Influence (GI). Geographical influence [12,16,22,24,25] is established according to the fact that an LBSN user tends to visit a location that is near to the venues where she has priorly visited. Methods such as *Power law distribution* [16,22], *Multi-Center Gaussian Model* [1], *Modelling the location instances using phrases* [8], and personalized *Kernel Density Estimation* [25] have been employed to integrate geographical influence. Our model comprises the geographical influence alongside with the social and temporal factors. Like [7], we have used Normal Equation to minimize the error and retrieve optimized parameters of distribution module.

Social Link. Social Influence [1,26] claims that a query user's mobility pattern is partially affected by her social links [15]. Accordingly, Ye et al. [16] compares the Jaccard similarity coefficient based on both locations and friends. Eventually, following the parameter tuning, they affirm that for a twin users in LBSN sphere, the number of shared locations gains a higher significance compared to the shared friends. [6] also studies the importance of the social link and the way it is propagated in a social network during a time constraint. While [1] confirms a minor significance for the social effect in location recommendation, some other work [26] jointly model the social effect with spatio-temporal evidence.

Temporal Influence. Temporal influence can be utilized to promote location recommendation in general or in a system to suggest new location to the query user at a specific time (Time-aware recommendation). Time has numerous properties. Some of the attributes such as long term/short term, periodicity, recency, consecutiveness, and non-uniformness have already been exploited in location recommendation. Long-term/short-term property [14] outlines that some locations are visited continuously, but some of them are visited during a short period of time. Non-uniformness [5] indicates that a user's check-in behavior drifts all the time (e.g. The places where people go during the week is work-oriented while it is amusement-related during the weekends). Periodicity [27] states that LBSN users perform periodic trips (e.g. home-work cycles) and the recency attribute [10] grants higher significance to the newly visited locations versus those visited while ago. Consecutiveness [2,26] claims that some places are visited after others successively (e.g. people go to drinking after the dinner). Intrinsically, the time entity is granular (e.g. minutes, quarters, hours, and etc.) with subset feature in between (e.g. $hour \subset day$). This model distinguishes itself from previous similar work [7,13,18–21] in the following aspects. First, we retrieve four-folded temporal similarity maps which deduce the data sparsity and reflect a multi-aspect temporal state of the user-location matrix; Second, rather than restricting the number of temporal dimensions, we propose a synthetic location recommendation model that comprises both heterogeneous temporal properties of recency and subdivision. Also, the latter property includes a four-folded set of homogeneous features ($hour \subset day \subset week \subset month$).

5   Conclusions

In this paper, we proposed a probabilistic location recommendation model which simultaneously employs two heterogeneous temporal properties of recency and the subset. On the one hand, as the time entity comprises numerous slots (e.g. $minute \subset hour \subset day$ and etc.), our model employs multiple latent temporal parameters where each one is dedicated to a specific time-related slot. On the other hand, we further utilize the recency feature in collaborative filtering module of the recommendation framework. The model reduces the sparsity in user-location matrices using the subset relation among temporal slots. It also employs a novel *Expectation-Maximization* method to find the best values for latent parameters and maximize the temporal correlation between the query user and the proposed locations. We perform two sets of experiments in this paper. Firstly, we apply multiple parameter adjustments to maximize the effectiveness of all competitive models. Finally, we compare all the models through another experiment which proves the supremacy of our method versus various temporal and non-temporal state-of-the-art location recommendation systems.

For the future work, we need to include a further diverse set of heterogeneous temporal attributes in location recommendation. Moreover, we need to include the fact that each temporal slot is affected by its parent factor. For instance, people's hourly mobility behaviors vary on different days of the week.

Acknowledgement. Meihui Zhang was supported by SUTD Start-up Research Grant under Project No. SRG ISTD 2014 084. The work is also partially supported by ARC Discovery Early Career Researcher Award (DE160100308) and ARC Discovery Project (DP170103954).

References

1. Cheng, C., Yang, H., King, I., Lyu, M.R.: Fused matrix factorization with geographical and social influence in location-based social networks. In: Twenty-Sixth AAAI Conference on Artificial Intelligence (2012)
2. Cheng, C., Yang, H., Lyu, M.R., King, I.: Where you like to go next: successive point-of-interest recommendation. IJCAI **13**, 2605–2611 (2013)
3. Das, M., Thirumuruganathan, S., Amer-Yahia, S., Das, G., Yu, C.: An expressive framework and efficient algorithms for the analysis of collaborative tagging. VLDB J. **23**(2), 201–226 (2014)
4. Ding, Y., Li, X.: Time weight collaborative filtering. In: Proceedings of the 14th ACM International Conference on Information and Knowledge Management, pp. 485–492. ACM (2005)
5. Gao, H., Tang, J., Hu, X., Liu, H.: Exploring temporal effects for location recommendation on location-based social networks. In: Proceedings of the 7th ACM Conference on Recommender Systems, pp. 93–100. ACM (2013)
6. Goyal, A., Bonchi, F., Lakshmanan, L.V.: Learning influence probabilities in social networks. In: Proceedings of the Third ACM International Conference on Web Search and Data Mining, pp. 241–250. ACM (2010)
7. Hosseini, S., Li, L.T.: Point-of-interest recommendation using temporal orientations of users and locations. In: Navathe, S.B., Wu, W., Shekhar, S., Du, X., Wang, X.S., Xiong, H. (eds.) DASFAA 2016. LNCS, vol. 9642, pp. 330–347. Springer, Cham (2016). doi:10.1007/978-3-319-32025-0_21
8. Hosseini, S., Unankard, S., Zhou, X., Sadiq, S.: Location oriented phrase detection in microblogs. In: Bhowmick, S.S., Dyreson, C.E., Jensen, C.S., Lee, M.L., Muliantara, A., Thalheim, B. (eds.) DASFAA 2014. LNCS, vol. 8421, pp. 495–509. Springer, Cham (2014). doi:10.1007/978-3-319-05810-8_33
9. Levandoski, J.J., Sarwat, M., Eldawy, A., Mokbel, M.F.: Lars: a location-aware recommender system. In: 2012 IEEE 28th International Conference on Data Engineering, pp. 450–461. IEEE (2012)
10. Li, X., Xu, G., Chen, E., Zong, Y.: Learning recency based comparative choice towards point-of-interest recommendation. Expert Syst. Appl. **42**(9), 4274–4283 (2015)
11. Liu, B., Xiong, H., Papadimitriou, S., Fu, Y., Yao, Z.: A general geographical probabilistic factor model for point of interest recommendation. IEEE Trans. Knowl. Data Eng. **27**(5), 1167–1179 (2015)
12. Liu, X., Liu, Y., Aberer, K., Miao, C.: Personalized point-of-interest recommendation by mining users' preference transition. In: Proceedings of the 22nd ACM International Conference on Conference on Information & Knowledge Management, pp. 733–738. ACM (2013)
13. Wang, W., Yin, H., Chen, L., Sun, Y., Sadiq, S., Zhou, X.: Geo-sage: a geographical sparse additive generative model for spatial item recommendation. In: Proceedings of the 21th ACM SIGKDD International Conference on Knowledge Discovery and Data Mining, pp. 1255–1264. ACM (2015)
14. Xiang, L., Yuan, Q., Zhao, S., Chen, L., Zhang, X., Yang, Q., Sun, J.: Temporal recommendation on graphs via long-and short-term preference fusion. In: Proceedings of the 16th ACM SIGKDD International Conference on Knowledge Discovery and Data Mining, pp. 723–732. ACM (2010)

15. Ye, M., Liu, X., Lee, W.-C.: Exploring social influence for recommendation: a generative model approach. In: Proceedings of the 35th International ACM SIGIR Conference on Research and Development in Information Retrieval, pp. 671–680. ACM (2012)

16. Ye, M., Yin, P., Lee, W.-C., Lee, D.-L.: Exploiting geographical influence for collaborative point-of-interest recommendation. In: Proceedings of the 34th International ACM SIGIR Conference on Research and Development in Information Retrieval, pp. 325–334. ACM (2011)

17. Yin, H., Cui, B.: Location-based and real-time recommendation. In: Spatio-Temporal Recommendation in Social Media, pp. 65–98. Springer, Singapore (2016)

18. Yin, H., Cui, B., Huang, Z., Wang, W., Wu, X., Zhou, X.: Joint modeling of users' interests and mobility patterns for point-of-interest recommendation. In: Proceedings of the 23rd ACM International Conference on Multimedia, pp. 819–822. ACM (2015)

19. Yin, H., Cui, B., Sun, Y., Hu, Z., Chen, L.: Lcars: a spatial item recommender system. ACM Trans. Inf. Syst. (TOIS) 32(3), 11 (2014)

20. Yin, H., Sun, Y., Cui, B., Hu, Z., Chen, L.: Lcars: a location-content-aware recommender system. In: Proceedings of the 19th ACM SIGKDD International Conference on Knowledge Discovery and Data Mining, pp. 221–229. ACM (2013)

21. Yin, H., Zhou, X., Shao, Y., Wang, H., Sadiq, S.: Joint modeling of user check-in behaviors for point-of-interest recommendation. In: Proceedings of the 24th ACM International on Conference on Information and Knowledge Management, pp. 1631–1640. ACM (2015)

22. Yuan, Q., Cong, G., Ma, Z., Sun, A., Thalmann, N.M.: Time-aware point-of-interest recommendation. In: Proceedings of the 36th International ACM SIGIR Conference on Research and Development in Information Retrieval, pp. 363–372. ACM (2013)

23. Yuan, Q., Cong, G., Sun, A.: Graph-based point-of-interest recommendation with geographical and temporal influences. In: Proceedings of the 23rd ACM International Conference on Conference on Information and Knowledge Management, pp. 659–668. ACM (2014)

24. Zhang, J.-D., Chow, C.-Y.: Ticrec: a probabilistic framework to utilize temporal influence correlations for time-aware location recommendations

25. Zhang, J.-D., Chow, C.-Y., Li, Y.: igeorec: a personalized and efficient geographical location recommendation framework. IEEE Trans. Serv. Comput. 8(5), 701–714 (2015)

26. Zhang, W., Wang, J.: Location and time aware social collaborative retrieval for new successive point-of-interest recommendation. In: Proceedings of the 24th ACM International on Conference on Information and Knowledge Management, pp. 1221–1230. ACM (2015)

27. Zhang, Y., Zhang, M., Zhang, Y., Lai, G., Liu, Y., Zhang, H., Ma, S.: Daily-aware personalized recommendation based on feature-level time series analysis. In: Proceedings of the 24th International Conference on World Wide Web, pp. 1373–1383. International World Wide Web Conferences Steering Committee (2015)

28. Zhao, S., Zhao, T., King, I., Lyu, M.R.: Gt-seer: geo-temporal sequential embedding rank for point-of-interest recommendation. arXiv preprint arXiv:1606.05859 (2016)

Location-Aware News Recommendation Using Deep Localized Semantic Analysis

Cheng Chen[1,2], Thomas Lukasiewicz[3], Xiangwu Meng[1,2(✉)], and Zhenghua Xu[3]

[1] Beijing Key Laboratory of Intelligent Telecommunications Software and Multimedia, Beijing University of Posts and Telecommunications, Beijing, China
[2] School of Computer Science, Beijing University of Posts and Telecommunications, Beijing, China
{ccbupt,mengxw}@bupt.edu.cn
[3] Department of Computer Science, University of Oxford, Oxford, UK
{thomas.lukasiewicz,zhenghua.xu}@cs.ox.ac.uk

Abstract. With the popularity of mobile devices and the quick growth of the mobile Web, users can now browse news wherever they want, so their news preferences are usually strongly correlated with their geographical contexts. Consequently, many research efforts have been put on location-aware news recommendation; the explored approaches can mainly be divided into physical distance-based and geographical topic-based ones. As for geographical topic-based location-aware news recommendation, ELSA is the state-of-the-art geographical topic model: it has been reported to outperform many other topic models, e.g., BOW, LDA, and ESA. However, the Wikipedia-based topic space in ELSA suffers from the problems of high dimensionality, sparsity, and redundancy, which greatly degrade the recommendation performance of ELSA. Therefore, to overcome these problems, in this work, we propose three novel geographical topic feature models, CLSA, ALSA, and DLSA, which integrate clustering, autoencoders, and recommendation-oriented deep neural networks, respectively, with ELSA to extract dense, abstract, low dimensional, and effective topic features from the Wikipedia-based topic space for the representation of news and locations. Experimental results show that (i) CLSA, ALSA, and DLSA all greatly outperform the state-of-the-art geographical topic model, ELSA, in location-aware news recommendation in terms of both the recommendation effectiveness and efficiency; (ii) Deep Localized Semantic Analysis (DLSA) achieves the most significant improvements: its precision, recall, MRR, and MAP are all about 3 times better than those of ELSA; while its recommendation time-cost is only about 1/29 of that of ELSA; and (iii) DLSA, ALSA, and CLSA can also remedy the "cold-start" problem by uncovering users' latent news preferences at new locations.

Keywords: Location-aware news recommendation · Explicit semantic analysis · Autoencoders · Deep neural networks

© Springer International Publishing AG 2017
S. Candan et al. (Eds.): DASFAA 2017, Part I, LNCS 10177, pp. 507–524, 2017.
DOI: 10.1007/978-3-319-55753-3_32

1   Introduction

Nowadays, news reading is an indispensable daily activity of many people. With the recent popularity of smart mobiles and the rapid development of the mobile Web, more and more people tend to read news online via their mobiles or other handheld devices, e.g., tablets. However, due to the huge volume of news articles generated everyday, readers cannot afford to go through all the news online. So, news recommendation systems, which aim to filter out irrelevant online information and recommend to users their preferred news, have been widely studied [1,8,11,22].

In typical news recommendation systems, a user's news preferences are usually learned from his/her news reading history or other online activity history; so his/her news preferences are (almost) static in these systems. However, in real-world contexts, users' news preferences usually evolve with the change of their locations; e.g., people may prefer economic or political news, when they are working in the office; but they may like to read entertainment or sports news, when they are at home. As the users' news preferences are strongly correlated with their geographical contexts, location-aware news recommendation systems that recommend news based on the geographical contexts of users have recently attracted many research efforts. There are mainly two research directions: physical distance-based and geographical topic-based approaches.

Specifically, physical distance-based news recommendation [2,3,14,16] aims to offer users with news happening nearest to them; so, the relevance of a news article to a user is measured by the physical distance between their locations based on GPS coordinates. However, the descriptions of event locations in many news articles are very vague and general (mentioning only a city or suburb) in practice; so, obtaining accurate GPS information for this kind of news is very difficult and sometimes even impossible, which greatly limits the application of physical distance-based methods.

Given this status quo, geographical topic-based methods [12,19,23] are proposed to achieve a more generic location-aware news recommendation, where, instead of using GPS coordinates, the locations are described using topic vectors, and the relevance of a news article to a user is measured by the similarity between the topic vectors of the news and the current location of the user. Therefore, the topic representations of locations are crucial for geographical topic-based location-aware news recommendation, and a range of topic models (such as Latent Dirichlet Allocation (LDA) [5], Explicit Semantic Analysis (ESA) [10], Probabilistic Latent Semantic Analysis (PLSA) [24], and their improved models [12,19,23]) have been used.

The state-of-the-art geographical topic model is *Explicit Localized Semantic Analysis (ELSA)* [19], which is reported to outperform many other geographical topic models (e.g., BOW, LDA, and ESA) in geographical topic-based location-aware news recommendation. The recommendation process of ELSA is briefly as follows: it first uses collections of documents with geo-tags (called *geo-tagged documents*) as the descriptions of corresponding locations; then, it projects both the geo-tagged documents and the news articles onto a topic space using Explicit

Semantic Analysis (ESA) [10], where Wikipedia concepts are regarded as topics; consequently, by considering link information between the corresponding concepts of local topics, both locations (e.g., country, city, or venue) and news are represented as topic vectors (called *localized location profiles* and *localized news profiles*, respectively) and the relevance score between a user and a candidate news article is estimated by the similarity between the corresponding localized location and news profiles; finally, the news with top-k highest relevance scores are recommended to the user.

However, since the volume of concepts is enormous (millions) on Wikipedia, the resulting Wikipedia-based topic space in ELSA is very high dimensional. Consequently, the process of online news recommendation in ELSA is very time-consuming, which is unacceptable for the need of real-time online responses in practice. In addition, the Wikipedia-based topic space also suffers from the problems of sparsity and redundancy, which degrade the news recommendation effectiveness of ELSA to a great extent.

Therefore, to achieve better recommendation performance, in this work, we first propose two geographical topic feature models, *Clustering-based Localized Semantic Analysis (CLSA)* and *Autoencoder-based Localized Semantic Analysis (ALSA)* to address these problems by *topic feature modeling*. Generally, CLSA and ALSA integrate clustering and autoencoders (neural networks), respectively, with ELSA to extract denser, more abstract and lower dimensional topic features from the Wikipedia-based topic space in ELSA for the representations of news and locations. Our experimental studies show that CLSA and ALSA both improve the performance of ELSA in location-aware news recommendations. However, these two solutions still suffer from the following drawback: the learning objectives of clustering in CLSA and autoencoders in ALSA are to minimize the within-cluster distances and the reconstruction errors, respectively, which are not directly correlated to the objective of news recommendation, i.e., distinguishing the user's local target news from the irrelevant ones; so the resulting cluster-based or autoencoder-based topic feature representations of news and locations may not be very effective in news recommendation.

Motivated by this observation, we further propose another novel geographical topic feature model, called *Deep Localized Semantic Analysis (DLSA)* model, to address the high dimensionality, sparsity, and redundancy problems in ELSA. DLSA utilizes deep neural networks to map the Wikipedia-based topic space in ELSA to an abstract, dense, and low dimensional topic feature space, where the localized similarities between the locations and users' local target news are maximized, and those with the users' irrelevant news are minimized. DLSA has the following advantage: the deep neural networks in DLSA are trained with a recommendation-oriented learning objective, i.e., to differentiate the users' local target news from the irrelevant ones, so the resulting deep topic feature representations of news and locations are more effective for location-aware news recommendations than ALSA and CLSA. Consequently, the performance of DLSA is superior to those of CLSA and ALSA in location-aware news recommendation. Although we only investigate their applications in ELSA, the proposed models,

CLSA, ALSA, and DLSA, can easily be used to tackle similar problems in other topic models.

In summary, the contributions of this paper are briefly as follows:

- We identify the high dimensionality, sparsity, and redundancy problems in the Wikipedia-based topic space of ELSA, which greatly degrades ELSA's location-aware news recommendation performance.
- We thus propose three novel geographical topic feature models (CLSA, ALSA, and DLSA) to address these problems by topic feature modeling. These three models integrate clustering, autoencoders, and recommendation-oriented deep neural networks, respectively, with ELSA to obtain an abstract, dense, low dimensional, and effective topic feature representation for locations and news.
- Extensive experiments are conducted using a public real-world dataset. The results show that: (i) the proposed CLSA, ALSA, and DLSA all greatly outperform the state-of-the-art geographic topic model, ELSA, in location-aware news recommendation in both the recommendation effectiveness and efficiency; (ii) DLSA achieves the most significant improvements: its precision, recall, MRR, and MAP are all about 3 times better than those of ELSA, while its recommendation time-cost is only about 1/29 of that of ELSA; (iii) DLSA, ALSA, and CLSA can remedy "cold-start" problems by uncovering users' latent news preferences at new locations.

2   Related Work

2.1   News Recommendation

Typical news recommendation systems aim to recommend to users the news that match their personal interests best [8,11]. Users' interests in news are usually modeled by their explicit ratings or browsing histories (e.g., visited pages, reading times, and downloads). Both heuristic [1] and model-based methods [22] are proposed for news recommendations: the former are mainly based on mathematical or statistical solutions (e.g., cosine similarity and Euclidean distance), while the latter make use of machine learning techniques or mathematical models (e.g., Bayesian networks and decision trees). Specifically, Abel et al. [1] proposed to combine news with information on social media (tweets) to construct three kinds of user profiles, and then compute the cosine similarity between user profiles and news articles for personalized news recommendation. Yeung et al. [22] used Bayesian networks to predict levels of interesting news categories for users and to provide real-time personalized news recommendation.

2.2   Location-Based News Recommendation

However, in the era of mobile and wireless networks, users' news preferences are also influenced by their geographical contexts, i.e., people usually pay more attention to the news happening nearby than those far away from them. Therefore, more and more research efforts have been put into location-aware news recommendations, which mainly focus on two research directions: physical distance-based and geographical topic-based.

As for physical distance-based news recommendation, GeoFeed [2] and GeoRank [3] recommend to users some news happening at the users' current locations or within a given range, where GeoRank uses only static location points of both users and news, while GeoFeed allows news with spatial extent; Pedro et al. [16] utilized the Euclidean distance between the locations of users and news articles to measure the importance of news articles; LocaNews [14] keeps three versions of news and offers to users the most suitable ones according to their different distances to the locations of news; Wen et al. [21] proposed a news stream recommendation framework, called MobiFeed, to further investigate news recommendation based on users' moving tracks.

However, in the real-world context, the descriptions of event locations in many news articles are very vague and general (mentioning only a city or suburb); so, obtaining accurate GPS information for this kind of news is very difficult and sometimes even impossible. Therefore, the application of physical distance-based methods is limited.

As for more generic location-aware news recommendations, geographical topic-based methods are proposed. Instead of using GPS coordinates, the locations are described using topic vectors, and the relevance of a news article to a user is measured by the similarity between the topic vectors of the news and the current location of the user. Therefore, the topic representations of locations are crucial for geographical topic-based location-aware news recommendation, and a range of topic models (such as Latent Dirichlet Allocation (LDA) [5], Explicit Semantic Analysis (ESA) [10], Probabilistic Latent Semantic Analysis (PLSA) [24], and their improved models [12,19,23]) have been used. The state-of-the-art geographical topic model in topic-based location-aware news recommendation is *Explicit Localized Semantic Analysis (ELSA)* [19], which is reported to outperform many other topic models, e.g., BOW, LDA [5], and ESA [10]. However, the Wikipedia-based topic space in ELSA suffers from the problems of high dimensionality, sparsity, and redundancy, which greatly degrade the recommendation performance of ELSA. Therefore, in this work, we propose three novel geographical topic feature models (CLSA, ALSA, and DLSA) to overcome these problems by topic feature modeling and to achieve better recommendation performance.

2.3   Recommendation Using Deep Learning

Due to its capability to extract effective representations [4], deep learning has already been successfully applied in many online recommendation applications, such as music recommendation [20], movie recommendation [17], tag-aware recommendation [25], and multi-view item recommendation [9].

Similarly to our work, the recommendation system proposed in [9] is also based on deep neural networks with a recommendation-oriented training objective. But [9] is very different from the DLSA proposed here: (i) [9] is not a location-aware model, so its recommendation is not sensitive to the changes of users' geographical contexts; (ii) [9] does not aim to solve the huge dimensionality, sparsity, and redundancy problems in the Wikipedia-based topic space; and (iii) [9] has to train own parameters for each neural network; while, in our work, shared parameters are applied in deep neural networks, as all input topic vectors in our work share the same Wikipedia-based topic space; consequently, the time needed for model training in our work is greatly reduced.

3   Explicit Localized Semantic Analysis in News Recommendation

Explicit localized semantic analysis (ELSA) [19] is the state-of-the-art topic model in geographical topic-based location-aware news recommendation, which is reported to outperform many other topic models, e.g., BOW, LDA, and ESA. Due to its close relation to our work, we briefly review ELSA in this section.

ELSA is an ESA-based [10] solution, where each Wikipedia concept is considered as a potential topic, and each location and news article is represented as a Wikipedia-based topic vector. Figure 1 shows the overall process of ELSA. First of all, ELSA collects for each location a set of documents with the corresponding geo-tags (denoted as D_l) as the description of this location. Then, these geo-tagged documents and the candidate set of news articles V are mapped onto a Wikipedia-based topic space (denoted as Z) to generate for each location

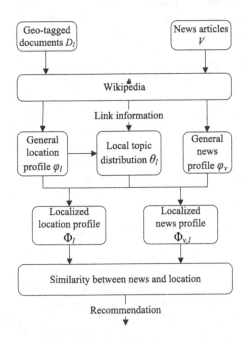

Fig. 1. Overall process of ELSA

or news article a topic vector, which is represented as a probability distribution over topics, called *general location profile* (φ_l) or *general news profile* (φ_v), respectively. Consequently, the topics related to a location are the ones with non-zero probability values in this location's general profile. Since these topics are generally dependent on one another, ELSA further uses the link information within the corresponding Wikipedia concepts to construct a topic dependency graph and then applies PageRank [13] to estimate a *local topic distribution* (θ_l). With the help of the local topic distribution, the location, and the news, general profiles are localized to obtain the local topic representations of locations and news, which are called *localized location profile* (Φ_l) and *localized news profile* $(\Phi_{v,l})$. Finally, ELSA estimates the similarities between news and locations based on their localized profiles, and makes recommendations by offering to users the news articles with top-k similarity scores to their current locations. For the detailed inferences of φ_l, φ_v, θ_l, Φ_l, and $\Phi_{v,l}$, please refer to [19].

4 Topic Feature Modeling

Although ELSA benefits from using Wikipedia-concept-based topics for semantic enrichment, due to the huge volume of concepts (millions) on Wikipedia, the resulting topic space in ELSA suffers from the problems of high dimensionality, sparsity, and redundancy, which greatly degrade ELSA's recommendation effectiveness and efficiency.

Therefore, we propose to apply *topic feature modeling* to address these problems and to achieve a better performance in location-aware news recommendation. Generally, the process of topic feature modeling takes the general location profile (φ_l), general news profile (φ_v), and local topic distribution (θ_l) in ELSA as inputs, and exploits either clustering or deep learning techniques to extract dense, abstract, low dimensional, and effective topic features from the Wikipedia-based topic space for the representations of news and locations. The solutions that utilize clustering, autoencoders, and recommendation-oriented deep neural networks for topic feature modeling are presented in the rest of this section.

4.1 Clustering-Based Localized Semantic Analysis

Due to its capability in extracting abstract and low dimensional features [18], in this work, we adopt hierarchical clustering as the first solution for the topic feature modelling in ELSA; the resulting geographical topic feature model that integrates clustering with ELSA is called *Clustering-based Localized Semantic Analysis (CLSA)*.

As shown in Fig. 2, the clustering-based topic feature modeling in CLSA first represents each Wikipedia topic $z \in Z$ as a vector of weights over the set of resources, i.e., the locations and news articles, where the weight on each dimension is measured by the probability of the resource generated from the corresponding topic. Then, hierarchical clustering [18] groups all topics into a number of clusters based on the distances between their corresponding resource vectors.

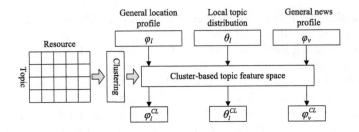

Fig. 2. Clustering-based topic feature modeling

Finally, with the resulting well-learned clusters, CLSA converts the Wikipedia-based topic space to a cluster-based topic feature space; so the representations of general location profile φ_l, local topic distribution θ_l, and general news profile φ_v are converted from Wikipedia-based topic vectors to cluster-based topic feature vectors, denoted φ_l^{CL}, θ_l^{CL}, and φ_v^{CL}, respectively.

CLSA tackles the high dimensionality, sparsity, and redundancy problems in ELSA, because: (i) the cluster-based topic feature space is lower dimensional than the Wikipedia-based topic space in ELSA; (ii) each cluster contains several topics, so sparsity is diminished; and (iii) redundant topics are aggregated to a cluster to reduce redundancy.

As for location-aware recommendation, similarly to ELSA, given φ_l^{CL}, θ_l^{CL}, and φ_v^{CL}, CLSA first obtains the cluster-based localized location and news profiles (denoted Φ_l^{CL} and $\Phi_{v,l}^{CL}$, respectively), which are formally defined as

$$\Phi_l^{CL} = (\varphi_l^{CL})^T \cdot \theta_l^{CL}, \qquad \Phi_{v,l}^{CL} = (\varphi_v^{CL})^T \cdot \theta_l^{CL}. \tag{1}$$

Then, given a user at a location l, CLSA generates news recommendations by ranking all news $v \in V$ according to their relevance to l (denoted $R_{l,v}^{CL}$), where the relevance is estimated by the cosine similarity between Φ_l^{CL} and $\Phi_{v,l}^{CL}$. Formally,

$$R_{l,v}^{CL} = Sim(\Phi_l^{CL}, \Phi_{v,l}^{CL}) = \frac{\Phi_l^{CL} \cdot \Phi_{v,l}^{CL}}{\|\Phi_l^{CL}\| \cdot \|\Phi_{v,l}^{CL}\|}. \tag{2}$$

4.2   Autoencoder-Based Localized Semantic Analysis

Besides clustering, autoencoders are another method to model low dimensional, dense, and abstract representations of raw data [25]. So, in this work, we also employ autoencoders as another solution for the high dimensionality, sparsity, and redundancy problems in ELSA; the resulting geographical topic feature model that integrates autoencoders with ELSA is called *Autoencoder-based Localized Semantic Analysis (ALSA)*.

Autoencoders are neural networks consisting of two parts: an encoder and a decoder. As shown in Fig. 3, to conduct topic feature modeling, autoencoders first take φ_l, θ_l, and φ_v as inputs, which are passed through multiple hidden

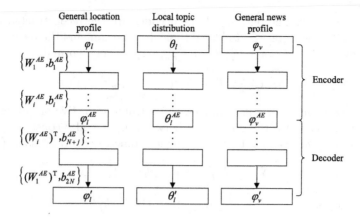

Fig. 3. Autoencoder-based topic feature modeling

layers in encoders. The intermediate outputs $f_i(l)$, $f_i(\theta)$, and $f_i(v)$ of the i-th hidden layers for location, local topics distribution, and news can be formally defined as

$$f_i(l) = \sigma(W_i^{AE} f_{i-1}(l) + b_i^{AE}), \tag{3}$$

$$f_i(\theta) = \sigma(W_i^{AE} f_{i-1}(\theta) + b_i^{AE}), \tag{4}$$

$$f_i(v) = \sigma(W_i^{AE} f_{i-1}(v) + b_i^{AE}), \tag{5}$$

where $i = 1, \ldots, N$; N is the total number of hidden layers in the encoder (decoder); W_i^{AE} and b_i^{AE} are the weight matrix and bias vector for the i-th hidden layer; $\sigma(\cdot)$ is the sigmoid activation function; and $f_0(l) = \varphi_l$, $f_0(\theta) = \theta_l$, $f_0(v) = \varphi_v$.

The outputs of the final layers of encoders ($f_N(l)$, $f_N(\theta)$, and $f_N(v)$) are the autoencoder-based topic feature representations for location (denoted φ_l^{AE}), local topic distribution (denoted θ_l^{AE}), and news (denoted φ_v^{AE}), respectively. Formally

$$\varphi_l^{AE} = f_N(l), \qquad \theta_l^{AE} = f_N(\theta), \qquad \varphi_v^{AE} = f_N(v).$$

Furthermore, the decoders in ALSA take φ_l^{AE}, θ_l^{AE}, and φ_v^{AE} as inputs and pass them through another N layers. Since we use tied-weights autoencoders, the weight matrices in the decoder are the transposes of those in the encoder. Formally, we have

$$f_{N+j}(l) = \sigma((W_{N-(j-1)}^{AE})^T f_{N+(j-1)}(l) + b_{N+j}^{AE}), \tag{6}$$

$$f_{N+j}(\theta) = \sigma((W_{N-(j-1)}^{AE})^T f_{N+(j-1)}(\theta) + b_{N+j}^{AE}), \tag{7}$$

$$f_{N+j}(v) = \sigma((W_{N-(j-1)}^{AE})^T f_{N+(j-1)}(v) + b_{N+j}^{AE}), \tag{8}$$

where $j = 1, \ldots, N$. The outputs of the decoder are the reconstructed general location profile, the reconstructed local topic distribution, and the reconstructed general news profile, denoted φ'_l, θ'_l, and φ'_v, respectively. Formally,

$$\varphi'_l = f_{2N}(l), \qquad \theta'_l = f_{2N}(\theta), \qquad \varphi'_v = f_{2N}(v).$$

As for the training of autoencoders, the learning objective of autoencoders in ALSA is to minimize the differences between input and reconstructed data, called *reconstruction errors*. Therefore, the loss function of ALSA is as follows:

$$L^{AE}(\Theta) = \tfrac{1}{2} \sum_{(l,v)} (\|\varphi'_l - \varphi_l\| + \|\theta'_l - \theta_l\| + \|\varphi'_v - \varphi_v\|), \qquad (9)$$

where Θ represents the set of parameters $\{W_i^{AE}, b_j^{AE}\}$ ($i = 1, \ldots, N; j = 1, \ldots, 2N$) in autoencoders.

After training, given the well-modeled autoencoder-based topic feature representations φ_l^{AE}, θ_l^{AE}, and φ_v^{AE}, ALSA first generates the autoencoder-based localized location and news profiles by

$$\Phi_l^{AE} = (\varphi_l^{AE})^T \cdot \theta_l^{AE}, \qquad (10)$$

$$\Phi_{v,l}^{AE} = (\varphi_v^{AE})^T \cdot \theta_l^{AE}. \qquad (11)$$

Then, given a user at a location l, ALSA generates the location-aware recommendations based on the relevance of all news $v \in V$ to l, which is computed by the cosine similarity between Φ_l^{AE} and $\Phi_{v,l}^{AE}$. Formally,

$$R_{l,v}^{AE} = Sim(\Phi_l^{AE}, \Phi_{v,l}^{AE}) = \frac{\Phi_l^{AE} \cdot \Phi_{v,l}^{AE}}{\|\Phi_l^{AE}\| \cdot \|\Phi_{v,l}^{AE}\|}. \qquad (12)$$

4.3   Deep Localized Semantic Analysis

However, the learning objectives of clustering in CLSA and autoencoders in ALSA are to minimize the within-cluster distances and minimize the reconstruction errors, respectively, which are not directly correlated to the objective of the news recommendation, i.e., distinguishing the users' local target news from the irrelevant ones; so the resulting clustering-based and autoencoder-based topic feature representations of news and locations may not be very effective in news recommendation.

Therefore, we further propose another novel geographical topic feature model, called *Deep Localized Semantic Analysis (DLSA)*, to address the high dimensionality, sparsity, and redundancy problems in ELSA. DLSA also applies deep neural networks for topic feature modeling; however, instead of using autoencoders, DLSA newly integrates ELSA with recommendation-oriented deep neural networks, which maps the Wikipedia-based topic space to an abstract, dense, and low dimensional topic feature space, where the localized similarities between the locations and the users' local target (resp., irrelevant) news are maximized (resp., minimized). Since the deep neural networks in DLSA are trained with a

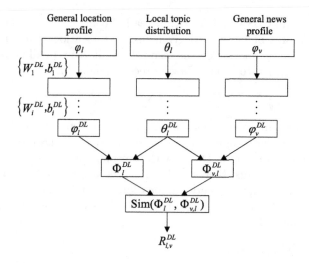

Fig. 4. Topic feature modeling based on recommendation-oriented deep neural networks

recommendation-oriented learning objective, i.e., to differentiate the users' local target news from the irrelevant ones, the resulting deep topic feature representations of news and locations in DLSA are more effective for location-aware news recommendation than CLSA and ALSA.

As shown in Fig. 4, similarly to the encoders in ALSA, the three deep neural networks in DLSA also take φ_l, θ_l, and φ_v as inputs, and the intermediate outputs h_i of the i-th hidden layers are formally defined as follows:

$$h_i(l) = \tan(W_i^{DL} h_{i-1}(l) + b_i^{DL}), \tag{13}$$

$$h_i(\theta) = \tan(W_i^{DL} h_{i-1}(\theta) + b_i^{DL}), \tag{14}$$

$$h_i(v) = \tan(W_i^{DL} h_{i-1}(v) + b_i^{DL}), \tag{15}$$

where $i = 1, \ldots, N$; tan is used as the activation function; and $h_0(l) = \varphi_l$, $h_0(\theta) = \theta_l$, $h_0(v) = \varphi_v$. Furthermore, the intermediate outputs in the N-th hidden layers are the deep topic feature representations for the general location profile (denoted φ_l^{DL}), local topics distribution (denoted θ_l^{DL}), and general news profile (denoted φ_v^{DL}); formally,

$$\varphi_l^{DL} = h_N(l), \qquad \theta_l^{DL} = h_N(\theta), \qquad \varphi_v^{DL} = h_N(v).$$

Given φ_l^{DL}, θ_l^{DL}, and φ_v^{DL}, the deep localized location and news profiles are defined as

$$\Phi_l^{DL} = (\varphi_l^{DL})^T \cdot \theta_l^{DL}, \qquad \Phi_{v,l}^{DL} = (\varphi_v^{DL})^T \cdot \theta_l^{DL}. \tag{16}$$

Then, for a user at a location l, the similarity between l and a news article v is measured via the cosine similarity between their deep localized profiles at l (i.e., Φ_l^{DL} and $\Phi_{v,l}^{DL}$):

$$Sim(\Phi_l^{DL}, \Phi_{v,l}^{DL}) = \frac{\Phi_l^{DL} \cdot \Phi_{v,l}^{DL}}{\|\Phi_l^{DL}\| \cdot \|\Phi_{v,l}^{DL}\|}. \tag{17}$$

Differently from CLSA and ALSA, instead of using cosine similarity directly, the relevance scores of news v to given users at locations l are measured by applying the softmax function on the resulting similarities of all news at l, which are then used to generate location-aware recommendation lists. Formally,

$$R_{l,v}^{DL} = e^{Sim(\Phi_l^{DL}, \Phi_{v,l}^{DL})} \Big/ \sum_{v' \in V} e^{Sim(\Phi_l^{DL}, \Phi_{v',l}^{DL})}. \tag{18}$$

Intuitively, to achieve good location-aware news recommendations, the local target news should have higher relevance scores than irrelevant ones. We thus conduct the model training in DLSA with a recommendation-oriented objective to maximize the relevance scores of local target news; equivalently, we maximize the localized similarities between locations and their local target news and minimize those with irrelevant ones. Formally, it is equivalent to minimize the following loss function:

$$
\begin{aligned}
L^{DL}(\Theta) &= -\sum_{(l,v^*)} \log(R_{l,v^*}^{DL}) \\
&= -\sum_{(l,v^*)} [\log(e^{Sim(\Phi_l^{DL}, \Phi_{v^*,l}^{DL})}) - \log(\sum_{v' \in V} e^{Sim(\Phi_l^{DL}, \Phi_{v',l}^{DL})})], \quad (19)
\end{aligned}
$$

where Θ is the set of parameters $\{W_i^{DL}, b_i^{DL}\}$ ($i = 1, \ldots, N$) in DLSA; tuple (l, v^*) is a training sample, indicating that v^* is a local target news to the user at location l.

As for the training of DLSA (resp., ALSA), we first initialize the weight matrices W_i^{DL} (resp., W_i^{AE}) using the random normal distribution and initialize the bias vectors b_i^{DL} (resp., b_i^{AE}) to be zero vectors; the model is then trained via stochastic gradient descent [7], which is a gradient-based optimization algorithm; finally, the training stops when the model converges or reaches the maximum training iterations. Two optimization solutions are used to enhance the training efficiency of DLSA and ensure its scalability in practice: (i) Rather than training own parameters for each neural network, networks in DLSA share parameters; and (ii) negative sampling is used to further reduce the training cost of DLSA. As shown in our prior work [26], sharing parameters is reasonable and negative sampling can greatly enhance the model's training efficiency by hundreds of times while maintaining almost the same training effectiveness.

In summary, applying deep neural networks for topic feature modeling in DLSA and ALSA is capable to overcome the high dimensionality, sparsity, and redundancy problems in ELSA, because (i) the number of nodes in the hidden layer is much smaller than that in the input layer, so the dimensionality of the resulting deep (or autoencoder-based) topic feature space is much lower than that of the Wikipedia-based topic space in ELSA; and (ii) deep neural networks in DLSA and ALSA extract more abstract and denser features layer-by-layer, so sparsity and redundancy problems are addressed.

5   Experiments

We evaluate the performances of ELSA, CLSA, ALSA, and DLSA using a publicly available Twitter dataset [1], which consists of 2,316,204 tweets posted by

Table 1. Statistic information of the dataset

Tweets	Users	News	Locations	Samples
2,316,204	1,619	63,485	2,366	98,321

Table 2. Details of training set and test (sub)sets

	Users	News	Locations	Samples
Training set	1,558	51,399	1,089	86,086
Test set	1077	11,965	1805	12,235
Old City test subset	941	10,734	581	11,000
New City test subset	527	1,231	1,224	1,235

1,619 users ($|U| = 1,619$). About half of these tweets explicitly contain URLs to the news articles; by using these URLs to download the corresponding news articles, we get 63,485 news articles, which are used as the candidate news articles for recommendation, i.e., $|V| = 63,485$. Then, we apply a Web service tool[1] to extract city names from the news articles, resulting in 2,366 locations ($|L| = 2,366$). Finally, we consider these city names as geo-tags and use the titles and keywords of the news articles, from which the city names are extracted, as the descriptions of these locations, i.e., geo-tagged documents D_l. The statistic information of the dataset is summarized in Table 1.

We assume that a user is specified only by his/her location. Then, if a user posts a tweet containing an URL to a news article v^* with a city name (location) l in its content, the user is believed to be interested in v^* at l, from which a sample (l, v^*) is generated, indicating v^* is a local target news to location l. Consequently, a total of 98,321 samples are obtained from the dataset; we randomly select 85% of the samples as the training set and the remaining 15% as the test set. To evaluate the different recommendation performance on "old" locations ("old" cities), which have appeared in the training set, and on new locations (new cities), which do not exist in the training set, we further divide the test set into two subsets: for each sample (l, v^*) in the test set, if the training set also contains some samples related to location l, l is seen as an "old" city (location), so (l, v^*) is added to Old City test subset; otherwise, l is a new city (location), and the sample (l, v^*) is added to New City test subset. The details are summarized in Table 2.

Finally, a Wikipedia snapshot of August 11, 2014 is used for the Wikipe-dia-based semantic enrichment [10] in ELSA, CLSA, ALSA, and DLSA, resulting in 1,301, 900 concepts with 1,618,970 distinct terms. To cut down the calculation and memory cost, we select 8,000 most frequent concepts as the Wikipedia-based topic space Z.

---

[1] OpenCalais at https://opencalais.com/.

All methods are implemented using Python and Theano and run on a server of Oxford University's ARC facility [15] with an NVIDIA Tesla K40 GPU and 12 GB GPU memory. ELSA is implemented based on [19]; # of clusters in CLSA is empirically set to 1024; and the parameters of ALSA and DLSA are empirically set as follows: (i) # of hidden layers in DLSA and in the encoder of ALSA: $N = 3$; (ii) # of neurons in the 1st, 2nd, and 3rd hidden layer: 1024, 512, and 256, respectively; (iii) learning rate for model training: 0.0001; additionally, (iv) ALSA adds another two hidden layers for the decoder, and # of neurons in the 4th and 5th layer are 512 and 1024, respectively.

The most popular metrics for the evaluation of recommendation systems are precision and recall [6]. Since users usually only browse the topmost recommended news, we apply these metrics at a given cut-off rank k, i.e., considering only the top-k results on the recommendation list, called *precision at k (P@k)* and *recall at k (R@k)*. Since users always prefer to have their target news ranked in the front of the recommendation list, we also use *mean average precision (MAP)* and *mean reciprocal rank (MRR)* as evaluation metrics, which give greater importance to news ranked higher.

5.1 Main Results

Figure 5 depicts the news recommendation performance of DLSA, ALSA, CLSA, and ELSA on three test (sub)sets in terms of precision at k $(P@k)$ and recall at k $(R@k)$, where k varies from 1 to 50. In addition, Table 3 shows the performance of DLSA, ALSA, CLSA, and ELSA on three test (sub)sets in terms of MRR and MAP.

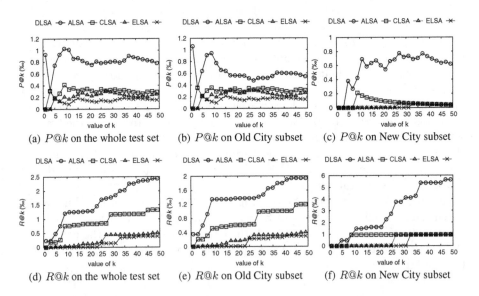

Fig. 5. Performance of DLSA, ALSA, CLSA and ELSA in terms of $P@k$ and $R@k$

Table 3. Performance of DLSA, ALSA, CLSA, and ELSA in terms of MRR and MAP

	The whole test set		Old City test subset		New City test subset	
	MRR(‰)	MAP(‰)	MRR(‰)	MAP(‰)	MRR(‰)	MAP(‰)
ELSA	0.1243	0.2483	0.1301	0.2799	0.0726	0.0951
CLSA	0.1734	0.3530	0.1796	0.3714	0.1185	0.1434
ALSA	0.2192	0.4514	0.2174	0.4441	0.2356	0.2532
DLSA	**0.4737**	**0.8491**	**0.4001**	**0.9132**	**1.130**	**0.7217**

Generally, as shown in both Fig. 5 and Table 3, the proposed three geographical topic feature models, DLSA, ALSA, and CLSA, all greatly outperform the Wikipedia-based topic model, ELSA, in location-aware news recommendation in terms of all evaluation metrics on all three test (sub)sets. This finding demonstrates that applying clustering or deep learning techniques for topic feature modeling can address the high dimensionality, sparsity, and redundancy problems in ELSA and greatly enhances the location-aware news recommendation effectiveness.

Furthermore, with the help of the recommendation-oriented deep neural networks, DLSA achieves a much better recommendation performance than ALSA and CLSA. For example, on the whole test set, the precision, recall, MRR, and MAP of DLSA are all roughly 3 times better than those of ELSA, about 2 times better than those of CLSA, and about double of ALSA. The superior performance of DLSA is mainly because DLSA uses a recommendation-oriented learning objective, which is directly correlated with distinguishing the user's local target news from the irrelevant ones; so, the resulting deep topic features for the representations of news and location profiles in DLSA are much more effective for news recommendations than those generated by clustering and autoencoders, whose learning objectives are not directly related to recommendations.

We also note that DLSA, ALSA, and CLSA generally achieve more significant improvements to ELSA at new locations than those at "old" locations. For example, the MAP's of DLSA, ALSA, and CLSA are 63.2%, 224.5%, and 14.6 times, respectively, better than that of ELSA on the New City test subset, while the corresponding improvements are only 38.0%, 67.1%, and 3.08 times on the Old City test subset. It may be that due to the lack of users' history data on new locations, the inferred localized location profiles on the New City test subset are less accurate than those on the Old City test subset, so ELSA is less likely to recommend users' local target news to the top positions of recommendation lists at new locations. But, by using clustering or deep learning techniques for topic feature modeling, the proposed geographical topic feature models, CLSA, ALSA, and DLSA, map the Wikipedia-based topic space to a more abstract feature topic space, where the correlations among similar topics (e.g., topics within a category) are strengthened. Thus, CLSA, ALSA, and DLSA can uncover users' latent localized news preferences to make the inferred abstract localized location profiles more accurate. So, DLSA, ALSA, and CLSA also remedy the "cold-start" problem.

5.2   Efficiency and Scalability

News recommendation requests real-time responses in practice; so the recommendation efficiency is crucial for online location-aware news recommendation. However, due to the high volume of concepts in Wikipedia, the Wikipedia-based topic space in ELSA is very huge. Consequently, the online news recommendation process that requests to compute the similarities between the localized location and news profiles in ELSA is usually computationally expensive. As shown in Table 4, even if we have limited the Wikipedia topic space to contain only 8,000 most frequent topics and speeded up computation using a GPU server, the total (resp., average) time-costs for the online recommendation processes in ELSA are still up to 142.1, 15.67, and 160.4 (resp., 0.0129, 0.0127, and 0.0131) min on Old City, New City, and the whole test (sub)sets, respectively, which are usually unacceptable in practice.

   Therefore, to ensure scalability in the real-world context, the proposed DLSA (resp., CLSA and ALSA) solve the high dimensionality problem by mapping the Wikipedia-based topic space to a deep (resp., clustering- and autoencoder-based) topic feature space with much lower dimensionality. As shown in Table 4, although DLSA and ALSA have to pass data through the well-trained neural networks prior to compute similarities between deep or autoencoder-based localized location and news profiles, the total online recommendation time-costs of DLSA and ALSA on the whole test set are still only about 1/29 of that of ELSA. In addition, although the total time-cost of CLSA is higher than those of DLSA and ALSA, it is still much lower than that of ELSA. Overall, these findings prove that, with the help of topic feature modeling, the proposed geographical topic feature models, CLSA, ALSA, and DLSA, can achieve much higher recommendation efficiency than ELSA; so, it is more scalable in practice.

Table 4. Total time-costs for online recommendations (in min)

	ELSA	CLSA	ALSA	DLSA
The whole test set	160.4	18.59	**5.552**	5.699
Old City test subset	142.1	16.89	**5.006**	5.160
New City test subset	15.67	3.997	0.5961	**0.5906**

6   Summary and Outlook

In this work, we proposed three novel geographical topic feature models, CLSA, ALSA, and DLSA, which integrate clustering, autoencoders, and recommendation-oriented deep neural networks, respectively, with ELSA to address the high dimensionality, sparsity, and redundancy problems existing in ELSA's Wikipedia-based topic space. Experimental studies showed that CLSA, ALSA, and DLSA all significantly outperform the state-of-the-art geographical topic model, ELSA, in location-aware news recommendation in terms of both effectiveness and efficiency, while DLSA achieves the best performance: it offers more effective (about 3 times better) location-aware news recommendation with much lower online recommendation time cost (about 28 times quicker) than ELSA. In addition, DLSA, ALSA, and CLSA can also remedy the "cold-start" problem by uncovering users' latent news preferences at new locations.

In the future, it would be interesting to consider user personal preferences and more contextual information, such as timeliness of news and the social relationships of users, to achieve better personalized context-aware news recommendation. In addition, hybrid learning signals, e.g., combining reconstruction errors with deep-semantic similarities, and more sophisticated neural networks (e.g., convolutional or long short-term memory (LSTM) neural networks) may be applied to learn a more effective abstract topic feature space, and so to further improve the performance of DLSA.

Acknowledgments. This work is supported by the Mutual Project of Beijing Municipal Education Commission. Thomas Lukasiewicz and Zhenghua Xu are supported by the UK EPSRC Grants EP/J008346/1, EP/L012138/1, and EP/M025268/1, and by The Alan Turing Institute under the EPSRC Grant EP/N510129/1.

References

1. Abel, F., Gao, Q., Houben, G.-J., Tao, K.: Analyzing user modeling on Twitter for personalized news recommendations. In: Konstan, J.A., Conejo, R., Marzo, J.L., Oliver, N. (eds.) UMAP 2011. LNCS, vol. 6787, pp. 1–12. Springer, Heidelberg (2011). doi:10.1007/978-3-642-22362-4_1

2. Bao, J., Mokbel, M., Chow, C.-Y.: GeoFeed: a location-aware news feed system. In: Proceedings of ICDE, pp. 54–65 (2012)

3. Bao, J., Mokbel, M.F.: GeoRank: an efficient location-aware news feed ranking system. In: Proceedings of SIGSPATIAL, pp. 184–193 (2013)

4. Bengio, Y., Courville, A., Vincent, P.: Representation learning: a review and new perspectives. TPAMI **35**(8), 1798–1828 (2013)

5. Blei, D.M., Ng, A.Y., Jordan, M.I.: Latent Dirichlet allocation. JMLR **3**, 993–1022 (2003)

6. Bobadilla, J., Ortega, F., Hernando, A., Gutiérrez, A.: Recommender systems survey. Knowl. Based Syst. **46**, 109–132 (2013)

7. Bottou, L.: Stochastic learning. In: Proceedings of ALML, pp. 146–168 (2003)

8. Chu, W., Park, S.-T.: Personalized recommendation on dynamic content using predictive bilinear models. In: Proceedings of WWW, pp. 691–700 (2009)

9. Elkahky, A.M., Song, Y., He, X.: A multi-view deep learning approach for cross domain user modeling in recommendation systems. In: Proceedings of WWW, pp. 278–288 (2015)

10. Gabrilovich, E., Markovitch, S.: Wikipedia-based semantic interpretation for natural language processing. JAIR **34**(1), 443–498 (2009)

11. Li, L., Chu, W., Langford, J., Schapire, R.E.: A contextual-bandit approach to personalized news article recommendation. In: Proceedings of WWW, pp. 661–670 (2010)

12. Noh, Y., Oh, Y.-H., Park, S.-B.: A location-based personalized news recommendation. In: Proceedings of BigComp, pp. 99–104 (2014)

13. Page, L., Brin, S., Motwani, R., Winograd, T.: The PageRank citation ranking: bringing order to the web (1999)
14. Qie, K.V.: Sensing the news: user experiences when reading locative news. Future Internet **4**, 161–178 (2012)
15. Richards, A.: University of Oxford Advanced Research Computing (2015)
16. Rosa, P.M.P., Rodrigues, J.J.P.C., Basso, F.: A weight-aware recommendation algorithm for mobile multimedia systems. Mob. Info. Sys. **9**(2), 139–155 (2013)
17. Salakhutdinov, R., Mnih, A., Hinton, G.: Restricted Boltzmann machines for collaborative filtering. In: Proceedings of ICML, pp. 791–798 (2007)
18. Shepitsen, A., Gemmell, J., Mobasher, B., Burke, R.: Personalized recommendation in social tagging systems using hierarchical clustering. In: Proceedings of RecSys, pp. 259–266 (2008)
19. Son, J.-W., Kim, A.-Y., Park, S.-B.: A location-based news article recommendation with explicit localized semantic analysis. In: Proceedings of SIGIR, pp. 293–302 (2013)
20. Van den Oord, A., Dieleman, S., Schrauwen, B.: Deep content-based music recommendation. In: Proceedings of NIPS, pp. 2643–2651 (2013)
21. Xu, W., Chow, C.-Y., Yiu, M.L., Li, Q., Poon, C.K.: MobiFeed: a location-aware news feed framework for moving users. GeoInformatica **19**(3), 633–669 (2015)
22. Yeung, K.F., Yang, Y.: A proactive personalized mobile news recommendation system. In: Proceedings of DeSE, pp. 207–212 (2010)
23. Yin, H., Cui, B., Sun, Y., Hu, Z., Chen, L.: LCARS: a spatial item recommender system. ACM TOIS **32**(3), 1–37 (2014)
24. Zhou, Y., Luo, J.: Geo-location inference on news articles via multimodal pLSA. In: Proceedings of ACM MM, pp. 741–744 (2012)
25. Zuo, Y., Zeng, J., Gong, M., Jiao, L.: Tag-aware recommender systems based on deep neural networks. Neurocomputing **204**, 51–60 (2016)
26. Xu, Z., Chen, C., Lukasiewicz, T., Miao, Y., Meng, X.: Tag-aware personalized recommendation using a deep-semantic similarity model with negative sampling. In: Proceedings of CIKM, pp. 1921–1924 (2016)

Review-Based Cross-Domain Recommendation Through Joint Tensor Factorization

Tianhang Song[1], Zhaohui Peng[1(✉)], Senzhang Wang[2], Wenjing Fu[1],
Xiaoguang Hong[1], and Philip S. Yu[3,4]

[1] School of Computer Science and Technology,
Shandong University, Jinan, China
{sth1202, fuwenjing}@mail.sdu.edu.cn,
{pzh, hxg}@sdu.edu.cn
[2] College of Computer Science and Technology, Nanjing University
of Aeronautics and Astronautics, Nanjing, China
szwang@nuaa.edu.cn
[3] Department of Computer Science,
University of Illinois at Chicago, Chicago, USA
psyu@uic.edu
[4] Institute for Data Science, Tsinghua University, Beijing, China

Abstract. Cross domain recommendation which aims to transfer knowledge from auxiliary domains to target domains has become an important way to solve the problems of data sparsity and cold start in recommendation systems. However, most existing works only consider ratings and tags, but ignore the text information like reviews. In reality, review text in some way well explains the reason why a product could gain such high or low ratings and reflect users' sentiment towards different aspects of an item. For instance, reviews can be taken advantage to obtain users' attitudes towards the specific aspect "screen" or "battery" of a "cell phone". Taking these aspect factors into cross domain recommendation will bring us more about user preference, and thus could potentially improve the performance of recommendation. In this paper, we for the first time study how to fully exploit the aspect factors extracted from the review text to improve the performance of cross domain recommendation. Specifically, we first model each user's sentiment orientation and concern degree towards different aspects of items extracted from reviews as tensors. To effectively transfer the aspect-level preferences of users towards items, we propose a joint tensor factorization model on auxiliary domain and target domain together. Experimental results on real data sets show the superior performance of the proposed method especially in the cold-start users in target domain by comparison with several state-of-the-arts cross domain recommendation methods.

Keywords: Cross-domain recommendation · Joint tensor factorization · Review text

S. Candan et al. (Eds.): DASFAA 2017, Part I, LNCS 10177, pp. 525–540, 2017.
DOI: 10.1007/978-3-319-55753-3_33

1 Introduction

Personalized recommendation systems have been widely applied on the Internet. However, until now, the problem of cold start for new users is still a challenge that cannot be neglected in recommendation systems [25, 26]. An effective solution worth considering is cross domain recommendation, which has gained increasing attention of many researchers [4, 7, 8, 16]. In many product review websites, users' feedback is "domain-unbalanced", which means that there may be sufficient user feedbacks in some domains, such as "Clothes" or "Books", but very few or even no user feedback in other domains such as "Shoes" or "Movies". This is also called "unacquainted world" phenomenon [4]. To cope with this problem, users' preferences in one domain can be utilized as auxiliary information to assist alleviates the cold start issue in another domain.

Most existing works about cross domain recommendation transfer users' preference among different domains only based on their ratings on items [6, 7, 13–15, 17] or tags [3, 21]. However, knowledge transfer merely based on ratings has rather great limitation in that ratings just give a sweeping statement about users' attitude towards an item, but cannot show us the detailed attitude about specific aspects of the item, for example, aspects "screen" and "battery" of a "cell phone". Luckily, to some extent, these detailed information which could make a contribution to describe users' preferences on different aspects could be extracted from text reviews, but it is largely ignored by previous works [6, 7, 16].

Table 1 shows three reviews given by one user, for two movies and one book. From Movie 1 we can see, though the movie gains a low rating from the user, the user also shows a positive attitude towards the aspects about "picture" and "stunt" of the movie. Also we can see the disappointment about the aspect about "politics" mainly leads to the low rating. So, instead of drawing a simple conclusion that the user "dislikes the movie",

Table 1. Three review texts of one user in movie domain and book domain

Item	Rating	Review text
Movie 1	2.0	Good parts: new Bond is OK, very physical and relatively **pleasant**, the **pictures** and **supporting cast** are **great**, **stunts** are simply **great**. The Obanno's **attack is a very refreshing move** and in my view the **best episode** of the movie. Bad parts:1. Bond girls are ... **looks very unnatural**. 2. ... **governments** do not care that much about $15 mln. anymore. 3. The **plot** is **annoyingly weak**: e.g. the **bad guy** does not have his own crew capable of minor bombing etc.
Movie 2	3.0	... **characters are very good**, and **pictures are beautiful**. ... There is one thing I **admire** about **Hollywood**: even when running for her life **girl** would never ditch her **high heels**. We have plenty of **high heel races** here.
Book 1	2.0	It has some **entertaining value**, but ... we all know that to ... **World** ...; there is no need to use **nukes** for the purpose. **CIA** participation line is even kookier, ... **world** would ever participate in ... **conspiracy**? I am **tired of** ....

learning the aspect level attitudes of users towards items is necessary. Furthermore, in Table 1, the user shows her/his attitude towards the aspect about "politics" both in Movie 1 and Book 1, and shows attitude towards the aspect about "entertaining" both in Movie 2 and Book 1. This inspires us that the aspect level attitudes may have internal relationships between different domains. Consequently, we introduce this fine-grained preference information into cross domain recommendation by exploiting review text. In our framework, instead of transferring the overall user preference to items, we transfer the more fine-grained user preference to different aspects of the items.

For example, if a user gave a high rating to the book "Harry Potter", previous works usually tend to make a prediction that the user will also be fond of the movie "Harry Potter". But in our model, a user shows a positive sentiment towards the "child" aspect of the book "Harry Potter" may lead to a positive predicted attitude on the "child" aspect of the movie "Harry Potter". Furthermore, the attitude towards other aspects such as "fantasy" and "music" may not be necessarily positive.

What's more, for an item, different users may place emphasis on different aspects. The sentiment orientation on the aspects which gain more attention usually play a more significant role in the overall rating in that they'd better be allocated greater weight. Therefore, users' concern degree on different aspects of an item is also needed to be extracted and taken into our cross domain recommendation framework.

In extracting aspects, obtaining aspect-level user ratings and concern degrees, in this paper, we employ the state-of-the-art method AIRS [9]. AIRS outputs the ratings and concern degree on each aspect for every review, these are the input of our framework. To solve the challenge of fully capturing implicit relationship between users, items and aspects, 3-order tensors are effectively utilized. To achieve the goal of transferring knowledge, extending traditional cross domain matrix collective factorization models [10, 16, 18, 27], we propose a novel tensor collective factorization framework, called Review Based Joint Tensor Factorization (RB-JTF). Through transferring latent factors of users and aspects in factorization process, the RB-JTF model gets a superior performance in cross domain recommendation.

The main contributions of this paper are summarized as follows:

- We for the first time propose a joint tensor factorization model to effectively transfer knowledge in cross-domain recommendation. How to transfer cross-domain knowledge by decomposing two different tensors jointly is challenging and not well studied.
- We propose to capture the latent relations of aspect-level sentiment of users towards items on both domains in cross domain recommendation, which is different from previous works. And we verify its rationality.
- We demonstrate the effectiveness of our method on real world datasets comparing with current state-of-the-art methods.

The rest of this paper is organized as follows. Section 2 presents some notations and gives the problem formulation. Section 3 describes the unified framework. Experimental results and some analysis are given in Sect. 4. In Sect. 5, we discuss related works and conclude the paper in Sect. 6.

2 Problem Formulation

Given two domains of data for recommendation, let $Rev^{(A)}$ and $Rev^{(T)}$ denote the set of reviews in auxiliary and target domain respectively. Correspondingly, let $Rat^{(A)}$ and $Rat^{(T)}$ denote the set of observed ratings in the two domains. We obtain users' sentiments towards different aspects of items from $Rev^{(A)}$, $Rat^{(A)}$ in auxiliary domain and $Rev^{(T)}$, $Rat^{(T)}$ in target domain and organize them into two tensors. Hence, 3-order tensors are employed to organize the ternary relationships of user, item and aspect. For the two domains, we correspondingly construct the rating tensors $\mathcal{A}^{RAT}$ and $\mathcal{T}^{RAT}$, $\mathcal{A}^{RAT} \in \mathbb{R}^{N_U \times N_{I^{(A)}} \times N_{AS^{(A)}}}$, $\mathcal{T}^{RAT} \in \mathbb{R}^{N_U \times N_{I^{(T)}} \times N_{AS^{(T)}}}$. The element a^{rat}_{xyz} of $\mathcal{A}^{RAT}$, records the rating preference of user x on aspect z of item y. If user x wrote any review about item y, the corresponding elements in $\mathcal{A}^{RAT}$(x, y, :) are unknown, and so as to tensor $\mathcal{T}^{RAT}$. Similarly, to model the concern degree of users on items towards different aspects, we also construct tensors $\mathcal{A}^{CON}$ and $\mathcal{T}^{CON}$ to organize them. The sizes and coordinates of $\mathcal{A}^{CON}$ and $\mathcal{T}^{CON}$ are corresponding to $\mathcal{A}^{RAT}$ and $\mathcal{T}^{RAT}$ respectively. If no data is observed for an element, the element is unknown. Table 2 lists some frequently used symbols.

Problem Definition. Given $Rev^{(A)}$, $Rev^{(T)}$, $Rat^{(A)}$, $Rat^{(T)}$, U, $I^{(A)}$, $I^{(T)}$, $N_{AS^{(A)}}$ and $N_{AS^{(T)}}$, to get the predicted ratings $\hat{r}_{ui}$, where $u \in$ U, $i \in I^{(T)}$ and $r_{ui} \notin Rat^{(T)}$.

Table 2. Summary of notations

Symbol	Description
U	The set of users with size $|U| = N_U$
$I^{(A)}$, $I^{(T)}$	The set of items in auxiliary or target domain with sizes $\left|I^{(A)}\right| = N_{I^{(A)}}$ and $\left|I^{(T)}\right| = N_{I^{(T)}}$
$N_{AS^{(A)}}$, $N_{AS^{(T)}}$	The number of aspects in auxiliary or target domain
$\mathcal{A}^{RAT}$, $\mathcal{T}^{RAT}$	The aspect rating tensor of auxiliary or target domain
$\mathcal{A}^{CON}$, $\mathcal{T}^{CON}$	The concern degree tensor of auxiliary or target domain
$a^{rat}_{u,i^{(a)},as^{(a)}}$, $t^{rat}_{u,i^{(t)},as^{(t)}}$	The element in $\mathcal{A}^{RAT}$ or $\mathcal{T}^{RAT}$, denotes the aspect rating of user u to item i on aspect as in auxiliary domain or target domain
$a^{con}_{u,i^{(a)},as^{(a)}}$, $t^{con}_{u,i^{(t)},as^{(t)}}$	The element in $\mathcal{A}^{CON}$ or $\mathcal{T}^{CON}$, denotes the concern degree of user u on item i towards aspect as in auxiliary domain or target domain

3 Review-Based Joint Tensor Factorization

In this section, we first describe how to model users' aspect-level preferences. Taking these as inputs, we propose our joint tensor factorization model to transfer knowledge among two domains. Finally we will present the method to predict overall ratings.

3.1   Aspect Rating and Concern Degree Modeling

Some techniques can be utilized to extract aspects and users' sentiment and concerns from review texts, such as the LARAM [19] and the AIRS [9]. We employ the AIRS model in our framework. In AIRS, each review is represented by a mixture over K topics (seen as K aspects) and the distribution of each review's topic θ_i is sampled from a Dirichlet distribution for review i, accordingly, $\theta_{i,k}$ is the weight of the k-th topic for review i. And the aspect ratings Ω_i are sampled from a Beta distribution with the prior that if an aspect is talked more in a review, the rating on this aspect would be affected more by the overall rating. Three Dirichlet distributions over words under neutral (φ_k^0), positive (φ_k^1) and negative (φ_k^2) sentiments are used to characterize each topic. EM for MAP is used to estimate parameters.

Case in the target domain, given $Rat^{(T)}$, $Rev^{(T)}$ and $N_{AS(T)}$, AIRS model generates a number of $N_{AS(T)}$ aspects. Meanwhile, for each review $Rev_{u,i^{(t)}}$, AIRS gets the aspect distribution $\theta_{u,i^{(t)}}$, and the ratings on each aspect $\Omega_{u,i^{(t)},as^{(t)}}$. Let $\theta_{u,i^{(t)},as^{(t)}}$ denotes the weight of the $as^{(t)}$-th aspect in review $Rev_{u,i^{(t)}}$; $\Omega_{u,i^{(t)},as^{(t)}}$ denotes the aspect rating of the $as^{(t)}$-th aspect in review $Rev_{u,i^{(t)}}$. Tensor $\mathcal{T}^{CON}$ is constructed as:

$$t^{con}_{u,i^{(t)},as^{(t)}} = \begin{cases} \theta_{u,i^{(t)},as^{(t)}}, & \text{if } u \text{ has review to } i^{(t)}. \\ unknown, & else. \end{cases} \quad (1)$$

$\mathcal{T}^{RAT}$ is constructed as:

$$t^{rat}_{u,i^{(t)},as^{(t)}} = \begin{cases} \Omega_{u,i^{(t)},as^{(t)}}, & \text{if } t^{con}_{u,i^{(t)},as^{(t)}} \text{ it not unknown} \\ & \text{and } t^{con}_{u,i^{(t)},as^{(t)}} > 0. \\ unknown, & else. \end{cases} \quad (2)$$

In auxiliary domain, $\mathcal{A}^{CON}$ and $\mathcal{A}^{RAT}$ are constructed in the similar way.

3.2   Joint Tensor Factorization

Now we have four tensors: $\mathcal{A}^{RAT}$, $\mathcal{T}^{RAT}$, $\mathcal{A}^{CON}$, $\mathcal{T}^{CON}$. In this subsection, we describe the proposed joint tensor factorization model with the four tensors. The overview framework is shown in Fig. 1. We jointly factorize the two tensors in different domains, each into three factor matrices. The knowledge is transferred in two parts: first, the sharing of user latent factors; second, the transferring of aspect latent factors.

Aspect Rating Tensor Joint Factorization. Our model is based on CANDECOMP/ PARAFAC (CP) model [5, 11, 29]. Given a 3-order tensor $x \in \mathbb{R}^{I \times J \times K}$, CP factorizes it as $x \approx [\![L, M, N]\!] = \sum_{r=1}^{R} L_{:,r} \circ M_{:,r} \circ N_{:,r}$, where $L \in \mathbb{R}^{I \times R}$, $M \in \mathbb{R}^{J \times R}$, $N \in \mathbb{R}^{K \times R}$ and R is the number of latent features. $\bigcirc$ denotes the outer product. Since our work is to predict the missing values in incomplete tensors, we use the weighted version of CP:

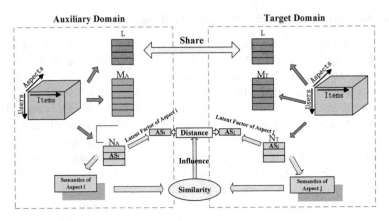

Fig. 1. Overview of the joint tensor factorization model.

$$f(L, M, N) = \sum_{i=1}^{I} \sum_{j=1}^{J} \sum_{k=1}^{K} \left[w_{ijk} \left(x_{ijk} - \sum_{r=1}^{R} L_{i,r} \cdot M_{j,r} \cdot N_{k,r} \right) \right]^2 \quad (3)$$

where $\mathcal{W}$ is an indicator tensor whose size is the same as $\mathbf{x}$.

Through weighted CP model, an aspect rating tensor is factorized into three matrices, which represent the latent features of users, items and aspects respectively. The objective functions of $\mathcal{A}^{RAT}$ is as follows:

$$fa^{rat}\left(L_A^{rat}, M_A^{rat}, N_A^{rat} \right) =$$
$$\sum_{u=1}^{N_U} \sum_{i^{(a)}=1}^{N_{I(A)}} \sum_{as^{(a)}=1}^{N_{AS(A)}} \left\{ W^{rat}_{a\,u,i^{(a)},as^{(a)}} \left(a^{rat}_{u,i^{(a)},as^{(a)}} - \sum_{r=1}^{R} l^{rat}_{a\,u,r} \cdot m^{rat}_{a\,i^{(a)},r} \cdot n^{rat}_{a\,as^{(a)},r} \right) \right\}^2 \quad (4)$$

In Eq. (4), $L_A^{rat} \in \mathbb{R}^{N_U \times R}$, $M_A^{rat} \in \mathbb{R}^{N_{I(A)} \times R}$ and $N_A^{rat} \in \mathbb{R}^{N_{AS(A)} \times R}$ is the latent feature matrices of users, items and aspects in auxiliary domain. R is the number of latent features. $\mathcal{W}_A^{RAT}$ is the indicator tensor of $\mathcal{A}^{RAT}$ with following definition:

$$\mathcal{W}_A^{RAT}\left(u, i^{(a)}, as^{(a)} \right) = \begin{cases} \mathcal{A}^{CON}\left(u, i^{(a)}, as^{(a)} \right), & \text{if } \mathcal{A}^{CON}\left(u, i^{(a)}, as^{(a)} \right) \text{ is not unknown} \\ & \text{and } \mathcal{A}^{CON}\left(u, i^{(a)}, as^{(a)} \right) > 0 \\ 0, & \text{else} \end{cases}$$

The objective functions of $\mathcal{T}^{RAT}$ is similar.

Equation (4) can be rewritten in the tensor and matrix form as:

$$fa^{rat}\left(L_A^{rat}, M_A^{rat}, N_A^{rat} \right) = \left\| \mathcal{W}_A^{RAT} * \left(\mathcal{A}^{RAT} - [\![L_A^{rat}, M_A^{rat}, N_A^{rat}]\!] \right) \right\|_F^2 \quad (5)$$

where $*$ denotes the Hadamard (element-wise) product, and $\|\cdot\|_F$ denotes the Frobenius norm.

To transfer knowledge from auxiliary domain to target domain, our main idea is mapping the users, items and aspects of the two domains into a shared latent feature space. First we jointly factorize the two tensors $\mathcal{A}^{RAT}$ and $\mathcal{T}^{RAT}$ with sharing users' latent factors L^{rat} in the two domains. We have:

$$f^{rat}\left(L^{rat}, M_A^{rat}, N_A^{rat}, M_T^{rat}, N_T^{rat}\right) =$$
$$\lambda \cdot \left\|\mathcal{W}_A^{RAT} * \left(\mathcal{A}^{RAT} - [[L^{rat}, M_A^{rat}, N_A^{rat}]]\right)\right\|_F^2 + \left\|\mathcal{W}_T^{RAT} * \left(\mathcal{T}^{RAT} - [[L^{rat}, M_T^{rat}, N_T^{rat}]]\right)\right\|_F^2$$

(6)

λ is a weight parameter that controls the proportion of the auxiliary domain.

More importantly, the aspects' factors should be mapping into a shared space simultaneously. However, the aspect factors cannot be directly equal in two domains. The strategy we adopt is, the more similar two aspects in semantic, the closer they should in the shared space of factor. We first define the similarity s_{ij} of two aspects i and j. An aspect corresponds to a set of words, top N words of each aspect are chosen to calculate the similarity, denoted as $wordset_i$ and $wordset_j$ respectively. Their Jaccard similarity $jsim_{ij}$ is:

$$jsim_{ij} = \frac{|wordset_i \cap wordset_j|}{|wordset_i \cup wordset_j|}$$

(7)

The scope of $jsim_{ij}$ is in [0, 1], thus we employ such a function to map $jsim_{ij}$ to the scope of [0, +∞), to have a large enough range to adjust the influence to the distance of aspect latent factors:

$$s_{ij} = \gamma \cdot \frac{jsim_{ij}}{1 - jsim_{ij}}$$

(8)

Finally we obtain the objective function:

$$\min f^{rat}\left(L^{rat}, M_A^{rat}, N_A^{rat}, M_T^{rat}, N_T^{rat}\right) =$$
$$\frac{\lambda}{2}\left\|\mathcal{W}_A^{RAT} * \left(\mathcal{A}^{RAT} - [[L^{rat}, M_A^{rat}, N_A^{rat}]]\right)\right\|_F^2 + \frac{1}{2}\left\|\mathcal{W}_T^{RAT} * \left(\mathcal{T}^{RAT} - [[L^{rat}, M_T^{rat}, N_T^{rat}]]\right)\right\|_F^2$$
$$+ \frac{\mu}{2} \sum_{i=1}^{N_{AS(A)}} \sum_{j=1}^{N_{AS(T)}} s_{ij} \cdot \left\|N_A^{rat}(i,:) - N_T^{rat}(j,:)\right\|_F^2$$

(9)

where $\left\|N_A^{rat}(i,:) - N_T^{rat}(j,:)\right\|_F^2$ is the Euclidean distance between two vectors of aspect factors respectively from the auxiliary and target domains. This achieves that the more similar in semantic between aspects i and j, the nearer their latent features in the shared space. We apply this constraint on each pair of aspects from different domains, thus there are $N_{AS(A)} \times N_{AS(T)}$ such terms. μ controls the overall weight of this kind of constraint. The implication of the last group of terms in Eq. (9) is ensuring that user's preferences on similar aspects in different domains tend to be similar.

Concern Degree Tensor Joint Factorization. We use a similar strategy, jointly decompose the two tensors $\mathcal{A}^{CON}$ and $\mathcal{T}^{CON}$, to transfer the concern degree. The objective function is:

$$\min f^{con}\left(L^{con}, M_A^{con}, N_A^{con}, M_T^{con}, N_T^{con}\right) =$$

$$\frac{\lambda}{2}\left\|\mathcal{W}_A^{CON} * \left(\mathcal{A}^{CON} - [\![L^{con}, M_A^{con}, N_A^{con}]\!]\right)\right\|_F^2 + \frac{1}{2}\left\|\mathcal{W}_T^{CON} * \left(\mathcal{T}^{CON} - [\![L^{con}, M_T^{con}, N_T^{con}]\!]\right)\right\|_F^2$$

$$+ \frac{\mu}{2}\sum_{i=1}^{N_{AS^{(A)}}}\sum_{j=1}^{N_{AS^{(T)}}} s_{ij} \cdot \left\|N_A^{con}(i,:) - N_T^{con}(j,:)\right\|_F^2$$

$$(10)$$

where $\mathcal{W}_A^{CON}$ is defined as:

$$\mathcal{W}_A^{CON}\left(u, i^{(a)}, as^{(a)}\right) = \begin{cases} 1, \mathcal{A}^{CON}\left(u, i^{(a)}, as^{(a)}\right) \text{ is not unknown} \\ 0, \mathcal{A}^{CON}\left(u, i^{(a)}, as^{(a)}\right) \text{ is unknown} \end{cases}$$

$\mathcal{W}_T^{CON}$ is similar to $\mathcal{W}_A^{CON}$.

Factor Matrices Learning. To learn factor matrices $L^{rat}, M_A^{rat}, N_A^{rat}, M_T^{rat}, N_T^{rat}$ in Eq. (9) and $L^{con}, M_A^{con}, N_A^{con}, M_T^{con}, N_T^{con}$ in Eq. (10), we extend the learning method in CP-WOPT [1, 2], which is proved to be effective in factorizing scalable tensors with unknown elements. Nonlinear conjugate gradient (NCG) method is adopted in learning factor matrices. The learning process is shown in Algorithm 1.

Algorithm 1 Factor matrices learning with NCG

Inputs: $\mathcal{A}, \mathcal{T}, \mathcal{W}_A, \mathcal{W}_T, S, \lambda, \mu$

Outputs: L, M_T, N_T, M_A, N_A

1: Initialize L, M_T, N_T, M_A, N_A, and reshape them into a vector x_0, set k = 0.

2: Calculate gradients of L, M_T, N_T, M_A, N_A at x_0 use Eq. (11) – Eq. (15) respectively, and reshape them into a vector g_0.

   Do Steps 3-7 until convergence

3: Calculate search direction d_k use:
$$d_k = \begin{cases} -g_k, & k = 0 \\ -g_k + \beta_{k-1}d_{k-1}, & k \geq 1 \end{cases} \text{ where } \beta_{k-1} = \frac{g_k^T g_k}{g_{k-1}^T g_{k-1}}.$$

4: Search the step size α_k.

5: Calculate $x_{k+1} = x_k + \alpha_k d_k$.

6: Calculate gradients of L, M_T, N_T, M_A, N_A at x_{k+1} use Eq. (11) – Eq. (15) respectively, and reshape them into a vector g_{k+1}.

7: k=k+1.

   end

8: Reshape x_k to L, M_T, N_T, M_A, N_A.

9: Return L, M_T, N_T, M_A, N_A.

The gradient of L^{rat} is:

$$\frac{\partial f^{rat}}{\partial L^{rat}} = \lambda \cdot \left[W_A^{RAT}{}_{(1)} * \left(Z_A^{RAT}{}_{(1)} - Y_A^{RAT}{}_{(1)} \right) \right] \left(N_A^{rat} \odot M_A^{rat} \right) \\ + \left[W_T^{RAT}{}_{(1)} * \left(Z_T^{RAT}{}_{(1)} - Y_T^{RAT}{}_{(1)} \right) \right] \left(N_T^{rat} \odot M_T^{rat} \right) \tag{11}$$

where $\odot$ denotes the Khatri-Rao product. $Z_A^{RAT} = W_A^{RAT} * [\![L^{rat}, M_A^{rat}, N_A^{rat}]\!]$, $Y_A^{RAT} = W_A^{RAT} * A^{RAT} \cdot Z_T^{RAT}$ and Y_T^{RAT} are similar to Z_A^{RAT} and Y_A^{RAT}. $Z_{(n)}$, $Y_{(n)}$ denote the mode-n matricization [5, 11] of tensor Z, Y respectively.

The gradients of M_A^{rat} and M_T^{rat} are:

$$\frac{\partial f^{rat}}{\partial M_A^{rat}} = \lambda \cdot \left[W_A^{RAT}{}_{(2)} * \left(Z_A^{RAT}{}_{(2)} - Y_A^{RAT}{}_{(2)} \right) \right] \left(N_A^{rat} \odot L^{rat} \right) \tag{12}$$

$$\frac{\partial f^{rat}}{\partial M_T^{rat}} = \left[W_T^{RAT}{}_{(2)} * \left(Z_T^{RAT}{}_{(2)} - Y_T^{RAT}{}_{(2)} \right) \right] \left(N_T^{rat} \odot L^{rat} \right) \tag{13}$$

The gradient of N_A^{rat} is:

$$\frac{\partial f^{rat}}{\partial N_A^{rat}} = \lambda \cdot \left[W_A^{RAT}{}_{(3)} * \left(Z_A^{RAT}{}_{(3)} - Y_A^{RAT}{}_{(3)} \right) \right] \left(M_A^{rat} \odot L^{rat} \right) \\ + \mu \cdot \begin{bmatrix} S(1,:) \cdot (N_A^{rat(1)} - N_T^{rat}) \\ S(2,:) \cdot (N_A^{rat(2)} - N_T^{rat}) \\ \vdots \\ S(m,:) \cdot (N_A^{rat(m)} - N_T^{rat}) \\ \vdots \\ S(N_{AS(A)},:) \cdot (N_A^{rat(N_{AS(A)})} - N_T^{rat}) \end{bmatrix} \tag{14}$$

where S is a $N_{AS(A)} \times N_{AS(T)}$ matrix which is composed of $S_{ij} \cdot N_A^{rat(m)}$ is a matrix with size of $N_{AS(T)} \times R$, each row of it is the same as the m-th row of N_A^{rat}, i.e., for each row i in $N_A^{rat(m)}$, $N_A^{rat(m)}(i,:) = N_A^{rat}(m,:)$, $1 \le m \le N_{AS(A)}$, $1 \le i \le N_{AS(T)}$; m, $i \in N^+$. Similarly, the gradient of N_T^{rat} is:

$$\frac{\partial f^{rat}}{\partial N_T^{rat}} = \left[W_T^{RAT}{}_{(3)} * \left(Z_T^{RAT}{}_{(3)} - Y_T^{RAT}{}_{(3)} \right) \right] \left(M_T^{rat} \odot L^{rat} \right) \\ + \mu \cdot \begin{bmatrix} S(:,)^T \cdot (N_T^{rat(1)} - N_A^{rat}) \\ S(:,2)^T \cdot (N_T^{rat(2)} - N_A^{rat}) \\ \vdots \\ S(:,m)^T \cdot (N_T^{rat(m)} - N_A^{rat}) \\ \vdots \\ S(:,N_{AS(T)})^T \cdot (N_T^{rat(N_{AS(T)})} - N_A^{rat}) \end{bmatrix} \tag{15}$$

where $S(:,m)^T$ is the transpose of $S(:,m) \cdot N_T^{rat(m)}$ is a matrix with size of $N_{AS(A)} \times R$, $N_T^{rat(m)}(i,:) = N_T^{rat}(m,:)$, $1 \le m \le N_{AS(T)}$, $1 \le i \le N_{AS(A)}$; m, $i \in N^+$.

The gradients of $L^{con}, M_A^{con}, N_A^{con}, M_T^{con}, N_T^{con}$ can be obtained similarly as $L^{rat}, M_A^{rat}, N_A^{rat}, M_T^{rat}, N_T^{rat}$. In the computation of the gradients, tensor y and its unfold matrices, the aspects similarity matrix S are all pre-calculated and never changed in the iterations. So the complexity of Algorithm 1 mainly depends on the computation of gradients. The complexity of Eq. (11) is $O(N_U N_I^2 N_{AS}^2 R)$, that of Eqs. (12) and (13) are all $O(N_I N_U^2 N_{AS}^2 R)$, the complexity of Eq. (14) is $O(N_{AS^{(A)}} N_U^2 N_{I^{(A)}}^2 R + N_{AS^{(T)}}^2 R N_{AS^{(A)}})$, Eq. (15) is similar to Eq. (14).

3.3   Overall Ratings Prediction

The predicted target domain tensors $\hat{\mathcal{T}}^{RAT}$ and $\hat{\mathcal{T}}^{CON}$ are reconstructed by $L^{rat}, M_T^{rat}, N_T^{rat}$ and $L^{con}, M_T^{con}, N_T^{con}$ respectively, with $\hat{\mathcal{T}}^{RAT} = [\![L^{rat}, M_T^{rat}, N_T^{rat}]\!]$ and $\hat{\mathcal{T}}^{CON} = [\![L^{con}, M_T^{con}, N_T^{con}]\!]$. The predicted rating between user u and item i in target domain is calculated as:

$$\hat{r}_{ui} = \frac{\sum_{as \in AS_{ui}} \mathcal{T}^{RAT}(u, i, as) \cdot \mathcal{T}^{CON}(u, i, as)}{\sum_{as \in AS_{ui}} \mathcal{T}^{CON}(u, i, as)}$$

where AS_{ui} is the set of aspects which are related to i and $\mathcal{T}^{CON}(u, i, AS_{ui}) > 0$. For items with reviews, we extract related aspects from reviews. For new items without reviews, their related aspects are extracted from meta-data like item descriptions.

4   Experiments

4.1   Experiment Setup

Dataset. We evaluate our method on the publicly available Amazon[1] dataset[2] [12]. We apply our model on the two pair of domains "Books - Movies" and "Movies – Music CDs" respectively. In order to better validate the superiority of our method's performance, we first select the items which have more than 120 reviews in "Books" and "Movies", more than 30 reviews in "Music CDs". Then, in each of the two sub-datasets, only the users who have more than 10 feedbacks in each of the two domains are kept. The ratings range from 1 star to 5 stars. The review texts are pre-processed by removing stop words and lemmatizing with corenlp[3]. The characteristics of dataset are summarized in Table 3. The domain "Books" and "Music CDs" are chosen as the target domain in each pair of domains because they are comparatively sparse. In each pair of domains,

---

[1] https://www.amazon.com/.

[2] http://jmcauley.ucsd.edu/data/amazon/.

[3] https://github.com/stanfordnlp/CoreNLP.

Table 3. Summary of dataset

	Books - Movies		Movies - Music CDs	
Domains	Books	Movies	Movies	Music CDs
Users	717		810	
Items	2,285	2,563	2,500	3,185
Feedbacks	20,047	38,964	38,768	27,057
Density	1.22%	2.12%	1.91%	1.05%
Avg. # words in each review	318.2	324.9	322.7	330.1

the last 20% feedbacks ordered by time stamp in the target domain are used as the test data, while the rest are used as the training data.

Evaluation Metrics. We adopt the metrics of Root Mean Square Error (RMSE) and Mean Absolute Error (MAE) to evaluate our method. They are defined as:

$$\text{RMSE} = \sqrt{\sum_{r_{ui} \in \mathcal{O}_{test}} (\hat{r}_{ui} - r_{ui})^2 / |\mathcal{O}_{test}|}, \quad \text{MAE} = \frac{1}{|\mathcal{O}_{test}|} \sum_{r_{ui} \in \mathcal{O}_{test}} |\hat{r}_{ui} - r_{ui}|$$

where $\mathcal{O}_{test}$ is the set of test ratings, r_{ui} represents an observed rating in test data, $\hat{r}_{ui}$ is the predictive value of r_{ui}. $|\mathcal{O}_{test}|$ is the number of test ratings.

Baselines. We compare our method with the following baselines:

1. Single Domain Matrix Factorization (SDMF) [24]. SDMF applies the basic MF model only on the rating matrix of target domain, minimizing the squared error by stochastic gradient descent.
2. Review Based Single Domain Tensor Factorization (RB-SDTF). RB-SDTF only considers the target domain, without transferring, i.e., set $\lambda = \mu = 0$ in our model.
3. Joint Tensor Factorization with Aspect Rating only (AR-JTF). AR-JTF removes the concern degree factors from our method. Use the average of aspect ratings.
4. CDTF [4]. CDTF is a state-of-the-art model in cross domain recommendation with triadic factorization of multi-domain rating tensor.
5. CTR-RBF [20]. CTR-RBF is also a state-of-the-art model in cross domain recommendation which utilizes review text.

4.2 Performance Comparison

In the experiments, a topic (aspect) is represented with around 20 words that have the largest probability belonging to the topic. Too many or too few words can harm the performance. 20 words is a reasonable choice in our experiment. In each pair of domains, the numbers of aspects are equally set to 20 by preliminary tests, and the rank of tensors R is set to 10. We repeat 5 times of parameter learning process for each result and average them as the final results. For all methods, we search parameters to achieve the best performance as far as possible. To test the performance of the methods in addressing the cold start issue in the target domain, we respectively select users whose existing interactions in target domain in training data are less than α and calculate their RMSE and MAE. Consider of the average number of reviews of a user in target domain in the dataset, α is chosen in 1, 3, 5, 10, and the set of all users is also chosen, denoted as α = 'all'. For every value of α, we search and choose satisfactory values of λ and μ. The discussion of λ and μ are in Sect. 4.3. The RMSE and MAE results in each pair of domains are shown in Tables 4 and 5.

Table 4. Performance comparison in target domain in "Books – Movies"

Methods	RMSE					MAE				
	$\alpha = 1$	$\alpha = 3$	$\alpha = 5$	$\alpha = 10$	α = 'all'	$\alpha = 1$	$\alpha = 3$	$\alpha = 5$	$\alpha = 10$	α = 'all'
SDMF	–	1.4997	1.2834	1.1192	1.0768	–	1.1364	0.9855	0.8598	0.8177
RB-SDTF	–	1.3024	1.0507	0.9421	0.9202	–	0.9721	0.8217	0.7146	0.6838
AR-JTF	0.9518	0.9334	0.9253	0.9097	0.9046	0.7326	0.7129	0.6974	0.6971	0.6803
CDTF	0.9877	0.9620	0.9470	0.9400	0.9233	0.7562	0.7307	0.7203	0.7086	0.6850
CTR-RBF	0.9698	0.9572	0.9433	0.9316	0.9138	0.7400	0.7253	0.7164	0.7057	0.6818
RB-JTF	**0.9430**	**0.9252**	**0.9127**	**0.9033**	**0.8973**	**0.7235**	**0.6970**	**0.6898**	**0.6804**	**0.6741**

Table 5. Performance comparison in target domain in "Movies – Music CDs"

Methods	RMSE					MAE				
	$\alpha = 1$	$\alpha = 3$	$\alpha = 5$	$\alpha = 10$	α = 'all'	$\alpha = 1$	$\alpha = 3$	$\alpha = 5$	$\alpha = 10$	α = 'all'
SDMF	–	1.5161	1.2936	1.1150	1.1023	–	1.1490	1.0466	0.8407	0.8336
RB-SDTF	–	1.3297	1.1055	0.9661	0.9317	–	0.9803	0.8442	0.7408	0.7205
AR-JTF	1.0142	0.9872	0.9573	0.9405	0.9341	0.7800	0.7386	0.7281	0.7087	0.7067
CDTF	1.0390	1.0216	0.9761	0.9587	0.9460	0.8043	0.7634	0.7471	0.7325	0.7232
CTR-RBF	1.0284	1.0089	0.9690	0.9519	0.9405	0.7857	0.7480	0.7349	0.7264	0.7197
RB-JTF	**0.9918**	**0.9701**	**0.9467**	**0.9311**	**0.9237**	**0.7643**	**0.7263**	**0.7097**	**0.7004**	**0.6959**

Table 4 shows the performance of different methods in "Books - Movies", and Table 5 shows the performance of different methods in "Movies - Music CDs". We respectively evaluate all the methods under different α in both pair of domains. The proposed method (RB-JTF) achieves the best performance, especially under a lower α, i.e., to those cold-start users in target domain. With α becoming lower, the performance of single domain methods becomes progressively worse while the cross domain methods keep satisfactory results, which shows the effectiveness of knowledge transferring. Compared with CTR-RBF and CDTF, our method (RB-JTF) gets an improvement of 2% to 4% both in RMSE and MAE. Moreover, RB-JTF performs better than RB-SDTF especially in lower α, which demonstrates that the knowledge transferred from the auxiliary domain is useful. And RB-JTF outperforms AR-JTF, indicating that the concern degree transfer can work.

4.3 Sensitivity Analysis

Parameters. We examine the effect of parameters λ and μ on the performance of our model. In the examination, same settings are used in the joint factorization of a pair of aspect rating tensors and their corresponding pair of concern degree tensors.

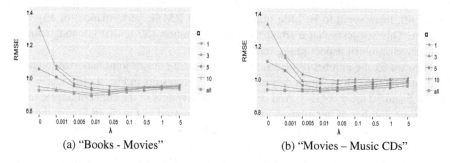

(a) "Books - Movies"   (b) "Movies – Music CDs"

Fig. 2. The influence of the proportion of two domains

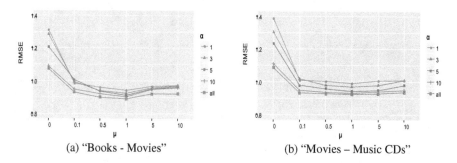

(a) "Books - Movies"   (b) "Movies – Music CDs"

Fig. 3. The influence of constraint of aspect similarity in transfer

λ controls the mixture of the two domains in the model. We fix the constraint of aspect similarity μ on 1 and search the λ from 0 to 5 on different α, the result is reported in Fig. 2. When λ is very small, it is close to the result of just exploiting the target domain. With λ increases, we can achieve better results. It is worth noting that to the cold start users (small α), the improvement of performance is more pronounced. Otherwise, when adopting a large λ, it is tend to just exploit the auxiliary domain, the performance might degrade. Figure 2 shows that the RMSE performance achieves the best when λ is in the near of 0.01 to 0.1 in these cases, which provides an evidence that our transfer framework is useful. We also find that when α is smaller, the best λ tends to be bigger.

μ controls the constraint scale of aspect similarity. We fix λ on 0.05 and search the value of μ from 0 to 10 on different α, the result is reported in Fig. 3. In Fig. 3, we can see that the RMSE performance reaches the best when the value of μ is in the vicinity of 1. This clearly demonstrates the effectiveness of the transferring of aspect factors.

Iterations. The influence of the number of iterations on the performance in the joint factorization of aspect rating tensor and concern degree tensor should be studied respectively. In the study, we observe the impact of number of iterations on RMSE when α varies in 1, 5 and 'all' in "Books - Movies". We set λ = 0.05 and μ = 1 here.

Figure 4 shows the impact of number of iterations on the RMSE in the aspect rating tensor joint factorization. It is equivalent to the method AR-JTF. We vary the number of iterations from 5 to 70. We observe that when the iteration increases to be between 30 and 40, there seem to be little improvement on RMSE performance for any large iteration. This suggests that a small number of iteration (30 to 40) is enough for the joint factorization of aspect rating tensors.

Then we fix the number of iterations of aspect rating tensor joint factorization under different α on the values which they achieves the best results in Fig. 4 (40 for α = 1, 35 for α = 5 and 30 for α = 'all'). And study the effectiveness of the concern degree tensor on different number of iterations. The results are reported in Fig. 5. In Fig. 5, the three horizontal lines are the best RMSE results in Fig. 4 of different α. After adding the concern degree tensor, the results improve. 20 iterations are enough for the joint factorization of concern degree tensors. This clearly shows that the joint factorization of concern degree tensor can work. This is the reason why RB-JTF outperforms AR-JTF.

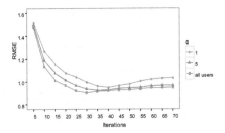

Fig. 4. The impact of number of iterations on the RMSE in the aspect rating tensor joint factorization in "Books - Movies"

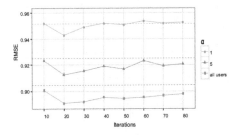

Fig. 5. The effectiveness of the concern degree tensor under different number of iterations of it in "Books - Movies"

5 Related Work

Works about cross domain recommendation can be roughly categorized into two types of models: adaptive models and collective models [22].

In adaptive models, transferring user rating patterns between domains technique is widely used. CBT [6] learns a rating pattern called codebook from a domain and transfer it into another domain; TALMUD [14] considers the possible interactions between the source domains and the different degrees of relatedness between the sources and the target domain; CLFM [23] considers the domain-specific factors in learning patterns. Coordinate system transfer (CST) [15] is another effective model. Tags are also exploited in some works in knowledge transfer [3, 21, 28].

Collective model contains sharing user or item latent feature matrix in factorization [10, 18]; sharing a latent cluster-level rating model [7, 13]; sharing user and item latent features while capturing the data-dependent effect via learning inner matrices in SVD [16]; employing a tensor triadic factorization for knowledge transfer [4], et al. They all achieve promising results in their applications, but they all only use the overall ratings. Our model is a collective model, we introduce users' sentiment which hidden in review text in the transfer. CTR-RBF [20] also exploits review text, it learns a non-linear mapping on users' preferences on different topics across domains, but it only consider which topics (i.e. aspects) a user refers in review texts, not exploit users' aspect-level sentiment towards different items.

6   Conclusions

In this paper, based on the fully exploiting of review text, aspect factor is introduced into the cross domain framework. We propose a novel Joint Tensor Factorization model to respectively transfer aspects rating knowledge and concern degree knowledge among domains. In the factorization, the user latent factors are shared in the two domains, and a constraint is added on the aspect factors to transfer the aspect features. Experimental results show that our method outperforms the baselines especially in the set of cold start users in the target domain. Additionally, optimizing the efficiency of the model for very large data sets is our future work.

Acknowledgements. This work is supported by NSF of China (No. 61602237), 973 Program (No. 2015CB352501), NSF of Shandong, China (No. ZR2013FQ009), the Science and Technology Development Plan of Shandong, China (No. 2014GGX101047, No. 2014GGX101019). This work is also supported by US NSF grants III-1526499, and CNS-1115234.

References

1. Acar, E., Dunlavy, D.M., Kolda, T.G., Morup, M.: Scalable tensor factorizations for incomplete data. Chemometr. Intell. Lab. Syst. **106**(2011), 41–56 (2011)
2. Acar, E., Dunlavy, D.M., Kolda, T.G.: A scalable optimization approach for fitting canonical tensor decompositions. Chemometrics **25**(2011), 67–86 (2011)
3. Chen, W., Hsu, W., Lee, M.L.: Making recommendations from multiple domains. In: KDD 2013, pp. 892–900 (2013)
4. Hu, L., Cao, J., Xu, G., Cao, L., Gu, Z., Zhu, C.: Personalized recommendation via cross-domain triadic factorization. In: WWW 2013, pp. 595–605 (2013)
5. Kolda, T.G., Bader, B.W.: Tensor decompositions and application. SIAM Rev. **51**(3), 455–500 (2009)

6. Li, B., Yang, Q., Xue, X.: Can movies and books collaborate? Cross-domain collaborative filtering for sparsity reduction. In: IJCAI 2009, pp. 2052–2057 (2009)
7. Li, B., Yang, Q., Xue, X.: Transfer learning for collaborative filtering via a rating-matrix generative model. In: ICML 2009, pp. 617–624 (2009)
8. Li, B., Zhu, X., Li, R., Zhang, C.: Rating knowledge sharing in cross-domain collaborative filtering. IEEE Trans. Cybern. **45**(5), 2015 (2015)
9. Li, H., Lin, R., Hong, R., Ge, Y.: Generative models for mining latent aspects and their ratings from short reviews. In: ICDM 2015, pp. 241–250 (2015)
10. Ma, H., Yang, H., Lyu, M.R., King, I.: Sorec: social recommendation using probabilistic matrix factorization. In: CIKM 2008, pp. 931–940 (2008)
11. Morup, M.: Applications of tensor (multiway array) factorizations and decompositions in · data mining. WIREs Data Min. Knowl. Discov. **1**, 24–40 (2011)
12. McAuley, J., Pandey, R., Leskovec, J.: Inferring networks of substitutable and complementary products. In: KDD 2015, pp. 785–794 (2015)
13. Mirbakhsh, N., Ling, C.X.: Improving Top-N recommendation for cold-start users via cross-domain information. ACM TKDD **9**(4), 33 (2015)
14. Orly, M., Bracha, S., Lior, R., Guy, S.: TALMUD – transfer learning for multiple domains. In: CIKM 2012, pp. 425–434 (2012)
15. Pan, W., Evan, W.X., Nathan, N.L., Yang, Q.: Transfer learning in collaborative filtering for sparsity reduction. In: AAAI 2010, pp. 230–235 (2010)
16. Pan, W., Yang, Q.: Transfer learning in heterogeneous collaborative filtering domains. Artif. Intell. **197**(2013), 39–55 (2013)
17. Pan, W., Xiang, E.W., Yang, Q.: Transfer learning in collaborative filtering with uncertain ratings. In: AAAI 2012, pp. 662–668 (2012)
18. Singh, A.P., Gordon, G.J.: Relational learning via collective matrix factorization. In: SIGKDD 2008, pp. 650–658 (2008)
19. Wang, H., Lu, Y., Zhai, C.: Latent aspect rating analysis without aspect keyword supervision. In: KDD 2011, pp. 618–626 (2011)
20. Xin, X., Liu, Z., Lin, C., Huang, H., Wei, X., Guo, P.: Cross-domain collaborative filtering with review text. In: IJCAI 2015, pp. 1827–1833 (2015)
21. Yang, D., He, J., Qin, H., Xiao, Y., Wang, W.: A graph-based recommendation across heterogeneous domains. In: CIKM 2015, pp. 463–472 (2015)
22. Iván, C., Ignacio, F., Shlomo, B., Paolo, C.: Cross-domain recommender systems. In: Recommender Systems Handbook, pp. 919–959 (2015)
23. Sheng, G., Hao, L., Da, C., Shantao, L., Patrick, G., Guo, J.: Cross-domain recommendation via cluster-level latent factor model. In: ECML-PKDD 2013, pp. 161–176 (2013)
24. Koren, Y., Bell, R., Volinsky, C.: Matrix factorization techniques for recommender systems. Computer **42**, 30–37 (2009)
25. Rong, Y., Wen, X., Cheng, H.: A monte carlo algorithm for cold start recommendation. In: WWW 2014, pp. 327–336 (2014)
26. Zhou, X., Chen, L., Zhang, Y., Cao, L., Huang, G., Wang, C.: Online video recommendation in sharing community. In: SIGMOD 2015, pp. 149–152 (2015)
27. Wang, S., He, L., Stenneth, L., Yu, P.S., Li, Z.: Citywide traffic congestion estimation with social media. SIGSPATIAL, Article No. 34 (2015)
28. Wang, S., Zhang, H., Zhang, J., Zhang, X., Yu, P.S., Li, Z.: Inferring diffusion networks with sparse cascades by structure transfer. In: DASFAA 2015, pp. 405–421 (2015)
29. Wang, Y., Liu, Y., Yu, X.: Collaborative filtering with aspect-based opinion mining: a tensor factorization approach. In: ICDM 2012, pp. 1152–1157 (2012)

Security, Privacy, Senor and Cloud

A Local-Clustering-Based Personalized Differential Privacy Framework for User-Based Collaborative Filtering

Yongkai Li[1], Shubo Liu[1($\boxtimes$)], Jun Wang[1], and Mengjun Liu[2]

[1] School of Computer, Key Laboratory of Aerospace Information Security
and Trusted Computing, Ministry of Education, Wuhan University, Wuhan, China
{yongkai.li,liu.shubo,jameswang}@whu.edu.cn
[2] School of Education, Hubei University, Wuhan, China
mengjun@hawaii.edu

Abstract. The Collaborative Filtering (CF) algorithm plays an essential role in recommender systems. However, the CF algorithm relies on the user's direct information to provide good recommendations, which may cause major privacy issues. To address these problems, Differential Privacy (DP) has been introduced into CF recommendation algorithms. In this paper, we propose a novel framework called Local-clustering-based Personalized Differential Privacy (LPDP) as an extension of DP. In LPDP, we take the privacy requirements specified at the item-level into consideration instead of employing the same level of privacy guarantees for all users. Moreover, we introduce a local-similarity-based item clustering process into the LPDP scheme, which leads to the result that any items within the same local cluster are hidden. We conduct a theoretical analysis of the privacy guarantees provided within the proposed LPDP scheme. We experimentally evaluate the LPDP scheme on real datasets and demonstrate the superior performance in recommendation quality.

Keywords: Differential privacy · Recommender system · Collaborative filtering · Privacy preferences

1 Introduction

The rapid development of web-based applications has created a large amount of information and makes it impossible for an individual to explore all the web content to extract relevant data. In turn, this clearly leads to the advent of recommender systems (RS) [1]. One core technology of these recommendation systems is the Collaborative Filtering (CF) [2] algorithm, which makes predictions about user preferences by learning patterns from similar users (user-based) or finding similar items (item-based) based on the neighborhood. The CF algorithm can provide good recommendations based on a user's direct information (or profile). However, the user profile contains large amounts of individualized information, such as the purchase history of a customer [3].

© Springer International Publishing AG 2017
S. Candan et al. (Eds.): DASFAA 2017, Part I, LNCS 10177, pp. 543–558, 2017.
DOI: 10.1007/978-3-319-55753-3_34

Widespread attention has been paid to the associated privacy risks in such RS. The ϵ-differential privacy(DP) [4] framework was proposed to hide any individual information in the output by perturbing the data prior to the release. Recently, DP method has been introduced into CF recommendation algorithms [5,6]. It is worth noting that a novel distance-based differential privacy (DDP) that ensures a strong form of DP was proposed in [7]. DDP in CF recommenders guarantees that any adversary is not only prevented from guessing whether a profile contains some item but also whether this profile contains any item within some distance λ from I. However, the distance metric (λ) in DDP is hard to quantify and the value of λ also seems to have no explicit physical or substantial meaning. Moreover, DDP aims to hide the items within a distance λ from the global perspective. When it comes to the procedure of a recommendation, only items in the same category referring to the local neighbors are actually used.

It has been noted that traditional DP affords the same level of privacy guarantees for all users. This "one size fits all" approach ignores the reality that different individuals may have very different expectations for various items [8,9]. When faced with a dataset with different privacy preferences, both the traditional DP and DDP employ the highest privacy level (the smallest ϵ) among the determined privacy preferences [9]. This is likely to introduce an unacceptable amount of noise into the outputs, resulting in poor utility.

In this paper, we introduce a novel framework called Local-clustering-based Personalized Differential Privacy (LPDP) as an extension of DP in a user-based CF. In LPDP, the adversary is not only prevented from guessing whether user X has item I in his profile but also whether the profile of X contains any item I_0 within the local cluster of I. LPDP allows privacy requirements to be specified at the item-level and guarantees precisely the required level of privacy to each user for each item. The main contributions of this paper are as follows:

(1) We present the LPDP scheme for a privacy-preserving user-based CF recommender, which hides not only the item I in profile X but also any item I_0 within the local cluster of I. LPDP guarantees item-specific privacy requirements.
(2) We conduct a theoretical analysis of the privacy guarantees provided within the proposed LPDP scheme and implement the scheme on real datasets to show that our proposed scheme exhibits high utility.

The paper is organized as follows: Sect. 2 presents the related work. Preliminaries of our paper, including some notations and definitions, are shown in Sect. 3. Section 4 discusses the privacy goal. Section 5 presents our proposed LPDP recommendation mechanism. The privacy analysis is reported in Sect. 6. Section 7 provides the experimental results. Section 8 draws some conclusions of this research.

2   Related Work

The framework of differential privacy, introduced by Dwork [4], was proposed to hide any individual information in the output by perturbing the data prior to the release. A limitation of the model is that the same level of privacy protection is afforded for all individuals. To addressing this problem, Jorgensen et al. [9] extended differential privacy to personalized differential privacy (PDP) in which users specify a personal privacy requirement for their data.

In this paper, we extend differential privacy to the context of recommenders. In fact, while there has been a lot of research work related to privacy in recommenders [10,11] and differential privacy [4,12–14], only a few have combined these two notions [5–7,15]. McSherry et al. [5] addressed the privacy issue in RS by using Laplace noise. They added Laplace noise to the movie average rating, user average rating and covariance matrix. Then, the noisy matrices were released and used in the current recommender algorithms. McSherry et al. did not consider updates to the covariance matrix, and hence their method was not applicable to a dynamic system without jeopardizing the privacy guarantee. Moritz Hardt et al. [6] converted the recommendation problem into the Matrix Completion problem and they presented an (ϵ, δ)-differential privacy approach to compute the low rank approximations of large matrices that contain sensitive information about individuals. Zhu et al. [15] proposed two differentially private recommender algorithms with sampling, named DP-IR and DP-UR. Both DP-IR and DP-UR are based on the exponential mechanism with a carefully designed quality function.

Guerraoui et al. [7] extended differential privacy to the context of recommenders by appending the original definition with a distance metric (λ). This work is the most relevant to ours. The proposed distance-based differential privacy (DDP) provides a stronger form of classical differential privacy in the context of RS. However, the distance metric (λ) in DDP is hard to qualify, and the value of λ seems to have no explicit physical or substantial meaning. Moreover, DDP aims to hide the items within the pre-defined distance λ from the *global* perspective. When it comes to the procedure of recommendation, only items in the same category with regard to the *local K-Nearest Neighbors (KNNs)* [2] are actually used. Therefore, it is more important to hide the items in the same category from the *local* context. In addition, all the research work above assumes that users have the same privacy preferences for all items. Obviously, each private user may independently specify the privacy requirements for their item ratings.

3 Preliminaries

In this section, we introduce some notations and initial definitions, and review the definition of differential privacy and the Laplace mechanism to achieve differential privacy, upon which our work is based.

3.1 Underlying Scheme

We model a recommender scheme as a tuple $\mathcal{G} = (\mathcal{U}, \mathcal{I})$ where $\mathcal{U}$ is the user set and $\mathcal{I}$ is the item set. We use $|\mathcal{U}|$ and $|\mathcal{I}|$ to denote the number of users and items, respectively. The user profiles and item profiles are stored in the recommender scheme $\mathcal{G}$. The profile of a user u, denoted by S_u, consists of all the items rated (alternatively shared or liked) by u along with the ratings, and we denote $S_u(i) = r(u, i)$, where $r(u, i)$ is a rating pattern of user u toward item i. In our implementation, we convert the numerical ratings into binary ratings: a like (1) or a dislike (0).[1] An item profile (v_i) consists of item features or types that item has (e.g., in the case of sports, whether the item is a ball game or not).

3.2   Locality Sensitive Hashing (LSH)

Locality sensitive hashing (LSH) is a widely used technique for searching for the approximate nearest-neighbor. We use LSH to allow the system to prune the less similar neighbors of targeted user. We briefly review the LSH in this section.

Definition 1 (Locality Sensitive Hashing [16]). *Let V be a set of vectors and d be the distance measure between vectors. Given two distances r_1, $r_2(r_1 < r_2)$, and two probabilities $p_1, p_2(p_1 > p_2)$, a family of hash functions $H = \{h : V \mapsto U\}$ is (r_1, r_2, p_1, p_2)-sensitive if for any $v_i, v_j \in V$ if $d(v_i, v_j) < r_1$ then $Pr[h(v_i) = h(v_j)] \geq p_1$; if $d(v_i, v_j) > r_2$ then $Pr[h(v_i) = h(v_j)] \leq p_2$.*

We use the LSH defined over cosine similarity, introduced in [17], in this paper. To calculate the LSH of a user profile S_u, we pick k random vectors $r_m(1 \leq m \leq k)$ whose components are $+1$ and -1 at first, and then we calculate the dot product of S_u with each random vector r_m, i.e., $S_u \cdot r_m$. The jth bit of $LSH(S_u)$ is defined as 1 if $S_u \cdot v_j > 0$ and 0 otherwise.

3.3   Differential Privacy

The profile set D in the recommender system is set to $D = (S_{u_0}, S_{u_1}, ..., S_{u_{|u|}})^T$ and $D(u)(i) = r(u, i) = S_u(i)$. We identify two profile sets D, D' as neighboring if D and D' only differ in one user-item rating pattern (or element). We use $D \xrightarrow{r(u,i)} D'$ to denote that D and D' are neighbors and that $D = D' \wedge r(u, i)$ or $D' = D \wedge r(u, i)$, where $r(u, i)$ is a rating pattern of user u toward item i.

Definition 2 (ϵ-differential privacy [4]). *A randomized algorithm $\mathcal{A}$ is ϵ-differentially private if for any two neighboring datasets D_1 and D_2, and for all $O \in Range(\mathcal{A})$, $Pr[\mathcal{A}(D_1) \in O] \leq e^{\epsilon} \cdot Pr[\mathcal{A}(D_2) \in O]$.*

Sequential composition of differentially private algorithms also provides differential privacy:

Sequential composition ([12], Theorem 3). *Let $\mathcal{A}_i$ each provides ϵ_i-differential privacy. The sequence of $\mathcal{A}_i(X)$ provides $(\sum_i \epsilon_i)$-differential privacy.*

Post-Processing reserves differentially private if the latter procedure does not access the input dataset D and is just applied on the output $\mathcal{A}$ derived from the former procedure. Formally,

---

[1] Binary ratings are considered for the sake of simplicity: this scheme can be generalized to numerical ratings.

Theorem 1. *[18] Let $\mathcal{A} : D \rightarrow R^d$ provide ϵ-differential privacy and let $g : R^d \rightarrow I^d$ be an arbitrary function. Then: $g \circ \mathcal{A} : D \rightarrow I^d$ is ϵ-differential privacy.*

The Laplace mechanism [19] is one of the most-widely used approaches to achieving differential privacy. For counting functions, i.e., $f : \mathcal{D} \rightarrow R^d$, injecting Laplace noise with a magnitude proportional to the global sensitivity of f into the output can achieve differential privacy. Formally,

Definition 3 (Global Sensitivity [19]). *The global sensitivity of the function $f : \mathcal{D} \rightarrow R^d$ is $\Delta(f) = \max_{d(G,G') \leq 1} \|f(D) - f(D')\|$ for all neighboring $D, D' \in \mathcal{G}$, where $\| \cdot \|$ denotes the L_1 norm.*

The Laplace Mechanism through which ϵ-differential privacy is achieved is outlined in the following theorem:

Theorem 2. *[19] Let $f : \mathcal{D} \rightarrow R^d$. A mechanism M that adds independently generated noise from a zero-mean Laplace distribution with scale $\lambda = \Delta(f)/\epsilon$ to each of the d output values $f(D)$, i.e., which produces $O = f(D) + \langle Lap(\Delta(f)/\epsilon) \rangle^d$, satisfies ϵ-differential privacy.*

4   Privacy Goal

4.1   Item-Level Privacy Specification

In general, we assume that each private user independently specifies the privacy requirement for their data. More formally, the Privacy Specification of private users is defined as follows:

Definition 4 (Item-Level Privacy Specification). *A privacy specification is a mapping $\mathcal{P} : \mathcal{U} \times \mathcal{I} \rightarrow R_+^{|\mathcal{U}| \times |\mathcal{I}|}$ from users to personal privacy preferences for all items, where a smaller value represents a stronger privacy preference. The notation $P^u(i)$ denotes the privacy specification of user $u \in \mathcal{U}$ for item i.*

We can also describe a specific instance of a privacy specification as a set of ordered pairs, e.g. $\mathcal{P} := \{(u_1, i_1, \varepsilon_1), (u_1, i_2, \varepsilon_2), \ldots, (u_j, i_k, \varepsilon_l), \ldots\}$ where $u_j \in \mathcal{U}, i_k \in \mathcal{I}$ and $\varepsilon_l \in R^+$. It is assumed that every user $u_j \in \mathcal{U}$ maintains an individual privacy preference for item $i_k \in \mathcal{I}$, or that a default privacy level is assigned.

4.2   Privacy Goal

The goal of this paper is to propose a privacy-preserving CF system under a strengthened notation of Personalized Differential Privacy (PDP) [9] named Local-clustering-based PDP (LPDP). LPDP ensures the item-level privacy specification, where each user independently requires an individual privacy specification for each item.

Given one target user u, the KNNs of u, denoted by $KNN(u)$, can be computed by a KNNs algorithm based on some distance metric. We use the measurement $dis(u,v) = \frac{\|S_u - S_v\|_2}{\sqrt{|\mathcal{I}|}}$ to quantify the distance between user u and user v, where $\|\cdot\|_2$ is the Euclidean distance. Furthermore, we define the local community of user u as $C(u) = u \cup KNN(u)$. On the basis of $C(u)$, the notion of Clustering-based Group for item and set are introduced as follows:

Definition 5 (Clustering-based Group for Item). *For every item $i \in \mathcal{I}$ and a local similarity metric $\theta(u)$ in $C(u)$, the Clustering-based Group $G(i, u)$ for i is defined as the collection of all items that belong to the same cluster containing i. More specifically: $G(i, u) = \{y \in \mathcal{I} | y$ and i belong to the same cluster under matrix $\theta(u)\}$.*

We extend this notion of groups to a set of items where each item in the set has a Group as defined by Definition 5.

Definition 6 (Clustering-based Group for Set). *For a set of items $\mathcal{S}$, $G(\mathcal{S}, u)$ is the union of all the groups: $G(\mathcal{S}, u)$ for each element $s \in \mathcal{S}$. More specifically: $G(\mathcal{S}, u) = \underset{s \in \mathcal{S}}{\cup} G(s, u)$.*

More formally, the definition of LPDP is shown in Definition 7.

Definition 7 (Local-clustering based Personalized Differential Privacy (LPDP)). *In the context of a privacy specification $\mathcal{P}$ and a target private user u_0, a recommendation mechanism M satisfies $\mathcal{P}$-local-clustering-based personalized differential privacy (or $\mathcal{P}$-LPDP), if for any two neighboring profile datasets D_1 and D_2, with $D_1 \xrightarrow{r(u,i)} D_2$, and for any possible subset of items $\mathcal{S}, Pr[M(D_1, u_0) \in G(\mathcal{S}, u_0)] \leq e^{P^u(i)} \cdot Pr[M(D_2, u_0) \in G(\mathcal{S}, u_0)]$.*

Intuitively, LPDP offers the similar strong, semantic notion of privacy that traditional differential privacy provides, but the privacy guarantee for LPDP is personalized to the needs of every user for every item. In the context of LPDP in CF, the adversary is not only prevented from guessing whether user X has item I in his profile but also whether the profile of X contains any item I_0 within the local cluster of I.

To achieve LPDP, we refer to the sampling mechanism proposed in [9]. The sample mechanism works by introducing two independent types of randomness: (1) non-uniform random sampling at the user-item rating pattern level, and (2) additional uniform randomness introduced by invoking a traditional differentially private mechanism on the sampled input.

Theorem 3 (The Sample Mechanism). *Consider a recommendation mechanism M, a profile set D, a target private user u_0, a configurable sampling threshold t and a privacy specification $\mathcal{P}$. Let $RS(D, \mathcal{P}, t)$ denote the procedure that independently samples each rating pattern $r(u, i) \in D$ with probability*

$$\pi(r(u,i), t) = \begin{cases} \frac{e^{P^u(i)} - 1}{e^t - 1} & \text{if } P^u(i) < t, \\ 1 & \text{otherwise.} \end{cases}$$

where $min_{u,i}P^u(i) \leq t \leq max_{u,i}P^u(i)$. The sample mechanism is defined as $S_M(D, \mathcal{P}, t, u_0) = DP_M^t(RS(D, \mathcal{P}, t), u_0)$ where DP_M^t is any t-differentially private mechanism M. Then, the sample mechanism $S_M(D, \mathcal{P}, t, u_0)$ achieves $\mathcal{P}$-LPDP.

The mechanism DP_M^t could be the Laplace or other mechanisms, or even a private mechanism containing the composition of several differentially private algorithms. As it is suggested in [9], the threshold t is always set to $t = max_{u,i}P^u(i)$ or $t = \frac{1}{|U||\mathcal{I}|}\sum_{u,i} P^u(i)$.

5 Proposed LPDP Recommendation Mechanism

Our proposed recommender scheme named LPDP consists of three phases (M_1, M_2, M_3) as shown in Fig. 1. Given the profile set D, a target private user u_0 and the corresponding privacy preference $\mathcal{P}$, we first use these inputs to sample a privacy specification aware profile set D_s. Then, we modify D_s by replacing items with those items in their local item clusters of user u_0. This step is the most important part in our scheme and consists of three substeps, including sampled-profile-based KNNs selection, local-similarity-based item clustering and cluster-based profile modification. The last step of our scheme is modified-profile-based recommendation and it is a general step in RS. The details of the aforementioned steps will be given in the following section.

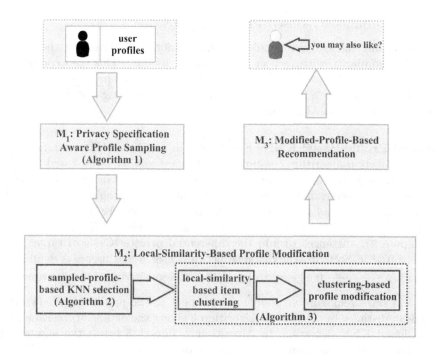

Fig. 1. LPDP recommendation scheme.

Algorithm 1

Input: user set $\mathcal{U}$, item set $\mathcal{I}$, profile set D, the privacy preference $\mathcal{P}$, sampling threshold t, random vector $r_m (1 \leq m \leq k)$
Output: sampled profile set D_s and the LSH of sampled profile set $LSH(D_s)$
1: **For** each $u \in \mathcal{U}$ **do**
2:     Set $D_s(u) = \phi$
3:     **For** each $i \in \mathcal{I}$ **do**
4:         Sample each rating pattern $D_s(u)(i) = RS(r(u,i), \mathcal{P}, t)$ with probability

$$\pi(r(u,i),t) = \begin{cases} \frac{e^{P^u(i)} - 1}{e^t - 1} & \text{if } P^u(i) < t, \\ 1 & \text{otherwise} \end{cases}.$$

5:     **End for**
6:     **For** each $m (1 \leq m \leq k)$ **do**
7:         Calculate $LSH(D_s(u))[m] = \begin{cases} 1, & \text{if } S_u \cdot v_m > 0 \\ 0, & \text{otherwise} \end{cases}$
8:     **End for**
9: **End for**
10: **Return** D_s and $LSH(D_s)$

5.1   Privacy Specification Aware Profile Sampling

Privacy Specification Aware Profile Sampling is used to sample the user-item rating patterns non-uniformly. The inclusion probability for a user-item rating depends on the personal privacy preference for the item and a sampling threshold t. The sampling procedure is shown in Algorithm 1. Lines 2–5 in Algorithm 1 can be denoted as $RS(D, \mathcal{P}, t)$ in the Sample Mechanism. The LSH of the sampled profile for each user is calculated in Lines 6–8.

5.2   Local-Similarity-Based Profile Modification

After obtaining the sampled profile set D_s, we modify D_s based on the predefined local similarity with respect to the target user u_0. This step consists of three substeps, and we will show them as follows: sampled-profile-based KNNs selection in Algorithm 2, local-similarity-based item clustering and cluster-based profile modification in Algorithm 3.

Sampled-Profile-Based KNNs Selection. In user-based CF recommenders, a KNNs Selection algorithm computes the K most similar users based on predetermined similarity metric. However, calculating the distances for users in $\mathcal{U}$ may be too time-consuming, and we should prune the profile set D_s regarding the target user u_0 by employing the LSH method. We assume that the pruned set D'_s has $2K$ profiles. With the pruned profile set D'_s, we only need to compute $2K$ distances to find the differential private KNNs of target user u_0. As stated in Sect. 3.2, the distance between user u and user v is defined as $dis(u,v) = \frac{\|S_u - S_v\|_2}{\sqrt{|\mathcal{I}|}}$. It is easy to prove that the sensitivity of the dis function is $1/\sqrt{|\mathcal{I}|}$ since removing or adding one element in the profile can change the Euclidean distance by at most 1. Algorithm 2 shows the details of the sampled-profile-based KNNs selection. In Algorithm 2, lines 1–5 prune the number of candidate neighboring users to $2K$, and lines 6–9 calculate the noisy KNNs of u_0.

Algorithm 2

Input: user set $\mathcal{U}$, sampled profile set D_s, the LSH of sampled profile set $LSH(D_s)$, sampling threshold t, the target user u_0
Output: noisy KNNs of the target user KNN(u_0)
1: Initialize an empty set $\mathcal{U}_{2K}$
2: **For** each $u \in \mathcal{U}$ and $u \neq u_0$ **do**
3:    Calculate $d_L(u, u_0) = \|LSH(D_s(u)) - LSH(D_s(u_0))\|$
4: **End for**
5: Sort $d_L(u, u_0)$ in decreasing order and pick the first $2K$ users as $\mathcal{U}_{2K}$
6: **For** each $u \in \mathcal{U}_{2K}$ **do**
7:    Calculate $d(u, u_0) = dis(u, u_0) + Lap(\frac{2}{\sqrt{|\mathcal{I}|t}}) = \frac{\|D_s(u_0) - D_s(u)\|_2}{\sqrt{|\mathcal{I}|}} + Lap(\frac{2}{\sqrt{|\mathcal{I}|t}})$
8: **End for**
9: Sort $d(u, u_0)$ in decreasing order and pick the first K users as KNN(u_0)
10: **Return** KNN(u_0)

Clustering-Based Profile Modification. The local community $C(u_0) = u_0 \cup KNN(u_0)$ can be obtained by Algorithm 2. We then focus on differentially private modification of the user-item rating patterns of $C(u_0)$. In $C(u_0)$, we define the local similarity for item i and item j as

$$sim(i, j, C(u_0)) =$$

$$\frac{\sum\limits_{u \in C(u_0), r(u,i)=r(u,j)=1} r(u, i)}{2\max\{\left|\sum\limits_{u \in C(u_0)} r(u, i) + \sum\limits_{u \in C(u_0)} r(u, j) - \sum\limits_{u \in C(u_0), r(u,i)=r(u,j)=1} r(u, i)\right|, \frac{K+1}{4}\}} + \frac{v_i \cdot v_j}{2\|v_i\|_2 \|v_j\|_2}.$$

The first part of the formula measures the similarity of item i and item j in $C(u_0)$, and the latter one measures the cosine similarity of item features. The items will then be clustered by the Affinity Propagation (AP) algorithm [20] based on the defined local similarity. The AP algorithm can find clusters with low error and does not need the initial selection of exemplars. We refer the readers to [20] for a full description of such clustering method. The profile of user u in $C(u_0)$ will be modified after the clusters are determined. Algorithm 3 gives the details of clustering-based profile modification. Lines 1–7 show the local-similarity-based clustering using the AP algorithm and lines 8–16 give the detailed procedure for modification of the sampled profiles.

5.3   Modified-Profile-Based Recommendation

Modified-profile-based recommendation is a general step in RS. Based on the modified profiles $D_m(C(u_0))$, the preference function $q(D_m(C(u_0)), u_0, i)$ for the target user u_0 is defined as follows: $q(D_m(C(u_0)), u_0, i) = \frac{\sum\limits_{u \in KNN(u_0)} (1 - dis(u, u_0)) D_m(u)(i)}{K}$. The preference function evaluates how much the user u_0 will prefer an item. Generally, a higher score indicates a higher probability that item i is recommended.

6   Privacy Guarantees

The proofs of privacy guarantees for the proposed mechanism are formally provided in this section. We show the proposed Algorithm 3 satisfies $t/2$- differential privacy first.

Algorithm 3

Input: the target user u_0, u_0's local community $C(u_0)$, sampled profiles for u_0's local community $D_s(C(u_0))$, sampling threshold t, item replacement parameter α_m
Output: modified profiles for u_0's local community $D_m(C(u_0))$
1: Initialize an $|K+1| \times |\mathcal{I}|$ profile matrix $D_m(C(u_0))$ and an $|\mathcal{I}| \times |\mathcal{I}|$ similarity matrix SIM
2: **For** each $i \in \mathcal{I}$ **do**
3:     **For** each $j \in \mathcal{I}$ and $j \neq i$ **do**
4:         Calculate $SIM(i)(j) = sim(i, j, C(u_0))$
5:     **End for**
6: **End for**
7: Cluster item set $\mathcal{I}$ using AP method based on SIM and denote all the k clusters as $c_m, m = 1, \ldots, k$
8: **For** each $u \in C(u_0)$ **do**
9:     **For** each $i \in \mathcal{I}$ **do**
10:        **If** $D_s(u)(i) = 1$ **do**
11:            Find the cluster c_m which contains i
12:            Select and item j from c_m with probability

$$p(i,j) = \begin{cases} \dfrac{e^{\alpha_m t/2}}{|\mathcal{I}| + e^{\alpha_m t/2} + |c_m|e^{\alpha_m t/4} - e^{\alpha_m t/4} - |c_m|} & \text{if } i = j \\[2ex] \dfrac{e^{\alpha_m t/4}}{|\mathcal{I}| + e^{\alpha_m t/2} + |c_m|e^{\alpha_m t/4} - e^{\alpha_m t/4} - |c_m|} & \text{if } i \neq j \text{ and } j \in c_m \\[2ex] \dfrac{1}{|\mathcal{I}| + e^{\alpha_m t/2} + |c_m|e^{\alpha_m t/4} - e^{\alpha_m t/4} - |c_m|} & \text{otherwise} \end{cases}$$

            where $0 < \alpha_m \leq 1$, and $|c_m|$ is the number of items in c_m
13:        Set $D_m(u)(j) = 1$
14:        **End if**
15:    **End for**
16: **End for**
17: **Return** $D_m(C(u_0))$

Lemma 1. *Let MF be the privacy preserving mechanism in Algorithm 3, which creates the modified profiles set D_m for the sampled profiles set D_s and c_{max} denote $\max_m(|c_m|)$. If for any α_m and $|c_m|$, we have $\frac{|\mathcal{I}| + e^{t/2} + c_{max}e^{t/4} - e^{t/4} - c_{max}e^{\alpha_m t/2}}{|\mathcal{I}| + e^{\alpha_m t/2} + |c_m|e^{\alpha_m t/4} - e^{\alpha_m t/4} - |c_m|} \leq e^{t/2}$, and then $Pr[MF(D_s) = S_m \in D_m] \leq e^{t/2} \cdot Pr[MF(D'_s) = S_m \in D_m]$, i.e., MF is $t/2$-differential privacy.*

Lemma 1 shows that Algorithm 3 satisfies $t/2$-differential privacy and its proof is omitted. We then show our proposed recommender scheme is $\mathcal{P}$-LPDP.

Theorem 4. *The proposed recommender scheme is $\mathcal{P}$-LPDP.*

The proof of Theorem 4 is omitted.

7   Experimental Evaluation

This section presents an experimental evaluation of our LPDP scheme. In particular, we compare the recommendation quality of LPDP with that of the Random

Recommendation (RR) method. We also provide a comparison with [7], denoted as DDP, one of the closest methods to our work. We implemented all methods in Matlab 2010b. The experiments were conducted on an Intel Core i7 machine at 3.40 GHz with 4 GB RAM running Windows 7.

7.1   Experimental Setup

Datasets. We evaluate our LPDP scheme with two datasets: the Jester dataset[2] [21] and the Hetrec2011-MovieLens-2k (HML)[3] [22]. The Jester dataset contains 4.1 million ratings of 100 jokes from 73,421 users. We use a subset of the Jester dataset with 100,000 ratings given by the first 1,000 users who rated 100 jokes. The item features of all 100 jokes are given by 5 randomly selected rating patterns where all 100 jokes are rated. The HML dataset consists of 855,598 ratings given by 2,113 users over 10,197 movies. We use a subset of the HML dataset with 1,541 users who rated at least 100 movies among 3,009 movies. All the selected 3,009 movies have at least 40 movie tags. The item profiles of these 3,009 movies are given by their tag data. We divide the dataset into a training set and a test set. For each rating in the test set, a set of top recommendations is selected as the Recommendation Set (RS).

Evaluation Metrics. To measure the recommendation quality, we set the size of the RS as $N=20$. We use Precision and Recall as classification accuracy metrics, which are conventionally used in top-k recommenders [23]. Precision is the ratio of the number of relevant recommended items to the total number of recommended items, i.e., $Precision = \frac{\# \ of \ relevant \ recommended \ items}{k}$. Recall is the ratio of the number of relevant recommended items to the total number of relevant items, i.e., $Recall = \frac{\# \ of \ relevant \ recommended \ items}{\# \ of \ relevant \ items}$. The F-Score is used to access precision and recall simultaneously. Mathematically, it is the harmonic mean of Precision and Recall, i.e., F-Score $= 2 \cdot \frac{Precision \cdot Recall}{Precision + Recall}$.

Parameter Selection. In our experimental evaluations for both LPDP and DDP schemes, we set the number of nearest neighbors $K = 100$. There are three important parameters in our LPDP scheme: privacy preference $\mathcal{P}$, sampling threshold t and item replacement parameter α_m. To generate the privacy preference $\mathcal{P}$ for our experiments, we randomly divided the privacy specifications for items into three groups: conservative, representing items with high privacy concern; moderate, representing items with medium concern; and liberal, representing items with low concern. The fraction of each item type is 20%, 30% and 50%, respectively. First, the privacy preference $\mathcal{P}$ for the items in the conservative, moderate and liberal groups received a privacy specification of $\epsilon_c = 0.2, \epsilon_m = 0.6$ and $\epsilon_l = 1$ respectively. As a result, the average privacy preference of all users is equal to 0.72. We then vary the privacy specifications

---

[2] http://eigentaste.berkeley.edu/dataset/.
[3] http://grouplens.org/datasets/hetrec-2011/.

to $\epsilon_c = 0.2 \times 5 = 1, \epsilon_m = 0.6 \times 5 = 3$ and $\epsilon_l = 1 \times 5 = 5$ for the conservative, moderate and liberal groups, respectively. We denote the first setting as $\mathcal{P}_1$ and the latter one as $\mathcal{P}_2$. For the sampling threshold t, we set its value to 0.72 or 0.4 for privacy preference $\mathcal{P}_1$ and 3.6 or 2 for privacy preference $\mathcal{P}_2$. The item replacement parameter α_m is set to 0.4 (low item replacement probability in the same cluster) or 0.8 (high item replacement probability in the same cluster) for comparison. For DDP scheme, we set the distance parameter $\lambda = 10$. To serve as control of $\mathcal{P}_1$, the privacy parameter ϵ_1 of DDP is set to the minimum privacy preference, i.e., $\epsilon_1 = \epsilon_c = 0.2$ (i.e., $p = 0.99999$, $p* = 0.00006$). Similarly, the privacy parameter ϵ_2 of DDP is set to 1 (i.e., $p = 0.9999$, $p* = 0.0005$) as a control to $\mathcal{P}_2$. The parameter sets, defined as LPDP$_1$, LPDP$_2$, LPDP$_3$, LPDP$_4$, LPDP$_5$, LPDP$_6$, LPDP$_7$, LPDP$_8$, DDP$_1$ and DDP$_2$ are shown in Table 1.

Table 1. Parameter sets

Parameter set	Parameter value	Parameter set	Parameter value
LPDP$_1$	$\mathcal{P}_1$, $t = 0.4$, $\alpha_m = 0.4$	LPDP$_5$	$\mathcal{P}_2$, $t = 2$, $\alpha_m = 0.4$
LPDP$_2$	$\mathcal{P}_1$, $t = 0.4$, $\alpha_m = 0.8$	LPDP$_6$	$\mathcal{P}_2$, $t = 2$, $\alpha_m = 0.8$
LPDP$_3$	$\mathcal{P}_1$, $t = 0.72$, $\alpha_m = 0.4$	LPDP$_7$	$\mathcal{P}_2$, $t = 3.6$, $\alpha_m = 0.4$
LPDP$_4$	$\mathcal{P}_1$, $t = 0.72$, $\alpha_m = 0.8$	LPDP$_8$	$\mathcal{P}_2$, $t = 3.6$, $\alpha_m = 0.8$
DDP$_1$	$\lambda = 10$, $\epsilon = 0.2$	DDP$_2$	$\lambda = 10$, $\epsilon = 1$

7.2   Results

We vary the recommender parameter k of the top-k recommenders from 1 to 10. Figure 2 demonstrates the performances of LPDP and DDP schemes under privacy preference $\mathcal{P}_1$ for the Jester dataset. Figure 2(A) shows the dependence of Precision and k, and Fig. 2(B) presents the dependence of Precision and Recall. Figure 2(C) illustrates that the F-score increases with the increasing of parameter k. In Fig. 2(A), we observe that the RR method outperforms the DDP scheme for small $k(k \leq 4)$. Our LPDP scheme under parameter set LPDP$_3$ performs best, outperforming the DDP scheme by up to 13.27% and the RR method by up to 8.98% in Precision. Simulations under privacy preference $\mathcal{P}_1$ indicate that both the sampling threshold t and item replacement parameter α_m have insignificant influence on the recommendation quality when k is greater than or equal to 4. Figure 3 shows the performances of LPDP and DDP schemes under privacy preference $\mathcal{P}_2$ for the Jester dataset. In Fig. 3, we observe that both the LPDP and DDP schemes outperform the RR method under privacy preference $\mathcal{P}_2$, as the privacy requirement for $\mathcal{P}_2$ is weaker than $\mathcal{P}_1$. Figure 3 shows that our LPDP scheme under parameter set LPDP$_8$ performs best, outperforming the DDP scheme by up to 40.54% and the RR method by up to 47.04% in Precision. Under the large privacy preference, the item replacement parameter α_m has a more significant influence on the recommendation quality than that of sampling threshold t. For item replacement parameter α_m, its most significant influence on the Precision is exemplified by LPDP$_8$ and LPDP$_7$, and LPDP$_8$ outperforms LPDP$_7$ by up to 30.10% in Precision at $k = 2$.

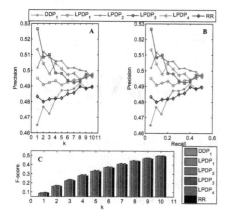

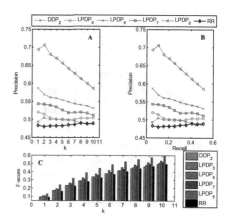

Fig. 2. The recommendation quality for the Jester dataset under preference $\mathcal{P}_1$. (A) Precision v.s. k Comparison. (B) Precision v.s. Recall Comparison. (C) F-score v.s. k Comparison.

Fig. 3. The recommendation quality for the Jester dataset under preference $\mathcal{P}_2$. (A) Precision v.s. k Comparison. (B) Precision v.s. Recall Comparison. (C) F-score v.s. k Comparison.

Figure 4 shows the performances of LPDP and DDP schemes under privacy preference $\mathcal{P}_1$ for the HML dataset. In Fig. 4, we observe that the DDP scheme and RR method performs almost the same for different ks. Figure 4 shows that our LPDP scheme under parameter set LPDP$_2$ performs best, outperforming the DDP scheme by up to 13.85% and the RR method by up to 13.25% in Precision. Similar to Fig. 2, we also find that both the sampling threshold t and item replacement parameter α_m have insignificant influence on the recommendation quality when k is greater than or equal to 3. Figure 5 shows the performances of the LPDP and DDP schemes under privacy preference $\mathcal{P}_2$ for the HML dataset. In Fig. 5, we observe that RR method outperforms DDP scheme for almost all k from 1 to 10. Figure 5 indicates that our LPDP scheme under parameter set LPDP$_8$ performs the best, and it outperforms the DDP scheme by up to 42.89% and the RR method by up to 41.94% in Precision. Moreover, it also indicates that the sampling threshold t has less significant influence on the recommendation quality than that of item replacement parameter α_m. For item replacement parameter α_m, its most significant influence on the Precision is exemplified by LPDP$_8$ and LPDP$_7$ and LPDP$_8$ outperforms LPDP$_7$ by up to 30.43% in Precision at $k = 2$.

We also evaluate the computational overhead of our proposed LPDP mechanism and compare it to the DDP scheme. The comparison results of

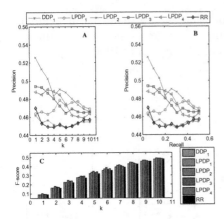

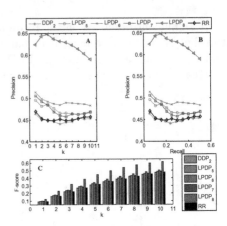

Fig. 4. The recommendation quality for the HML dataset under preference $\mathcal{P}_1$. (A) Precision v.s. k Comparison. (B) Precision v.s. Recall Comparison. (C) F-score v.s. k Comparison.

Fig. 5. The recommendation quality for the HML dataset under preference $\mathcal{P}_2$. (A) Precision v.s. k Comparison. (B) Precision v.s. Recall Comparison. (C) F-score v.s. k Comparison.

Table 2. Computational overheads

	LPDP overhead	DDP overhead
Jester dataset	0.52 s	2.25 s
HML dataset	28.19 s	21.81 s

computational overhead are shown in Table 2. As shown in Table 2, the computation overhead of our proposed LPDP for the Jester dataset is approximately 4.33 times smaller than that of DDP, while approximately 1.29 times larger for the HML dataset. The larger computational overhead of our LPDP mechanism for HML Dataset comes from the clustering process of a great many items in the KNNs.

8   Conclusion

Leveraging item-specified privacy preference acts as an important factor for personalized recommendations. In this paper, we present a mechanism named LPDP as the extension notion of differential privacy to the context of user-based CF recommenders. In LPDP, an adversary is not only prevented from guessing whether user profile X has item I but also whether the profile X contains any item I_0 within the local cluster of I. The LPDP scheme guarantees precisely the required level of privacy to each user for each item. The implementations show that the LPDP scheme is superior in recommendation quality. One limitation of LPDP stems from the fact that clustering large numbers of items in the local community of a target user is time-consuming, which results in heavy computational overhead for LPDP. For future work, we would like to elaborate more on the computational overhead and make it scalable for larger numbers of items.

Acknowledgement. This work was supported by the National Natural Science Foundation of China (Grant no. 41671443); the Fundamental Research Funds for the Central Universities under Grant no. 2015211020201.

References

1. Konstan, J.A., Riedl, J.: Recommender systems: from algorithms to user experience. User Model. User-Adap. Interact. **22**(1–2), 101–123 (2012)
2. Su, X., Khoshgoftaar, T.M.: A survey of collaborative filtering techniques. Adv. Artif. Intell. **4**, 1–19 (2009). doi:10.1155/2009/421425
3. Dwork, C.: Differential privacy. In: Encyclopedia of Cryptography and Security, pp. 338–340 (2011)
4. Dwork, C.: Differential privacy. In: Bugliesi, M., Preneel, B., Sassone, V., Wegener, I. (eds.) ICALP 2006. LNCS, vol. 4052, pp. 1–12. Springer, Heidelberg (2006). doi:10.1007/11787006_1
5. McSherry, F., Mironov, I.: Differentially private recommender systems: building privacy into the net. In: Proceedings of the 15th ACM SIGKDD International Conference on Knowledge Discovery and Data Mining, pp. 627–636 (2009)
6. Hardt, M., Roth, A.: Beating randomized response on incoherent matrices. In: Proceedings of the Forty-fourth Annual ACM Symposium on Theory of Computing, pp. 1255–1268 (2012)
7. Guerraoui, R., Kermarrec, A.M., Patra, R., et al.: D2P: distance-based differential privacy in recommenders. Proc. VLDB Endowment **8**(8), 862–873 (2015)
8. Berendt, B., Günther, O., Spiekermann, S.: Privacy in e-commerce: stated preferences vs. actual behavior. Commun. ACM **48**(4), 101–106 (2005)
9. Jorgensen, Z., Yu, T., Cormode, G.: Conservative or liberal? Personalized differential privacy. In: IEEE International Conference on Data Engineering, pp. 1023–1034 (2015)
10. Luo, Y., Le, J., Chen, H.: A privacy-preserving book recommendation model based on multi-agent. In: Computer Science and Engineering, pp. 323–327 (2009)
11. Zhao, Y., Chow, S.S.M.: Privacy preserving collaborative filtering from asymmetric randomized encoding. In: International Conference on Financial Cryptography and Data Security, pp. 459–477 (2015)
12. Mcsherry, F.D.: Privacy integrated queries: an extensible platform for privacy-preserving data analysis. Commun. ACM **53**(9), 89–97 (2010)
13. Friedman, A., Schuster, A.: Data mining with differential privacy. In: SIGKDD, pp. 493–502 (2010)
14. Xiao, X., Wang, G., Gehrke, J.: Differential privacy via wavelet transforms. IEEE Trans. Knowl. Data Eng. **23**(8), 1200–1214 (2011)
15. Zhu, X., Sun, Y.: Differential privacy for collaborative filtering recommender algorithm. In: ACM on International Workshop on Security and Privacy Analytics (2016)
16. Datar, M., Immorlica, N., Indyk, P., et al.: Locality-sensitive hashing scheme based on p-stable distributions. In: Twentieth Symposium on Computational Geometry, pp. 253–262 (2004)

17. Charikar, M.S.: Similarity estimation techniques from rounding algorithms. In: Applied and Computational Harmonic Analysis, pp. 380–388 (2002)
18. Li, C., Hay, M., Rastogi, V., et al.: Optimizing linear counting queries under differential privacy. In: Proceedings of the Twenty-ninth ACM SIGMOD-SIGACT-SIGART Symposium on Principles of Database Systems, pp. 123–134 (2010)
19. Dwork, C., McSherry, F., Nissim, K., Smith, A.: Calibrating noise to sensitivity in private data analysis. In: Halevi, S., Rabin, T. (eds.) TCC 2006. LNCS, vol. 3876, pp. 265–284. Springer, Heidelberg (2006). doi:10.1007/11681878_14
20. Frey, B.J., Dueck, D.: Clustering by passing messages between data points. Science 315(5814), 972–6 (2007)
21. Goldberg, K., Roeder, T., Gupta, D., et al.: Eigentaste: a constant time collaborative filtering algorithm. Inf. Retrieval J. 4(2), 133–151 (2001)
22. Cantador, I., Brusilovsky, P., Kuflik, T.: HetRec 2011: Second Workshop on Information Heterogeneity and Fusion in Recommender Systems, Chicago (2011)
23. Cremonesi, P., Koren, Y., Turrin, R.: Performance of recommender algorithms on top-n recommendation tasks. In: RecSys, pp. 39–46 (2010)

Fast Multi-dimensional Range Queries
on Encrypted Cloud Databases

Jialin Chi[1,3](✉), Cheng Hong[1], Min Zhang[1,2], and Zhenfeng Zhang[1,2]

[1] Trusted Computing and Information Assurance Laboratory, Institute of Software,
Chinese Academy of Sciences, Beijing, China
{chijialin,hongcheng,mzhang,zfzhang}@tca.iscas.ac.cn
[2] State Key Laboratory of Computer Science, Institute of Software, Chinese
Academy of Sciences, Beijing, China
[3] University of Chinese Academy of Sciences, Beijing, China

Abstract. With the adoption of cloud computing, data owners can store
their datasets on clouds for lower cost and better performance. However,
privacy issues compel sensitive data to be encrypted before outsourcing,
which inevitably introduces challenges in terms of search functionalities.
This paper considers the issue of multi-dimensional range queries on
encrypted cloud databases. Prior schemes focusing on this issue are weak
in either security or efficiency. In this paper, using our improved asymmetric scalar-product-preserving encryption, we present an innovative
technique for the encrypted rectangle intersection problem. Based on this
technique, we propose a tree-based method to handle multi-dimensional
range queries in encrypted form. Thorough analysis demonstrates that
our method is secure under the honest-but-curious model and the known-
plaintext attack model. Experimental results on both real-life and artificial datasets and comprehensive comparisons with other schemes show
the high efficiency of our proposed approach.

Keywords: Searchable encryption · Multi-dimensional range queries ·
Cloud computing · Rectangle intersection problem

1 Introduction

Cloud computing has become a great solution for providing scalable and flexible
storage and computation resources for individuals and organizations. One of
the most promising cloud computing services is Database-as-a-Service (DBaaS)
[1]. In the DBaaS model, outsourcing large datasets to cloud providers, such as
Amazon Web Services [2] and Microsoft Azure [3], allows cloud customers to
be relieved of the burden of maintaining their own databases. However, privacy
concerns from data owners have been the key roadblocks to cloud computing,
since their sensitive data, e.g., health records, financial transactions and location
information, may be revealed by service providers. For instance, an engineer in
Google's Seattle offices broke into the Gmail and Google Voice accounts of several
children in 2010 [4]. To protect privacy, the simplest solution is to encrypt data

© Springer International Publishing AG 2017
S. Candan et al. (Eds.): DASFAA 2017, Part I, LNCS 10177, pp. 559–575, 2017.
DOI: 10.1007/978-3-319-55753-3_35

using traditional encryption algorithms (e.g., block ciphers) before uploading. This in turn posts challenges in terms of search functionalities on clouds.

To enable users to search their encrypted data without disclosing privacy information, many searchable encryption techniques have been proposed. Most of the existing works [5–10] focus on supporting keyword queries or single-dimensional range queries, while they are limited and insufficient in executing range queries over multiple numeric attributes. The problem addressed in this paper is multi-dimensional range search in a secure and efficient manner, which is a major type of database queries. Specifically, data tuples (also called records or rows) are denoted as numeric values in several dimensions (i.e., points such as $(2.3, 5.1, 6.9)$), and queries are described as conjunctions of ranges of interest over these dimensions (i.e., hyper-rectangles such as $([1, 5], [3, 8], [-2, 7])$). The search results are the data points that fall into the query hyper-rectangles. Notice that equality and greater-than (smaller-than) tests can be viewed as special cases of range queries.

Both security and efficiency are important to multi-dimensional range queries on outsourced data. However, current schemes designed for this issue offer either security or efficiency, but not both. In [11,12], two searchable encryption methods are proposed to support multi-dimensional range queries without loss of privacy, but the computation overheads of these two methods are linear with the number of data records. To provide sub-linear search, some predicate-encryption based approaches [13–15] employ tree structures, such as the k-d tree [18] and the R-tree [19], to index data. However, their search complexities increase significantly with the precision required. The above schemes offer provable security, but they are not suitable for handling large-scale and high-dimensional databases with real-valued attributes. Besides, the scheme introduced in [13] reveals the ordering information of data tuples due to the inherent weakness of the k-d tree. In [16], Wang and Ravishankar present an efficient multi-dimensional range search technique which applies the R-tree to create indices and the asymmetric scalar-product-preserving encryption [20] to encrypt nodes. However, this method suffers from privacy leakage, since it discloses the relative positions between nodes and queries. The bucketing scheme developed in [17] also provides fast range queries, but it requires indices to be stored and searched at clients. Hence, it is of great significance to study how to perform multi-dimensional range queries securely and efficiently on clouds.

This paper makes the following contributions.

(1) Utilizing our enhanced asymmetric scalar-product-preserving encryption, we design an innovative mechanism for the encrypted rectangle intersection problem. This mechanism can tell whether two rectangles intersect without disclosing their relative position.

(2) Based on the above technique, we propose a tree-based approach that can securely and efficiently support multi-dimensional range queries. In particular, we leverage the R-tree to index data and encrypt each node separately, while parent-children relationships are preserved.

(3) We demonstrate that our scheme achieves index, trapdoor and ordering privacy altogether under the honest-but-curious model as well as the known-plaintext attack model.

(4) We implement and evaluate the proposed scheme on both real-life and artificial datasets. Experimental results and comparisons with competing methods show the efficiency and scalability of our scheme.

The remainder of this paper is organized as follows. Section 2 reviews related works and Sect. 3 introduces the preliminaries. In Sect. 4, we describe our system model, threat model and design goals. Section 5 presents the details of our scheme. The security and performance are discussed in Sects. 6 and 7, respectively. Finally, Sect. 8 concludes this paper.

2   Related Work

The issue addressed in this paper is range search on encrypted multi-dimensional data, which has been a hot topic in recent years.

In [11], Boneh and Waters propose a primitive called Hidden Vector Encryption and use it to execute multi-dimensional range queries on encrypted data. Meanwhile, Shi et al. [12] also design an encryption scheme for the same purpose. These two schemes both achieve provable security. Unfortunately, they are not practical to handle large-scale datasets, since their search efficiencies increase linearly with the number of data tuples.

To achieve faster-than-linear search, Lu [13] designs a multi-dimensional range search method, which utilizes the k-d tree [18] to index data and predicate encryption to encrypt nodes. Due to the adoption of predicate encryption, this approach provides provable security as well. However, it suffers from privacy loss, since the k-d tree reveals the ordering information of points. Particularly, each node in the k-d tree is specified by a k-dimensional point which splits the space into two parts through one of the k dimensions. Without decrypting nodes, the cloud servers can reconstruct the relative ordering of data along each dimension.

In order to overcome the inherent limitation of the k-d tree, some other predicate-encryption based schemes [14,15] leverage the R-tree [19] to build indices. The main idea of the R-tree is to group nearby objects (i.e., points or hyper-rectangles) on the same level and enclose them with a minimum bounding box in the higher level. Therefore, approaches using the R-tree achieve stronger security properties than [13]. Even though tree structures can provide sub-linear queries, efficiency and scalability are still major obstacles for the applications of these predicate-encryption based methods. On one hand, their computational costs increase significantly with the precision required. On the other hand, predicate encryption requires expensive bilinear-paring operations.

In [16], Wang and Ravishankar introduce a fast multi-dimensional range query scheme, which applies the R-tree to index data and the asymmetric scalar-product-preserving encryption (ASPE) [20] to encrypt nodes. The query processing of this method has been proved to be much more efficient than those of predicate-encryption based schemes, and it is able to handle high-precision values efficiently. However, since this method discloses the relative positions between nodes and queries, the ordering information of tuples is revealed from searches. Additionally, ASPE is proved to be not secure in [21].

Hore et al. [17] use the bucketization technique to perform multi-dimensional range queries on encrypted databases. The space is partitioned into multiple buckets and each bucket is identified by a unique tag. The bucket tag is stored on clouds as an index, together with the encrypted data in this bucket. To perform range queries, the tags of buckets overlapping with the search condition are sent to the cloud server and the results are tuples indexed by these tags. The search complexity of this method is linear in the number of buckets. Unfortunately, bucketing schemes are not true outsourcing schemes, since their indices have to be locally stored and searched by cloud users.

3   Preliminaries

Some ideas central to our scheme are briefly introduced in this section.

3.1   R-Tree

The R-tree [19] is a height-balanced tree proposed to achieve faster-than-linear search for multi-dimensional data. Each node in the R-tree denotes the minimum bounding hyper-rectangle (MBR) of its underlying points. Hence, the root node describes the MBR that covers the whole dataset. An example of an R-tree with fanout (i.e., node capacity) 2 is shown in Fig. 1. The search algorithm of the R-tree is conducted from the root node by recursively solving the hyper-rectangle intersection problem. In particular, if one non-leaf node intersects with the query hyper-rectangle, continue to search its children; otherwise, stop searching along this path. If one leaf node overlaps with the query hyper-rectangle, return this node; otherwise, ignore it. While traversing the R-tree, the search algorithm filters most of the unmatched data efficiently and hence achieves better search efficiency than linear search.

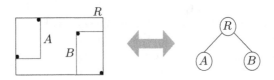

Fig. 1. The R-tree structure.

3.2   Asymmetric Scalar-Product-Preserving Encryption

The asymmetric scalar-product-preserving encryption (ASPE) proposed in [20] allows the scalar product computation of two vectors under encryption. In ASPE, data vectors and query vectors are encrypted differently. The details are presented as follows.

- **ASPE.Setup**$(d) \rightarrow \mathcal{SK}$: Given the security parameter d, output the secret key $\mathcal{SK}$, including two invertible matrices $M', M'' \in \mathbb{R}^{d \times d}$ and one bit string $S \in \{0, 1\}^d$.
- **ASPE.Data_Enc**$(\mathcal{SK}, P) \rightarrow \widehat{P}$: Given the secret key $\mathcal{SK}$ and a data vector $P \in \mathbb{R}^d$, split P into two vectors $\{P', P''\}$ following the rule: for $i = 1$ to d, if $S[i] = 1$, $P'[i]$ and $P''[i]$ are set to two random numbers such that $P'[i] + P''[i] = P[i]$; otherwise, $P'[i]$ and $P''[i]$ are the same as $P[i]$. Output the encrypted data vector $\widehat{P} = \{\widehat{P}' = M'^T P', \widehat{P}'' = M''^T P''\}$.
- **ASPE.Query_Enc**$(\mathcal{SK}, Q) \rightarrow \widehat{Q}$: Given the secret key $\mathcal{SK}$ and a query vector $Q \in \mathbb{R}^d$, split Q into two vectors $\{Q', Q''\}$ following the rule: for $i = 1$ to d, if $S[i] = 0$, $Q'[i]$ and $Q''[i]$ are set to two random numbers such that $Q'[i] + Q''[i] = Q[i]$; otherwise, $Q'[i]$ and $Q''[i]$ are both set to $Q[i]$. Output the encrypted query vector $\widehat{Q} = \{\widehat{Q}' = M'^{-1} Q', \widehat{Q}'' = M''^{-1} Q''\}$.

Now, the encrypted data vector $\widehat{P}$ and query vector $\widehat{Q}$ can be used to blindly evaluate the scalar product of their corresponding plaintext vectors, since

$$\widehat{P} \cdot \widehat{Q} = \widehat{P}' \cdot \widehat{Q}' + \widehat{P}'' \cdot \widehat{Q}'' = M'^T P' \cdot M'^{-1} Q' + M''^T P'' \cdot M''^{-1} Q'' = P \cdot Q.$$

4 Problem Formulation

We now describe the system and threat models considered in this paper, and discuss our design goals.

4.1 System Model

Our system model shown in Fig. 2 recognizes three fundamental entities: *cloud server*, *data owner* and *data user*. The cloud server hosted by service providers provides storage and search services. The data owner uploads the encrypted database to clouds, together with the encrypted index used to enable the searching

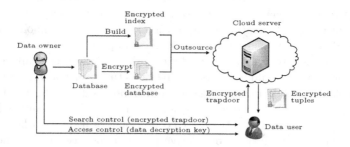

Fig. 2. The system architecture of queries on encrypted cloud databases.

capability of the cloud server. The encryption of the database is performed at the tuple level and treated as a black box in our discussion. To execute queries on the encrypted database, the data user acquires an encrypted trapdoor for ranges of interest through the search control mechanism, and sends it to the cloud server. The cloud server searches the encrypted index, and returns all matched encrypted tuples to the data user. Finally, the access control mechanism is applied to manage decryption capabilities. Here, the search control and access control mechanisms are separate issues and out of the scope of this paper.

4.2   Threat Model

In this paper, the cloud server is assumed to be honest-but-curious, which has been widely accepted in prior works [14–17]. In particular, the cloud server is scrupulous in following the designed algorithms, but may try to infer and analyze the meaningful information about users' databases and queries. We also adopt the known-plaintext attack model, which is consistent with [16,20]. In this model, the cloud server knows a set of plaintext-ciphertext pairs of index vectors, and attempts to solve the plaintexts corresponding to other ciphertexts. Furthermore, the cloud server possesses several data tuples and the background knowledge of the database, such as the value distribution of each attribute.

4.3   Design Goals

Security Goals. During query processing, the cloud server cannot be able to obtain useful information, including the plaintext contents of encrypted indices and trapdoors as well as the ordering information of tuples in the database. We are not interested in hiding the access pattern which is defined as the sequence of search results. Though the Oblivious RAM technique [22] can provide access pattern privacy, the complexity of this technique prevents its application in real scenarios. Our detailed security goals are presented as follows.

(1) Index and Trapdoor Privacy. The index and trapdoor privacy requires that the cloud server should not deduce the plaintext information about encrypted indices and trapdoors. This goal is dependent on the security of the encryption method used for protecting indices and trapdoors.

(2) Ordering Privacy. The ordering privacy requires that the cloud server should not infer the relative ordering of data points along each dimension. The leakage of ordering information permits the cloud server to exploit the values of encrypted tuples by using order statistic techniques [23]. There are two key technical challenges to protect the ordering privacy. The first challenge is the tree structure used to build index. Some tree structures, such as k-d tree applied in [13], disclose the ordering information of data points no matter what encryption schemes are used. The second challenge is the mechanism applied to test whether a node intersects with the search condition. For instance, when performing queries in [16], the leakage of relative positions between nodes and search conditions leads to that of ordering information between tuples.

Usability Goals. Considering the values of attributes are usually high-precision numbers in real-world datasets, our scheme should provide efficient queries for large-scale and high-dimensional databases with real-valued attributes.

5 Our Scheme

In this section, we first introduce an innovative solution for the rectangle intersection problem without encryption technique, and then solve this problem in the ciphertext domain by utilizing our improved ASPE. Next, we modify the R-tree, and design a privacy-preserving multi-dimensional range search scheme which is based on the modified tree structure and the method for solving the encrypted rectangle intersection problem.

5.1 Rectangle Intersection Problem

The 2-dimensional rectangle intersection problem is an important issue in computational geometry. The core idea of our solution here is to first transform the geometric relationship of two rectangles into an equivalent geometric relationship between a point and a rectangle, and then propose two conditions to solve the point inclusion problem.

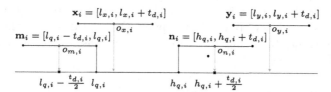

$\mathbf{m}_i, \mathbf{n}_i, \mathbf{x}_i, \mathbf{y}_i$: ranges of the same length with $\mathbf{d}_i$, i.e., $t_{d,i} = h_{d,i} - l_{d,i}$
$o_{m,i}, o_{n,i}, o_{x,i}, o_{y,i}$: mid-points of ranges

Fig. 3. Examples of whether ranges of the same length with $\mathbf{d}_i$ intersect with $\mathbf{q}_i$.

Let us consider a data rectangle $\mathbf{D} = (\mathbf{d}_1, \mathbf{d}_2)$ and a query rectangle $\mathbf{Q} = (\mathbf{q}_1, \mathbf{q}_2)$, where $\mathbf{d}_i = [l_{d,i}, h_{d,i}]$ and $\mathbf{q}_i = [l_{q,i}, h_{q,i}]$ are ranges in the i-th dimension. If the two rectangles intersect (i.e., $\mathbf{D} \cap \mathbf{Q} \neq \varnothing$), they intersect in each dimension (i.e., $\mathbf{d}_i \cap \mathbf{q}_i \neq \varnothing$, for $i \in \{1, 2\}$), and visa versa. As described in Fig. 3, we observe that $\mathbf{d}_i$ overlaps with $\mathbf{q}_i$, if and only if, the mid-point $o_{d,i}$ of $\mathbf{d}_i$ lies inside the extended range $\mathbf{r}_i = [l_{q,i} - \frac{t_{d,i}}{2}, h_{q,i} + \frac{t_{d,i}}{2}]$, where $o_{d,i} = \frac{l_{d,i} + h_{d,i}}{2}$ and $t_{d,i} = h_{d,i} - l_{d,i}$. Then, we can determine whether $\mathbf{D}$ intersects with $\mathbf{Q}$ by testing whether the point $O_d = (o_{d,1}, o_{d,2})$ is within the rectangle $\mathbf{R} = (\mathbf{r}_1, \mathbf{r}_2)$. The transformation from the rectangle intersection problem into the point inclusion problem is summarized as below

$$\mathbf{D} \cap \mathbf{Q} \neq \varnothing \Longleftrightarrow \begin{cases} \mathbf{d}_1 \cap \mathbf{q}_1 \neq \varnothing \\ \mathbf{d}_2 \cap \mathbf{q}_2 \neq \varnothing \end{cases} \Longleftrightarrow \begin{cases} o_{d,1} \in \mathbf{r}_1 \\ o_{d,2} \in \mathbf{r}_2 \end{cases} \Longleftrightarrow O_d \in \mathbf{R}. \qquad (1)$$

After the above transformation, we now need to check the geometric relationship between the point $O_d = (o_{d,1}, o_{d,2})$ and the rectangle $\mathbf{R} = (\mathbf{r}_1, \mathbf{r}_2)$. There have been several methods designed for the point inclusion problem in plaintext form, but most of them reveal privacy information when applied in our system. In this paper, we introduce two conditions to solve this problem.

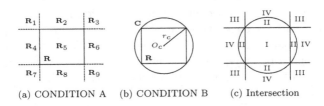

(a) CONDITION A   (b) CONDITION B   (c) Intersection

Fig. 4. The conditions used to solve the point inclusion problem.

CONDITION A. The first condition is applied to determine which region the point O_d lies within. As illustrated in Fig. 4(a), the four edges of the rectangle $\mathbf{R}$ divides the plane into nine regions, where the region $\mathbf{R}_5$ exactly corresponds to $\mathbf{R}$. We generate a formula and split it into an expression of the scalar product of two length-16 vectors V_A and W_A as Eq. (2), where V_A is transformed from the data rectangle $\mathbf{D}$ and W_A is transformed from the query rectangle $\mathbf{Q}$. If the scalar product satisfies $V_A \cdot W_A \geq 0$, the point O_d is inside one of the regions $\{\mathbf{R}_1, \mathbf{R}_3, \mathbf{R}_5, \mathbf{R}_7, \mathbf{R}_9\}$ (including on the edges); otherwise, it lies inside one of the regions $\{\mathbf{R}_2, \mathbf{R}_4, \mathbf{R}_6, \mathbf{R}_8\}$.

$$
\begin{aligned}
&(o_{d,1} - l_{r,1})(o_{d,2} - l_{r,2})(h_{r,1} - o_{d,1})(h_{r,2} - o_{d,2}) \\
&= 1 \cdot l_{q,1} l_{q,2} h_{q,1} h_{q,2} - l_{d,1} \cdot l_{q,1} l_{q,2} h_{q,2} - l_{d,2} \cdot l_{q,1} l_{q,2} h_{q,1} - h_{d,1} \cdot l_{q,2} h_{q,1} h_{q,2} \\
&\quad - h_{d,2} \cdot l_{q,1} h_{q,1} h_{q,2} + l_{d,1} l_{d,2} \cdot l_{q,1} l_{q,2} + h_{d,1} h_{d,2} \cdot h_{q,1} h_{q,2} + l_{d,1} h_{d,1} \cdot l_{q,2} h_{q,2} \\
&\quad + l_{d,1} h_{d,2} \cdot l_{q,1} h_{q,2} + l_{d,2} h_{d,1} \cdot l_{q,2} h_{q,1} + l_{d,2} h_{d,2} \cdot l_{q,1} h_{q,1} - l_{d,1} h_{d,1} h_{d,2} \cdot h_{q,2} \\
&\quad - l_{d,2} h_{d,1} h_{d,2} \cdot h_{q,1} - l_{d,1} l_{d,2} h_{d,1} \cdot l_{q,2} - l_{d,1} l_{d,2} h_{d,2} \cdot l_{q,1} + l_{d,1} l_{d,2} h_{d,1} h_{d,2} \cdot 1 \\
&= V_A \cdot W_A, \tag{2}
\end{aligned}
$$

$$
\begin{aligned}
V_A = (1,\ & l_{d,1},\ l_{d,2},\ h_{d,1},\ h_{d,2},\ l_{d,1} l_{d,2},\ h_{d,1} h_{d,2},\ l_{d,1} h_{d,1},\ l_{d,1} h_{d,2},\ l_{d,2} h_{d,1},\ l_{d,2} h_{d,2}, \\
& l_{d,1} h_{d,1} h_{d,2},\ l_{d,2} h_{d,1} h_{d,2},\ l_{d,1} l_{d,2} h_{d,1},\ l_{d,1} l_{d,2} h_{d,2},\ l_{d,1} l_{d,2} h_{d,1} h_{d,2})^T, \tag{3}
\end{aligned}
$$

$$
\begin{aligned}
W_A = (& l_{q,1} l_{q,2} h_{q,1} h_{q,2},\ -l_{q,1} l_{q,2} h_{q,2},\ -l_{q,1} l_{q,2} h_{q,1},\ -l_{q,2} h_{q,1} h_{q,2},\ -l_{q,1} h_{q,1} h_{q,2},\ l_{q,1} l_{q,2}, \\
& h_{q,1} h_{q,2},\ l_{q,2} h_{q,2},\ l_{q,1} h_{q,2},\ l_{q,2} h_{q,1},\ l_{q,1} h_{q,1},\ -h_{q,2},\ -h_{q,1},\ -l_{q,2},\ -l_{q,1},\ 1)^T. \tag{4}
\end{aligned}
$$

CONDITION B. The second condition is used to check whether the point O_d is within the circumcircle of the rectangle $\mathbf{R}$, since points that lie inside $\mathbf{R}$ must lie inside its circumcircle. Let the circle $\mathbf{C}$ be the circumcircle of $\mathbf{R}$ (see Fig. 4(b)), and its center $O_c(o_{c,1}, o_{c,2})$ and radius r_c are described as below

$$
o_{c,1} = \frac{l_{q,1} + h_{q,1}}{2}, o_{c,2} = \frac{l_{q,2} + h_{q,2}}{2},
$$

$$
r_c = \frac{1}{2} \sqrt{(h_{q,1} + h_{d,1} - l_{q,1} - l_{d,1})^2 + (h_{q,2} + h_{d,2} - l_{q,2} - l_{d,2})^2}.
$$

To decide the relationship of the point O_d and the circumcircle $\mathbf{C}$, we need to compare r_c with the distance between O_d and O_c. More specifically, we construct a polynomial and split it into an expression of the scalar product of two length-6 vectors V_B and W_B as Eq. (5), where V_B is transformed from the data rectangle $\mathbf{D}$ and W_B is transformed from the query rectangle $\mathbf{Q}$. If the scalar product satisfies $V_B \cdot W_B \geq 0$, the point O_d lies within the circumcircle $\mathbf{C}$ (including on the boundary); otherwise, it is outside $\mathbf{C}$.

$$r_c^2 - ((o_{d,1} - o_{c,1})^2 + (o_{d,2} - o_{c,2})^2)$$
$$= -1 \cdot (l_{q,1}h_{q,1} + l_{q,2}h_{q,2}) + l_{d,1} \cdot l_{q,1} + l_{d,2} \cdot l_{q,2} + h_{d,1} \cdot h_{q,1}$$
$$+ h_{d,2} \cdot h_{q,2} - (l_{d,1}h_{d,1} + l_{d,2}h_{d,2}) \cdot 1$$
$$= V_B \cdot W_B, \tag{5}$$

$$V_B = (1, \; l_{d,1}, \; l_{d,2}, \; h_{d,1}, \; h_{d,2}, \; l_{d,1}h_{d,1} + l_{d,2}h_{d,2})^T, \tag{6}$$

$$W_B = (-l_{q,1}h_{q,1} - l_{q,2}h_{q,2}, \; l_{q,1}, \; l_{q,2}, \; h_{q,1}, \; h_{q,2}, \; -1)^T. \tag{7}$$

Clearly, the intersection area of the regions $\{\mathbf{R}_1, \mathbf{R}_3, \mathbf{R}_5, \mathbf{R}_7, \mathbf{R}_9\}$ and the circumcircle $\mathbf{C}$ is exactly equivalent to the rectangle $\mathbf{R}$ (see Fig. 4(c)). Hence, if $V_A \cdot W_A \geq 0$ and $V_B \cdot W_B \geq 0$, we can derive that the point O_d is within the rectangle $\mathbf{R}$, and further conclude that the data rectangle $\mathbf{D}$ and the query rectangle $\mathbf{Q}$ intersect with each other because of Eq. (1).

Our method utilizes the following algorithms to transform the data rectangle and the query rectangle into vectors.

• **DRect_Vec(D)** $\rightarrow \widetilde{\mathbf{D}}$: This function accepts the data rectangle $\mathbf{D}$ and outputs the vectors $\widetilde{\mathbf{D}} = \{V_A, V_B\}$, where V_A, V_B are evaluated as Eqs. (3) and (6), respectively.

• **QRect_Vec(Q)** $\rightarrow \widetilde{\mathbf{Q}}$: This function accepts the query rectangle $\mathbf{Q}$ and outputs the vectors $\widetilde{\mathbf{Q}} = \{W_A, W_B\}$, where W_A, W_B are computed as Eqs. (4) and (7), respectively.

5.2 Intersection Problem Under Encryption

Wong et al. [20] propose that the invertible matrices used for encryption in ASPE are secure against known-plaintext attacks if the attacker cannot derive the bit string S. However, contrary to their claims, Chunsheng et al. [21] provide a polynomial-time algorithm for directly solving a secret key which can decrypt arbitrary encrypted data vectors. Additionally, ASPE cannot protect query vectors due to its feature of scalar product preservation, i.e., the scalar product of two encrypted vectors is equal to that of their corresponding plaintext vectors. We assume that an attacker obtains the plaintext-ciphertext pairs of several data vectors $\mathcal{H} = \{(P_1, \widehat{P}_1), (P_2, \widehat{P}_2), \ldots, (P_w, \widehat{P}_w)\}$, where $P_i \in \mathbb{R}^d$ and $\widehat{P}_i$ represents its encrypted form. Given an encrypted query vector $\widehat{Q}$, the attacker can set up the equations $\{P_i \cdot Q = \widehat{P}_i \cdot \widehat{Q}\}$. The number of unknowns on the left side of equal signs is d, while the values on the right side of equal signs are constants.

If there exist at least d pairs in $\mathcal{H}$ such that they are linearly independent, the attacker can recover Q from $\widehat{Q}$. Hence, ASPE is not suitable to be directly used.

Fortunately, when determining the geometric relationship of two rectangles, we only need to check the signs of $V_A \cdot W_A$ and $V_B \cdot W_B$ instead of their exact values. So we can introduce random factors to improve the security of ASPE. In particular, for the data rectangle, first create two positive random numbers r_A, r_B and scale V_A, V_B by r_A, r_B. Then utilize **ASPE.Data_Enc** to encrypt $r_A V_A$ and $r_B V_B$, separately. For the query rectangle, first generate two positive random numbers r'_A, r'_B and scale W_A, W_B by r'_A, r'_B. Then apply **ASPE.Query_Enc** to encrypt $r'_A W_A$ and $r'_B W_B$, independently.

The algorithms used to solve the rectangle intersection problem in encrypted form can be summarized as follows.

- **SK_Gen**(16, 6) → $\{\mathcal{SK}_A, \mathcal{SK}_B\}$: Given the parameters 16 and 6, where 16 is the length of vectors V_A, W_A and 6 is the length of vectors V_B, W_B, output two secret keys $\mathcal{SK}_A$ and $\mathcal{SK}_B$ by invoking **ASPE.Setup**(16) and **ASPE.Setup**(6) respectively.
- **DRect_Enc**($\mathcal{SK}_A$, $\mathcal{SK}_B$, **D**) → $\widehat{\mathbf{D}}$: This algorithm accepts the secret keys $\mathcal{SK}_A, \mathcal{SK}_B$ and the data rectangle **D**. It first invokes **DRect_Vec(D)** to obtain the vectors $\widetilde{\mathbf{D}} = \{V_A, V_B\}$. It then generates two random numbers $r_A, r_B > 0$, and encrypts V_A as $\widehat{V}_A$ by using **ASPE.Data_Enc**($\mathcal{SK}_A$, $r_A V_A$) and V_B as $\widehat{V}_B$ by using **ASPE.Data_Enc**($\mathcal{SK}_B$, $r_B V_B$). Finally, output the encrypted data rectangle $\widehat{\mathbf{D}} = \{\widehat{V}_A, \widehat{V}_B\}$.
- **QRect_Enc**($\mathcal{SK}_A$, $\mathcal{SK}_B$, **Q**) → $\widehat{\mathbf{Q}}$: This algorithm accepts the secret keys $\mathcal{SK}_A, \mathcal{SK}_B$ and the query rectangle **Q**. It first invokes **QRect_Vec(Q)** to get the vectors $\widetilde{\mathbf{Q}} = \{W_A, W_B\}$. It then creates two random numbers $r'_A, r'_B > 0$, and encrypts W_A as $\widehat{W}_A$ by using **ASPE.Query_Enc**($\mathcal{SK}_A$, $r'_A W_A$) and W_B as $\widehat{W}_B$ by using **ASPE.Query_Enc**($\mathcal{SK}_B$, $r'_B W_B$). Finally, output the encrypted query rectangle $\widehat{\mathbf{Q}} = \{\widehat{W}_A, \widehat{W}_B\}$.
- **Xsect_Rect**($\widehat{\mathbf{D}}$, $\widehat{\mathbf{Q}}$) → $\{0, 1\}$: Given the encrypted data rectangle $\widehat{\mathbf{D}} = \{\widehat{V}_A, \widehat{V}_B\}$ and the encrypted query rectangle $\widehat{\mathbf{Q}} = \{\widehat{W}_A, \widehat{W}_B\}$, if $\widehat{V}_A \cdot \widehat{W}_A \geq 0$ and $\widehat{V}_B \cdot \widehat{W}_B \geq 0$, this algorithm outputs 1, and 0 otherwise.

5.3 Modified R-Tree

To provide better search efficiency than linear search, we would like to apply R-trees to index multi-dimensional data. While searching R-trees, the basic operation is to test whether a node (i.e., hyper-rectangle) intersects with the query hyper-rectangle. However, the geometric relationship of two objects in high-dimensional spaces is difficult to decide without privacy leakage. To solve this problem, we modify the R-tree by converting each node into a set of rectangles.

Let us consider a node $\mathbf{HD} = (\mathbf{d}_1, \mathbf{d}_2, ..., \mathbf{d}_n)$ and a query hyper-rectangle $\mathbf{HQ} = (\mathbf{q}_1, \mathbf{q}_2, ..., \mathbf{q}_n)$, where $\mathbf{d}_i = [l_{d,i}, h_{d,i}]$ and $\mathbf{q}_i = [l_{q,i}, h_{q,i}]$ are ranges in the i-th dimension and the number of dimensions is n. For the node $\mathbf{HD}$, generate $\lceil \frac{n}{2} \rceil$ (i.e., the smallest integer greater than or equal to $\frac{n}{2}$) rectangles $\overline{\mathbf{HD}} = \{\mathbf{D}_1,$

$\mathbf{D}_2, \ldots, \mathbf{D}_{\lceil \frac{n}{2} \rceil}\}$ following the rule: if n is an even integer, $\lceil \frac{n}{2} \rceil$ equals $\frac{n}{2}$ and for $i = 1$ to $\lceil \frac{n}{2} \rceil$, $\mathbf{D}_i = (\mathbf{d}_{2i-1}, \mathbf{d}_{2i})$; otherwise, $\lceil \frac{n}{2} \rceil$ equals $\frac{n+1}{2}$ and for $i = 1$ to $\lceil \frac{n}{2} \rceil - 1$, $\mathbf{D}_i = (\mathbf{d}_{2i-1}, \mathbf{d}_{2i})$, while $\mathbf{D}_{\lceil \frac{n}{2} \rceil} = (\mathbf{d}_1, \mathbf{d}_n)$. Correspondingly, the query hyper-rectangle $\mathbf{HQ}$ is also converted into $\lceil \frac{n}{2} \rceil$ rectangles $\overline{\mathbf{HQ}} = \{\mathbf{Q}_1, \mathbf{Q}_2, \ldots, \mathbf{Q}_{\lceil \frac{n}{2} \rceil}\}$, where $\mathbf{Q}_i$ lies on the same plane with $\mathbf{D}_i$. Now, if $\mathbf{D}_i \cap \mathbf{Q}_i \neq \varnothing$ for every i, then $\mathbf{HD} \cap \mathbf{HQ} \neq \varnothing$, and visa versa. The correctness of this transformation is summarized as follows.

$$\mathbf{HD} \cap \mathbf{HQ} \neq \varnothing \Longleftrightarrow \begin{cases} \mathbf{d}_1 \cap \mathbf{q}_1 \neq \varnothing \\ \mathbf{d}_2 \cap \mathbf{q}_2 \neq \varnothing \\ \ldots \end{cases} \Longleftrightarrow \begin{cases} \mathbf{d}_1 \cap \mathbf{q}_1 \neq \varnothing \wedge \mathbf{d}_2 \cap \mathbf{q}_2 \neq \varnothing \\ \mathbf{d}_3 \cap \mathbf{q}_3 \neq \varnothing \wedge \mathbf{d}_4 \cap \mathbf{q}_4 \neq \varnothing \\ \ldots \end{cases} \Longleftrightarrow \begin{cases} \mathbf{D}_1 \cap \mathbf{Q}_1 \neq \varnothing \\ \mathbf{D}_2 \cap \mathbf{Q}_2 \neq \varnothing \\ \ldots \end{cases} \quad (8)$$

The algorithm used to transform nodes and queries into a set of rectangles is described as below

- **HRect** $(\mathbf{HX}) \rightarrow \overline{\mathbf{HX}}$: This function takes as input a hyper-rectangle $\mathbf{HX} = (\mathbf{x}_1, \mathbf{x}_2, \ldots, \mathbf{x}_n)$. It generates $\lceil \frac{n}{2} \rceil$ rectangles $\overline{\mathbf{HX}} = \{\mathbf{X}_1, \mathbf{X}_2, \ldots, \mathbf{X}_{\lceil \frac{n}{2} \rceil}\}$ following the rule: if n is an even integer, for $i = 1$ to $\lceil \frac{n}{2} \rceil$, $\mathbf{X}_i = (\mathbf{x}_{2i-1}, \mathbf{x}_{2i})$; otherwise, for $i = 1$ to $\lceil \frac{n}{2} \rceil - 1$, $\mathbf{X}_i = (\mathbf{x}_{2i-1}, \mathbf{x}_{2i})$, while $\mathbf{X}_{\lceil \frac{n}{2} \rceil} = (\mathbf{x}_1, \mathbf{x}_n)$. Output the set of rectangles $\overline{\mathbf{HX}}$.

5.4   Multi-Dimensional Range Queries

To construct the encrypted index $\widehat{\mathrm{R}}$-tree, we build a regular R-tree for the given dataset and encrypt its nodes, while the parent-children relationships are preserved. Particularly, for the node $\mathbf{HD}$, first transform it into several rectangles by invoking **HRect**, and then encrypt each rectangle by using **DRect_Enc**, independently. Each internal node in the encrypted index has the form $(\widehat{\mathbf{HD}}, ptrs)$ where $\widehat{\mathbf{HD}}$ denotes the set of encrypted rectangles and $ptrs$ are the pointers to the node's children. Each leaf node is of the form $(\widehat{\mathbf{HD}}, ids)$ where $\widehat{\mathbf{HD}}$ represents the set of encrypted rectangles and ids are the identifiers of data points contained in this node. Similarly, the encrypted trapdoor is generated from the query in the same way, except that each rectangle is encrypted by **QRect_Enc**. The algorithms applied to encrypt nodes and queries in our method are summarized as follows.

- **Setup**$(16, 6) \rightarrow \{\mathcal{SK}_A, \mathcal{SK}_B\}$: This algorithm outputs the secret keys $\mathcal{SK}_A$ and $\mathcal{SK}_B$ that **SK_Gen**$(16, 6)$ outputs.
- **Node_Enc**$(\mathcal{SK}_A, \mathcal{SK}_B, \mathbf{HD}) \rightarrow \widehat{\mathbf{HD}}$: Given the secret keys $\mathcal{SK}_A, \mathcal{SK}_B$ and the node $\mathbf{HD}$, this function first invokes **HRect**$(\mathbf{HD})$ to obtain a set of rectangles $\overline{\mathbf{HD}} = \{\mathbf{D}_1, \mathbf{D}_2, \ldots, \mathbf{D}_{\lceil \frac{n}{2} \rceil}\}$. Then, it encrypts each $\mathbf{D}_i$ as $\widehat{\mathbf{D}}_i$ by using **DRect_Enc**$(\mathcal{SK}_A, \mathcal{SK}_B, \mathbf{D}_i)$. Finally, return the encrypted node $\widehat{\mathbf{HD}} = \{\widehat{\mathbf{D}}_1, \widehat{\mathbf{D}}_2, \ldots, \widehat{\mathbf{D}}_{\lceil \frac{n}{2} \rceil}\}$.
- **Query_Enc**$(\mathcal{SK}_A, \mathcal{SK}_B, \mathbf{HQ}) \rightarrow \widehat{\mathbf{HQ}}$: Given the secret keys $\mathcal{SK}_A, \mathcal{SK}_B$ and the query $\mathbf{HQ}$, this function first invokes **HRect**$(\mathbf{HQ})$ to obtain its rectangle set $\overline{\mathbf{HQ}} = \{\mathbf{Q}_1, \mathbf{Q}_2, \ldots, \mathbf{Q}_{\lceil \frac{n}{2} \rceil}\}$. Next, it encrypts each $\mathbf{Q}_i$ as $\widehat{\mathbf{Q}}_i$ by using

QRect_Enc$(\mathcal{SK}_A, \mathcal{SK}_B, \mathbf{Q}_i)$. Finally, return the encrypted trapdoor $\widehat{\mathbf{HQ}} = \{\widehat{\mathbf{Q}}_1, \widehat{\mathbf{Q}}_2, \ldots, \widehat{\mathbf{Q}}_{\lceil \frac{n}{2} \rceil}\}$.

Given the encrypted index $\widehat{\mathbf{R}}$-tree and trapdoor, multi-dimensional range queries can be executed securely and efficiently on encrypted cloud databases. The search algorithm of $\widehat{\mathbf{R}}$-tree is similar to that of R-tree. While traversing the $\widehat{\mathbf{R}}$-tree, its basic operation is to test whether an encrypted node intersects with the encrypted trapdoor. The function used for intersection determination is described as below

- **Xsect_NodeQuery**$(\widehat{\mathbf{HD}}, \widehat{\mathbf{HQ}}) \rightarrow \{0, 1\}$: Given the encrypted node $\widehat{\mathbf{HD}} = \{\widehat{\mathbf{D}}_1, \widehat{\mathbf{D}}_2, \ldots, \widehat{\mathbf{D}}_{\lceil \frac{n}{2} \rceil}\}$ and the encrypted trapdoor $\widehat{\mathbf{HQ}} = \{\widehat{\mathbf{Q}}_1, \widehat{\mathbf{Q}}_2, \ldots, \widehat{\mathbf{Q}}_{\lceil \frac{n}{2} \rceil}\}$. For each $i \in \mathbb{N}_{\lceil \frac{n}{2} \rceil}$, if there exists some i such that **Xsect_Rect**$(\widehat{\mathbf{D}}_i, \widehat{\mathbf{Q}}_i)$ returns 0, this algorithm outputs 0 and 1 otherwise.

6   Security Analysis

We prove that our scheme can provide index and trapdoor privacy as well as ordering privacy under the honest-but-curious model and the known-plaintext attack model.

Theorem 1. The proposed scheme can provide index and trapdoor privacy, if the bit string used for splitting cannot be known to the adversary.

Proof. In this paper, each hyper-rectangle is converted into a set of 2-dimensional rectangles, and each rectangle is encrypted with the improved ASPE separately. Therefore, the index and trapdoor privacy in our scheme is based on the security of the encryption functions for protecting rectangles. Without loss of generality, we assume that the number of dimensions equals 2, i.e., both nodes and queries are denoted as rectangles. To launch the known-plaintext attack, suppose that the adversary (i.e. the cloud server) obtains w plaintext-ciphertext pairs of nodes $\{(\widetilde{\mathbf{D}}_i, \widehat{\mathbf{D}}_i)\}$, where $\widetilde{\mathbf{D}}_i = \{V_{i,A}, V_{i,B}\}$ and $\widehat{\mathbf{D}}_i = \{\widehat{V}_{i,A}, \widehat{V}_{i,B}\}$.

Using the linear relationship between the input vectors and output vectors of **ASPE.Data_Enc**, Chunsheng et al. [21] propose a polynomial-time algorithm to solve a secret key which can decrypt arbitrary encrypted index vectors. One basic requirement of their attack is that the adversary obtains the input data vector of ASPE, i.e., the parameter P in **ASPE.Data_Enc**. However, before encrypting $\widetilde{\mathbf{D}}_i$, our scheme scales $V_{i,A}, V_{i,B}$ as $r_{i,A}V_{i,A}, r_{i,B}V_{i,B}$, respectively. The linear relationship between $V_{i,A}, V_{i,B}$ and $\widehat{V}_{i,A}, \widehat{V}_{i,B}$ is different from that between $r_{i,A}V_{i,A}, r_{i,B}V_{i,B}$ and $\widehat{V}_{i,A}, \widehat{V}_{i,B}$. In other words, the adversary no longer possesses the exact input of **ASPE.Data_Enc**. So our encryption functions used to protect rectangles are secure against the attack proposed in [21].

Given an encrypted trapdoor $\widehat{\mathbf{Q}} = \{\widehat{W}_A, \widehat{W}_B\}$, the attacker can set up the equations $\{r_{i,A}V_{i,A} \cdot r'_A W_A = \widehat{V}_{i,A} \cdot \widehat{W}_A\}$ to infer W_A. We view $r_{i,A}r'_A$ as an entirety. Then, there are $4 + w$ unknowns on the left side of the equal signs. Considering there are only w equations, which are less than the number of unknowns, the adversary can not get W_A. Similarly, W_B can also be protected. In addition, since our encryption method extends ASPE, it retains the security properties of ASPE and artificial dimensions still work in our scheme, which can further improve the security. Hence, the privacy of indices and trapdoors are provided. □

Theorem 2. The proposed scheme can provide ordering privacy.

Proof. Our scheme transforms each hyper-rectangle into $\lceil \frac{n}{2} \rceil$ rectangles on different planes, where n denotes the number of dimensions. When checking whether we have to continue to search one internal node's children and whether one leaf node should be returned, the search algorithm needs to determine the relationship of the two rectangles on the same plane. So the security of our method is dependant on that of our technique used for solving encrypted rectangle intersection problem. For the ease of description, we start to first assume the number of dimensions is 2. To decide whether **D** overlaps with **Q**, the cloud server needs to compute the scalar products $\widehat{V}_A \cdot \widehat{W}_A$ and $\widehat{V}_B \cdot \widehat{W}_B$, where $\{\widehat{V}_A, \widehat{V}_B\}$ and $\{\widehat{W}_A, \widehat{W}_B\}$ are the encrypted versions of **D** and **Q**, respectively. According to the computation results, the cloud server only obtains which region the center $O_d(o_{d,1}, o_{d,2})$ of **D** is within. As shown in Fig. 4(c), if $\widehat{V}_A \cdot \widehat{W}_A \geq 0$ and $\widehat{V}_B \cdot \widehat{W}_B \geq 0$, the center is in region I; if $\widehat{V}_A \cdot \widehat{W}_A < 0$ and $\widehat{V}_B \cdot \widehat{W}_B \geq 0$, the center lies inside region II; if $\widehat{V}_A \cdot \widehat{W}_A \geq 0$ and $\widehat{V}_B \cdot \widehat{W}_B < 0$, the center is within region III; otherwise, it is inside region IV. Observe that, the distribution of the four regions are symmetric about the center of the rectangle **R** and the lines $x = (l_{r,1} + h_{r,1})/2$, $y = (l_{r,2} + h_{r,2})/2$. So the adversary can not determine the exact relationship between $O(o_{d,1}, o_{d,2})$ and **R**, which means that the ordering information between points in node **D** and query **Q** is protected. When $n > 2$, the rectangles transformed from each node and each query are on the different planes. So the security of these cases relies on that of the case when $n = 2$. Hence, the ordering privacy can be protected in our scheme.   □

For all tree-based methods, if the fanout (i.e., node capacity) of the tree is small, the points contained in the same leaf node are "close" in space. Then the known tuples in the databases will disclose the unknown tuples. Obviously, increasing the fanout will mitigate this problem, but this incurs more false positive results. In this paper, we use the variance of the underlying points of each leaf node to reflect the level of security. Since variance measures how far a set of numbers is spread out, a large variance corresponds to a high level of security.

7   Performance Evaluation

In this section, the effectiveness of our scheme is demonstrated on both real and artificial datasets. We focus on evaluating the average query time, false positive rate and variance of leaf nodes in our experiments. We also present comparisons with competing methods.

Implementation Details. Our scheme is implemented in C on a desktop PC running Windows 7 Professional with 3.40 GHz Intel(R) Core(TM) i7-3770 processor and 4 GB RAM inside. Since double precision in C cannot meet our requirements due to precision loss, the GMP Library [24] is used to handle high-precision operations. To ensure security, artificial dimensions are introduced in ASPE and the parameter d' is set to 80 as recommended in [20].

We use three datasets in our experiments, including the check-in dataset for Gowalla (GO) [25], the spatial network dataset for North Jutland, Denmark (DE) [26] and an artificial dataset (AR). The details are given in Table 1. We normalize the data range along every dimension to $[0, 1]$. For each dataset, our scheme builds an R-tree and then encrypts each node separately to obtain the encrypted index $\hat{R}$-tree. Regarding the query distribution, we generate 100 queries as randomly distributed hyper-squares with side lengths varying from 0.001 to 0.1 with a scaling factor of 0.001. Every query result is not empty.

Table 1. Datasets.

Dataset	Attributes	Number of tuples
GO	latitude, longitude	6,442,890
DE	osm_id, latitude, longitude, altitude	434,874
AR	30 attributes following the uniform distribution	1,000,000

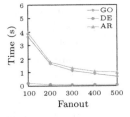

Fig. 5. Average query time.

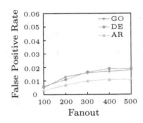

Fig. 6. Average false positive rate.

Query Time. Figure 5 plots the average query times for 100 random queries over the datasets GO, DE and AR. As the fanout of R-tree increases, the average query time drops significantly. Our approach provides efficient queries for large-scale and high-dimensional datasets. In particular, the average query times over the real dataset GO and the artificial dataset AR are both less than 4 seconds when the fanout is set to 100, where GO contains 6 million-plus total tuples and AR consists of 30 dimensions.

False Positive Rate. The false positive rate is defined as the fraction of negative records that are returned as positive. Figure 6 describes that the average false positive rate increases with the fanout of R-tree. To ensure security, a high false positive rate is not a huge problem as mentioned in [16].

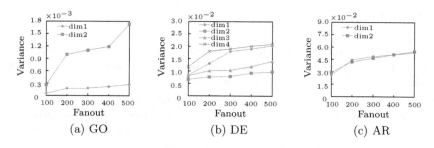

Fig. 7. Average variance of leaf nodes.

Variance of Leaf Nodes. Figure 7 shows the average variance of leaf nodes with respect to the fanout of R-tree. Specifically, for each leaf node, we measure the variance of its underlying points along each dimension and then average this value over all leaf nodes. Considering all attributes of the dataset AR follow the uniform distribution, we only evaluate the first two of these attributes. As shown in Figs. 5, 6 and 7, a larger fanout leads to the decrease of query time and the increases of false positive rate and variance, where a large variance corresponds to a high level of security. So data owners can tune the balance between efficiency and security by using different fanouts.

Comparisons with Competing Methods. Both [14,16] leverage R-tree to index data and encrypt each node separately, which are similar to our scheme. The encryption technique used in [16] and our scheme is ASPE, while [14] uses predicate encryption to encrypt nodes. In our experiments, we implement the algorithms in [14] using PBC Library [27] and each value has a precision of 10^{-3}. Table 2 gives the ratios of running times and storage costs for [14,16] to those for ours, respectively. Our scheme provides the best efficiency in both query encryption and intersection test. Even though we require more overheads in node encryption and node storage than [16], the performance of our scheme is much better than [14]. Furthermore, the data users pay more attention to query encryption and search time in real-life applications.

Table 2. Comparisons with other methods.

Number of attributes	Encryption time per node		Encryption time per query		Search time per node		Storage cost per node	
	[14]/ours	[16]/ours	[14]/ours	[16]/ours	[14]/ours	[16]/ours	[14]/ours	[16]/ours
2	80.11	1.00	2.52	2.00	3761.25	2.89	26.21	1.00
3	65.03	0.50	1.93	1.50	3305.13	2.33	19.66	0.50
4	88.12	0.50	2.38	2.00	4226.52	3.12	26.03	0.50
5	69.35	0.33	1.85	1.67	3512.03	2.57	20.01	0.33
6	86.81	0.33	2.57	2.00	4180.60	3.06	26.19	0.33

8   Conclusion

In this paper, we propose a novel technique for encrypted rectangle intersection problem using our enhanced asymmetric scalar-product-preserving encryption. Based on this technique, we present a multi-dimensional range query scheme over encrypted cloud databases. The proposed scheme is secure under the honest-but-curious model and the known-plaintext attack model. By evaluating the performance of our scheme and comparing it with other methods, we show that our system is highly performant to be used in real scenarios.

Acknowledgments. This work was supported by the National Natural Science Foundation of China under Grant No. U1636216, No. 61232005 and No. 61402456.

References

1. Hacigumus, H., Iyer, B., Mehrotra, S.: Providing database as a service. In: ICDE, pp. 29–38 (2002)
2. Amazon Web Services. https://aws.amazon.com
3. Microsoft Azure. https://azure.microsoft.com
4. Google Fires Engineer for Privacy Breach. http://edition.cnn.com/2010/TECH/web/09/15/google.privacy.firing
5. Song, D.X., Wagner, D., Perrig, A.: Practical techniques for searches on encrypted data. In: S&P, pp. 44–55 (2000)
6. Cao, N., Wang, C., Li, M., Ren, K., Lou, W.: Privacy-preserving multi-keyword ranked search over encrypted cloud data. IEEE Trans. Parallel Distrib. Syst. **25**(1), 222–233 (2014)
7. Hong, C., Li, Y., Zhang, M., Feng, D.: Fast multi-keywords search over encrypted cloud data. In: Cellary, W., Mokbel, M.F., Wang, J., Wang, H., Zhou, R., Zhang, Y. (eds.) WISE 2016. LNCS, vol. 10041, pp. 433–446. Springer, Cham (2016). doi:10.1007/978-3-319-48740-3_32
8. Hacigumus, H., Iyer, B., Li, C., Mehrotra, S.: Executing SQL over encrypted data in the database-service-provider model. In: SIGMOD, pp. 216–227 (2002)
9. Li, R., Liu, A.X., Wang, A.L., Bruhadeshwar, B.: Fast range query processing with strong privacy protection for cloud computing. In: VLDB, pp. 1953–1964 (2014)
10. Chi, J., Hong, C., Zhang, M., Zhang, Z.: Privacy-enhancing range query processing over encrypted cloud databases. In: Wang, J., Cellary, W., Wang, D., Wang, H., Chen, S.-C., Li, T., Zhang, Y. (eds.) WISE 2015. LNCS, vol. 9419, pp. 63–77. Springer, Cham (2015). doi:10.1007/978-3-319-26187-4_5
11. Boneh, D., Waters, B.: Conjunctive, subset, and range queries on encrypted data. In: Vadhan, S.P. (ed.) TCC 2007. LNCS, vol. 4392, pp. 535–554. Springer, Heidelberg (2007). doi:10.1007/978-3-540-70936-7_29
12. Shi, E., Bethencourt, J., Chan, T.H.H., Song, D., Perrig, A.: Multi-dimensional range query over encrypted data. In: S&P, pp. 350–364 (2007)
13. Lu, Y.: Privacy-preserving logarithmic-time search on encrypted data in cloud. In: NDSS (2012)
14. Wang, B., Hou, Y., Li, M., Wang, H., Li, H.: Maple: scalable multi-dimensional range search over encrypted cloud data with tree-based index. In: AsiaCCS, pp. 111–122 (2014)

15. Wang, B., Li, M., Wang, H.: Geometric range search on encrypted spatial data. IEEE Trans. Inf. Forensics Secur. **11**(4), 704–719 (2016)
16. Wang, P., Ravishankar, C.V.: Secure and efficient range queries on outsourced databases using R-trees. In: ICDE, pp. 314–325 (2013)
17. Hore, B., Mehrotra, S., Canim, M., Kantarcioglu, M.: Secure multidimensional range queries over outsourced data. VLDB J. **21**(3), 333–358 (2012)
18. Bentley, J.L.: Multidimensional binary search trees used for associative searching. Commun. ACM **18**(9), 509–517 (1975)
19. Guttman, A.: R-trees: a dynamic index structure for spatial searching. In: SIGMOD, pp. 47–57 (1984)
20. Wong, W.K., Cheung, D.W.L., Kao, B., Mamoulis, N.: Secure kNN computation on encrypted databases. In: SIGMOD, pp. 139–152 (2009)
21. Chunsheng, G., Jixing, G.: Known-plaintext attack on secure kNN computation on encrypted databases. Secur. Commun. Netw. **7**(12), 2432–2441 (2014)
22. Goldreich, O., Ostrovsky, R.: Software protection and simulation on oblivious rams. J. ACM **43**(3), 431–473 (1996)
23. David, H.A., Nagaraja, H.N.: Order Statistics, 3rd edn. Wiley, New York (2003)
24. GMP: The GNU Multiple Precision Arithmetic Library. http://gmplib.org
25. Cho, E., Myers, S.A., Leskovec, J.: Friendship and mobility: user movement in location-based social networks. In: KDD, pp. 1082–1090 (2011)
26. Kaul, M., Yang, B., Jensen, C.S.: Building accurate 3D spatial networks to enable next generation intelligent transportation systems. In: MDM, pp. 137–146 (2013)
27. PBC: The Pairing-Based Cryptography Library. https://crypto.stanford.edu/pbc

When Differential Privacy Meets Randomized Perturbation: A Hybrid Approach for Privacy-Preserving Recommender System

Xiao Liu[1], An Liu[1,2(✉)], Xiangliang Zhang[2], Zhixu Li[1], Guanfeng Liu[1],
Lei Zhao[1], and Xiaofang Zhou[3]

[1] Soochow University, Suzhou, China
anliu@suda.edu.cn
[2] King Abdullah University of Science and Technology,
Thuwal, Kingdom of Saudi Arabia
[3] University of Queensland, Brisbane, Australia

Abstract. Privacy risks of recommender systems have caused increasing attention. Users' private data is often collected by probably untrusted recommender system in order to provide high-quality recommendation. Meanwhile, malicious attackers may utilize recommendation results to make inferences about other users' private data. Existing approaches focus either on keeping users' private data protected during recommendation computation or on preventing the inference of any single user's data from the recommendation result. However, none is designed for both hiding users' private data and preventing privacy inference. To achieve this goal, we propose in this paper a hybrid approach for privacy-preserving recommender systems by combining differential privacy (DP) with randomized perturbation (RP). We theoretically show the noise added by RP has limited effect on recommendation accuracy and the noise added by DP can be well controlled based on the sensitivity analysis of functions on the perturbed data. Extensive experiments on three large-scale real world datasets show that the hybrid approach generally provides more privacy protection with acceptable recommendation accuracy loss, and surprisingly sometimes achieves better privacy without sacrificing accuracy, thus validating its feasibility in practice.

Keywords: Recommender systems · Privacy-preserving · Differential privacy · Randomized perturbation

1 Introduction

During the last few decades we have witnessed the increasing use of recommender systems in various domains to solve the problem of information seeking in an extremely large volume of content. With the help of recommender systems, customers can quickly find things that are interesting or new by narrowing down the set of choices. Meanwhile, service providers using recommender systems can

S. Candan et al. (Eds.): DASFAA 2017, Part I, LNCS 10177, pp. 576–591, 2017.
DOI: 10.1007/978-3-319-55753-3_36

increase sales or click-through rate (CTR) by providing personalized service for customers. For example, McKinsey[1] reported that "35% of what consumers purchase on Amazon and 75% of what they watch on Netflix come from product recommendations".

The benefits brought by recommender systems are significant. However, the use of recommender systems introduces privacy threats and concerns. In order to provide high quality recommendations, recommender systems need to collect customers' private data, such as history data (e.g., the books bought last month or the movies watched last week) and rating data (e.g., the rate for a book or a movie). However, recommender systems may not be trustable. It is common for customers to raise privacy concerns as the collected data may be shared with, rent or sold to third parties. According to a survey done by PewResearch[2], "86% of Internet users have taken steps online to remove or mask their digital footprints" and "68% of Internet users believe current laws are not good enough in protecting people's privacy online". It is thus crucial to develop technologies that can keep users' private data protected while enabling personalized recommendation, which is a necessary and beneficial complement to the efforts made in the non-technical domain such as privacy policies and related laws.

1.1   Related Work

Cryptography is one of the most important technologies to realize privacy-preserving recommender systems. Using some well-known encryption algorithms, users can transform their private data from meaningful plaintext to meaningless ciphertext, thus achieving privacy preservation. To enable recommender systems to carry out computation over ciphertext directly, the encryption algorithms to be used have to be homomorphic, that is, the result of operations performed on ciphertext, when decrypted, matches the result of operations performed on the corresponding plaintext. For example, Paillier cryptosystem [14] was employed by Erkin et al. [6] and Ma et al. [10], ElGamal cryptosystem [5] was used by Zhan et al. [17] and Badsha et al. [1], to realize privacy-preserving recommender systems. However, homomorphic encryption is built on expensive public-key cryptography, which is theoretical in nature and cannot be applied in practice due to the prohibitive computation cost. In addition, Nikolaenko et al. [13] and Liu et al. [9] built privacy-preserving recommender systems based on another renowned cryptographic tool, Yao's garbled circuits [8,16]. However, these approaches require the existence of a trusted third party, which also hinders their application in practice.

To overcome the weakness of cryptography based techniques, Polat and Du [15] proposed a *Randomized Perturbation* (RP) technique which adds noise to users' private data before releasing the data to recommender systems. RP is much faster than cryptography based techniques, but this is at the cost of sacrificing recommendation accuracy and privacy protection degree. In particular,

---

[1] http://www.mckinsey.com/industries/retail/our-insights/
    how-retailers-can-keep-up-with-consumers.

[2] http://www.pewinternet.org/2013/09/05/anonymity-privacy-and-security-online/.

the smaller the random noise added, the more accurate the predicted rating. However, the smaller noise also results in weak privacy guarantee. For example, the data reconstruction methods proposed by Zhang et al. [18] can derive more original data when smaller noise is injected. Therefore, a trade-off between accuracy and privacy should be made when applying RP.

The above work aims at keeping users' private data secret during recommendation computation. However, the output of recommender systems can be also utilized by malicious users to make inferences about other users' private data [12], that is, based on the recommendation she gets, a malicious user can guess whether someone else has, for example, bought some book or seen some movie. To avoid this kind of information leakage, *Differential Privacy* (DP) [3,4] has been introduced into recommender systems recently [7,11,12]. By adding noise, DP guarantees the distribution of the recommendation is insensitive to any individual user's data, thus preventing the inference of any single user's data from the recommendation. Due to the injected noise, DP also needs to strike a balance between recommendation accuracy and privacy protection degree. In addition, DP does not protect users' data from recommender systems, as the latter has full access to users' data in the clear.

1.2   Contributions

From the above discussion, it is expected that a privacy-friendly recommender system should respect user privacy at two stages: (1) does not ask users to submit their original data in the data collection stage; and (2) can prevent the inference of any single user's data from the final recommendation result in the normal execution stage. To the best of our knowledge, however, these two aspects have not been considered simultaneously. In this paper, we aim at designing an approach that can hide users' private data and prevent privacy inference simultaneously. At first glance, an intuitive solution is to integrate the techniques mentioned above. Nevertheless, there are some interesting issues worthy of investigation but largely overlooked by recent studies. For example, the amount of noise injected by DP is based on the sensitivity of a query function, which sometimes is not easy to estimate, especially when considering that the underlying data will be disguised by RP or encryption. For another, since DP and RP both introduce noise to original data, can we be certain that the recommendation accuracy will inevitably become worse? Or what is the trade-off between accuracy and privacy in this new context?

As the initial step towards more privacy-friendly recommender systems, we propose in this paper a hybrid approach which combines RP and DP. Specifically, users mask their original data through RP and send the disguised values to the recommender system, which injects calibrated noise again to the perturbed data to achieve DP. Our contributions are summarized as follows:

– We design a hybrid approach for privacy-preserving recommender systems by combining DP with RP. Compared with existing works, our approach provides more privacy guarantee as users' private data is kept secret and no one can infer any single user's data from the recommendation result.

- We theoretically show the noise added by RP has limited effect on recommendation accuracy and the noise added by DP can be well controlled based on the sensitivity analysis of functions on the perturbed data.
- We conduct extensive experiments to evaluate the performance of our hybrid approach on three large-scale real world datasets. The results show that the combination of DP and RP is feasible in practice. Generally it provides more privacy protection with acceptable accuracy loss, and surprisingly sometimes it achieves better privacy and accuracy at the same time.

The rest of the paper is organized as follows. Section 2 introduces a representative non-private recommendation algorithm and some background knowledge. Section 3 presents the detailed design of the hybrid approach. Section 4 discusses the experimental results and Sect. 5 concludes the paper.

2 Preliminaries

2.1 Recommendation Algorithm Without Privacy Guarantee

We first describe a recommendation algorithm [12] without privacy guarantee. Suppose there are n users and m items. Based on the data provided by n users, the recommender system has two matrices in hand. One is a rating matrix $R_{n \times m}$ that contains the ratings of n users for m items where r_{ui} indicates the rating of user u for item i. The other auxiliary (binary) matrix $E_{n \times m}$ indicates the presence of ratings, where $e_{ui} = 1$ means u has rated for i and $e_{ui} = 0$ means u does not. The two matrices are the input to the recommendation algorithm, while the output is predicted ratings of items that users have not rated.

Some users tend to give higher ratings than other users, and some items tend to receive higher ratings than others. This difference will make the recommendation result disappointing, so it is necessary to subtract user effects and item effects from ratings. We first compute the global average of $R_{n \times m}$:

$$GAvg = \frac{\sum_R r_{ui}}{\sum_E e_{ui}}$$

Then, we center ratings by computing and subtracting average ratings for items and users:

$$r'_{ui} = r_{ui} - UAvg(u)$$

$$UAvg(u) = \frac{\sum_i (r_{ui} - IAvg_i) + \beta_u \cdot GAvg}{\sum_i e_{ui} + \beta_u}, \quad IAvg_i = \frac{\sum_u r_{ui} + \beta_m \cdot GAvg}{\sum_u e_{ui} + \beta_m}$$

where $IAvg$ and $UAvg$ are dampened by β_m and β_u fictitious ratings of the global average, respectively. Here, β_m is the average number of ratings for item m, and β_u is the average number of rating items for user u.

Finally, we use the centered ratings to calculate the covariance matrix, which indicates the relationships between items:

$$Cov(ij) = \frac{\sum_u w_u r'_{ui} r'_{uj}}{\sum_u w_u e_{ui} e_{uj}}$$

where w_u is per-user weights equaling to the reciprocal of $||e_u||$. The final recommendation result can be made by passing this covariance matrix to a large number of advanced learning and prediction algorithms, such as the k-nearest neighbor (kNN) method proposed by Bell and Koren [2].

2.2   Differential Privacy (DP)

Intuitively, differential privacy means the probability an attacker who is able to observe the computation's output learns any record's presence in or absence from the computation's input should be indistinguishable [3,4]. The formal definition is as follows:

Definition 1. *A randomized function f provides ϵ-differential privacy if for any neighboring data bases A and B ($A \triangle B = 1$), and any subset S of possible outcomes Range(f),*

$$Pr[f(A) \in S] \leq exp(\epsilon) \times Pr[f(B) \in S]$$

Two datasets A and B are adjacent if there is only one individual record difference between them ($A \triangle B = 1$). The parameter ϵ is the privacy budget, which can be used to control the level of privacy protection. The smaller the value of ϵ is, the stronger privacy protection it provides. DP guarantees the output is insensitive to any individual record. The probability that an attacker can correctly guess whether or not an individual record is in the dataset is at most $exp(\epsilon)$ based on the outputs of calculations. It satisfies a *composability property* defined as follows: The sequence of $f_i(A)$ provides $(\sum_i \epsilon_i)$-differential privacy, where f_i each provides ϵ_i-differential privacy. Therefore, the ϵ parameter can be considered as an accumulative privacy cost as more steps are executed. These costs keep accumulating until they reach an allotted privacy budget.

A common way to obtain differential privacy is by applying random noise to the measurement. The amount of noise added depends on the L_1-sensitivity of the evaluated function, which is the largest possible change in the measurement given a change in a single record in the dataset. In general, the L_k-sensitivity of a function f is given by:

$$S_k(f) = \max_{(A \triangle B = 1)} ||f(A) - f(B)||_k$$

where $|| \cdot ||_k$ denotes the L_k-norm.

Given a function $f : D \rightarrow \mathbb{R}^d$, Laplace mechanism obtains ϵ-differential privacy by adding noise sampled from Laplace distribution, with a calibrated scale $b = S_1(f)/\epsilon$. The following computation maintains ϵ-differential privacy:

$$K(x) = f(x) + (Laplace(S_1(f)/\epsilon))^d$$

2.3   Randomized Perturbation

The basic idea of randomized perturbation is to perturb the data in such a way that certain computations can be done while preserving users' privacy. Although data from each user is scrambled, if the number of users is significant large, the aggregate information of these users can be estimated with decent accuracy. Such property is very useful for computations that are based on aggregate information. Scaler product and sum are among such computations.

Let r^a and r^b be the original vectors, where $r^a = (r_1^a, \ldots, r_i^a)$ and $r^b = (r_1^b, \ldots, r_i^b)$. r^a is disguised by $v^a = (v_1^a, \ldots, v_i^a)$, and r^b by $v^b = (v_1^b, \ldots, v_i^b)$, where v^a and v^b are uniformly distributed in domain $[-\gamma, \gamma]$. Let $r'^a = r^a + v^a$ and $r'^b = r^b + v^b$ be disguised data that are known. Because v^a and v^b are uniformly distributed, the scalar product of r^a and r^b can be estimated from r'^a and r'^b and the sum of the values of r^a can be estimated from r'^a as follows:

$$\sum_{i=1}^{n}(r_i + v_i) = \sum_{i=1}^{n} r_i + \sum_{i=1}^{n} v_i \approx \sum_{i=1}^{n} r_i \tag{1}$$

$$r'^a \cdot r'^b = \sum_{i=1}^{n}(r_i^a r_i^b + r_i^b v_i^a + r_i^a v_i^b + v_i^a v_i^b) \approx \sum_{i=1}^{n} r_i^a r_i^b \tag{2}$$

3   The Hybrid Approach

Figure 1 shows the whole life-cycle of a typical recommender system armed with our hybrid privacy-preserving approach. Three stages are involved in the process of recommendation: *data collection*, *data publication* and *data prediction*. In the first stage, users' original data are disguised through randomized perturbation, resulting in perturbed rating matrix R and auxiliary matrix E. Based on the two perturbed matrices, the recommender system computes global average, item averages, user averages, and finally the covariance matrix for data publication. All these data are masked with particular amount of noise to guarantee differential privacy. With the added noise, the covariance matrix is ready for publication and can be fed into an existing learning and prediction algorithm (e.g., the kNN method [2]) with no changes. As mentioned earlier, the challenge here is how to ensure recommendation accuracy and realize differential privacy on the perturbed data effectively.

3.1   Methodology

In the data collection stage, the recommender system decides on a range $[-\gamma, \gamma]$ and let each user know. Then, each user u disguises her ratings r_{ui} by adding noise that is uniformly distributed in the domain $[-\gamma, \gamma]$. The recommender system collects these disguised data r'_{ui} to form two perturbed matrices.

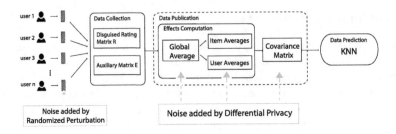

Fig. 1. Overview of the hybrid approach for privacy-preserving recommender systems

In the stage of data publication, the recommender system injects noise to three different values: the global average, the per-item average and the covariance matrix. Note that, the noises for these values are different and depend on the sensitivity of the underlying functions of computing these values. For global average, Laplace distributed noise is added to guarantee its privacy:

$$GAvg = \frac{\sum_R r'_{ui} + Laplace(\Delta r_1/\epsilon_1)}{\sum_E e_{ui} + Laplace(\Delta r'_1/\epsilon_1)}$$

where Δr_1 and $\Delta r'_1$ are the sensitivity of function $\sum_R r'_{ui}$ and $\sum_E e_{ui}$, respectively. Their exact values will be discussed in the next subsection. We then use the global average to produce a stabilized per-item average rating by β_m at value GAvg for each item:

$$IAvg_i = \frac{(\sum_u r'_{ui} + Laplace(\Delta r_2/\epsilon_2)) + \beta_m \cdot GAvg}{(\sum_u e_{ui} + Laplace(\Delta r'_2/\epsilon_2)) + \beta_m}$$

where Δr_2 and $\Delta r'_2$ are the sensitivity of $\sum_u r'_{ui}$ and $\sum_u e_{ui}$, respectively.

Having published the average rating for each item, we center the ratings for each user as follows, taking shrinking parameter β_u at global average, where c_u is the number of ratings by user u:

$$UAvg(u) = \frac{\sum_i (r'_{ui} - IAvg_i) + \beta_u \cdot GAvg}{c_u + \beta_u}$$

We subtract user effects average from the appropriate ratings and clamp the resulting centered ratings to the intervals $[-B, B]$, to lower the sensitivity of the measurements at the expense of the relatively few remaining large entries:

$$\hat{r_{ui}} = \begin{cases} -B & \text{if } r'_{ui} - UAvg(u) < -B \\ r'_{ui} - UAvg(u) & \text{if } -B \geq r'_{ui} - UAvg(u) < B \\ B & \text{if } r'_{ui} - UAvg(u) \geq B \end{cases}$$

The final measurement we make of the private data is the covariance of the perturbed and clamped user ratings vectors. To retain the difference between users, we take the non-uniform averages by using per-user weights w_u which equals to the reciprocal of $||e_u||$. Then, the covariance will be published as:

$$Cov(ij) = \frac{\sum_u w_u \hat{r}_{ui} \hat{r}_{uj} + Laplace(\Delta r_3/\epsilon_3)}{\sum_u w_u e_{ui} e_{uj} + Laplace(\Delta r'_3/\epsilon_3)}$$

where Δr_3 and $\Delta r'_3$ are the sensitivity of $\sum_u w_u \hat{r}_{ui} \hat{r}_{uj}$ and $\sum_u w_u e_{ui} e_{uj}$, respectively.

3.2   Theoretical Analysis

As mentioned earlier, different functions have different sensitivities, which determines the amount of noise needed for differential privacy. In this subsection, we analyze the sensitivities of different functions involved in the data publication. The sensitivity values Δr_1 and Δr_2 are both $\tau + 2\gamma$, where τ is the maximum possible difference in raw ratings, and γ is the parameter of RP. For example, if the range of rating is from 1 to 5, the τ then equals to 4. From Theorem 2, the sensitivity value Δr_3 is $2B(\tau + 2\gamma) + 3B^2$. For $\Delta r'_1$ and $\Delta r'_2$, their values are both 1, because the maximum possible difference is 1 in the auxiliary matrix when e^a and e^b differ on only one value. The value of $\Delta r'_3$ is 3 which is clear from Theorem 3.

Theorem 1. *Let r^a and r^b differ on one rating, τ be the maximum possible difference in raw ratings. Considering the randomized perturbation before collecting the data, the maximum possible difference in the processed ratings is $\tau + 2\gamma$. For centered and clamped ratings $\hat{r}^a$ and $\hat{r}^b$, we have*

$$||\hat{r}^a - \hat{r}^b||_1 \leq \tau + 2\gamma + B$$

Proof: If r^a and r^b are two sets of ratings which differ on one rating, present in r^b at r^b_{ui}, others are everywhere equal, except for the ratings of user u. For the ratings in common between r^a and r^b, the difference is at most the difference in the subtracted averages:

$$|UAvg(u)^b - UAvg(u)^a| = \frac{|r_{ui} - UAvg(u)^a|}{c^b_u + \beta_p} \leq \frac{\tau + 2\gamma}{c^b_u + \beta_p}$$

For the new rating r_{ui}, its previous contribution of zero is replaced with the new centered and clamped rating, at most B in magnitude. Hence, we have

$$||\hat{r}^a - \hat{r}^b||_1 \leq c^a_u \times \frac{\tau + 2\gamma}{c^b_u + \beta_p} + B$$

Note that $c^b_u = c^a_u + 1$ and the maximal value of c^a_u is $\beta_p + 1$. Therefore, the upper bound of $||\hat{r}^a - \hat{r}^b||_1$ is $\tau + 2\gamma + B$.   □

Theorem 2. *Let r^a and r^b differ on one rating. Taking $w_u = 1/||e_u||_1$, we have*

$$||w_u^a r_{ui}'^a r_{uj}'^a - w_u^b r_{ui}'^b r_{uj}'^b|| \leq 2B(\tau + 2\gamma) + 3B^2$$

Proof: For the difference $w_u^a r_{ui}'^a r_{uj}'^a - w_u^b r_{ui}'^b r_{uj}'^b$, we can rewrite it as $w_u^a r_{ui}'^a (r_{uj}'^a - r_{uj}'^b) + w_u^b (r_{ui}'^a - r_{ui}'^b) r_{uj}'^b + (w_u^a - w_u^b) r_{ui}'^a r_{uj}'^b$, as $||e_u^b - ||e_u^a|| \leq 1$, we have that

$$w_u^a - w_u^b = \frac{1}{||e_u^a||} - \frac{1}{||e_u^b||} \leq \frac{1}{||e_u^a|| ||e_u^b||}$$

The original matrix difference is bounded by

$$\left(\frac{||r_i'^a||}{||e_i^a||} + \frac{||r_i'^b||}{||e_i^b||} \right) ||r_i'^a - r_i'^b|| + \frac{||r_i'^a|| ||r_i'^b||}{||e_i^r|| ||e_i^b||}$$

Giving Theorem 2, we have the upper bound $2B(\tau + 2\gamma) + 3B^2$. □

Theorem 3. *Let e^a and e^b differ on one rating presence or absence. Taking $w_u = 1/||e_u||_1$, we have*

$$||w_u^a e_{ui}^a e_{uj}^a - w_u^b e_{ui}^b e_{uj}^b|| \leq 3$$

Proof: Between the two weight matrices, similarly, we can rewrite it as $w_u^a e_{ui}^a (e_{uj}^a - e_{uj}^b) + w_u^b (e_{ui}^a - e_{ui}^b) e_{uj}^b + (w_u^a - w_u^b) e_{ui}^a e_{uj}^b$. Then, we have the bound as follows:

$$||w_u^a e_{ui}^a e_{uj}^a - w_u^b e_{ui}^b e_{uj}^b|| \leq \left(\frac{||e_i^a||}{||e_i^a||} + \frac{||e_i^b||}{||e_i^b||} \right) ||e_i^a - e_i^b|| + \frac{||e_i^a|| ||e_i^b||}{||e_i^r|| ||e_i^b||} = 3$$

□

The above theorems show that the hybrid approach takes into account the effect of noise introduced by RP on the noise injected by DP. If we directly use DP without considering the noise of RP, we will obtain weaker privacy protection. This result is guaranteed by the following theorem.

Theorem 4. *The hybrid approach can provide stronger privacy protection than DP when raw rating data are disguised by RP.*

Proof: First note that the sensitivity of $\sum_R r_{ui}'$ and $\sum_R r_{ui}$ are $\tau + 2\gamma$ and τ, respectively. To provide ϵ_1-differential privacy for the global average, the hybrid approach injects noise v based on $Laplace(\frac{\tau + 2\gamma}{\epsilon_1})$. As DP does not consider the noise introduced by RP, it will inject noise v' based on $Laplace(\frac{\tau}{\epsilon_1})$. The noise v' on the disguised raw data can actually provide ϵ_1'-differential privacy where $\frac{\tau + 2\gamma}{\epsilon_1'} = \frac{\tau}{\epsilon_1}$. Thus we have $\epsilon_1' = \frac{\tau + 2\gamma}{\tau} \epsilon_1 > \epsilon_1$. Likewise, we can have $\epsilon_2' > \epsilon_2$ and $\epsilon_3' > \epsilon_3$ based on the sensitivity values given in Theorem 2, where ϵ_2' and ϵ_3' are the actual privacy budget DP can provide for item average and covariance, respectively. According to the composability property of differential privacy, we have $\epsilon = \epsilon_1 + \epsilon_2 + \epsilon_3$ and $\epsilon' = \epsilon_1' + \epsilon_2' + \epsilon_3'$. Clearly, $\epsilon < \epsilon'$, which completes the proof. □

We now examine the effect of the noise added by RP in the hybrid approach on recommendation accuracy. As mentioned earlier, though raw ratings from each user are perturbed, the aggregate information of these ratings can be estimated with decent accuracy if the number of users is significant large. Suppose $GAvg$ is the global average of the perturbed raw data collected in the hybrid approach and $GAvg^*$ is the global average of the raw data collected in DP. Clearly, we have

$$GAvg = \frac{\sum_R r'_{ui} + Laplace(\Delta r_1/\epsilon_1)}{\sum_E e_{ui} + Laplace(\Delta r'_1/\epsilon_1)}, \quad GAvg^* = \frac{\sum_R r_{ui} + Laplace(\Delta r^*_1/\epsilon_1)}{\sum_E e_{ui} + Laplace(\Delta r'_1/\epsilon_1)}$$

where Δr_1, $\Delta r'_1$, and Δr^*_1 are the sensitivity of function $\sum_R r'_{ui}$, $\sum_E e_{ui}$, and $\sum_R r_{ui}$, respectively. If R is sufficiently large, we have $\sum_R r'_{ui} \approx \sum_R r_{ui}$ as the noise injected into r_{ui} is uniformly sampled from $[-\gamma, \gamma]$. Besides, it is important to notice that $\sum_R r'_{ui} \gg Laplace(\Delta r_1/\epsilon_1)$. Thus, we have: $GAvg \approx GAvg^*$. Likewise, we can conclude that $Cov(ij) \approx Cov^*(ij)$, which indicates that the noise added by RP in the hybrid approach has limited effect on recommendation accuracy.

4   Experiments

4.1   Experimental Setting

In this section, we evaluate our hybrid approach for privacy-preserving recommender systems. As discussed earlier, both RP and DP introduce noise into recommendation computation, so it is worth studying the prediction accuracy when combining the two techniques. Therefore, we examined each of the three methods (i.e., RP, DP, and the hybrid one) in turn to see its effect on recommendation accuracy. All experiments were conducted on three real world datasets: Netflix[3] consists of roughly 100 M ratings of 17770 movies contributed by 480 K users; MovieLens[4] consists of 100 K ratings of 1682 movies contributed by 943 users; Yahoo[5] consists of 23 M ratings of 11915 movies contributed by 7742 users. The rating of the three datasets are all from 1 to 5.

By adjusting the parameters of the noise distributions we use (i.e., γ of RP and ϵ of DP), our approach provides different randomized perturbation and differential privacy guarantees, and consequently, the recommendation outputs have different accuracy values. In our experiment, the recommendation accuracy is measured by the *root mean squared error* (RMSE) on the test datasets: $RMSE = \sqrt{\frac{\sum_X (x-x')^2}{|X|}}$ where X consists of all values needs to be predicted in the test set and $|X|$ is the size of X, x' is the predicted value and x is the original value in the test set. A smaller RMSE value indicates a more accurate recommendation result. Regarding the training set and test set, the MovieLens

---

[3] http://www.netflixprize.com.

[4] http://grouplens.org/datasets/movielens.

[5] https://webscope.sandbox.yahoo.com.

data is divided into two parts, 80% for the training set and 20% for the test set. For Netflix data, the test set is the Probe set. For Yahoo dataset, the training set contains 7642 users and the test set has 2309 users. The test set is gathered chronologically after the training set.

We applied the kNN method [2] to the covariance matrix for the final recommendation. The value of k is fixed at 20, and the clamping parameter B is set to 1. Following the work in [12], for any ϵ, we set the respective ϵ_i as follows: $\epsilon_1 = 0.02 \times \epsilon, \epsilon_2 = 0.19 \times \epsilon, \epsilon_3 = 0.79 \times \epsilon$. All experiments were conducted on a Dell PowerEdge R930 server which is equipped with 2.2 GHz CPU and 2 TB RAM. Each experiment was run 10 times and the average results were reported.

4.2 Experimental Results

4.2.1 RP's Effect on Accuracy.
Figure 2 shows the recommendation accuracy when γ increases from 0.5 to 3.5 with a step of 0.5. Clearly, the accuracy decreases on all three datasets. This is because the noise added by RP is determined by the parameter γ. In particular, the larger the γ is, the wider range the random noise is in. Therefore, more randomness is likely to be added into the original data, resulting in less accurate recommendation.

4.2.2 DP's Effect on Accuracy.
Figure 3 shows the recommendation accuracy when ϵ increases from 0.1 to 10. From the results, we can see that the accuracy decreases rapidly when DP provides strong privacy guarantee (i.e.,

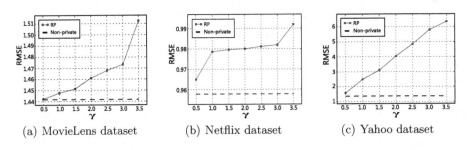

(a) MovieLens dataset   (b) Netflix dataset   (c) Yahoo dataset

Fig. 2. RP's effect on recommendation accuracy

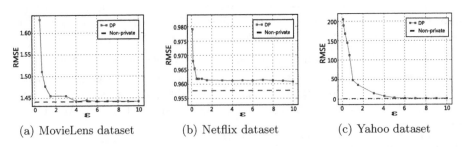

(a) MovieLens dataset   (b) Netflix dataset   (c) Yahoo dataset

Fig. 3. DP's effect on recommendation accuracy

when $\epsilon < 1$). When providing a weak privacy protection (i.e., when $\epsilon > 1$), the accuracy approaches to a constant. Such observations imply that a large ϵ contributes little to the accuracy but weakens the privacy protection.

4.2.3 Effect of the Hybrid Approach on Accuracy.

We first examine the hybrid approach by taking DP as the baseline. Figure 4 depicts how the recommendation accuracy of DP is affected by RP. It is clear that no matter which γ is used in RP, the overall trend of DP remains the same, that is, the accuracy decreases as ϵ approaches to 0, indicating a stronger privacy guarantee. Besides, when DP and RP work together, larger γ often leads to less accuracy. For example, DP plus RP with $\gamma = 0.5$ is more accurate than DP plus RP with $\gamma = 3.5$ on MovieLens dataset. Finally, it is worth noting that in most cases the combination of DP and RP makes recommendation less accurate, which coincides with our common sense as both of them introduce noise into the original data. However, their combination sometimes results in a win-win situation where both the accuracy and the privacy becomes better, as seen in Fig. 4(b). To make this clear, we draw in Fig. 5 the RMSE ratio between the hybrid approach and DP. We can see that for Netflix dataset, the combination of DP and RP sometimes outperforms DP only, especially when γ is small, for example, 0.5. Besides, the accuracy loss of DP plus RP is acceptable on MovieLens and Netflix datasets, but is not satisfactory on Yahoo dataset. A possible reason might be that MoveiLens and Netflix datasets have similar rating distribution, which is different from Yahoo dataset. We then examine the hybrid approach by taking RP as the baseline.

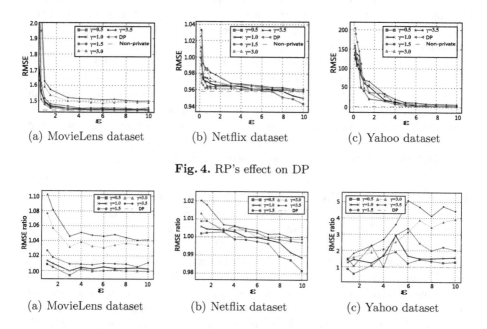

(a) MovieLens dataset   (b) Netflix dataset   (c) Yahoo dataset

Fig. 4. RP's effect on DP

(a) MovieLens dataset   (b) Netflix dataset   (c) Yahoo dataset

Fig. 5. RMSE ratio between the hybrid approach and DP

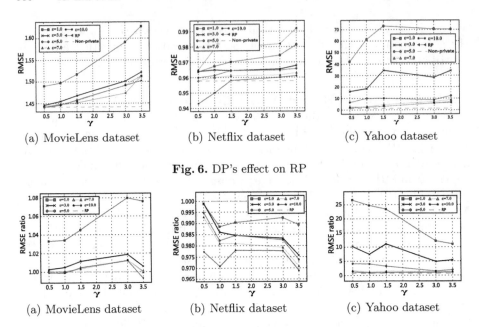

Fig. 6. DP's effect on RP

Fig. 7. RMSE ratio between the hybrid approach and RP

Figure 6 depicts how the recommendation accuracy of RP is affected by DP. We can again see that the overall trend of RP remains the same (i.e. the accuracy decreases as γ increases) no matter which ϵ is used in DP. Besides, smaller ϵ is more likely to make the recommendation less accurate, as stronger privacy guarantee is provided in DP through injecting more noise into the data. We also notice that, for MovieLens dataset, the combination of RP and DP makes recommendation less accurate, but in an acceptable range. For the Netflix dataset, however, their combination is indeed a good choice as we can obtain additional privacy guarantee while not sacrificing recommendation accuracy. Further note that this is true for any combination of γ and ϵ in our experiments, as shown in Fig. 7(b). The accuracy loss is still unsatisfactory on Yahoo dataset, especially when DP provides strong privacy guarantee, as depicted in Fig. 7(c).

4.2.4 Efficiency of the Hybrid Approach. Figure 8 shows the running time of the hybrid approach. It is clear that the running time increases when the rating matrix becomes large, but the total computation cost is acceptable even on a moderate server. In particular, for the MovieLens dataset where the size of rating matrix is about 1000 * 1700, the hybrid approach only needs 33 s. Even for large Netflix dataset whose rating matrix is 380 K * 500, the hybrid approach can be completed within less than 2.5 h. The computation cost of the hybrid approach mainly comes from DP, as RP only requires few simple operations and is done at user side. Thus, the hybrid approach has the same computation complexity as DP, but can provide stronger privacy guarantee than DP as shown in Theorem 4.

4.2.5   The Hybrid Approach Vs DP. Figure 9 depicts the accuracy comparison of the hybrid approach and DP when the raw rating data are disguised by RP with different γ. The privacy budget ϵ is set to 1 in both methods. In all datasets, we can see that DP has a better performance than the hybrid approach. This, however, exactly shows DP cannot provide sufficient privacy guarantee over the perturbed data, as it underestimates the sensitivity of the functions on the perturbed data. This result coincides with Theorem 4, which says the hybrid approach can provide stronger privacy protection than DP over the data disguised by RP. Further, the RMSE difference of the two methods is small, which means the hybrid approach has an acceptable accuracy loss while providing stronger privacy guarantee.

4.2.6   Summary. From the above discussion, we can see that, by carefully injecting appropriate noise into the perturbed data based the sensitivity analysis of different functions involved in the recommendation computation, the hybrid approach can provide more privacy protection with acceptable accuracy loss. More interestingly, the hybrid approach will not necessarily lead to less recommendation accuracy, which initially contradicts our common sense but has been validated subsequently by experiments on Netflix dataset, which is the largest dataset in our experiments. Besides, the integration of DP and RP does not affect their original trend of the relation between accuracy and privacy, which is also appealing as we still have control of the balance between accuracy and privacy in the hybrid approach.

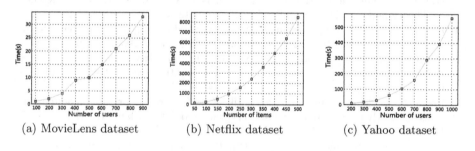

(a) MovieLens dataset   (b) Netflix dataset   (c) Yahoo dataset

Fig. 8. Efficiency of the hybrid approach

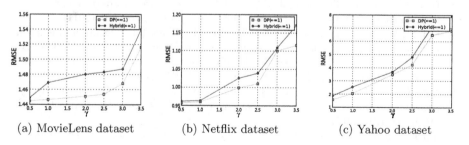

(a) MovieLens dataset   (b) Netflix dataset   (c) Yahoo dataset

Fig. 9. RMSE of the hybrid approach and DP over perturbed data

5 Conclusion

We have presented a hybrid approach for privacy-preserving recommender systems by combining randomized perturbation (RP) and differential privacy (DP), which is more privacy-friendly than existing works as the user's private data are protected by randomized perturbation and no one can infer any single user's data from the normal recommendation output thanks to differential privacy. We have theoretically shown the noise added by RP has limited effect on recommendation accuracy and the noise added by DP can be well controlled based on the sensitivity analysis of functions on the perturbed data. We have conducted extensive experiments on real datasets and concluded that the combination of DP and RP is feasible not only in theory but also in practice.

Acknowledgment. This work was done while the first author was a visiting student at King Abdullah University of Science and Technology (KAUST). Research reported in this publication was partially supported by KAUST and Natural Science Foundation of China (Grant Nos. 61572336, 61572335, 61632016, 61402313).

References

1. Badsha, S., Yi, X., Khalil, I.: A practical privacy-preserving recommender system. Data Sci. Eng. **1**(3), 161–177 (2016)
2. Bell, R.M., Koren, Y.: Scalable collaborative filtering with jointly derived neighborhood interpolation weights. In: ICDM, pp. 43–52 (2007)
3. Dwork, C.: Differential privacy: a survey of results. In: International Conference on Theory and Applications of Models of Computation, pp. 1–19 (2008)
4. Dwork, C., McSherry, F., Nissim, K., Smith, A.: Calibrating noise to sensitivity in private data analysis. In: Halevi, S., Rabin, T. (eds.) TCC 2006. LNCS, vol. 3876, pp. 265–284. Springer, Heidelberg (2006). doi:10.1007/11681878_14
5. Elgamal, T.: A public key cryptosystem and a signature scheme based on discrete logarithms. IEEE Trans. Inf. Theory **31**(4), 469–472 (1985)
6. Erkin, Z., Veugen, T., Toft, T., Lagendijk, R.L.: Generating private recommendations efficiently using homomorphic encryption and data packing. IEEE Trans. Inf. Forensics Secur. **7**(3), 1053–1066 (2012)
7. Guerraoui, R., Kermarrec, A.M., Patra, R., Taziki, M.: D 2 p: distance-based differential privacy in recommenders. VLDB **8**(8), 862–873 (2015)
8. Huang, Y., Evans, D., Katz, J., Malka, L.: Faster secure two-party computation using garbled circuits. In: USENIX Security Symposium, vol. 201 (2011)
9. Liu, S., Liu, A., Liu, G., Li, Z., Xu, J., Zhao, P., Zhao, L.: A secure and efficient framework for privacy preserving social recommendation. In: Cheng, R., Cui, B., Zhang, Z., Cai, R., Xu, J. (eds.) APWeb 2015. LNCS, vol. 9313, pp. 781–792. Springer, Cham (2015). doi:10.1007/978-3-319-25255-1_64
10. Ma, X., Li, H., Ma, J., Jiang, Q., Gao, S., Xi, N., Lu, D.: Applet: a privacy-preserving framework for location-aware recommender system. Sci. China Inf. Sci. **60**(9), 092101 (2017)
11. Machanavajjhala, A., Korolova, A., Sarma, A.D.: Personalized social recommendations: accurate or private. VLDB **4**(7), 440–450 (2011)

12. McSherry, F., Mironov, I.: Differentially private recommender systems: building privacy into the netflix prize contenders. In: KDD, pp. 627–636 (2009)
13. Nikolaenko, V., Ioannidis, S., Weinsberg, U., Joye, M., Taft, N., Boneh, D.: Privacy-preserving matrix factorization. In: CCS, pp. 801–812 (2013)
14. Paillier, P.: Public-key cryptosystems based on composite degree residuosity classes. In: Stern, J. (ed.) EUROCRYPT 1999. LNCS, vol. 1592, pp. 223–238. Springer, Heidelberg (1999). doi:10.1007/3-540-48910-X_16
15. Polat, H., Du, W.: Privacy-preserving collaborative filtering using randomized perturbation techniques. In: ICDM, pp. 625–628 (2003)
16. Yao, A.C.C.: How to generate and exchange secrets. In: FOCS, pp. 162–167 (1986)
17. Zhan, J., Hsieh, C.L., Wang, I.C., Hsu, T.S., Liau, C.J., Wang, D.W.: Privacy-preserving collaborative recommender systems. IEEE Trans. Syst. Man Cybern. Part C (Appl. Rev.) **40**(4), 472–476 (2010)
18. Zhang, S., Ford, J., Makedon, F.: Deriving private information from randomly perturbed ratings. In: SDM, pp. 59–69 (2006)

Supporting Cost-Efficient Multi-tenant Database Services with Service Level Objectives (SLOs)

Yifeng Luo[1], Junshi Guo[1], Jiaye Zhu[1], Jihong Guan[2], and Shuigeng Zhou[1(✉)]

[1] Shanghai Key Lab of Intelligent Information Processing,
School of Computer Science, Fudan University, Shanghai, China
{luoyf,jsguo14,zhujy14,sgzhou}@fudan.edu.cn
[2] Department of Computer Science and Technology, Tongji University,
Shanghai, China
jhguan@tongji.edu.cn

Abstract. Quality of Service (QoS) is at the core of the vision of Database as a Service (DBaaS). Traditional approaches in DBaaS often reserve computation resources (e.g. CPU and memory) to satiate tenants' QoS guarantees under various circumstances, which inevitably results in poor resource utilization, as the tenants' actual workloads are usually below their expectations described by their Service Level Objectives (SLOs). In this paper, we propose a novel scheme FrugalDB to enhance resource utilization for DBaaS systems with QoS guarantees. FrugalDB accommodates two independent database engines, an in-memory engine for heavy workloads with tight SLOs, and a disk-based engine for light workloads with loose SLOs. By allocating each tenant' workload to an appropriate engine via workload migration, this dual-engine scheme can substantially save computation resources, and thus consolidate more tenants on a single database server. FrugalDB tries to minimize workload migration cost incurred in moving workloads between the two engines. By an effective workload estimation method and an efficient migration schedule algorithm, FrugalDB responds quickly to workload changes and executes workload migrations with minimal overhead. We evaluate FrugalDB with extensive experiments, which show that it achieves high tenant consolidation rate yet with few performance SLO violations.

Keywords: Cloud computing · Database-as-a-Service · Multi-tenancy · Workload consolidation · Workload migration

1 Introduction

Service providers of DBaaS [1,8,9,11,13,21] typically consolidate multiple tenants into the same hardware/software stack to reduce operational cost. When tenants require QoS guarantees, service providers usually reserve enough resources for tenants according to their performance SLOs, so that all tenants'

S. Candan et al. (Eds.): DASFAA 2017, Part I, LNCS 10177, pp. 592–606, 2017.
DOI: 10.1007/978-3-319-55753-3_37

workloads can be handled properly, no matter what the real workload condition is. However, service providers can only achieve modest resource utilization via resource reservation, as multi-tenancy tends to possess low overall tenant activeness (OTA) [22,23], thus the actual resource demands often fall below the amount of resources that tenants subscribe. To improve resource utilization, service providers usually consolidates more tenants onto a server than it can serve via resource reservation. But it is challenging to guarantee that enough resources are dispatched for each tenant. Take memory buffer resource as an example, when massive memory is consumed on light workloads, there would not be enough memory for heavy workloads.

Targeting to boost resource utilization for multi-tenancy, where a large number of small tenants with changing workloads are served with QoS guarantees, this paper proposes a cost-efficient scheme FrugalDB. Here, small tenants refer to theses that have relatively light-weight databases with tens of or a few hundred megabytes data. For tenants with gigabyte-scale databases and strict QoS requirements as well as stable workloads, it is more appropriate to encapsulate them in independent properly-configured virtual machines (VMs), which is out of the scope of this paper. One typical application of our work is CRM SaaS aiming at small/medium enterprises, where databases are usually at the size of tens of megabytes with varying workloads [9]. However, it is difficult to satiate the performance SLOs of tenants with similar database schemas via shared-table DBaaS workload consolidation [21].

The proposed FrugalDB scheme employs two independent database engines in the same framework: an in-memory engine for heavy workloads, and a disk-based engine for light workloads. Here, heavy workloads refer to workloads approaching the tenants' tight SLOs, meaning that plentiful amounts of resources will be consumed to process such workloads, while light workload indicates the workloads of tenants with loose SLOs, and falling far behind the tenants' tight SLOs, meaning that only moderate amount of resources will be consumed to process such workloads. By assigning each tenant's workload to an appropriate engine for query processing via workload migration, FrgualDB can consolidate more tenants with performance SLOs on a single database server.

Though this paper focuses on a single server, the proposed method can be applied to multiple-node computing environments. Concretely, for an existing DBaaS system, a FrugalDB server is first launched to handle the workloads in the system. When workloads outnumber the capacity of the first FrugalDB and SLO-violations begin to occur, a new FrugalDB server will be initiated to handle the workloads jointly with the first one, and so on. Each FrugalDB server functions independently according to the workload consolidation policy proposed in this paper. As for how to coordinate multiple FrugalDB servers in a system is left for future work. Major contributions of this paper include: (1) proposing the FrugalDB approach with a workload migration mechanism to handle mixed workloads with performance SLOs; (2) formulating and solving the workload migration problem as an optimization problem; (3) implementing a prototype and evaluating its performance with extensive experiments.

The rest of this paper is organized as follows: we first introduce the proposed scheme in Sect. 2, and formally define the workload migration problem and present the algorithms in Sect. 3. We provide experimental results in Sect. 4, and review the related work in Sect. 5. Finally, we conclude the paper in Sect. 6.

2 The Framework

Computation resources are usually heavily under-utilized, while memory and disk bandwidth are scarce resources for OLTP database systems. So the key for satiating performance SLOs of an OLTP application is whether its working data set can be completely hosted in memory. If so, the system can process most database operations efficiently without accessing the underlying disks, and thus satiate much higher performance SLOs; Otherwise, the system has to frequently visit disk pages, and achieves deteriorated overall performance.

With this in mind, we propose the FrugalDB scheme. Our basic idea is to seamlessly integrate a disk-based database engine and an in-memory database engine, to dynamically separate heavy workloads from light workloads, and guarantee heavy workloads with enough critical memory resources. Here, we define a *workload burst* as the circumstance where the overall workloads submitted to the system exceed the processing capacity of the disk-based database engine. When a workload burst is to happen, FrugalDB migrates heavy workloads to the in-memory database engine; when workloads ebb, workloads on the in-memory database engine are migrated back to the disk-based database engine. Thus the tremendous data-serving pressure caused by heavy workloads can be relieved from the disk-based database engine. In such a way, FrugalDB can get through the bursty workload periods and alleviate or even avoid SLO violations. Figure 1 illustrates the following major functional modules of FrugalDB:

1. *Access Controller* (AC): this module routes tenants' requests to the right database engine according to the optimized workload assignments generated by the *Workload Migration Engine* (WME), and then returns the request results to the tenants. Besides, it keeps log records of all tenants' requests.
2. *Log Analyzer* (LA): this module analyzes tenants' request logs to extract tenants' workload characteristics and to generate the workload description for each tenant.
3. *Workload Monitor* (WM): this module benchmarks the processing capacity of the disk-based database engine with various configurations of write request percentage and concurrent tenant number, and then leverages the benchmarking results to make predictions on whether a workload burst will occur in the next time interval, and initiates necessary workload migrations.
4. *Workload Migration Engine* (WME): this module executes workload migration plans based on workload information and a workload prediction model. It consists of three submodules: *Migration Benefit Evaluator* (MBE), *Data Migrator* (DM) and *Data Reloader* (DR). When a workload migration process needs initiating, *Migration Benefit Evaluator* evaluates the workload migration benefits of all active tenants, and computes the optimized workload

assignments by combining the tenants' workload descriptions, current workload metrics and tenants' SLOs. Then, *Data Migrator* migrates data of the selected tenants from the disk-based database into the in-memory database. When the workload burst fades out, *Data Reloader* reloads data into the disk-based database to release occupied memory resources of the in-memory database engine.

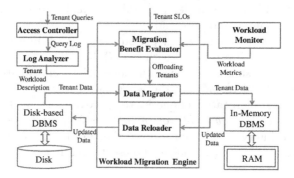

Fig. 1. The FrugalDB architecture.

As data migration takes time, each tenant to be migrated keeps two copies of its data in the system during migration. Therefore, we have to handle data requests properly during workload migration, especially when the requests involve data changes, including insert, update and delete operations, to make sure that data consistency is not damaged. When FrugalDB migrates a tenant's data into the in-memory database, FrugalDB collects all rows that have been changed during migration, and replays all operations recorded on these collected rows in the in-memory database. When FrugalDB migrates back a tenant's workload into the disk-based database, operations that change the tenant's data are executed on the disk-based database.

3   Migration Scheduling

To optimize the workload migration process, we build a proper workload model and solve the problem of scheduling workload migration between the two database engines as an optimization problem, to determine which active tenants' workloads should be migrated with optimal migration cost.

3.1   Workload Modeling

Here, we have a basic assumption on tenants' workloads: *a tenant follows some repeatable and predictable pattern to generate its workload.* This assumption is reasonable because a tenant tends to perform similar business logic periodically, and thus follows some statistical pattern to generate workloads conforming to

the tenant's business logics. Existing works [5,7,16] validated this assumption
by experiments with real-life workloads, and revealed that tenants' workloads
exhibit obvious patterns. We can find the patterns by analyzing historic data of
workloads. For this end, we first split the time span of historic data into a number
of *periods*, which indicate the time cycles of workload (e.g. one day/week/month)
and can be inferred by mining the historic data. We then further divide each
period into a series of equal-size time slots, which are called *intervals* and used as
the finest time granularity for modeling workload. Finally, we estimate workload
distribution in each interval of a period.

Let the size of a period be T_w, a period is split to k intervals of size t_w,
which are denoted as $\{t_i|i=1,2,...,k\}$ with $k = \lceil T_w/t_w \rceil$. The workload of a tenant
consists of the queries submitted at various time points, and we use the average
workload in an interval to indicate the workload of this interval. That is, we
employ interval-wise aggregation of queries to approximate tenant workload. We
call this method *piecewise aggregate approximation*. Concretely, we first sums up
the queries in an interval and then average the sum by the interval duration.
Figure 2 shows an example of workload representation in a period.

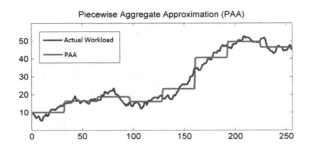

Fig. 2. Workload representation by PAA. Here, the whole time span is a period, which
is split to eight intervals. The blue curve shows the actual workload, and the red line
segments indicate the average workloads in different intervals. (Color figure online)

For each tenant, we estimate its workload distribution by considering historic
workload data of a number of periods. For simplicity, we discretize the workload
amount of each interval into l *subranges* or *levels*, denoted as S_1, ..., S_l. We
estimate the distribution of workload over different subranges. Formally, let w_{ij}
be the workload in the ith interval of the jth period ($1 \leq i \leq k$, $1 \leq j \leq n$), the
probability of a tenant's workload in the ith interval falling in subrange $S_m(1 \leq m \leq l)$ is estimated by $Pr(i,m) = \frac{Fr(S_m)}{n}$, where $Fr(S_m) = \sum_{j=1}^{n} H(w_{ij})$, and
$H(w_{ij}) = 1$ if w_{ij} falls in S_m, and 0 otherwise. For each tenant, we estimate
$Pr(i,m)$ for $1 \leq i \leq k$ and $1 \leq m \leq l$.

Figure 3 shows an example of workload modeling. Here, each period contains
five intervals, and the workload in each interval is discretized into three subranges, which are denoted conceptually as *Low*, *Middle* and *High* respectively.
The workload distribution of the fifth interval is 10%, 30% and 60% for the *Low*,
Middle and *High* levels, respectively.

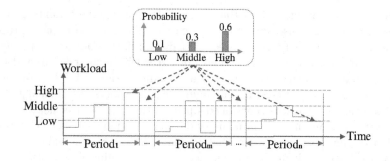

Fig. 3. An example of workload modeling.

3.2 Problem Formulation and Optimization

Here, we formally present the workload migration problem. Let U denote the set of all tenants, and each tenant $u_i \in U$ is described as a triple $\{O_i, S_i, W_{i,t}\}$, where O_i denotes u_i's performance SLO (namely the maximum workload pressure that u_i will generate), S_i denotes u_i's data size, and $W_{i,t}$ denotes u_i's workload pressure at time t. The set of all active tenants at time t is denoted as UA_t, and we assume that the ratio of UA_t over all tenants is nearly a constant. We denote the average percentage of write requests over all requests submitted to the DBaaS system at time t as $\overline{W_t}$, the maximum request processing capacity of the disk-based database and the in-memory database at time t as $WDISK_t$ and $WMEM_t$ respectively. We also denote $MEMLIMIT$ as the maximum size of memory configured for data storage in the in-memory database engine.

When a workload burst occurs at time t, our goal is to assign each active tenant to exactly one of the three disjoint tenant sets: $UDISK_t$, $UMEM_t$ and $UVIO_t$, so as to minimize the total performance SLO violations. Here, $UDISK_t$ is the set of tenants whose performance SLOs will be met by the disk-based engine at time t, $UMEM_t$ is the set of tenants whose performance SLOs will be met by the in-memory engine, and $UVIO_t$ is the set of tenants whose data serving requests will be rejected. Our problem is formulated as

$$\sum_{i \in UVIO_t} W_{i,t}, \tag{1}$$

subjected to:

$$\sum_{i \in UDISK_t} W_{i,t} \leq WDISK_t, \tag{2}$$

$$\sum_{i \in UMEM_t} S_i \leq MEMLIMIT, \tag{3}$$

$$\sum_{i \in UMEM_t} W_{i,t} \leq WMEM_t. \tag{4}$$

Constraint (2) above guarantees that workloads of all active tenants assigned to the disk-based database engine do not exceed its maximum request processing capacity; Constraint (3) guarantees that overall data of all tenants assigned to the in-memory database engine does not exceed its memory size, and Constraint (4) guarantees that workloads of all tenants assigned to the in-memory database engine do not exceed its maximum request processing capacity.

Our solution to the fore-mentioned optimization problem is heuristic. For simplicity, we assume that the in-memory database engine can efficiently process all workloads that its memory can accommodate. This is reasonable because the performance bottleneck of the in-memory database engine lies in its memory capacity, rather than its query processing capacity. Thus, Constraint (4) can be ignored as it surely holds. Note that the total workload at time t is fixed, we denote it as follows:

$$WTOT_t = \sum_{i \in UDISK_t} W_{i,t} + \sum_{i \in UMEM_t} W_{i,t} + \sum_{i \in UVIO_t} W_{i,t}. \qquad (5)$$

Consequently, our problem turns to maximizing the sum of the first part and the second part of Eq. (5), subjected to Constraints (2) and (3). Then we try to maximize the second part of Eq. (5) subjected to Constraints (3). As we can see later, it can be formulated as a knapsack problem. Now, we assume the second part is maximized, and maximizing the first part means minimizing the third part. Let's consider the tenants not belonging to $UMEM_t$. We need to decide which tenants should be assigned to $UDISK_t$ subjected to Constraint (2), so that the object function Eq. (1) is minimized. For this end, we assign all tenants not belonging to $UMEM_t$ to $UDISK_t$, and then selectively assign tenants to $UVIO_t$ until Constraint (2) is satisfied. We first choose tenants with highest workloads and then choose tenants with lower workloads, based on greedy strategy. Then our remaining work is to maximize the second part of Eq. (5) subjected to Constraint (3), which is a classical 0/1 knapsack problem.

We define function $F(n, m)$ as the maximum of the second term in Eq. (5), when we consider the first n tenants in UA_t, subjected to the following condition:

$$\sum_{i \in UMEM_t} S_i \leq m. \qquad (6)$$

Constraint (6) is different from Constraint (3), where $MEMLIMIT$ is replaced by the value m. Consequently, $F(N, MEMLIMIT)$ is the optimal value we are to compute, assuming there are N tenants in total. Obviously,

$$F(0, m) = 0, 0 \leq m \leq MEMLIMIT. \qquad (7)$$

For $n > 0$, we have:

$$F(n, m) = \begin{cases} min\{F(n-1, m), F(n-1, m - S_n) + W_{n,t}\}, \\ \qquad\qquad\qquad S_n \leq m \leq MEMLIMIT \\ F(n-1, m), 0 \leq m < S_n \end{cases} \qquad (8)$$

We also define a function $P(n, m)$ as:

$$P(n, m) = \begin{cases} -1, & n = 0 \\ 0, & (n > 0) \wedge (m < S_n \vee F(n, m) \neq \\ & \qquad F(n - 1, m - S_n) + W_{n,t}) \\ 1, & n > 0 \wedge m \geq S_n \wedge F(n, m) = \\ & \qquad F(n - 1, m - S_n) + W_{n,t} \end{cases} \qquad (9)$$

We iterate every possible value n from 0 to N and every possible value m from 0 to $MEMLIMIT$ to compute the function F. Also, we use function P to find out which tenants should be assigned to $UMEM_t$, where $P(n, m) = 1$ means the nth tenant should be assigned to $UMEM_t$. If $P(n, m) = 1$, then we check $P(n - 1, m - S_n)$ to see whether $(n - 1)$th tenant should be assigned to $UMEM_t$; otherwise we check $P(n - 1, m)$. We continue this process until $n = 0$, and then we will figure out the $UMEM_t$ set, just as described in Algorithm 1.

Algorithm 1. Maximizing workloads in the in-memory database engine

Input: UA_t: The set of active tenants

$MEMLIMIT$: the memory size for data storage in the in-memory database engine

Output: $UMEM_t$: tenants assigned to the in-memory database engine
1: $UMEM_t \leftarrow \{\}$
2: **for** $m \leftarrow 0$ to $MEMLIMIT$ **do**
3:     $F(0, m) \leftarrow 0$
4:     $P(0, m) \leftarrow -1$
5: **end for**
6: **for** $n \leftarrow 1$ to N **do** /*the 0/1 knapsack problem solver*/
7:     **for** $m \leftarrow 0$ to $MEMLIMIT$ **do**
8:         **if** $S_n \leq m$ **and** $F(n - 1, m) \leq F(n - 1, m - S_n) + W_{n,t}$ **then**
9:             $F(n, m) \leftarrow F(n - 1, m - S_n) + W_{n,t}$
10:            $P(n, m) \leftarrow 1$
11:         **else**
12:            $F(n, m) \leftarrow F(n - 1, m)$
13:            $P(n, m) \leftarrow 0$
14:         **end if**
15:     **end for**
16: **end for**
17: $id \leftarrow N$
18: $capacity \leftarrow MEMLIMIT$
19: **while** $id \neq 0$ **do**
20:     **if** $P(id, capacity) = 1$ **then**
21:         $UMEM_t \leftarrow UMEM_t \cup \{id_{th}\ \text{tenant}\}$
22:         $capacity \leftarrow capacity - S_{id}$
23:     **end if**
24:     $id \leftarrow id - 1$
25: **end while**

---

4   Performance Evaluation

4.1   Experimental Setting

We conduct experiments on two interconnected servers. As there are few real-life multi-tenant OLTP workloads available, we generate synthetic workloads based on YCSB [4] to benchmark the performance of FrugalDB and corresponding baseline settings. We extract all read and update operations contained in YCSB queries, and leverage them to generate the workloads, where each tenant request contains only one extracted operation. The aggregate benchmark workloads are composed of tenants with corresponding performance SLOs, different data sizes and different read/write percentages, as Table 1 shows. The value of performance SLO specifies the request upper bound that a tenant can generate per minute. For simplicity, we set three levels of tenant performance SLOs: Low, Medium and High, and set 5–50–500 as the Low-Medium-High combination to test FrugalDB, representing scenarios where tenants have diverse performance SLOs.

Table 1. Workload statistics.

Feature	Setting		
SLO	L(30%)	M(50%)	H(20%)
TPC-C data size (MB)	6.9(50%)	16.3(30%)	35.2(20%)
YCSB data size (MB)	10(50%)	20(30%)	29(20%)
Update/Read percentage	U(20%)	R(80%)	—

Note that tenants are not simultaneously active in real-world multi-tenant environments, we set the percentage of active tenants over all tenants to 30% by default, and randomly choose a part of active/inactive tenants and let them change their inactive/active status, but maintain the overall number of active tenants unchanged. In each interval, we first assign workloads to active tenants by following the Poisson distribution, requiring that the workload assigned to a tenant does not exceed its performance SLO, and then maintain the workload throughout the interval. If the assigned workload exceeds a tenant's performance SLO, we set the tenant's workload to its performance SLO. Besides, we set the third, the fourth and the sixth out of the seven intervals as workload-bursty intervals, and in these bursty intervals all active tenants' workloads are set to their performance SLOs. Tenants' workloads in non-workload-bursty intervals are set proportionally to their performance SLOs, and the proportion is set to 20% in our experiments. We set the default memory storage capacity of the memory-based VoltDB database to 2000 MB, which determines the number of tenants whose workloads can be migrated into the VoltDB database.

4.2   Results and Analysis

We compare FrugalDB's performance on satisfying tenants' performance SLOs and responsiveness to tenant queries with the baseline settings. We define whether or not a tenant's performance SLO is satisfied as follows: If a tenant's performance SLO is A, that is, the tenant can get at most A queries processed by the system in a minute. Now the tenant actually issues B queries but gets only C query responses in a minute, then we say that the tenant has $min(A, B)$-C queries not being fulfilled. If $min(A, B)$-C is 0, we say that this tenant's performance SLO is satisfied in a minute; Otherwise, we say this tenant's performance SLO is violated and $min(A, B)$-C queries are not finished for this tenant in a minute. We report the number of tenants whose performance SLOs are violated in each minute as the metric for evaluating FrugalDB's performance on satisfying tenants' performance SLOs, and the percentage of query latency as the metric for evaluating FrugalDB's responsiveness to tenant queries.

Comparison with Static-SLO. We firstly compare FrugalDB by simulation with the method proposed by Lang et al. [12], which aims to minimize the number of servers with different hardware configurations, to satisfy the performance SLOs of a group of tenants. They considered only tenants' SLOs and did not take dynamic workloads into consideration, so their method belongs to static workload consolidation with SLO guarantees, and we call it Static-SLO.

As FrugalDB combines disk-based and in-memory databases, we consider that we have both disk-based database servers and in-memory database servers, and we implement the Static-SLO scheme based on these two types of servers. We first benchmark the tenant consolidation that FrugalDB can yield, and then we compute the minimal number of the two types of database servers that the Static-SLO scheme needs to satiate these tenants' SLOs. According to the Static-SLO method, we need at least **three** MySQL and **two** VoltDB servers to satiate all tenants' SLOs, while FrugalDB needs only one server. We mainly benchmark the query latency achieved on FrugalDB and Static-SLO, and the results are presented in Fig. 4. We can see that FrugalDB and Static-SLO provide similar latency for 85% queries, while FrugalDB provides much higher yet tolerable latency for the other 15% queries. The major reason is that FrugalDB consolidates far more tenants onto the same server, and most tenants' queries are answered by the disk-based database in FrugalDB with relatively higher latencies.

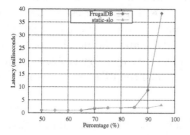

Fig. 4. Query latency.

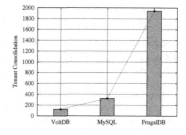

Fig. 5. Tenant consolidation.

Tenant Consolidation. We measure the number of tenants that can be served without performance SLO violations, namely the capacity for tenant consolidation, that different multi-tenant implementations can support for YCSB workloads. We compare FrugalDB with the implementation purely on VoltDB and the implementation purely on MySQL of the resource-reservation scheme, based on the non-deterministic workload model. We evaluate the number of tenants whose workloads amount to their performance SLOs that the MySQL database can support, and take this number as the capacity of tenant consolidation that the MySQL database can provide under the resource reservation scheme.

Figure 5 presents the results of the tenant consolidation capacity obtained on various multi-tenant database implementations, and we can see that: the implementation purely based on the VoltDB database provides the lowest tenant consolidation; the resource-reservation implementation purely based on the MySQL database provides two times higher tenant consolidation than the VoltDB-based implementation; while FrugalDB could achieve almost ten times higher tenant consolidation than the VoltDB-based implementation, which is four to five times higher than the resource-reservation MySQL-based implementation. The VoltDB-based implementation provides the lowest tenant consolidation, as the memory storage capacity bottlenecks its tenant consolidation capacity. The resource-reservation MySQL-based implementation is bottlenecked by the storage subsystem. FrugalDB could provide much higher tenant consolidation, mainly because of two reasons: (1) massive numbers of non-active tenants reside in the MySQL database to share database resources; (2) most workloads of highly active tenants are guaranteed with enough crucial memory resources.

Satisfaction of Tenants' SLOs. We benchmark the satisfaction of tenants' performance SLOs obtained both on the multi-tenant database implementation purely based on the MySQL database and on FrugalDB with various memory storage capacity. As we need to guarantee that comparison experiments between FrugalDB and the MySQL-based implementation are performed with the same configurations, we do not follow the resource-reservation fashion to benchmark the MySQL-based implementation, which could only support far less tenants than FrugalDB, and we follow the best-effort fashion without any control mechanism for guaranteeing tenants' performance SLOs. Figure 6 present the evaluation results. We can see that the number of tenants whose performance SLOs are violated on MySQL far exceeds that on FrugalDB, and the number of tenants whose SLOs are violated on FrugalDB decreases as its memory storage capacity increases, while all tenants' SLOs are satisfied on FrugalDB when its memory storage capacity is set to 2000 MB.

Besides, we benchmark the satisfaction of tenants' performance SLOs with various tenant active ratios, and the evaluation results are presented in Fig. 7. We can see that: on the MySQL-based implementation, the number of tenants whose SLOs are violated increases by 120–140 as tenants active ratio increases by 5%; on FrugalDB, all tenants' SLOs are satisfied until tenant active ratio increases to 30%, when a moderate portion of tenants' SLOs could not be guaranteed as the

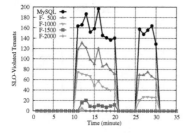
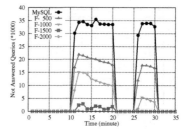

Fig. 6. Satisfaction of tenants' performance SLOs for various implementations.

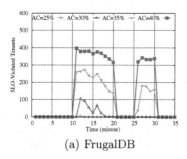
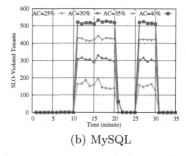

(a) FrugalDB   (b) MySQL

Fig. 7. Satisfaction of tenants' performance SLOs, with various tenant active ratios.

overall number of active tenants exceeds FrugalDB's processing capacity, and the number of tenants whose SLOs are violated increases to around 380, when the tenant active ratio increases to 40%.

Memory Consumption and Data Migration Cost. At last, we report statistics about memory consumption of the VoltDB database contained in FrugalDB and the temporal cost of migrating data into/from the VoltDB database from/into the MySQL database throughout the whole experiments, and Figs. 8 and 9 present the evaluation results. We can see that: the aggregate memory consumption of the VoltDB database increases rapidly as its memory storage capacity increases, and increasing the memory storage capacity by 500 MB makes the aggregate memory consumption increase 1000 MB–1500 MB, because VoltDB needs more memory buffer resources to process increased workloads; the aggregate memory consumption decreases after the workload bursts, because there are few tenants still residing in the VoltDB database, and memory resources consumed could be released to the operating system; the temporal cost of loading tenants' data into VoltDB of course increases linearly as its memory storage capacity increases, and it takes about 130 s to load 500 MB tenants' data, and this is the reason why we should initiate the workload migration process in advance; while migrating updated data from VoltDB into MySQL for data persistence only takes 1/4 of the time spent on data loading, as only a part of the whole data is updated and needs to be persisted into the MySQL database.

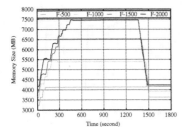

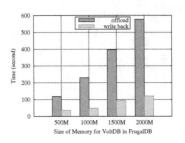

Fig. 8. Memory consumption.   **Fig. 9.** Cost of data migration.

5   Related Work

There have been extensive research efforts devoted to improving services for multi-tenancy to support mixed workloads in cloud environments w/o performance guarantees. Previous works on DBaaS without performance guarantees were mainly devoted to achieving extremely high levels of consolidation for shared-schema multi-tenancy. Aulbach et al. [2] proposed Chunk Folding, which vertically partitions tenant logical tables into chunks and folds them into shared physical tables, which are joined on request for query processing. Hui et al. [10] proposed MStore to improve scalability and efficiency of consolidating tuples from different tenants into shared tables. Schiller et al. [17] implemented tenant-aware multi-tenancy in databases to gain flexibility on database extensibility. These works seldom consider tenants' resource demands on the per-tenant basis.

Some existing works on DBaaS with performance guarantees mainly focused on resource provisioning in virtualized cloud environments [20,24], configuring, provisioning and consolidating virtual machines (VMs) to avoid resource over-provisioning. Soror et al. [19] proposed a method to automatically configure VMs for different database workloads consolidated on the same hardware; Soundararajan et al. [20] presented a system to determine resource partitions for VMs; Shen et al. [18] introduced CloudScale for online predicting resource demands of VM-based applications, and automating fine-grained elastic resource scaling, as well as resolving scaling conflicts via VM migration. Cecchet et al. [3] presented Dolly that explores VM cloning techniques to spawn database replicas to address the challenges of provisioning shared-nothing replicated databases in clouds. Nathuji et al. [15] introduced Q-Clouds as a QoS-aware control framework to dynamically tune resource allocations to mitigate performance interference between applications consolidated on the same server. The VM-based consolidation approaches could achieve only moderate consolidation, and thus are applicable only when hard isolation between databases is more important than cost/performance.

Instead of packing resources in the form of VMs, Narasayya et al. [6,14,26] presented SQLVM to provide isolated performance for multi-tenant DBaaS by reserving resources specified by tenants within database server process. Curino et al. [5] and Lang et al. [12] presented mechanisms for consolidating multiple database workloads into the same database process, aiming to minimize the number of servers needed to satisfy the performance goals of a large number of tenants. All of these approaches followed the resource-reservation fashion, which are suitable for scenarios with relatively stable workloads.

Our work focuses on properly consolidating numerous tenants with highly-dynamic workloads on the same database server, by dynamically migrating shortly-sustaining heavy workloads to the in-memory database engine back and forth. Some existing methods can complement our method to better handle some issues, for example, employing admission control [25] and workload scheduling [23] mechanisms to avoid workload overloads.

6   Conclusion

Targeting scenarios where a moderate portion of a huge number of tenants are simultaneously active, we propose the FrugalDB DBaaS approach that includes a dual-engine framework, a workload prediction model and the corresponding workload migration scheduling algorithm. Our experiments show that it can achieve impressively high tenant consolidation with few SLO violations. To the best of our knowledge, this is the first work that tries to serve a massive number of tenants by employing in-memory database techniques for multi-tenancy.

Acknowledgement. This work was supported by National Natural Science Foundation of China (No. 61572141 and No. 61373036), the Key Projects of Fundamental Research Program of Shanghai Municipal Commission of Science and Technology (No. 14JC1400300), the Program of Shanghai Subject Chief Scientist (No. 15XD1503600), and Project funded by China Postdoctoral Science Foundation.

References

1. Agrawal, D., Abbadi, A., Emekci, F., Metwally, A.: Database management as a service: challenges and opportunities. In: Proceedings of ICDE 2009, pp. 1709–1716 (2009)
2. Aulbach, S., Grust, T., Jacobs, D., Kemper, A., Rittinger, J.: Multi-tenant database for software as a service: schema-mapping techniques. In: Proceedings of SIGMOD 2008, pp. 1195–1206 (2008)
3. Cecchet, E., Singh, R., Sharma, U., Shenoy, P.: Dolly: Virtualization-driven database provisioning for the cloud. In: Proceedings of VEE 2011, pp. 51–62 (2011)
4. Cooper, B.F., Silberstein, A., Tam, E., Ramakrishnan, R., Sears, R.: Benchmarking cloud serving systems with YCSB. In: Proceedings of SOCC 2010, pp. 143–154 (2010)
5. Curino, C., Jones, E.P., Madden, S., Balakrishnan, H.: Workload-aware database monitoring and consolidation. In: Proceedings of SIGMOD 2011, pp. 832–843 (2011)
6. Das, S., Narasayya, V.R., Li, F., Syamala, M.: CPU sharing techniques for performance isolation in multi-tenant relational database-as-a-service. In: Proceedings of VLDB 2014, vol. 7, no. 1, pp. 37–48 (2014)

7. Elmore, A.J., Das, S., Pucher, A., Agrawal, D., Abbadi, A.E., Yan, X.: Characterizing tenant behavior for placement and crisis mitigation in multitenant DBMSS. In: Proceedings of SIGMOD 2013, pp. 517–528 (2013)
8. Hacigumus, H., Iyer, B., Mehrotra, S.: Providing database as a service. In: Proceedings of ICDE 2002, pp. 29–39 (2002)
9. http://www.salesforce.com
10. Hui, M., Jiang, D., Li, G., Zhou, Y.: Supporting database applications as a service. In: Proceedings of ICDE 2009, pp. 832–843 (2009)
11. Jacobs, D., Aulbach, S.: Ruminations on multi-tenant databases. In: Proceedings of BTW 2007, pp. 514–521 (2007)
12. Lang, W., Shankar, S., Patel, J.M., Kalhan, A.: Towards multi-tenant performance SLOs. In: Proceedings of ICDE 2012, pp. 702–713 (2012)
13. Lehner, W., Sattler, K.: Database as a service (DBaaS). In: Proceedings of ICDE 2010, pp. 1216–1217 (2010)
14. Narasayya, V., Das, S., Syamala, M., Chandramouli, B., Chaudhuri, S.: SQLVM: performance isolation in multi-tenant relational database-as-a-service. In: Proceedings of CIDR 2013 (2013)
15. Nathuji, R., Kansal, A., Ghaffarkhah, A.: Q-Clouds: managing performance interference effects for QoS-aware clouds. In: Proceedings of EuroSys 2010, pp. 237–250 (2010)
16. Schaffner, J., Januschowski, T.: Realistic tenant traces for enterprise DBaaS. In: Workshops Proceedings of ICDE 2013, pp. 29–35 (2013)
17. Schiller, O., Schiller, B., Brodt, A., Mitschang, B.: Native support of multi-tenancy in RDBMS for software as a service. In: Proceedings of EDBT 2011, pp. 117–128 (2011)
18. Shen, Z., Subbiah, S., Gu, X., Wilkes, J.: Cloudscale: elastic resource scaling for multi-tenant cloud systems. In: Proceedings of SOCC 2011 (2011)
19. Soror, A.A., Minhas, U.F., Aboulnaga, A., Salem, K., Kokosielis, P., Kamath, S.: Automatic virtual machine configuration for database workloads. In: Proceedings of SIGMOD 2008, pp. 953–966 (2008)
20. Soundararajan, G., Lupei, D., Ghanbari, S., Popescu, A.D., Chen, J., Amza, C.: Dynamic resource allocation for database servers running on virtual storage. In: Proceedings of FAST 2012, pp. 71–84 (2012)
21. Weissman, C., Bobrowski, S.: The design of the force.com multitenant internet application development platform. In: Proceedings of SIGMOD 2009, pp. 889–896 (2009)
22. Reinwald, B.: Multitenancy. UW MSR Summer Institute (2010)
23. Wong, P., He, Z., Lo, E.: Parallel analytics as a service. In: Proceedings of SIGMOD 2013, pp. 25–36 (2013)
24. Xiong, P., Chi, Y., Zhu, S., Moon, H.J., Pu, C., Hacigms, H.: Intelligent management of virtualized resources for database systems in cloud environment. In: Proceedings of ICDE 2011, pp. 87–98 (2011)
25. Xiong, P., Chi, Y., Zhu, S., Tatemura, J., Pu, C., Hacigms, H.: ActiveSLA: a profit-oriented admission control framework for database-as-a-service providers. In: Proceedings of SOCC 2011 (2011)
26. Narasayya, V., Menache, I., Singh, M., Li, F., Syamala, M., Chaudhuri, S.: Sharing buffer pool memory in multitenant relational database-as-a-service. In: Proceedings of VLDB 2015, pp. 726–737 (2015)

Recovering Missing Values from Corrupted Spatio-Temporal Sensory Data via Robust Low-Rank Tensor Completion

Wenjie Ruan[1]([⊠]), Peipei Xu[1,2], Quan Z. Sheng[3], Nickolas J.G. Falkner[1],
Xue Li[4], and Wei Emma Zhang[1]

[1] School of Computer Science, The University of Adelaide,
Adelaide, SA 5005, Australia
{wenjie.ruan,nickolas.falkner,wei.zhang01}@adelaide.edu.au
[2] School of Electronic Engineering, UESTC, Chengdu, China
peipei.xu6@gmail.com
[3] Department of Computing, Macquarie University, Sydney, NSW 2109, Australia
michael.sheng@mq.edu.au
[4] School of ITEE, University of Queensland, Brisbane, QLD 4072, Australia
xueli@itee.uq.edu.au

Abstract. With the booming of the Internet of Things, tremendous amount of sensors have been installed in different geographic locations, generating massive sensory data with both time-stamps and geo-tags. Such type of data usually have shown complex spatio-temporal correlation and are easily missing in practice due to communication failure or data corruption. In this paper, we aim to tackle the challenge – how to accurately and efficiently recover the missing values for corrupted spatio-temporal sensory data. Specifically, we first formulate such sensor data as a high-dimensional tensor that can naturally preserve sensors' both geographical and time information, thus we call *spatio-temporal Tensor*. Then we model the sensor data recovery as a low-rank robust tensor completion problem by exploiting its latent low-rank structure and sparse noise property. To solve this optimization problem, we design a highly efficient optimization method that combines the alternating direction method of multipliers and accelerated proximal gradient to minimize the tensor's convex surrogate and noise's ℓ_1-norm. In addition to testing our method by a synthetic dataset, we also use passive RFID (radio-frequency identification) sensors to build a real-world sensor-array testbed, which generates overall 115,200 sensor readings for model evaluation. The experimental results demonstrate the accuracy and robustness of our approach.

1 Introduction

With the rapid development of sensor technology, enormous numbers of smart devices or sensors have been deployed in our planet and thus served as a basic yet essential component of IoT (Internet of Things) [9,26]. Such tremendous

© Springer International Publishing AG 2017
S. Candan et al. (Eds.): DASFAA 2017, Part I, LNCS 10177, pp. 607–622, 2017.
DOI: 10.1007/978-3-319-55753-3_38

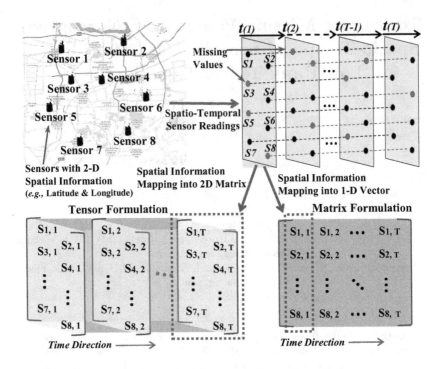

Fig. 1. Matrix formulation *vs.* Tensor formulation

smart devices enable ease access, retrieval and monitoring of our surrounding environment in a real-time manner. For instance, fine-particles (*e.g.,* PM 2.5) sensors are deployed in different locations within a city to continuously and cooperatively monitor the air quality [14]. Usually, many sensory data in real world share a common character that they are not only related to the time dimension (*e.g.,* time series data) but also have a two-dimensional (*e.g.,* sensors deployed in different latitudes and longitudes) or even three-dimensional spatial attribute (*e.g.,* sensors placed in various latitudes, longitudes and altitudes), which we thus call multi-dimensional spatio-temporal sensory data.

However, in practice, sensors easily experience an issue of missing readings due to the unexpected hardware failures (*e.g.,* power outages) or communication interruptions [14,24]. Those missing values will not only decrease the real-time monitoring performance, but also compromise the accuracy of back-end data analysis such as data predication, inference or visualization. Besides the data loss, the observed sensory data are also easily polluted by the environmental noise, making an accurate data recovery even more difficult [14]. Therefore, accurately yet efficiently interpolating the missing sensory data is a non-trivial and challenging task, especially for the multi-dimensional spatio-temporal sensory data with noise pollution.

The key to tackle this challenge lies on how to accurately model the quantitative dependencies of the missing readings with the known ones. The most widely

used and straightforward technique is various filtering or regression algorithms
that estimate the missing values according to their local temporal/spatial inter-
dependence, such as median filtering, exponential moving averaging [19], Kriging,
Kalman filtering [12], or regression methods with different complexities including ARMA/SARIMA (AutoRegressive Moving Average/Seasonal AutoRegressive Integrated Moving Average), SVR (Support Vector Regression) [22], kNN
(k-Nearest Neighbors) *etc.* However, such intuitive approaches suffer from two
shortcomings: *(i)* it only learns either spatial or temporal dependencies among
readings, and is hard to capture both features; *(ii)* it unavoidably ignores the
global correlations of data (*e.g.,* for some occasions, the missing readings may
depend on some far-away entries instead of those nearby values), leading to an
inaccurate estimation in some circumstances.

To solve this issue, some researchers treat the sensory data as a matrix and
propose various matrix completion/recovery methods to estimate the missing
values by capturing their inherent low-rank structure[1] [5,20]. Those methods
usually model time-dependent sensory data as a matrix $S = [s_1, s_2, ..., s_T]$ where
vector $s_k \in \mathbb{R}^N$ represents readings of N different sensors at time-stamp k. In
this regard, matrix completion based methods are, in principle, to recover the
unknown entries by solving an optimization problem: $\min_M\{\text{rank}(M)|P_\Omega(M) = P_\Omega(S)\}$ where $M \in \mathbb{R}^{N \times T}$ indicates recovered data matrix and P_Ω is the project
operator that means only entries in Ω are observed [20]. Also, several robust low-
rank matrix completion methods are recently proposed to deal with a case that
the observed data matrix S are polluted by noise [4]. Although matrix-based
methods can well take advantage of the temporal information, still they are limited to capturing *one-dimensional* spatial structure due to a fact that, in matrix
formulation, sensors with two-dimensional spatial coordinates are mapped into
a one-dimensional vector, unavoidably resulting in the spatial information loss,
as illustrate in Fig. 1.

As a result, to resolve those unsatisfied issues, this paper formates the spatio-
temporal sensor data as a *multi-dimensional tensor* - a natural high-order extension of a matrix. Figure 1 illustrates our general idea, for spatio-temporal sensory
data, compared to the matrix formulation that only preserves one-dimensional
spatial similarity, a tensor-based method naturally captures the two-dimensional
geographic dependency among sensors. Nevertheless, applying this high-level
idea into the practical still requires to address several challenges. Similar to a
matrix-based method [4], low-rank tensor-based data recovery can be formu-
lated as an optimization problem: $\min_{\mathcal{M},\mathcal{N}}\{\text{rank}(\mathcal{M}) + \lambda\|\mathcal{N}\|_0 \mid P_\Omega(\mathcal{M}+\mathcal{N}) = P_\Omega(\mathcal{S})\}$ where $\mathcal{M}, \mathcal{N}, \mathcal{S}$ represent the recovered data tensor, additive noise tensor
and observed data tensor respectively, P_Ω means the known entries of tensor.
To this end, the first challenge is that, the above optimization is a NP-hard
untraceable problem [21]. For the matrix version, we can replace rank($\mathcal{M}$) by
its tightest convex surrogate (*i.e.,* trace norm), enabling it solvable (*i.e.,* a convex optimization problem). But how to define a convex surrogate for a tensor

---

[1] The rank of a matrix is often linked to the order, complexity, or dimensionality of
the underlying system, which tends to be much smaller than the data size.

rank needs some careful design. Secondly, even we can define an effective convex surrogate and make the problem traceable, how to efficiently solve the convex optimization problem with a convergence guarantee also deserves an elaborative consideration.

To address above challenges, we generalize the idea of trace norm in matrix completion into the tensor, replacing the rank regularization term by the sum of tensor unfoldings' trace norms under all modes (see details in Sect. 3). Moreover, to optimize the objective function, we first apply a variable-splitting trick by introducing auxiliary tensor variables to decouple the interdependency of different tensor-modes, then we design an efficient optimization method with a strict convergence guarantee by drawing upon recent advances of Alternating Direction Method of Multipliers (ADMM) [30] and Accelerate Proximal Gradient (APG) [29] (see details in Sect. 3 and Algorithm 1). In a nutshell, our main contributions are summarized as follows:

- We propose a robust low-rank tensor completion method to accurately recover the missing sensor readings under a circumstance of noise pollution by exploiting the latent spatio-temporal structures and sparse noise property. We also introduce an efficient ADMM based optimization scheme to solve the robust tensor completion problem with a theoretical guarantee of convergence to a global optimum.
- We design a real-world sensor-array testbed consisting of 4 × 4 passive radio-frequency identification sensor-tags, generating overall 115,200 received signal strength indicator (RSSI) readings for the model evaluations. The experiments in both synthetic and real-world datasets demonstrate that our approach outperforms the state-of-the-art approaches in terms of accuracy and robustness.

The rest of the paper is organized as follows. Sect. 2 introduces problem formulation and notations of our solution. We present our model and optimization method in Sect. 3 and Sect. 4 presents experimental results and analysis. In Sect. 5, we review related work and Sect. 6 offers some concluding remarks.

2   Problem Formulation

First, we mathematically define our target problem. Assuming that we have $I_1 \times I_2$ sensors deployed in different spatial areas and collect (noisy) sensor readings[2] for T timestamps (see the example in Fig. 1), we then can formulate it as a 3-order tensor $\mathcal{O} \in \mathbb{R}^{I_1 \times I_2 \times T}$ and $\mathcal{O} = \mathcal{M} + \mathcal{N}$ where $\mathcal{M}$ represents the true sensor readings (without noise) and $\mathcal{N}$ means the added noise. We use the projection operator $P_\Omega(\mathcal{O}) : \mathbb{R}^{I_1 \times I_2 \times T} \rightarrow \mathbb{R}^K$ that indicates the K observed sensor readings $o_{i,j,t}$ where the index $(i,j,t) \in \Omega$, mapping a tensor to a vector. Formally, this paper therefore aims to solve the following *Corrupted Sensor Value Recovery* problem:

---

[2] We assume the additive noises are sufficiently sparse relative to the data tensor $\mathcal{O}$.

Problem 1 (Corrupted Sensor Value Recovery). *Given a partially observed data tensor $\mathcal{O}_\Omega$, our task is to accurately recover the true sensor readings $\mathcal{M}$ and additive noise $\mathcal{N}$, where $\mathcal{M}, \mathcal{N}, \mathcal{O} \in \mathbb{R}^{I_1 \times I_2 \times \cdots \times I_d}$.*

Throughout this paper, we represent scalars, vectors and matrices by lowercase letters *e.g.*, x, bold lowercase letters such as $\mathbf{x}$, and upper letters X. Tensors of d-order/dimension are written by calligraphic letters like $\mathcal{X} \in \mathbb{R}^{I_1 \times I_2 \times \cdots \times I_d}$, whose elements are represented by $x_{i_1 \cdots i_k \cdots i_d} \in \mathbb{R}$ and $1 \le i_k \le I_k, 1 \le k \le d$. Thus a vector can be seen as a 1-order tensor and a matrix can be seen as a 2-order tensor.

Definition 1. *Unfolding Operator: the mode-k unfolding of $\mathcal{X} \in \mathbb{R}^{I_1 \times I_2 \times \cdots \times I_d}$ is denoted by $unfold(\mathcal{X}, k) = X_{(k)} \in \mathbb{R}^{I_k \times \prod_{i \ne k} I_i}$, i.e., the row of the matrix $X_{(k)}$ are determined by the k-th component of the tensor $\mathcal{X}$, whereas all the remaining components form its columns.*

This operation transforms a tensor into a matrix, i.e. matricization or flattening.

Definition 2. *Folding Operator: the mode-k folding of a matrix $X_{(k)}$ is defined as $fold(X_{(k)}, k) = \mathcal{X}$.*

Definition 3. *Inner Product of Tensor: the inner product of two tensors with identical size $\mathcal{X}, \mathcal{Y} \in \mathbb{R}^{I_1 \times I_2 \times \cdots \times I_d}$ is computed by $\langle \mathcal{X}, \mathcal{Y} \rangle := \sum_{i_1, i_2, \cdots, i_d} x_{i_1 i_2 \cdots i_d} y_{i_1 i_2 \cdots i_d}$.*

Definition 4. *Frobenius Norm: Frobenius norm of $\mathcal{X}$ is defined as $\|\mathcal{X}\|_F := \left(\sum_{i_1, i_2, \cdots, i_d} |x_{i_1 i_2 \cdots i_d}|^2 \right)^{\frac{1}{2}}$.*

Thus, for any $k \in \{1, ..., d\}$, we have $\|\mathcal{X}\|_F = \|X_{(k)}\|_F$, and $\langle \mathcal{X}, \mathcal{Y} \rangle = \langle X_{(k)}, Y_{(k)} \rangle$.

Definition 5. *Tensor-matrix multiplication: the multiplication of a d-order tensor $\mathcal{X} \in \mathbb{R}^{I_1 \times I_2 \times \cdots \times I_d}$ with a matrix $A \in \mathbb{R}^{J \times I_k}$ in mode-k is mathematically defined as $\mathcal{X} \times_k A \in \mathbb{R}^{I_1 \times \cdots \times I_{k-1} \times J \times I_{k+1} \times \cdots \times I_d}$.*

Definition 6. *Tucker decomposition: Given an input tensor, Tucker decomposition uses a smaller/core tensor multiplied by a matrix along each mode to describe the original tensor. A tensor $\mathcal{X} \in \mathbb{R}^{I_1 \times I_2 \times \cdots \times I_d}$ decomposes as*

$$\mathcal{X} = \mathcal{G} \times_1 A^{(1)} \times_2 A^{(2)} \cdots \times_d A^{(d)} = [\![\mathcal{G}, A^{(1)}, ..., A^{(d)}]\!]$$

$$= \sum_{r_1=1}^{R_1} \sum_{R_2}^{r_2=1} \cdots \sum_{R_d}^{r_d=1} g_{r_1 r_2 \cdots r_d} \mathbf{a}_{r_1}^{(1)} \circ \cdots \circ \mathbf{a}_{r_d}^{(d)} \tag{1}$$

where $\{A^{(k)}\}_{k=1}^d \in \mathbb{R}^{I_k \times R_k}$ are a set of factor matrices and $R_1, R_2, ..., R_d$ is defined as the Tucker rank. 'o' denotes the outer product. When $d = 3$, Tucker decomposition of tensor is $\mathcal{X} = \mathcal{G} \times_1 A \times_2 B \times_3 C = [\![\mathcal{G}, A, B, C]\!]$.

3   Robust Low-Rank Spatio-Temporal Tensor Recovery

Being similar to matrix completion, Problem (1) can be formulated as solving a low-rank minimization problem.

$$\min_{\mathcal{M},\mathcal{N}} \; \text{rank}_{Tucker}(\mathcal{M}) + \lambda \|\mathcal{N}\|_0$$

$$\text{s.t. } P_\Omega(\mathcal{M} + \mathcal{N}) = P_\Omega(\mathcal{O}) \tag{2}$$

where $\text{rank}_{Tucker}(\mathcal{M})$ is the Tucker-rank of a tensor [11]. Similar to matrix completion, this problem is NP-hard. Thus, to make it tractable, we replace Tucker rank by its *convex surrogate* and use ℓ_1-norm instead of ℓ_0-norm as $\min_{\mathcal{M},\mathcal{N}}\{\text{ConSurro}(\mathcal{M}) + \lambda\|\mathcal{N}\|_1 \mid \text{s.t. } P_\Omega(\mathcal{M}+\mathcal{N}) = P_\Omega(\mathcal{O})\}$.

Then the first issue is how to define the convex surrogate of a tensor. For a matrix, the trace norm $\|.\|_*$ is the tightest convex envelop of its rank, used as the convex surrogate. Thus, the idea can be generalized into the high-order tensor, defining its trace norm as the sum of the trace norms [21] of the mode-i unfolding in tensor $\mathcal{M}$, *i.e.*, $\text{ConSurro}(\mathcal{M}) = \sum_i \|M_{(i)}\|_*$. Equation (2) can be therefore transformed into a convex problem:

$$\min_{\mathcal{M},\mathcal{N}} \; \sum_{i=1}^{d} \|M_{(i)}\|_* + \lambda\|\mathcal{N}\|_1$$

$$\text{s.t. } P_\Omega(\mathcal{M} + \mathcal{N}) = P_\Omega(\mathcal{O}) \tag{3}$$

To solve Eq. (3), we introduce an Alternating Direction Method of Multipliers [30] that is very efficient in dealing with convex optimization problems by breaking them into smaller pieces, each of which is then easier to handle. However, the trace norm of each mode unfolding $\|M_{(i)}\|_*, (i = 1, ..., d)$ shares the same values in data tensor $\mathcal{M}$ and cannot be optimized independently so that existing ADMM cannot directly be applied to our problem. Hence, we split these interdependent terms by introducing auxiliary variables $\mathcal{M}_1, \mathcal{M}_2, ..., \mathcal{M}_d$, so that they can be solved independently. Specifically, we reformulate Eq. (3) as

$$\min_{\mathcal{M}_1,...,\mathcal{M}_d,\mathcal{N}} \; \sum_{i=1}^{d} \|M_{i,(i)}\|_* + \lambda\|\mathcal{N}\|_1$$

$$\text{s.t. } \quad P_\Omega(\mathcal{M}_i + \mathcal{N}) = P_\Omega(\mathcal{O}), \quad i = 1, ..., d. \tag{4}$$

We hence define its *augmented Lagrangian function* as

$$\mathcal{L}_\mu(\mathcal{M}_1, ..., \mathcal{M}_d, \mathcal{N}, \mathcal{Y}_1, ..., \mathcal{Y}_d) = \sum_{i=1}^{d} \left(\frac{1}{2\mu} \|P_\Omega(\mathcal{M}_i + \mathcal{N}) - P_\Omega(\mathcal{O})\|^2 \right.$$

$$- \langle \mathcal{Y}_i, P_\Omega(\mathcal{M}_i + \mathcal{N}) - P_\Omega(\mathcal{O}) \rangle \right) + \sum_{i=1}^{d} \|M_{i,(i)}\|_* + \lambda\|\mathcal{N}\|_1. \tag{5}$$

According to ADMM, we first fix $\mathcal{N}$ to optimize $\mathcal{M}_i$ $(i = 1, ..., d)$ by solving

$$\min_{\mathcal{M}_1, ..., \mathcal{M}_d} \sum_{i=1}^{d} (\frac{1}{2\mu} \| P_\Omega(\mathcal{M}_i + \mathcal{N}) - P_\Omega(\mathcal{O}) \|^2 - \langle \mathcal{Y}_i, P_\Omega(\mathcal{M}_i + \mathcal{N})$$

$$- P_\Omega(\mathcal{O}) \rangle + \| M_{i,(i)} \|_* \equiv \sum_{i=1}^{d} (\mu \| M_{i,(i)} \|_* + \frac{1}{2} \| P_\Omega(\mathcal{M}_i) - \mathcal{A}_i \|^2)$$

(6)

where $\mathcal{A}_i = P_\Omega(\mathcal{O}) - P_\Omega(\mathcal{N}) + \mu \mathcal{Y}_i$. We define the function $f(\mathcal{M}_i) = \frac{1}{2} \| P_\Omega(\mathcal{M}_i) - \mathcal{A}_i \|^2$ and calculate the gradient $\nabla f(\mathcal{M}_i) = P_\Omega^*(P_\Omega(\mathcal{M}_i) - \mathcal{A}_i)$, where $P_\Omega^*(\cdot)$ means the adjoint operation of $P_\Omega(\cdot)$ such as $P_\Omega^*(\mathcal{O}) : \mathbb{R}^K \rightarrow \mathbb{R}^{I_1 \times I_2 \times \cdots \times T}$. According to Accelerated Proximal Gradient (APG) method [29], we can independently minimize $\mathcal{M}_i$ through iterative optimization to make the final sum minimal. Specifically, we get optimal $\mathcal{M}_i^{(k+1)}$ given $\mathcal{M}_i^{(k)}$ until it converges by solving

$$\min_{\mathcal{M}_i^{(k+1)}} f(\mathcal{M}_i^{(k)}) + \nabla f(\mathcal{M}_i^{(k)})(\mathcal{M}_i^{(k+1)} - \mathcal{M}_i^{(k)}) + \frac{1}{2\eta} \| \mathcal{M}_i^{(k+1)} - \mathcal{M}_i^{(k)} \|^2 +$$

$$\mu \| M_{i,(i)}^{(k+1)} \|_* = \frac{1}{2\eta} \| \mathcal{M}_i^{(k+1)} - \mathcal{M}_i^{(k)} + \eta \nabla f(\mathcal{M}_i^{(k)}) \|^2 + \mu \| M_{i,(i)}^{(k+1)} \|_* \quad (7)$$

$$\propto \frac{1}{2} \| \mathcal{M}_i^{(k+1)} - \mathcal{M}_i^{(k)} + \eta \nabla f(\mathcal{M}_i^{(k)}) \|^2 + \eta \mu \| M_{i,(i)}^{(k+1)} \|_*$$

To solve Eq. (7), we first need to define *singular value thresholding operator* for tensor.

Theorem 1. *For matrix, the singular value threshold operator is defined as* $T_\mu(M) := U \, diag(\bar{\sigma}) V^\intercal$, *where* $M = U \, diag(\sigma) V^T$ *is the singular value decomposition (SVD) and* $\bar{\sigma} := max(\sigma - \mu, 0)$.

Similarly, we define the singular value threshold [3] operator for tensor as $T_{i,\mu}(\mathcal{M}) := \text{fold}(T_\mu(M_{(i)}), i)$. We thus can calculate the closed-form solution of Eq. (7) as follows:

$$\mathcal{M}_i^{(k+1)} = T_{i,\eta\mu}(\mathcal{M}_i^{(k)} - \eta \nabla f(\mathcal{M}_i^{(k)})) \quad (8)$$

In the next, we will optimize $\mathcal{N}$ when fixed $\mathcal{M}_i$ $(i = 1, ..., d)$ by solving the following problem:

$$\min_{\mathcal{N}} \mathcal{Y} \| \mathcal{N} \|_1 + \sum_{i=1}^{d} (\frac{1}{2\mu} \| P_\Omega(\mathcal{M}_i + \mathcal{N}) - P_\Omega(\mathcal{O}) - \mu \mathcal{Y}_i \|^2)$$

(9)

$$\propto \mu\lambda \| \mathcal{N} \|_1 + \frac{1}{2} \sum_{i=1}^{d} \| P_\Omega(\mathcal{N}) - \mathcal{B}_i \|^2$$

Algorithm 1. ADMM based Robust Tensor Completion

Input: Observed Sensory Data Tensor: $\mathcal{O}$, Set of Missing Value Indexes: Ω
Model Parameters: λ, η, μ
Initialization: $\mathcal{M}_i^{(0)} = \mathcal{N}^{(0)} = \mathcal{Y}_i^{(0)} = 0$ $(i = 1, ..., d)$
Output: Recovered Sensory Data Tensor: $\mathcal{M}$, Estimated Noise Tensor: $\mathcal{N}$

1 **while** $k > 0$ **do**
2 | **for** $i = 1 : d$ **do**
3 | | $\mathcal{M}_i^{(k+1)} = \mathcal{T}_{i,\eta\mu}(\mathcal{M}_i^{(k)} - \eta\nabla f(\mathcal{M}_i^{(k)}))$;
                     /* Optimize $\mathcal{M}_i$ using singular value threshold */
4 | **end**
5 | $\mathcal{N}^{(k+1)} = \mathcal{S}_{\frac{\mu\lambda}{d}}(\frac{1}{d}\sum_{i=1}^{d}(P_\Omega(\mathcal{O}) + \mu\mathcal{Y}_i^{(k)} - P_\Omega(\mathcal{M}_i^{(k+1)})))$;
                     /* Optimize $\mathcal{N}$ using shrinkage thresholding operator */
6 | **for** $i = 1 : d$ **do**
7 | | $\mathcal{Y}_i^{(k+1)} = \mathcal{Y}_i^{(k)} - \frac{1}{\mu}(P_\Omega(\mathcal{M}_i^{(k+1)} + \mathcal{N}^{(k+1)}) - P_\Omega(\mathcal{O}))$;
                     /* Update Lagrangian multiplier parameters $\mathcal{Y}_i$ */
8 | **end**
9 | **if** $StoppingCondition == TRUE$ **then**
10 | | Break;   /* Ending loop when stop condition satisfied */
11 | **end**
12 **end**
13 **return** $\mathcal{M} = \frac{1}{d}\sum_{i=1}^{d}\mathcal{M}_i^{(k+1)}$ and $\mathcal{N}^{(k+1)}$;   /* Return results */

---

where $\mathcal{B}_i = P_\Omega(\mathcal{O}) + \mu\mathcal{Y}_i - P_\Omega(\mathcal{M}_i)$. To solve Eq. (9), we define *Homogeneous Tensor Array* [11] by introducing an operator that combines the component tensors with the same size along the tensor mode-1 as:

$$\bar{\mathcal{M}} := (\mathcal{M}_1, ..., \mathcal{M}_d)^\mathsf{T} \in \mathbb{R}^{dI_1 \times I_2 \times \cdots \times I_d}, \tag{10}$$

which is written as TArray$(\mathcal{M}_1, ..., \mathcal{M}_d)$ and its linear operator $\mathcal{C}$: $\mathbb{R}^{I_1 \times I_2 \times \cdots \times I_d} \to \mathbb{R}^{dI_1 \times I_2 \times \cdots \times I_d}$, *i.e.*, $\bar{\mathcal{M}} = \mathcal{C}(\mathcal{M}) \in \mathbb{R}^{dI_1 \times I_2 \times \cdots \times I_d}$.

Then, we can attain its adjoint operator $\mathcal{C}^* : \mathbb{R}^{dI_1 \times I_2 \times \cdots \times I_d} \to \mathbb{R}^{I_1 \times I_2 \times \cdots \times I_d}$, such that $\mathcal{M} = \mathcal{C}^*(\bar{\mathcal{M}}) = \sum_{i=1}^{d}\mathcal{M}_i$.

According to this definition, we can rewrite Eq. (9) as

$$\min_{\mathcal{N}} \; \mu\lambda\|\mathcal{N}\|_1 + \frac{1}{2}\|\mathcal{C}(P_\Omega(\mathcal{N})) - \bar{\mathcal{B}}\|^2$$

$$\propto \frac{\mu\lambda}{d}\|\mathcal{N}\|_1 + \frac{1}{2}\|P_\Omega(\mathcal{N}) - \frac{\mathcal{C}^*(\bar{\mathcal{B}})}{d}\|^2 \tag{11}$$

$$= \frac{\mu\lambda}{d}\|\mathcal{N}\|_1 + \frac{1}{2}\|P_\Omega(\mathcal{N}) - \frac{1}{d}\sum_{i=1}^{d}(P_\Omega(\mathcal{O}) + \mu\mathcal{Y}_i - P_\Omega(\mathcal{M}_i))\|^2$$

where $\bar{\mathcal{B}} = (\mathcal{B}_1, ..., \mathcal{B}_d)^\mathsf{T}$ and $\mathcal{C}(P_\Omega(\mathcal{N})) = (P_\Omega(\mathcal{N}), ..., P_\Omega(\mathcal{N}))^\mathsf{T}$.
Before solving the Eq. (11), we need the following Theorem.

Theorem 2. *When* $\min_{\mathbf{y}} \{\frac{1}{2}\|\mathbf{y} - \mathbf{x}\|_2^2 + \mu\|\mathbf{y}\|_1\}$, *it has a closed form solution* $\mathbf{y} = \mathcal{S}_\mu(\mathbf{x}) := sign(\mathbf{x})max(|\mathbf{x}| - \mu, 0)$, *where* $\mathcal{S}_\mu(\mathbf{x})$ *is the shrinkage operator, where all the operations are element-wise.*

According to the shrinkage thresholding operator [2], we can define $\mathcal{S}_\mu(\mathcal{M})$ on the $vec(\mathcal{M}) = \mathbf{m}$. As a result, we can solve Eq. (11) to get $\mathcal{N} = \mathcal{S}_{\frac{\mu\lambda}{d}}(\frac{1}{d}\sum_{i=1}^{d}(P_\Omega(\mathcal{O}) + \mu\mathcal{Y}_i - P_\Omega(\mathcal{M}_i)))$. We thus have $\mathcal{C}^*\mathcal{C}(\mathcal{N}) = d\,\mathcal{N}$, then the optimal condition of Eq. (11) is $0 \in d\,\mathcal{N} + \mu\lambda_1\partial\|\mathcal{N}\|_1 \Leftrightarrow 0 \in \mathcal{N} + \frac{\mathcal{C}^*(\mathcal{B})}{d} + \frac{\mu\lambda_1}{d}\partial\|\mathcal{N}\|_1$.

Finally, given $\mathcal{M}_i^{(k+1)}$ and $\mathcal{N}_i^{(k+1)}$, we can update the Lagrangian multiplier parameter by $\mathcal{Y}_i^{(k+1)} = \mathcal{Y}_i^{(k)} - \frac{1}{\mu}(P_\Omega(\mathcal{M}_i^{(k+1)} + \mathcal{N}^{(k+1)}) - P_\Omega(\mathcal{O}))$. Algorithm 1 shows the pseudo-code of our optimization method.

Essentially, when $d = 3$, Algorithm 1 alternatively optimizes two blocks of variables $\{\mathcal{M}_1, \mathcal{M}_2, \mathcal{M}_3\}$ and $\mathcal{N}$. By defining $f(\mathcal{M}_1, \mathcal{M}_2, \mathcal{M}_3) := \sum_{i=1}^{3}\|X_{i,(i)}\|_*$ and $g(\mathcal{N}) := \lambda_1\|\mathcal{N}\|_1$, it is easy to verify that Algorithm 1 meets the convergence condition of ADMM. Briefly, the sequence $\{\mathcal{M}_1^{(k)}, \mathcal{M}_2^{(k)}, \mathcal{M}_3^{(k)}, \mathcal{N}^{(k)}\}$ obtained from Algorithm 1 can converge to optimal tensors as $(\mathcal{M}_1^{(*)}, \mathcal{M}_2^{(*)}, \mathcal{M}_3^{(*)}, \mathcal{N}^{(*)})$ for Eq. (4). Hence, the sequence $\{\frac{1}{3}(\sum_{i=1}^{3}\mathcal{M}_i^{(k)}), \mathcal{N}^{(k)}\}$ can reach optimal values. Due to the page limitation, we make the proof details available online[3].

4   Experiments

In this section, we first conduct a simulation experiment using synthetic data to compare with the state-of-the-art data recovery methods under different data loss percentages and additive noise ratios. Then we design a real-world experimental testbed using passive RFID (Radio-frequency identification) sensors to generate geo-tagged time-series RSSI readings, which are used to test the practical performance of our method. We run the experiments on a laptop computer (CORE i7-4710HQ 2.50GHz CPU and 16GB RAM) using MATLAB R2015b. We use the Tensor Toolbox[4] and for tensor operations and decompositions and PROPACK Toolbox for SVD (Singular Value Decomposition) calculation[5].

Similar to other data recovery works [20, 21], we adopt the *relative error* $\frac{\|\mathcal{M} - \mathcal{M}_0\|}{\|\mathcal{M}_0\|}$, where $\mathcal{M}, \mathcal{M}_0$ mean the recovered and original data tensor respectively, to evaluate the recovery performance.

4.1   Comparison Methods

We compare our method with the following typical data recovery methods.

- *MAF* means the moving averaging interpolation that is the most widely-used method to fill in missing values in time-series sensory data[6].

---

[3] www.dropbox.com/s/mcqqpxc6m0b5jyn/Appendix.pdf?dl=0.

[4] Available in www.sandia.gov/~tgkolda/TensorToolbox/index-2.6.html.

[5] Available in http://sun.stanford.edu/~rmunk/PROPACK/.

[6] Available in au.mathworks.com/help/curvefit/smoothing-data.html.

- *IAL-MC* [20] represents the inexact augmented Lagrangian method that can recover a data matrix of being arbitrarily corrupted, it is a matrix-based robust completion method and greatly motivates our work[7].
- *LR-TC* [21] is the earliest yet very effective tensor completion method using block coordinate descent optimization but it cannot deal with the corrupted data (not robust version)[8].
- *ADMM-TC* [30] also utilizes ADMM for solving the tensor completion problem. It provides valuable intuitions for the optimization part in this paper[9].

4.2 Evaluations on Synthetic Data

Similar to the works in [21,30], we generate a $50 \times 50 \times 20$ data tensor with Tucker rank-$(5,5,5)$. We randomly choose a fraction ρ_n of the tensor entries that are polluted by an additive *i.i.d.* (*i.e.*, independent and identically distributed) noise following uniform distribution $\mathsf{unif}(-a, a)$. Then a fraction ρ_o of the corrupted tensor elements are randomly picked as observed values in $\mathcal{O}_\Omega$. In the experiments, we set μ and η as constants for simplicity, *i.e.*, $\mu = 5 \times \mathsf{std}(\mathsf{vec}(\mathcal{O}))$ and $\eta = 0.91$. We set parameter as $\lambda = \alpha r \lambda_*$, where $\lambda_* = 1, r = \dfrac{1}{\sqrt{max(I_1, I_2, T)}}$ and α are tuned in $1 < \alpha < 2$.

Recovery Accuracy: Figure 2 compares the relative errors of different methods under an observation percentage from 5% to 100% and a noise ($\rho_n = 0.1, a = 1$). We can see that our method has a better recovery accuracy than the other three algorithms. Especially, during the interval between 30% and 60%, our method reveals significantly higher recovery capability. We also observe that, from 5% to 40% the recovery performance dramatically increases, while it does not show significant improvement from 40% to 100% observation. We then add more polluted data ($\rho_n = 0.25, a = 1$) to test these approaches. As Fig. 3, all four methods perform similarly under a circumstance that the missing data are less than 40%, while the proposed method achieves a smaller relative error of the data recovery when more than 50% data are polluted. Combining both Figs. 2 and 3, our method appears an obvious "thresholding phenomena" that the recovery accuracy is continuously improved when the observed data increases below a certain threshold (*e.g.*, 40% in Fig. 2 and 70% in Fig. 3) and the threshold is bigger when adding more corrupted data. In the next, we set the observation $\rho_o = 1$ and investigate the recovery performance under different corruption percentages (from 0% to 40%). The results are shown in Fig. 4. Similarly, our method shows a relatively higher recovery accuracy (*e.g.*, improving around 3 times under 30% noise pollution) and the other three methods reveal a similar performance. It is worth mentioning that the recovering performance greatly degenerates when more than 35% data are polluted.

---

[7] Available in www.cis.pku.edu.cn/faculty/vision/zlin/RPCA+MC_codes.zip.

[8] Available in www.cs.rochester.edu/u/jliu/code/TensorCompletion.zip.

[9] Available in https://github.com/ryotat/tensor.

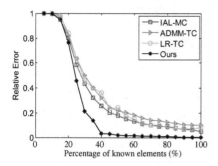

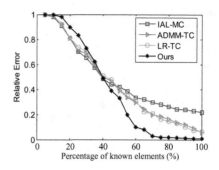

Fig. 2. Relative errors for different known elements ($\rho_n = 0.1$, $a = 1$)

Fig. 3. Relative errors for different known elements ($\rho_n = 0.25$, $a = 1$)

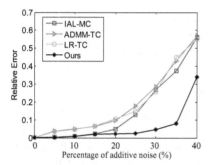

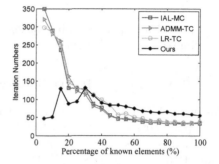

Fig. 4. Relative errors for different corruption percentages ($\rho_o = 1$, $a = 1$)

Fig. 5. Iteration numbers for different known elements ($\rho_n = 0.15$, $a = 1$)

Iteration Number: Figures 5, 6 and 7 compare the computation time of different methods in terms of iteration numbers[10]. In details, Fig. 5 illustrates the iteration times needed for different percentages of known elements. We observe that the proposed method is super fast when a few data are observed (*e.g.,* from 5% to 30%) comparing to other solutions, however it requires a similar or slightly higher iteration times when observing more data (*e.g.,* from 40% to 100%). A similar result applies to Fig. 6 where the proposed method only needs around 50–100 iteration times under 5% to 30% known data comparing to other methods that requires 180–370 iterations. Figure 6 shows the experimental results of iteration numbers for different data corruption percentages with no missing values. Our method overall reveals a slightly better computation efficiency.

In summary, the experimental results on the synthetic data suggest that our model can achieve better recovery performance and computation efficiency with

---

[10] In each iteration, all the tensor completion based methods require to calculate 3 times SVD that is most time-consuming computation task so we use iteration number as evaluation metric for computation time.

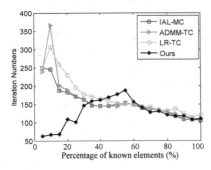
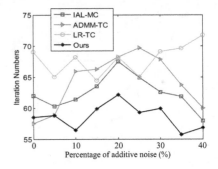

Fig. 6. Iteration numbers for different known elements ($\rho_n = 0.3$, $a = 1$)

Fig. 7. Iteration numbers for different corruption percentages ($\rho_o = 1$, $a = 1$)

both partial and polluted observations, especially for the scenarios that only very limited data are known such as less than 30%.

4.3 Evaluations on RFID Sensory Data

Passive RFID tags are one of the most frequently-used sensors due to its cheap price (<5 cents each) and battery-free features [25,33]. It is widely used to identify and track objects through remotely accessing the electronically stored data. However, since passive RFID tags are powered by radio signals and deliver the data via the weak backscatter signal, they experience severe RSSI reading loss, especially with a high sampling rate or when tag/reader is moving [23]. As a result, how to accurately recover missing RSSI readings is still a challenge, especially for a large-scale RFID usage.

To deal with this practical issue as well as to test our method, we design a testbed consisting geo-tagged 4×4 RFID sensor array (see Fig. 8a) and collected overall 115,200 RSSI readings[11]. To be more practical, we formulate the readings of RFID sensor array it into a tensor with different sizes to simulate various real-world application scenarios (*e.g.*, 4×4, 20×20, 20×40 and 40×40 sensor array). Similarly, we add noise ($\rho_n = 0.1$, $a = 10$) with uniform distribution and randomly choose 20% elements as the unknown in our experiments. Figure 8b shows the recovery results of our method and MAF (most frequently used in practical RFID system) as well as other matrix-tensor completion methods. For small-scaled deployment (*i.e.*, 4×4 sensor array), our method achieves similar performance to MAF. However, the tensor-based methods perform significantly better than MAF in a large-scale deployment (*e.g.*, 20×20 sensor array). The lack of performance improvement in 4×4 array mainly lies in the fact that the low-rank structure only exists in mode-3 of data tensor which conflicts

---

[11] We collect over all one hour's RSS readings, the sampling rate is 2 Hz. During the data collection, a participant is doing various activities between the RFID sensor-array and reader, including walking, sitting, standing, lying down as well as falling down etc. By doing so, the collected RSSI reading will reveal different patterns.

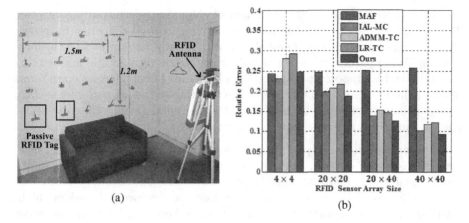

Fig. 8. (a) Experimental testbed of RFID sensor array; (b) Relative errors for different tag-array size with 20% missing values

our assumption that requires low-rank in all tensor modes. We will analyze this issue and point out a possible solution in Sect. 6.

5   Related Work

Imputing/estimating the missing values from a partially observed data have attracted many interest in the past decades such as signal processing, data mining, computer vision [13,28]. Generally, we categorize the techniques of recovering missing values into three types - *regression* based methods, *matrix completion* and *tensor completion* based methods. In this paper, we will concentrate on discussing the latter two categories that are more related to our model.

Matrix Completion Technique: To capture the global information of targeted dataset, the "rank" of the matrix is a powerful tool and many matrix completion/recovery based on the inherent low-rank structure assumption have drawn significant interest. Massive optimization models and efficient algorithms are proposed [29]. Some researchers [5] have shown that under some mild conditions, most low-rank matrices can be perfectly recovered from an incomplete set of entries by solving a simple convex optimization program, namely, solving $\min_M\{\text{rank}(M)|P_\Omega(X) = P_\Omega(T)\}$, where M indicates recovered data matrix and P_Ω means only entries in Ω are observed. Although *low-rank matrix completion* has drawn significant interest and have played an important role in missing data recovery, however, such methods cannot work or fail to recover the data matrix under some circumstances that a subset of its entries may be corrupted or polluted by various sparse noises [6].

As a result, many robust versions of matrix completion that can recover the low-rank matrix from both noisy and partial observations of data are proposed lately [15,31]. For example, Chen *et al.* [7] investigate the problem of low-rank matrix completion where a large number of columns are arbitrarily corrupted. They show that only a small fraction of the entries are needed in order to recover the low-rank matrix with high probability, without any assumptions on the location nor the amplitude of the corrupted entries. Recently, a multi-view learning based method is proposed to capture both local and global information in terms of spatial and temporal perspective, achieving state-of-the-art performance [32]. It also demonstrates that both local and global spatial/temporal correlations play an important role in sensor data reconstruction.

Tensor Completion Technique: Though promising of matrix-based models, however the recovered dataset, in many practical applications, has complex multi-dimensional spatio-temporal correlations, which can be naturally treated as a tensor instead of a matrix [21]. Therefore data recovery based on high-dimensional tensor or multi-way data analysis is becoming prevalent in recent several years. Generally, there are two state-of-the-art techniques used for tensor completion. One is the nuclear norm minimization, many pioneering similar works are emerged [10,17] since Liu *et al.* [21] first extended the nuclear norm of matrix (*i.e.,* the sum of all the singular values) to tensor. Later on, Gandy *et al.* [10] and Signoretto *et al.* [27] consider a tractable and unconstrained optimization problem of low-n-rank tensor recovery $\min_{\mathcal{M}} \sum_{i=1}^{N} \|M_{(i)}\|_* \|P_{\Omega}(\mathcal{M}) - P_{\Omega}(\mathcal{T})\|^2$ and adopt the Douglas-Rachford splitting method and ADMM method. Another popular technique is to utilize the tensor decomposition [18], *i.e.,* decomposing the Nth-order tensor into another smaller Nth-order tensor (*i.e.,* core tensor) and N factor matrices. Generally, Tucker and CANDECOMP/PARAFAC are the two most popular tensor decomposition frameworks [16]. For example, Acar *et al.* [1] develop an algorithm called CP-WOPT (CP Weighted OPTimization) used a first-order optimization approach for dealing with missing value and has been testified to provide a good imputation performance. More recently, Da Silva *et al.* [8] and Kressner *et al.* [18] have proposed a nonlinear conjugate gradient method for Riemannian optimization based on the hierarchical Tucker decomposition and Tucker decomposition separately. However, those tensor completion methods are neither applied into recovering spatio-temporal sensory data, nor can deal with a circumstance that the known sensor readings are corrupted by noise. Our ADMM based robust tensor completion method, on the contrary, can fill both two gaps and recover the missing sensor values with a high accuracy and robustness.

6   Conclusion

In summery, we propose a method for recovering the missing data by using the robust tensor completion. The proposed method can accurately recover the missing values given partial observed corrupted data. In the future, we will investigate how to accurately recover the sensor readings when the low-rank only exists in certain mode of a data tensor and the type of pollution noise is unknown.

References

1. Acar, E., Dunlavy, D.M., Kolda, T.G., Mørup, M.: Scalable tensor factorizations for incomplete data. Chemometr. Intell. Lab. Syst. **106**(1), 41–56 (2011)
2. Beck, A., Teboulle, M.: A fast iterative shrinkage-thresholding algorithm for linear inverse problems. SIAM J. Imaging Sci. **2**(1), 183–202 (2009)
3. Cai, J.F., Candès, E.J., Shen, Z.: A singular value thresholding algorithm for matrix completion. SIAM J. Optim. **20**(4), 1956–1982 (2010)
4. Candès, E.J., Li, X., Ma, Y., Wright, J.: Robust principal component analysis? J. ACM (JACM) **58**(3), 11 (2011)
5. Candès, E.J., Recht, B.: Exact matrix completion via convex optimization. Found. Comput. Math. **9**(6), 717–772 (2009)
6. Chen, Y., Jalali, A., Sanghavi, S., Caramanis, C.: Low-rank matrix recovery from errors and erasures. IEEE Trans. Info. Theor. **59**(7), 4324–4337 (2013)
7. Chen, Y., Xu, H., Caramanis, C., Sanghavi, S.: Robust matrix completion with corrupted columns. arXiv preprint (2011). arXiv:1102.2254
8. Da Silva, C., Herrmann, F.J.: Hierarchical tucker tensor optimization-applications to tensor completion. In: Proceedings of 10th International Conference on Sampling Theory and Applications (2013)
9. Da Xu, L., He, W., Li, S.: Internet of things in industries: a survey. IEEE Trans. Industr. Inf. **10**(4), 2233–2243 (2014)
10. Gandy, S., Recht, B., Yamada, I.: Tensor completion and low-n-rank tensor recovery via convex optimization. Inverse Prob. **27**(2), 025010 (2011)
11. Goldfarb, D., Qin, Z.: Robust low-rank tensor recovery: models and algorithms. SIAM J. Matrix Anal. Appl. **35**(1), 225–253 (2014)
12. Grewal, M.S.: Kalman Filtering. Springer, Heidelberg (2011)
13. Hazan, T., Polak, S., Shashua, A.: Sparse image coding using a 3D non-negative tensor factorization. In: Tenth IEEE International Conference on Computer Vision, ICCV 2005, vol. 1, pp. 50–57. IEEE (2005)
14. Hsieh, H.P., Lin, S.D., Zheng, Y.: Inferring air quality for station location recommendation based on urban big data. In: Proceedings of the 21th ACM SIGKDD International Conference on Knowledge Discovery and Data Mining, pp. 437–446 (2015)
15. Ji, H., Huang, S., Shen, Z., Xu, Y.: Robust video restoration by joint sparse and low rank matrix approximation. SIAM J. Imaging Sci. **4**(4), 1122–1142 (2011)
16. Kolda, T.G., Bader, B.W.: Tensor decompositions and applications. SIAM Rev. **51**(3), 455–500 (2009)
17. Kreimer, N., Stanton, A., Sacchi, M.D.: Tensor completion based on nuclear norm minimization for 5D seismic data reconstruction. Geophysics **78**(6), V273–V284 (2013)
18. Kressner, D., Steinlechner, M., Vandereycken, B.: Low-rank tensor completion by Riemannian optimization. BIT Numer. Math. **54**(2), 447–468 (2014)
19. Lawrance, A., Lewis, P.: An exponential moving-average sequence and point process. J. Appl. Probab. **14**, 98–113 (1977)
20. Lin, Z., Chen, M., Ma, Y.: The augmented lagrange multiplier method for exact recovery of corrupted low-rank matrices. arXiv preprint (2010). arXiv:1009.5055
21. Liu, J., Musialski, P., Wonka, P., Ye, J.: Tensor completion for estimating missing values in visual data. In: 2009 IEEE 12th International Conference on Computer Vision, pp. 2114–2121. IEEE (2009)

22. Norcia, A.M., Clarke, M., Tyler, C.W.: Digital filtering and robust regression techniques for estimating sensory thresholds from the evoked potential. IEEE Eng. Med. Biol. Mag. **4**(4), 26–32 (1985)

23. Ruan, W., Sheng, Q.Z., Yao, L., Gu, T., Ruta, M., Shangguan, L.: Device-free indoor localization and tracking through human-object interactions. In: 2016 IEEE 17th International Symposium on A World of Wireless, Mobile and Multimedia Networks (WoWMoM), pp. 1–9. IEEE (2016)

24. Ruan, W., Xu, P., Sheng, Q.Z., Tran, N.K., Falkner, N.J., Li, X., Zhang, W.E.: When sensor meets tensor: filling missing sensor values through a tensor approach. In: Proceedings of the 25th ACM International on Conference on Information and Knowledge Management, pp. 2025–2028. ACM (2016)

25. Ruan, W., Yao, L., Sheng, Q.Z., Falkner, N., Li, X., Gu, T.: TagFall: towards unobstructive fine-grained fall detection based on UHF passive RFID tags. In: Proceedings of the 12th International Conference on Mobile and Ubiquitous Systems: Computing, Networking and Services, pp. 140–149 (2015)

26. Sakurai, Y., Matsubara, Y., Faloutsos, C.: Mining and forecasting of big time-series data. In: Proceedings of the 2015 ACM SIGMOD International Conference on Management of Data, pp. 919–922. ACM (2015)

27. Signoretto, M., De Lathauwer, L., Suykens, J.A.: Nuclear norms for tensors and their use for convex multilinear estimation. Linear Algebra Appl. **43** (2010)

28. Sun, J., Tao, D., Faloutsos, C.: Beyond streams and graphs: dynamic tensor analysis. In: Proceedings of the 12th ACM SIGKDD International Conference on Knowledge Discovery and Data Mining, pp. 374–383. ACM (2006)

29. Toh, K.C., Yun, S.: An accelerated proximal gradient algorithm for nuclear norm regularized linear least squares problems. Pac. J. Optim. **6**(615–640), 15 (2010)

30. Tomioka, R., Hayashi, K., Kashima, H.: Estimation of low-rank tensors via convex optimization. arXiv preprint (2010). arXiv:1010.0789

31. Wright, J., Ganesh, A., Rao, S., Peng, Y., Ma, Y.: Robust principal component analysis: exact recovery of corrupted low-rank matrices via convex optimization. In: Advances in Neural Information Processing Systems, pp. 2080–2088 (2009)

32. Xiuwen Yi, Y., Zheng, J.: ST-MVL: filling missing values in geo-sensory time series data. In: IJCAI 2016 (2016)

33. Yao, L., Ruan, W., Sheng, Q.Z., Li, X., Falkner, N.J.: Exploring tag-free RFID-based passive localization and tracking via learning-based probabilistic approaches. In: Proceedings of the 23rd ACM International Conference on Conference on Information and Knowledge Management, pp. 1799–1802. ACM (2014)

Social Network Analytics (I)

Group-Level Influence Maximization
with Budget Constraint

Qian Yan[1], Hao Huang[1]($\boxtimes$), Yunjun Gao[2], Wei Lu[3], and Qinming He[2]

[1] State Key Laboratory of Software Engineering, Wuhan University, Wuhan, China
{qy,haohuang}@whu.edu.cn
[2] College of Computer Science and Technology, Zhejiang University,
Hangzhou, China
{gaoyj,hqm}@zju.edu.cn
[3] DEKE, School of Information, MOE, Renmin University of China, Beijing, China
lu-wei@ruc.edu.cn

Abstract. Influence maximization aims at finding a set of seed nodes in a social network that could influence the largest number of nodes. Existing work often focuses on the influence of individual nodes, ignoring that infecting different seeds may require different costs. Nonetheless, in many real-world applications such as advertising, advertisers care more about the influence of groups (e.g., crowds in the same areas or communities) rather than specific individuals, and are very concerned about how to maximize the influence with a limited budget. In this paper, we investigate the problem of group-level influence maximization with budget constraint. Towards this, we introduce a statistical method to reveal the influence relationship between the groups, based on which we propose a propagation model that can dynamically calculate the influence spread scope of seed groups, following by presenting a greedy algorithm called GLIMB to maximize the influence spread scope with a limited cost budget via the optimization of the seed-group portfolio. Theoretical analysis shows that GLIMB can guarantee an approximation ratio of at least $(1 - 1/\sqrt{e})$. Experimental results on both synthetic and real-world data sets verify the effectiveness and efficiency of our approach.

1 Introduction

Given a social network, influence maximization aims at finding a subset of nodes (refer to as seeds) that could influence the largest number of nodes [7]. Over the last decade, this problem has received considerable attention due to its key importance in applications such as epidemic prevention, public opinion monitoring and viral marketing, in which local influence relationships between people may lead to an unexpectedly wide spread of disease, ideas, and product adoption [1,2].

Existing studies on influence maximization mostly focus on the influence of individuals [11], ignoring different costs required for infecting different seeds [13]. Nonetheless, analyzing the influence of specific individuals is trivial in many

S. Candan et al. (Eds.): DASFAA 2017, Part I, LNCS 10177, pp. 625–641, 2017.
DOI: 10.1007/978-3-319-55753-3_39

real-world scenarios, and cost budgets for infecting seeds are usually limited in practice. For example, to prevent and control epidemic diseases, establishing epidemic prevention stations is a common method. The siting of the stations depends on, firstly, the influence of crowds grouped in the same areas rather than the influence of individuals, and secondly, the cost of establishing a station in each area. The final goal, as a rule, is to make the best of the cost budget and prevent the diseases as much as possible. Analogous considerations also exist in advertising and promotional activities.

Driven by the practical applications above, in this paper, we study the problem of group-level influence maximization with budget constraint. A straightforward solution could be extending the existing individual-level approaches to solve this problem, since the influence of a group can be considered to be the sum of the influence of each group member [11]. Nonetheless, this solution has three drawbacks. (1) As the number of individuals is much greater than that of groups, analysing the influence of individuals instead of groups is much more computationally expensive. (2) Exact and clear influence relationship between individuals are hard to obtain. For example, although an epidemic prevention station can identify an infected person, it is difficult to find out which one infected him or her. (3) Cost estimation is based mostly on groups, e.g., crowds in the same geographical regions or human social communities, but few based on individuals.

To avoid the drawbacks of individual-level approaches, we propose to analyse the influence relationship at the level of groups. To this end, we adopt association probability to describe the influence relationship between a group pair. With historical infection data sets, we learn the association probability by checking the conditional independence between the infection statuses of each two groups, and construct an influence relationship graph. With the graph, we present an influence propagation model which can calculate the influence spread scope of any group or group set.

Based on the aforementioned group-level influence relationship graph and influence propagation model, we propose GLIMB (**G**roup-**L**evel **I**nfluence **M**aximization with **B**udget constraint) algorithm to approximate the optimal seed groups that maximize the influence spread scope with a limited cost budget. Towards this, GLIMB keeps searching the group that maximizes the incremental spread scope over cost ratio. Before the ending of searching, GLIMB checks whether there is an alternative group or group portfolio that can bring a greater incremental spread scope. Theoretical analysis shows that GLIMB provides at least a $(1 - 1/\sqrt{e})$-approximation. Experimental results on synthetic and real-world data sets verify the effectiveness and efficiency of our approach.

The remaining sections are organized as follows. In Sect. 2, we review the related work. In Sect. 3, we introduce how to construct the influence relationship graph and model the influence propagation at the level of groups, followed by presenting our GLIMB algorithm in Sect. 4. Experimental results and our findings are reported in Sect. 5 before concluding the paper in Sect. 6.

2   Related Work

The related work to group-level influence maximization with budget constraint can be classified into three categories, i.e., (1) influence relationship modeling, (2) individual-level influence maximization, and (3) influence maximization with budget constraint.

Influence relationship modeling aims at inferring the influence relationship between entities. Existing work focuses on inferring individual-level influence relationship based on historical infection statuses and infection time. Individuals that are sequentially infected within a time interval are regarded to have influence relationship [4,12]. Nonetheless, this idea is not appropriate to inferring group-level influence relationship. Because that in a group, the infection time of different individuals often vary a lot so that the time interval between the infections of two groups is hard to be determined. To model the group-level influence relationship, only a few approaches have been proposed. COLD model [5] carries out this work via subject analysis, but cannot construct an influence relationship graph for influence maximization. CSI model [11] regards the individual-level influence relationship across two groups as the group-level influence relationship, but requires the influence relationship between individuals, which is hard to obtain in practice. As CSI needs to calculate the strength of influence relationship between each two individuals, its time complexity is $O(n^2)$, where n is the number of individuals.

Individual-level influence maximization tries to find top k individuals that can maximize the expected influence spread scope. The existing approaches to this problem can be divided into two types, namely (1) greedy searching approaches [1,7,15,16], which utilize the submodularity of influence propagation model and keep selecting individuals that maximize current incremental spread scope, and (2) heuristic searching approaches, which efficiently identify the candidate seeds satisfying some heuristic rules, e.g., having the highest degree [3], influence ranking [6] or local influence [2]. Nonetheless, these approaches mostly still require the influence relationship between individuals.

Influence maximization with budget constraint considers the cost of each seed, and tries to make the best of a cost budget to maximize the influence spread scope. Only a few literature [10,13] addresses this problem with a basic idea of iteratively selecting a candidate seed that maximizes the incremental spread scope over cost ratio. Among the existing approaches, the best guarantee for the approximation ratio is $(1 - 1/\sqrt{e})$ [13], while our proposed GLIMB algorithm provides at least a $(1 - 1/\sqrt{e})$-approximation.

3   Group-Level Influence Propagation Model

In this paper, we assume that the underlying influence propagation between the individuals follows the IC (Independent Cascade) model, which is one of the most commonly used influence propagation models for individuals. Table 1 summarizes the notations. Given a set $\mathbf{S} \subseteq \mathbf{V}$ ($\mathbf{V} = \{v_1, \ldots, v_n\}$ refers to the set of

n individuals) of seed individuals, IC model works as follows: Let $\mathbf{S}_t$ be the set of individuals newly infected at time t, with $\mathbf{S}_0 = \mathbf{S}$ and $\mathbf{S}_t \cap \mathbf{S}_{t+1} = \emptyset$. In round $t+1$, each infected individual $u \in \mathbf{S}_t$ tries to infect its uninfected neighbors (i.e., uninfected individuals having influence relationship with u) in $\mathbf{V} \backslash \bigcup_{0 \leq j \leq t} \mathbf{S}_j$ independently with probability $p_{u,v}$. When there are multiple infected individuals trying to infect the same uninfected individual simultaneously, the infections can be carried out in any order. If the influence relationship between individuals is given, IC model can help calculate the influence spread scope $\sigma(\mathbf{S})$ of $\mathbf{S}$, which is the expected number of infected individuals given seed set $\mathbf{S}$.

Nevertheless, in many real-world application scenarios of influence maximization, the influence relationship between individuals is hard to obtain. It is often the case that available data resources only include the historical infection state $s_i^k \in \{1,0\}$ of each individual $v_i \in \mathbf{V}$ in the k-th ($k \in \{1, \ldots, \kappa\}$) outbreak of an infection event, forming a historical infection data set $\mathbf{D} = \left\{ s_i^k \in \{1,0\} \mid 1 \leq i \leq n, 1 \leq k \leq \kappa \right\}$. In order to find out which seeds can bring the greatest influence spread scope, the influence propagation model should be constructed in advance. In this paper, we propose to carry out this work at the granularity

Table 1. Notations

Symbol	Description	Symbol	Description
$\mathbf{V}$	The set of individuals	$\mathbf{M}$	The set of groups of individuals in $\mathbf{V}$
n	The number of individuals in $\mathbf{V}$	m	The number of groups in $\mathbf{M}$
v_i	The i-th individual in $\mathbf{V}$	M_i	The i-th group in $\mathbf{M}$
$\mathbf{S}$	The set of seed groups, $\mathbf{S} \subseteq \mathbf{M}$	κ	The number of infection outbreaks
s_i^k	The historical infection state of v_i in the k-th infection outbreak	$\mathbf{D}$	The historical infection data set that records each s_i^k, where $k \in \{1, \ldots, \kappa\}$
X	Any group in $\mathbf{M}$	$\lvert X \rvert$	The number of individuals in X
x	Infection status value of X, $x \in \{0,1\}$	$G(\mathbf{M}, \mathbf{E}, \mathbf{W})$	The group-level influence relationship graph
$\mathbf{E}$	The set of directed edges in $G(\mathbf{M}, \mathbf{E}, \mathbf{W})$	$\mathbf{W}$	The set of edge weights of edges in $\mathbf{E}$
W_{ij}	The edge weight of directed edge (M_i, M_j)	p_{ij}	The probability that M_i can influence M_j
$\mathbf{N}'_{M_i}$	The set of groups that can directly influence M_i, $\mathbf{N}'_{M_i} = \{M_j \mid (M_j, M_i) \in \mathbf{E}\}$	$\mathbf{N}_{M_i}$	The set of groups that can be directly influenced by M_i, $\mathbf{N}_{M_i} = \{M_j \mid (M_i, M_j) \in \mathbf{E}\}$
θ_i	The infection ratio of M_i	θ_i^t	The infection ratio of M_i at time t
$\theta_{i \to j}$	The pass ratio of influence on a directed edge (M_i, M_j) from M_i to M_j	$\theta_{i \to j}^t$	The pass ratio of influence on a directed edge (M_i, M_j) from M_i to M_j at time t
$\theta_{\mathbf{S} \to j}$	The pass ratio of influence from $\mathbf{S}$ to M_j	$\sigma(\mathbf{S})$	The influence spread scope of $\mathbf{S}$
$\mathbf{M}_{\mathbf{S} \to j}$	The set of groups in all the path from any seed group in $\mathbf{S}$ to M_j	$\mathbf{\Gamma}_{\mathbf{S} \to j}$	The set of groups in $\mathbf{M}_{\mathbf{S} \to j}$ that can directly influence M_j
$I_{i \to j}$	The set of individuals in M_j that are infected by M_i	$I_{\mathbf{S} \to j}$	The set of individuals in M_j that are infected by $\mathbf{S}$
$c(M_i)$	The cost of infecting M_i	b	The cost budget
$\triangle\sigma(\mathbf{S}, M_i)$	The incremental spread scope caused by adding M_i in to $\mathbf{S}$	$\delta(\mathbf{S}, M_i)$	The ratio of incremental spread scope $\triangle\sigma(\mathbf{S}, M_i)$ over cost $c(M_i)$
$\mathbf{L}$	The set of candidate seed groups	$\mathbf{P}$	The set of candidate seed group pairs

of groups via the following two steps, namely (1) revealing influence relationship between groups, and (2) modeling influence propagation between groups.

3.1 Revealing Influence Relationship Between Groups

Statistically speaking, if a group of individuals can influence another group, there should be an association relationship between the two groups. The strength of the influence can be reflected by an association probability. To check association relationship, statistical independence testing is a commonly used approach. Formally, for any infection status values $x \in \{0,1\}$ and $y \in \{0,1\}$ of groups X and Y, if relationship $p(x,y) = p(x)p(y)$ always holds, then X and Y are called independent to each other, denoted as $(X \perp\!\!\!\perp Y)$. Here, probabilities $p(x)$, $p(y)$ and $p(x,y)$ can be estimated based on the historical infection data set $\mathbf{D}$ under the assumption that individuals in a same group are homogeneous [12]. With this homogeneous assumption, the infection probability of each individual in a group is equal to each other, and can be regarded as the infection probability of the group. In other words, if we observed that 20% of individuals in a group were infected by a disease, it indicates that the disease has a 20 percent chance of infecting each individual of this group. Hence, we utilize $\mathbf{D}$ to estimate the infection probability of each group. For example, in the k-th outbreak of an infection event, $p_k(x = 1) = \sum_{v_i \in X} s_i^k / |X|$, where $|X|$ refers to the number of individuals in X; then, considering all κ outbreaks of this infection event, we have $p(x = 1) = \sum_{k=1}^{\kappa} p_k(x = 1)/\kappa$ and $p(x = 0) = 1 - p(x = 1)$. Moreover, according to the definitions of joint probability and conditional probability, we can also estimate probabilities $p(x,y)$ and $p(x|y)$ based on $\mathbf{D}$.

Nonetheless, sometimes, the independence between X and Y is not enough to comprehensively express the association relationship between groups X and Y. Because X may influence Y directly, or X may influence Y through group Z even if X cannot directly influence Y. Both cases result in $(X \not\perp\!\!\!\perp Y)$. To avoid this ambiguity, we adopt conditional independence to reveal the direct association relationship between groups X and Y. For any infection status values $x \in \{0,1\}$, $y \in \{0,1\}$ and $z \in \{0,1\}$ of groups X, Y and Z, if $p(x,y \mid z) = p(y \mid z)p(x|z)$ always holds, then X is independent of Y conditioned on Z, denoted as $(X \perp\!\!\!\perp Y \mid Z)$. The physical interpretation of $(X \perp\!\!\!\perp Y \mid Z)$ is the independence between X and Y when the mediating effect of Z is excluded.

In information theory, conditional mutual information is commonly used to quantify the conditional independence. Formally, given Z, the conditional mutual information of X and Y, denoted by $Inf(X, Y \mid Z)$, is calculated as

$$Inf(X, Y \mid Z) = \sum_{z \in \{0,1\}} p(z) \sum_{x \in \{0,1\}} \sum_{y \in \{0,1\}} p(x,y \mid z) log_2 \frac{p(x,y \mid z)}{p(x \mid z)p(y|z)}. \tag{1}$$

$Inf(X, Y \mid Z) = 0$ indicates that $(X \perp\!\!\!\perp Y \mid Z)$. A higher $Inf(X, Y \mid Z)$ indicates a stronger direct association relationship between X and Y given Z. Moreover, conditional mutual information has the following properties.

Theorem 1. *For any variable sets X, Y and Z, if the mutual information $Inf(X,Y)$ of X and Y is equal to 0, then relationship $Inf(X,Y \mid Z) = 0$ always holds.*

Proof. Since $Inf(X,Y) = Inf(X,Y \mid \emptyset)$, when $Inf(X,Y) = 0$, we have $Inf(X,Y \mid \emptyset) = 0$, indicating that X is independent of Y conditioned on $\emptyset$, i.e. $(X \perp\!\!\!\perp Y \mid \emptyset)$. Then, we can have $(X \perp\!\!\!\perp Y \mid \emptyset \cup Z)$ which is equal to $(X \perp\!\!\!\perp Y \mid Z)$, due to strong union property of conditional independence relation, indicating that $Inf(X,Y \mid Z) = 0$. ∎

Theorem 2. *For any variable sets X, Y, Z and W, if $Inf(X,Y \mid Z) = 0$, then relationship $Inf(X,Y \mid Z \cup W) = 0$ always holds.*

Proof. $Inf(X,Y \mid Z) = 0$ indicates that $(X \perp\!\!\!\perp Y \mid Z)$. Then, due to strong union property of conditional independence relation, relationship $(X \perp\!\!\!\perp Y \mid Z \cup W)$ also holds, indicating that $Inf(X,Y \mid Z \cup W) = 0$. ∎

With the help of conditional mutual information, if group X has a strong direct association relationship with group Z, then we can add a directed edge from X to Z and add Z into the neighbor set $\mathbf{N}_X$ of X, indicating that X can directly influence Z. When we check the direct association relationship between X and group $Y \notin \mathbf{N}_X$, the mediating effect of groups in $\mathbf{N}_X$ should be excluded by calculating $Inf(X,Y \mid \mathbf{N}_X)$. Based on the above basic ideas, Algorithm 1 provides a construction approach for the group-level influence relationship graph $G(\mathbf{M}, \mathbf{E}, \mathbf{W})$, in which $\mathbf{M}$ is the set of groups, $\mathbf{E}$ is the set of directed edges, and $\mathbf{W}$ is the set of edge weights.

Algorithm 1 takes as inputs the given group set $\mathbf{M}$, and the historical infection data set $\mathbf{D}$, which is used for calculating probabilities required in the computation of conditional mutual information. The algorithm first initializes $\mathbf{E}$, $\mathbf{W}$, and the neighbor set $\mathbf{N}_{M_i}$ for each group $M_i \in \mathbf{M}$ as empty sets (line 1), and calculates the mutual information $Inf(M_i, M_j)$ of each two groups in $\mathbf{M}$ (line 2), which is equal to the conditional mutual information $Inf(M_i, M_j \mid \emptyset)$ of the

---

Algorithm 1. Construction of Group-Level Influence Relationship Graph

Input : Group set $\mathbf{M}$; historical infection data set $\mathbf{D}$.
Output: Influence relationship graph $G(\mathbf{M}, \mathbf{E}, \mathbf{W})$.
1 Initial $\mathbf{E} \leftarrow \emptyset$, $\mathbf{W} \leftarrow \emptyset$, and an empty neighbor set $\mathbf{N}_{M_i}$ for each group $M_i \in \mathbf{M}$;
2 Calculate $Inf(M_i, M_j)$ for each two groups $M_i, M_j \in \mathbf{M}$ $(i \neq j)$;
3 **for** each $M_i \in \mathbf{M}$ **do**
4   **for** each $M_j \in \mathbf{M}$ *($j \neq i$)* having $Inf(M_i, M_j) > 0$ **do**
5   **if** $Inf(M_i, M_j \mid \mathbf{N}_{M_i}) > \varepsilon$ **then**
6              $\mathbf{N}_{M_i} \leftarrow \mathbf{N}_{M_i} \cup \{M_j\}$;

7   **for** each $M_j \in \mathbf{N}_{M_i}$ **do**
8          $\mathbf{E} \leftarrow \mathbf{E} \cup \{(M_i, M_j)\}$;   //$(M_i, M_j)$ is a directed edge from M_i to M_j
9          $W_{ij} \leftarrow Inf(M_i, M_j \mid \mathbf{N}_{M_i} \backslash \{M_j\})$;   //$W_{ij}$ is the weight of edge (M_i, M_j)
10         $\mathbf{W} \leftarrow \mathbf{W} \cup \{W_{ij}\}$;

two groups conditioned on $\emptyset$. Then, for each group $M_i \in \mathbf{M}$, it identifies which of other groups $\{M_j \in \mathbf{M} \mid i \neq j\}$ have strong direct association relationship with group M_i by checking whether $Inf(M_i, M_j \mid \mathbf{N}_{M_i})$ is greater than a threshold ε (line 5). Instead of adopting a user-specified ε, we suggest to determine the ε based on mutual information $Inf(M_i, M_j)$ of each two groups in $\mathbf{M}$. Specifically, by performing K-means with $K = 2$, the non-zero values of mutual information can be classified into two parts. Let ε be the minimal value in the part containing greater mutual information values. Then, condition $Inf(M_i, M_j \mid \mathbf{N}_{M_i}) > \varepsilon$ helps find direct association relationship that are strong enough. Groups that satisfy the condition above will be added into the neighbor set $\mathbf{N}_{M_i}$ of M_i (line 6). Finally, for each $M_j \in \mathbf{N}_{M_i}$, we add the directed edge (M_i, M_j) into the edge set $\mathbf{E}$ (line 8), calculate the weight W_{ij} of this edge (line 9), followed by adding W_{ij} into the edge weight $\mathbf{W}$ (line 10).

The overall time complexity of Algorithm 1 is $O(mn\kappa + m^2 + 2^\alpha m^2)$, where m is the number of groups in $\mathbf{M}$, n is the number of individuals in all groups (usually $m \ll n$), κ is the number of historical infection outbreaks recorded in $\mathbf{D}$, and α is the maximal number of groups that can be directly influenced by each $M_i \in \mathbf{M}$ (i.e., $\alpha = \max_{1 \leq i \leq m} |\mathbf{N}_{M_i}|$). To be specific, statistics for the probabilities used for calculating conditional mutual information take $O(mn\kappa)$ time. Then, calculating mutual information in line 2 requires $O(m^2)$ time. The time complexity of the loop of line 3 is $O(2^\alpha m^2)$. In the loop of line 3, the most computationally expensive step is in line 5, i.e., the computation of conditional mutual information. For each M_i and M_j, it takes $O(2^{|\mathbf{N}_{M_i}|})$ time, where $|\mathbf{N}_{M_i}| \leq \alpha$ refers to the number of groups that can be directly influenced by group M_i. In practice, the influence of each group is usually limited, and only a few groups can be directly influenced by each group ($\alpha \ll m$). Furthermore, to reduce the time complexity of line 5, users can adopt a greater threshold ε to reduce the cardinality $|\mathbf{N}_{M_i}|$ of each set $\mathbf{N}_{M_i}$ and obtain a smaller α. Though, the compensation of a greater ε is that the constructed graph will have less edges which only capture the strongest direct association relationship. Besides, to avoid unnecessary calculation in line 5, we carry out a pruning in line 4. According to Theorems 1 and 2, if $Inf(M_i, M_j) = 0$, then $Inf(M_i, M_j \mid \mathbf{N}_{M_i}) = 0$ always holds. Thus, it is not necessary to calculate $Inf(M_i, M_j \mid \mathbf{N}_{M_i})$ for each $M_j \in \mathbf{M}$ ($j \neq i$) having $Inf(M_i, M_j) = 0$.

With the influence relationship graph constructed by Algorithm 1, in what follows, we introduce how the influence is propagated between the groups on the graph.

3.2 Modeling Influence Propagation Between Groups

In this section, we first introduce (1) the pass ratio of influence on each directed edge, based on which we elaborate (2) the rules of influence propagation on the influence relationship graph, followed by presenting (3) the function of influence spread scope, which calculates the expected number of of infected individuals given seed groups.

Pass Ratio of Influence. For groups M_i, $M_j \in \mathbf{M}$ in the influence relationship graph $G(\mathbf{M}, \mathbf{E}, \mathbf{W})$, if there is a directed edge $(M_i, M_j) \in \mathbf{E}$ from M_i to M_j, then M_i can directly influence M_j, and the strength of influence is proportional to the edge weight $W_{ij} \in \mathbf{W}$. Among all the groups $\mathbf{N}'_{M_j} = \{M_\ell \in \mathbf{M} \mid (M_\ell, M_j) \in \mathbf{E}\}$ that can directly influence M_j, group $M_i \in \mathbf{N}'_{M_j}$ can influence group M_j with a probability $p_{ij} = W_{ij} / \sum_{M_\ell \in \mathbf{N}'_{M_j}} W_{\ell j}$.

Furthermore, according to IC model, uninfected individuals can only be influenced by infected ones. Thus, when the infection ratio θ_j of M_j is less than 1, M_i with a higher infection ratio θ_i (i.e., more infected individuals) has more chances to influence M_j.

In brief, the pass ratio of influence on a directed edge $(M_i, M_j) \in \mathbf{E}$, denoted by $\theta_{i \to j}$, is affected by infection ratio θ_i of the influence source M_i, the influence probability p_{ij} from M_i to M_j, and the infection ratio θ_j of the target group M_j. Formally, $\theta_{i \to j}$ can be calculated as $\theta_{i \to j} = \theta_i \times p_{ij} \times (1 - \theta_j)$. The physical interpretation of $\theta_{i \to j}$ is the newly-increased infection ratio of M_j caused by the influence from M_i.

Rules of Influence Propagation. According to IC model, infected individuals at time t (or round t) only have infectivity at time $t + 1$. Analogously, in the process of group-level influence propagation, we calculate and record the infection ratio θ_i^t of each $M_i \in \mathbf{M}$ at time t, based on which we can deduce each infection ratio θ_i^{t+1} at time $t + 1$.

At time $t = 0$, if the set $\mathbf{S}$ of seed groups is given, the expected infection ratio θ_j^0 of $M_j \in \mathbf{S}$ can be predicted based on historical infection data.

At time $t > 0$, if the infection ratio of a group $M_i \in \mathbf{M}$ has increased at time $t - 1$, then M_i will try to influence its neighbors $M_\ell \in \mathbf{N}_{M_i}$ through the directed edge $(M_i, M_\ell) \in \mathbf{E}$. For a target group $M_\ell \in \mathbf{N}_{M_i}$, (1) if $\mathbf{N}'_{M_\ell} = \{M_i\}$, i.e., M_ℓ will be only influenced by M_i, then the infection ratio θ_ℓ^t of M_ℓ at time t can be calculated as $\theta_\ell^t = \theta_\ell^{t-1} + \theta_{i \to \ell}^t$, where $\theta_{i \to \ell}^t = \theta_i^{t-1} \times p_{i\ell} \times (1 - \theta_\ell^{t-1})$; (2) if $\mathbf{N}'_{M_\ell} = \{M_i, M_k\}$ and infection ratio of M_k has increased at time $t - 1$, then M_i and M_k will influence M_ℓ simultaneously at time t. Following the rules of IC model, the influences can be carried out in any order. Let M_i execute the influence first, we can calculate $\theta_{i \to \ell}^t$ in the same way, and exclude this newly-increased infection ratio of M_ℓ in the calculation of $\theta_{k \to \ell}^t$ to avoid repeatedly infecting the same part in M_ℓ. Specifically, $\theta_{k \to \ell}^t = \theta_k^{t-1} \times p_{k\ell} \times (1 - \theta_\ell^{t-1} - \theta_{i \to \ell}^t)$, and the infection ratio of M_ℓ at time t can be updated by $\theta_\ell^t = \theta_\ell^{t-1} + \theta_{i \to \ell}^t + \theta_{k \to \ell}^t$. The above calculation rules can be easily extended to the cases that more groups influence M_ℓ simultaneously.

Moreover, we consider that the process of influence propagation is acyclic, i.e., once M_i pass its influence to M_j, M_j will not pass back its influence to M_i any more. This consideration is commonly used in influence maximization [13]. When the infection ratio of each group does not increase any more, the influence propagation will end.

Function of Influence Spread Scope. According to the rules of influence propagation, each $M_j \in \mathbf{M} \backslash \mathbf{S}$ can be influenced by $\mathbf{S}$ iff there is at least one directed path from any seed group to M_j. Let $\mathbf{M}_{\mathbf{S}\to j}$ denote the union of groups in all the paths from any seed group in $\mathbf{S}$ to M_j, group set $\mathbf{\Gamma}_{\mathbf{S}\to j} = \mathbf{M}_{\mathbf{S}\to j} \cap \mathbf{N}'_{M_j}$ refers to the groups that can directly influence M_j in $\mathbf{M}_{\mathbf{S}\to j}$. By combining the influence from each $M_k \in \mathbf{\Gamma}_{\mathbf{S}\to j}$ to M_j, we have the recursion formula for the pass ratio of influence from $\mathbf{S}$ to M_j, i.e.,

$$
\theta_{\mathbf{S}\to j} = \begin{cases} 1 - \prod_{M_k \in \mathbf{\Gamma}_{\mathbf{S}\to j}}(1 - \theta_{\mathbf{S}\to k} \times p_{kj}), & M_j \notin \mathbf{S} \\ \theta_j^0, & M_j \in \mathbf{S} \end{cases} \tag{2}
$$

Let $|M_j|$ denote the number of individuals in group M_j, the function $\sigma(\mathbf{S})$ of influence spread scope can be be formulated as follows.

$$
\sigma(\mathbf{S}) = \sum_{M_j \in \mathbf{M}} |M_j| \times \theta_{\mathbf{S}\to j}. \tag{3}
$$

Function $\sigma(\mathbf{S})$ has the following properties.

Theorem 3. *Function $\sigma(\mathbf{S})$ is monotone.*

Proof. We first proof that $\theta_{\mathbf{S}\to j}$ is monotone increasing, i.e., given $G(\mathbf{M}, \mathbf{E}, \mathbf{W})$ and $\mathbf{S} \subseteq \mathbf{T} \subseteq \mathbf{M}$, for any $M_j \in \mathbf{M}$, relationship $\theta_{\mathbf{T}\to j} \geq \theta_{\mathbf{S}\to j}$ always holds. When $M_j \in \mathbf{S}$, according to the definition of $\theta_{\mathbf{S}\to j}$, we have $\theta_{\mathbf{S}\to j} = \theta_{\mathbf{T}\to j} = \theta_j^0$. Hence, relationship $\theta_{\mathbf{S}\to j} \leq \theta_{\mathbf{T}\to j}$ holds for $M_j \in \mathbf{S}$. When $M_j \notin \mathbf{S}$, we can proof relationship $\theta_{\mathbf{S}\to j} \leq \theta_{\mathbf{T}\to j}$ by induction. (1) For each $M_k \in \mathbf{M}_{\mathbf{S}\to j}$ which is directed influenced by $\mathbf{S}$, if there is a directed path from $\mathbf{T}\backslash\mathbf{S}$ to M_k, i.e., $\theta_{\mathbf{T}\backslash\mathbf{S}\to k} > 0$, then $\theta_{\mathbf{T}\to k} > \theta_{\mathbf{S}\to k}$; otherwise $\theta_{\mathbf{T}\to k} = \theta_{\mathbf{S}\to k}$. (2) For each $M_\ell \in \mathbf{M}_{\mathbf{S}\to j}$ which is directed influenced by M_k, since $\theta_{\mathbf{T}\to k} \geq \theta_{\mathbf{S}\to k}$, we have relationship $1 - \prod_{M_k \in \mathbf{\Gamma}_{\mathbf{S}\to\ell}}(1 - \theta_{\mathbf{T}\to k} \times p_{k\ell}) \geq 1 - \prod_{M_k \in \mathbf{\Gamma}_{\mathbf{S}\to\ell}}(1 - \theta_{\mathbf{S}\to k} \times p_{k\ell}) = \theta_{\mathbf{S}\to\ell}$. Moreover, since $\mathbf{S} \subseteq \mathbf{T} \subseteq \mathbf{M}$, we have $\mathbf{M}_{\mathbf{S}\to\ell} \subseteq \mathbf{M}_{\mathbf{T}\to\ell}$, and thus $\mathbf{\Gamma}_{\mathbf{S}\to\ell} \subseteq \mathbf{\Gamma}_{\mathbf{T}\to\ell}$. Then, relationship $\theta_{\mathbf{T}\to\ell} = 1 - \prod_{M_k \in \mathbf{\Gamma}_{\mathbf{T}\to\ell}}(1 - \theta_{\mathbf{T}\to k} \times p_{k\ell}) \geq 1 - \prod_{M_k \in \mathbf{\Gamma}_{\mathbf{S}\to\ell}}(1 - \theta_{\mathbf{T}\to k} \times p_{k\ell})$ holds, and hence relationship $\theta_{\mathbf{T}\to\ell} \geq \theta_{\mathbf{S}\to\ell}$ holds. (3) By induction, we can proof that for each group $M_i \in \mathbf{M}_{\mathbf{S}\to j}$, relationship $\theta_{\mathbf{T}\to i} \geq \theta_{\mathbf{S}\to i}$ holds. (4) If there is at least one path from $\mathbf{S}$ to M_j, then relationship $\theta_{\mathbf{T}\to j} \geq \theta_{\mathbf{S}\to j}$ holds; otherwise, $\theta_{\mathbf{S}\to j} = 0$, and relationship $\theta_{\mathbf{T}\to j} \geq \theta_{\mathbf{S}\to j}$ also holds since $\theta_{\mathbf{T}\to j} \geq 0$. In summary, $\theta_{\mathbf{S}\to j}$ is monotone increasing.

Since a non-negative linear combination of monotone increasing functions is still a monotone increasing function, function $\sigma(\mathbf{S})$ is a monotone increasing function. ∎

Theorem 4. *Function $\sigma(\mathbf{S})$ is submodular.*

Proof. We first proof the submodularity of $\theta_{\mathbf{S}\to j}$, i.e., given $G(\mathbf{M}, \mathbf{E}, \mathbf{W})$ and $\mathbf{S} \subseteq \mathbf{T} \subseteq \mathbf{M}$, for any $M_i, M_j \in \mathbf{M}$, relationship $\theta_{\mathbf{S}\cup\{M_i\}\to j} - \theta_{\mathbf{S}\to j} \geq \theta_{\mathbf{T}\cup\{M_i\}\to j} - \theta_{\mathbf{T}\to j}$ always holds. (1) When $M_i \in \mathbf{S}$, we have $\theta_{\mathbf{S}\cup\{M_i\}\to j} - \theta_{\mathbf{S}\to j} = \theta_{\mathbf{T}\cup\{M_i\}\to j} - \theta_{\mathbf{T}\to j} = 0$. Hence, the relationship $\theta_{\mathbf{S}\cup\{M_i\}\to j} - \theta_{\mathbf{S}\to j} \geq \theta_{\mathbf{T}\cup\{M_i\}\to j} - \theta_{\mathbf{T}\to j}$ holds for $M_i \in \mathbf{S}$. (2) When $M_i \in \mathbf{T}\backslash\mathbf{S}$, we have $\theta_{\mathbf{T}\cup\{M_i\}\to j} - \theta_{\mathbf{T}\to j} = 0$. Since $\theta_{\mathbf{S}\cup\{M_i\}\to j} - \theta_{\mathbf{S}\to j} \geq 0$, the relationship $\theta_{\mathbf{S}\cup\{M_i\}\to j} - \theta_{\mathbf{S}\to j} \geq \theta_{\mathbf{T}\cup\{M_i\}\to j} - \theta_{\mathbf{T}\to j}$ holds for $M_i \in \mathbf{T}\backslash\mathbf{S}$. (3) When $M_i \notin \mathbf{T}$, we proof the submodularity of $\theta_{\mathbf{S}\to j}$

at the granularity of individuals. Let $\theta_{i\rightarrow j}$ denotes the newly-increased infection ratio of M_j caused by M_i regardless of the influence of any other seed groups. As $M_j \notin \mathbf{S} \subseteq \mathbf{T}$, $\theta_j^0 = 0$. Hence, $|M_j| \times \theta_{i\rightarrow j}$ is the expected number of individuals (directly and indirectly) infected by M_i. We denote these individuals by set $I_{i\rightarrow j}$. Similarly, sets $I_{\mathbf{S}\rightarrow j}$ and $I_{\mathbf{T}\rightarrow j}$ refer to the individuals (directly and indirectly) infected by $\mathbf{S}$ and $\mathbf{T}$, respectively, regardless of the influence of any other seed groups. $I_{\mathbf{S}\rightarrow j} \subseteq I_{\mathbf{T}\rightarrow j}$ since $\mathbf{S} \subseteq \mathbf{T}$. With sets $I_{i\rightarrow j}$, $I_{\mathbf{S}\rightarrow j}$ and $I_{\mathbf{T}\rightarrow j}$, we have $\theta_{\mathbf{S}\cup\{M_i\}\rightarrow j} - \theta_{\mathbf{S}\rightarrow j} = (|I_{i\rightarrow j}| - |I_{i\rightarrow j} \cap I_{\mathbf{S}\rightarrow j}|)/|M_j|$, and $\theta_{\mathbf{T}\cup\{M_i\}\rightarrow j} - \theta_{\mathbf{T}\rightarrow j} = (|I_{i\rightarrow j}| - |I_{i\rightarrow j} \cap I_{\mathbf{T}\rightarrow j}|)/|M_j|$. As $I_{\mathbf{S}\rightarrow j} \subseteq I_{\mathbf{T}\rightarrow j}$, we have $|I_{i\rightarrow j} \cap I_{\mathbf{S}\rightarrow j}| \leq |I_{i\rightarrow j} \cap I_{\mathbf{T}\rightarrow j}|$, and thus $(|I_{i\rightarrow j}| - |I_{i\rightarrow j} \cap I_{\mathbf{S}\rightarrow j}|)/|M_j| \geq (|I_{i\rightarrow j}| - |I_{i\rightarrow j} \cap I_{\mathbf{T}\rightarrow j}|)/|M_j|$, indicating that relationship $\theta_{\mathbf{S}\cup\{M_i\}\rightarrow j} - \theta_{\mathbf{S}\rightarrow j} \geq \theta_{\mathbf{T}\cup\{M_i\}\rightarrow j} - \theta_{\mathbf{T}\rightarrow j}$ holds for $M_i \notin \mathbf{T}$, $M_j \notin \mathbf{S}$.

In brief, $\theta_{\mathbf{S}\rightarrow j}$ is submodular. Since a non-negative linear combination of submodular functions is still a submodular function, function $\sigma(\mathbf{S})$ is also submodular. ∎

Corollary 1. *Given* $G(\mathbf{M}, \mathbf{E}, \mathbf{W})$, *relationship* $\sigma(\{M_i\}) \geq \sigma(\mathbf{S} \cup \{M_i\}) - \sigma(\mathbf{S})$ *holds for any group set* $\mathbf{S} \subset \mathbf{M}$ *and group* $M_i \in \mathbf{M}\backslash\mathbf{S}$.

Proof. Since $\sigma(\mathbf{S})$ is submodular, i.e., given $G(\mathbf{M}, \mathbf{E}, \mathbf{W})$ and $\mathbf{T} \subseteq \mathbf{S} \subseteq \mathbf{M}$, for any $M_i \in \mathbf{M}$, relationship $\sigma(\mathbf{T} \cup \{M_i\}) - \sigma(\mathbf{T}) \geq \sigma(\mathbf{S} \cup \{M_i\}) - \sigma(\mathbf{S})$ always holds, we have $\sigma(\{M_i\}) = \sigma(\emptyset \cup \{M_i\}) - \sigma(\emptyset) \geq \sigma(\mathbf{S} \cup \{M_i\}) - \sigma(\mathbf{S})$. ∎

4   The GLIMB Algorithm

Given the constructed influence relationship graph $G(\mathbf{M}, \mathbf{E}, \mathbf{W})$, the function $\sigma(\mathbf{S})$ of influence spread scope, and the cost $c(M_i)$ for each group $M_i \in \mathbf{M}$, in this section, we address the problem of finding a set $\mathbf{S}$ of groups that maximizes $\sigma(\mathbf{S})$ under a cost budget constraint b, i.e., $\max_{\mathbf{S}\subseteq\mathbf{M}} \sigma(\mathbf{S})$, s.t. $\sum_{M_i \in \mathbf{S}} c(M_i) \leq b$.

An intuitive strategy, which can be denoted as `NaiveGreedy`, is to select at each step a group M_i that maximizes the incremental spread scope over cost ratio $\delta(\mathbf{S}, M_i) = (\sigma(\mathbf{S}\cup\{M_i\})-\sigma(\mathbf{S}))/c(M_i)$ if $c(M_i)$ is less than the remaining budget [10,13]. However, this strategy is easy to plunge into local optima. For example, assume there are three equal-sized groups M_1, M_2 and M_3, $\theta_1^0 = \theta_2^0 = \theta_3^0$, $c(M_1) = 0.9$, $c(M_2) = c(M_3) = 2$, and cost budget $b = 2$. Let M_1 be an isolated group, while M_2 and M_3 are connected with influence probability one to each other. Then, although M_2 or M_3 can maximize the spread scope under the cost budget constraint, `NaiveGreedy` will select M_1 as the seed group and stop, since $\theta_1^0|M_1|/c(M_1) > (\theta_2^0|M_2| + \theta_3^0|M_3|)/c(M_2) = (\theta_2^0|M_2| + \theta_3^0|M_3|)/c(M_3)$.

To avoid the drawbacks of `NaiveGreedy` method, an improved solution known as `ImprovedGreedy` records the set $\mathbf{S}_{naive}$ of seed groups selected by `NaiveGreedy`, and identifies the group M_{max} having the largest influence scope and a cost no more than b, followed by returning set $\mathbf{S} = \arg\max (\sigma(\mathbf{S}_{naive}), \sigma(\{M_{max}\}))$ as the seed group set. It has been proven that `ImprovedGreedy` provides an approximation ratio of $(1 - 1/\sqrt{e})$, when the function of influence spread scope is monotone and submodular [13].

Algorithm 2. The GLIMB Algorithm

Input : $G(\mathbf{M}, \mathbf{E}, \mathbf{W})$; budget b; cost $c(M_i)$ and infection ratio θ_i^0 for each $M_i \in \mathbf{M}$.

Output: Set $\mathbf{S}$ of seed groups.

1 Initial $\mathbf{S} \leftarrow \emptyset$, $\mathbf{L} \leftarrow \mathbf{M}$;   //$\mathbf{L}$ records the candidates for seed groups
2 Calculate $\sigma(\{M_i\})$ for each $M_i \in \mathbf{M}$, $M_\lambda \leftarrow \arg\max_{M_i \in \mathbf{M}} \sigma(\{M_i\})$;
3 **while** $\mathbf{L} \neq \emptyset$ **do**
4      $\mathbf{L} \leftarrow \mathbf{L} \backslash \{M_i \in \mathbf{L} \mid c(M_i) > b - \sum_{M_j \in \mathbf{S}} c(M_j)\}$;
5      $\ell \leftarrow 0$, $MaxRatio \leftarrow 0$;
6   **for** $M_i \in \mathbf{L}$ **do**
7   **if** $\frac{\sigma(\{M_i\})}{c(M_i)} > MaxRatio$ **then**
8              $\delta(\mathbf{S}, M_i) \leftarrow \triangle\sigma(\mathbf{S}, M_i)/c(M_i)$; //$\triangle\sigma(\mathbf{S}, M_i) = \sigma(\mathbf{S} \cup \{M_i\}) - \sigma(\mathbf{S})$
9   **if** $\delta(\mathbf{S}, M_i) > MaxRatio$ **then**
10                  $MaxRatio \leftarrow \delta(\mathbf{S}, M_i)$, $\ell \leftarrow i$;

11   **if** $\forall M_k \in \mathbf{L} \backslash \{M_\ell\}$, $c(M_k) > b - \sum_{M_j \in \mathbf{S} \cup \{M_\ell\}} c(M_j)$ **then**
12   **if** $c(M_k) \geq c(M_\ell)$ **then**
13              $M_\ell \leftarrow \arg\max_{M_i \in \mathbf{L}, b \geq \sum_{M_j \in \mathbf{S} \cup \{M_i\}} c(M_j)} \triangle\sigma(\mathbf{S}, M_i)$;
14              $\mathbf{S} \leftarrow \mathbf{S} \cup \{M_\ell\}$, $\mathbf{L} \leftarrow \mathbf{L} \backslash \{M_\ell\}$;
15   **else**
16              $\mathbf{P} \leftarrow \{< M_i, M_j > \mid M_i, M_j \in \mathbf{L} \backslash \{M_\ell\},\ i \neq j,\ \triangle\sigma(\mathbf{S}, M_i) +$
                 $\triangle\sigma(\mathbf{S}, M_j) > \triangle\sigma(\mathbf{S}, M_\ell),\ c(M_i) + c(M_j) \leq b - \sum_{M_j \in \mathbf{S}} c(M_j)\}$;
17   **while** $\mathbf{P} \neq \emptyset$ **do**
18                  $< M_{i*}, M_{j*} > \leftarrow \arg\max_{< M_i, M_j > \in \mathbf{P}} \left(\triangle\sigma(\mathbf{S}, M_i) + \triangle\sigma(\mathbf{S}, M_j) \right)$;
19   **if** $\triangle\sigma\left(\mathbf{S} \cup \{M_{i*}\}, M_{j*}\right) > \triangle\sigma(\mathbf{S}, M_\ell)$ **then**
20                      $\mathbf{S} \leftarrow \mathbf{S} \cup \{M_{i*} \cup M_{j*}\}$, $\mathbf{L} \leftarrow \mathbf{L} \backslash \{M_{i*} \cup M_{j*}\}$;
21   break;
22   **else**
23                      $\mathbf{P} \leftarrow \mathbf{P} \backslash \{< M_{i*}, M_{j*} >\}$;

24   **else**
25          $\mathbf{S} \leftarrow \mathbf{S} \cup \{M_\ell\}$, $\mathbf{L} \leftarrow \mathbf{L} \backslash \{M_\ell\}$;
26   **if** $\mathbf{L} = \emptyset$ and $\sigma(\{M_\lambda\}) \geq \sigma(\mathbf{S})$ **then**
27          $\mathbf{S} \leftarrow \{M_\lambda\}$, $\mathbf{L} \leftarrow \mathbf{M} \backslash \{M_\lambda\}$;

Nevertheless, both NaiveGreedy and ImprovedGreedy have a high risk of waste budge. For example, (1) Case 1: when the remaining budget is 4, and there are still two candidates M_1 and M_2, of which the costs are 2 and 3, respectively, assume that $\delta(\mathbf{S}, M_1) = 1$ while $\delta(\mathbf{S}, M_2) = 0.8$, NaiveGreedy and ImprovedGreedy will select M_1 and stop, although candidate M_2 can bring a greater incremental spread scope, which is 2.4 (it is 2 for M_1); (2) Case 2: when the remaining budget is 4, and there are still two candidates M_1, M_2 and M_3, of which the costs are 3, 2 and 2, respectively, assume that $\delta(\mathbf{S}, M_1) = 1$ while

$\delta(\mathbf{S}, M_2) = \delta(\mathbf{S}, M_3) = 0.8$, `NaiveGreedy` and `ImprovedGreedy` will select M_1 and stop, although selecting the candidate portfolio of M_2 and M_3 can bring a greater incremental spread scope, which is 3.2 (it is 3 for M_1).

To avoid the above waste-budget cases, in Algorithm 2, we propose a novel algorithm called GLIMB for the problem of influence maximization with budget constraint.

The GLIMB algorithm takes as inputs the influence relationship graph $G(\mathbf{M}, \mathbf{E}, \mathbf{W})$, the cost budget b, the cost $c(M_i)$ for each $M_i \in \mathbf{M}$ (assume that $\forall M_i \in \mathbf{M}$, $c(M_i) \leq b$), and the expected infection ratio θ_i^0 for each $M_i \in \mathbf{M}$ which is used in the calculation of influence spread scope. It first initializes an empty set $\mathbf{S}$ to record the seed groups, and a set $\mathbf{L}$ which is initially set as $\mathbf{M}$ to record which candidate groups are left for $\mathbf{S}$ (line 1), followed by calculating $\sigma(\{M_i\})$ of each $M_i \in \mathbf{M}$ (line 2). Then, it iteratively searches the currently best candidate or candidate portfolio for seed groups. Each iteration has two *routine phases*, namely (1) removing each group having a cost greater than current remaining budget from set $\mathbf{L}$ (line 4), and (2) finding the group $M_\ell \in \mathbf{L}$ that can maximize the incremental spread scope over cost ratio $\delta(\mathbf{S}, M_\ell)$ for current $\mathbf{S}$ (lines 5–10), and two *extra phases* which are carried out before ending, namely (3) searching an alternative candidate or candidate portfolio (if any) that can bring a greater incremental spread scope (lines 11–23), and (4) checking whether the algorithm plunges into local optima with the strategy used by `ImprovedGreedy` (lines 26–27).

The first *extra phase* (lines 11–23) plays the central role to help avoid waste-budget cases. It works as follows. If each remaining candidate has a cost no less than $c(M_\ell)$, it is easy to prove that the remaining cost budget cannot afford more than one alternative candidate. Thus, in the groups that can be afforded by the remaining cost budget, we select the one that can bring the greatest incremental spread scope (line 13) as the latest seed group (line 14). Otherwise, we find out the set $\mathbf{P}$ of alternative candidate portfolios (a portfolio consists of two candidates) that can be afforded by the remaining cost budget (line 16). In $\mathbf{P}$, if there is a candidate portfolio that can bring a greater incremental spread scope than current $\triangle\sigma(\mathbf{S}, M_\ell)$ (line 19), we add the two candidates in this portfolio into $\mathbf{S}$ (line 20), and go to the next iteration of the loop of line 3 (line 21).

The time complexity of GLIMB is dominated by the time complexity of the second *routine phase* (lines 4–10) and the first *extra phase* (lines 11–23). Let τ be the maximum time required by calculating $\delta(\mathbf{S}, M_i)$ for each candidate group $M_i \in \mathbf{L}$. The second *routine phase* requires $O(m^2\tau)$ time since there are at most $m \times (m - 1)$ times of calculation on $\delta(\mathbf{S}, M_i)$, where m is the number of groups in $\mathbf{M}$. To avoid redundant computations in the second *routine phase*, we adopt a pruning method based on Corollary 1, which indicates that the incremental spread scope $\triangle\sigma(\mathbf{S}, M_i)$ of M_i will not be greater than $\sigma(\{M_i\})$. In other words, for each incremental spread scope over cost ratio $\delta(\mathbf{S}, M_i)$, the upper bound is $\sigma(\{M_i\})/c(M_i)$. Hence, if this upper bound $\sigma(\{M_i\})/c(M_i)$ is less than $MaxRatio$ which records current maximal ratio of incremental spread scope over cost, then this M_i definitely cannot maximize the incremental spread scope over cost ratio, and thus can be excluded to calculate the $\delta(\mathbf{S}, M_i)$. The first *extra phase* requires $O\big((|\mathbf{L}| - 1)^2 \tau\big)$ time since there are at most $|\mathbf{L}| - 1$ candidate groups for the calculation of line 13 and $(|\mathbf{L}| - 1)^2$ candidate portfolios for the calculation of line 19 ($|\mathbf{L}| \leq m$).

Moreover, GLIMB algorithm has the following performance guarantees.

Theorem 5. *GLIMB provides at least a $(1 - 1/\sqrt{e})$-approximation.*

Proof. Without the execution of the first *extra phase*, the result of GLIMB is equal to that of `ImprovedGreedy`. If (1) the first *extra phase* is executed and finds an alternative candidate or candidate portfolio that can bring a greater incremental spread scope, and (2) after this execution of the first *extra phase*, the second *extra phase* is not executed till ending, then $\sigma(\mathbf{S}) > \sigma(\mathbf{S}')$; otherwise, $\sigma(\mathbf{S}) = \sigma(\mathbf{S}')$. In brief, we have $\sigma(\mathbf{S}) \geq \sigma(\mathbf{S}')$.

With the monotone and submodular function $\sigma(\cdot)$, `ImprovedGreedy` provides a $(1 - 1/\sqrt{e})$-approximation. Thus, GLIMB can provide at least a $(1 - 1/\sqrt{e})$-approximation. ∎

5   Experimental Evaluation

In this section, we first describe the data sets used for experiments, and then verify the efficacy of the two algorithms proposed in this paper.

5.1   Experimental Setup

We adopt (1) LFR benchmark graphs [8] and (2) Amazon product co-purchasing network [9], respectively, as the underlying individual-level influence relationship graphs. A LFR benchmark graph can be generated by setting the number n of nodes, the number m of communities (groups), the average size *avg-s* of communities, and the average degree *avg-d* of each node. If there is an edge between two nodes, we regard that these two nodes can directly influence each other. We generate three series of LFR benchmark graphs with properties summarized in Table 2. Amazon product co-purchasing network was crawled from Amazon website. It contains 262111 nodes, each of which refers to a product, and 1234877 directed edges. A directed edge from node i to node j indicates that the i-th product is frequently co-purchased with the j-th product. Community detection approaches [14] can help divide the nodes into different number of groups.

Table 2. Properties of LFR benchmark graphs used for experiments

Graph data sets	Group number m	Average group size *avg-s*	Average degree *avg-d*
LFR1.1–1.5	50, 100, 150, 200, 250	200	3
LFR2.1–2.5	200	50, 100, 150, 200, 250	3
LFR3.1–3.5	200	150	3, 5, 7, 9, 11

The historical infection data set $\mathbf{D}$ can be obtained by simulating κ times of infection outbreaks on each underlying individual-level influence relationship graph with randomly selected seed nodes in each simulation. In each infection outbreak, each infected node tries to activate its uninfected neighbors with probability p. In all the experiments, κ is set to 50, the proportion of the seed nodes is set to 10%, and p is set to 0.2.

5.2 Performance Study of Influence Relationship Graph Construction

In this experiment, we carry out performance study on the construction of group-level influence relationship graph by comparing our proposed approach, i.e., Algorithm 1, with the existing algorithm known as CSI [11] in terms of (1) runtime for construction, and (2) effect to influence maximization. Since CSI requires the influence relationship between individuals to estimate group-level influence relationship, we give it a privilege that the underlying individual-level influence relationship graphs are available for CSI.

Runtime for Graph Construction. To evaluate the effects of (1) group number m, (2) average group size $avg\text{-}s$, and (3) the compactness of individual-level influence relationship (reflected by average degree $avg\text{-}d$) to the runtime for the construction of group-level influence relationship graph, we carry out runtime comparisons on graphs LFR1.1–1.5 that have varying m, graphs LFR2.1–2.5 that have varying $avg\text{-}s$, and graphs LFR3.1–3.5 that have varying $avg\text{-}d$, respectively. For the Amazon product co-purchasing network that has a fixed number of nodes and a fixed degree for each node, we only vary the number m of groups returned by community detection from 50 to 250 with an interval of 50 (the corresponding $avg\text{-}s$ will also vary with the varying m).

Figures 1(a)–(d) illustrate the runtime comparison result on each graph data set, from which we can have the following observations. (1) Our approach is significantly faster than CSI on the graph construction. This is because the time complexity of Algorithm 1 is $O(mn\kappa + m^2 + 2^\alpha m^2)$, which is linear to the number $n = m \times avg\text{-}s$ of nodes (individuals), while the time complexity of CSI is quadratic to n. (2) The gradients of runtime curves of Algorithm 1 in Figs. 1(a) and (d) are greater than that in Figs. 1(b) and (c), indicating that the number of groups is dominant affecting factor for the efficiency performance of Algorithm 1. (3) The compactness of individual-level influence relationship (i.e., $avg\text{-}d$) can also slightly affect the runtime of Algorithm 1. This is because with a higher average degree, each node is expected to spread its influence to more neighbors, and thus more groups may have significant influence relationship, resulting in that Algorithm 1 needs to execute more computation of conditional mutual information.

Effect to Influence Maximization. On the group-level influence relationship graphs constructed by Algorithm 1 and CSI, we perform our GLIMB algorithm with the same paraments. Figures 1(e)–(h) illustrate the corresponding influence spread scopes of the seed groups selected by GLIMB. From the figures, we can

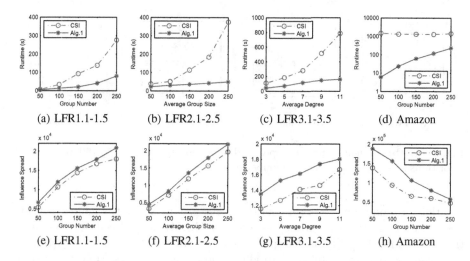

Fig. 1. (a)–(d): Runtime for the construction of group-level influence relationship graph on different graph data sets. (e)–(h): The influence spread scopes of seed groups selected by GLIMB on the corresponding constructed group-level influence relationship graphs.

observe that our proposed Algorithm 1 can help our GLIMB algorithm to find better seed groups that bring larger influence spread scopes.

In brief, compared with CSI which learns the influence relationship between groups based on individuals, our proposed group-level approach can not only achieve a better efficiency performance, but also help improve the results of influence maximization.

5.3 Performance Study of GLIMB Algorithm

In this experiment, we verify the effectiveness and efficiency of our GLIMB algorithm for the problem of influence maximization with budget constraint. For the purpose of comparison, we modify the existing individual-level influence maximization approaches, including a state-of-the-art greedy searching method TIM [15] and two canonical heuristic searching methods DegreeDiscount [3] and IRIE [6], to search optimal seed groups with budget constraint by (1) considering the number of individuals in each group as a weight during their estimation of influence spread scope, and (2) adopting an `ImprovedGreedy`-like strategy. Moreover, the `ImprovedGreedy` approach [13] (denoted as Greedy in short) is also involved in the comparison.

Comparison on Influence Spread Scope. We compare our GLIMB algorithm with the other tested algorithms on all the LFR benchmark graph data sets listed in Table 2 and all the Amazon graph data sets used in Sect. 5.2, and record the influence spread scope when different number of seed groups are selected. Figures 2(a) and (b) illustrate the comparison results on LFR1.4 (in which $m = 200$)

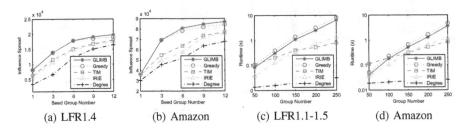

Fig. 2. (a)–(b): Influence spread scopes of the seed groups selected by different tested algorithms. (c)–(d): Scalability to number of groups.

and the Amazon graph data set containing 200 groups. From the figures, we can observe that (1) the seed groups selected by GLIMB and Greedy always have the significantly larger influence spread scopes than the other tested algorithms; (2) when the budget is sufficient to afford more seed group portfolios, the seed groups selected by GLIMB can bring a larger influence spread scope than the seed groups selected by Greedy. Similar observations can be observed on the rest of tested data sets.

Scalability Study. To investigate the scalability of GLIMB to the number m of groups, in Figs. 2(c) and (d), we report the runtimes of GLIMB and the other tested algorithms on LFR1.1 to 1.5 and the Amazon graph data sets, in which the number m of groups varies from 50 to 250 with an interval of 50. From the figures, we can have the following observations. (1) GLIMB is slightly more efficient than Greedy. This is due to the pruning method adopted in GLIMB, which reduces some redundant computation. (2) The gradient of IRIE's runtime curve is close to that of GLIMB's and Greedy's runtime curves, while IRIE's runtime is less than the runtimes of GLIMB and Greedy. This advantage in runtime of IRIE may be from a lower coefficient of the dominant item in its time complexity (3) The gradient of TIM's runtime curve is slightly smaller than that of GLIMB's and Greedy's runtime curves, since it has a $O(m \log m)$ time complexity. (4) DegreeDiscount is the most efficient, although its performance on influence maximization is often not comparable to GLIMB and Greedy.

In summary, with a compensation of more runtime, our GLIMB algorithm can make a better use of cost budget to achieve a larger influence spread scope, compared against the other tested approaches.

6   Conclusion

In this paper, we have studied the problem of group-level influence maximization with budget constraint. Towards this, we have proposed an efficient construction approach for group-level influence relationship graphs, introduced how to model influence propagation on the graph, and presented the GLIMB algorithm to search the optimal seed groups with at least a $(1 - 1/\sqrt{e})$-approximation. Experimental results on both synthetic and real-world data sets have demonstrated the efficacy of our approaches.

Acknowledgements. This work was supported in part by the NSFC Grants (61502347, 61522208, 61502504, and 61472359), and the Nature Science Foundation of Hubei Province (2016CFB384).

References

1. Borgs, C., Brautbar, M., Chayes, J., Lucier, B.: Maximizing social influence in nearly optimal time. In: SODA, pp. 946–957 (2014)
2. Chen, W., Wang, C., Wang, Y.: Scalable influence maximization for prevalent viral marketing in large-scale social networks. In: KDD, pp. 1029–1038 (2010)
3. Chen, W., Wang, Y., Yang, S.: Efficient influence maximization in social networks. In: KDD, pp. 199–208 (2009)
4. Gomez-Rodriguez, M., Leskovec, J., Krause, A.: Inferring networks of diffusion and influence. ACM Trans. Knowl. Disc. Data **5**(4), 1019–1028 (2012)
5. Hu, Z., Yao, J., Cui, B., Xing, E.: Community level diffusion extraction. In: SIG-MOD, pp. 1555–1569 (2015)
6. Jung, K., Heo, W., Chen, W.: IRIE: scalable and robust influence maximization in social networks. In: ICDM, pp. 918–923 (2012)
7. Kempe, D., Kleinberg, J., Tardos, É.: Maximizing the spread of influence through a social network. In: KDD, pp. 137–146 (2003)
8. Lancichinetti, A., Fortunato, S., Radicchi, F.: Benchmark graphs for testing community detection algorithms. Phys. Rev. E **78**(4), 046110 (2008)
9. Leskovec, J., Adamic, L., Adamic, B.: The dynamics of viral marketing. ACM Trans. Web **1**(1), 5 (2007)
10. Leskovec, J., Krause, A., Guestrin, C., Faloutsos, C., VanBriesen, J., Glance, N.: Cost-effective outbreak detection in networks. In: KDD, pp. 420–429 (2007)
11. Mehmood, Y., Barbieri, N., Bonchi, F., Ukkonen, A.: CSI: community-level social influence analysis. In: Blockeel, H., Kersting, K., Nijssen, S., Železný, F. (eds.) ECML PKDD 2013. LNCS (LNAI), vol. 8189, pp. 48–63. Springer, Heidelberg (2013). doi:10.1007/978-3-642-40991-2_4
12. Myers, S., Leskovec, J.: On the convexity of latent social network inference. In: NIPS, pp. 1741–1749 (2010)
13. Nguyen, H., Zheng, R.: On budgeted influence maximization in social networks. IEEE J. Sel. Areas Commun. **31**(6), 1084–1094 (2013)
14. Shang, R., Luo, S., Li, Y., Jiao, L., Stolkin, R.: Large-scale community detection based on node membership grade and sub-communities integration. Phys. A Stat. Mech. Appl. **428**, 279–294 (2015)
15. Tang, Y., Xiao, X., Shi, Y.: Influence maximization: near-optimal time complexity meets practical efficiency. In: SIGMOD, pp. 75–86 (2014)
16. Wang, Y., Cong, G., Song, G., Xie, K.: Community-based greedy algorithm for mining top-k influential nodes in mobile social networks. In: KDD, pp. 1039–1048 (2010)

Correlating Stressor Events for Social Network Based Adolescent Stress Prediction

Qi Li[1], Liang Zhao[2], Yuanyuan Xue[1], Li Jin[1], Mostafa Alli[1], and Ling Feng[1(✉)]

[1] Department of Computer Science and Technology,
Centre for Computational Mental Healthcare Research, Institute of Data Science,
Tsinghua University, Beijing 100084, China
{liqi13,xue-yy12,l-jin12,allim10}@mails.tsinghua.edu.cn,
fengling@tsinghua.edu.cn
[2] Institute of Social Psychology, Xi'an Jiaotong University, Xi'an 710049, China
zhaoliang0415@gmail.com

Abstract. The increasingly severe psychological stress damages our mental health in this highly competitive society, especially for immature teenagers who cannot settle stress well. It is of great significance to predict teenagers' psychological stress in advance and prepare targeting help in time. Due to the fact that stressor events are the source of stress and impact the stress progression, in this paper, we give a novel insight into the correlation between stressor events and stress series (stressor-stress correlation, denotes as SSC) and propose a SSC-based stress prediction model upon microblog platform. Considering both linguistic and temporal correlations between stressor series and stress series, we first quantify the stressor-stress correlation with KNN method. Afterward, a dynamic NARX recurrent neural network is constructed to integrate such impact of stressor events for teens' stress prediction in future episode. Experiment results on the real data set of 124 high school students verify that our prediction framework achieves promising performance and outperforms baseline methods. Integrating the correlation of stressor events is proved to be effective in stress prediction, significantly improving the average prediction accuracy.

Keywords: Stress · Stressor events · Correlation · Prediction · Microblog

1 Introduction

Motivation. The increasingly severe psychological stress threatens our mental health in this highly competitive society, especially for immature teenagers who cannot settle stress well. According to China's Center for Disease Control and Prevention, suicide has become the top cause of death among Chinese youth, and excessive stress is considered to be a major factor of suicide[1].

---

[1] http://theweek.com/articles/457373/rise-youth-suicide-china.

© Springer International Publishing AG 2017
S. Candan et al. (Eds.): DASFAA 2017, Part I, LNCS 10177, pp. 642–658, 2017.
DOI: 10.1007/978-3-319-55753-3_40

Social network based psychological stress analysis attracts much attention in recent years, and offers a new channel to sense users' psychological status in a low-cost and timely manner [7,14,19]. Compared to posteriori stress detection, it is of greater significance to predict teenagers' psychological stress in advance such that in-time targeting psychological help can be prepared to prevent further stress deterioration. Actually, stress is triggered by various stressor events. Different teenagers perform diverse sensitivity to different stressors. Even for the same teenager, various stressors leave distinguishing impacts upon his/her stress progression as well. Mining such personalized hidden impact of stressors can provide powerful cues for future stress prediction.

Challenges. However, the state-of-the-art stress prediction in social network seldom incorporates the stressor impact as guidance. Only [13] considers the stressor impact which is simply measured by the historically mean stress value and directly added to the stress prediction results. Since psychological stress is a highly personalized and subjective feeling, it is difficult to quantitatively measure the personalized impact of stressor events. As a widely-used psychological scale, the Social Readjustment Rating Scale (SRRS) [5] introduces 43 daily-life stressor events with corresponding stress score for each event as impact description. However, the stress score in the scale is the average impact deduced from general groups which compares the relative impacts among different stressors, not individually quantified. Besides, the scale highly depends on manpower and lacks timeliness and flexibility. On the other hand, social network based stressor analysis so far only focuses on qualified stressor detection and extraction [10,15], while quantified stressor impact has not been investigated yet. Thus, the first and the most significant challenge is *how to define and quantify the impact of stressor events upon a specific individual*. The second challenge then becomes *how to integrate such stressor impact into stress prediction*.

Our Work. In this paper, we give a novel insight into the correlation between stressor events and stress series (stressor-stress correlation, denotes as SSC) to describe the stressor impact and propose an SSC-based stress prediction model upon teenagers' microblog. Considering both linguistic and temporal correlations between stressor series and stress series, we first quantify the stressor-stress correlation as a two sample problem with KNN method. Then a dynamic NARX (Nonlinear Autoregressive model process with eXogenous input) recurrent neural network is constructed to integrate such impact of stressor events for teens' stress prediction in future episode. Experiment results on the real data set of 124 high school students verify that our prediction framework achieves promising performance (MAPE 0.156, MSE 0.192) and outperforms baseline methods. Integrating the correlation of stressor events as stressor impact is proved to be effective in stress prediction, significantly improving the average prediction accuracy (with MAPE, MSE reduced by 18.8%, 30.4%, respectively).

Our contributions lie in the following two aspects:

- We formulate the correlation between stressor events and stress series to quantitatively measure the stressor impact upon individuals.
- Integrating the stressor impact, we propose a stressor-stress correlation based prediction model upon social network to predict teenagers' future stress.

To the best of our knowledge, this is the first work exploring the underlying mechanisms of the stressor impact upon adolescent stress, and further incorporating it into social network based adolescent stress prediction.

2   Related Work

Stressor Analysis from Social Network. Traditional stressor event analysis in the field of psychology is based on scales. The Social Readjustment Rating Scale (SRRS) [5] is one of the most widely used psychological scale. It introduces 43 daily-life stressor events, and illustrates corresponding stress score for each event as its impact. However, the stress score in the scale is the average impact deduced from questionnaires of general groups, which generally compares the relative impacts among different stressors, not personal targeting stressor impact in detail. While the scale-based method depends highly on manpower and lacks timeliness, recent researches explore to qualitatively detect stressor events from social network. [15] identifies stressor subjects and categories from tweets using a hybrid multi-task model, and maps the detected events into SRRS scale to get the corresponding stress level. [10] analyzes stressful periods from microblogs using poisson-process-based statistical model based on teens' stressful posting behaviours, and extracts typical stressor events in hierarchy based on common stress dimensions. So far, either in psychology field or computer-aided social network based stress analysis, quantified stressor impact upon individual user has not been investigated yet.

Stress Prediction from Social Network. In recent years, social network offers a low-cost and timely channel to sense individual's daily stress [7,14,19]. Utilizing social media for psychological stress prediction has also been proved to be feasible [2,4,6]. Focusing on teenagers, [12] gets inspiration from stock analysis and predicts change of stress level using candlestick charts upon teenagers' stress series deduced from microblog. [11] further considers the fuzzy theory and proposes a fuzzy candlestick line based approach to predict teenagers' future stress trend when the stress level comes to the peak/valley. The most closest work is [13], which predicts teens' stress in the next time point from microblog using the linear seasonal Autoregressive Integrated Moving Average (SVARIMA) time series prediction method. Influence of forthcoming events is considered in the prediction model. However, the event impact here is simply involved by superimposing the mean stress value during historical events on the prediction result, which ignores abundant linguistic information from tweet content as well as the hidden relationship between stressor events and stress tendency.

From the above, there are no literal publications integrating individually quantified stressor impact into stress prediction in social network.

3   Problem Formulation

3.1   Problem Definition

As illustrated in Fig. 1, given a sequence of stressor events (*exams* in this example), once the event happens, the teen's stress dramatically increases. According to daily life experience, stressor events usually start to influence a teen's psychological status before it happens and continue the impact in the following days after it happens. The stress trends in the two cases may also be different. For example, a teen increasingly feels stressful when it approaches final exam and may decreasingly keep the anxiety waiting for the marks after the exam. Considering such continuity of stressor events, for a stressor event e happening at time t, we extend its lasting interval into two conjunctive phases, the pre-event phase and post-event phase, denoted as $\ell^{pre}(e) = [t - a, t)$, $\ell^{post}(e) = (t, t + b]$, with length a and b respectively, as shown in Fig. 1. Thus, the continuously lasting interval of the event e becomes the combination of pre-event phase, happening time t, and post-event phase.

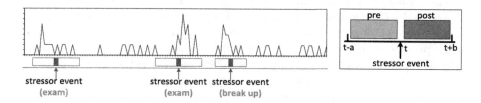

Fig. 1. Illustration of stress series and stressor event series.

Definition 1. *Stressor Event:* *Let* $\mathbb{E} = \{\texttt{school}, \texttt{family}, \texttt{peer}, \texttt{self-cognition}, \texttt{romantic}\}$ *are types of teens' stressor events, representing 'school life', 'family life', 'peer relationship', 'self-cognition' and 'romantic relationship', respectively (as in [10]). A stressor event of type* $y \in \mathbb{E}$ *happened at time* t *is denoted as a four dimensional tuple* $e = (t, y, \ell^{pre}(e), \ell^{post}(e))$, *where* $\ell^{pre}(e) = [t - a)$, $\ell^{post}(e) = (t + b]$ *denote the pre-/post-event phase, describing the continuously lasting interval of the event. For stressor event type* $y \in \mathbb{E}$, *the type-specific sequence of stressor events up to time* t *is represented as* $E_y^t = (e_1, e_2, \ldots, e_k)$, *where* e_i *is the i-th stressor event under type* y *happened before time* t. □

Based on the tweet-wise stress detection model of [19], we are able to detect the stress from a single tweet and a teen's microblog series can then be mapped to stress series.

Definition 2. *Stress Series:* *Let* $\mathbb{L} = \{0, 1, 2, 3, 4, 5\}$ *be the set of tweet-wise stress level, denoting non-stress, tiny, light, moderate, heavy, and severe stress, respectively. Given a series of tweets* $W^t = (w_1, w_2, \cdots, w_n)$ *posted up to time* t, *the corresponding stress series is denoted as* $(s_1, s_2, \cdots, s_n)$, *where* $s_i \in \mathbb{L}$

corresponds to the stress level detected from tweet w_i. Aggregated by equal time unit day, the stress series is then normalized as $S^t = (s^1, s^2, \ldots, s^t)$, where s^k is the stress amount in day k, $s^k \geq 0$.   □

Obviously, the bigger value of s^k denotes the teenager is more stressful in day k. Multiple methods can be used for time series value aggregation, and in this study we adopt the maximal, average and accumulated stress.

Problem Definition: Given a teen's tweet sequence W^t up to time t, the corresponding stress series S^t and stressor event sequence E^t, the problem investigated in this paper becomes predicting the teen's stress in the next episode $[t+1, t+c]$, denoted as $F : (W^t, S^t, E^t) \rightarrow S^{t+c}$. F is the function/model receiving tweet sequence, stress series and stressor event sequence as input, and return the stress predicted in the next c days as output.

3.2   Framework Overview

To solve the problem defined above, in this paper, we propose a stressor-stress correlation based prediction framework to predict teens' future stress. Figure 2 illustrates the overview of the framework. The stress prediction is performed by three stages: (1) preprocessing, (2) model construction, and (3) stress prediction. In the preprocessing stage, a teen's microblog sequence is mapped to two time series by preprocessing, the stress series and the stressor event series. Considering the fact that stressor events closely impact users' stress progression, we go deep into the correlation between stressor event sequence E^t and stress series S^t and propose the stressor-stress correlation (SSC) to quantify personalized stressor impact. Comprehensively integrating four kinds of microblog features (i.e., SSC correlation, event distribution, stress tendency and tweeting behavior features), an SSC-based NARX neural network is then constructed for stress prediction.

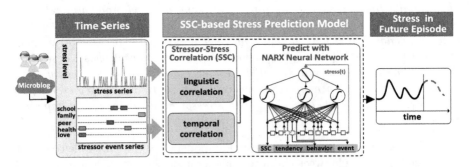

Fig. 2. Model illustration of teens' stress prediction under the impact of stressor events.

4 Stressor-Stress Correlation (SSC)

In this section, we firstly formulate Stressor-Stress Correlation as linguistic correlation and temporal correlation in detail, and then model it as a multivariate two-sample problem, which is widely applied to judge whether two samples obey the same underlying distribution. The nearest neighbour based method is used to quantify the correlation.

4.1 Temporal Correlation and Linguistic Correlation

Based on Definitions 1 and 2, we formulate the Stressor-Stress Correlation from two perspectives: (1) the *temporal correlation* describes the impact of stressor events on stress changing patterns in time line; (2) the *linguistic correlation* represents semantic characters the teen talking about under stressor events. To formulate our problem, we use bold capital letters (e.g., $\mathbf{X}$) and bold lowercase letters (e.g.,$\mathbf{x}$) to denote matrices and vectors, respectively.

Temporal Correlation (C^T). As human's stress is stimulated by stressor events and lasts for a period, a teen's stress changes in this process and forms hidden temporal changing patterns. For a stressor event e and its corresponding interval $\ell(e) = \{\ell^{pre}(e), \ell^{post}(e)\}$ with n unit time slots, in each unit time slot i, its stress pattern is described as:

- **Stress level.** Let a vector $\mathbf{l}_i = [l_i^{acc}, l_i^{avg}, l_i^{max}]$ denote the accumulated, average and maximal stress level of tweets posted during unit time slot i.
- **Stress deviation.** Let $\mathbf{v}_i = [v^{acc}, v^{avg}, v^{max}] = [(l_i - \frac{1}{n}\sum_{1 \leq t \leq n} l_t)^2 | l_i = l_i^{acc}, l_i^{avg}, l_i^{max}]$ denote the stress deviation degree in time slot i compared with average stress level during $l(e)$.
- **Stress trend.** Let $\mathbf{r}_i = [r_i^{acc}, r_i^{avg}, r_i^{max}]$ denote stress changing trend on time slot i, computed as $\mathbf{r}_i = \mathbf{l}_i - \mathbf{l}_{i-1}$, with $\{-1, 0, 1\}$ indicating increase, decrease and steady state, respectively.

Thus we formally describe the stress pattern of stressor event e during interval $\ell(e)$ as a matrix $\mathbf{P}$ composed of three parts: $\mathbf{P} = [\mathbf{P}^l, \mathbf{P}^r, \mathbf{P}^v] \in \mathbb{R}^{N \times 9}$, where $\mathbf{P}^l = [\mathbf{l}_1, \mathbf{l}_2, \cdots, \mathbf{l}_n]^T \in \mathbb{R}^{N \times 3}$, $\mathbf{P}^v = [\mathbf{v}_1, \mathbf{v}_2, \cdots, \mathbf{v}_n]^T \in \mathbb{R}^{N \times 3}$, and $\mathbf{P}^r = [\mathbf{r}_1, \mathbf{r}_2, \cdots, \mathbf{r}_n]^T \in \mathbb{R}^{N \times 3}$, denoting series of stress level, stress deviation and stress trend during $\ell(e)$, respectively.

Linguistic Correlation (C^L). The linguistic distribution in interval $\ell(e)$ indicates the topic centralisation degree under the impact of stressor event e, which measures the confidence that the teen's stress is caused by stressor e. From the perspective of natural language comprehension, we represent a stressor event as $e = \{role, act, emotion, type\}$. Based on this representation, we describe the linguistic distribution during $\ell(e)$ from three aspects:

- **Stressor type word frequency:** Stressor type word frequency is used to measure the proportion of each type of stressor events. We follow work [10] to categorize teens stressor events into five dimensions as $\mathbb{E} = \{$'school life', 'family life', 'peer relation', 'romantic relation' and 'self-cognition'$\}$, containing 69 types of daily hassles[2] and major life events[3] that teens likely encounter. Five corresponding lexicons containing 1,952 stressor event words/phrases based on SC-LIWC [18] are adopted. For each tweet, we apply the word segmentation module of Chinese natural language processing tool LTP[4] for word segmentation, and count the word occurrence frequency in five dimensions during stressor interval $\ell(e)$ respectively, denoted as $\mathbf{d}^{event} = (d_1^e, \cdots, d_5^e)$, $d_i \in \mathbb{E}$.

- **Stressor role word frequency:** We follow work [10] and define the *role* of a stressor event e as the doer that conducts the *act* and the object on which the action is performed. To identify the set of roles for a stressor event e, the parser model of LTP is deployed to linguistically localize the central verb/verb-phase (i.e., the *act*) and its associated semantic roles, based on a lexicon containing 152 role words extended from SC-LIWC [18]. We categorize the set of roles during interval $\ell(e)$ from two aspects:
  - **Role type distribution:** The distribution of role words during $\ell(e)$ reflects the occurrence of each type of stressor event, thus we count the frequency of *role* words in five dimension respectively, and denote the role type distribution as $\mathbf{d}^{type} = (d_1^t, \cdots, d_5^t)$.
  - **First person distribution:** Inspired by work [9], we record the occurrence frequency of role words in first person (e.g.,'I'), second person (e.g.,'your') or third person (e.g.,'her','Mom') respectively, to indicate whether the stressor event e is closely related to the author, denoted as $\mathbf{d}^{per} = (d_1^f, d_2^f, d_3^f)$. Thus the distribution of stressor role word frequency is denoted as $\mathbf{d}^{role} = (\mathbf{d}^{type}, \mathbf{d}^{per})$ in 8 dimensions.
- **Stressor press word frequency:** The occurrence of press word during $\ell(e)$ indicates teen's general stressful *emotion*. We indicate the frequency of press word as d^{press}, and process word matching based on the press word lexicon, which contains 1,113 stressful adjectives, nouns, verbs and adverbs.

Combining the stressor type word frequency $\mathbf{d}^{event}$, the stressor role word frequency $\mathbf{d}^{role}$, and the stressor press word frequency d^{press}, the linguistic distribution of a stressor event e during interval $\ell(e)$ is denoted as a 14-dimension vector: $\mathbf{d} = (\mathbf{d}^{event}, \mathbf{d}^{role}, d^{press}) \in \mathbb{R}^{14}$.

4.2 Modeling Correlation as a Two-Sample Problem

We model the correlation between stressor event sequence E and stress series S as a multivariate two sample hypothesis test[5] problem, which is widely applied to judge whether two samples obey the same underlying distribution (*e.g, Gaussian Distribution*). In our problem, we set the two sample sets as (1) the set of stressor

---

[2] http://aspsychologyblackpoolsixth.weebly.com/daily-hassles.html.

[3] https://en.wikipedia.org/wiki/Holmes_and_Rahe_stress_scale.

[4] http://www.ltp-cloud.com.

[5] https://en.wikipedia.org/wiki/Statistical_hypothesis_testing.

intervals, and (2) the set of randomly chosen stress series in time line, represented as follows:

Definition 3. *Set of stressor intervals:* *For the set of stressor events $E_y = \{e_1, \cdots, e_n\}$ with a specific type $y \in \mathbb{E}$, the set of stressor event intervals is denoted as $S_y^{phase} = \{\ell^{phase}(e_i) | e_i \in E_y\}$, phase $\in \{pre, post\}$, where $\ell^{pre}(e_i)$ and $\ell^{post}(e_i)$ denote the sub-interval before and after stressor event e_i, respectively.*

In this section, we take S_y^{pre} as the example to model the correlation. The length of sub-intervals in S_y^{pre} for each type of stressor event is closely related with the correlation result. For simply and generality, let $\kappa = \frac{1}{n} \sum_{1 \leq i \leq n} \ell^{pre}(e_i)$ denote the general length of sub-interval ℓ^{pre} caused by stressor events with type y, where $\ell^{pre}(e_i)$ is identified using method in [10].

Definition 4. *Set of randomly selected stress series:* *To denote the common state of a teens' stress in timeline, a set of stress series with length κ are randomly picked out, denoted as $S^r = \{\ell_j^r | 1 \leq j \leq n\}$, where $|\ell^r| = |\ell^{pre}(e)| = \kappa$.*

Let ℓ_x denote a point in either of the two sets, which contains κ unit time slots, i.e., $\ell_x \in S_y^{pre} \cup S^r, |\ell_x| = k$. Each ℓ_x is composed of two parts: a matrix $\mathbf{P} = [\mathbf{P}^l, \mathbf{P}^r, \mathbf{P}^v] \in \mathbb{R}^{k \times 9}$ and a vector $\mathbf{d} = (\mathbf{d}^{event}, \mathbf{d}^{role}, d^{press}) \in \mathbb{R}^{14}$, describing the temporal pattern and linguistic distribution during ℓ_x, respectively.

Assuming that S_y^{pre} and S^r are randomly generated from two distribution F and G with densities f and g, the hypotheses of two-sample test is denoted as:

$$\begin{cases} H_0 : F = G \\ H_1 : F \neq G \end{cases} \tag{1}$$

The hypotheses measures whether stressor events of type y (i.e., E_y) have obvious impact on the teen. If H_1 is true, the character distribution of intervals in S_y^{pre} is different from S_r, showing the high correlation between E_y and stress series. Otherwise, if H_0 is true, stressor events E_y have little correlation.

4.3 K-Nearest Neighbor Method

To measure the distance between κ-dimension points in two sample sets S_y^{pre} and S_r, we adopt the nearest neighbor based method here for our two-sample test problem [17]. The basic idea is that for each point ℓ_x in the two sets, we hope its nearest neighbors (*most similar points*) belonging to the same set of ℓ_x.

Given the two sample sets composed of stress intervals $A_1 = S_y^{pre}$ and $A_2 = S^r$, for each point $\ell x \in A = A_1 \bigcup A_2$, we define function $NN_r(\ell_x, A)$ as the r−th nearest neighbor of ℓ_x. Specifically, let $SNN_r(\ell_x, A)$ and $LNN_r(\ell_x, A)$ be two sub-functions of $NN_r(\ell_x, A)$ based on temporal pattern and linguistic distribution, respectively. For point ℓ_x with temporal pattern matrix $\mathbf{P}_x$ and linguistic distribution vector $\mathbf{d}_x$, the r−th temporal-based neighbor is:

$$SNN_r(\ell_x, A) = \{y | min\{||\mathbf{P}_x - \mathbf{P}_y||_2 | y \in (A/\ell_x)\}\} \tag{2}$$

The $r-$th linguistic-based neighbor is:

$$LNN_r(\ell_x, A) = \{y|min\{||\mathbf{d}_x - \mathbf{d}_y||_2|y \in (A/\ell_x)\}\} \tag{3}$$

We define function $I_r(\ell_x, A1, A2)$ to indicate whether the r-th nearest neighbor $NN_r(\ell_x, A)$ belongs to the same sub-set with ℓ_x:

$$I_r(\ell_x, A_1, A_2) = \begin{cases} 1, \ if \ \ell_x \in A_i \&\& NN_r(\ell_x, A) \in A_i, \\ 0, \ otherwise \end{cases} \tag{4}$$

We use $T_{r,n}$ to indicate the proportion that pairs containing two points from the same set among all pairs formed by $\ell_x \in A$ and its nearest neighbors:

$$T_{r,n} = \frac{1}{nr} \sum_{i=1}^{n} \sum_{j=1}^{r} I_j(x, A_1, A_2) \tag{5}$$

The ratio of above neighbors indicates the difference of the two sample sets. We expect that $T_{r,n}$ is close to 1 thus the two underlying distribution F and G for S_y^{pre} and S^r is significantly different, meaning stressor events of type y have obvious impact on teens' stress series.

According to the hypothesis test theory [8], when the size of S_y^{pre} and S_r are large enough, the statistic value $Z = (nr)^{1/2}(T_{r,n} - \mu_r)/\sigma_r$ obeys a standard Gaussian distribution, with the expectation $\mu_r = (\lambda_1)^2 + (\lambda_2)^2$ and variance $\sigma_r^2 = \lambda_1\lambda_2 + 4\lambda_1^2\lambda_2^2$, where $\lambda_1 = |A_1|$ and $\lambda_2 = |A_2|$. Thus we judge whether stressor events of type y has intensive correlation with teens' stress series as: if $C = (nr)^{1/2}(T_{r,n} - \mu_r)/\mu_r^2 > \alpha$ (here $\alpha = 1.96$ for $P = 0.025$), the stressor events E_y has intensive correlation with stress series, where $C^L = C$ if $NN_r = LNN_r(.)$, $C^T = C$ if $NN_r = SNN_r(.)$.

The linguistic correlation based on $LNN_r(\ell_x, A)$ and the temporal correlation based on $TNN_r(\ell_x, A)$ are identified according to above hypotheses test method, and denoted as the pair $\{C_y^L, C_y^T\}$ for stressor events belonging to E_y.

5   SSC-based Stress Prediction Model (SPM)

To capture the impact of stressor events for better stress prediction, in this section, we build a Nonlinear AutoRegressive with Exogenous inputs (NARX) recurrent neural network, and handle the problem as an issue of dynamic multi-step-ahead nonlinear time series prediction. The basic idea of NARX RNN is providing a weighted feedback connection between layers of neurons, and assigning time significance to entire network. It has been proved that NARX network with gradient-descending learning algorithm has more effective learning ability, faster convergency and better generalization compared with other networks (i.e. conventional recurrent neural network) [16].

Table 1. Feature space for teens' stressor-impact-based stress prediction

u_1: stressor event feature space (16 dimensions)	
Role word distribution (8 dimensions)	$R_p = \{firstperson, secondperson, thirdperson\}$ (e.g., 'I', 'you', 'Mom') $R_c = \{family, self\text{-}cognition, peer, academic, romantic\}$
Type word distribution (5 dimensions)	Frequency of lexicon words for each type of stressors
Press word distribution (1 dimension)	Frequency of general press words (e.g., 'hopeless', 'sad')
Event phase (2 dimensions)	① $p(e) = $ 'pre' ② $p(e) = $ 'post'
u_2: stress tendency feature space (6 dimensions)	
Stress trend (3 dimensions)	$Tr_t^{max} = s_t^{max} - s_{t-1}^{max}$ (trend of maximal stress) $Tr_t^{sum} = s_t^{sum} - s_{t-1}^{sum}$ (trend of accumulated stress) $Tr_t^{avg} = s_t^{avg} - s_{t-1}^{avg}$ (trend of average stress)
Stress deviation (3 dimensions)	$D_j^{max} = \|s_j^{max} - \frac{\sum_1^k s_i^{max}}{k}\| / \frac{\sum_1^k s_i^{max}}{k}$ (deviation of maximal stress) $D_j^{sum} = \|s_j^{sum} - \frac{\sum_1^k s_i^{sum}}{k}\| / \frac{\sum_1^k s_i^{sum}}{k}$ (deviation of accumulated stress) $D_j^{avg} = \|s_j^{avg} - \frac{\sum_1^k s_i^{avg}}{k}\| / \frac{\sum_1^k s_i^{avg}}{k}$ (deviation of average stress)
u_3: tweeting behavior feature space (30 dimensions)	
Tweeting frequency (3 dimensions)	Number of tweets Number of stressful tweets Ratio of stressful tweets
Tweeting type (3 dimensions)	Number of original tweets Number of re-tweet tweets Ratio of original tweets
Tweeting time (24 dimensions)	Distribution of tweeting time in 24 h
u_4: correlation feature space (10 dimensions)	
Temporal correlation (5 dimensions)	Whether each type of stressor events have temporal correlation with stress series $(C_y^T, E_y \in \mathbb{E})$
Linguistic correlation (5 dimensions)	Whether each type of stressor events have linguistic correlation with stress series $(C_y^L, E_y \in \mathbb{E})$

5.1   Feature Space

For each time slot, we extract four groups of features for teens' stress prediction: (1) the *stressor event features* (stressor type words, role words, press words and event phase) describe the probability of occurrence for each type of stressor events from linguistic perspective; (2) the *stress tendency features* (stress

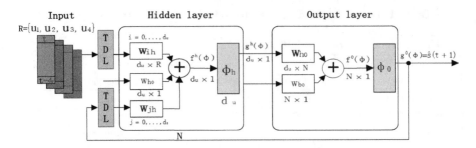

Fig. 3. NARX recurrent network topology for teens' stress prediction.

trend and stress deviation) describe stress changing patterns in time series continuously considering the decay effect of stressor events, aggregated in accumulated, average and maximal format, respectively; 3) the *tweeting behavior features* capture teens stressful and abnormal posting behaviors; and 4) the *linguistic* and *temporal correlation features* (C_y^T, C_y^L) of each type of stressor events $E_y \in \mathbb{E}, y = 1, 2, 3, 4, 5$ learned from Stressor Event Correlation (SSC), which are set to $(0, 0)$ if $E_y = \varnothing$. Table 1 shows details of the feature space for teens' stressor-impact-based stress prediction.

5.2 NARX Recurrent Neural Network

We construct the NARX model with a two-layers (*input layer* and *output layer*) feedforward network for our future episode stress prediction problem, as shown in Fig. 3. For simplicity, we indicate the series of value $\{s^t, s^{t-1}, \cdots, s^{t-d_s}\}$ as $s^{t, \cdots, t-d_s}$. The prediction equation is denoted as:

$$s^{t+1} = \Phi(s^{t, \cdots, t-d_s}, \{\mathbf{u}_1, \mathbf{u}_2, \mathbf{u}_3, \mathbf{u}_4\}^{t, \cdots, t-d_u}) \quad (6)$$

where the stress value in next time slot s^{t+1} is determined by past outputs $(s^{t, \cdots, t-d_s})$ and past observations of the multidimensional input features $(\{\mathbf{u}_1, \mathbf{u}_2, \mathbf{u}_3, \mathbf{u}_4\}^{t, \cdots, t-d_u}$ in Table 1). Here d_s and d_u indicate the history time window for prediction and the length of future episode, respectively. Considering the connexion and transfer between each node in the network, the function Φ is further represented as:

$$s^{t+1} = \Phi_0(w_{b0} + \sum_{h=1}^{N} w_{ho}\Phi_h(w_{ho} + \sum_{k=1}^{4}\sum_{i_k=0}^{d_u} w_{i_k h}u_k(t - i_k) + \sum_{j=0}^{d_s} w_{jh}s(t - j))) \quad (7)$$

Model Learning: The multi-step-ahead stress prediction is a dynamic process, since the model's output are fed back to the input regressor for fixed time plots. For the approximation of function Φ in gradient-based training process, to avoid the problem of vanishing gradients, we adopt the embedded memory [3] strategy in the feed-forward network, represented as the Time Delay Line (DTL) in Fig. 3. The sigmoid function is used as the activation function, denoted as:

$$Sig(x) = \frac{1}{1 + exp(-x)} \tag{8}$$

6   Experiment

In this section, we investigate both the effectiveness and efficiency of our SSC-based stress prediction model. We first compare the performance of our stress prediction model (SPM) with baseline prediction methods. Furthermore, we test the importance of stressor-stress correlation (SSC) of each type of stressor events in teens' stress prediction (Table 2).

Table 2. Statistics of dataset collected from 2012/1/1 to 2015/2/1.

	Description	Data scale
Students	#Students in Taicang High School	124
Posts	#Tencent microblog	29,232
Time slots	Aggregated by days	114,506
Stressor events	#School scheduled events	273
	#Stressful study-related events	122

6.1   Experimental Settings

Data Description. The data set contains 29,232 microblogs from Tencent Weibo[6] of 124 high school students in Taicang High School in Zhejiang Province of China. The school publishes detailed description of scheduled school events (e.g., *monthly exam*) on the official web site[7]. From 2012/1/1 to 2015/2/1, 273 school events are published, among which 122 study-related stressor events are filtered out, covering 114,506 time slots for 124 students. Each of the students posted over 100 posts, with an average of 236 posts and maximal posts of 1,387. Three volunteers manually labeled the stress level of each post from 0 to 5.

Evaluation Metrics. As we mentioned previously, we measure the effectiveness and efficiency of our proposed framework against the baselines. For the **effectiveness**, we adopt two widely used evaluation metrics MSE (Mean Squared Error) and MAPE (Mean Absolute Percentage Error) to measure the absolute and relative error of predictive value, denoted as $MSE = \frac{1}{n}\sum_{i=1}^{n}(x_i - \hat{x_i})^2$, $MAPE = \frac{1}{n}\sum_{i=1}^{n}|x_i - \hat{x_i}|/x_i$, where $\hat{x_i}$ is the predicted value and x_i is the

---

[6] http://t.qq.com/.
[7] http://www.tcsyz.com/col/col11201/index.html.

observed value. To test the **efficiency**, we take the CPU time as the measurement for the efficiency of training each model. We conduct the experiments using an server with two 2.3 GHz Intel Xeon E5 CPU and 256 GB RAM.

Baselines. We choose following baselines to compare with our SSC correlation method and SPM stress prediction method, respectively. For SSC correlation comparison, Pearson Correlation and the average based method are adopted:

- **Pearson Correlation.** Pearson correlation is widely adopted for mining correlation between two time series data, calculated as $\rho_{X,Y} = \frac{cov(X,Y)}{\sigma_X \sigma_Y}$, where $cov(X,Y) = E[(X - \mu_x)(Y - \mu_Y)]$ is the covariance, σ is the standard deviation, μ is the mean value, and E denotes the expectation.
- **Average Based Method.** [13] proposes a simple average-based method to measure the impact of stressor events by calculating the mean stress level in each phase of historical stressor events. We compare our SSC method with this average based method for stress prediction.

For SPM model validation, LR, SVR and SVARIMA methods are used:

- **Logistic Regression (LR).** LR is a statistical liner method to estimate the simultaneous effect of a set of predictors. We use LR to train a linear model on aggregated time slot.
- **Support Vector Regression (SVR).** The SVR method learns a non-linear function and is mapped into high dimensional kernel induced feature space. We choose SVR with RBF kernel to handle the nonlinear problem.
- **Autoregressive Integrated Moving Average (SVARIMA).** The seasonal SVARIMA is a time series analysis method known to be efficient for handling data seasonality and non-stationarity.

6.2   Experiment Results

General Performance of SPM. We conduct the SPM stress prediction model on the real data set of 124 high school students. Figure 4(a) shows the general stress prediction performance integrating SSC correlation. When the correlation of all types of stressors are taken into stress prediction, the model achieves the best performance with MAPE 15.6% and MSE 0.192 (average stress level), respectively. Specifically, the test on correlating each type of stressor events for prediction shows that the SSC correlation of stressors in 'school life' and 'self-cognition' contribute to the prediction significantly. This can be explained with statistical results illustrated in Fig. 4(b) and (c)): the average linguistic and temporal correlation of stressor events from 'school life' and 'self-cognition' in 124 students are higher than other stressor types; and the number of students both showing obvious linguistic and temporal correlations in above two types are highest. This is consistent with the investigation result that problems in school life are usually accompanied with teens inner cognition problems [1].

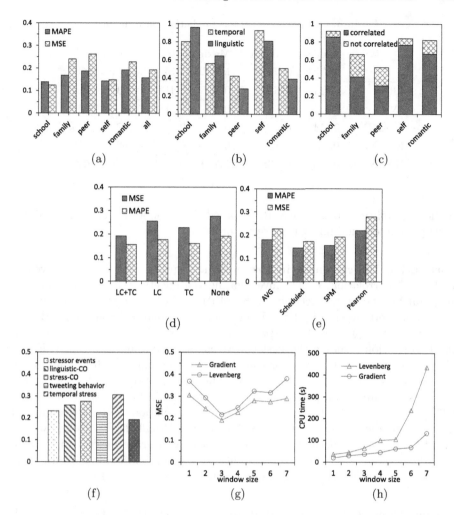

Fig. 4. (a) Predict performance, (b) Correlation values, (c) Correlated students, (d) Correlation contribution, (e) Comparing with baselines, (f) Feature contribution, (g) Window size and (h) CPU time

Comparing Different Prediction Models. In this part, we compare our SPM prediction model with baseline prediction methods. Two back propagation training algorithms (Gradient descent and Levenberg-Marquardt) are tested to train the NARX recurrent neural network. Table 3 shows the prediction performance using different prediction models. The NARX RNN method with Gradient descent algorithm (denoted as 'NARX-2') achieves the best performance (e.g., average stress) both on MAPE (reduced by 46.8%, 30.8% and 16.6%) and MSE performance (reduced by 38.9%, 23.2% and 21.6%) compared with baseline LR, SVR and SVARIMA methods, respectively, demonstrating the strong

Table 3. Comparison of stress prediction performance.

	Accumulated stress		Maximal stress		Average stress	
	MAPE	MSE	MAPE	MSE	MAPE	MSE
LR	0.411	0.76	0.367	0.413	0.293	0.314
SVR	0.331	0.666	0.266	0.318	0.225	0.25
SVARIMA	0.351	0.63	0.253	0.292	0.187	0.245
NARX-1	**0.276**	0.481	0.192	0.233	0.169	0.216
NARX-2	0.289	**0.422**	**0.181**	**0.203**	**0.156**	**0.192**

predicting ability of our NARX RNN based SPM model in handling complex multi-parameters non-linear time series problems.

Correlation Effectiveness in Prediction. Figure 4 shows the effectiveness of the SSC correlation, by (1) investigating the effectiveness of linguistic and temporal correlation respectively (in Fig. 4(d)) and (2) comparing with baseline correlation methods (in Fig. 4(e)). Experiments are conducted on the 'academic' stressor event sequences. The experiment group with both linguistic and temporal correlation achieves the best effectiveness, decreasing 18.8% MAPE and 30.4% MSE compared to non-correlation group. Compared with the automatically extracted stressor event sequences, using the list of scheduled school events improves the prediction performance (with MSE, MAPE reduced by 9.8% and 6.4%, respectively). Furthermore, the baseline correlation methods are implied into stress prediction based on NARX RNN model. Results in Fig. 4(e) show that our SSC correlation method outperforms both the average based and Pearson baseline methods, showing the effectiveness of SSC correlation in capturing the complex relationship between stressor events and teens' stress series.

Feature Contribution in Prediction. To evaluate the contribution of each feature group, we conduct tests on the SPM prediction model by removing one of the four groups of features respectively: stressor event features (F-SE), stressor correlation features (F-CO), tweeting behaviour features (F-BE) and stress tendency (F-SS) features. Figure 4(f) illustrates that the prediction performance without stress tendency features and temporal correlation features decreases with MSE increased by 37.3% and 30.4%, respectively, followed by linguistic correlation which decreases MSE by 25.6% during stress prediction.

Size of Prediction Window. We test our model with different size of prediction windows on the aforementioned server to test the training efficiency. Figure 4(g) and (h) show the efficiency and prediction accuracy changing with the window size. The MSE value decreases first with the window size expending, and then increases. The best performance (MSE 0.192 and 0.216 with Gradient descent and Levenberg-Marquardt training algorithms respectively) is achieved when the window size equals 3. This is because a too small window losts important information for future prediction, while a large window contains more noisy data not relating to current prediction.

7   Conclusion

In this paper, we investigate the problem of predicting teens' psychological stress based on the impact of stressor events in social network. From both the linguistic and temporal perspective, the correlation between stressor event sequence and stress series (stressor-stress correlation, short for SSC) is defined to quantify the stressor impact upon individuals. An SSC-based dynamic NARX recurrent neural network model is then proposed to predict teens' stress in the future episode. Experiments on the real microblog data set of 124 high school students validate that the proposed SSC-based prediction model achieves a promising performance (MAPE 0.156 and MSE 0.192). Integrating the correlation of stressor events as stressor impact is proved to be effective, significantly improving the average stress prediction accuracy (with MAPE, MSE reduced by 18.8%, 30.4%, respectively).

Acknowledgement. The work is supported by National Natural Science Foundation of China (61373022, 61532015, 61521002, 71473146, 2016ZD102) and Chinese Major State Basic Research Development 973 Program (2015CB352301).

References

1. Byrne, D.G., Davenport, S.C., Mazanov, J.: Profiles of adolescent stress: the development of the adolescent stress questionnaire (ASQ). J. Adolesc. **30**(3), 393–416 (2007)
2. Choudhury, M.D., Counts, S., Horvitz, E.: Predicting postpartum changes in emotion and behavior via social media. In: Proceedings of CHI, pp. 3267–3276 (2013)
3. Diaconescu, E.: The use of NARX neural networks to predict chaotic time series. WSEAS Trans. Comput. Res. **3**(3), 182–191 (2008)
4. Gamon, M., Choudhury, M.D., Counts, S., Horvitz, E.: Predicting depression via social media. In: Proceedings of AAAI, pp. 128–137 (2013)
5. Holmes, T.H., Rahe, R.H.: The social readjustment rating scale. J. Psychosom. Res. **11**(2), 213–218 (1967)
6. Hu, Q., Li, A., Heng, F., Li, J., Zhu, T.: Predicting depression of social media user on different observation windows. In: Proceedings of WI-IAT, pp. 361–364 (2015)
7. Jin, L., Xue, Y., Li, Q., Feng, L.: Integrating human mobility and social media for adolescent psychological stress detection. In: Navathe, S.B., Wu, W., Shekhar, S., Du, X., Wang, X.S., Xiong, H. (eds.) DASFAA 2016. LNCS, vol. 9643, pp. 367–382. Springer, Cham (2016). doi:10.1007/978-3-319-32049-6_23
8. Johnson, R.A., Wichern, D.W.: Applied Multivariate Statistical Analysis. Springer, Heidelberg (2003)
9. Li, J., Ritter, A., Cardie, C., Hovy, E.: Major life event extraction from Twitter based on congratulations/condolences speech acts. In: Proceedings of EMNLP (2014)

10. Li, Q., Xue, Y., Zhao, L., Jia, J., Feng, L.: Analyzing and identifying teens stressful periods and stressor events from a microblog. IEEE J. Biomed. Health Inform. (2016)
11. Li, Y., Feng, Z., Feng, L.: When a teen's stress level comes to the top/bottom: a fuzzy candlestick line based approach on micro-blog. In: Proceedings of ICSH (2015)
12. Li, Y., Feng, Z., Feng, L.: Using candlestick charts to predict adolescent stress trend on micro-blog. In: Proceedings of EUSPN, pp. 221–228 (2015)
13. Li, Y., Huang, J., Wang, H., Feng, L.: Predicting teenager's future stress level from micro-blog. In: Proceedings of CBMS, pp. 208–213 (2015)
14. Lin, H., Jia, J., Guo, Q., et al.: Psychological stress detection from cross-media microblog data using deep sparse neural network. In: Proceedings of ICME (2014)
15. Lin, H., Jia, J., Nie, L., Shen, G., Chua, T.S.: What does social media say about your stress? In: Proceedings of IJCAI (2016)
16. Lin, T., Horne, B.G., Tino, P., Giles, C.L.: Learning long-term dependencies in NARX recurrent neural networks. IEEE Trans. Neural Netw. **7**(6), 1329–1338 (1996)
17. Schilling, M.F.: Multivariate two-sample tests based on nearest neighbors. J. Am. Stat. Assoc. **81**(395), 799–806 (1986)
18. Tausczik, Y.R., Pennebaker, J.W.: The psychological meaning of words: LIWC and computerized text analysis methods. Proc. JLSP **29**(1), 24–54 (2010)
19. Xue, Y., Li, Q., Jin, L., Feng, L., Clifton, D.A., Clifford, G.D.: Detecting adolescent psychological pressures from micro-blog. In: Zhang, Y., Yao, G., He, J., Wang, L., Smalheiser, N.R., Yin, X. (eds.) HIS 2014. LNCS, vol. 8423, pp. 83–94. Springer, Cham (2014). doi:10.1007/978-3-319-06269-3_10

Emotion Detection in Online Social Network Based on Multi-label Learning

Xiao Zhang[1], Wenzhong Li[1,2,3(✉)], and Sanglu Lu[1,2,3]

[1] State Key Laboratory for Novel Software Technology,
Nanjing University, Nanjing, China
tobexiao1@dislab.nju.edu.cn
[2] Sino-German Institutes of Social Computing, Nanjing University, Nanjing, China
[3] Collaborative Innovation Center of Novel Software Technology
and Industrialization, Nanjing, China
{lwz,sanglu}@nju.edu.cn

Abstract. Emotion detection in online social networks benefits many applications such as recommendation systems, personalized advertisement services, etc. Traditional sentiment or emotion analysis mainly address polarity prediction or single label classification, while ignore the co-existence of emotion labels in one instance. In this paper, we address the multiple emotion detection problem in online social networks, and formulate it as a multi-label learning problem. By making observations to an annotated Twitter dataset, we discover that multiple emotion labels are correlated and influenced by social network relationships. Based on the observations, we propose a factor graph model to incorporate emotion labels and social correlations into a unified framework, and solve the emotion detection problem by a multi-label learning algorithm. Performance evaluation shows that the proposed approach outperforms the existing baseline algorithms.

Keywords: Emotion detection · Online social network · Factor graph · Multi-label learning

1 Introduction

With the increasing popularity of online social media, people like expressing their emotions or sharing meaningful events with their friends in the online social network (OSN) platforms such as Twitter, Facebook etc. Exploring the rich source of published content in OSNs provides the opportunity to understand users' emotions, therefore enables the emerging emotion-aware applications. For example, a recommendation system could recommend personalized songs, movies or products according to users' current emotions; a personalized advertisement and activity management can make suggestions and schedule in a way of emotion awareness.

This work was partially supported by the National Natural Science Foundation of China (Grant Nos. 61672278, 61373128, 61321491), and the Sino-German Institutes of Social Computing.

S. Candan et al. (Eds.): DASFAA 2017, Part I, LNCS 10177, pp. 659–674, 2017.
DOI: 10.1007/978-3-319-55753-3_41

Table 1. Example of tweets with multiple emotions

Tweets	Emotions
Although I am more excited because I get to see my friends who moved to England two and a half years ago!	Happy, Surprise
Lost exams! I hate homework! I hate studying for tests! I just hate school!	Sad, Anger, Disgust
Talking to Mouse on MSN! Its amazing that even with the time difference, we somehow get to talk! :)	Happy, Surprise

Emotion detection in OSNs has drawn researchers' attentions in many fields in the recent years [1,2,18,30]. Existing approaches for inferring users' emotions in online social media could be classified into two major categories: lexicon based approach [14] and machine learning based approach [1]. Lexicon based approach is to extract emotional keywords based on the dictionaries, such as Linguistic Inquiry and Word Count (LIWC) [14]. As for machine learning based approaches, Go et al. found that machine learning could perform well in classifying sentiment of tweets [8]. Vo et al. analyzed Twitter emotions in earthquake situations [27]. Hu et al. predicted polarity sentiment of one single tweet considering social relations [13]. However, most of the existing approaches usually consider single emotion classification. In fact, multiple emotions could co-exist in one tweet or even one sentence. For example, Table 1 provides several examples of tweets with multiple emotions, where multiple emotion labels such as "Happy" and "Surprise" co-appear in one tweet. To the best of our knowledge, the problem of multiple emotions detection problem in OSNs has not been addressed, and the existing single emotion classification approaches can not deal with this problem well.

In this paper, we study the problem of multiple emotion detection in OSNs based on a multi-label learning approach. We first adopt the Ekman's emotion model [5] to express emotion by six basic categories: *Happy, Surprise, Anger, Disgust, Sad,* and *Fear.* Then we study the influential factors of user's emotion by making observations to an annotated Twitter dataset, where two significant correlations are found: the *emotion label correlation* and the *social correlation.* Emotion label correlation means that some emotion pairs such as *happy* and *surprise* are more likely to co-exist than other emotion label pairs such as *happy* and *fear,* while social correlation means that neighboring users in OSNs are more likely to have similar emotions.

Based on the observed correlations, we formulate the multi-emotion detection problem as a multiple label classification problem, and propose a factor graph model to tackle the problem by addressing both emotion label correlation and social correlation. The factor graph regards each variable as a node in a graph, and the edges represent the correlation between variables, which is called factor function that can be used to express the emotion label correlation and social correlation naturally. We propose a learning algorithm to solve the problem by maximizing the joint probability of the factor functions. Performance evaluation shows that the proposed approach outperforms the baseline algorithms such as logistic regression, decision tree and support vector machine.

The main contributions of our paper are as follows:

- We present the multiple emotion detection problem in online social networks, and formulate it as a multi-label learning problem.
- We make observations to an annotated Twitter dataset and discover the correlations between emotion labels and social relationships, which can be used as features for the task of multi-emotion detection.
- We propose a factor graph model to incorporate emotion labels and social correlations into a unified framework, and solve the emotion detection problem by a learning algorithm, which achieve better performance than the state-of-the-art.

2   Related Work

2.1   Emotion Model

Emotion models have been widely studied in the past. The researchers have proposed various kinds of models to describe emotion. The PANAS model [4] used a checklist more then 20 items to measure positive and negative affect of various aspects. The Circumplex mood model [19] employed a two-dimensional circumplex to represent the emotion state: the pleasure dimension and the activeness dimension. The pleasure dimension measures the degree of positive and negative feelings, and the activeness dimension measures the likelihood for a use to take action. The discrete category model [5,25] described emotion through a set of categories. One of the most popular model is Ekman's six basic categories [5]: *Happy, Surprise, Anger, Disgust, Sad,* and *Fear.* The Ekman's model is intuitive and understandable to normal users, and it allows the co-existence of multiple emotions. With its high applicability, the Ekman's emotion model was widely adopted by many studies [33,34], and it is used in our work of emotion detection in OSNs.

2.2   Emotion Analysis

Sentiment analysis or emotion analysis has drawn many researchers' interests in recent years. Sentiment analysis could be regraded as polarity prediction (positive, negative or neutral), which is widely studied in analyzing product views [12], movie reviews [17], etc. Emotion analysis is a fine-grained sentiment analysis, which could contain more categories than traditional sentiment analysis. As for the emotion analysis in human social network, Fowler et al. found that happiness [6], depression [20] or loneliness [3] could spread from person to person based on clinical trial. With the fast development of online social media, much more data could be provided for researchers to analyze. For example, Wang et al. quantified users' emotion influence and proposed an emotion prediction model in online image social network such as Flicker [28]. Yang et al. modeled the visual features and comment information of images jointly to further improve the accuracy of detecting users' emotions from image [30]. Moodcast [24,32] predicted users' emotions by taking the location information, social influence and temporal correlation as features.

Particularly, recognizing emotions form online social media such as Twitter of FaceBook is always a hot topic [2,9,27]. Generally, there exists two major approaches: lexicon based approach and machine learning based approach. The lexicon based approach tends to extract emotions from text using Linguistic Inquiry and Word Count (LIWC) dictionaries [14]. For example, Bollen et al. extracted six mood states (tension, depression, anger, vigor, fatigue, confusion) from the Twitter and compared the results to some important events [2]. However, the lexicon based approach is heavily dependent on the quality and quantity of words in the dictionary. The machine learning based approach recognizes emotions by extracting features from Tweet content and then predicting sentiment based on various kinds of classification models [13,16,27]. For example, Vo et al. analyzed Twitter emotions in earthquake situations [27], and proposed several emotion categories for the earthquake situations including calm, unpleasantness, sadness, anxiety, fear and relief. However, most of the existing works consider single emotion detection problem while ignore the co-existence of multiple emotions. Different from the literature, our work focus on the multiple emotion detection problem in OSNs, which has not been well addressed in the past.

3   Problem

Let $G = (V, E)$ represents a social network in Twitter, where V denotes a set of users, $E \subset V \times V$ is the relationships between users, such as following each other in Twitter. For a particular user $i \in V$, we denote his (her) published tweets in the t-th time interval as a set S_t. Let X_t be the textual feature vector extracting from S_t, the goal of our work is to learn a function that maps X_t to multiple emotional states represented by a vector Y_t of the user in the t-th time interval:

$$f : (G = (V, E), X_t) \rightarrow Y_t. \tag{1}$$

Here Y_t is a vector with multiple dimensions each of which takes binary values "0" or "1". Therefore the proposed multiple emotions detection problem is considered as a multi-label learning problem.

4   Data

4.1   Dataset Description

The Tweet dataset we use in the study is from an openly available dataset [29]. The dataset was collected starting from the most popular 20 users as the seed users. The snow-ball crawling method was used to crawl the tweet data. The crawling period lasted for one month. All the tweets published by the users were up to Apr. 2, 2010.

4.2   Data Preprocessing

Since tweets have a limit number of words and there are many noises among them, preprocessing the data is important. To improve the quality of textual features, we process the dataset as follows. Firstly, we remove the retweeted tweets since they are from other users which may not represent their own opinions. Secondly, due to the fact that there are many non-English tweets in the dataset, so we utilize LangDetect [21] to remove the non-English Tweets.

4.3   Data Annotation

In order to obtain the users' emotional states, we annotate a small dataset firstly. We randomly select 100 active users (who publish tweets daily) from the original dataset. The 100 users published 16,424 tweets in total. The average number of days each user has posted tweets is 56.

We utilize the popular Ekman's discrete category model [5] to describe users' emotion states. The Ekman's model categories the emotions into six basic classes: *Happy, Surprise, Anger, Disgust, Sad,* and *Fear.* We annotate the tweets using these 6 emotions, one emotion labeled with "1" indicates that this emotion exists in the tweets, and "0" is the opposite.

4.4   Data Observation

We make observations on the annotated Twitter dataset as follows:

Number of Emotion Labels: The distribution of number of emotions is shown in Fig. 1. As in the figure, about 54% instances have single kind of emotion; about 28% instances have two kinds of emotions; and about 18% have at least 3 kinds of emotions. In another words, about 46% instances are multi-labeled. This justify our motivation of multiple emotion detection for OSNs.

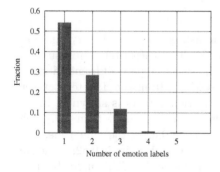

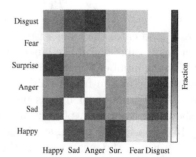

Fig. 1. Statistics of the number of emotion label.

Fig. 2. Correlation of different emotions.

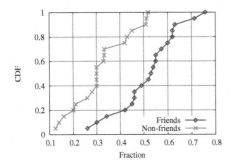
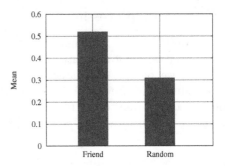

Fig. 3. CDF of fraction of common emotions.

Fig. 4. Mean fraction of common emotions.

Emotion Label Correlation: Then we study the correlation between emotion labels in each instance. We count the fraction that two emotions co-exist in one instance, which is shown in Fig. 2, where darker color corresponds to higher frequency of co-existence. As shown in the figure, several emotion pairs, such as *sadness&disgust, sadness&anger, happyness&surprise*, appear in the same instance frequently, while other pairs co-occurs less common. And some pairs such as *happyness&fear, surprise&fear* rarely appear together.

Social Correlation: Sociologists have found that emotions like happiness [6], depression [20] and loneliness [3] could spread in the social network. This means that an individual's emotional states are influenced by his/her social neighbors, which is referred to as *social correlation* in our paper. To show the significance of social influence, we study emotion correlation between friends and between randomly chosen pairs. Similar to the work of [16], we use the *fraction of common emotions* to show the correlation, which is defined as follows.

$$P(Y_i, Y_j) = \frac{\sum_{t=1}^{N} I(Y_i^t, Y_j^t)}{N} \tag{2}$$

where Y_i^t indicates the label set of user i's tth instance, $I(\cdot)$ is a binary function, which has a value of 1 if the label set Y_i^t and Y_j^t has at least one common label and N is the number of instances. We randomly select 20 pairs of friends and the same number of non-friend pairs randomly selected from the Twitter dataset. The CDF of fraction of friends pairs and non-friends pairs is shown in Fig. 3. As shown in the result, 60% friends pairs have fraction larger than 0.6, while only 15% non-friends pairs have fraction larger than 0.6. At the same time, as is shown in Fig. 4, the mean value of the fraction of friends pairs is 0.52, while the mean value of the fraction of the non-friends pairs is 0.31. This indicates that friends could have higher correlation and more similar emotions in Twitter.

In summary, our observations indicates two significant correlations: (1) Emotion label correlation: Some emotion label pairs are more likely to coexist than other emotion label pairs; (2) Social correlation: friends in OSNs are more likely to have common emotions. In the next section, we will exploit such correlations to devise a model to solve the multiple emotion detection problem.

5   Solution

Based on the observed correlations, we adopt the factor graph model [15] to depict the emotion label correlation and social correlation, and propose a learning algorithm to achieve multi-label emotion detection. The factor graph regards each variable as a node in a graph, and the edges represent the correlation between variables, which is called factor function. Hence, the factor graph could model the emotion label correlation and social correlation as factor functions very naturally. Apart from this, due to the fact that the number of textual features is large, features selection is used to reduce the complexity of emotion detection. The details are discussed in the following.

Textual Features: We adopt the Unigram model [10] to extract textual features from Twitter, which has been show to be effective in [8,11]. At the same time, similar with [13], we keep stemming or stop-words because there may contain emotional information among these words. With the Unigram model, each tweet is represented by a multiple dimensional textual feature vector. Normally the dimensional of the obtained feature vector is very high, and dimension reduction is needed to reduce the computational complexity.

5.1   Feature Selection

Feature selection is the process to find a subset of features $X^* \subseteq X$ to describe the dataset as well as X does, where X is the feature space. When learning from the high-dimensional data, feature selection provides a good way to reduce the dimensions.

Considering the multi-label learning problem, we use the *problem transformation approach* from [22] to select features. The steps are as follows:

- (1) First, binary relevance [26] as the transformation approach is used to transform the multi-label dataset to k single label datasets, where k is the number of labels.
- (2) Then, ReliefF [23] is applied to measure the importance of the features in each single label dataset. ReliefF outputs a value w for every feature, which ranges from -1 to 1. The larger w is, the more important the feature is.
- (3) For one feature in q single label datasets, sum up the q importance outputs as the importance measure of the feature.
- (4) Finally, sort all the features and select the top K features according to their importance measurements.

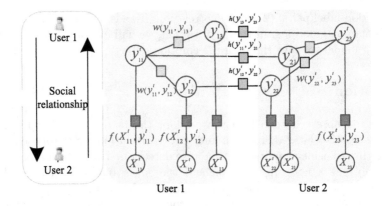

Fig. 5. Factor graph model for multiple emotions detection.

5.2 Factor Graph Model

We propose a factor graph model for learning emotion labels taking into account various of correlations, as is described in Fig. 5. Specially, we consider a feature vector X_i^t for the i-th user's published tweets in the t-th time period, which contains N feature attributes after feature selection. We also denote the emotion labels at the t-th time interval as $y_i^t = (y_{i1}^t, ..., y_{iK}^t)$, where K is the number of different emotion states. To account for the correlations between features and emotion labels, the correlations across a user's different emotion labels and the correlations between two different users' emotion labels, three factor functions are introduced in our factor graph model:

(1) **Textual feature-Label Factor:** $f(X_i^t, y_{ik}^t)$ represents the correlation between the emotion label y_{ik}^t and the feature vector X_i^t. The feature-label factor measures the correlation between the feature vector and the emotion labels. The red rectangles in Fig. 5 represent the feature-label factors.

Just like the traditional multi-label learning problem, we replicate the feature vector X_i^t by k copies (denoted as $\{X_{ik}^t\}_{k=1}^K$), such that we associate the copy X_{ik}^t with the emotion label y_{ik}^t (see Fig. 5 for an illustration). Then, we can define the feature-label factor as

$$f(X_{ik}^t, y_{ik}^t) = \frac{1}{Z_1} \exp\{\sum_{n=1}^N \alpha_{ikn} \Phi(X_{ikn}^t, y_{ik}^t)\}, \tag{3}$$

where $\Phi(X_{ikn}^t, y_{ik}^t)$ is a binary indicator function. For example, $\Phi(X_{ikn}^t = $ "true", $y_{ik}^t = $ "1") means that if the user i has the kth emotion state and the n-th attribute in feature vector X_{ikn}^t is true, then the indicator function value is 1, otherwise 0. α_{ikn} is the weight of the indicator function which describes how strong the correlation between X_{ikn} and y_{ik}^t is. They are also the parameters we need to learn later. Z_1 is a normalization term.

Algorithm 1. Construct label colleration tree

Input: Label matirx M, Number of labels K

Output: Label correlation graph $G = (V, E)$

1: Count the times t_{ij} that label pair $< l_i, l_j >$ coexist in each row of M,
    $T = \{< l_i, l_j >, t_{ij}\}_{i,j=1 \& i \neq j}^{K}$
2: Sort T according to the times t_{ij}
3: **for** each label pair $< l_i, l_j >$ in T **do**
4:     **if** $|V| < K$ **then**
5:         **if** $l_i \in V \& l_j \in V$ **then**
6:             continue;
7:         **else**
8:             $V = V \cup \{l_i, l_j\}$;
9:             $E = E \cup (l_i, l_j)$;
10:        **end if**
11:    **else**
12:        break;
13:    **end if**
14: **end for**

(2) **Label Correlation Factor:** Let $w(y_{ik}^t, y_{id}^t)$ represent the correlation between the user i's kth emotion label y_{ik}^t and the user i's dth emotion label y_{id}^t. The yellow rectangles in Fig. 5 represent the label correlation factors.

Since we have multiple emotion labels, if we take into account all the possible correlations among any two labels, the exact inference in the factor graph model would be difficult. Thus, based on the observation in Sect. 4, we then design an Algorithm 1 to elicit the strong correlation label pairs and meanwhile avoid the cyclic structure in the factor graph.

First, we sort the times that two emotion labels coexist in the same time period in a descending order. Then we select the emotion label pairs in accordance with the order until the *label correlation tree* is formed. We denote the set of user i's emotion labels that have the correlation with the label y_{ik}^t as $W(y_{ik}^t)$ on the tree. Similar to the idea above, we define the factor function $w(y_{ik}^t, y_{id}^t)$ for any $y_{id}^t \in W(y_i k^t)$ as

$$w(y_{ik}^t, y_{id}^t) = \frac{1}{Z_2} \exp\{\beta_{kd} \Phi(y_{ik}^t, y_{id}^t)\}, \qquad (4)$$

where β_{kd} is the wight and quantifies the influence degree between two different labels and Z_2 is a normalization term. For simplicity we further define the factor function $w(y_{ik}^t, W(y_{ik}^t))$ that describes the collective influence between the label y_{ik}^t and the label set $W(y_{ik}^t)$ as

$$w(y_{ik}^t, W(y_{ik}^t))) = \frac{1}{Z_3} \exp\{ \sum_{y_{id}^t \in W(y_i k^t)} \beta_{kd} \Phi(y_{ik}^t, y_{id}^t)\}. \qquad (5)$$

(3) **Social Correlation Factor**: Let $h(y_{ik}^t, y_{jk}^t)$ represent the correlation between user i's kth emotion label y_{ik}^t and user j's kth emotion label y_{jk}^t. To tackle the social correlation, similar to the label correlation factor, $h(y_{ik}^t, y_{jk}^t)$ can be defined as follows. The green rectangles in Fig. 5 represent the social correlation factors.

$$h(y_{ik}^t, y_{jk}^t) = \frac{1}{Z_4} \exp\{\delta_{ij} \Phi(y_{ik}^t, y_{jk}^t)\}, \tag{6}$$

$$h(y_{ik}^t, H(y_{ik}^t)) = \frac{1}{Z_5} \exp\{ \sum_{y_{jk}^t \in H(y_{ik}^t)} \delta_{ij} \Phi(y_{ik}^t, y_{jk}^t)\}, \tag{7}$$

where $H(y_{ik}^t)$ denotes the set of user i's friends' kth emotion labels.

Based on the factor functions above, the objective function can be defined as follows:

$$P(Y|X, \theta) = \prod_{t=1}^{T} \prod_{k=1}^{K} \prod_{i=1}^{M} f(y_{ik}^t, X_{ik}^t) w(y_{ik}^t, W(y_{ik}^t)) h(y_{ik}^t, H(y_{ik}^t)). \tag{8}$$

5.3   Learning Algorithm

We would like to obtain the optimal parameter configuration $\theta = \{\alpha, \beta, \delta\}$ to minimize the following negative log-likelihood objective function together with a L_2-regularization penalty to prevent overfitting:

$$L(\theta) = -\log P(Y|X, \theta) + \frac{\lambda}{2} \sum_{n=1}^{N} \sum_{k=1}^{K} \sum_{i=1}^{M} {\alpha^2}_{ikn} + \frac{\lambda}{2} \sum_{k=1}^{K} \sum_{d=1}^{K} {\beta_{kd}}^2 + \frac{\lambda}{2} \sum_{i=1}^{M} \sum_{j=1}^{M} {\delta_{ij}}^2. \tag{9}$$

We apply the gradient decent method to learn the parameters θ. We obtain the gradients for our factor graph model as follows:

$$\frac{\partial L(\theta)}{\partial \alpha_{ikn}} = E_\theta[\Phi(X_{ikn}^t, y_{ik}^t)] - E_D[\Phi(x_{ikn}^t, y_{ik}^t)] + \lambda\alpha_{ikn}; \tag{10}$$

$$\frac{\partial L(\theta)}{\partial \beta_{kd}} = E_\theta[\Phi(y_{ik}^t, y_{id}^t)] - E_D[\Phi(y_{ik}^t, y_{id}^t)] + \lambda\beta_{kd}; \tag{11}$$

$$\frac{\partial L(\theta)}{\partial \delta_{ij}} = E_\theta[\Phi(y_{ik}^t, y_{jk}^t)] - E_D[\Phi(y_{ik}^t, y_{jk}^t)] + \lambda\delta_{ij}; \tag{12}$$

where $E_\theta[\cdot]$ is the expectation of feature values with respect to the model parameters θ and $E_D[\cdot]$ is the average value by counting the given pattern over the given training dataset. The learning algorithm is summerized in Algorithm 2, wherein we apply the belief propagation (BP) algorithm [7] to infer the expectation value $E_\theta[\cdot]$.

Algorithm 2. Learning algorithm based on factor graph model

Input: Learning rate η
Output: Model parameters θ
1: Initialize $\theta \leftarrow 0$
2: **repeat**
3:   Calculate $E_D[\Phi(X_{ikn}^t, y_{ik}^t)], E_D[\Phi(y_{ik}^t, y_{id}^t)]$ $E_D[\Phi(y_{ik}^t, y_{jk}^t)]$using the training dataset
4:   Calculate $E_\theta[\Phi(X_{ikn}^t, y_{ik}^t)], E_\theta[\Phi(y_{ik}^t, y_{id}^t)]$ $E_\theta[\Phi(y_{ik}^t, y_{jk}^t)]$ using BP algorithm
5:   Calculate the gradients $\frac{\partial L(\theta)}{\partial \theta}$ according to Eqs.(10), (11) and (12)
6:   Update parameter θ as $\theta = \theta + \eta \frac{\partial L(\theta)}{\partial \theta}$
7: **until** Convergence

Given the learned parameter θ, we can obtain user's multiple emotions based on the principle of Maximum a Posterriori (MAP), i.e., finding the emotion levels that maximizes the likelihood given the learned parameters θ as

$$\arg \max P(Y = y|X, \theta). \tag{13}$$

Similarly, we use the BP algorithm to calculate the marginal probabilities. Hence, the emotion labels with the highest probabilities will be identified.

6   Performance Evaluation

6.1   Experiment Setup

In the experiments, we build a personalized model for each annotated user in the Twitter dataset. For each user, 70% data are used as training set to train the model, and the remaining 30% data are used as test set for performance evaluation.

Baseline: We compare the performance of the proposed algorithm with three widely used classification algorithms: Decision Tree (DT), Support Vector Machine (SVM), and Logistic Regression (LR). Since these algorithms only work for single-label classification, we transform the multi-label dataset to k single label dataset, apply the algorithms on each dataset separately, and combine the results to form the prediction on multiple labels.

6.2   Performance Metrics

We adopt the four widely used metrics for performance evaluation for multi-label learning [31], which includes accuracy, precision, recall and F1-score.

- **Accuracy:** It is the proportion of the number of correctly predicted emotion labels to the total number (predicted and the ground truth) of labels for one instance. The overall accuracy is the average on the all instances.

$$Accuracy = \frac{1}{p} \sum_{i=1}^{p} \frac{|Y_i \cap P_i|}{|Y_i \cup P_i|} \tag{14}$$

- **Precision:** It is the proportion of the number of correctly predicted emotion labels to the number of predicted emotion labels for one instance. The overall precision is the average on the all instances.

$$Precision = \frac{1}{p} \sum_{i=1}^{p} \frac{|Y_i \cap P_i|}{|P_i|} \quad (15)$$

- **Recall:** It is the proportion of the number of correctly predicted emotion labels to the number of true emotion labels for one instance. The overall recall is the average on the all instances.

$$Recall = \frac{1}{p} \sum_{i=1}^{p} \frac{|Y_i \cap P_i|}{|Y_i|} \quad (16)$$

- **F1 score:** It is a weighted harmonic mean between the precision and the recall.

$$F1 = \frac{2 * Recall * Precision}{Recall + Precision} \quad (17)$$

where Y_i is the ground-truth label set, P_i is the predicted label set and p is the number of instances.

Table 2. Evaluation results of 15 randomly selected users

ID	Factor graph (%)				DT (%)				SVM (%)				LR (%)			
	Acc.	Pre.	Rec.	F1	Acc.	Pre.	Rec.	F1	Acc.	Pre.	Rec.	F1	Acc.	Pre.	Rec.	F1
227037	**58**	**61**	**58**	**60**	58	61	58	60	58	61	58	60	58	61	58	60
550855	**43**	**47**	**43**	**45**	43	47	43	45	43	47	43	45	43	47	43	45
142614	**82**	**92**	**82**	**87**	82	92	82	87	82	92	82	87	82	92	82	87
223269	**71**	**79**	**71**	75	66	74	71	72	68	76	**76**	**76**	71	79	71	75
192100	**56**	**59**	**56**	**57**	32	35	38	37	35	35	35	35	32	35	32	34
162499	**39**	**50**	41	**45**	15	17	21	19	37	47	**44**	45	37	47	37	41
186507	**61**	**72**	**81**	**76**	56	71	73	72	55	71	72	72	56	71	73	72
159589	51	**63**	52	57	**52**	62	**56**	**59**	52	63	54	58	49	59	49	53
143990	49	70	58	**63**	45	70	49	57	45	70	49	58	45	59	**59**	59
184106	**39**	65	40	**50**	39	58	**44**	50	37	55	42	48	38	**66**	38	48
150811	**46**	**63**	**46**	**53**	46	63	46	53	44	61	46	52	46	63	46	53
606965	**58**	**67**	**58**	**62**	58	67	58	62	58	67	58	62	58	67	58	62
128310	**56**	**67**	**78**	**72**	41	62	43	51	41	62	43	51	56	67	78	72
717973	**35**	**50**	**35**	**41**	22	30	33	31	21	28	31	29	21	33	21	26
179256	**39**	**52**	39	**44**	30	40	33	36	30	37	35	36	38	48	**46**	**47**

Table 3. Mean value of evaluation results of all users

ID	Factor graph (%)				DT (%)				SVM (%)				LR (%)			
	Acc.	Pre.	Rec.	F1	Acc.	Pre.	Rec.	F1	Acc.	Pre.	Rec.	F1	Acc.	Pre.	Rec.	F1
Ave.	49	62	53	56	43	55	47	50	44	56	49	52	45	57	50	53

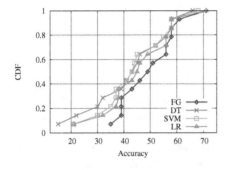

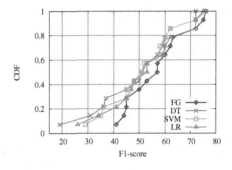

Fig. 6. CDF of accuracy of different emotion detection models.

Fig. 7. CDF of F1-score of different emotion detection models.

6.3   Experiment Results

Table 2 compares the accuracy, precision, recall, F1 score of proposed factor graph model with the baseline methods for 15 randomly chosen users in the dataset. As shown in the result, 93% individuals' factor graph models can achieve the highest accuracy, 93% factor graph models have the highest precision, 67% models have the highest recall, 80% models have the highest F1-score separately. The CDF of accuracy and F1-score are shown in Figs. 6 and 7. As shown in Fig. 6, 53% factor graph models can achieve accuracy larger than 50%, while 40% DT models, 40% SVM models, 40% LR models have accuracy larger than 50%. Similarly in Fig. 7, 100% factor graph models obtain F1-score larger than 40%, 73% factor graph models achieve F1-score larger than 50%, while 73% DT models, 80% SVM models, 86% LR models have F1-score larger than 40% and 67% DT models, 60% SVM models, 60% LR models have F1-score larger than 50%.

Particularly, for user 192100, the factor graph can gain at most +24% accuracy, +24% precision, +18% recall, +20% F1-score compared with DT(decision tree). Similarly, compared with SVM, factor graph model can achieve +21%accuracy, +24% precision, +21% recall and +22% F1-score improvement. As well as +24%accuracy, +24% precision, +24% recall and +23% F1-score improvement compared with LR (logistic regression). As for the performance metrics that factor graph model performs worse than the baseline, the difference is at a scale of 7%-1% compared with the best results. The reason why factor graph behaves differently in users' personalized models is that there is no abundant emotion label correlation information in the dataset.

On average, as shown in Table 3, our proposed factor graph model can achieve +6% accuracy, +7% precision, +6% recall and +6% F1-score improvement compared with DT (decision tree). Similarly, compared with SVM, factor graph model can achieve +5%accuracy, +6% precision, +4% recall and +4% F1-score improvement. As well as +4%accuracy, +5% precision, +3% recall and +3% F1-score improvement compared with LR (logistic regression). This proves the effectiveness of our proposed factor graph model.

7 Conclusion

In this paper, we study the multiple emotion detection problem in online social networks (OSNs) and formulate it as a multi-label learning problem. Based on the observations to the annotated users in a Tweet dataset, we discover the emotion label correlation and social correlation in OSNs, based on which a factor graph model is introduced to incorporate such correlations into a unified framework for emotion detection. A multi-label learning algorithm is proposed to solve the problem. Performance evaluation shows that the proposed approach outperforms the existing baseline algorithms.

References

1. Alm, C.O., Roth, D., Sproat, R.: Emotions from text: machine learning for text-based emotion prediction. In: Proceedings of the Conference on Human Language Technology and Empirical Methods in Natural Language Processing, pp. 579–586. Association for Computational Linguistics (2005)
2. Bollen, J., Mao, H., Pepe, A.: Modeling public mood and emotion: twitter sentiment and socio-economic phenomena. ICWSM **11**, 450–453 (2011)
3. Cacioppo, J.T., Fowler, J.H., Christakis, N.A.: Alone in the crowd: the structure and spread of loneliness in a large social network. J. Pers. Soc. Psychol. **97**(6), 977 (2009)
4. Crawford, J.R., Henry, J.D.: The positive and negative affect schedule (panas): construct validity, measurement properties and normative data in a large non-clinical sample. Br. J. Clin. Psychol. **43**(3), 245–265 (2004)
5. Ekman, P., Friesen, W.V., O'Sullivan, M., Chan, A., Diacoyanni-Tarlatzis, I., Heider, K., Krause, R., LeCompte, W.A., Pitcairn, T., Ricci-Bitti, P.E., et al.: Universals and cultural differences in the judgments of facial expressions of emotion. J. Pers. Soc. Psychol. **53**(4), 712 (1987)
6. Fowler, J.H., Christakis, N.A., et al.: Dynamic spread of happiness in a large social network: longitudinal analysis over 20 years in the framingham heart study. Br. Med. J. **337**, a2338 (2008)
7. Frey, B.J., MacKay, D.J.: A revolution: belief propagation in graphs with cycles. In: Advances in Neural Information Processing Systems, pp. 479–485 (1998)
8. Go, A., Bhayani, R., Huang, L.: Twitter sentiment classification using distant supervision. CS224N Project Report, Stanford, vol. 1, no. 12 (2009)
9. Guillory, J., Spiegel, J., Drislane, M., Weiss, B., Donner, W., Hancock, J.: Upset now? emotion contagion in distributed groups. In: Proceedings of the SIGCHI Conference on Human Factors in Computing Systems, pp. 745–748. ACM (2011)

10. Guthrie, D., Allison, B., Liu, W., Guthrie, L., Wilks, Y.: A closer look at skip-gram modelling. In: Proceedings of the 5th International Conference on Language Resources and Evaluation (LREC-2006), pp. 1–4 (2006)
11. He, H., Jin, J., Xiong, Y., Chen, B., Sun, W., Zhao, L.: Language feature mining for music emotion classification via supervised learning from lyrics. In: Kang, L., Cai, Z., Yan, X., Liu, Y. (eds.) ISICA 2008. LNCS, vol. 5370, pp. 426–435. Springer, Heidelberg (2008). doi:10.1007/978-3-540-92137-0_47
12. Hu, M., Liu, B.: Mining and summarizing customer reviews. In: Proceedings of the Tenth ACM SIGKDD International Conference on Knowledge Discovery and Data Mining, pp. 168–177. ACM (2004)
13. Hu, X., Tang, L., Tang, J., Liu, H.: Exploiting social relations for sentiment analysis in microblogging. In: Proceedings of the Sixth ACM International Conference on Web Search and Data Mining, pp. 537–546. ACM (2013)
14. Kramer, A.D.: An unobtrusive behavioral model of gross national happiness. In: Proceedings of the SIGCHI Conference on Human Factors in Computing Systems, pp. 287–290. ACM (2010)
15. Kschischang, F.R., Frey, B.J., Loeliger, H.-A.: Factor graphs and the sum-product algorithm. IEEE Trans. Inf. Theory $47(2)$, 498–519 (2001)
16. Li, S., Huang, L., Wang, R., Zhou, G.: Sentence-level emotion classification with label and context dependence. In: Proceedings of ACL-2015, pp. 1045–1053 (2013)
17. Pang, B., Lee, L.: A sentimental education: sentiment analysis using subjectivity summarization based on minimum cuts. In: Proceedings of the 42nd Annual Meeting on Association for Computational Linguistics, p. 271. Association for Computational Linguistics (2004)
18. Pang, B., Lee, L., Vaithyanathan, S.: Thumbs up? sentiment classification using machine learning techniques. In: Proceedings of the ACL-02 Conference on Empirical Methods in Natural Language Processing, vol. 10, pp. 79–86. Association for Computational Linguistics (2002)
19. Posner, J., Russell, J.A., Peterson, B.S.: The circumplex model of affect: an integrative approach to affective neuroscience, cognitive development, and psychopathology. Dev. Psychopathol. $17(03)$, 715–734 (2005)
20. Rosenquist, J.N., Fowler, J.H., Christakis, N.A.: Social network determinants of depression. Mol. Psychiatry $16(3)$, 273–281 (2011)
21. Shuyo, N.: Language detection library for java (2010)
22. Spolaôr, N., Cherman, E.A., Monard, M.C., Lee, H.D.: A comparison of multi-label feature selection methods using the problem transformation approach. Electron. Notes Theor. Comput. Sci. 292, 135–151 (2013)
23. Spolaôr, N., Cherman, E.A., Monard, M.C., Lee, H.D.: Relieff for multi-label feature selection. In: 2013 Brazilian Conference on Intelligent Systems (BRACIS), pp. 6–11. IEEE (2013)
24. Tang, J., Zhang, Y., Sun, J., Rao, J., Yu, W., Chen, Y., Fong, A.M.: Quantitative study of individual emotional states in social networks. IEEE Trans. Affect. Comput. $3(2)$, 132–144 (2012)
25. Tomkins, S.S.: Affect, Imagery, Consciousness: Vol. I The Positive Affects (1962)
26. Tsoumakas, G., Katakis, I., Vlahavas, I.: Mining multi-label data. In: Data Mining and Knowledge Discovery Handbook, pp. 667–685. Springer (2009)
27. Vo, B., Collier, N.: Twitter emotion analysis in earthquake situations. Intl. J. Comput. Linguist. Appl. $4(1)$, 159–173 (2013)
28. Wang, X., Jia, J., Tang, J., Wu, B., Cai, L., Xie, L.: Modeling emotion influence in image social networks. IEEE Trans. Affect. Comput. $6(3)$, 286–297 (2015)

29. Xu, T., Chen, Y., Jiao, L., Zhao, B.Y., Hui, P., Fu, X.: Scaling microblogging services with divergent traffic demands. In: Proceedings of the 12th International Middleware Conference, pp. 20–39. International Federation for Information Processing (2011)

30. Yang, Y., Jia, J., Zhang, S., Wu, B., Chen, Q., Li, J., Xing, C., Tang, J.: How do your friends on social media disclose your emotions? AAAI **14**, 1–7 (2014)

31. Zhang, M.-L., Zhou, Z.-H.: A review on multi-label learning algorithms. IEEE Trans. Knowl. Data Eng. **26**(8), 1819–1837 (2014)

32. Zhang, Y., Tang, J., Sun, J., Chen, Y., Rao, J.: Moodcast: emotion prediction via dynamic continuous factor graph model. In: Proceedings of the 10th International Conference on Data Mining, pp. 1193–1198. IEEE (2010)

33. Zhao, S., Yao, H., Jiang, X.: Predicting continuous probability distribution of image emotions in valence-arousal space. In: Proceedings of the 23rd ACM International Conference on Multimedia, pp. 879–882. ACM (2015)

34. Zhou, Y., Xue, H., Geng, X.: Emotion distribution recognition from facial expressions. In: Proceedings of the 23rd ACM International Conference on Multimedia, pp. 1247–1250. ACM (2015)

Tutorials

Urban Computing: Enabling Urban Intelligence with Big Data

Yu Zheng[⊠]

Microsoft Research, Building 2 No. 5 Danling Street,
Haidian District, Beijing, China
yuzheng@microsoft.com

Urban computing is a process of acquisition, integration, and analysis of big and heterogeneous data generated by a diversity of sources in cities to tackle urban challenges, e.g. air pollution, energy consumption and traffic congestion. Urban computing connects unobtrusive and ubiquitous sensing technologies, advanced data management and analytics models, and novel visualization methods, to create win-win-win solutions that improve urban environment, human life quality, and city operation systems. Urban computing is an inter-disciplinary field where computer science meets urban planning, transportation, economy, the environment, sociology, and energy, etc., in the context of urban spaces [1]. The vision of urban computing has been leading to better cities that matter to billions of people.

Though the concept of urban computing has been proposed for a few years, there are still quite a few questions open. For example, what are the core research problems of urban computing? What are the challenges of the research theme? What are the key methodologies for urban computing? What are the representative applications in this domain, and how does an urban computing system work?

Outline

In this tutorial, I will overview the framework of urban computing, discussing its key challenges and methodologies from computer science's perspective. This tutorial will also present a diversity of urban computing applications, ranging from big data-driven environmental protection [10, 13] to transportation [11], from urban planning to urban economy [12]. Here after is an outline of this tutorial.

- Urban sensing
    - Challenges and key techniques
    - Filling missing values in geo-sensory data [3]
    - Resource allocation in urban sensing [2, 5]
- Urban data management
    - Urban big data platform [6]
    - Trajectory data management on the cloud [4]

© Springer International Publishing AG 2017
S. Candan et al. (Eds.): DASFAA 2017, Part I, LNCS 10177, pp. 677–679, 2017.
DOI: 10.1007/978-3-319-55753-3

- Urban data analytics
    - Key challenges and techniques
    - Cross-domain data fusion [7]
        Stage-based methods
        - Feature-based data fusion: Feature concatenation + regularization [12]; Deep learning-based methods [8]
        Semantic meaning-based data fusion: Multi-view-based, probabilistic dependency-based, similarity-based, and transfer learning-based fusion methods [9]

References

Publications Related to the Framework of Urban Computing

1. Zheng, Y., Capra, L., Wolfson, O., Yang, H.: Urban computing: concepts, methodologies, and applications. ACM Trans. Intell. Syst. Technol.

Publications Related to Urban Sensing

2. Ji, S., Zheng, Y., Li, T.: Urban sensing based on human mobility. In: UbiComp (2016)
3. Yi, X., Zheng, Y. Zhang, J., Li, T.: ST-MVL: filling missing values in geo-sensory time series data. In: IJCAI (2016)

Publications Related to Urban Data Management

4. Bao, J., Li, R., Yi, X., Zheng, Y.: Managing massive trajectories on the cloud. In: ACM SIGSPATIAL (2016)
5. Li, Y., Bao, J., Li, Y., Wu, Y., Gong, Z., Zheng, Y.: Mining the most influential k-location set from massive trajectories. In: ACM SIGSPATIAL (2016)
6. Zheng, Y.: Trajectory data mining: an overview. ACM Trans. Intell. Syst. Technol. (ACM TIST)

Publications Related to Urban Data Analytics

7. Zheng, Y.: Methodologies for cross-domain data fusion: an overview. IEEE Trans. Big Data
8. Zhang, J., Zheng, Y., Qi, D.: Deep spatio-temporal residual networks for citywide crowd flows prediction. In: AAAI (2017)
9. Wei, Y., Zheng, Y., Yang, Q.: Transfer knowledge between cities. In: KDD (2016)
10. Zheng, Y., Yi, X., Li, M., Li, R., Shan, Z., Chang, E., Li, T.: Forecasting fine-grained air quality based on big data. In: KDD (2015)

11. Shang, J., Zheng, Y., Tong, W., Chang, E., Yu, Y.: Inferring gas consumption and pollution emission of vehicles throughout a city. In: KDD (2014)
12. Fu, Y., Xiong, H., Ge, Y., Yao, Z., Zheng, Y., Zhou, Z.-H.: Exploiting geographic dependencies for real estate appraisal: a mutual perspective of ranking and clustering. In: KDD (2014)
13. Zheng, Y., Liu, F., Hsieh, H.-P.: U-Air: when urban air quality inference meets big data. In: KDD (2013)

Incentive-Based Dynamic Content Management in Mobile Crowdsourcing for Smart City Applications

Sanjay K. Madria(⊠)

Department of Computer Science,
Missouri University of Science and Technology, Rolla, USA
madrias@mst.edu

Motivation, Objective and Scope

Ever-increasing prevalence of social networking using mobile devices has catalyzed the growth of interesting and innovative new-age mobile crowdsourcing applications, which work at the intersection of human-centric computation (e.g., economic incentive management and social computing) and dynamic management of information as well as content in mobile networks. The prevalence and proliferation of mobile devices coupled with popularity of social media and increasingly technology-savvy users have fuelled the growth of mobile crowdsourcing and participatory sensing. In particular, participatory sensing can occur in various ways by means of devices (e.g., mobile phones, PDAs, laptops and various types of sensors) or by including humans in the loop or both. Notably, participatory sensing can also potentially act as a key enabling technology for various applications involving smarter cities initiatives.

Incidentally, large-scale collection of city-related event data is crucial to effective planning and decision-making for improving city management. Examples of city-related event data include traffic congestion [1], illegal parking, accidents, dysfunctional streetlights, broken pavements, potholes, planned road construction works, and waterlogging, and garbage collection. Notably, existing sensor-based data collection mechanisms cannot always take human judgment and the context of the event into consideration, and the costs of deploying them across all city locations would be prohibitively expensive. Hence, event data collection can be used to complement sensor-based data collection. Since mobile devices often come equipped with various kinds of sensors, resident-driven data collection is also well aligned with current technological trends. However, incentives [2–4] need to be provided to users for encouraging them to contribute better quality event data.

I will discuss in this tutorial incentives & economic models for crowd participation, mobile data management and resource discovery, mobile replica allocation and consistency, indexing & query processing as well as analytics. These issues have generated a significant amount of interest in academia as well as in industry recently. This tutorial

---

Author is supported by NSF Grant CNS-1461914.

S. Candan et al. (Eds.): DASFAA 2017, Part I, LNCS 10177, pp. 680–681, 2017.
DOI: 10.1007/978-3-319-55753-3

thus intends to foster discussions on the important research challenges as well as design issues of key enabling technologies that need to be addressed to make next-generation mobile crowdsourcing [5] and human computation effective in the real-world.

Acknowledgement: I would like to thank Anirban Mondal for the discussion and in contributing some of the tutorial slides.

Biography: Sanjay Kumar Madria is a full tenured Professor, Department of Computer Science at the Missouri University of Science and Technology, USA. His research interests are in mobile computing, cloud computing, sensor networks and security. He received several awards for his research including faculty research excellence, IEEE best papers, and ACM Distinguished Scientist.

References

1. Ilarri, S., Wolfson, O., Delot, T.: Collaborative sensing for urban transportation. IEEE Data Eng. Bull. **37**, 3–14 (2014)
2. Padhariya., N., Mondal., A., Madria, S.: Efficient processing of mobile crowdsourcing queries with multiple sub-tasks for facilitating smart cities. In: SmartCities workshop with ACM Middleware, Italy (2016)
3. Padhariya, N., Mondal, A., Madria, S.K.: Top-k query processing in mobile-P2P networks using economic incentive schemes. Peer-to-Peer Networking Appl. **9**(4), 731–751 (2016)
4. Padhariya, N., Wolfson, O., Mondal, A., Gandhi, V., Madria, S.K.: E-VeT: economic reward/penalty-based system for vehicular traffic management. In: MDM, vol. 1, pp. 99–102 (2014)
5. Mondal, A., Rao, P., Madria, S.K.: Mobile computing, internet of things, and big data for urban informatics. In: IEEE MDM, pp. 8–11 (2016)

Tutorial on Data Analytics
in Multi-engine Environments

Verena Kantere[✉] and Maxim Filatov

University of Geneva, Geneva, Switzerland
{verena.kantere,maxim.filatov}@unige.ch

Motivation for the Tutorial. The performance of analytics on Big Data collections is the focus of a lot of research and is becoming a leading requirement in many business domains and scientific disciplines. Data analytics includes techniques, algorithms and tools for the inspection of data collections in order to extract patterns, generalizations and other useful information. Much of the recent research is focused on the employment of Hadoop for efficient data analysis, which takes advantage of the MapReduce paradigm and distributed storage, and, furthermore, Spark mostly for real-time analysis, which takes advantage of in-memory processing. The success and effectiveness of such analysis depend on numerous challenges related to the data itself, the nature of the analytics tasks, as well as the data processing platforms over which the analysis is performed.

Data analytics may vary significantly in terms of the properties of data, type of analysis and processing system and may require cross-platform analysis with migration of both data and processing. For example, data may be structured, semi-structured, or unstructured residing in files, and may have different interdependencies, such as relational constraints, tree-like or graph dependencies. The type of analysis may include stream processing, information retrieval, query processing, mining, clustering, integration, and other. The underlying processing systems may be traditional relational DBMSs, but also RDF stores, NoSQL databases, graph databases etc.

The above diversity across data, processing and systems has recently spawned an interest in the research community for the creation of transparent all-inclusive solutions for data analytics in multi-engine environments, which enable the interoperability of the involved systems, and focus in the construction of inter-system optimizers. The challenges for the creation of such solutions lie in the consolidation of different engine capabilities for data processing, such that the integrated system adapts its operation depending on the input workload and the data location and type, and the processing load of the engines. Research to tackle this challenge includes the definition of versatile programming models, engine performance modeling and monitoring, planning and optimization techniques, parallel deployment and execution on multiple engines, workflow management and visualization techniques.

Brief Outline of the Tutorial Content. The proposed tutorial will present the current picture in research for the realization of multi-engine data analytics. Towards this end, the tutorial will start with the discussion of modeling

© Springer International Publishing AG 2017
S. Candan et al. (Eds.): DASFAA 2017, Part I, LNCS 10177, pp. 682–684, 2017.
DOI: 10.1007/978-3-319-55753-3

of analytics, including new models for the representation of complex tasks and languages for programming scalable processing. The tutorial will focus on the execution of analytics: planning, optimizing and executing complex or multiple workflows especially on dynamic multi-engine and elastic environments. Finally, the tutorial will trigger the discussion for the development of tools for advanced analytics tasks, such as operators for regular and irregular computations.

The tutorial was presented at Data Analytics and Management in Data Intensive Domains conference (DAMDID) in October 2016.

Biographies of the Presenters. **Verena Kantere** is a Maître d'Enseignement et de Recherche (equivalent to Associate Professor) at the Centre Universitaire d'Informatique (CUI) of the University of Geneva (UniGe) and an Assistant Professor at the School of Electrical and Computer Engineering in the National Technical University of Athens, working towards the provision and exchange of data services in cloud environments, focusing on the management of Big Data and performance of Big Data analytics, by developing methods, algorithms and fully fledged systems. Before coming to the UniGe she was a tenure-track junior assistant professor at the Department of Electrical Engineering and Information Technology at the Cyprus University of Technology (CUT). She has received a Diploma and a Ph.D. from the National Technical University of Athens, (NTUA) and a M.Sc. from the Department of Computer Science at the University of Toronto (UofT), where she also started her Ph.D. studies. After the completion of her Ph.D. studies she worked as a postdoctoral researcher at the Ecole Polytechnique Federale de Lausanne (EPFL). She is currently leading the research on Adaptive Analytics in the ASAP (www.asap-fp7.eu) EU project. She has created and co-chaired several workshops, the most recent being the Workshop on Multi-Engine Data Analytics (collocated with EDBT 2016), she has given 30 invited talks in international conferences, universities and research institutes and has served in the PC of 40 international conferences and workshops.

Maxim Filatov is a Ph.D. student working in CUI of the University of Geneva (UniGe), under the supervision of Prof. Verena Kantere. Prior to Geneva, Maxim obtained his Masters degree in Mechanics at the Computational Mechanics department of Lomonosov Moscow State University, where he also started his Ph.D. studies. During this period he also worked as a researcher in RD of Roxar and TimeZYX on the development of reservoir simulation software. He developed methods, algorithms and fully fledged systems for modeling of multiphase flow, focusing on the performance aspect. His Ph.D. research is focused on the management of Big Data and performance of Big Data analytics. He is currently working on Adaptive Analytics in the ASAP (www.asap-fp7.eu) FP7 EU project.

References

1. Alexandrov, A., Volker, M., Salzmann, A.: Emma in action: declarative dataflows for scalable data analysis. In: SIGMOD (2016)
2. Agrawal, D.: Rheem: enabling multi-platform task execution. In: SIGMOD (2016)
3. Duggan, J., Elmore, A.: The BigDAWG polystore system. In: SIGMOD (2015)
4. Alexandrov, A., Bergmann, R.: The stratosphere platform for big data analytics. VLDB J. (2014)
5. Simitsis, A., Wilkinson, K.: HFMS: managing the lifecycle and complexity of hybrid analytic data flows. In: ICDE (2013)
6. Deelman, E., Singh, G.: Pegasus: a framework for mapping complex scientific workflows onto distributed systems. Sci. Program J. (2005)
7. Oinn, T., Addis, M.: Taverna: a tool for the composition and enactment of bioinformatics workflows. Bioinform. J. (2004)

Author Index

in the United States
masters